DARK AGES

(A.D. 450 to 1120)

A sterile period for all learning in Western Europe

Slender thread of Greek and Latin learning preserved in monasteries

PERIOD OF TRANSMISSION

(A.D. 950 to 1500)

Learning preserved by the Arabs slowly transmitted to Western Europe

Translation of Arabic works
(Plato of Tivoli, A.D. 1120; Robert of Chester, A.D. 1140; Adelhard of Bath, A.D. 1142; Gherardo of Cremona, A.D. 1150; Campanus, A.D. 1260)

Advocacy of Hindu-Arabic numeral system
(Fibonacci, A.D. 1202)

Fourteenth century, the century of the Black Death

First mathematics book printed in the Western World
(*Treviso Arithmetic*, 1478)

First printed edition of Euclid's *Elements*
(Campanus' translation, A.D. 1482)

MODERN

(first half, A.D. 1450 to 1700)

Early trigonometry
(Regiomontanus, 1464; Copernicus, 1530; Rhaeticus, 1550)

Early arithmetics
(Borghi, 1484; Widman, 1489; Pacioli, 1494; Köbel,1512; Riese, 1518; Tonstall, 1522; Buteo, 1525)

Beginnings of algebraic symbolism
(Recorde, 1557; Bombelli, 1572; Viète, 1579; Oughtred, 1631)

Algebraic solution of cubic and quartic equations
(Tartaglia, Cardano, Ferrari, 1545)

Development of classical algebra
(Viète, 1580; Harriot, 1631)

Decimal fractions
(Stevin, 1585)

Boost from science
(Galileo, 1600; Kepler, 1609)

Logarithms
(Napier, 1614; Briggs, 1615)

Modern number theory
(Fermat, 1635)

Analytic geometry
(Fermat, 1629; Descartes, 1637)

Start of projective geometry
(Desargues, 1639; Pascal, 1648)

Mathematical probability
(Fermat and Pascal, 1654)

Calculus
(Fermat, 1629; Cavalieri, 1635; Barrow, 1669; Leibniz, 1684; Newton, 1687)

Applied calculus
(Jakob and Johann Bernoulli, 1700; Clairaut, 1743; d'Alembert, 1743; Euler, 1750; Lagrange, 1788; Laplace, 1805; Fourier, 1822; Legendre, 1825; Green, 1828; Poisson, 1831)

Infinite series
(Taylor, 1715; Maclaurin, 1742; Fourier, 1822)

Non-Euclidean geometry
(Saccheri, 1733; Lambert, 1770; Legendre, 1794; Gauss, 1800; Lobachevski, 1829; J. Bolyai, 1832)

Topology
(Euler, 1736; Gauss, 1799; Listing, 1847; Riemann, 1851; Möbius, 1865; Poincaré, 1895)

Advanced analytic geometry
(Monge, 1795; Plücker, 1826; Möbius, 1827)

Analysis
(Lagrange, 1797; Abel, 1826; Cauchy, 1827; Riemann, 1851; Dedekind, 1872; Weierstrass, 1874; Lebesgue, 1903)

Projective geometry
(Poncelet, 1822; Gergonne, 1826; Steiner, 1834; Von Staudt, 1847; Clifford, 1878)

Modern computing machines
(Babbage, 1823; ASCC, 1944; ENIAC, 1945; SSEC; EDVAC; MANIAC; UNIVAC)

Rise of abstract algebra
(Galois, 1832; Hamilton, 1843; Grassmann, 1844; Cayley, 1857)

Mathematical logic
(Boole, 1847; De Morgan, 1847; Schröder, 1890; Peano, 1894; Whitehead and Russell, 1910; Lukasiewicz, 1921)

Set theory
(Cantor, 1874; Hausdorff, 1914)

Foundations and philosophy of mathematics
(Frege, 1884–1903; Hilbert, 1899; Brouwer, 1907; Whitehead and Russell, 1910; Gödel, 1931)

Abstract spaces
(Fréchet, 1906; Hausdorff, 1914; Banach, 1923)

HOWARD EVES
University of Maine

AN INTRODUCTION TO THE HISTORY OF MATHEMATICS

with Cultural Connections
by
Jamie H. Eves

Sixth Edition

Australia • Canada • Mexico • Singapore • Spain
United Kingdom • United States

Text Typeface: Times Roman
Acquisitions Editor: Robert Stern
Production Manager: Charlene Squibb
Project Editor: Rebecca Gruliow
Text Designer: Lawrence R. Didona, Doris Bruey
Art Director: Christine Schueler
Art and Design Coordinator: Doris Bruey
Copy Editor: Diane M. Lamsback
Illustrator: Larry Ward
Cover Designer: Lawrence R. Didona
Cover Printer: R. R. Donnelley, Crawfordsville
Compositor: TechBooks
Printer: R. R. Donnelley, Crawfordsville

Printed in the United States of America
18 19 20 07

AN INTRODUCTION TO THE HISTORY OF MATHEMATICS, Sixth Edition

Library of Congress Control Number: 89-043140

ISBN-13: 978-0-03-029558-4
ISBN-10: 0-03-029558-0

Brooks/Cole—Thomson Learning
511 Forest Lodge Road
Pacific Grove, CA 93950
USA

Asia
Thomson Learning
5 Shenton Way #01-01
UIC Building
Singapore 068808

Australia
Nelson Thomson Learning
102 Dodds Street
South Melbourne, Victoria 3205
Australia

Canada
Nelson Thomson Learning
1120 Birchmount Road
Toronto, Ontario M1K 5G4
Canada

Europe/Middle East/Africa
Thomson Learning
High Holborn House
50/51 Bedford Row
London WC1R 4LR
United Kingdom

Latin America
Thomson Learning
Seneca, 53
Colonia Polanco
11560 Mexico D.F.
Mexico

Spain
Paraninfo Thomson Learning
Calle/Magallanes, 25
28015 Madrid, Spain

TO MIMSE

in fond recollection of countless such things
as eating ice cream together
in the middle of a lake during the rain

PREFACE

Advantage has been taken in this sixth edition to include a large number of improvements, ranging from historical amplifications and updatings to the introduction of some new sections and the expansion of some old ones. Much new illustrative material has been added and women in mathematics have been given a more deserving attention.

There is scarcely a section of the 15 chapters of the book that has not undergone some amplification and/or updating—these improvements are far too numerous to list here. Among the more major changes are a considerable expansion of the discussion of the contents of Euclid's *Elements* in Chapter 5, the entire treatment of Chinese mathematics in Chapter 7, the treatment of logarithms in Chapter 9, an entirely new section on Maria Agnesi and the Marquise du Châtelet in Chapter 12, a consideration of the contributions of Argand and Wessel to the geometric representation of complex numbers in Chapter 13, a new section in Chapter 13 devoted to Sophie Germain and Mary Somerville, another new section in Chapter 13 devoted to Bolzano, a considerable expansion in Chapter 13 of the material on the liberation of geometry in the early nineteenth century, a complete rewriting and expansion of the section on differential geometry in Chapter 14, the addition of material on Grace Chisholm and Charlotte Scott in Chapter 14, and a new concluding section of the book devoted to a prognostication of the future of mathematics.

A very significant addition to the book are the Cultural Connections written by Jamie Eves. These have been supplied at the request of those earlier users of the book who have felt that a more in-depth cultural setting of the various eras and times of the history of mathematics would be beneficial to the student. A wise student will peruse each Cultural Connection before embarking upon the historical material of the associated chapter.

Ten new pieces of pictorial material have been added to the book and 16 new portraits of mathematicians have been added (bringing the total number of such portraits to 76). Finally, the Bibliography has been significantly updated.

One desiring a more detailed description of many of the features of the book may consult the Introduction that immediately precedes Chapter 1.

As with the previous editions, it is a pleasure once again to express my appreciation of the very warm reception given to the book by both school teachers and college professors. I especially want to thank all who took the time and trouble to write me encouraging words and to send me suggestions for further betterment of the book. It is largely from a carefully filed collection of these suggestions that each new edition has been fashioned.

There are many others who have been particularly helpful. Among these are Duane E. Deal of Ball State University, Florence D. Fasanelli of Sidwell

Friends School, David E. Kullman of Miami University, and Gregorio Fuentes of the University of Maine, each of whom made valuable suggestions that have led to improvement of the text. Of these reviewers, I want to extend special thanks to Professor Deal, who unstintingly gave so much of his time toward supplying me with excellent and scholarly material enhancing many parts of the book. Ouyang Jiang and Zhang Liangjin of P.R. China furnished helpful advice and valuable material concerning the mathematics of ancient China. The Bookstore and Library of the University of Maine at Machias and the Article Retrieval Service of the University of Maine at Orono were very helpful.

It gives me special pleasure to thank my son Jamie H. Eves for embellishing the book with his Cultural Connections. It has been a great advantage to benefit from his wide, deep, and enthusiastic scholarship in the field of history.

And, finally, thanks go to the efficient folks of Saunders College Publishing for their splendid help and cooperation.

Fox Hollow, Lubec, Maine
Summer, 1989

H. E.

CONTENTS

AN INTRODUCTION TO THE HISTORY OF MATHEMATICS

INTRODUCTION

This book differs from many existing histories of mathematics in that it is not primarily a work for the reference shelf, but an attempt to *introduce* the history of mathematics to undergraduate college mathematics students. Therefore, in addition to the historical narrative, there are also pedagogical devices designed to assist, interest, and involve the student. Let us describe some of these devices and comment on other characteristics of this work.

1. In the belief that a college course in the history of mathematics should be primarily a *mathematics* course, an effort has been made to inject a considerable amount of genuine mathematics into this book. It is hoped that a student using this book will learn much mathematics, as well as history.

2. Perhaps chief among the pedagogical devices of the book are the Problem Studies listed at the conclusion of each chapter. Each Problem Study contains a number of related problems and questions concerning some part of the material of the associated chapter. It is felt that by discussing a number of these Problem Studies in class and assigning others to be worked out at home, the course will become more concrete and meaningful for the student, and the student's grasp of a number of historically important concepts will crystallize. For example, the student can gain a better appreciation and understanding of numeral systems by actually working with the systems. Again rather than just reading that the ancient Greeks solved quadratic equations geometrically, the student can solve some by the Greek method and, in doing so, attain a deeper appreciation of Greek mathematical achievement. Some of the Problem Studies concern themselves with historically important problems and procedures, others furnish valuable material for the future teacher of either high school or college mathematics, still others are purely recreational, and many are designed to lead to short "junior" research papers by the students. A large number of instructors in both high schools and colleges have used material from these Problem Studies to enliven and augment various courses that they teach. The Problem Studies have been extensively employed by college mathematics clubs, and many school students have used them in high-school mathematics fairs.

3. There are many more Problem Studies than can be covered in either one or two semesters, and they are of varying degrees of difficulty. This permits the instructor to select problems that fit his or her students' abilities and to vary assignments from year to year.

4. At the end of the book is a collection of hints and suggestions for the solution of many of the Problem Studies. It is hoped that these hints and

suggestions are not so broad as to "spoil" the problems. A good problem should be more than a mere exercise; it should be challenging and not too easily solved by the student, and it should require some "dreaming" time.

5. It is of interest that, on the grounds that problems constitute the heart of mathematics,[1] problem courses have been given in some colleges based solely upon the Problem Studies of this book.

6. Many instructors of the history of mathematics like to assign essay papers; therefore, at the end of each chapter, immediately following the Problem Studies, some Essay Topics are listed that relate to material covered in that chapter. These topics are merely suggested; an instructor can easily devise an extended list of his or her own. An assigned Essay Topic should require the student to read more than the textbook; the student should find it necessary to delve into some of the literature listed in the chapter's Bibliography. A number of these Essay Topics have led to excellent term papers, many to masters theses, and several to student papers that received publication in mathematics and pedagogical journals.

7. It is axiomatic that the history of a subject cannot be properly appreciated without at least a fair acquaintance with the subject itself.[2] Accordingly, an attempt has been made to explain the material under consideration, especially in the later chapters, where the subject matter is more advanced. This is one of the ways a beginning student can learn a considerable amount of mathematics, as well as history, from a study of this book.

8. One will notice that terms that are defined in the text are made prominent by appearing in boldface type.

9. The historical material is presented in essentially chronological order, with occasional departures motivated either by pedagogical and logical considerations or by the desires of some readers and instructors. A couple of places where a more direct chronological development may be desired are clearly marked, and instructions are given for carrying out the rearrangement.

10. The reader will find that a knowledge of simple arithmetic, high school algebra, geometry, and trigonometry is generally sufficient for a proper understanding of the first nine chapters. A knowledge of the rudiments of plane analytic geometry is needed for Chapter 10, and a knowledge of the basic concepts of the calculus is required for the remaining chapters (11 through 15). Any concepts or developments of a more advanced nature appearing in the book are, it is hoped, sufficiently explained at the points where they are introduced. A certain amount of mathematical maturity is desirable, and whether

---

[1] See P. R. Halmos, "The heart of mathematics," *The American Mathematical Monthly* 87 (1980): 519–524.

[2] It is interesting and pertinent that, conversely, a true appreciation of a branch of mathematics is impossible without some acquaintance with the history of that branch, for mathematics is largely a study of ideas, and a genuine understanding of ideas is not possible without an analysis of origins. A particularly obvious example of this observation is the study of non-Euclidean geometry. It was J. W. L. Glaisher who aptly said, "I am sure that no subject loses more than mathematics by any attempt to dissociate it from its history."

nine, ten, eleven, or all fifteen chapters are to be covered depends upon class time and the students' previous preparation. Here the Problem Studies form an elastic element, for one can include or omit as many problems as convenience and time dictate.

11. Frankly, it is not easy to cover the history of mathematics from antiquity up through modern times in a one-semester course that meets three hours a week; to do so requires too much reading on the part of the student and almost complete neglect of the problem material. An ideal situation is to offer a one-year course in the subject, covering Part 1 (the first eight chapters), or Part 1 along with selections from Chapters 9, 10, and 11 in the first semester, and Part 2, or the remaining material, in the second semester. The advanced students and the mathematics majors would enroll for both semesters; the elementary students and the prospective teachers of high-school mathematics might enroll for only the first semester.

12. The history of mathematics is so vast that only an *introduction* to the subject is possible at the undergraduate college level, even in a two-semester course. Accordingly, a Bibliography has been appended to each chapter that deals with the material of that chapter. A General Bibliography, which immediately follows the final chapter, applies to all, or almost all, chapters. It must be realized that the General Bibliography, extensive as it is, makes no pretense to completeness and is intended merely to serve as a starting point for any search for further material. Many periodical references are furnished in the book at appropriate places in footnotes. An excellent source of periodical references appears near the end of the General Bibliography; important references of this sort are very numerous and will soon be encountered by an inquiring student. To accommodate the general undergraduate student, the references in the Bibliographies are generally accessible and in English.

13. A great pitfall in writing a book such as this one lies in including more material than can be covered and/or digested within the time limits of the course; a writer simply knows too much about his or her subject. The delicate balance between a too brief and a too voluminous treatment is not easy to maintain, and is perhaps achieved more through teaching experience than anything else. No one is more aware than the author of the many topics that, because of the purpose and clientele of the book, had to be slighted or omitted. If an instructor feels deeply that certain omitted material should be included in his course, by all means introduce it if it can be managed. A textbook is never meant to replace an instructor or to interfere with creative teaching; it is merely offered as an aid.

14. The Cultural Connections supplied by Jamie H. Eves may be included or omitted at the discretion of the instructor. They have been inserted for those who feel that such connections are important because mathematics did not develop in a vacuum. Some of the material presented in the Cultural Connections has been repeated in the text proper since some instructors may elect to omit the Cultural Connections.

Part 1

BEFORE THE SEVENTEENTH CENTURY

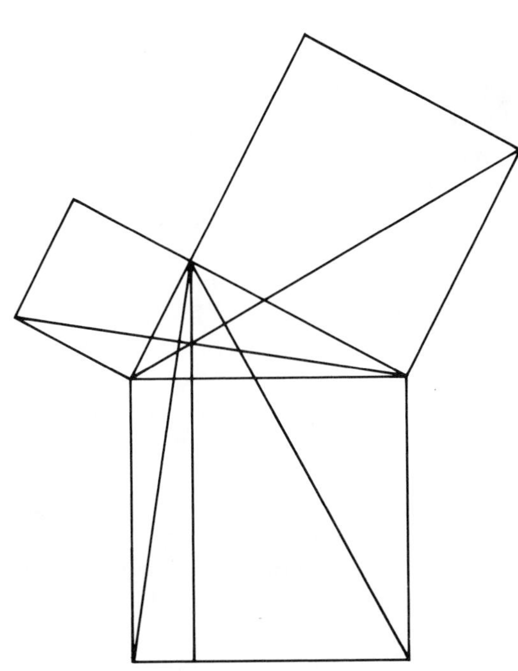

Cultural Connection

THE HUNTERS OF THE SAVANNA

The Stone Age—ca. 5,000,000–3000 B.C.
(to accompany Chapter One)

The first people were hunters of small game and gatherers of fruits, nuts, and roots. They lived, for the most part, in the open spaces of the savannas, seas of tall grass that covered most of the habitable portions of Africa, southern Europe, southern Asia, and Middle America. They were a migratory people, constantly moving from place to place in search of food and in response to climatic flux. Their culture was forged in the crucible of a harsh, hostile world where life was short and the search for food an unabating constant. Everything was geared toward the hunt: their tools of stone, wood, bone, and shell were designed for either hunting or food preparation; the fire they tamed was used for cooking as well as keeping warm; their art depicted scenes of hunts; their religion was a fearful attempt to understand and control the raw wilderness around them and only dimly grappled with the concept of ultimate destiny.

We cannot say with certainty when the Stone Age began. It was perhaps as early as 5,000,000 B.C., when *Australopithecus,* a four-foot-tall ancestor of humanity who lived in Africa, might have made crude stone choppers and cutting flakes by striking one pebble against another. Certainly by approximately 400,000 B.C., *Homo erectus* in China routinely constructed stone chopping tools, flakes, and scrapers. *Homo erectus* also moved out of the storms of the open savanna into caves near modern Peking, an innovation continued by their cousins, *Homo neanderthalensis,* who lived in Europe and the Middle East between approximately 110,000 B.C. and 35,000 B.C. *Homo neanderthalensis* heated their caves with fire and cooked the game they had captured on the savanna. They preserved records of their hunts in detailed, elegant wall paintings. By 30,000 B.C., *Homo sapiens* (modern people) replaced cave dwellings with mobile structures, wooden lean-tos and huts of animal skins, that they could take with them on the hunt. At about the same time, they also began to carve stone fertility figurines and other religious icons.

We cannot precisely fix the end of the Stone Age. Stone Age cultures persisted in some parts of the world into the nineteenth and twentieth centuries. Most of southern Africa, Australia, and the Americas were peopled with Stone Age hunter/gatherers when encountered by European explorers in the sixteenth and seventeenth centuries. In the mid-twentieth century, lumberjacks chanced upon the hitherto "undiscovered" Tasadays, a Stone Age forest tribe living deep in the interior of one of the Philippine islands. Historical convention, however, ends the Stone Age in approximately 3000 B.C., when metal-smelting city cultures emerged in the Middle East, India, and China.

Like all historical epochs, the Stone Age was not static. Society and culture changed over time to adapt to a changing world. Historians diagram this change by dividing the Stone Age into three periods. During the Paleolithic period, or Old Stone Age (ca. 5,000,000–10,000 B.C.), *Homo sapiens* evolved from smaller, slighter creatures and developed the basic socioeconomic structures of the Stone Age. In the Mesolithic period, or Middle Stone Age (ca. 10,000–7000 B.C.), the hunter/gatherer economy of the Stone Age crystallized. In the Neolithic period, or New Stone Age (ca. 7000–3000 B.C.), the Stone Age began to fade into the Bronze and Iron Ages, as people began to turn away from a hunter/gatherer society to one involving early forms of agriculture and the domestication of animals. The Paleolithic era was a transition from a prehuman world to a society of human hunters. The Neolithic period was also a transition, from a society of hunters to one of farmers.

Because it was a time when almost all people were migratory hunters, the Stone Age was a period of limited scientific and intellectual advancement. This is not because the people of the Stone Age lacked intelligence. By 20,000 B.C., the hunters of the savanna had developed a complex culture that included tool making, language, religion, art, music, and commerce. Progress in mathematics and science, however, was hampered by the social and economic structures of those early times. Because Stone Age people were hunters rather than farmers, they had to move with the seasons, following the migrations of animals and the seasonal availability of naturally growing fruits and nuts. They were able to carry along with them only small, easily transported tools, clothing, and personal items. There was no room in a hunter/gatherer society for the bulky equipment needed to forge metals, or for voluminous libraries; hence, Stone Age people did not develop metal tools or a written language. There were no cities, as the savannas could provide enough food to support only about forty persons per hundred square miles. In the busy, often short life of a hunter, there was no leisure time to ponder questions of philosophy or science. To be sure, some very basic scientific progress took place during the Stone Age. Stone Age people traded with one another, and they needed to keep track of each family's share of the hunt; both activities required the necessity of counting, a prelude to scientific thinking. Some Stone Age people, like the Sioux Indians, had pictographic calendars that recorded several decades of history. Anything beyond the most primitive counting systems, however, had to wait until the development of full-scale, intensive agriculture, which required more sophisticated arithmetic.

In the final millennia of the Stone Age, during the Neolithic era, humanity moved from simply gathering naturally growing wild fruits, nuts, roots, and vegetables to actually planting seeds and cultivating crops. Neolithic men and women were still primarily hunters and gatherers, however, and their small, tangled fields would have resembled unweeded vegetable gardens more than farms. These late-Stone Age gardens probably looked very much like the cornfields planted by American Indians and described by European explorers in the sixteenth century, with several different crops planted haphazardly in the same field.

To recapitulate, the Stone Age lasted several million years, from perhaps as early as 5,000,000 B.C. until about 3000 B.C. In a world of vast grasslands and savannas where wild game was abundant, people were chiefly hunters and gatherers. Their lives were harsh and difficult, so early people were too busy staying alive to develop scientific traditions. After 3000 B.C., densely populated farming communities emerged along the Nile River in Africa, along the Tigris and Euphrates Rivers in the Middle East, and along the Yellow River in China. These communities developed cultures in which science and mathematics could begin to develop.

Chapter 1

NUMERAL SYSTEMS

1–1 Primitive Counting

In giving a chronological account of the development of mathematics, one is beset with the problem of where to begin. Should one start with the first methodical deductions in geometry traditionally credited to Thales of Miletus around 600 B.C.? Or should one go back further and start with the empirical derivation of certain mensuration formulas made by the pre-Greek civilizations of Mesopotamia and Egypt? Or should one go back even further and start with the first groping efforts made by prehistoric man to systematize size, shape, and number? Or can one say mathematics originated in prehuman times in the meager number sense and pattern recognition of certain animals, birds, and insects? Or even before this, in the number and spatial relations of plants? Or still earlier, in the spiral nebulae, the courses of planets and comets, and the crystallization of minerals in preorganic times? Or was mathematics, as Plato believed, *always* in existence, merely awaiting discovery? Each of these possible origins can be defended.[1]

Since it is man's primal efforts to systematize the concepts of size, shape, and number that are popularly regarded as the earliest mathematics, we shall commence there, and begin with the emergence in primitive man of the concept of *number* and the process of *counting*.

The number concept and the counting process developed so long before the time of recorded history (there is archeological evidence that counting was employed by man as far back as 50,000 years ago) that the manner of this development is largely conjectural. It is not difficult, though, to imagine how it probably came about. It seems fair to argue that humans, even in most primitive times, had some number sense, at least to the extent of recognizing *more* and *less* when some objects were added to or taken from a small group, for studies have shown that some animals possess such a sense. With the gradual evolution of society, simple counting became imperative. A tribe had to know how many members it had and how many enemies, and a man found it necessary to know if his flock of sheep was decreasing in size. Probably the earliest way of keeping a count was by some simple tally method, employing the principle of one-to-one correspondence. In keeping a count on sheep, for exam-

[1] For a start, see D. E. Smith, *History of Mathematics,* vol. 1, chap. 1, and Howard Eves, *In Mathematical Circles* (Items 1°, 2°, 3°, 4°), which are cited in the General Bibliography at the end of the book.

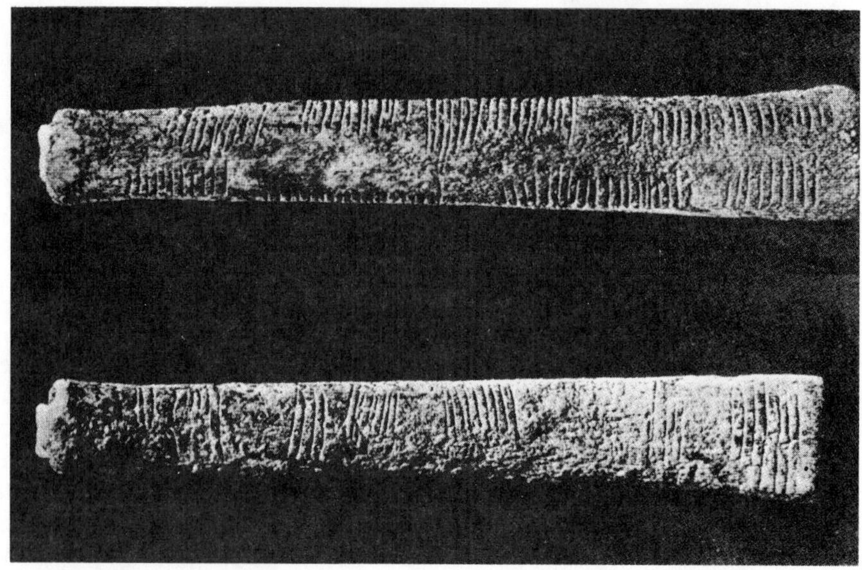

Two views of the Ishango bone, over 8000 years old and found at Ishango, on the shore of Lake Edward in Zaire (Congo), showing numbers preserved by notches cut in the bone.
(Dr. de Heinzelin.)

A Peruvian Indian census quipu, showing numbers recorded by knots in cord. Larger knots are multiples of smaller ones, and cord color may distinguish male from female.
(Collection Musée de L'Homme, Paris.)

ple, one finger per sheep could be turned under. Counts could also be maintained by making collections of pebbles or sticks, by making scratches in the dirt or on a stone, by cutting notches in a piece of wood, or by tying knots in a string. Then, perhaps later, an assortment of vocal sounds was developed as a word tally against the number of objects in a small group. And still later, with the refinement of writing, an assortment of symbols was devised to stand for these numbers. Such an imagined development is supported by reports of anthropologists in their studies of present-day primitive peoples.

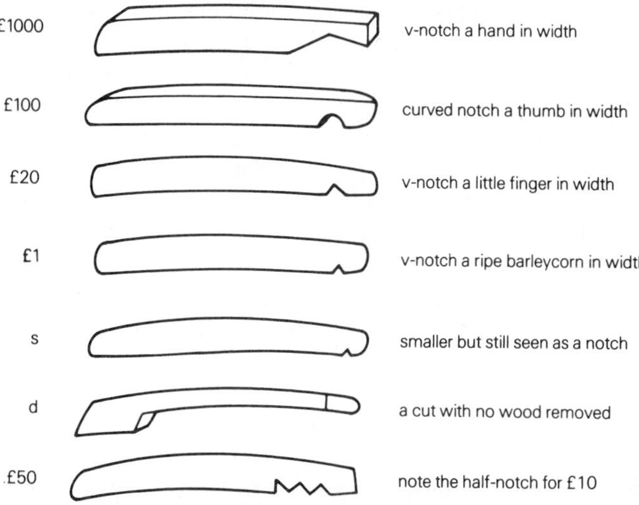

Drawing showing the official system of notching used on twelfth-century exchequer tallies of the British Royal Treasury. Such tallies continued in use until 1826.

In the earlier stages of the period of vocal counting, different sounds (words) were used, for example, for *two* sheep and *two* men. (Consider, for example, in English: *team* of horses, *span* of mules, *yoke* of oxen, *brace* of partridge, *pair* of shoes, *couple* of days.) The abstraction of the common property of *two,* represented by some sound considered independently of any concrete association, probably was a long time in arriving. Our present number words in all likelihood originally referred to sets of certain concrete objects, but these associations, except for that perhaps relating five and hand, are now lost to us.[2]

[2] For an interesting alternative to the classical evolutionary view of nonliterate peoples, see Marcia and Robert Ascher, "Euthomathematics," *History of Science* 24, no. 2 (June 1980): 125–144.

1–2 Number Bases

When it became necessary to make more extensive counts, the counting process had to be systematized. This was done by arranging the numbers into convenient basic groups, the size of the groups being largely determined by the matching process employed. Essentially, the method was like this. Some number b was selected as a base (also called **radix** or **scale**) for counting, and names were assigned to the numbers 1, 2, . . . , b. Names for numbers larger than b were then given by combinations of the number names already selected.

Since fingers furnished such a convenient matching device, it is not surprising that 10 was ultimately chosen far more often than not for the number base b. Consider, for example, our present number words, which are formed on 10 as a base. We have the special names *one, two, . . . , ten* for the numbers 1, 2, . . . , 10. When we come to 11, we say *eleven,* which, the philologists tell us, derives from *ein lifon,* meaning "one left over," or one over ten. Similarly, *twelve* is from *twe lif* ("two over ten"). Then we have *thirteen* ("three and ten"), *fourteen* ("four and ten"), up through *nineteen* ("nine and ten"). Then comes *twenty* (*twe-tig*, or "two tens"), *twenty-one* ("two tens and one"), and so on. The word *hundred,* we are told, comes originally from a term meaning "ten times" (ten).

There is evidence that 2, 3, and 4 have served as primitive number bases. For example, there are natives of Queensland who count "one, two, two and one, two twos, much," and some African pygmies count "*a, oa, ua, oa-oa, oa-oa-a,* and *oa-oa-oa*" for 1, 2, 3, 4, 5, and 6. A certain tribe of Tierra del Fuego has its first few number names based on 3, and some South American tribes similarly use 4.

As might be expected, the **quinary scale,** or number system based on 5, was the first scale to be used extensively. To this day, some South American tribes count by hands: "one, two, three, four, hand, hand and one," and so on. The Yukaghirs of Siberia use a mixed scale by counting "one, two, three, three and one, five, two threes, one more, two three-and-ones, ten with one missing, ten." German peasant calendars used a quinary scale as late as 1800.

There is evidence that the **duodecimal scale,** or number system based on 12, may have been used in some societies during prehistoric times, chiefly in relation to measurements. Such a base may have been suggested by the approximate number of lunations in a year, or perhaps because 12 has so many integral fractional parts. At any rate, we have 12 as the number of inches in a foot, ounces in the ancient pound, pence in a shilling, lines in an inch, hours about the clock, months in a year, and the words *dozen* and *gross* used as higher units.

The **vigesimal scale,** or number system based on 20, has been widely used, and recalls man's barefoot days. This scale was used by American Indian peoples, and is best known in the well-developed Mayan number system. Celtic traces of a base 20 are found in the French *quatre-vingt* instead of *huitante,* and *quatre-vingt-dix* instead of *nonante*. Traces are also found in Gaelic, Danish, and Welsh. The Greenlanders use "one man" for 20, "two men" for 40, and so on. In English we have the frequently used word *score*.

The **sexagesimal scale,** or number system based on 60, was used by the ancient Babylonians, and is still used when measuring time and angles in minutes and seconds.

1–3 Finger Numbers and Written Numbers

In addition to spoken numbers, **finger numbers** were at one time widely used. Indeed, the expression of numbers by various positions of the fingers and hands probably predates the use of either number symbols or number names. Thus, the early written symbols for 1, 2, 3, and 4 were invariably the suitable number of vertical or horizontal strokes, representing the corresponding number of raised or extended fingers, and the word *digit* (that is, "finger") for the numbers 1 through 9 can be traced to the same source.

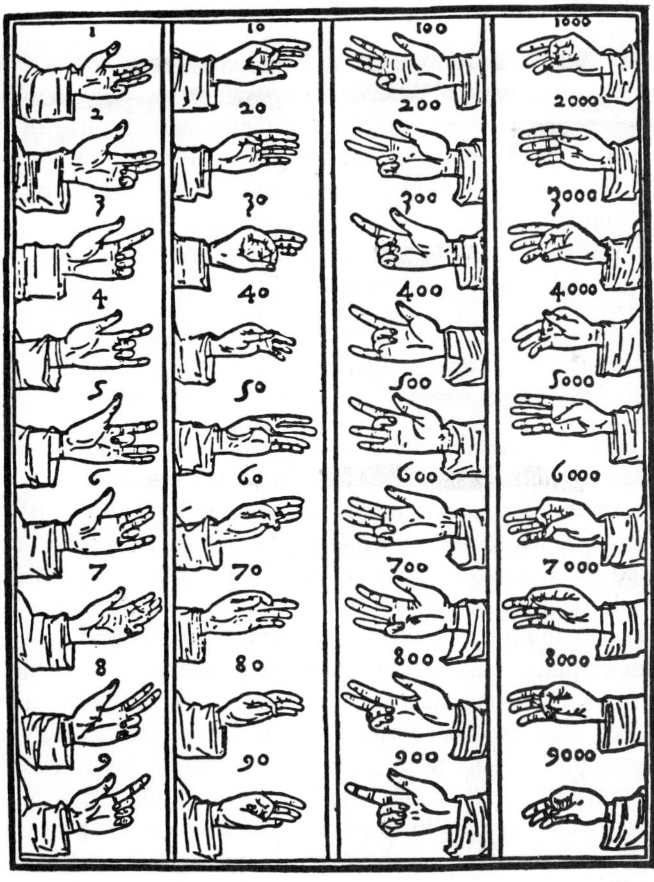

Finger numbers from Pacioli's *Sūma* of 1494. The first two columns represent the left hand, the other two the right hand.

In time, finger numbers were extended to include the largest numbers occurring in commercial transactions; by the Middle Ages, they had become international. In the ultimate development, the numbers 1, 2, . . . , 9 and 10, 20, . . . , 90 were represented on the left hand, and the numbers 100, 200, . . . , 900 and 1000, 2000, . . . , 9000 on the right hand. In this way, any number up to 10,000 was representable by the use of the two hands. Pictures of the finger numbers were given in Renaissance arithmetic books. For example, using the left hand, 1 was represented by partially folding down the little finger; 2 by partially folding down the little and ring fingers; 3 by partially folding down the little, ring, and middle fingers; 4 by folding down the middle and ring fingers; 5 by folding down the middle finger; 6 by folding down the ring finger; 7 by completely folding down the little finger; 8 by completely folding down the little and ring fingers; and 9 by completely folding down the little, ring, and middle fingers.

Although finger numbers originated in very early times, they are still used today by some primitive races of Africa, by Arabs, and by Persians. In North and South America, some native Indian and Eskimo tribes still employ the fingers.

Finger numbers had the advantage of transcending language differences but, like the vocal numbers, lacked permanence and were not suitable for performing calculations. We have already mentioned the use of marks and notches as early ways of recording numbers. In such devices, we probably have the first attempt at writing. At any rate, various written number systems gradually evolved from these primitive efforts to make permanent number records. A written number is called a **numeral,** and we now turn our attention to a simple classification of early numeral systems.

1–4 Simple Grouping Systems

Perhaps the earliest type of numeral system that was developed is that which has been called a **simple grouping system.** In such a system, some number b is selected for number base, and symbols are adopted for 1, b, b^2, b^3, and so on. Then any number is expressed by using these symbols *additively,* each symbol being repeated the required number of times. The following illustrations will clarify the underlying principle.

A very early example of a simple grouping system is that furnished by the Egyptian hieroglyphics, employed as far back as 3400 B.C. and chiefly used by the Egyptians when making inscriptions on stone. Although the hieroglyphics were sometimes used on other writing media than stone, the Egyptians early developed two considerably more rapid writing forms for work on papyrus, wood, and pottery. The earlier of these forms was a running script, known as the **hieratic,** derived from the hieroglyphic and used by the priesthood. From the hieratic, there later evolved the **demotic** writing, which was adopted for general use. The hieratic and demotic numeral systems are not of the simple grouping type.

The Egyptian hieroglyphic numeral system is based on the scale of 10. The symbols adopted for 1 and the first few powers of 10 are

1 | a vertical staff, or stroke

10 ∩ a heel bone, or hobble, or yoke

10^2 a scroll, or coil of rope

10^3 a lotus flower

10^4 a pointing finger

10^5 a burbot fish, or tadpole

10^6 a man in astonishment, or a god holding up the universe

Any number is now expressed by using these symbols additively, each symbol being repeated the required number of times. Thus,

$$13015 = 1(10^4) + 3(10^3) + 1(10) + 5 =$$

We have written this number from left to right, although it was more customary for the Egyptians to write from right to left.

The early Babylonians, lacking papyrus and having little access to suitable stone, resorted principally to clay as a writing medium. The inscription was pressed into a wet clay tablet by a stylus, the writing end of which may have been a sharp isosceles triangle. By tilting the stylus slightly from the perpendicular, one could press either the vertex angle or a base angle of the isosceles triangle into the clay, producing two forms of wedge-shaped **(cuneiform)** characters. The finished tablet was then baked in an oven to a time-resisting hardness that resulted in a permanent record. On cuneiform tablets dating from 2000 to 200 B.C., numbers less than 60 are expressed by a simple grouping system to base 10, and it is interesting that the writing is often simplified by using a subtractive symbol. The subtractive symbol and the symbols for 1 and 10 are

respectively, where the symbol for 1 and the two parts making up the subtractive symbol are obtained by using the vertex angle of the isosceles triangle, and the symbol for 10 is obtained by using one of the base angles. As examples of written numbers employing these symbols, we have

25 = 2(10) + 5 =

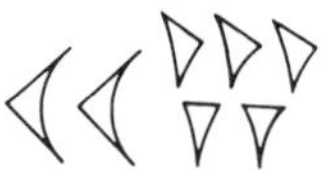

and

38 = 40 − 2 =

The method employed by the Babylonians for writing larger numbers will be considered in Section 1-7.

The Attic, or Herodianic, Greek numerals were developed some time prior to the third century B.C. and constitute a simple grouping system to base 10 formed from initial letters of number names. In addition to the symbols I, Δ, H, X, M for 1, 10, 10^2, 10^3, 10^4, there is a special symbol for 5. This special symbol is an old form of Π, the initial of the Greek *pente* ("five"), and Δ, H, X, and M are the initial letters of the Greek *deka* (ten), *hekaton* (hundred), *kilo* (thousand), and *myriad* (ten thousand). The symbol for 5 was frequently used both alone and in combination with other symbols in order to shorten number representations. As an example, in this numeral system we have

2857 =

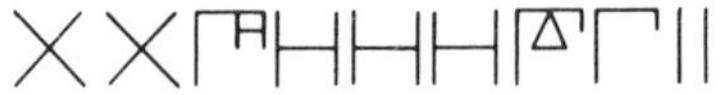

in which one can note the special symbol for 5 appearing once alone and twice in combination with other symbols.

As a final example of a simple grouping system, again to base 10, we have the familiar Roman numerals. Here the basic symbols I, X, C, M for 1, 10, 10^2, 10^3 are augmented by V, L, D for 5, 50, and 500. The subtractive principle, in which a symbol for a smaller unit placed before a symbol for a larger unit means the difference of the two units, was used only sparingly in ancient and medieval times. The fuller use of this principle was introduced in modern times. As an example, in this system we have

1944 = MDCCCCXXXXIIII,

or, in more modern times, when the subtractive principle became common,

1944 = MCMXLIV.

In using the subtractive principle, however, one is to abide by the following rule: I can precede only V or X, X can precede only L or C, C can precede only D or M.

There has been no lack of imagination in the attempts to account for the origins of the Roman number symbols. Among the more plausible explanations, acceptable to many authorities on Latin history and epigraphy, is that I, II, III, IIII were derived from the raised fingers of the hand. The symbol X may be a compound of two V's, or may have been suggested by crossed hands or thumbs, or may have originated from the common practice of crossing groups of ten when counting by strokes. There is some evidence that the original symbols for 50, 100, and 1000 may have been the Greek aspirates Ψ (*psi*), θ (*theta*), and Φ (*phi*). Older forms for *psi* were

$$\downarrow, \ \perp, \ \perp, \ \text{L},$$

all of which were used for 50 in early inscriptions. The symbol θ for 100 probably later developed into the somewhat similar symbol C, influenced by the fact that C is the initial letter of the Latin word *centum* ("hundred"). A commonly used early symbol for 1000 is $\subset\!|\!\supset$, which could be a variant of Φ. The symbol for 1000 became an M, influenced by the fact that M is the initial letter of the Latin word *mille* ("thousand"). Five hundred, being half of 1000, was represented by $|\!\supset$, which later became a D. The symbols $\subset\!|\!\supset$ and $|\!\supset$ for 1000 and 500 are found as late as 1715.

1–5 Multiplicative Grouping Systems

There are instances in which a simple grouping system developed into what may be called a **multiplicative grouping system.** In such a system, after a base b has been selected, symbols are adopted for $1, 2, \ldots, b - 1$, and a second set of symbols for $b, b^2, b^3, \ldots$. The symbols of the two sets are employed *multiplicatively* to show how many units of the higher groups are needed. Thus, if we should designate the first nine numbers by the usual symbols, but designate 10, 100, and 1000 by a, b, c, say, then in a multiplicative grouping system we would write

$$5625 = 5c6b2a5.$$

The traditional Chinese-Japanese numeral system is a multiplicative grouping system to base 10. Writing vertically, the symbols of the two basic groups and of the number 5625 are as shown on p. 18.

Lacking a paperlike writing material, the early Chinese and Japanese recorded their findings on bamboo slips. The piece of a bamboo stalk between two knots was split lengthwise into thin strips. After these strips were dried and scraped, they were laid side by side and tied together by four crosswise cords. The narrowness of the strips necessitated that the characters written on them be arranged vertically from top to bottom, giving rise to a custom of writing that persisted into more modern times, when bamboo slips were replaced by silk and paper as more convenient writing materials.

Example: 5625

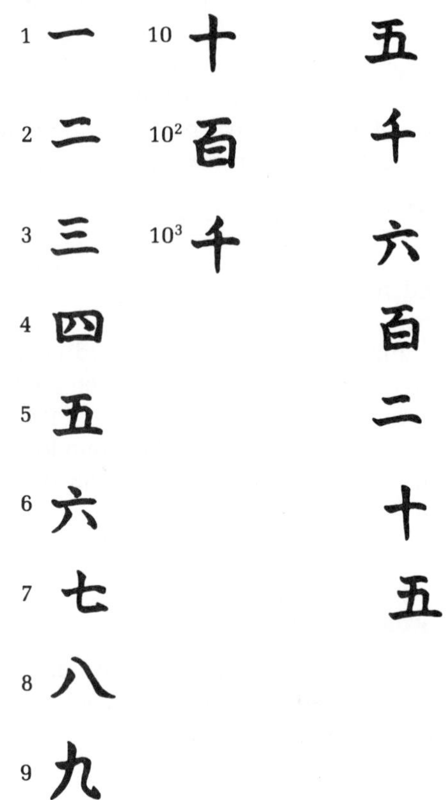

1–6 Ciphered Numeral Systems

In a **ciphered numeral system,** after a base b has been selected, sets of symbols are adopted for $1, 2, \ldots, b - 1; b, 2b, \ldots, (b - 1)b; b^2, 2b^2, \ldots, (b - 1)b^2$; and so on. Although many symbols must be memorized in such a system, the representation of numbers is compact.

The so-called Ionic, or alphabetic, Greek numeral system is of the ciphered type and can be traced as far back as about 450 B.C. It is a system that is based on 10 and employs twenty-seven characters—the twenty-four letters of the Greek alphabet together with the symbols for the obsolete *digamma, koppa,* and *sampi.* Although the capital letters were used (the small letters were substituted much later), we shall now illustrate the system with the small letters. The following equivalents had to be memorized:

1	α	alpha	10	ι	iota	100	ρ	rho
2	β	beta	20	κ	kappa	200	σ	sigma
3	γ	gamma	30	λ	lambda	300	τ	tau
4	δ	delta	40	μ	mu	400	υ	upsilon
5	ε	epsilon	50	ν	nu	500	ϕ	phi
6	obsolete	digamma	60	ξ	xi	600	χ	chi
7	ζ	zeta	70	o	omicron	700	ψ	psi
8	η	eta	80	π	pi	800	ω	omega
9	θ	theta	90	obsolete	koppa	900	obsolete	sampi

As examples of the use of these symbols, we have

$$12 = \iota\beta, \qquad 21 = \kappa\alpha, \qquad 247 = \sigma\mu\zeta.$$

Accompanying bars or accents were used for larger numbers (see Problem Study 1.3 (b)).

Symbols for the obsolete digamma, koppa, and sampi are

ϛ, ϙ, ϡ.

Other ciphered numeral systems are the Egyptian hieratic and demotic, Coptic, Hindu Brahmi, Hebrew, Syrian, and early Arabic. The last three, like the Ionic Greek, are *alphabetic* ciphered numeral systems.

1–7 Positional Numeral Systems

Our own numeral system is an example of a **positional numeral system** with base 10. For such a system, after the base b has been selected, basic symbols are adopted for $0, 1, 2, \ldots, b-1$. Thus, there are b basic symbols, frequently called the **digits** of the system. Now any (whole) number N can be written uniquely in the form

$$N = a_n b^n + a_{n-1} b^{n-1} + \cdots + a_2 b^2 + a_1 b + a_0,$$

where $0 \leqq a_i < b$, $i = 0, 1, \ldots, n$. We then represent the number N to base b by the sequence of basic symbols

$$a_n a_{n-1} \ldots a_2 a_1 a_0 .$$

Thus, a basic symbol in any given numeral represents a multiple of some power of the base, the power depending on the position in which the basic symbol occurs. In our own *Hindu-Arabic* numeral system, for example, the 2 in 206 stands for $2(10^2)$, or 200, whereas in 27, the 2 stands for 2(10), or 20. Note that for complete clarity some symbol for zero is needed to indicate any possible missing powers of the base. A positional numeral system is a logical, although not necessarily historical, outgrowth of a multiplicative grouping system.

Sometime between 3000 and 2000 B.C., the ancient Babylonians evolved a sexagesimal system employing the principle of position. The numeral system, however, is really a mixed one in that, although numbers exceeding 60 are written according to the positional principle, numbers within the basic 60 group are written by a simple grouping system to base 10, as explained in Section 1–4. As an illustration we have

$$524{,}551 = 2(60^3) + 25(60^2) + 42(60) + 31 =$$

Until after 300 B.C., this positional numeral system suffered from the lack of a zero symbol to stand for any missing powers of 60, thus leading to possible misinterpretations of given number expressions. The symbol that was finally introduced consisted of two small, slanted wedges, but this symbol was used only to indicate a missing power of the base 60 *within* a number, and not for any missing power of the base 60 occurring at the *end* of a number. Thus, the symbol was only a partial *zero,* for a true zero serves for missing powers of the base both within and at the end of numbers, as in our 304 and 340. In the Babylonian numeral system, then, 10,804 would appear as

$$10{,}804 = 3(60^2) + 0(60) + 4 =$$

and 11,040 as

$$11{,}040 = 3(60^2) + 4(60) =$$

rather than as

.

The Mayan numeral system is very interesting. Of remote but unknown date of origin, it was uncovered by the early sixteenth-century Spanish expeditions into Yucatán. This system is essentially a vigesimal one, except that the second number group is $(18)(20) = 360$ instead of $20^2 = 400$. The higher groups are of the form $(18)(20^n)$. The explanation of this discrepancy probably lies in the fact that the official Mayan year consisted of 360 days. The symbol for zero given in the table below, or some variant of this symbol, is consistently used. The numbers within the basic 20 group are written very simply by dots and dashes (pebbles and sticks) according to the following simple grouping scheme, the dot representing 1 and the dash 5.

1		6		11		16	
2		7		12		17	

3	8	13	18
4	9	14	19
5	10	15	0

An example of a larger number, written in the vertical Mayan manner, is shown below.

$$43{,}487 = 6(18)(20^2) + 0(18)(20) + 14(20) + 7 =$$

The mixed-base system we have described was used by the priest class. There are reports of a pure vigesimal system that was used by the common people but which has not survived in written form.

1–8 Early Computing

Many of the computing patterns used today in elementary arithmetic, such as those for performing long multiplications and divisions, were developed as late as the fifteenth century. Two reasons are usually advanced to account for this tardy development; namely, the mental difficulties and the physical difficulties encountered in such work.

The first reason, mental difficulties, must be somewhat discounted. The impression that the ancient numeral systems are not amenable to even the simplest calculations is largely based on lack of familiarity with these systems. It is clear that addition and subtraction in a simple grouping system require only the ability to count the number symbols of each kind and then to convert to higher units. No memorization of number combinations is needed.[3] In a ciphered numeral system, if sufficient addition and multiplication tables have been memorized, the work can proceed much as we do it today. The French mathematician Paul Tannery attained considerable skill in multiplication with the Greek Ionic numeral system and even concluded that that system has some advantages over our present one.

The physical difficulties encountered, however, were quite real. Without a plentiful and convenient supply of some suitable writing medium, any very extended development of arithmetic processes was bound to be hampered. It must be remembered that our common machine-made pulp paper is little more than a hundred years old. The older rag paper was made by hand; conse-

[3] For the performance of long multiplications and divisions with Roman numerals, see, for example, James G. Kennedy, "Arithmetic with Roman numerals," *The American Mathematical Monthly* 88 (1981): 29–33.

quently, it was expensive and scarce. It was not introduced into Europe until the twelfth century, although it is likely that the Chinese knew how to make it a thousand years before.

An early paperlike writing material, called **papyrus,** was invented by the ancient Egyptians, and by 650 B.C. had been introduced into Greece. It was made from a water reed called *papu,* which is found in abundance in the Nile delta. The stems of the reed were cut into long, thin strips and laid side by side to form a sheet. Another layer of strips was laid crosswise on top and the whole soaked with water, after which the sheet was pressed out and dried in the sun. Probably because of a natural gum in the plant, the layers stuck together. After the sheets were dry, they were readied for writing by laboriously smoothing them with a hard, round object. Papyrus was too valuable to be used in any quantity as mere scratch paper.

Another early writing medium was parchment, which was made from the skins of animals, usually sheep and lambs. Naturally, this was scarce and hard to get. Even more valuable was vellum, a parchment made from the skin of calves. In fact, parchment was so costly that the custom arose in the Middle Ages of washing the ink off old parchment manuscripts and using them over again. Such manuscripts are called **palimpsests** (*palin,* "again"; *psao,* "rub smooth"). In some instances, after the passage of years, the original writing of a palimpsest reappeared faintly beneath the later treatment. Some interesting restorations have been made in this manner.

Small boards bearing a thin coat of wax, along with a stylus, formed a writing medium for the Romans of about 2000 years ago. Before and during the Roman Empire, sand trays were frequently used for simple counting and for the drawing of geometrical figures. Of course, stone and clay were used very early for making written records.

The way around these mental and physical difficulties was the invention of the **abacus** (Greek *abax,* "sand tray"), which can be called the earliest mechanical computing device used by man. It appeared in many forms in parts of the ancient and medieval world. Let us describe a rudimentary form of abacus and illustrate its use in the addition and subtraction of some Roman numbers. Draw four vertical parallel lines and label them from left to right by M, C, X, and I, and obtain a collection of convenient counters, like checkers, pennies, or pebbles. A counter will represent 1, 10, 100, or 1000 units according to its position on the I, X, C, or M line. To reduce the number of counters that may subsequently appear on a line, we agree to replace any five counters on a line by one counter in the space just to the left of that line. Any number less than 10,000 may then be represented on our frame of lines by placing not more than four counters on any line, and not more than one counter in the space just to the left of that line.

Let us now add

MDCCLXIX and MXXXVII.

Represent the first of the two numbers by counters on the frame, as illustrated at the left in Figure 1. We now proceed to add the second number, working

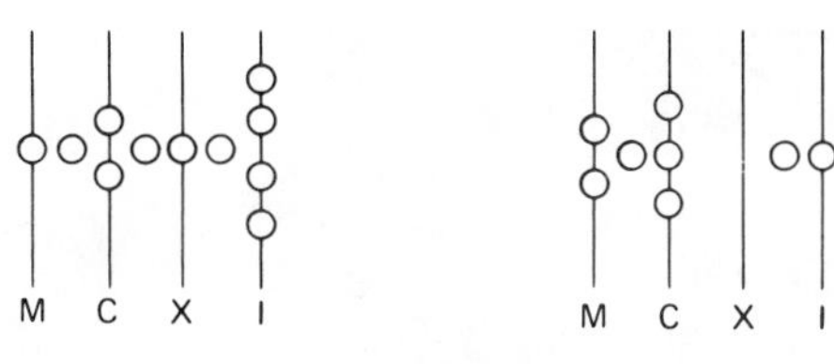

FIGURE 1

from right to left. To add the VII, put another counter between the X and I lines and two more counters on the I line. The I line now has six counters on it. We remove five of them and instead put another counter between the X and I lines. Of the three counters now between the X and I lines, we "carry over" two of them as a single counter on the X line. We now add the XXX by putting three more counters on the X line. Since we now have a total of five counters on the X line, they are replaced by a single counter between the C and X lines, and the two counters now found there are "carried over" as a single counter on the C line. We finally add the M by putting another counter on the M line. The final appearance of our frame is illustrated at the right in Figure 1, and the sum can be read off as MMDCCCVI. We have obtained the sum of the two numbers by simple mechanical operations and without requiring any scratch paper or recourse to memorization of any addition tables.

Subtraction is similarly carried out, except that now, instead of "carrying over" *to* the left, we may find it necessary to "borrow" *from* the left.

The Hindu-Arabic positional numeral system represents a number very simply by recording in order the number of counters belonging to the various lines of the abacus. The symbol 0 stands for a line with no counters on it. Our present addition and subtraction patterns, along with the concepts of "carrying over" and "borrowing" may have originated in the processes for carrying out these operations on the abacus. With the Hindu-Arabic numeral system, we are working with symbols instead of the actual counters, so it becomes necessary either to commit the simple number combinations to memory or to have recourse to an elementary addition table.

1–9 The Hindu-Arabic Numeral System

The Hindu-Arabic numeral system is named after the Hindus, who may have invented it, and after the Arabs, who transmitted it to western Europe. The earliest preserved examples of our present number symbols are found on some stone columns erected in India about 250 B.C. by King Aśoka. Other early examples in India, if correctly interpreted, are found among records cut about 100 B.C. on the walls of a cave in a hill near Poona and in some inscriptions of about A.D. 200 carved in the caves at Nasik. These early specimens contain no zero and do not employ positional notation. Positional value and a zero must have been introduced in India sometime before A.D. 800, because the Persian mathematician al-Khowârizmî describes such a completed Hindu system in a book of A.D. 825.

The abacist versus the algorist.
(From Gregor Reisch, *Margarita Philosophica,* Strassbourg, 1504.)

How and when the new numeral symbols first entered Europe is not settled. In all likelihood, they were carried by traders and travelers of the Mediterranean coast. They are found in a tenth-century Spanish manuscript and may have been introduced into Spain by the Arabs, who invaded the peninsula in A.D. 711 and remained there until A.D. 1492. The completed system was more widely disseminated by a twelfth-century Latin translation of al-Khowârizmî's treatise and by subsequent European works on the subject.

The next 400 years saw the battle between the abacists and the algorists, as the advocates of the new system were called, and by A.D. 1500 our present rules in computing won supremacy. In another hundred years, the abacists were almost forgotten, and by the eighteenth century no trace of an abacus was found in western Europe. Its reappearance, as a curiosity, was due to the French geometer Poncelet, who brought back a specimen to France after his release as a Russian prisoner of war following the Napoleonic Russian campaign.

Considerable variation was found in the number symbols until these symbols became stabilized by the development of printing. Our word *zero* probably comes from the Latinized form *zephirum* of the Arabic *sifr,* which in turn is a translation of the Hindu *sunya,* meaning "void" or "empty." The Arabic *sifr* was introduced into Germany in the thirteenth century by Nemorarius, as *cifra,* from which we have obtained our present word *cipher.*

1–10 Arbitrary Bases

We recall that to represent a number in a positional numeral system with base b we need basic symbols for the integers zero up through $b - 1$. Even though the base $b = 10$ is such an important part of our culture, the choice of 10 is really quite arbitrary, and other bases have great practical and theoretical importance. If $b \leqq 10$, we may use our ordinary digit symbols; thus, for example, we may consider 3012 as a number expressed to base 4 with the basic symbols 0, 1, 2, 3. To make clear that the number is considered as expressed to base 4, we shall write it as $(3012)_4$. When no subscript is written, it will be understood in this treatment that the number is expressed to the ordinary base 10. If $b > 10$, we must augment our digit symbols by some new basic symbols, for we always need b basic symbols. If $b = 12$, therefore, we may take 0, 1, 2, 3, 4, 5, 6, 7, 8, 9, t, e for our basic symbols, where t and e are symbols for *ten* and *eleven*; for example, we might have $(3t1e)_{12}$.

It is easy to convert a number from a given base to the ordinary base 10. Thus, we have

$$(3012)_4 = 3(4^3) + 0(4^2) + 1(4) + 2 = 198$$

and

$$(3t1e)_{12} = 3(12^3) + 10(12^2) + 1(12) + 11 = 6647.$$

If we have a number expressed in the ordinary scale, we may express it to base b as follows. Letting N be the number, we have to determine the integers $a_n, a_{n-1}, \ldots, a_0$ in the expression

$$N = a_n b^n + a_{n-1} b^{n-1} + \cdots + a_2 b^2 + a_1 b + a_0,$$

where $0 \leqq a_i < b$. Dividing the above equation by b, we have

$$N/b = a_n b^{n-1} + a_{n-1} b^{n-2} + \ldots + a_2 b + a_1 + a_0/b = N' + a_0/b.$$

That is, the remainder a_0 of this division is the last digit in the desired representation. Dividing N' by b, we obtain

$$N'/b = a_n b^{n-2} + a_{n-1} b^{n-3} + \cdots + a_2 + a_1/b,$$

and the remainder of this division is the next to the last digit in the desired representation. Proceeding in this way, we obtain all the digits $a_0, a_1, \ldots, a_n$. This procedure can be systematized quite conveniently, as shown below. Suppose, for example, we wish to express 198 to the base 4. We find

$$\begin{array}{rl} 4\,\underline{|198} & \\ 4\,\underline{|49} & \text{remainder } 2 \\ 4\,\underline{|12} & \text{remainder } 1 \\ 4\,\underline{|3} & \text{remainder } 0 \\ 0 & \text{remainder } 3 \end{array}$$

The desired representation is $(3012)_4$. Again, suppose we wish to express 6647 to the base 12, where *t* and *e* are employed to represent *ten* and *eleven*, respectively. We find

$$\begin{array}{rl} 12\,\underline{|6647} & \\ 12\,\underline{|553} & \text{remainder } e \\ 12\,\underline{|46} & \text{remainder } 1 \\ 12\,\underline{|3} & \text{remainder } t \\ 0 & \text{remainder } 3 \end{array}$$

The desired representation is $(3t1e)_{12}$.

One is apt to forget, when adding or multiplying numbers in our ordinary system, that the actual work is accomplished mentally and that the number symbols are used merely to retain a record of the mental results. Our success and efficiency in carrying out such arithmetic operations depend on how well we know the addition and multiplication tables, the learning of which absorbed so much of our time in the primary grades. With corresponding tables constructed for a given base *b*, we can similarly perform additions and multiplications within the new system, without spending any time reverting to the ordinary system.

Let us illustrate with base 4. We first construct the following addition and multiplication tables for base 4.

Addition

	0	1	2	3
0	0	1	2	3
1	1	2	3	10
2	2	3	10	11
3	3	10	11	12

Multiplication

	0	1	2	3
0	0	0	0	0
1	0	1	2	3
2	0	2	10	12
3	0	3	12	21

The addition of 2 and 3, therefore, by reference to the table, is 11, and the multiplication of 2 and 3 is 12. Using these tables, exactly as we are accustomed to using the corresponding tables for base 10, we can now perform additions and multiplications. As an example, for the multiplication of $(3012)_4$ by $(233)_4$ we have, omitting the subscript 4,

$$
\begin{array}{r}
3012 \\
\underline{233} \\
21102 \\
21102 \\
\underline{12030} \\
2101122
\end{array}
$$

Considerable familiarity with the tables will be needed in order to perform the inverse operations of subtraction and division. This, of course, is also true for the base 10 and is the reason for much of the difficulty encountered in teaching the inverse operations in the elementary grades.

Problem Studies

1.1 Number Words

Furnish explanations of the following primitive number words.

(a) For a Papuan tribe in southeast New Guinea, it was found necessary to translate the Bible passage (John 5:5): "And a certain man was there, which had an infirmity 30 and 8 years" into "A man lay ill one man, both hands, 5 and 3 years."

(b) In (British) New Guinea, the number 99 comes out as "four men die, two hands come to an end, one foot ends, and four."

(c) The South American Kamayura tribe uses the word *peak-finger* as their word for 3, and "3 days" comes out as "peak-finger days."

(d) The Zulus of South Africa use the following equivalents: 6 ("taking the thumb"), 7 ("he pointed").

(e) The Malinké of West Sudan use the word *dibi* for 40. The word literally means "a mattress."

(f) The Mandingo tribe of West Africa use the word *kononto* for 9. The word literally means "to the one in the belly."

1.2 Written Numbers

Write 574 and 475 in **(a)** Egyptian hieroglyphics, **(b)** Roman numerals, **(c)** Attic Greek numerals, **(d)** Babylonian cuneiform, **(e)** traditional Chinese-Japanese, **(f)** alphabetic Greek, **(g)** Mayan numerals.

Record in Roman numerals: **(h)** $\frac{1}{4}$ of MCXXVIII, **(i)** 4 times XCIV.

Record in alphabetic Greek: **(j)** $\frac{1}{8}$ of $\tau\delta$, **(k)** 8 times $\rho\kappa\alpha$.

1.3 Alphabetic Greek Numeral System

(a) How many different symbols must one memorize in order to write numbers less than 1000 in alphabetic Greek? In Egyptian hieroglyphics? In Babylonian cuneiform?

(b) In the alphabetic Greek numeral system, the numbers 1000, 2000, . . . , 9000 were often represented by priming the symbols for 1, 2, . . . , 9. Thus, 1000 might appear as α'. The number 10,000, or **myriad,** was denoted by M. The multiplication principle was used for multiples of 10,000. Thus, 20,000, 300,000, and 4,000,000 appeared as βM, λM, and υM. Write, in alphabetic Greek, the numbers 5780, 72,803, 450,082, 3,257,888.

(c) Make an addition table up through $10 + 10$ and a multiplication table up through 10×10 for the alphabetic Greek numeral system.

1.4 Old and Hypothetical Numeral Systems

(a) As an alternative to the cuneiform, or wedge-shaped, numeral symbols, the ancient Babylonians sometimes used *circular* numeral symbols, so named because they were formed by *circular-shaped* imprints in clay tablets, made with a *round*-ended stylus instead of a *triangular*-ended one. Here the symbols for 1 and 10 are ☽ and ○. Write, with circular Babylonian numerals, the numbers 5780, 72,803, 450,082, 3,257,888.

(b) State a simple rule for multiplying by 10 a number expressed in Egyptian hieroglyphics.

(c) An interesting numeral system is the **Chinese scientific** (or **rod**) **numeral system,** which is probably 2000 or more years old. The system is essentially positional, with base 10. Figure 2 shows how the digits 1, 2, 3, 4, 5, 6, 7, 8, 9 are represented when they appear in an odd (units, hundreds, and so forth) position. But when they appear in an even (tens, thousands, and so forth) position, they are represented as shown in Figure 3. In this system, a circle, ○, was used for zero in the Sung Dynasty (960–1279) and later. Write, with rod numerals, the numbers 5780, 72,803, 450,082, 3,257,888.

(d) In a simple grouping system to base 5, let 1, 5, 5^2, 5^3 be represented by /, *,), (. Express the numbers 360, 252, 78, 33 in this system.

(e) In a positional numeral system to base 5, let 0, 1, 2, 3, 4 be represented by #, /, *,), (. Express the numbers 360, 252, 78, 33 in this system.

1.5 Finger Numbers

(a) Finger numbers were widely used for many centuries; from this use, finger processes were developed for some simple computations. One of

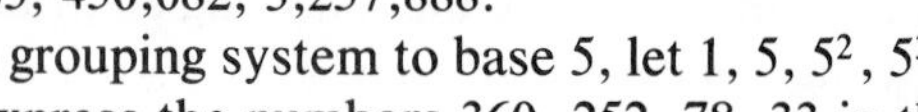

FIGURE 2

FIGURE 3

these processes, by giving the product of two numbers, each between 5 and 10, served to reduce the memory work connected with the multiplication tables. To multiply 7 by 9, for example, raise $7 - 5 = 2$ fingers on one hand and $9 - 5 = 4$ fingers on the other hand. Now add the raised fingers, $2 + 4 = 6$, for the tens digit of the product, and multiply the closed fingers, $3 \times 1 = 3$, for the units digit of the product, giving the result 63. This process is still used by some European peasants. Prove that the method gives correct results.

(b) Explain the ninth-century riddle that is sometimes attributed to Alcuin (ca. 775): "I saw a man holding 8 in his hand, and from the 8 he took 7, and 6 remained."

(c) Explain the following, found in Juvenal's tenth satire: "Happy is he indeed who has postponed the hour of his death so long and finally numbers his years upon his right hand."

1.6 Radix Fractions

Fractional numbers can be expressed, in the ordinary scale, by digits following a decimal point. The same notation is also used for other bases; therefore, just as the expression .3012 stands for

$$3/10 + 0/10^2 + 1/10^3 + 2/10^4,$$

the expression $(.3012)_b$ stands for

$$3/b + 0/b^2 + 1/b^3 + 2/b^4.$$

An expression like $(.3012)_b$ is called a **radix fraction** for base b. A radix fraction for base 10 is commonly called a **decimal fraction.**

(a) Show how to convert a radix fraction for base b into a decimal fraction.

(b) Show how to convert a decimal fraction into a radix fraction for base b.

(c) Approximate to four places $(.3012)_4$ and $(.3t1e)_{12}$ as decimal fractions.

(d) Approximate to four places .4402 as a radix fraction, first for base 7, and then for base 12.

1.7 Arithmetic in Other Scales

(a) Construct addition and multiplication tables for bases 7 and 12.

(b) Add and then multiply $(3406)_7$ and $(251)_7$, first using the tables of (a) and then by converting to base 10. Similarly, add and then multiply $(3t04e)_{12}$ and $(51tt)_{12}$.

(c) We may apply the tables for base 12 to simple mensuration problems involving feet and inches. For example, if we take 1 foot as a unit, then

3 feet 7 inches become $(3.7)_{12}$. To find, to the nearest square inch, the area of a rectangle 3 feet 7 inches long by 2 feet 4 inches wide, we may multiply $(3.7)_{12}$ by $(2.4)_{12}$ and then convert the result to square feet and square inches. Complete this example.

1.8 Problems in Scales of Notation

(a) Express $(3012)_5$ in base 8.

(b) For what base is $3 \times 3 = 10$? For what base is $3 \times 3 = 11$? For what base is $3 \times 3 = 12$?

(c) Can 27 represent an even number in some scale? Can 37? Can 72 represent an odd number in some scale? Can 82?

(d) Find b such that $79 = (142)_b$. Find b such that $72 = (2200)_b$.

(e) A 3-digit number in the scale of 7 has its digits reversed when expressed in the scale of 9. Find the 3 digits.

(f) What is the smallest base for which 301 represents a square integer?

(g) If $b > 2$, show that $(121)_b$ is a square integer. If $b > 4$, show that $(40{,}001)_b$ is divisible by $(221)_b$.

1.9 Some Recreational Aspects of the Binary Scale

The positional number system with base 2 has applications in various branches of mathematics. Also, there are many games and puzzles, like the well-known game of *Nim* and the puzzle of the *Chinese rings,* that have solutions that depend on this system. Following are two easy puzzles of this sort.

(a) Show how to weigh, on a simple equal-arm balance, any weight w of a whole number of pounds, using a set of weights of 1 pound, 2 pounds, 2^2 pounds, 2^3 pounds, and so forth, there being only one weight of each kind.

(b) Consider the following four cards containing numbers from 1 through 15.

Card 1		Card 2		Card 3		Card 4	
1	9	2	10	4	12	8	12
3	11	3	11	5	13	9	13
5	13	6	14	6	14	10	14
7	15	7	15	7	15	11	15

On the first card are all those numbers whose last digit in the binary system is 1; the second contains all those numbers whose second digit from the right is 1; the third contains all those whose third digit from the right is 1; the fourth contains all those whose fourth digit from the right is 1. Now someone is asked to think of a number N from 1 through 15 and to tell on which cards N can be found. It is then easy to announce the number N by merely adding the top left numbers on the cards where it appears. Make a similar set of 6 cards for detecting any number from 1 through 63. It has been noted that if the numbers are written on cards weighing 1, 2, 4, . . . units, then an automaton in the form of a postal scale could express the number N.

1.10 Some Number Tricks

Many simple number tricks, in which one is to "guess a selected number," have explanations depending on our own positional scale. Expose the following tricks of this kind.

(a) Someone is asked to think of a 2-digit number. He is then requested to multiply the tens digit by 5, add 7, double, add the units digit of the original number, and announce the final result. From this result, the conjurer secretly subtracts 14 and obtains the original number.

(b) Someone is asked to think of a 3-digit number. He is then requested to multiply the hundreds digits by 2, add 3, multiply by 5, add 7, add the tens digit, multiply by 2, add 3, multiply by 5, add the units digit, and announce the result. From this result, the conjurer secretly subtracts 235 and obtains the original number.

(c) Someone is asked to think of a 3-digit number whose first and third digits are different. He is then requested to find the difference between this number and that obtained by reversing the 3 digits. Upon disclosing only the last digit of this difference, the conjurer announces the entire difference. How does the trickster do this?

Essay Topics

1/1 Possible number sense in the animal world.
1/2 Linguistic evidence of the use at some time of bases other than ten.
1/3 Advantages and disadvantages of bases other than ten.
1/4 The history of writing materials.
1/5 The battle of the abacists and the algorists.
1/6 Finger numbers and finger arithmetic.
1/7 Arithmetic on an abacus.
1/8 The ancient quipu.
1/9 Mayan arithmetic.
1/10 Tally sticks.
1/11 The Babylonian zero.
1/12 Three for plural.
1/13 Number taboos.
1/14 The Kensington stone mystery.
1/15 Some fancied origins of our numeral symbols.
1/16 Contests between the abacus and an electronic desk calculator.

Bibliography[4]

ANDREWS, F. E. *New Numbers*. New York: Harcourt, Brace & World, 1935.

ASCHER, MARCIA, and ROBERT ASCHER *Code of the Quipu: A Study in Media, Mathematics, and Culture*. Ann Arbor, Mich.: University of Michigan Press, 1981.

BALL, W. W. R., and H. S. M. COXETER *Mathematical Recreations and Essays*. 12th ed. Toronto: University of Toronto Press, 1974. Reprinted by Dover, New York.

CAJORI, FLORIAN *A History of Mathematical Notations*. 2 vols. Chicago: Open Court Publishing, 1928–1929.

CLOSS, M. P., ed. *Native American Mathematics*. Austin, Tex.: University of Texas Press, 1986.

COHEN, P. C. *A Calculating People: The Spread of Numeracy in North America*. Chicago: University of Chicago Press, 1983.

CONANT, L. L. *The Number Concept: Its Origin and Development*. New York: Macmillan, 1923.

DANTZIG, TOBIAS *Number: The Language of Science*. New York: Macmillan, 1946.

FLEGG, GRAHAM *Numbers: Their History and Meaning*. Glasgow: Andrew Deutch Ltd., or New York: Schocken Books, 1983.

FREITAG, H. T., and A. H. FREITAG *The Number Story*. Washington, D.C.: National Council of Teachers of Mathematics, 1960.

GALLENKAMP, CHARLES *Maya*. New York: McKay, 1959.

GATES, W. E. *Yucatan Before and After the Conquest, by Friar Diego de Landa, etc*. Translated with notes. Maya Society Publication No. 20, Baltimore: The Maya Society, 1937.

GLASER, ANTON *History of Binary and Other Nondecimal Numeration*. Published by Anton Glaser, 1237 Whitney Road, Southampton, Pennsylvania 18966, 1971.

HILL, G. F. *The Development of Arabic Numerals in Europe*. New York: Oxford University Press, 1915.

IFRAH, GEORGES *From One to Zero: A Universal History of Numbers*. Translated by Lowell Bair. New York: Viking Press, 1985.

KARPINSKI, LOUIS CHARLES *The History of Arithmetic*. New York: Russell & Russell, 1965.

KRAITCHIK, MAURICE *Mathematical Recreations*. New York: W. W. Norton, 1942.

LARSON, H. D. *Arithmetic for Colleges*. New York: Macmillan, 1950.

LOCKE, L. LELAND *The Ancient Quipu or Peruvian Knot Record*. New York: American Museum of Natural History, 1923.

MENNINGER, KARL *Number Words and Number Symbols, A Cultural History of Numbers*. Cambridge, Mass.: The M.I.T. Press, 1969.

MORLEY, S. G. *An Introduction to the Study of the Maya Hieroglyphs*. Washington, D.C.: Government Printing Office, 1915.

_____ *The Ancient Maya*. Stanford, Calif.: Stanford University Press, 1956.

ORE, OYSTEIN *Number Theory and Its History*. New York: McGraw-Hill, 1948.

[4] See the General Bibliography (p. 670) as a supplement to this and the bibliographies in the following chapters.

PULLAN, J. M. *The History of the Abacus*. New York: Praeger, 1968.

RINGENBERG, L. A. *A Portrait of 2*. Washington, D.C.: National Council of Teachers of Mathematics, 1956.

SCRIBA, CHRISTOPHER *The Concept of Number*. Mannheim: Bibliographisches Institut, 1968.

SMELTZER, DONALD *Man and Number*. New York: Emerson Books, 1958.

SMITH, D. E. *Number Stories of Long Ago*. Washington, D.C.: National Council of Teachers of Mathematics, 1958.

_____, and JEKUTHIEL GINSBURG *Numbers and Numerals*. Washington, D.C.: National Council of Teachers of Mathematics, 1958.

_____, and L. C. KARPINSKI *The Hindu-Arabic Numerals*. Boston: Ginn, 1911.

SWAIN, R. L. *Understanding Arithmetic*. New York: Holt, Rinehart and Winston, 1957.

TERRY, G. S. *The Dozen System*. London: Longmans, Green, 1941.

_____ *Duodecimal Arithmetic*. London: Longmans, Green, 1938.

THOMPSON, J. E. S. "Maya arithmetic." *Contributions to American Anthropology and History*. vol. 36: 37–62. Washington, D.C.: Carnegie Institution of Washington Publication No. 528, 1941.

_____ *The Rise and Fall of Maya Civilization*. Norman, Okla.: University of Oklahoma Press, 1954.

VON HAGEN, VICTOR M. *Maya: Land of the Turkey and the Deer*. Cleveland: World Publishing, 1960.

YOSHINO, Y. *The Japanese Abacus Explained*. New York: Dover, 1963.

ZASLAVSKY, CLAUDIA *Africa Counts: Number and Pattern in African Culture*. Boston: Prindle, Weber & Schmidt, 1973. Republished by Lawrence-Hill & Company, Westport, Conn.

Cultural Connection 

THE AGRICULTURAL REVOLUTION

The Cradles of Civilization—ca. 3000–525 B.C.
(to accompany Chapter Two)

Near the end of the Stone Age, in certain parts of the world, people were compelled toward a full-scale, intensive agriculture by changes in the world's climate. The vast, grassy savannas where Stone Age hunters lived began to shrink late in the Neolithic period, as they continue to shrink today. In some places, the savannas were reclaimed by expanding forests; in others, they became arid and lifeless, turning into deserts. As their environment changed, people adapted as best they could. In Europe, southern Africa, southeast Asia, and eastern North and South America, people moved into the new forests and became woodland hunters, a relatively minor adaptation.

In the growing deserts of north Africa, the Middle East, and central Asia, however, the transformation was not so simple. As the grass withered and the streams dried up, as mammoth sand dunes marched out from the centers of the new deserts, the animals that once lived in these regions left, crowding into oases, and then moving on when the oases dried up. The people followed the animals in this flight before the advance of the towering dunes, eventually settling in oasis-like wetlands on the edges of the deserts. These new places acted as catch basins for all forms of life, including people, and large numbers of men and women came to live there after they fled the desert. In Africa, on the marches of the Sahara Desert, which had once been rolling prairie, the Nile River valley offered water to the migrating animals and their human hunters. In the Middle East, the Tigris and Euphrates Rivers, sharing a single valley, formed a catch basin for those fleeing the growing Arabian Desert. The Indus River valley at the periphery of India's Thar Desert and the Yellow River valley in China at the frontier of the Gobi Desert also served as catch basins. In the Americas, although at a later date, the Pacific coastal plain turned dry and sere, and people climbed the high peaks of the Sierra Madre Mountains in Mexico and Central America and the Andes in Peru and Colombia, where the loftiest mountains scraped the clouds and tore loose the rain. Today, a similar process of desertification is being played out on a terrifying scale in Africa, where the Sahara is again on the move and people from the withering grasslands are crowding into refugee camps along the Niger River and Upper Nile.

The civilizations that emerged in these catch basins were vastly different from the hunter/gatherer societies of the Stone Age. Human population densities in these wetlands were too high to permit everyone to continue to survive as hunters and gatherers. To keep from starving, people in these places had to find other ways to obtain food. Not surprisingly, they turned to intensive agriculture, which could support populations as thick as forty people to the square mile. This was a sort of "agricultural revolution," and it precipitated profound cultural changes.

One such change was the creation of written language. Farming meant irrigation in the largely rainless valleys of north Africa and the Middle East, and the seasonal flooding of the Yellow, Nile, and Tigris and Euphrates Rivers meant diking—activities that required not only cooperation and engineering skill, but record-keeping systems as well. The farmers needed to know when the floods, or the rainy season, would come, and that meant calendars and almanacs. Landowners kept written accounts of agricultural production and drew maps noting the locations of irrigation ditches. Farmers prayed to the gods to ensure that the floods and rains would come as scheduled and, in the process, watched the movements of the stars. All these activities gave rise to a new class of educated men: priests, scribes, and astrologers.

Along with literacy came a need for new technologies. Early engineers planned dikes and irrigation works. Metal plows were better than wooden ones; people learned to forge bronze around 3000 B.C. and iron about 1100 B.C. The need for specialized tools created a need for yet another new social class: skilled artisans.

Another important change was the adoption of a sedentary lifestyle. Unlike hunters and gatherers, farmers did not need to travel great distances searching for food. They built permanent villages and towns, and small cities grew up along the river banks. By 2500 B.C., the cities of Memphis and Thebes had emerged as the leading metropoli in Egypt; not long after, Pharoah Pepi II (?–ca. 2200 B.C.) built the city of Heracleopolis as his capital. In the valley of the Tigris and Euphrates, the city of Ur emerged earlier, in approximately 3000 B.C. Although small by modern standards, these early cities dwarfed Neolithic villages. Ur had 24,000 inhabitants and covered 150 acres. The cities provided central marketplaces where farmers and artisans could exchange goods, and a merchant class sprang into being to facilitate that process.

For the first time in history, some people had leisure time. While the farmers, who made up the majority of the population during the agricultural revolution, generally spent the entire day in toil, other people—kings, priests, merchants, scribes—found time at the end of the day that could be spent pondering the mysteries of nature and science. At last, all the ingredients for scientific progress were brought together: written languages, a need for new technologies, urban environments, and leisure time. It is little wonder, then, that historians refer to ancient Egypt, India, China, and the Middle East as "cradles of civilization." (The deserts in the Americas appeared later than those in the eastern hemisphere; hence, the agricultural revolution in the west was longer in coming. Historians now acknowledge, however, that Mexico and Peru during the days of the Mayas, Incas, and their forebears were also true "cradles of civilization.")

The agriculturalists developed new forms of political organization. In the

Stone Age, "the government" had been the tribe or clan—a small band of men and women bound together by kinship ties and under the nominal direction of a chief. The complex activities attendant to farming (planting common fields, building granaries, digging irrigation trenches, regulating marketplaces to protect unwary buyers, appeasing the gods) required more centralized systems of government. Tribes were replaced by city-states, kingdoms, and small empires, and the tribal chief was supplanted by extensive bureaucracies.

The city-state was the most common form of government in the cradles of civilization, consisting of a single city or town and the surrounding countryside. Tiny by modern standards, city-states were so small that, in their ideal form, which was described by the Chinese philosopher Confucius (551–479 B.C.), a citizen might hear roosters crowing in neighboring countries. Each of the cradles of civilization was, at one time or another, divided into city-states: Egypt between 2200 B.C. and 2050 B.C., and again between 1786 B.C. and 1575 B.C.; the valley of the Tigris and Euphrates Rivers between approximately 3000 B.C. and 2150 B.C.; and China from 600 B.C. (or earlier) until 221 B.C. Most often, a city-state was an oligarchy, governed by a small clique of wealthy citizens. A few were monarchies, however, and some were theocracies (that is, ruled by a class of priests). A very few were republics, with broad citizen participation in affairs of state. We shall visit some of these republics in Greece, Rome, and Carthage in Cultural Connections III and IV.

In each of the cradles of civilization, city-states eventually gave way to expanding empires. According to tradition, Egypt was united under a single pharoah in 3100 B.C., at the beginning of the agricultural revolution, although the kingdom seems to have broken up by 2200 B.C. into a collection of small principalities ruled by petty lords called *nomarchs*. In 1575 B.C., Egypt was reunited under a single, absolute ruler, and would remain so until conquered by Persia in 525 B.C. Like Egypt, tradition holds that the China of antiquity was a united country under the mysterious Hsia Dynasty, about which little is actually known. Between 1500 B.C. and 1027 B.C., the land along the Yellow River was governed from the city of An-Yang by the Shang Dynasty and, after that, by the Chou Dynasty. By 600 B.C., Chou power had declined, and China was actually a collection of city-states until unified in 221 B.C. by the Chin Dynasty. The Chins were supplanted by the Han Dynasty fifteen years later, which established an empire that would endure unil A.D. 221. In both Egypt and China, we cannot be sure whether the traditional early dynasties represented centralized empires or merely strong city-states that dominated their neighbors, but later dynasts ruled as powerful autocrats over large, cohesive empires. The Tigris and Euphrates River valley in the Middle East was first united into a single empire by the warrior Sargon I (ca. 2276–2221 B.C.), although his kingdom broke up soon after his death. Permanent unification did not come until Amorite invaders conquered the valley in approximately 2000 B.C. and forged the Babylonian Empire. It is not known what political systems existed in the Indus River valley.

The fruits of the new agricultural civilizations were not enjoyed equally by everyone. There were strict class divisions. Most of the people, probably in excess of ninety percent, were poor farmers. These folk could not read or write. They often did not own the land they tilled, which belonged instead to an overlord. They toiled constantly, with scant time to relax or enjoy life. They had little in the way of material wealth or comforts, although they did most of the work. Wealth was instead concentrated in the hands of a small upper class of lords, priests, warriors (the first recorded war in history was a battle over an irrigation ditch in the Middle East in ca. 2000 B.C.), merchants, and craftsmen. Below even the farmers on the social scale were slaves, usually the victims of conquest, and women, who with few exceptions were treated merely as workers and bearers of children and were not accorded opportunities for intellectual expression.

All the new agricultural societies were not identical. On the fringes of the cultivated valleys lived nomadic tribes of herders who periodically made war on their plant-cultivating neighbors. In India, Aryan nomads from Central Asia possibly wiped out the civilization of the Indus River. In the Middle East, invading armies came in a number of waves, either horsemen from the Arabian Desert or fierce warriors from the Zargos Mountains. Each new conqueror established itself as the new ruling class and adopted the customs and manners of those it had supplanted. Among these conquerors were the Amorites, who invaded the valley of the Tigris and Euphrates Rivers in approximately 2000 B.C., learned the local culture, and produced the legal code of Hammurabi. The Amorites built the city of Babylon, and from it ruled a great empire that persisted for a thousand years, when the Assyrians conquered the land between the two rivers. The Assyrians were in turn overthrown by a revolt in approximately 600 B.C., and the rebels instituted the Chaldean or Neo-Babylonian Empire of Nebuchadnezzar. In 550 B.C., the Persians massed out of the Zargos Mountains and conquered Babylon. China was threatened by invaders from the Gobi Desert but managed each time to drive the would-be conquerors back.

In summary, the period from 3000 to 525 B.C. witnessed the birth of a new human civilization that was sparked by an agricultural revolution. New societies based on farming economies emerged from the mists of the Stone Age in the valleys of the Nile, Yellow, Indus, and Tigris and Euphrates Rivers. These new societies created written languages; worked metals; built cities; empirically developed the fundamental mathematics of surveying, engineering, and commerce; and spawned upper classes who had enough leisure time to pause and consider the mysteries of nature. After millions of years, humanity was at last embarked on the road toward scientific achievement.

Chapter 2

BABYLONIAN AND EGYPTIAN MATHEMATICS

2–1 The Ancient Orient

Early mathematics required a practical basis for its development, and such a basis arose with the evolution of more advanced forms of society. It was along some of the great rivers of Africa and Asia that the new forms of society made their appearance: the Nile in Africa, the Tigris and Euphrates in western Asia, the Indus and then the Ganges in south-central Asia, and the Hwang Ho and then the Yangtze in eastern Asia. The rivers furnished convenient transportation, and with marsh drainage, flood control, and irrigation, it was possible to convert the lands along the rivers into rich agricultural regions. Extensive projects of this sort not only knit together previously separated localities, but the engineering, financing, and administration of both the projects and the purposes for which they were created required the development of considerable technical knowledge and its concomitant mathematics. Thus, early mathematics can be said to have originated in certain areas of the ancient Orient (the world east of Greece) primarily as a practical science to assist in agriculture, engineering, and business pursuits. These pursuits required the computation of a usable calendar; the development of systems of weights and measures to serve in the harvesting, storing, and apportioning of foods; the creation of surveying methods for canal and reservoir construction and for parceling land; and the evolution of financial and commercial practices for raising and collecting taxes and for purposes of trade.[1]

As we have seen, the initial emphasis of mathematics was on practical arithmetic and mensuration. A special craft came into being for the cultivation, application, and instruction of this practical science. In such a situation, however, tendencies toward abstraction were bound to develop, and to some extent, the science was then studied for its own sake. It was in this way that algebra ultimately evolved from arithmetic and the beginnings of theoretical geometry grew out of mensuration.

[1] There is an alternative thesis that finds the origin of mathematics in religious ritual—agriculture, trade, and surveying being later contributors. See A. Seidenberg, "The ritual origin of geometry," *Archive for History of Exact Sciences* 1 (1962): 488–527, and "The ritual origin of counting," *Archive for History of Exact Sciences* 2 (1962): 1–40. Another alternative thesis can be advanced claiming that mathematics originated in art, the universal language of man.

It should be noted, however, that in all ancient Oriental mathematics one cannot find even a single instance of what we today call a demonstration. In place of an argument, there is merely a description of a process. One is instructed, "Do thus and so." Moreover, except possibly for a few specimens, these instructions are not even given in the form of general rules, but are simply applied to sequences of specific cases. Thus, if the solution of quadratic equations is to be explained, we do not find a derivation of the process used, nor do we find the process described in general terms; instead we are offered a large number of specific quadratics, and we are told step by step how to solve each of these specific instances. It was expected that from a sufficient number of specific examples, the general process would become clear. Unsatisfactory as the "do-thus-and-so" procedure may seem to us, it should not seem strange, for it is the procedure we must frequently use in teaching portions of grade-school and high-school mathematics.

There are difficulties in dating discoveries made in the ancient Orient. One of these difficulties lies in the static nature of the social structure and the prolonged seclusion of certain areas. Another difficulty is due to the writing media upon which discoveries were preserved. The Babylonians used imperishable baked clay tablets, and the Egyptians used stone and papyrus, the latter fortunately being long lasting because of the unusually dry climate of the region. But the early Chinese and Indians used very perishable media like bark and bamboo. Thus, although a fair quantity of definite information is now known about the science and the mathematics of ancient Babylonia and Egypt, very little is known with any degree of certainty about these studies in ancient China and India. Accordingly, this chapter, which is largely devoted to the mathematics of the pre-Hellenic centuries, will be limited to Babylonia and Egypt.

BABYLONIA

2–2 Sources

Since the first half of the nineteenth century, archeologists working in Mesopotamia have systematically unearthed some half-million inscribed clay tablets. Over 50,000 tablets were excavated at the site of ancient Nippur alone. There are many excellent collections of these tablets, such as those in the great museums at Paris, Berlin, and London, and in the archeological exhibits at Yale, Columbia, and the University of Pennsylvania. The tablets vary in size from small ones of only a few square inches to ones approximately the size of the present textbook, the latter being about an inch and a half thick through their centers. Sometimes writing appears on only one side of the tablet, sometimes on both sides, and frequently on the rounded edges of the tablet.

Of the approximately half-million tablets, about 400 have been identified as strictly mathematical tablets, containing mathematical tables and lists of mathe-

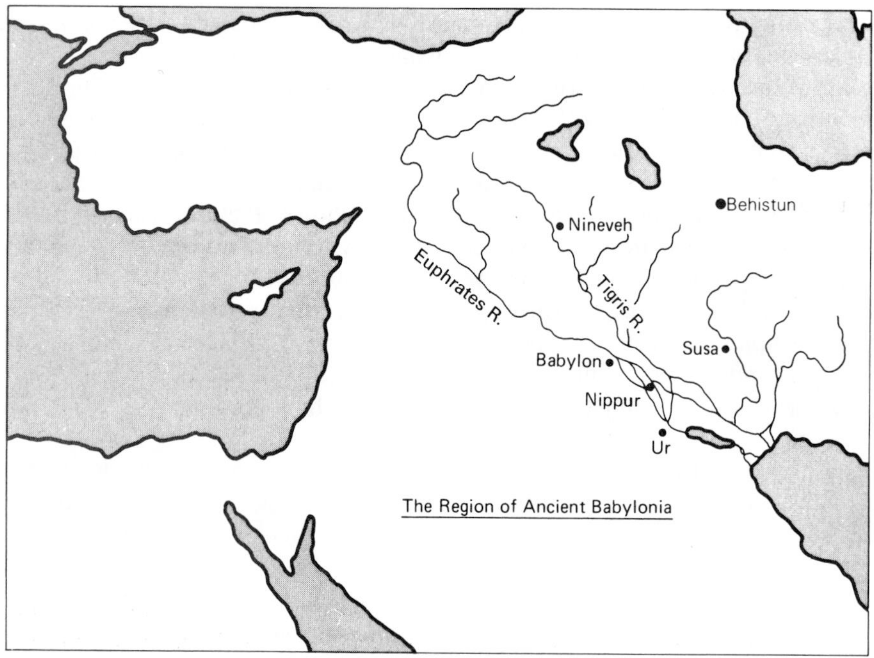

The Region of Ancient Babylonia

matical problems. We owe our knowledge of ancient Babylonian[2] mathematics to the scholarly deciphering and interpretation of many of these mathematical tablets.

Successful attempts at deciphering cuneiform writing did not occur until shortly before 1800, when European travelers noticed the inscriptions accompanying a monumental bas relief carved some 300 feet above ground on the great limestone cliff near the village of Behistun, in the northwestern part of present-day Iran. The puzzle of the inscriptions was finally solved in 1846 by the remarkable pertinacity of Sir Henry Creswicke Rawlinson (1810–1895), an English diplomat and Assyriologist who perfected a key earlier suggested by the German archeologist and philologist Georg Friedrich Grotefend (1775–1853). The inscriptions are engraved in thirteen panels on a smoothed surface measuring 150 feet by 100 feet and are in the three ancient languages of Old Persian, Elamite, and Akkadian, all of which employed cuneiform script. The relief and inscriptions were executed in 516 B.C. at the command of Darius the Great.

With the ability to read the cuneiform texts of the excavated Babylonian tablets, it was found that these tablets appear to bear upon all phases and

[2] It should be understood that the descriptive term *Babylonian* is used merely for convenience, and that many peoples, such as Sumerians, Akkadians, Chaldeans, Assyrians, and other early peoples who inhabited the area at one time or another, are subsumed under the general term.

interests of daily life and to range over many periods of Babylonian history. There are mathematical texts dating from the latest Sumerian period of perhaps 2100 B.C.; a second and very large group from the succeeding First Babylonian Dynasty of King Hammurabi's era, and on down to about 1600 B.C.; and a third generous group running from about 600 B.C. to A.D. 300, covering the New Babylonian Empire of Nebuchadnezzar and the following Persian and Seleucidan eras. The lacuna between the second and third groups coincides with an especially turbulent period of Babylonian history. Most of our knowledge of the contents of these mathematical tablets does not predate 1935 and is largely due to the remarkable findings of Otto Neugebauer and F. Thureau-Dangin. Since the work of interpreting these tablets is still proceeding, new and perhaps equally remarkable discoveries are quite probable in the near future.

2–3 Commercial and Agrarian Mathematics

Even the oldest tablets show a high level of computational ability and make it clear that the sexagesimal positional system was already long established. There are many texts of this early period dealing with farm deliveries and with arithmetical calculations based on these transactions. The tablets show that the ancient Sumerians were familiar with all kinds of legal and domestic contracts, like bills, receipts, promissory notes, accounts, both simple and compound interest, mortgages, deeds of sale, and guaranties. There are tablets that are records of business firms, and others that deal with systems of weights and measures.

Many arithmetic processes were carried out with the aid of various tables. Of the 400 mathematical tablets, a good half contain mathematical tables. These table tablets show multiplication tables, tables of reciprocals, tables of squares and cubes, and even tables of exponentials. These latter tables were probably used, along with interpolation, for problems on compound interest. The reciprocal tables were used to reduce division to multiplication.

The calendar used by the Babylonians was established ages earlier, as evidenced by the facts that their year started with the vernal equinox and that the first month was named after Taurus. Because the sun was in Taurus at this equinox around 4700 B.C., it seems safe to say that the Babylonians had some kind of arithmetic as far back as the fourth or fifth millennium B.C.

For examples concerning Babylonian table construction and Babylonian use of tables in business transactions, see Problem Studies 2.1 and 2.2.

2–4 Geometry

Babylonian geometry is intimately related to practical mensuration. From numerous concrete examples, the Babylonians of 2000 to 1600 B.C. must have been familiar with the general rules for the area of a rectangle, the areas of right and isosceles triangles (and perhaps the general triangle), the area of a trapezoid having one side perpendicular to the parallel sides, the volume of a rectangular parallelepiped, and, more generally, the volume of a right prism with a special trapezoidal base. The circumference of a circle was taken as three times the diameter and the area as one-twelfth the square of the circumference (both correct for $\pi = 3$), and the volume of a right circular cylinder was then obtained by finding the product of the base and the altitude. The volume of a frustum of a cone or of a square pyramid is incorrectly given as the product of the altitude and half the sum of the bases. The Babylonians also knew that corresponding sides of two similar right triangles are proportional, that the perpendicular through the vertex of an isosceles triangle bisects the base, and that an angle inscribed in a semicircle is a right angle. The Pythagorean theorem was also known. (In this connection, see Section 2–6.) There is a recently discovered tablet in which $3\frac{1}{8}$ is used as an estimate for π [see Problem Study 2.5(b)].

The chief feature of Babylonian geometry is its algebraic character. The more intricate problems that are expressed in geometric terminology are essentially nontrivial algebra problems. Typical examples may be found in Problem Studies 2.3 and 2.4. There are many problems concerning a transversal parallel to a side of a right triangle that lead to quadratic equations; there are others that lead to systems of simultaneous equations, one instance giving ten equations in ten unknowns. There is a Yale tablet, possibly from 1600 B.C., in which a general cubic equation arises in a discussion of volumes of frustrums of a pyramid, as the result of eliminating z from a system of equations of the type

$$z(x^2 + y^2) = A, \qquad z = ay + b, \qquad x = c.$$

We undoubtedly owe to the ancient Babylonians our present division of the circumference of a circle into 360 equal parts. Several explanations have been put forward to account for the choice of this number, but perhaps none is more plausible than the following, advocated by Otto Neugebauer. In early Sumerian times, there existed a large distance unit, a sort of *Babylonian mile,* equal to about seven of our miles. Since the Babylonian mile was used for measuring longer distances, it was natural that it should also become a time unit—namely, the time required to travel a Babylonian mile. Later, sometime in the first millennium B.C., when Babylonian astronomy reached the stage in which systematic records of celestial phenomena were kept, the Babylonian time-mile was adopted for measuring spans of time. Since a complete day was found to be equal to twelve time-miles, and one complete day is equivalent to one revolution of the sky, a complete circuit was divided into twelve equal parts. For convenience, however, the Babylonian mile was subdivided into thirty equal parts; thus, we arrive at (12)(30) = 360 equal parts in a complete circuit.

2–5 Algebra

By 2000 B.C. Babylonian arithmetic had evolved into a well-developed rhetorical, or prose, algebra. Not only were quadratic equations solved, both by the equivalent of substituting in a general formula and by completing the square, but some cubic (third degree) and biquadratic (fourth degree) equations were discussed. A tablet has been found giving a tabulation not only of the squares and the cubes of the integers from 1 to 30, but also of the combination $n^3 + n^2$ for this range. A number of problems are given that lead to cubics of the form $x^3 + x^2 = b$. These can be solved by using the $n^3 + n^2$ table. Problem Study 2.4 concerns itself with possible uses of this particular table.

There are some Yale tablets of about 1600 B.C. listing hundreds of unsolved problems involving simultaneous equations that lead to biquadratic equations for solution. As an example, we have

$$xy = 600, \qquad 150(x - y) - (x + y)^2 = -1000.$$

As another illustration from the same tablets, we have a pair of equations of the form

$$xy = a, \qquad bx^2/y + cy^2/x + d = 0,$$

that lead to an equation of the sixth degree in x, but that is quadratic in x^3.

Neugebauer has found two interesting series problems on a Louvre tablet of about 300 B.C. One of them states that

$$1 + 2 + 2^2 + \ldots + 2^9 = 2^9 + 2^9 - 1,$$

and the other one that

$$1^2 + 2^2 + 3^2 + \ldots + 10^2 = \left[1\left(\frac{1}{3}\right) + 10\left(\frac{2}{3}\right)\right]55 = 385.$$

One wonders if the Babylonians were familiar with the formulas

$$\sum_{i=0}^{n} r^i = \frac{r^{n+1} - 1}{r - 1}$$

and

$$\sum_{i=1}^{n} i^2 = \frac{2n + 1}{3}\sum_{i=1}^{n} i = \frac{n(n + 1)(2n + 1)}{6}.$$

The first of these was known to contemporary Greeks, and Archimedes found practically the equivalent of the second.

The Babylonians gave some interesting approximations to the square roots of nonsquare numbers, like 17/12 for $\sqrt{2}$ and 17/24 for $1/\sqrt{2}$. Perhaps the Babylonians used the approximation formula

$$(a^2 + h)^{1/2} \approx a + h/2a.$$

A very remarkable approximation for $\sqrt{2}$ is

$$1 + 24/60 + 51/60^2 + 10/60^3 = 1.41421296,$$

found on the Yale table tablet 7289 of about 1600 B.C. (see Problem Study 2.7).

There are astronomical tablets of the third century B.C. that make explicit use of the law of signs in multiplication.

In summary, we conclude that the ancient Babylonians were indefatigable table makers, computers of high skill, and definitely stronger in algebra than geometry. One is certainly struck by the depth and the diversity of the problems that they considered.

2–6 Plimpton 322

Perhaps the most remarkable of the Babylonian mathematical tablets yet analyzed is that known as *Plimpton 322,* meaning that it is the item with catalogue number 322 in the G. A. Plimpton collection at Columbia University. The tablet is written in Old Babylonian script, which dates it somewhere from 1900 to 1600 B.C., and it was first described by Neugebauer and Sachs in 1945.[3]

Figure 4 gives an idea of the shape of the tablet. Unfortunately, a missing piece has been broken from the entire left edge, and the tablet is further marred by a deep chip near the middle of the right edge and a flaked area in the top left corner. Upon examination, crystals of modern glue were found along the left broken edge of the tablet. This suggests that the tablet was probably complete when excavated, that it subsequently broke, that an attempt was made to glue the pieces back together, and that later the pieces again separated. Thus, the missing piece of the tablet may still be in existence but, like a needle in a haystack, lost somewhere among the collections of these ancient tablets. We shall shortly see that it would prove very interesting if this missing piece were to be found.

The tablet contains three essentially complete columns of figures that, for convenience, are reproduced on Figure 4 in our own decimal notation. There is a fourth and partly incomplete column of figures along the broken edge. We shall later reconstruct this column.

It is clear that the column on the extreme right merely serves to number the lines. The next two columns seem, at first glance, to be rather haphazard. With study, however, one discovers that corresponding numbers in these columns, with four unfortunate exceptions, constitute the hypotenuse and a leg of integral-sided right triangles. The four exceptions are noted in Figure 4 by placing the original readings in parentheses to the right of the corrected readings. The exception in the second line has received an involved explanation,[4]

[3] A detailed study of the tablet has more recently been done by Jöran Friberg. See "Methods and traditions of Babylonian mathematics," *Historia Mathematica* 8, no. 3 (August 1981): 277–318.

[4] See R. J. Gillings, *The Australian Journal of Science,* 16 (1953): 34–36, or Otto Neugebauer, *The Exact Sciences in Antiquity,* 2d ed., 1962.

119	169	1
3367	4825 (11521)	2
4601	6649	3
12709	18541	4
65	97	5
319	481	6
2291	3541	7
799	1249	8
481 (541)	769	9
4961	8161	10
45	75	11
1679	2929	12
161 (25921)	289	13
1771	3229	14
56	106 (53)	15

FIGURE 4

but the other three exceptions can easily be accounted for. Thus, in the ninth line, 481 and 541 appear as (8,1) and (9,1) in the sexagesimal system. Clearly the occurrence of 9 instead of 8 could be a mere slip of the stylus when writing these numbers in cuneiform script. The number in line 13 is the square of the corrected value, and the number in the last line is half of the corrected value.

A set of three positive integers, like (3,4,5), which can be the sides of a right triangle, is known as a **Pythagorean triple.** Again, if the triple contains no

Plimpton 322.
(Columbia University.)

common integral factor other than unity, it is known as a **primitive Pythagorean triple.** Thus, (3,4,5) is a primitive triple, whereas (6,8,10) is not. One of the mathematical achievements over a millenium after the date of the Plimpton tablet was to show that all primitive Pythagorean triples (a,b,c) are given parametrically by

$$a = 2uv, \qquad b = u^2 - v^2, \qquad c = u^2 + v^2,$$

where u and v are relatively prime, of different parity, and $u > v$; thus, if $u = 2$ and $v = 1$, we obtain the primitive triple $a = 4$, $b = 3$, $c = 5$.

Suppose we compute the other leg a of the integral-sided right triangles determined by the given hypotenuse c and leg b on the Plimpton tablet. We find the following Pythagorean triples:

a	b	c	u	v
120	119	169	12	5
3456	3367	4825	64	27
4800	4601	6649	75	32
13500	12709	18541	125	54
72	65	97	9	4
360	319	481	20	9
2700	2291	3541	54	25
960	799	1249	32	15
600	481	769	25	12
6480	4961	8161	81	40
60	45	75	2	1
2400	1679	2929	48	25
240	161	289	15	8
2700	1771	3229	50	27
90	56	106	9	5

One may notice that all of these triples, except the ones in lines 11 and 15, are primitive triples. For discussion, we have also listed the values of the parameters u and v leading to these Pythagorean triples. The evidence seems good that the Babylonians of this remote period were acquainted with the general parametric representation of primitive Pythagorean triples as given above. This evidence is strengthened when we notice that u and v, and hence a (since $a = 2uv$), are *regular* sexagesimal numbers (see Problem Study 2.1). It appears that the table on the tablet was constructed by deliberately choosing small regular numbers for the parameters u and v.

This choice of u and v must have been motivated by some subsequent process involving division, because regular numbers appear in tables of reciprocals and are used to reduce division to multiplication. An examination of the fourth, and partially destroyed, column gives the answer, for this column is found to contain the values of $(c/a)^2$ for the different triangles. To carry out the division, the side a and, hence, the numbers u and v, had to be regular.

It is worth examining the column of values for $(c/a)^2$ a little more closely. This column, of course, is a table giving the square of the secant of the angle B opposite side b of the right triangle. Because side a is regular, secant B has a finite sexagesimal expansion. Moreover, it turns out, with the particular choice of triangles as given, that the values of secant B form a surprisingly regular sequence that decreases by almost exactly 1/60 as we pass from one line of the table to the next, and the corresponding angle decreases from 45° to 31°. We thus have a secant table for angles from 45° to 31°, formed by means of integral-sided right triangles, in which there is a regular jump in the function, rather than in the corresponding angle. All this is truly remarkable. It seems highly probable that there were companion tables giving similar information for angles ranging from 30° to 16° and from 15° to 1°.

The analysis of Plimpton 322 shows the careful examination to which some of the Babylonian mathematical tablets must be subjected. Formerly, such a tablet might have been summarily dismissed as being merely a business list or record.

EGYPT

2–7 Sources and Dates

Ancient Babylonia and ancient Egypt differ considerably in their political histories. The former was open to invasion by neighboring peoples, with the result that there were periods of much turmoil when one empire succeeded another. Ancient Egypt, on the other hand, remained secluded and naturally protected from foreign invasion and was governed more peacefully and uninterruptedly by a succession of dynasties. Both societies were essentially theocracies that were ruled by rich and powerful bureaucrats hand-in-glove with the temple priests. Most of the manual labor was done by a large slave class, established in Babylonia mainly by the overthrowing of a present empire by a conquering invading one, and in Egypt by deliberate military importation from foreign lands. It was principally this slave class that dug and maintained the irrigation systems and built the ziggurats in Babylonia and erected the great temples and pyramids in Egypt. Basic surveying and engineering practices, with their concomitant mathematics, were created to assist in the design and construction of these works.

Contrary to popular opinion, the mathematics of ancient Egypt never reached the level attained by Babylonian mathematics. This may have been due to the more advanced economic development of Babylonia. Babylonia was located on a number of great caravan routes, whereas Egypt stood in semi-isolation. Also, the relatively peaceful Nile did not demand such extensive engineering and administrative efforts as did the more erratic Tigris and Euphrates.

Nevertheless, until the recent deciphering of so many Babylonian mathematical tablets, Egypt was long the richest field for ancient historical research. The reasons for this lie in the veneration that the Egyptians had for their dead

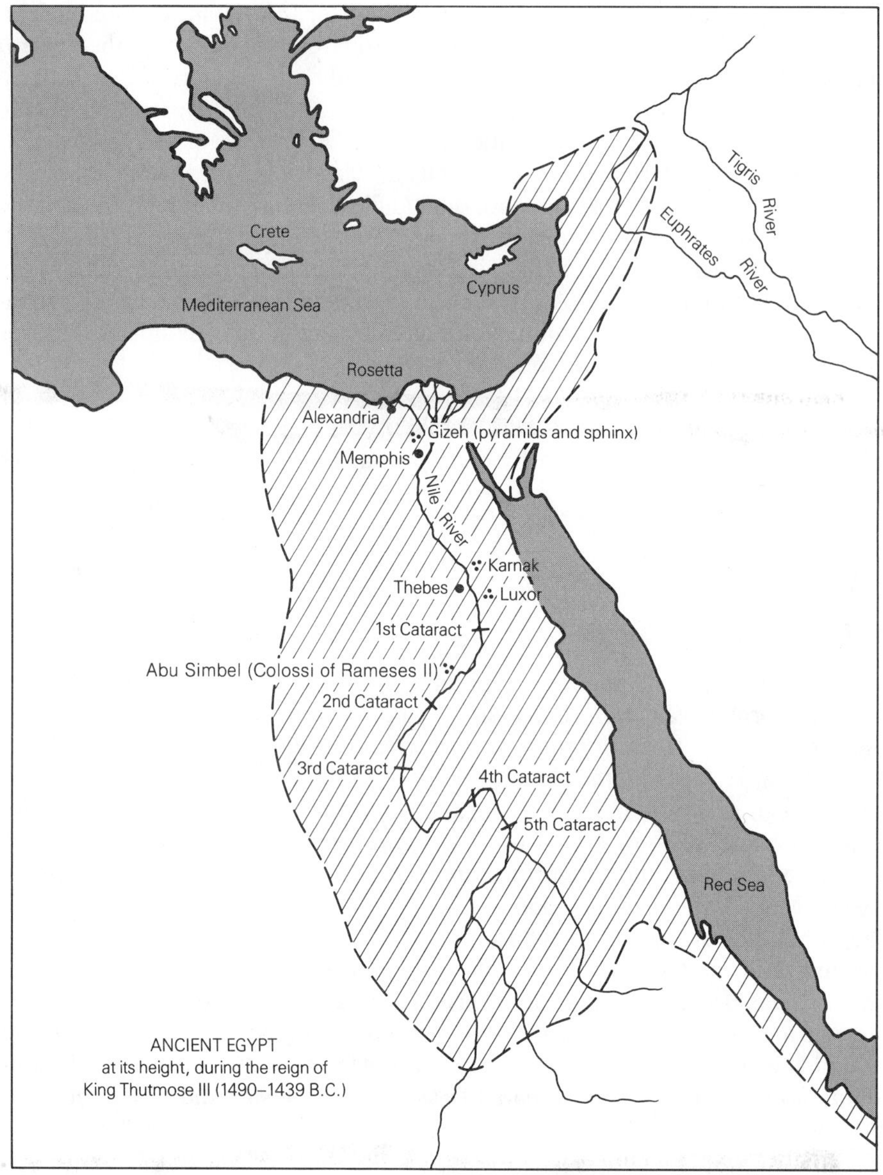

and in the unusually dry climate of the region. The former led to the erection of long-lasting tombs and temples with richly inscribed walls, and the latter preserved many papyri and objects that would otherwise have perished.

Following is a chronological list of some of the tangible items bearing on the mathematics of ancient Egypt. In addition to these items, there are numerous wall inscriptions and minor papyri that contribute to our knowledge.

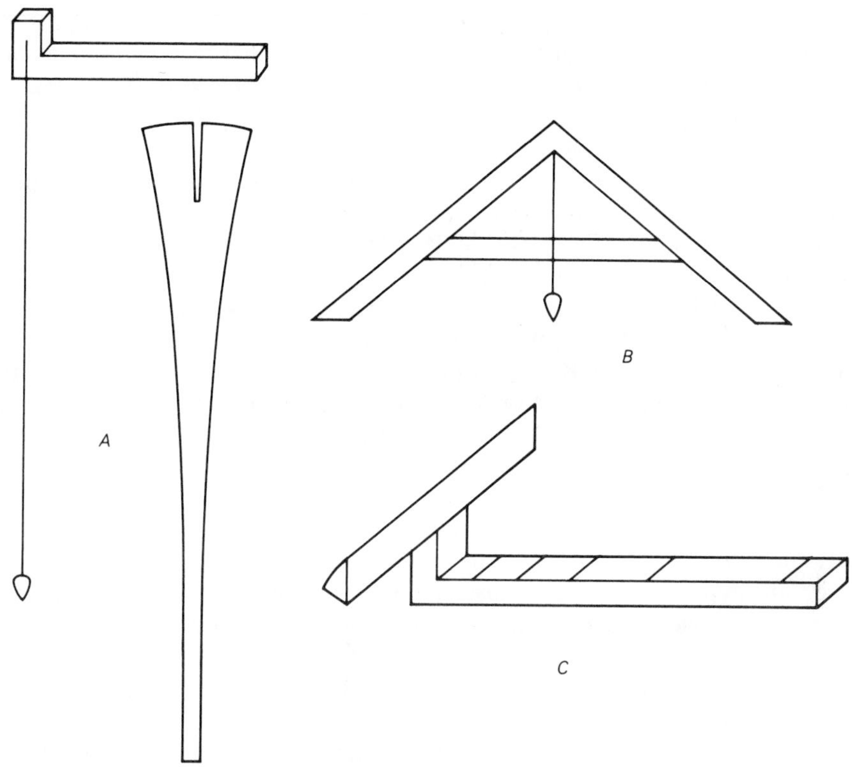

Sketches of some ancient Egyptian instruments
A. Oldest extant astronomical instrument (plumb line and sight rod) (in the Berlin Museum). With the aid of the plumb line, an observer could hold the rod vertically over a given point and sight through the slit to some object, such as the North Star.
B. A level (exhibited in the Museum at Cairo).
C. Oldest extant sundial (in the Berlin Museum). In the morning the crosspiece would be turned to the east and in the afternoon to the west.

1. ***3100 B.C.*** In a museum at Oxford is a royal Egyptian mace dating from this time. On the mace are several numbers in the millions and hundred thousands, written in Egyptian hieroglyphics, recording exaggerated results of a successful military campaign.

2. ***2600 B.C.*** The Great Pyramid at Gizeh was erected about 2600 B.C. and undoubtedly involved some mathematical and engineering problems. The structure covers thirteen acres and contains over 2,000,000 stone blocks, averaging 2.5 tons in weight, very carefully fitted together. These stone blocks were brought from sandstone quarries located on the other side of the Nile. Some chamber roofs are made of fifty-four-ton granite blocks, twenty-seven feet long and four feet thick, hauled from a quarry 600 miles away and set 200 feet above ground. It is reported that the sides of the square base involve a relative error of less than 1/14,000, and that the relative error in the right angles at the corners

does not exceed 1/27,000. The engineering skill implied by these impressive statistics is considerably diminished when we realize that the task was accomplished by an army of 100,000 laborers working for a period of thirty years.

The Great Pyramid is the largest of three pyramids sitting on the desert at Gizeh, a little south of the site of present-day Cairo. These huge structures were built as royal tombs. The Egyptians believed in an afterlife that depended upon the preservation of the deceased body. Embalming was accordingly developed, and valuables and objects of daily life were placed in the tombs for use in the afterlife.

The Great Pyramid (originally some 481 feet high) was built to house the body of Pharaoh Khufu (Cheops). The other two lesser pyramids at Gizeh were constructed as tombs for Khafre (Chephren) and Menkaure (Mycerinus), the two pharaohs who immediately succeeded Khufu. There are about eighty Egyptian pyramids still standing. The Great Pyramid became known as one of the Seven Wonders of the Ancient World.[4]

3. ***1850 B.C.*** This is the approximate date of the Golenischev, or Moscow, papyrus, a mathematical text containing twenty-five problems that were already old when the manuscript was compiled. The papyrus, which was purchased in Egypt in 1893 by the Russian collector Golenischev, now reposes in the Moscow Museum of Fine Arts. It was published with editorial comment in 1930. It is about eighteen feet long and about three inches high. For a sample of problems from the papyrus, see Problem Studies 2.14 and 2.15. The problem discussed in Problem Study 2.14 is particularly remarkable.

4. ***1850 B.C.*** The oldest extant astronomical instrument, a combination plumb line and sight rod, dates from this time and is preserved in the Berlin Museum.

5. ***1650 B.C.*** This is the approximate date of the Rhind (or Ahmes) papyrus, a mathematical text in the form of a practical handbook, which contains eighty-five problems copied in hieratic writing by the scribe Ahmes from an earlier work. The papyrus was purchased in 1858 in Egypt by the Scottish Egyptologist A. Henry Rhind and then later acquired by the British Museum. This and the Moscow papyrus are our chief sources of information concerning ancient Egyptian mathematics. The Rhind papyrus was published in 1927. It is about eighteen feet long and about thirteen inches high. When the papyrus arrived at the British Museum, however, it was shorter and in two pieces, with a central portion missing. About four years after Rhind purchased his papyrus, the American Egyptologist Edwin Smith (d. 1906) bought in Egypt what he thought was a medical papyrus. The Smith purchase was given to the New York Historical Society in 1932, where antiquarians discovered that it was a pasted-up deception, and that beneath the fraudulent covering lay the missing

---

[4] The Seven Wonders of the Ancient World are as follows: (1) The Great Pyramid of Egypt, (2) the Hanging Gardens of Babylon, (3) the Statue of Zeus at Olympia, (4) the Temple of Diana at Ephesus, (5) the Mausoleum at Halicarnassus, (6) the Colossus of Rhodes, and (7) the Pharos Lighthouse at Alexandria. Of the Seven Wonders, only the Great Pyramid is still standing.

piece of the Ahmes papyrus. The Society accordingly gave the scroll to the British Museum, thus completing the entire Ahmes work.

The Rhind papyrus is a rich primary source of ancient Egyptian mathematics, describing the Egyptian methods of multiplying and dividing, the Egyptian use of unit fractions, their employment of false position, their solution of the problem of finding the area of a circle, and many applications of mathematics to practical problems. The reader will find much of this material in the succeeding sections of the chapter and in Problem Studies 2.9, 2.11, 2.12, and 2.13.

6. ***1500 B.C.*** The largest existing obelisk, erected before the Temple of the Sun at Thebes, was quarried about this time. It is 105 feet long with a square base 10 feet to the side and weighs about 430 tons.

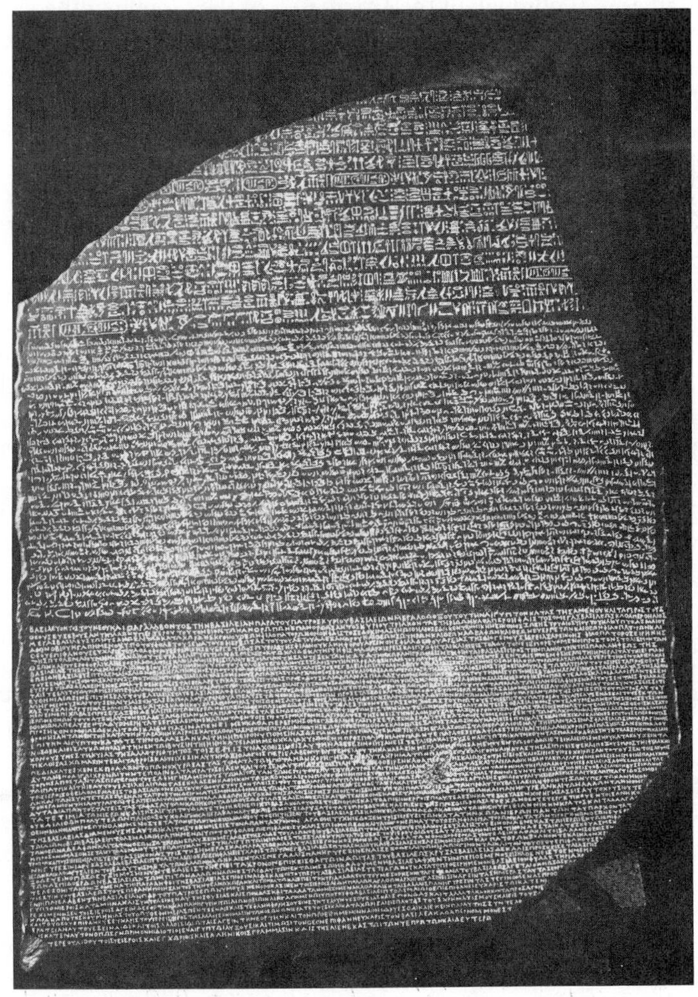

The Rosetta Stone (196 B.C.).
(Courtesy of the Trustees of the British Museum.)

7. ***1500 B.C.*** The Berlin Museum possesses an Egyptian sundial dating from this period. It is the oldest sundial extant.

8. ***1350 B.C.*** The Rollin papyrus of about 1350 B.C., now preserved in the Louvre, contains some elaborate bread accounts showing the practical use of large numbers at the time.

9. ***1167 B.C.*** This is the date of the Harris papyrus, a document prepared by Rameses IV when he ascended the throne. It sets forth the great works of his father, Rameses III. The listing of the temple wealth of the time furnishes the best example of practical accounts that has come to us from ancient Egypt.

Among other great structures of ancient Egypt that involved some engineering prowess are the Colossi of Rameses II at Abu Simbel, the Great Sphinx standing near the Great Pyramid at Gizeh, and the Temple of Amon-Re at Karnak. Rameses II completed the Great Hall of the temple in the 1200s B.C.; with columns seventy-eight feet tall, it was the largest columned hall ever built by man.

Ancient Egyptian sources more recent than those just listed show no appreciable gain in either mathematical knowledge or mathematical technique. In fact, there are instances showing definite regression.

Ability to read Egyptian hieroglyphic and demotic characters resulted from the successful decipherment by Jean François Champollion (1790–1832) of inscriptions on the Rosetta Stone, a polished basaltic slab that was found in 1799, during Napoleon's fateful Egyptian campaign, by French engineers while they were digging foundations for a fort near the Rosetta branch in the delta of the Nile. The stone measures three feet and seven inches by two feet and six inches, and the inscriptions on it give a common message repeated in Egyptian hieroglyphic, Egyptian demotic, and Greek. Since scholars were able to read the Greek, the stone furnished a clue to the decipherment of ancient Egyptian writing. The stone was engraved in 196 B.C., and as part of the treaty of capitulation when the French surrendered to the British, it went to England, where it now rests in the British Museum.

2–8 Arithmetic and Algebra

All of the 110 problems found in the Moscow and Rhind papyri are numerical, and many of them are very simple. Although most of the problems have a practical origin, there are some of a theoretical nature.

One consequence of the Egyptian numeral system is the additive character of the dependent arithmetic. Thus, multiplication and division were usually performed by a succession of doubling operations, based on the fact that any number can be represented as a sum of powers of 2. As an example of multiplication, let us find the product of 26 and 33. Since $26 = 16 + 8 + 2$, we have merely to add these multiples of 33. The work may be arranged as follows:

```
    1    33
  * 2    66
    4   132
```

* 8	264
*16	528
	858

Addition of the proper multiples of 33, that is, those indicated by an asterisk, gives the answer 858. Again, to divide 753 by 26, we successively double the divisor 26 up to the point where the next doubling would exceed the dividend 753. The procedure is shown below.

1	26
2	52
* 4	104
* 8	208
*16	416
28	

Now, since

$$\begin{aligned} 753 &= 416 + 337 \\ &= 416 + 208 + 129 \\ &= 416 + 208 + 104 + 25, \end{aligned}$$

we see, noting the starred items in the column above, that the quotient is 16 + 8 + 4 = 28, with a remainder of 25. This Egyptian process of multiplication and division not only eliminates the necessity of learning a multiplication table, but is so convenient on the abacus that it persisted as long as that instrument was in use, and even for some time beyond.

The Egyptians endeavored to avoid some of the computational difficulties encountered with fractions by representing all fractions, except $\frac{2}{3}$, as the sum of so-called **unit fractions,** or fractions with unit numerators. This reduction was made possible by tables so representing fractions of the form $2/n$, the only case necessary because of the dyadic nature of Egyptian multiplication. The problems of the Rhind papyrus are preceded by such a table for all odd n from 5 to 101. Thus, we find $\frac{2}{7}$ expressed as $\frac{1}{4} + \frac{1}{28}$; $\frac{2}{97}$ as $\frac{1}{56} + \frac{1}{679} + \frac{1}{776}$; and $\frac{2}{99}$ as $\frac{1}{66} + \frac{1}{198}$. Only one decomposition is offered for any particular case. The table is utilized in some of the problems of the papyrus.

Unit fractions were denoted in Egyptian hieroglyphics by placing an elliptical symbol above the denominator number. A special symbol was also used for the exceptional $\frac{2}{3}$, and another symbol sometimes appeared for $\frac{1}{2}$. These symbols are shown below.

$= \frac{1}{3}$, $= \frac{1}{4}$,

or $= \frac{1}{2}$,

$= \frac{2}{3}$.

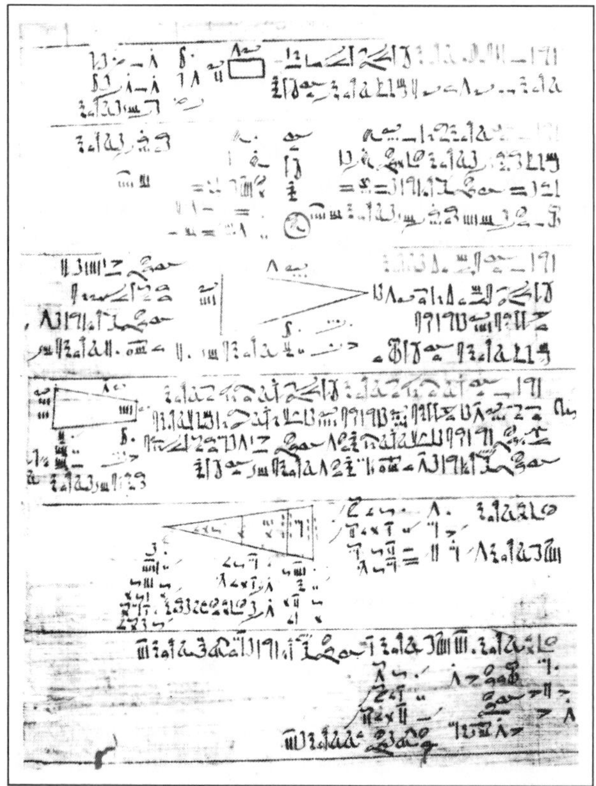

A portion of the Rhind papyrus.
(British Museum.)

There are interesting theories to explain how the Egyptians obtained their unit fraction decompositions (see Problem Study 2.9).

Many of the 110 problems in the Rhind and Moscow papyri show their practical origin by dealing with questions regarding the strength of bread and of beer, with feed mixtures for cattle and domestic fowl, and with the storage of grain. Many of these require nothing more than a simple linear equation and are generally solved by the method later known in Europe as the **rule of false position.** Thus, to solve

$$x + x/7 = 24$$

assume any convenient value for x, say $x = 7$. Then $x + x/7 = 8$, instead of 24. Since 8 must be multiplied by 3 to give the required 24, the correct x must be 3(7), or 21.

There are some theoretical problems involving arithmetic and geometric progressions. (See, for example, Problem Study 2.12**(c)** and Section 2–10.) A papyrus of about 1950 B.C., found at Kahun, contains the following problem:

"A given surface of one hundred units of area shall be represented as the sum of two squares whose sides are to each other as 1 : 3/4." Here we have $x^2 + y^2 = 100$ and $x = 3y/4$. Elimination of x yields a pure quadratic in y. We may, however, solve the problem by false position. Thus, take $y = 4$. Then $x = 3$, and $x^2 + y^2 = 25$, instead of 100. We must therefore correct x and y by doubling the initial values, obtaining $x = 6$, $y = 8$.

There is some symbolism in Egyptian algebra. In the Rhind papyrus, we find symbols for *plus* and *minus*. The first of these symbols represents a pair of legs walking from left to right and the other a pair of legs walking from right to left. Symbols, or ideograms, were also employed for *equals* and for the *unknown*.

2–9 Geometry

Twenty-six of the 110 problems in the Moscow and Rhind papyri are geometric. Most of these problems stem from mensuration formulas needed for computing land areas and granary volumes. The area of a circle is taken as equal to that of the square on $\frac{8}{9}$ of the diameter, and the volume of a right circular cylinder as the product of the area of the base by the length of the altitude. Recent investigations seem to show that the ancient Egyptians knew that the area of any triangle is given by half the product of base and altitude. Some of the problems concern themselves with the cotangent of the dihedral angle between the base and a face of a pyramid (see Problem Study 2.11), and others show an acquaintance with the elementary theory of proportion. Contradicting repeated and apparently unfounded stories, no documentary evidence has been found showing that the Egyptians were aware of even a particular case of the Pythagorean theorem. In later Egyptian sources, the incorrect formula $K = (a + c)(b + d)/4$ is used for finding the area of an arbitrary quadrilateral with successive sides of lengths a, b, c, d.

The existence, in the Moscow papyrus, of a numerical example of the correct formula for the volume of a frustum of a square pyramid is quite remarkable (see Problem Study 2.14(a)). No other unquestionably genuine example of this formula has been found in ancient Oriental mathematics, and several conjectures have been formulated to explain how it might have been discovered. E. T. Bell aptly refers to this early Egyptian example as the "greatest Egyptian pyramid."

2–10 A Curious Problem in the Rhind Papyrus

Although little difficulty was encountered in deciphering and then in interpreting most of the problems in the Rhind papyrus, there is one problem (Problem Number 79) for which the interpretation is not so certain. In it occurs the following curious set of data, here transcribed:

Estate	
Houses	7
Cats	49
Mice	343
Heads of wheat	2401
Hekat measures	16807
	19607

One easily recognizes the numbers as the first five powers of 7, along with their sum. Because of this, it was at first thought that perhaps the writer was here introducing the symbolic terminology *houses, cats,* and so on, for *first power, second power,* and so on.

A more plausible and interesting explanation, however, was given by the historian Moritz Cantor in 1907. He saw in this problem an ancient forerunner of a problem that was popular in the Middle Ages, and that was given by Leonardo Fibonacci in 1202 in his *Liber abaci.* Among the many problems occurring in this work is the following: "There are seven old women on the road to Rome. Each woman has seven mules; each mule carries seven sacks; each sack contains seven loaves; with each loaf are seven knives; and each knife is in seven sheaths. Women, mules, sacks, loaves, knives, and sheaths, how many are there in all on the road to Rome?" As a later and more familiar version of the same problem, we have the old English children's rhyme:

As I was going to St. Ives
I met a man with seven wives;
Every wife had seven sacks;
Every sack had seven cats;
Every cat had seven kits.
Kits, cats, sacks, and wives,
How many were going to St. Ives?

According to Cantor's interpretation, the original problem in the Rhind papyrus might then be formulated somewhat as follows: "An estate consisted of seven houses; each house had seven cats; each cat ate seven mice; each mouse ate seven heads of wheat; and each head of wheat was capable of yielding seven hekat measures of grain. Houses, cats, mice, heads of wheat, and hekat measures of grain, how many of these in all were in the estate?"

Here, then, may be a problem that has been preserved as part of the puzzle lore of the world. It was apparently already old when Ahmes copied it, and older by close to 3000 years when Fibonacci incorporated a version of it in his *Liber abaci.* More than 750 years later, we are reading another variant of it to our children. One cannot help wondering if a surprise twist such as occurs in the old English rhyme also occurred in the ancient Egyptian problem.

There are many puzzle problems popping up every now and then in our present-day magazines that have medieval counterparts. How much further back some of them go is now almost impossible to determine.[6]

---

[6] See D. E. Smith, "On the origin of certain typical problems," *The American Mathematical Monthly* 24 (February 1917): 64–71.

Problem Studies

2.1 Regular Numbers

A number is said to be (sexagesimally) **regular** if its reciprocal has a finite sexagesimal expansion (that is, a finite expansion when expressed as a radix fraction for base 60). With the exception of a single tablet in the Yale collection, all Babylonian tables of reciprocals contain only reciprocals of regular numbers. A Louvre tablet of about 300 B.C. contains a regular number of 7 sexagesimal places and its reciprocal of 17 sexagesimal places.

(a) Show that a necessary and sufficient condition for n to be regular is that $n = 2^a 3^b 5^c$, where a, b, c are nonnegative integers.

(b) Express, by finite sexagesimal expansions, the numbers $\frac{1}{2}$, $\frac{1}{3}$, $\frac{1}{5}$, $\frac{1}{15}$, $\frac{1}{360}$, $\frac{1}{3600}$.

(c) Generalize (a) to numbers having general base b.

(d) List all the sexagesimally regular numbers less than 100, and then list all the decimally regular numbers less than 100.

(e) Show that the decimal representation of $\frac{1}{7}$ has six-place periodicity. How many places are there in the periodicity of the sexagesimal representation of $\frac{1}{7}$?

2.2 Compound Interest

There are tablets in the Berlin, Yale, and Louvre collections containing problems in compound interest, and there are some Istanbul tablets that appear originally to have had tables of a^n for $n = 1$ to 10 and $a = 9, 16, 100$, and 225. With such tables, one can solve exponential equations of the type $a^x = b$.

(a) On a Louvre tablet of about 1700 B.C. occurs the problem: Find how long it will take for a certain sum of money to double itself at compound annual interest of 20 percent. Solve this problem by modern methods.

(b) Solve the problem of (a) by first finding $(1.2)^3$ and $(1.2)^4$ and then, by linear interpolation, x such that $(1.2)^x = 2$. Show that the result so obtained agrees with the Babylonian solution 3;47,13,20 (expressed sexagesimally) of this problem.[7]

2.3 Quadratic Equations

(a) A Babylonian problem asks for the side of a square if the area of the square diminished by the side of the square is the (sexagesimal) number 14,30. The solution of the problem is described as follows: "Take half of 1, which is 0;30; multiply 0;30 by 0;30, which is 0;15; add the 0;15 to 14,30 to obtain 14,30;15. This last is the square of 29;30. Now add 0;30 to 29;30; the result is 30, which is the side of the square." Show that this Babylonian solution is exactly equivalent to solving the quadratic

---

[7] As an illustration, the expression 9,20,8;30,10,23 means $9(60)^2 + 20(60) + 8 + 30/60 + 10/(60)^2 + 23/(60)^3$.

equation

$$x^2 - px = q$$

by substituting in the formula

$$x = \sqrt{(p/2)^2 + q} + p/2.$$

(b) Another Babylonian text solves the quadratic equation

$$11x^2 + 7x = 6;15$$

by first multiplying through by 11 to obtain

$$(11x)^2 + 7(11)x = 1,8;45,$$

which, by setting $y = 11x$, has the "normal form"

$$y^2 + py = q.$$

This is solved by substituting in the formula

$$y = \sqrt{(p/2)^2 + q} - p/2.$$

Finally, $x = y/11$.

Show that any quadratic equation $ax^2 + bx + c = 0$ can, by a similar transformation, be reduced to one of the normal forms

$$y^2 + py = q, \qquad y^2 = py + q, \qquad y^2 + q = py,$$

where p and q are both nonnegative. The solution of such three-term quadratic equations seems to have been beyond the capabilities of the ancient Egyptians.

2.4 Algebraic Geometry

(a) The algebraic character of Babylonian geometry problems is illustrated by the following, found on a Strassburg tablet of about 1800 B.C. "An area A, consisting of the sum of two squares is 1000. The side of one square is 10 less than $\frac{2}{3}$ of the side of the other square. What are the sides of the squares?" Solve this problem.

(b) On a Louvre tablet of about 300 B.C. are four problems concerning rectangles of unit area and given semiperimeter. Let the sides and semiperimeter be x, y, and a. Then we have

$$xy = 1, \qquad x + y = a.$$

Solve this system by eliminating y and thus obtaining a quadratic in x.

(c) Solve the system of (b) by using the identity

$$\left(\frac{x-y}{2}\right)^2 = \left(\frac{x+y}{2}\right)^2 - xy.$$

This is essentially the method used on the Louvre tablet. It is interesting that the identity appeared contemporaneously as Proposition 5 of Book II of Euclid's *Elements*.

(d) An Old Babylonian problem reads: "One leg of a right triangle is 50. A line parallel to the other leg and at distance 20 from that leg cuts off a right trapezoid of area 5,20. Find the lengths of the bases of the trapezoid." Solve this problem.

(e) Another Old Babylonian problem claims that an isosceles trapezoid with bases 14 and 50 and with sides 30 has area 12,48. Verify this.

(f) Still another Old Babylonian problem concerns a ladder of length 0;30 standing upright against a wall. The problem asks how far the lower end of the ladder will move out from the wall if the upper end slides down the wall a distance of 0;6. Solve this problem.

(g) A Seleucid tablet of 1500 years later proposes a problem similar to that of (f). Here a reed is given standing upright against a wall. The problem asks for the length of the reed if the top end of the reed slides down the wall 3 units when the lower end of the reed moves 9 units away from the wall. The answer is given as 15 units. Is this correct?

2.5 The Susa Tablets

(a) In 1936 a group of Old Babylonian tablets was lifted at Susa, about 200 miles from Babylon. One of the tablets compares the areas and the squares of the sides of regular polygons of 3, 4, 5, 6, and 7 sides. For the pentagon, hexagon, and heptagon, these ratios are given as 1;40, 2;37,30, and 3;41. Check these values for accuracy.

(b) On the same tablet considered in (a), the ratio of the perimeter of a regular hexagon to the circumference of the circumscribed circle is given as 0;57,36. Show that this leads to 3;7,30 or $3\frac{1}{8}$ as an approximation of π.

(c) On one of the Susa tablets appears the problem: "Find the circumradius of a triangle whose sides are 50, 50, and 60." Solve this problem.

(d) Another Susa tablet requests the sides x and y of a rectangle, given

$$xy = 20{,}0 \qquad \text{and} \qquad x^3d = 14{,}48{,}53{,}20,$$

where d is a diagonal of the rectangle. Solve this problem.

2.6 Cubics

(a) A Babylonian tablet has been discovered that gives the values of $n^3 + n^2$ for $n = 1$ to 30. Make such a table for $n = 1$ to 10.

(b) Find, by means of the above table, a root of the cubic equation $x^3 + 2x^2 - 3136 = 0$.

(c) A Babylonian problem of about 1800 B.C. seems to call for the solution of the simultaneous system $xyz + xy = 7/6$, $y = 2x/3$, $z = 12x$. Solve this system using the table of (a).

(d) Otto Neugebauer believes that the Babylonians were quite capable of reducing the general cubic equation to the "normal form" $n^3 + n^2 = c$, although there is as yet no evidence that they actually did do this. Show how such a reduction might be made.

(e) In connection with the table of (a), Neugebauer has noted that the Babylonians may well have observed the relation $\sum_{i=1}^{n} i^3 = \left(\sum_{i=1}^{n} i\right)^2$ for various values of n. Establish this relation by mathematical induction.

2.7 Square Root Approximations

It is known that the infinite series obtained by expanding $(a^2 + h)^{1/2}$ by the binomial theorem process converges to $(a^2 + h)^{1/2}$ if $-a^2 < h < a^2$.

(a) Establish the approximation formula

$$(a^2 + h)^{1/2} \approx a + \frac{h}{2a}, \qquad 0 < |h| < a^2.$$

(b) Take $a = \frac{4}{3}$ and $h = \frac{2}{9}$ in the approximation formula of (a), and thus find a Babylonian rational approximation for $\sqrt{2}$. Find a rational approximation for $\sqrt{5}$ by taking $a = 2$, $h = 1$.

(c) Establish the better approximation formula

$$(a^2 + h)^{1/2} \approx a + \frac{h}{2a} - \frac{h^2}{8a^3}, \qquad 0 < |h| < a^2,$$

and approximate $\sqrt{2}$ and $\sqrt{5}$ by using the same values for a and h as in (b).

(d) Take $a = \frac{3}{2}$ and $h = -\frac{1}{4}$ in the formula of (a) and find the ancient Babylonian approximation $\frac{17}{12}$ for $\sqrt{2}$.

(e) Take $a = \frac{17}{12}$ and $h = -\frac{1}{144}$ in the formula of (a) and find the value 1;24,51,10 for $\sqrt{2}$ as given on the Yale table tablet 7289.

2.8 Duplation and Mediation

The Egyptian process of multiplication later developed into a slightly improved method known as **duplation and mediation,** the purpose of which was mechanically to pick out the required multiples of one of the factors that have to be added in order to give the required product. Taking the example in the text, suppose we wish to multiply 26 by 33. We may successively halve the 26 and double the 33, thus

26	33
13	66*
6	132

3	264*
1	528*
	858

In the doubling column, we now add those multiples of 33 corresponding to the *odd* numbers in the halving column. Thus, we add 66, 264, and 528 to obtain the required product 858. The process of duplation and mediation is utilized by high-speed electronic computing machines.

(a) Multiply 424 by 137 using duplation and mediation.

(b) Prove that the duplation and mediation method of multiplication gives correct results.

(c) Find, by the Egyptian method, the quotient and remainder when 1043 is divided by 28.

2.9 Unit Fractions

(a) Show that $z/pq = 1/pr + 1/qr$, where $r = (p + q)/z$. This method for finding possible decompositions of a fraction into 2 unit fractions is indicated on a papyrus written in Greek probably sometime between A.D. 500 and 800, and found at Akhmim, a city on the Nile River.

(b) Take $z = 2, p = 1, q = 7$, and obtain the unit fraction decomposition of 2/7 as given in the Rhind papyrus.

(c) Represent 2/99 as the sum of two different unit fractions in 3 different ways.

(d) By taking $z = 1, p = 1, q = n$ in the relation of (a), obtain the more particular relation

$$1/n = 1/(n + 1) + 1/n(n + 1),$$

and show that when n is odd, this leads to a representation of $2/n$ as a sum of 2 unit fractions. Many of the entries in the Rhind papyrus can be obtained in this way.

(e) Show that if n is a multiple of 3, then $2/n$ can be broken into a sum of two unit fractions of which one is $1/(2n)$.

(f) Show that if n is a multiple of 5, then $2/n$ can be broken into a sum of two unit fractions of which one is $1/(3n)$.

(g) Show that for any positive integer n, $2/n$ can be expressed by the sum $1/n + 1/(2n) + 1/(3n) + 1/(6n)$. (In the $2/n$ table of the Rhind papyrus, only $\frac{2}{101}$ is expressed by this decomposition.)

(h) Show that if a rational number can be represented as a sum of unit fractions in one way, then it can be represented as a sum of unit fractions in an infinite number of ways.

2.10 The Sylvester Process

The British mathematician J. J. Sylvester (1814–1897) provided the following procedure for uniquely expressing any rational fraction between 0 and 1 as a sum of unit fractions:

1. Find the largest unit fraction (that is, the one with the smallest denominator) less than the given fraction.
2. Subtract this unit fraction from the given fraction.
3. Find the largest unit fraction less than the resulting difference.
4. Subtract again, and continue the process.
5. To find the largest unit fraction less than a given fraction, divide the denominator of the given fraction by the numerator of the fraction and take the next integer greater than the quotient as the denominator of the unit fraction sought.

(a) Express $\frac{2}{7}$ as a sum of unit fractions using the Sylvester process. Note that the decomposition is the same as that given in the $2/n$ table of the Rhind papyrus.

(b) Express $\frac{2}{97}$ as a sum of unit fractions using the Sylvester process. Note that the decomposition is different from that given in the $2/n$ table of the Rhind papyrus.

(c) Establish the rule given in the fifth step of the Sylvester process.

2.11 The Seqt of a Pyramid

(a) The Egyptians measured the steepness of a face of a pyramid by the ratio of the "run" to the "rise"—that is, by giving the horizontal departure of the oblique face from the vertical for each unit of height. The vertical unit was taken as the cubit and the horizontal unit as the hand; there were 7 hands in a cubit. Utilizing these units of measurement, the measure of steepness was called the **seqt** of the pyramid. Show that the seqt of a pyramid is 7 times the cotangent of the dihedral angle formed by the base and a face of the pyramid.

(b) In Problem 56 of the Rhind papyrus, one is asked to find the seqt of a pyramid 250 cubits high with a square base 360 cubits on a side. The answer is given as $5\frac{1}{25}$ hands per cubit. Is this correct?

(c) The great pyramid of Cheops has a square base 440 cubits on a side and a height of 280 cubits. What is the seqt of this pyramid?

(d) Problem 57 of the Rhind papyrus asks for the height of a square pyramid with a seqt of 5 hands and 1 finger per cubit and a base 140 cubits on a side. Solve this problem, where there are 5 fingers in a hand.

2.12 Egyptian Algebra

The following problems are found in the Rhind papyrus.

(a) "If you are asked, what is $\frac{2}{3}$ of $\frac{1}{5}$, take the double and the six-fold; that is $\frac{2}{3}$ of it. One must proceed likewise for any other fraction." Interpret this and prove the general statement.

(b) "A quantity, its $\frac{2}{3}$, its $\frac{1}{2}$, and its $\frac{1}{7}$, added together, become 33. What is the quantity?" Solve this problem by the rule of false position.

(c) "Divide 100 loaves among 5 men in such a way that the shares received shall be in arithmetic progression and that one-seventh of the sum of the largest 3 shares shall be equal to the sum of the smallest two." Solve this problem using modern methods.

2.13 Egyptian Geometry

(a) In the Rhind papyrus, the area of a circle is repeatedly taken as equal to that of the square of $\frac{8}{9}$ of the diameter. This leads to what value for π?

(b) Form an octagon from a square of side 9 units by trisecting the sides of the square and then cutting off the 4 triangular corners. The area of the octagon looks, by eye, to differ very little from the area of the circle inscribed in the square. Show that the area of the octagon is 63 square units, whence the area of the circle cannot be far from that of a square of 8 units on a side. There is evidence in Problem 48 of the Rhind papyrus that the formula for the area of a circle as given in (a) may have been arrived at in this way.

(c) Prove that of all triangles having a pair of given sides, the one in which these sides form a right angle is the maximum.

(d) Denote the lengths of the sides AB, BC, CD, DA of a quadrilateral $ABCD$ by a, b, c, d, and let K represent the area of the quadrilateral. Show that $K \leqq (ad + bc)/2$, equality holding if and only if angles A and C are right angles.

(e) For the hypothesis of (d) now show that $K \leqq (a + c)(b + d)/4$, equality holding if and only if $ABCD$ is a rectangle. Thus, the Egyptian formula for the area of a quadrilateral, cited in Section 2–9, gives too large an answer for all nonrectangular quadrilaterals.

(f) An extant deed from Edfu, dating some 1500 years after the Rhind papyrus, employs the inexact Egyptian formula for the area of a quadrilateral. From this formula, the author of the deed deduces, as a corollary, that the area of a triangle is half the sum of two sides multiplied by half the third side. Show how this corollary can be so deduced. Is the corollary correct?

(g) It looks to the eye that the area of a circle may be exactly halfway between those of an inscribed and a circumscribed square. Show that this is equivalent to taking $\pi = 3$.

2.14 The Greatest Egyptian Pyramid

(a) In Problem 14 of the Moscow papyrus, we find the following numerical example: "If you are told: A truncated pyramid of 6 for the vertical height by 4 on the base by 2 on the top. You are to square this 4, result 16. You are to double 4, result 8. You are to square 2, result 4. You are to add the 16, the 8, and the 4, result 28. You are to take one third of 6, result 2. You are to take 28 twice, result 56. See, it is 56. You will find it right." Show that this illustrates the general formula.

$$V = (\tfrac{1}{3})h(a^2 + ab + b^2)$$

giving the volume of a frustum of a square pyramid in terms of the height h and the sides a and b of the bases.

(b) If m and n are two positive numbers, $m \geqq n$, then we define the **arithmetic mean**, the **heronian mean**, and the **geometric mean** of m and

n to be $A = (m + n)/2$, $R = (m + \sqrt{mn} + n)/3$, $G = \sqrt{mn}$. Show that $A \geqq R \geqq G$, the equality signs holding if and only if $m = n$.

(c) Assuming the familiar formula for the volume of any pyramid (volume equals one-third the product of base and altitude), show that the volume of a frustum of the pyramid is given by the product of the height of the frustrum and the heronian mean of the bases of the frustum.

(d) Let a, b, and h denote the lengths of an edge of the lower base, an edge of the upper base, and the altitude of a frustum T of a regular square pyramid. Dissect T into: (1) A rectangular parallelepiped P of upper base b^2 and altitude h. (2) 4 right triangular prisms A, B, C, and D each of volume $b(a - b)h/4$, (3) 4 square pyramids E, F, G, H each of volume $(a - b)^2h/12$. Now obtain the formula of (a) for the volume of T.

(e) Consider the dissected frustum of (d). Horizontally slice P into 3 equal parts, each of altitude $h/3$, and designate one of these slices by J. Combine A, B, C, D into a rectangular parallelepiped Q of base $b(a - b)$ and altitude h, and horizontally slice Q into three equal parts, each of altitude $h/3$. Replace E, F, G, H by a rectangular parallelepiped R of base $(a - b)^2$ and altitude $h/3$. Combine one slice of P with one slice of Q to form a rectangular parallelepiped K of base ab and altitude $h/3$. Combine one slice of P, two slices of Q, and R to form a rectangular parallelepiped L of base a^2 and altitude $h/3$. The volume of T is then

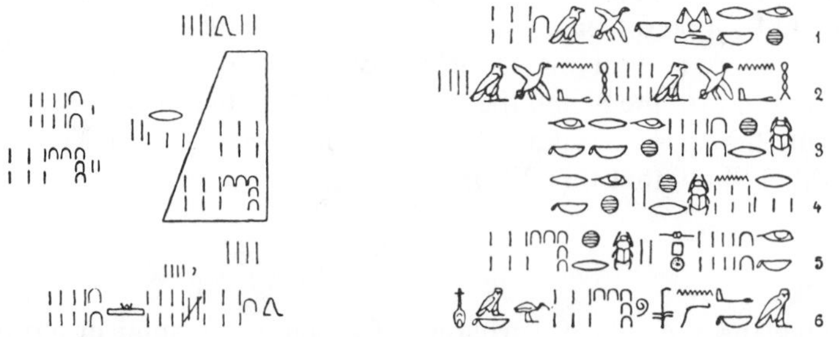

Problem 14 of the Moscow papyrus, with hieroglyphic transcription of the hieratic text.

equal to the sum of the volumes of the three rectangular parallelepipeds J, K, L. Using this fact, find the formula of (a) for the volume of T. It has been suggested that the Egyptian formula of (a) may have been obtained in this fashion. The procedure assumes familiarity with the formula for the volume of a (regular square) pyramid.

2.15 Some Problems from the Moscow Papyrus

Solve the following two problems found in the Moscow papyrus:

(a) The area of a rectangle is 12 and the width is $\frac{3}{4}$ the length. What are the dimensions?

(b) One leg of a right triangle is $2\frac{1}{2}$ times the other and the area is 20. What are the dimensions?

2.16 The 3,4,5 Triangle

There are reports that ancient Egyptian surveyors in the time of the pharaohs laid out right angles by constructing 3, 4, 5 triangles with a rope divided into 12 equal parts by 11 knots. Since there is no documentary evidence to the effect that these Egyptians were aware of even a particular case of the Pythagorean theorem, the following purely academic problem arises:[8] Show, without using the Pythagorean theorem, its converse, or any of its consequences, that the 3, 4, 5 triangle is a right triangle. Solve this problem by means of Figure 5, which appears in the *Chóu-peï,* the oldest known Chinese mathematical work, which may date back to the second millennium B.C.

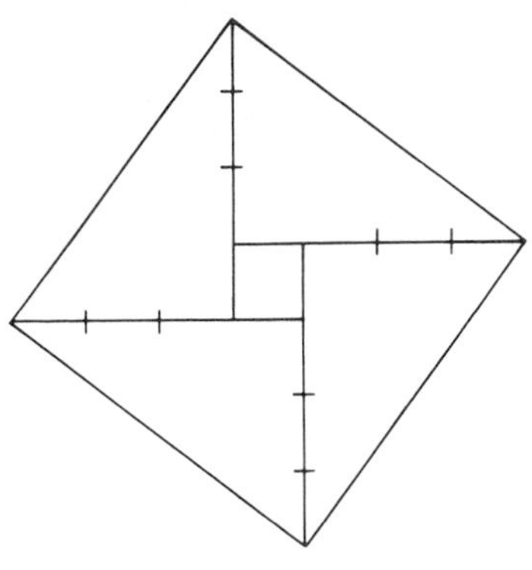

FIGURE 5

2.17 The Cairo Mathematical Papyrus

The so-called Cairo Mathematical papyrus was unearthed in 1938 and examined in 1962. Dating from about 300 B.C., this papyrus contains forty mathematical problems, nine of which deal exclusively with the Pythagorean theorem

[8] See Victor Thébault, "A note on the Pythagorean theorem," *The Mathematics Teacher* 43 (October 1950): 278.

and show that the Egyptians of that time not only knew that the 3,4,5 triangle is right angled, but that the 5,12,13 and 20,21,29 triangles were right angled as well. Solve the following problems found in the Cairo Mathematical papyrus.

(a) A ladder of 10 cubits has its foot 6 cubits from a wall. To what height does the ladder reach?

(b) A rectangle with an area of 60 square cubits has a diagonal of 13 cubits. Find the sides of the rectangle.

(c) A rectangle with an area of 60 square cubits has a diagonal of 15 cubits. Find the sides of the rectangle.

The scribe's method for solving (b) and (c) is as follows: Denoting the sides, diagonal, and area of a rectangle by x, y, d, and A, we have

$$x^2 + y^2 = d^2 \quad \text{and} \quad xy = A,$$

which yield

$$x^2 + 2xy + y^2 = d^2 + 2A, \qquad x^2 - 2xy + y^2 = d^2 - 2A,$$

or

$$(x + y)^2 = d^2 + 2A, \qquad (x - y)^2 = d^2 - 2A.$$

In (b), $d^2 + 2A$ and $d^2 - 2A$ are perfect squares, and one readily finds values for $x + y$ and $x - y$, and so forth. In (c), $d^2 + 2A$ and $d^2 - 2A$ are not perfect squares, and the scribe uses the approximation formula

$$\sqrt{a^2 + b} \approx a + b/2a,$$

arriving at

$$\sqrt{345} = \sqrt{18^2 + 21} \approx 18 + \tfrac{21}{36} = 18 + \tfrac{1}{2} + \tfrac{1}{12}$$

and

$$\sqrt{105} = \sqrt{10^2 + 5} \approx 10 + \tfrac{5}{20} = 10 + \tfrac{1}{4}.$$

Essay Topics

2/1 The "do-thus-and-so" procedure in teaching portions of elementary mathematics today.

2/2 Inductive (or empirical) mathematics versus deductive (or demonstrative) mathematics.

2/3 The pedagogical value of inductive mathematics.

2/4 The importance of inductive procedures in mathematical discovery.

2/5 The comparative influence in the rise of early geometry of an interest in astronomy and a need for surveying.

2/6 The importance of early religious ritual in the origin of geometry.
2/7 Grotefend, Rawlinson, and the Behistun Rock.
2/8 Napoleon, Champollion, and the Rosetta Stone.
2/9 The origin of certain typical problems.
2/10 Representation by unit fractions.
2/11 The Egyptian Mathematical Leather Roll.
2/12 The Babylonian tablet, Yale Babylonian Collection, 7289.
2/13 Pyramidology.

Bibliography

AABOE, ASGER *Episodes from the Early History of Mathematics*. New Mathematical Library no. 13. New York: Random House and L. W. Singer, 1964.

BALL, W. W. R., and H. S. M. COXETER *Mathematical Recreations and Essays*. 12th ed. Toronto: University of Toronto Press, 1974.

BRATTON, FRED *A History of Egyptian Archaeology*. New York: Thomas Y. Crowell, 1968.

BUDGE, E. A. W. *The Rosetta Stone*. Rev. ed. London: Harrison and Sons, 1950.

BUNT, L. N. H.; P. S. JONES; and J. D. BEDIENT *The Historical Roots of Elementary Mathematics*. Englewood Cliffs, N.J.: Prentice-Hall, 1976.

CHACE, A. B.; L. S. BULL; H. P. MANNING; and R. C. ARCHIBALD, eds. *The Rhind Mathematical Papyrus*. 2 vols. Buffalo, N.Y.: Mathematical Association of America, 1927–1929. Republished, in large part, by the National Council of Teachers of Mathematics, 1979.

CHIERA, EDWARD *They Wrote on Clay*. Chicago: University of Chicago Press, 1938.

COOLIDGE, J. L. *A History of Geometrical Methods*. New York: Oxford University Press, 1940.

GILLINGS, R. J. *Mathematics in the Time of the Pharaohs*. Cambridge, Mass.: The M.I.T. Press, 1972.

KRAITCHIK, MAURICE *Mathematical Recreations*. New York: W. W. Norton, 1942.

NEUGEBAUER, OTTO *The Exact Sciences in Antiquity*. 2d ed. New York: Harper and Row, 1962. Reprinted by Dover, New York.

———, and A. J. SACHS, eds. *Mathematical Cuneiform Texts*. American Oriental Series, vol. 29. New Haven: American Oriental Society, 1946.

ORE, OYSTEIN *Number Theory and Its History*. New York: McGraw-Hill, 1948.

PARKER, R. A. *The Calendars of Ancient Egypt*. Chicago: University of Chicago Press, 1950.

SANFORD, VERA *The History and Significance of Certain Standard Problems in Algebra*. New York: Teachers College, Columbia University, 1927.

VAN DER WAERDEN, B. L. *Science Awakening*. Translated by Arnold Dresden. New York: Oxford University Press, 1961. Paperback ed. New York: John Wiley, 1963.

——— *Geometry and Algebra in Ancient Civilizations*. New York: Springer-Verlag, 1983.

Cultural Connection III

THE PHILOSOPHERS OF THE AGORA

Hellenic Greece—ca. 800–336 B.C.
(to accompany Chapters Three and Four)

As we have seen, an agricultural revolution beginning in approximately 3000 B.C. sparked a long period of intellectual and scientific progress. In agricultural regions called "cradles of civilization" (the Middle East, China, and Egypt), people built the first cities, sprawling irrigation projects, and towering monuments like the pyramids, the Sphinx, and the Hanging Gardens of Babylon. These same people invented writing, early mathematics, astrology, and metallurgy. Complex systems of government, city-states and small empires, replaced the tribe as the principal forms of political organization. Probably the most impressive cultural achievements of the agricultural revolution took place in Greece during its Hellenic Age (ca. 800–336 B.C.) and in China during its early Classical period (ca. 600–221 B.C.). We will look at China in Cultural Connection V: The Asian Empires. In the following pages, we will explore the society and culture of the ancient Greeks.

Without doubt, the greatest scientists of the ancient world lived in tiny Greece, a collection of city-states perched atop a jumble of rocky islands and peninsulas at the eastern end of the Mediterranean Sea, on the very edge of Middle Eastern civilization. The agricultural revolution reached Greece from Egypt and the Middle East about 2000 B.C., shortly after the founding of the Babylonian Empire by the Amorites. Within 300 years, a mysterious, highly advanced, literate culture had evolved on the Greek island of Crete. This civilization, designated Minoan by historians, flourished between 1700 and 1200 B.C. The Greek mainland was populated by a less advanced, more warlike, but also literate people, the Myceneans, who, according to legend, fought the Trojan War. Between 1200 and 1150 B.C., these civilizations were destroyed abruptly by barbaric invaders from Asia, the Dorians, a tribe of herders closely related to the Aryans, whom we have met previously as the supplanters of the Indus River civilization in India. The Dorians settled on the lands they conquered and adopted much of the farming culture of the previous inhabitants. By 800 B.C., written language, which was lost after the collapse of the Minoan and Mycenean civilizations, was reintroduced by Phoenician merchants from the Middle East. The period of Greek history that followed (from ca. 800–336 B.C.), termed the *Hellenic Age* by historians, was an era of breathtaking intellectual and scientific progress—one of the most remarkable epochs of human achievement in history!

Hellenic Greece was a mosaic of city-states and small, scattered farms. Not a broad plain cleaved by great, muddy rivers, like Egypt and Babylonia, it was instead a country cut by steep mountain ranges and long, winding bays that bit deep inland from the sea. Its valleys were narrow and clotted with large stones, its rivers shallow, and its soil parched. Its city-states were separated from each other by rugged, steep-cliffed mountains; the farms in the little valleys were divided by rocky outcroppings and patches of infertile ground. Due in part to their isolation and in part to the small size of their neighbors, the small cities and farms of Hellenic Greece were somewhat protected against the designs of aggrandizers. To be sure, the Greeks fought numerous wars, but rarely was one city-state successful in annexing another. Some wealthy Greek farmers did succeed in assembling large estates, but never on the scale found in Egypt or Babylonia. In such a crucible, where wealth and power were dispersed, it was possible to create democratic republics; in the city of Athens, overlooking the island-dotted Saronic Gulf, the Greeks did just that.

Although several dozen Greek city-states existed, some were more prominent than others. Corinth and Argos, both seaports, were bustling commercial centers. Miletus and Smyrna were central market towns on the shores of Ionia, in present-day Turkey. Rhodes, Delos, and Samos were fishing and trading communities on islands. Delphi was home to the oracle of Apollo, the sun god. Syracuse was the largest of the Greek colonies in Italy. Aristocratic Thebes (not to be confused with the Thebes of Egypt) was an important agricultural center. Olympia hosted the famous quadrennial Olympic Games. The most important cities of Hellenic Greece, however, were commercial Athens and militaristic Sparta.

Sparta was located inland, away from the sea, in the small, confined valley of the Evrótas River, a place the Greeks called Laconia. At the beginning of the eighth century B.C., Sparta faced a food shortage; its population had grown too large and could no longer be supported by the meager crops produced by the poor, stony soil of the Evrótas valley. Driven by hunger, Sparta, in two bloody wars, invaded and conquered the neighboring, more populous city-state of Messene, which was located in the next valley, on the other side of the Taíyetos Mountains. The Spartans enslaved the Messenians, called *helots,* and put them to work in the fields growing food for the tables of their new overlords. From time to time, the *helots* staged rebellions, but each revolt was brutally suppressed. Outnumbered by its Messenian thralls, Spartans exerted control by maintaining a large standing army and staging periodic raids on *helot* villages. Spartan boys were taken from their parents at an early age and placed into military regiments, where they lived under a military discipline throughout most of their lives. The Spartan army was feared throughout Greece for its ferocity and fighting prowess, and Sparta was the preeminent Greek military power. However, although the Spartans unquestionably stood foremost among the Greeks as soldiers, the barracks proved infertile ground for scholarship, and Spartan intellectual accomplishments were negligible.

Although Sparta had the most powerful army in Hellenic Greece, the commercial and cultural center of the Greek world was the city-state of Athens. Located on a small, sere, rocky plain overlooking the sea, Athens, like Sparta, admitted to only a meager agriculture, and, before 600 B.C., it too faced chronic food shortages. The community was rocked by civil wars between rich and poor. In 594 B.C., the small Athenian middle class (merchants, artisans, and some farmers) engineered the election of the reform-minded Solon (639?–599? B.C.) as *archon,* or leader. Solon outlawed slavery for debt (although other forms of slavery persisted), granted citizenship to foreign craftspeople in the hope that they would teach their skills to native Athenians, encouraged farmers to abandon unprofitable wheat cultivation and grow olives and grapes instead, and instituted a popular assembly, or legislature. Despite such reforms, democracy did not come easily in Athens, and several times over the next century despots seized power in *coups d'état.* In 510 B.C., after one such *coup,* a new constitution was instituted. This constitution was even more democratic than Solon's and granted the right to vote to all adult male citizens. It was not a perfect democracy (women could not vote, and neither could slaves, who made up about a quarter of the city's population), but it was the closest thing to it in the ancient world.

Athens after Solon was as prosperous as it was democratic. Athenian olive oil and wine were considered the finest produced in the Mediterranean Sea region. They were sold widely in Greece and beyond and packaged in ornate vases crafted by the city's talented artisans. The city's marketplace, the *agora,* became the principal commercial nexus of the eastern Mediterranean. Intellectual life in Athens focused about the agora. There, farmers from the countryside, merchants and artisans from the city's shops, and traders and sailors just off the docks mingled and talked. Philosophers like Socrates (469?–399 B.C.) and Plato (427?–347 B.C.), scientists like Aristotle (384–322 B.C.), and playwrights like Aristophanes (445?–385? B.C.) sat in the shade of the marketplace, surrounded by students, admirers, and interested citizens, and exchanged ideas. Although the Athenian *agora* was the grandest in Hellenic Greece, marketplaces served a similar function in other commercial cities, such as Corinth, Rhodes, and Miletus. Furthermore, as the Greek population continued to grow, pioneers erected new city-states in far away Italy and Cyprus, and on the shores of the Black Sea. Such colonies, among them Syracuse and Neapolis (Naples—literally, "New City") in Italy, Massilla (Marseilles) on the French riviera, and Sinope in modern Turkey, had *agoras* too—smaller imitations of the one at Athens—where philosophers and scientists gathered.

In 432 B.C., Athens was at the height of its prestige and power and was led by its greatest statesman, Pericles (490?–429 B.C.). It had a powerful navy, built to repulse two earlier invasions by Persia, one in 490 B.C. and another a decade later. The city stood at the center of the Delian League, a political and commercial network that included a dozen or more other Greek city-states and controlled the League's treasury.

The prosperity did not last. Persia revenged itself for its defeats by annexing Miletus, Smyrna, and other Greek towns along the Ionian coast. Worse,

Sparta grew jealous of the powerful Athenian navy, and the two states quarrelled frequently. In 431 B.C., they were at war, a struggle that lasted until 404 B.C., ruined both countries, and involved most of the other Greek city-states. This war was followed by others that lasted until 336 B.C., when Alexander the Great (356–323 B.C.) united all of Greece in his Macedonian Empire.

Despite political disunity, chronic food shortages, overpopulation, and almost constant warfare, the Hellenic Age in Greece (ca. 800–336 B.C.) witnessed remarkable intellectual achievement. In the *agoras* of Athens and other city-states, philosophers taught students and advanced new ideas. It was a time that witnessed the writing of the first real histories: Herodotus' (484?–424? B.C.) optimistic account of Greece's glorious victories over the Persian invaders and Thucydides' (460?–400? B.C.) anguished telling of the fratricidal struggle between Sparta and Athens. It saw the application of deductive reasoning to mathematics by Thales of Miletus (640?–564? B.C.) and Pythagoras (586?–500? B.C.), the foundation of modern medicine by Hippocrates of Cos (460?–377? B.C.), who devised the famous Hippocratic Oath of physicians; and the systematization of logic by Aristotle. It was an age of great literature and theater, with playwrights like Sophocles (496?–406? B.C.) and Aristophanes (445?–385? B.C.). Here, in the small cities in the rocky valleys at the eastern end of the Mediterranean Sea, more than 2000 years ago, were laid the foundations of western society.

Chapter 3

PYTHAGOREAN MATHEMATICS

3–1 Birth of Demonstrative Mathematics

The last centuries of the second millennium B.C. witnessed many economic and political changes. Some civilizations disappeared, the power of Egypt and Babylonia waned, and new peoples, especially the Hebrews, Assyrians, Phoenicians, and Greeks, came to the fore. The Iron Age was ushered in and brought with it sweeping changes in warfare and in all pursuits that required tools. The alphabet was invented, and coins were introduced. Trade was increasingly stimulated, and geographical discoveries were made. The world was ready for a new type of civilization.

The new civilization made its appearance in the trading towns that sprang up along the coast of Asia Minor and later on the mainland of Greece, on Sicily, and on the Italian shore. The static outlook of the ancient Orient became impossible, and in a developing atmosphere of rationalism, men began to ask *why* as well as *how*.

For the first time, in mathematics, as in other fields, men began to ask fundamental questions such as "*Why* are the base angles of an isosceles triangle equal?" and "*Why* does a diameter of a circle bisect the circle?" The empirical processes of the ancient Orient, quite sufficient for the question *how,* no longer sufficed to answer these more scientific inquiries of *why*. Some attempt at demonstrative methods was bound to assert itself, and the deductive feature, which modern scholars regard as a fundamental characteristic of mathematics, came into prominence. Thus, mathematics, in the modern sense of the word, was born in this atmosphere of rationalism and in one of the new trading towns located on the west coast of Asia Minor. For tradition has it that demonstrative geometry began with Thales of Miletus, one of the "seven wise men" of antiquity, during the first half of the sixth century B.C.[1]

Thales seems to have spent the early part of his life as a merchant, becoming wealthy enough to devote the latter part of his life to study and some travel. It is said that he resided for a time in Egypt, and there evoked admiration by calculating the height of a pyramid by means of shadows (see Problem Study 3.1). Back in Miletus, his many-sided genius won him a reputation as a statesman, counselor, engineer, businessman, philosopher, mathematician, and as-

[1] There are some historians of ancient mathematics, in particular Otto Neugebauer, who disagree with the traditional *evolutionary* account of the origin of demonstrative mathematics and favor a more *revolutionary* account, wherein the change was probably brought on by the discovery of the irrationality of $\sqrt{2}$.

tronomer. Thales is the first known individual with whom mathematical discoveries are associated. In geometry, he is credited with the following elementary results:

1. A circle is bisected by any diameter.
2. The base angles of an isosceles triangle are equal.
3. The vertical angles formed by two intersecting lines are equal.
4. Two triangles are congruent if they have two angles and one side in each respectively equal. [Thales perhaps used this result in his determination of the distance of a ship from shore (see Problem Study 3.1).]
5. An angle inscribed in a semicircle is a right angle. (This was recognized by the Babylonians some 1400 years earlier.)

The value of these results is not to be measured by the theorems themselves, but rather by the belief that Thales supported them by some logical reasoning instead of intuition and experiment.

Take, for example, the matter of the equality of a pair of vertical angles formed by two intersecting lines. In Figure 6, we wish to show that angle a is equal to angle b. In pre-Hellenic times, the equality of these two angles probably would have been considered as quite obvious, and if anyone should have had doubts, that person would have been convinced by performing the simple experiment wherein the angles are cut out and then one applied to the other. Thales, on the other hand, preferred to establish the equality of angles a and b by logical reasoning, perhaps in much the same way as we do today in our elementary geometry texts. In Figure 6, angle a plus angle c equals a straight angle; also, angle b plus angle c equals a straight angle. Since all straight angles are equal, therefore, angle a equals angle b (if equals are subtracted from equals, the remainders are equal). The equality of angles a and b has been established by a short chain of deductive reasoning, starting from more basic principles.

As with other great men, many charming anecdotes are told about Thales that, if not true, are at least apposite. There was the occasion when he demonstrated how easy it is to get rich; foreseeing a heavy crop of olives coming, he obtained a monopoly on all the oil presses of the region and then later realized a fortune by renting them out. And there is the story, recounted by Aesop, of the recalcitrant mule that, when transporting salt, found that by rolling over in the stream he could dissolve the contents of his load and thus travel more lightly. Thales broke him of the troublesome habit by loading him with sponges. He answered Solon's query as to why he never married by having a runner appear next day with a fictitious message for Solon stating that Solon's favorite son

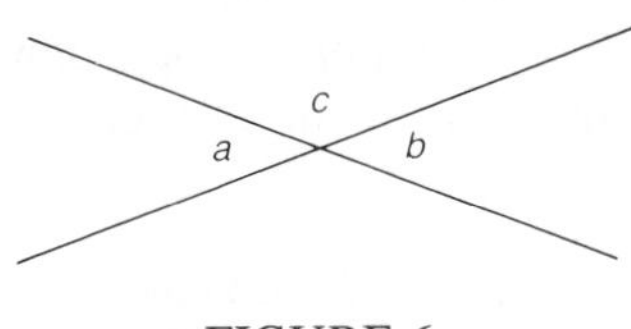

FIGURE 6

had been suddenly killed in an accident. Thales then calmed the grief-stricken father, explained everything, and said, "I merely wanted to tell you why I never married."

Recent research indicates that there is no evidence backing an often-repeated story that Thales predicted a solar eclipse that took place in 585 B.C.

3–2 Pythagoras and the Pythagoreans

The history of the first 300 years of Greek mathematics is obscured by the greatness of Euclid's *Elements,* written about 300 B.C., because this work so completely eclipsed so many preceding Greek writings on mathematics that those earlier works were thenceforth discarded and have become lost to us. As the eminent twentieth-century mathematician David Hilbert once remarked, one can measure the importance of a scientific work by the number of earlier publications rendered superfluous by it.

Consequently, unlike ancient Egyptian and Babylonian mathematics, there exist virtually no primary sources that throw much light upon early Greek mathematics. We are forced to rely upon manuscripts and accounts that are dated several hundred years after the original treatments were written. In spite of this difficulty, however, scholars of classicism have been able to build up a rather consistent, although somewhat hypothetical, account of the history of early Greek mathematics, and have even plausibly restored many of the original Greek texts. This work required amazing ingenuity and patience; it was carried through by painstaking comparisons of derived texts and by the examination of countless literary fragments and scattered remarks made by later authors, philosophers, and commentators.[2]

The debt of early Greek mathematics to ancient Oriental mathematics is difficult to evaluate, and the path of transmission from the one to the other has not yet been satisfactorily uncovered. That the debt is considerably greater than formerly believed became evident with twentieth-century research of Babylonian and Egyptian records. Greek writers themselves expressed respect for the wisdom of the East, and this wisdom was available to anyone who could travel to Egypt and Babylonia. There are also internal evidences of a connection with the East. Early Greek mysticism in mathematics smacks strongly of Oriental influence, and some Greek writings exhibit a Hellenic perpetuation of the more arithmetic tradition of the Orient. Also, there are strong links connecting Greek and Mesopotamian astronomy.

Our principal source of information concerning very early Greek mathematics is the so-called *Eudemian Summary* of Proclus. This summary constitutes the opening pages of Proclus' *Commentary on Euclid, Book I,* and is a very brief outline of the development of Greek geometry from the earliest times to Euclid. Although Proclus lived in the fifth century A.D., a good thousand

---

[2] A debt is owed, along these lines, to the profound and scholarly investigations of such men as Paul Tannery, T. L. Heath, H. G. Zeuthen, A. Rome, J. L. Heiberg, and E. Frank.

years after the inception of Greek mathematics, he still had access to a number of historical and critical works that are now lost to us, except for the fragments and allusions preserved by him and others. Among these lost works was a resumé of an apparently full history of Greek geometry, already lost in Proclus' time, covering the period prior to 335 B.C., written by Eudemus, a pupil of Aristotle. The *Eudemian Summary* is so named because it is based upon this earlier work. The account of the mathematical achievements of Thales, sketched in the preceding section, was furnished by the *Eudemian Summary*.

The next outstanding Greek mathematician mentioned in the *Eudemian Summary* is Pythagoras, whom his followers enveloped in such a mythical haze that very little is known about him with any degree of certainty. It seems that he was born about 572 B.C. on the Aegean island of Samos. Being about fifty years younger than Thales and living so near Thales' home city of Miletus, it may be that Pythagoras studied under the older man. He then appears to have sojourned in Egypt and may even have indulged in more extensive travel. Returning home, he found Samos under the tyranny of Polycrates and Ionia under the dominion of the Persians; accordingly, he migrated to the Greek seaport of Crotona, located in southern Italy. There he founded the famous Pythagorean school, which, in addition to being an academy for the study of philosophy, mathematics, and natural science, developed into a closely knit brotherhood with secret rites and observances. In time, the influence and aristocratic tendencies of the brotherhood became so great that the democratic forces of southern Italy destroyed the buildings of the school and caused the society to disperse. According to one report, Pythagoras fled to Metapontum where he died, maybe murdered, at an advanced age of seventy-five to eighty. The brotherhood, although scattered, continued to exist for at least two centuries more.

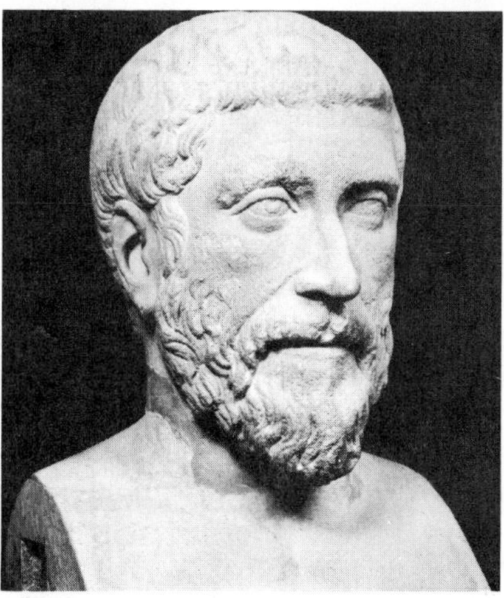

PYTHAGORAS
(David Smith Collection)

The Pythagorean philosophy rested on the assumption that whole number is the cause of the various qualities of man and matter. This led to an exaltation and study of number properties, and arithmetic (considered as the theory of numbers), along with geometry, music, and spherics (astronomy), constituted the fundamental liberal arts of the Pythagorean program of study. This group of subjects became known in the Middle Ages as the **quadrivium,** to which was added the **trivium** of grammar, logic, and rhetoric. These seven liberal arts came to be looked upon as the necessary equipment of an educated person.

Because Pythagoras' teaching was entirely oral, and because of the brotherhood's custom of referring all discoveries back to the revered founder, it is now difficult to know just which mathematical findings should be credited to Pythagoras himself and which to other members of the fraternity.

3–3 Pythagorean Arithmetic

The ancient Greeks made a distinction between the study of the abstract relationships connecting numbers and the practical art of computing with numbers. The former was known as **arithmetic** and the latter as **logistic.** This classification persisted through the Middle Ages until about the close of the fifteenth century, when texts appeared treating both the theoretical and practical aspects of number work under the single name *arithmetic*. It is interesting that today *arithmetic* has its original significance in continental Europe, while in England and America the popular meaning of *arithmetic* is synonymous with that of ancient *logistic*. In these two countries, the descriptive term **number theory** is used to denote the abstract side of number study.

It is generally conceded that Pythagoras and his followers, in conjunction with the fraternity's philosophy, took the first steps in the development of number theory, and at the same time laid much of the basis of future number mysticism. Thus, Iamblichus, an influential Neoplatonic philosopher of about A.D. 320, has ascribed to Pythagoras the discovery of **amicable,** or **friendly, numbers.** Two numbers are amicable if each is the sum of the proper divisors[3] of the other. For example, 284 and 220, constituting the pair ascribed to Pythagoras, are amicable, since the proper divisors of 220 are 1, 2, 4, 5, 10, 11, 20, 22, 44, 55, 110, and the sum of these is 284, whereas the proper divisors of 284 are 1, 2, 4, 71, 142, and the sum of these is 220. This pair of numbers attained a mystical aura, and superstition later maintained that two talismans bearing these numbers would seal perfect friendship between the wearers. The numbers came to play an important role in magic, sorcery, astrology, and the casting of horoscopes. It seemed that no new pair of amicable numbers was discovered until the great French number theorist Pierre de Fermat in 1636 announced 17,296 and 18,416 as another pair. It has recently been established,

[3] The **proper divisors** of a positive integer N are all the positive integral divisors of N except N itself. Note that 1 is a proper divisor of N. A somewhat antiquated synonym for proper divisor is **aliquot part.**

however, that this was a rediscovery, and that this pair of amicable numbers had been previously found by the Arab al-Banna (1256–1321) in the late thirteenth or early fourteenth century, perhaps by using the T̲âbit ibn Qorra formula. (For this formula, see Problem Study 7.11.) Two years after Fermat's announcement, the French mathematician and philosopher René Descartes gave a third pair. The Swiss mathematician Leonhard Euler undertook a systematic search for amicable numbers and, in 1747, gave a list of thirty pairs, which he later extended to more than sixty. A curiosity in the history of these numbers was the late discovery, by the sixteen-year-old Italian boy Nicolo Paganini[4] in 1866, of the overlooked and relatively small pair of amicable numbers, 1184 and 1210. All amicable number pairs below one billion have now been found.

Other numbers having mystical connections essential to numerological speculations, and sometimes ascribed to the Pythagoreans, are the **perfect, deficient,** and **abundant numbers.** A number is *perfect* if it is the sum of its proper divisors, *deficient* if it exceeds the sum of its proper divisors, and *abundant* if it is less than the sum of its proper divisors. So God created the world in six days, a perfect number, since $6 = 1 + 2 + 3$. On the other hand, as Alcuin (735–804) observed, the whole human race descended from the eight souls of Noah's ark, and this second creation was imperfect, for 8, being greater than $1 + 2 + 4$, is deficient. Until 1952, there were only twelve known perfect numbers, all of them even numbers, of which the first three are 6, 28, and 496. The last proposition of the ninth book of Euclid's *Elements* (ca. 300 B.C.) proves that *if* $2^n - 1$ *is a prime number,*[5] *then* $2^{n-1}(2^n - 1)$ *is a perfect number*. The perfect numbers given by Euclid's formula are even numbers, and Euler has shown that every even perfect number must be of this form. The existence or nonexistence of odd perfect numbers is one of the celebrated unsolved problems in number theory. There certainly is no number of this type having less than 200 digits.

In 1952, with the aid of the SWAC digital computer, five more perfect numbers were discovered, corresponding to $n = 521, 607, 1279, 2203$, and 2281 in Euclid's formula. In 1957, using the Swedish machine BESK, another was found, corresponding to $n = 3217$. In 1961, with an IBM 7090, two more were found, for $n = 4253$ and 4423. There are no other even perfect numbers for $n < 5000$. The values $n = 9689$, 9941, 11213, 19937, 21701, 23209, 44,497, 86243, 132049, and 216091 also yield perfect numbers, bringing the list of known perfect numbers to thirty. The last was found by scientists at Chevron in 1985 on a $10,000,000 Cray X-MP supercomputer.

The concept of perfect numbers has inspired certain generalizations by modern mathematicians. If we let $\sigma(n)$ represent the sum of *all* the divisors of n

[4] Not to be confused with Nicolo Paganini (1782–1840), the noted Italian violinist and composer.

[5] A **prime number** is a positive integer greater than 1 and having no positive integral divisors other than itself and unity. An integer greater than 1 that is not a prime number is called a **composite number,** thus, 7 is a prime number, whereas 12 is a composite number.

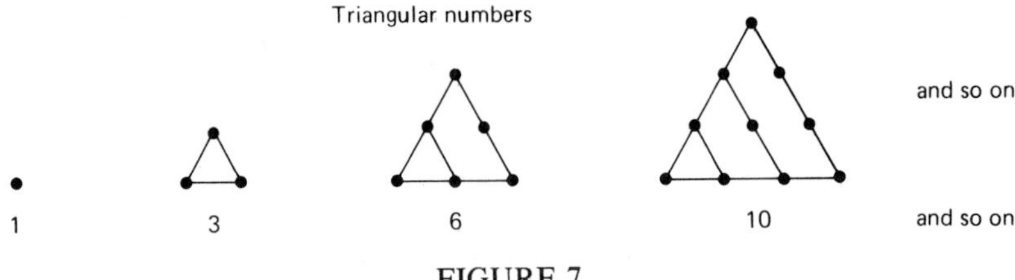

FIGURE 7

(including n itself), then n is perfect if and only if $\sigma(n) = 2n$. In general, if we should have $\sigma(n) = kn$, where k is a natural number, then n is said to be **k-tuply perfect.** One can show, for example, that 120 and 672 are triply perfect. It is not known if infinitely many multiply perfect numbers, let alone just perfect ones, exist. It is also not known if any odd multiply perfect number exists. In 1944, the concept of **superabundant numbers** was created. A natural number n is *superabundant* if and only if $\sigma(n)/n > \sigma(k)/k$ for all $k < n$. It is known that there are infinitely many superabundant numbers. Other numbers related to perfect, deficient, and abundant numbers that have been introduced in recent times are *practical numbers, quasiperfect numbers, semiperfect numbers,* and *weird numbers.* We merely mention these concepts to illustrate how ancient number work has inspired related modern investigations.

Although not all historians of mathematics feel that amicable and perfect numbers can be ascribed to the Pythagoreans, there seems to be universal agreement that the **figurate numbers** did originate with the earliest members of the society. These numbers, considered as the number of dots in certain geometrical configurations, represent a link between geometry and arithmetic. Figures 7, 8, and 9 account for the geometrical nomenclature of **triangular numbers, square numbers, pentagonal numbers,** and so on.

Many interesting theorems concerning figurate numbers can be established in purely geometric fashion. To show Theorem I (*any square number is the sum of two successive triangular numbers*), for example, we observe that a square number, in its geometric form, can be divided as in Figure 10. Again, Figure 11 illustrates Theorem II (*the nth pentagonal number is equal to n plus three times the* $(n - 1)$*th triangular number*). Theorem III (*the sum of any number of consecutive odd integers, starting with* 1, *is a perfect square*) is exhibited geometrically by Figure 12.

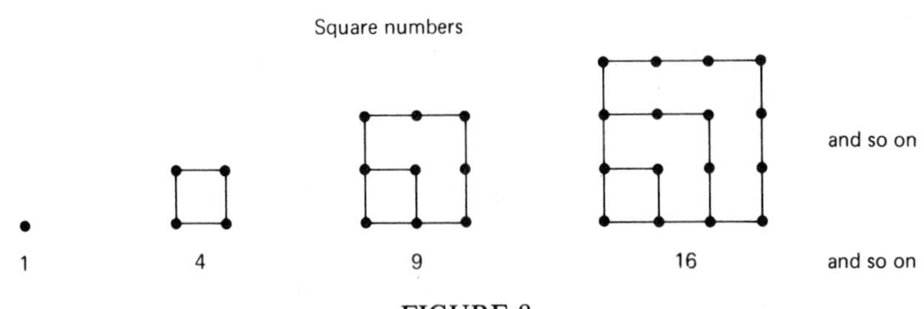

FIGURE 8

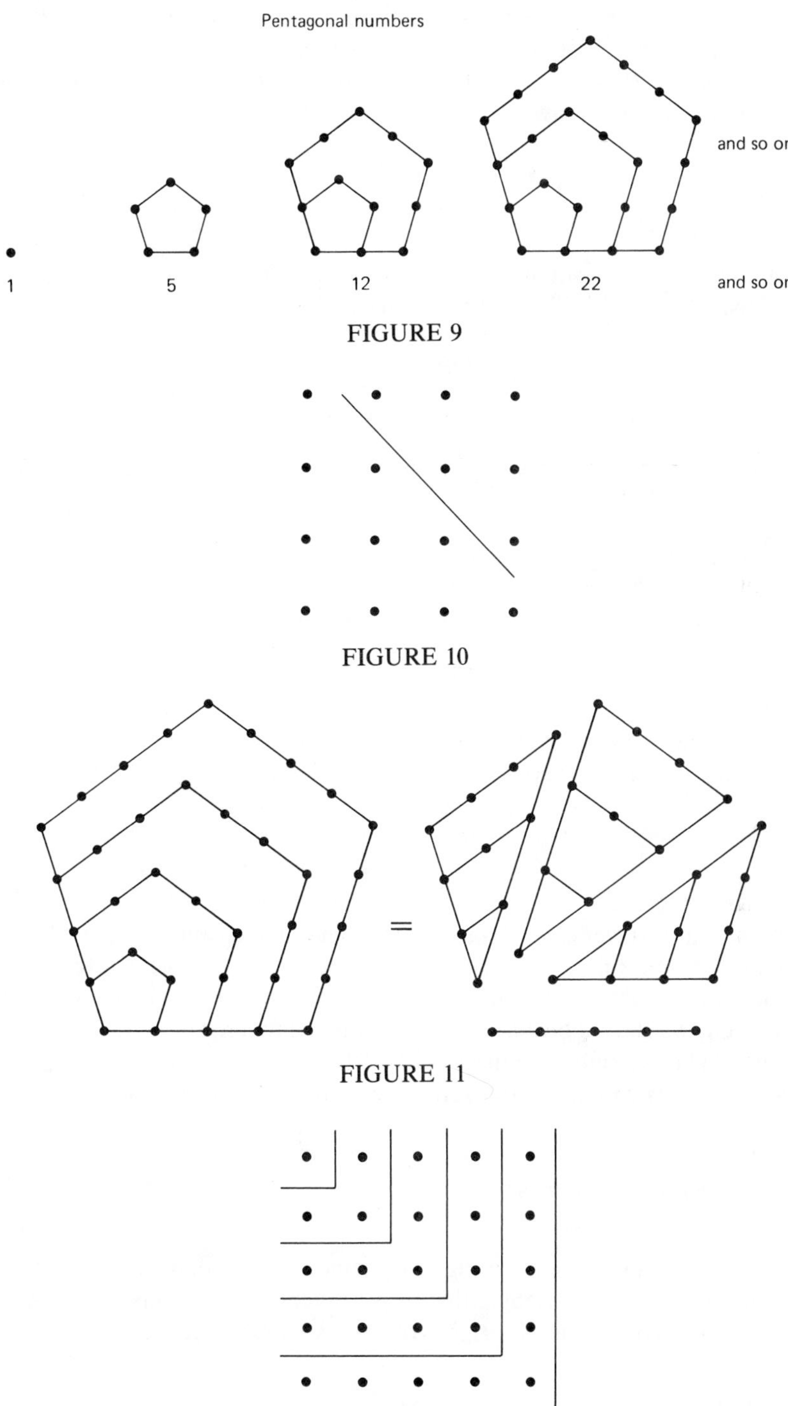

FIGURE 9

FIGURE 10

FIGURE 11

FIGURE 12

Of course, these theorems can also be established algebraically once we obtain the algebraic representations of the general triangular, square, and pentagonal numbers. It is clear that the nth triangular number, T_n, is given by the sum of an arithmetic series,[6]

$$T_n = 1 + 2 + 3 + \cdots + n = \frac{n(n+1)}{2},$$

and, of course, the nth square number, S_n, is n^2. Our first theorem may now be re-established algebraically by an identity as follows:

$$S_n = n^2 = \frac{n(n+1)}{2} + \frac{(n-1)n}{2} = T_n + T_{n-1}.$$

The nth pentagonal number, P_n, is also given by the sum of an arithmetic series.

$$\begin{aligned} P_n &= 1 + 4 + 7 + \cdots + (3n-2) \\ &= \frac{n(3n-1)}{2} = n + \frac{3n(n-1)}{2} \\ &= n + 3T_{n-1}. \end{aligned}$$

This proves the second theorem. The third theorem is obtained algebraically by summing the arithmetic series

$$1 + 3 + 5 + \ldots + (2n-1) = \frac{n(2n)}{2} = n^2.$$

As a last and very remarkable discovery about numbers, made by the Pythagoreans, we might mention the dependence of musical intervals upon numerical ratios. The Pythagoreans found that for strings under the same tension, the lengths should be 2 to 1 for the octave, 3 to 2 for the fifth, and 4 to 3 for the fourth. These results, the first recorded facts in mathematical physics, led the Pythagoreans to initiate the scientific study of musical scales.

3–4 Pythagorean Theorem and Pythagorean Triples

Tradition is unanimous in ascribing to Pythagoras the independent discovery of the theorem on the right triangle that now universally bears his name—that the square on the hypotenuse of a right triangle is equal to the sum of the squares

[6] The sum of an arithmetic series is equal to the product of the number of terms and half the sum of the two extreme terms.

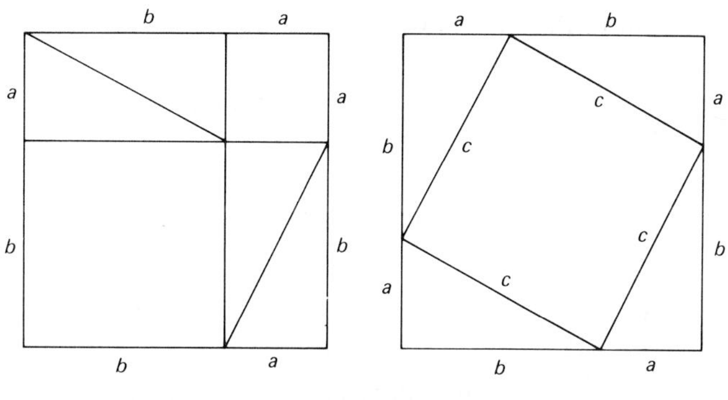

FIGURE 13

on the two legs. We have seen that this theorem was known to the Babylonians of Hammurabi's time, more than a thousand years earlier, but the first general proof of the theorem may well have been given by Pythagoras. There has been much conjecture as to the proof Pythagoras might have offered, and it is generally felt that it probably was a dissection type of proof[7] like the following, illustrated in Figure 13. Let a, b, c denote the legs and hypotenuse of the given right triangle, and consider the two squares in the accompanying figure, each having $a + b$ as its side. The first square is dissected into six pieces—namely, the two squares on the legs and four right triangles congruent to the given triangle. The second square is dissected into five pieces—namely, the square on the hypotenuse and four right triangles congruent to the given triangle. By subtracting equals from equals, it now follows that the square on the hypotenuse is equal to the sum of the squares on the legs.

To prove that the central piece of the second dissection is actually a square of side c, we need to employ the fact that the sum of the angles of a right triangle is equal to two right angles. But the *Eudemian Summary* attributes this theorem for the general triangle to the Pythagoreans. Because a proof of this theorem requires, in turn, a knowledge of some properties of parallels, the early Pythagoreans are also credited with the development of that theory.

Since Pythagoras' time, many different proofs of the Pythagorean theorem have been supplied. In the second edition of his book, *The Pythagorean Proposition,* E. S. Loomis has collected and classified 370 demonstrations of this famous theorem.

Closely allied to the Pythagorean theorem is the problem of finding integers a, b, c that can represent the legs and hypotenuse of a right triangle. A triple of numbers of this sort is known as a **Pythagorean triple** and, as we have seen in Section 2–6, the analysis of Plimpton 322 offers fairly convincing evidence that the ancient Babylonians knew how to calculate such triples. The

[7] See, however, Daniel Shanks, *Solved and Unsolved Problems in Number Theory,* vol. 1, pp. 124, 125.

Pythagoreans have been credited with the formula

$$m^2 + \left(\frac{m^2 - 1}{2}\right)^2 = \left(\frac{m^2 + 1}{2}\right)^2,$$

the three terms of which, for any odd m, yield a Pythagorean triple. The similar formula

$$(2m)^2 + (m^2 - 1)^2 = (m^2 + 1)^2,$$

where m may be even or odd, was devised for the same purpose and is attributed to Plato (ca. 380 B.C.). Neither of these formulas yields all Pythagorean triples.

3–5 Discovery of Irrational Magnitudes

The integers are abstractions arising from the process of counting finite collections of objects. The needs of daily life require us in addition to counting individual objects, to measure various quantities, such as length, weight, and time. To satisfy these simple measuring needs, fractions are required, for seldom will a length, as an example, appear to contain an exact integral number of linear units. Thus, if we define a **rational number** as the quotient of two integers p/q, $q \neq 0$, this system of rational numbers, since it contains all the integers and fractions, is sufficient for practical measuring purposes.

The rational numbers have a simple geometrical interpretation. Mark two distinct points O and I on a horizontal straight line (I to the right of O) and choose the segment OI as a unit of length. If we let O and I represent the numbers 0 and 1, respectively, then the positive and negative integers can be represented by a set of points on the line spaced at unit intervals apart, the positive integers being represented to the right of O and the negative integers to the left of O. The fractions with denominator q may then be represented by the points that divide each of the unit intervals into q equal parts. Then, for each rational number, there is a point on the line. To the early mathematicians, it seemed evident that all the points on the line would in this way be used up. It must have been something of a shock to learn that there are points on the line not corresponding to any rational number. This discovery was one of the greatest achievements of the Pythagoreans. In particular, the Pythagoreans showed that there is no rational number corresponding to the point P on the line where the distance OP is equal to the diagonal of a square having a unit side (see Figure 14). New numbers had to be invented to correspond to such points, and since these numbers cannot be rational numbers, they came to be called **irrational numbers** (meaning, nonrational numbers). Their discovery marks one of the great milestones in the history of mathematics.

To prove that the length of the diagonal of a square of unit side cannot be represented by a rational number, it suffices to show that $\sqrt{2}$ is irrational. To

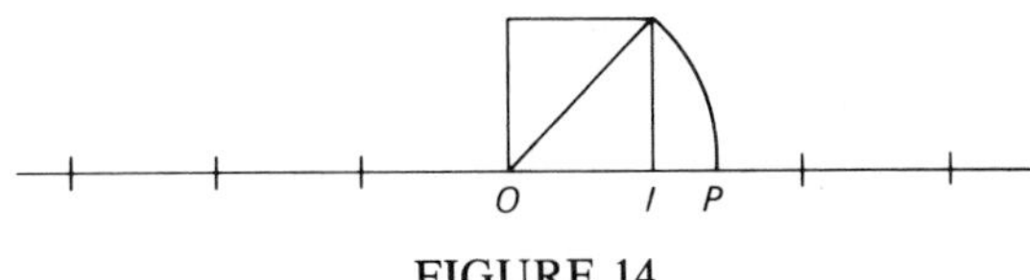

FIGURE 14

this end, we first observe that, for a positive integer s, s^2 is even if and only if s is even. Now, suppose, for the purpose of argument, that $\sqrt{2}$ is rational—that is, $\sqrt{2} = a/b$—where a and b are relatively prime integers.[8] Then

$$a = b\sqrt{2},$$

or

$$a^2 = 2b^2.$$

Since a^2 is twice an integer, we see that a^2 and, hence a, must be even. Put $a = 2c$; then the last equation becomes

$$4c^2 = 2b^2,$$

or

$$2c^2 = b^2,$$

from which we conclude that b^2 and, hence, b must be even. This is impossible, however, since a and b were assumed to be relatively prime. Thus, the assumption that $\sqrt{2}$ is rational, which led to this impossible situation, must be abandoned.

The discovery of the existence of irrational numbers was surprising and disturbing to the Pythagoreans. First of all, it seemed to deal a mortal blow to the Pythagorean philosophy that all depends upon the whole numbers. Next, it seemed contrary to common sense, for it was felt intuitively that any magnitude could be expressed by *some* rational number. The geometrical counterpart was equally startling, for who could doubt that for any two given line segments one is able to find some third line segment, perhaps very very small, that can be marked off a whole number of times into each of the two given segments? But take as the two segments a side s and a diagonal d of a square. Now if there exists a third segment t that can be marked off a whole number of times into s and d, we would have $s = bt$ and $d = at$, where a and b are positive integers. But $d = s\sqrt{2}$, whence $at = bt\sqrt{2}$—that is, $a = b\sqrt{2}$, or $\sqrt{2} = a/b$, a rational number. Contrary to intuition, then, there exist **incommensurable** line segments—that is, line segments having no common unit of measure.

---

[8] Two integers are **relatively prime** if they have no common positive integral factor other than unity. Thus, 5 and 18 are relatively prime, whereas 12 and 18 are not relatively prime.

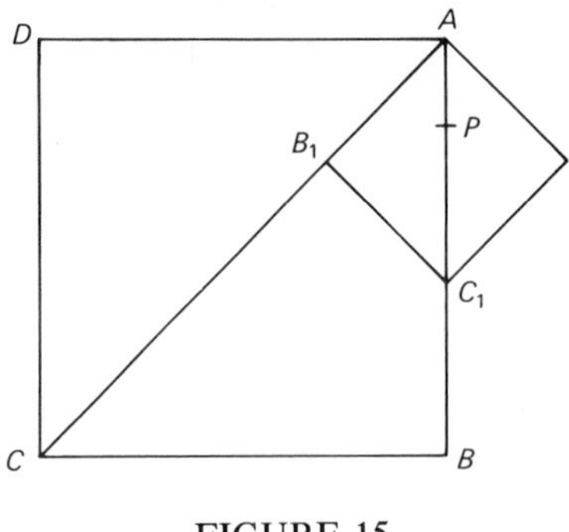

FIGURE 15

Let us sketch an alternative, geometrical demonstration of the irrationality of $\sqrt{2}$ by showing that a side and diagonal of a square are incommensurable. Suppose the contrary. According to this supposition, then, there exists a segment AP (see Figure 15) such that both the diagonal AC and side AB of a square $ABCD$ are integral multiples of AP; that is, AC and AB are commensurable with respect to AP. On AC, lay off $CB_1 = AB$ and draw B_1C_1 perpendicular to CA. One may easily prove that $C_1B = C_1B_1 = AB_1$. Then $AC_1 = AB - AB_1$ and AB_1 are commensurable with respect to AP. But AC_1 and AB_1 are a diagonal and a side of a square of dimensions less than half those of the original square. It follows that, by repeating the process, we may finally obtain a square whose diagonal AC_n and side AB_n are commensurable with respect to AP, and $AC_n < AP$. This absurdity proves the theorem.

The first proof is essentially the traditional one known to Aristotle (384–322 B.C.). This discovery of the irrationality of $\sqrt{2}$ caused some consternation in the Pythagorean ranks. Not only did it appear to upset the basic assumption that everything depends on the whole numbers, but because the Pythagorean definition of proportion assumed any two like magnitudes to be commensurable, all the propositions in the Pythagorean theory of proportion had to be limited to commensurable magnitudes, and their general theory of similar figures became invalid. So great was the "logical scandal" that efforts were made for a while to keep the matter secret. One legend has it that the Pythagorean Hippasus (or perhaps some other) perished at sea for his impiety in disclosing the secret to outsiders, or (according to another version) was banished from the Pythagorean community and a tomb was erected for him as though he was dead.

For some time, $\sqrt{2}$ was the only known irrational.[9] Later, according to Plato, Theodorus of Cyrene (ca. 425 B.C.) showed that $\sqrt{3}$, $\sqrt{5}$, $\sqrt{6}$, $\sqrt{7}$, $\sqrt{8}$, $\sqrt{10}$, $\sqrt{11}$, $\sqrt{12}$, $\sqrt{13}$, $\sqrt{14}$, $\sqrt{15}$, $\sqrt{17}$ are also irrational. About 370 B.C., the "scandal" was resolved by the brilliant Eudoxus, a pupil of Plato and of the

[9] There is some possibility that $(\sqrt{5} - 1)/2$, which is the ratio of a side to a diagonal of a regular pentagon, was the first known irrational.

Pythagorean Archytas who put forth a new definition of proportion. Eudoxus' masterful treatment of incommensurables appears in the fifth book of Euclid's *Elements,* and coincides essentially with the modern exposition of irrational numbers that was given by Richard Dedekind in 1872.

The treatments of ratio and proportion and similar triangles in early twentieth-century high school geometry texts reflect the difficulties and subtleties introduced by incommensurable magnitudes. In these treatments, two cases, depending upon the commensurability or incommensurability of certain magnitudes, are considered (see, for example, Section 5–5 and Problem Study 5.6). More recent texts circumvent the difficulties by the use of more sophisticated postulational bases.

3–6 Algebraic Identities

Imbued with the representation of a number by a length and completely lacking any adequate algebraic notation, the early Greeks devised ingenious geometrical processes for carrying out algebraic operations. Much of this geometrical algebra has been attributed to the Pythagoreans and can be found scattered through several of the earlier books of Euclid's *Elements*. Thus, Book II of the *Elements* contains a number of propositions that in reality are algebraic identities couched in geometric terminology. It seems quite certain that these propositions were developed, through means of a dissection method, by the early Pythagoreans. We may illustrate the method by considering a few of the propositions of Book II.

Proposition 4 of Book II establishes geometrically the identity

$$(a + b)^2 = a^2 + 2ab + b^2$$

by dissecting the square of side $a + b$ into two squares and two rectangles having areas a^2, b^2, ab, and ab, as indicated in Figure 16. Euclid's statement of the proposition is: *If a straight line is divided into any two parts, the square on the whole line is equal to the sum of the squares on the two parts together with twice the rectangle contained by the two parts.*

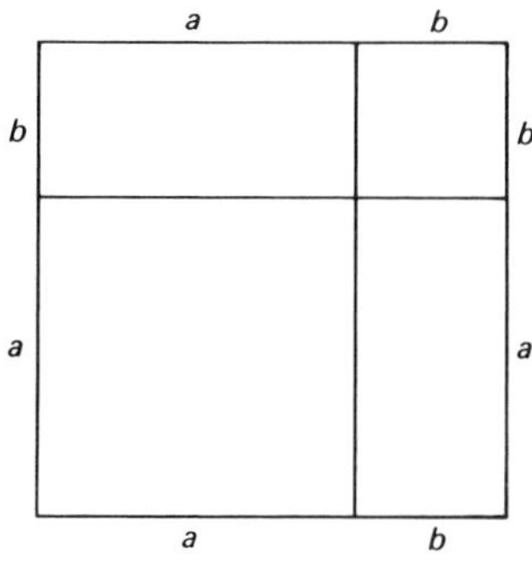

FIGURE 16

The statement of Proposition 5 of Book II is: *If a straight line is divided equally and also unequally, the rectangle contained by the unequal parts, together with the square on the line between the points of section, is equal to the square on half the line.* Let AB be the given straight-line segment, and let it be divided equally at P and unequally at Q. Then the proposition says that

$$(AQ)(QB) + (PQ)^2 = (PB)^2.$$

If we set $AQ = 2a$ and $QB = 2b$, this leads to the algebraic identity

$$4ab + (a - b)^2 = (a + b)^2,$$

or, if we set $AB = 2a$ and $PQ = b$, to the identity

$$(a + b)(a - b) = a^2 - b^2.$$

The dissection given in the *Elements* for establishing this theorem appears in Figure 17. It is more complicated than that for Proposition 4. In the figure, $PCDB$ and $QFLB$ are squares described on PB and QB as sides. Then

$$\begin{aligned}(AQ)(QB) + (PQ)^2 &= AGFQ + HCEF = AGHP + PHFQ + HCEF\\ &= PHLB + PHFQ + HCEF\\ &= PHLB + FEDL + HCEF = (PB)^2.\end{aligned}$$

The statement of Proposition 6 of Book II is: *If a straight line is bisected and produced to any point, the rectangle contained by the whole line thus produced and the part of it produced, together with the square on half the line bisected, is equal to the square on the straight line made up of the half and the part produced.* Here (see Figure 18), if the given straight-line segment AB with midpoint P is produced to Q, we are to show that

$$(AQ)(BQ) + (PB)^2 = (PQ)^2.$$

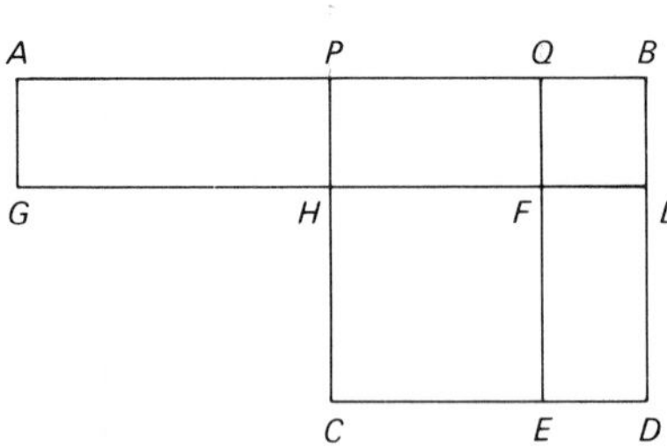

FIGURE 17

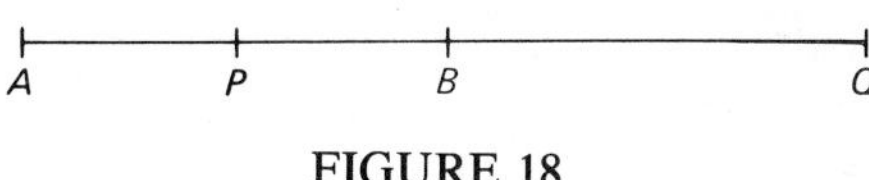

FIGURE 18

If we set $AQ = 2a$ and $BQ = 2b$, we are led again to the identity

$$4ab + (a - b)^2 = (a + b)^2,$$

and a similar dissection to that used for Proposition 5 may be used here.

Figure 19, with $AB = a$ and $BC = b$, suggests a less cumbrous proof of the identity

$$4ab + (a - b)^2 = (a + b)^2.$$

3–7 Geometric Solution of Quadratic Equations

In their geometric algebra, the Greeks employed two principal methods for solving certain simple equations—the method of proportions and the method of application of areas. There is evidence that both of these methods originated with the Pythagoreans.

The method of proportions permits one to construct (exactly as we do today in our high school geometry courses; see Figure 20) a line segment x given either by $a:b = c:x$ or by $a:x = x:b$, where a, b, c are given line segments. That is, the method of proportions furnishes geometrical solutions of the equations

$$ax = bc \qquad \text{and} \qquad x^2 = ab.$$

To explain the method of application of areas, consider (see Figure 21) a line segment AB and a parallelogram $AQRS$ having side AQ lying along the ray AB. If Q is not at B, take C so that $QBCR$ is a parallelogram. When Q is

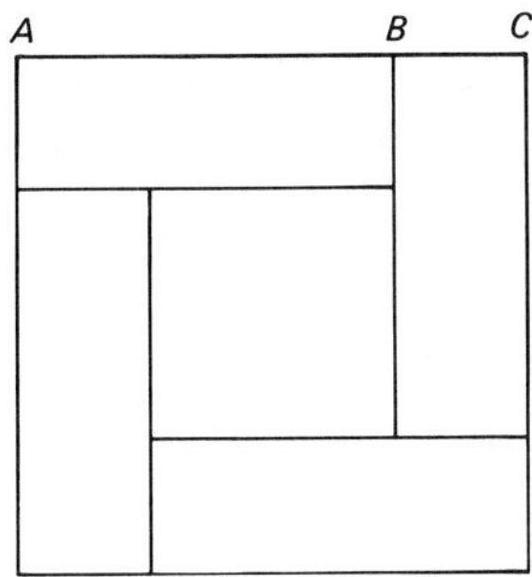

FIGURE 19

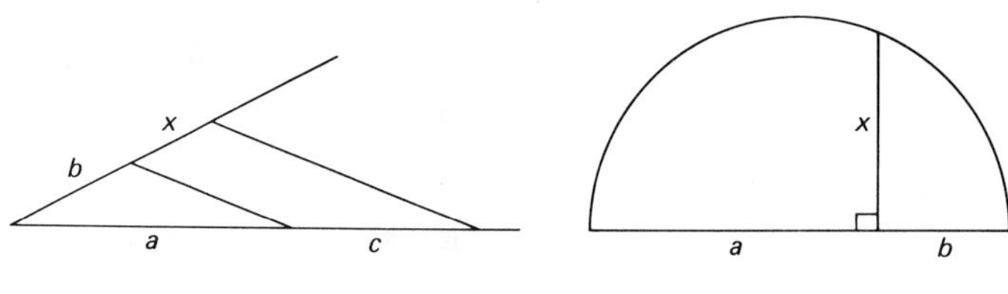

FIGURE 20

between A and B, parallelogram $AQRS$ is said to be *applied to segment AB, falling short by parallelogram QBCR;* when Q coincides with B, parallelogram $AQRS$ is said to be *applied to segment AB;* when Q lies on AB produced through B, parallelogram $AQRS$ is said to be *applied to segment AB, exceeding by parallelogram QBCR.*

Proposition 44 of Book I of Euclid's *Elements* solves the construction: *To apply to a given line segment AB a parallelogram of given area and given base angles.* Consider the special case in which the given base angles are right angles, so that the applied parallelogram is a rectangle. Denote the length of AB by a, the altitude of the applied rectangle by x, and the dimensions of a rectangle of area equal to that of the applied rectangle by b and c. Then

$$ax = bc \qquad \text{or} \qquad x = \frac{bc}{a}.$$

Proposition 28 of Book VI of the *Elements* solves the construction: *To apply to a given line segment AB a parallelogram AQRS equal in area to a given rectilinear figure F, and falling short by a parallelogram QBCR similar to a given parallelogram, the area of F not exceeding that of the parallelogram described on half of AB and similar to the defect QBCR.* Consider the special case in which the given parallelogram is a square. Denote the length of AB by a, the base AQ of the applied parallelogram (which is now a rectangle) by x, and the side of a square F equal in area to the applied rectangle by b. Then

$$x(a - x) = b^2 \qquad \text{or} \qquad x^2 - ax + b^2 = 0. \tag{1}$$

Proposition 29 of Book VI solves the construction: *To apply to a given line segment AB a parallelogram AQRS equal in area to a given rectilinear figure*

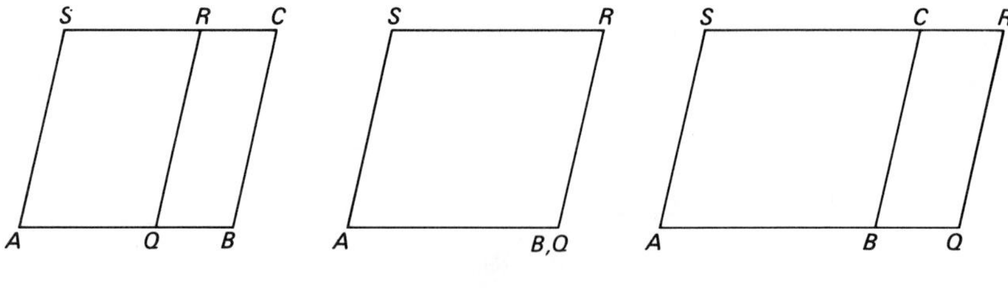

FIGURE 21

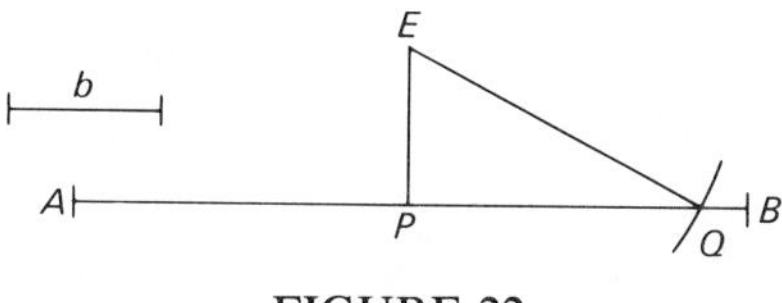

FIGURE 22

F, and exceeding by a parallelogram QBCR similar to a given parallelogram. Consider the special case in which the given parallelogram is a square. Denote the length of AB by a, the base AQ of the applied parallelogram (which is now a rectangle) by x, and the side of a square F equal in area to the applied rectangle by b. Then

$$x(x - a) = b^2 \qquad \text{or} \qquad x^2 - ax - b^2 = 0. \tag{2}$$

It follows that Proposition I 44 yields a geometric solution to the linear equation $ax = bc$, and Propositions VI 28 and 29 yield geometric solutions to the quadratic equations $x^2 - ax + b^2 = 0$ and $x^2 - ax - b^2 = 0$, respectively.

Constructions can easily be devised for the above special cases of Propositions VI 28 and 29 that are considerably simpler than the more general constructions given in the *Elements*.

Consider, for example, the special case of Proposition VI 28. Here we want to apply to a given line segment a rectangle that falls short by a square. From the first of equations (1), we see that we may restate the problem as follows: *To divide a given line segment so that the rectangle contained by its parts will equal a given square, the square not exceeding the square on half the given line segment.* To clarify the problem, let AB and b be two line segments, b not greater than half of AB. We are to divide AB by a point Q such that $(AQ)(QB) = b^2$. To accomplish this, we mark off $PE = b$ on the perpendicular to AB at its midpoint P, and with E as center and PB as radius, draw an arc cutting AB in the sought point Q, as in Figure 22. The proof is furnished by Proposition II 5 (which was probably devised by the Pythagoreans to serve here), for by that proposition

$$(AQ)(QB) = (PB)^2 - (PQ)^2 = (EQ)^2 - (PQ)^2 = (EP)^2 = b^2.$$

Denoting the length of AB by a and that of AQ by x, we have solved the quadratic equation $x^2 - ax + b^2 = 0$; the roots are represented by AQ and QB.[10] The roots of the quadratic equation

$$x^2 + ax + b^2 = 0$$

are represented by the negatives of the lengths of AQ and QB.

---

[10] If r and s are the roots of the quadratic equation $x^2 - ax + b^2 = 0$, we know from elementary algebra that $r + s = a$ and $rs = b^2$. But it is AQ and QB whose sum is AB, or a, and whose product is b^2.

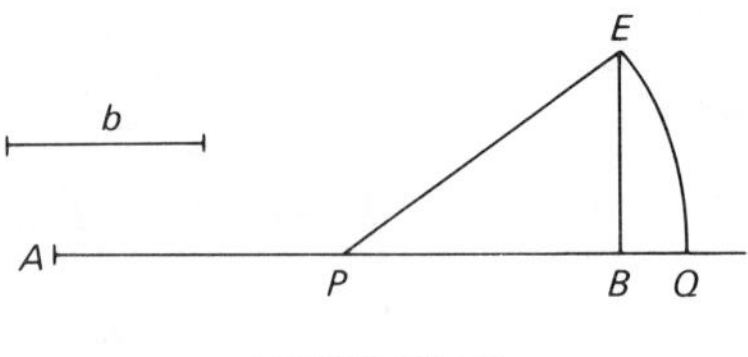

FIGURE 23

For the special case of Proposition VI 29, we want to apply to a given line segment a rectangle that exceeds by a square. From the first of equations (2) above, we see that we may restate the problem as follows: *To produce a given line segment so that the rectangle contained by the extended segment and the extension will equal a given square*. Again, let AB and b be two line segments. We are to produce AB to a point Q such that $(AQ)(BQ) = b^2$. To this end, we mark off $BE = b$ on the perpendicular to AB at B, and with P, the midpoint of AB, as center and PE as radius, draw an arc cutting AB produced in the sought point Q, as in Figure 23. This time, the proof is furnished by Proposition II 6, for by that proposition

$$(AQ)(BQ) = (PQ)^2 - (PB)^2 = (PE)^2 - (PB)^2 = (BE)^2 = b^2.$$

As before, we see that AQ and BQ, where we take the first one as positive and the second one as negative, are the roots of the quadratic equation

$$x^2 - ax - b^2 = 0,$$

a being the length of AB. The roots of

$$x^2 + ax - b^2 = 0$$

are the same as those of $x^2 - ax - b^2 = 0$, only with their signs changed.

The geometric algebra of the Pythagoreans, ingenious though it is, intensifies one's appreciation of the simplicity and convenience inherent in present-day algebraic notation.

3–8 Transformation of Areas

The Pythagoreans were interested in transforming an area from one rectilinear shape into another rectilinear shape. Their solution of the basic problem of constructing a square equal in area to that of a given polygon may be found in Propositions 42, 44, 45 of Book I and Proposition 14 of Book II of Euclid's *Elements*. A simpler solution, probably also known to the Pythagoreans, is the following. Consider any polygon $ABCD$ . . . (see Figure 24). Draw BR parallel to AC to cut DC in R. Then, since triangles ABC and ARC have a common base AC and equal altitudes on this common base, these triangles have equal areas.

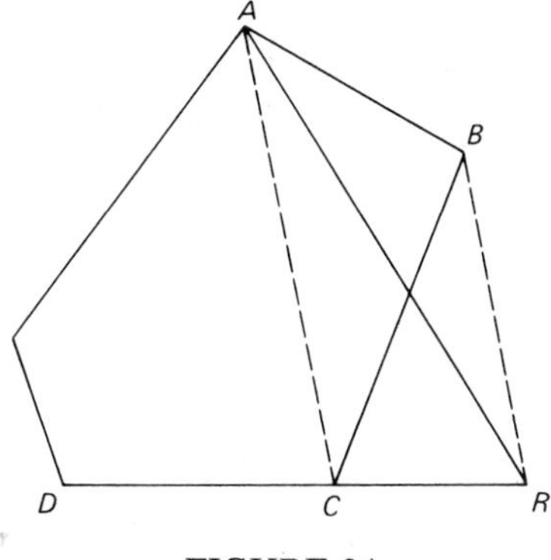

FIGURE 24

It follows that polygons $ABCD \ldots$ and $ARD \ldots$ have equal areas. But the derived polygon has one less side than the given polygon. By a repetition of this process, we finally obtain a triangle having the same area as the given polygon. Now if b is any side of this triangle and h the altitude on b, the side of an equivalent square is given by $\sqrt{(bh)/2}$—that is, by the mean proportional between b and $h/2$. Since this mean proportional is easily constructed with straightedge and compasses, the entire problem can be carried out with these tools.

Many interesting area problems can be solved by this simple process of drawing parallel lines (see Problem Study 3.11).

3–9 The Regular Solids

A polyhedron is said to be **regular** if its faces are congruent regular polygons and if its polyhedral angles are all congruent. Although there are regular polygons of all orders, it turns out that there are only five different regular polyhedra (see Problem Study 3.12). The regular polyhedra are named according to the number of faces each possesses. Thus, there is the tetrahedron with four triangular faces, the hexahedron, or cube, with six square faces, the octahedron with eight triangular faces, the dodecahedron with twelve pentagonal faces, and the icosahedron with twenty triangular faces (see Figure 25).

The early history of these regular polyhedra is lost in the dimness of the past. A mathematical treatment of them is initiated in Book XIII of Euclid's

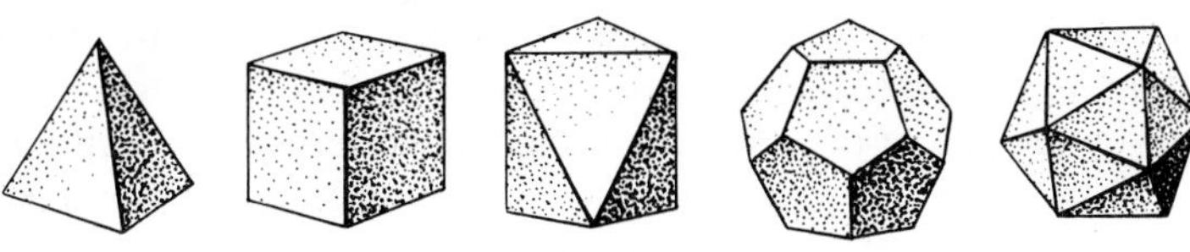

FIGURE 25

Elements. The first scholium of this book remarks that the book "will treat of the so-called Platonic solids, incorrectly named, because three of them, the tetrahedron, cube, and dodecahedron are due to the Pythagoreans, while the octahedron and icosahedron are due to Theaetetus." This could well be the case.

In any event, a description of all five regular polyhedra was given by Plato, who, in his *Timaeus,* shows how to construct models of the solids by putting triangles, squares, and pentagons together to form their faces. Plato's Timaeus is the Pythagorean Timaeus of Locri, whom Plato presumably met when he visited Italy. In Plato's work, Timaeus mystically associates the four easily constructed solids—the tetrahedron, octahedron, icosahedron, and cube—with the four Empedoclean primal "elements" of all material bodies—fire, air, water, and earth. The disturbing difficulty of accounting for the fifth solid, the dodecahedron, is taken care of by associating it with the enveloping universe.

Johann Kepler (1571–1630), master astronomer, mathematician, and numerologist, gave an ingenious explanation of the Timaeus associations. Of the regular solids, he intuitively assumed that the tetrahedron encloses the smallest volume for its surface, while the icosahedron encloses the largest. Now these volume-surface relations are qualities of dryness and wetness, respectively, and since fire is the driest of the four "elements" and water the wettest, the tetrahedron must represent fire and the icosahedron water. The cube is associated with earth, since the cube, resting foursquare on one of its square faces, has the greatest stability. The octahedron, held lightly by two of its opposite vertices between a forefinger and thumb, easily spins and has the instability of air. Finally, the dodecahedron is associated with the universe because the dodecahedron has twelve faces and the zodiac has twelve signs.

The tetrahedron, cube, and octahedron can be found in nature as crystals, for example, of sodium sulphantimoniate, common salt, and chrome alum, respectively. The other two cannot occur in crystal form, but have been observed as skeletons of microscopic sea animals called *radiolaria*. In 1885, a toy regular dodecahedron of Etruscan origin, believed to date back to about 500 B.C., was unearthed on Monte Loffa, near Padua.

3–10 Postulational Thinking

Sometime between Thales in 600 B.C. and Euclid in 300 B.C., the notion was perfected of a logical discourse as a sequence of rigorous deductions from some initial and explicitly stated assumptions. This process, the so-called **postulational method,** has become the very core of modern mathematics; undoubtedly, much of the development of geometry along this pattern is due to the Pythagoreans. Certainly one of the greatest contributions of the early Greeks was the development of this postulational method of thinking. We shall return to a fuller discussion of the subject in Sections 5–7 and 15–2.

Problem Studies

3.1 The Practical Problems of Thales

(a) There are two versions of how Thales calculated the height of an Egyptian pyramid by shadows. The earlier account, given by Hieronymus, a pupil of Aristotle, says that Thales noted the length of the shadow of the pyramid at the moment when his shadow was the same length as himself. The later version, given by Plutarch, says that he set up a stick and then made use of similar triangles. Both versions fail to mention the difficulty, in either case, of obtaining the length of the shadow of the pyramid—that is, the distance from the apex of the shadow to the center of the base of the pyramid.

Devise a method, based on similar triangles and independent of latitude and time of year, for determining the height of a pyramid *from two shadow observations.*

(b) We are told that Thales measured the distance of a ship from shore, using the fact that 2 triangles are congruent if 2 angles and the included side of one are equal to 2 angles and the included side of the other. Heath has conjectured that this was probably done by an instrument consisting of 2 rods AC and AD, hinged together at A, as shown in Figure 26. The rod AD was held vertically over point B on shore, while rod AC was pointed toward the ship P. Then, without changing the angle DAC, the instrument was revolved about AD, and point Q noted on the ground at which arm AC was directed. What distance must be measured in order to find the distance from B to the inaccessible point P?

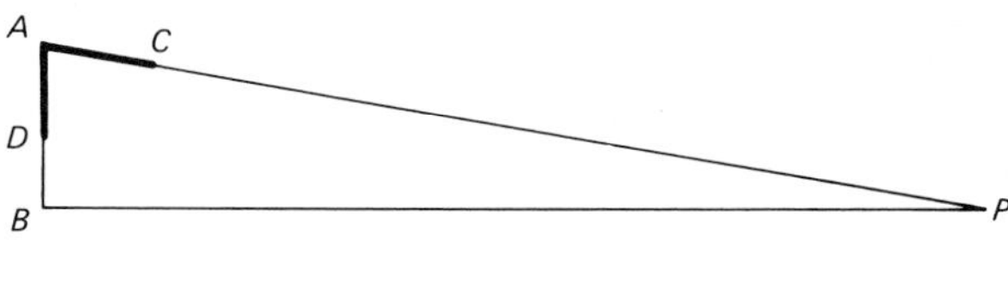

FIGURE 26

3.2 Perfect and Amicable Numbers

(a) Show that in Euclid's formula for perfect numbers, n must be prime.
(b) What is the fourth perfect number furnished by Euclid's formula?
(c) Prove that the sum of the reciprocals of *all* the divisors of a perfect number is equal to 2.
(d) Show that if p is a prime, then p^n is deficient.
(e) Show that Nicolo Paganini's numbers, 1184 and 1210, are amicable.
(f) Show that any multiple of an abundant or perfect number is abundant.

(g) Find the 21 abundant numbers less than 100. It will be noticed that they are all even numbers. To show that all abundant numbers are not even, show that $945 = 3^3 \cdot 5 \cdot 7$ is abundant. This is the first odd abundant number.

(h) Estimate the number of digits in the perfect numbers corresponding to (1) $n = 7$, (2) $n = 127$.

(i) A cyclic sequence of three or more numbers such that the sum of the proper divisors of each is equal to the next in the sequence is known as a **sociable chain** of numbers. Only two sociable chains involving numbers below 1,000,000 are known: one of 5 "links" (found by the Frenchman P. Poulet) starting with 12,496, and one of 28 links starting with 14,316. Find the first of these sociable chains. A sociable chain of exactly 3 links is called a **crowd;** no crowds have yet been found.

(j) Show that 120 is triply perfect.

(k) Is 12 superabundant?

3.3 Figurate Numbers

(a) List the first four hexagonal numbers.

(b) An **oblong number** is the number of dots in a rectangular array having one more column than rows. Show, geometrically and algebraically, that the sum of the first n positive even integers is an oblong number.

(c) Show, both geometrically and algebraically, that any oblong number is twice a triangular number.

(d) Show, geometrically and algebraically, that 8 times any triangular number, plus 1, is a square number.

(e) Show, geometrically and algebraically, that the nth pentagonal number equals the nth square number plus the $(n - 1)$th triangular number—that is, that $P_n = S_n + T_{n-1}$.

(f) Denoting the oblong number $n(n + 1)$ by O_n, show, geometrically and algebraically, that $O_n + S_n = T_{2n}$ and $O_n - S_n = n$.

(g) Prove that every even perfect number is also a triangular number.

(h) Prove that the sequence of m-gonal numbers is given by

$$an^2 + bn, \qquad n = 1, 2, \ldots,$$

for a certain fixed pair of rational numbers a and b.

(i) Find a and b of (h) when $m = 7$.

3.4 Means

The *Eudemian Summary* says that in Pythagoras' time there were three means, the **arithmetic,** the **geometric,** and the **subcontrary,** the last name being later changed to **harmonic** by Archytas and Hippasus. We may define these three means of two positive numbers a and b as

$$A = \frac{a + b}{2}, \qquad G = \sqrt{ab}, \qquad H = \frac{2ab}{a + b},$$

respectively.

(a) Show that $A \geq G \geq H$, equality holding if and only if $a = b$.

(b) Show that $a : A = H : b$. This was known as the "musical" proportion.

(c) Show that H is the harmonic mean between a and b if there exists a number n such that $a = H + a/n$ and $H = b + b/n$. This was the Pythagorean definition of the harmonic mean of a and b.

(d) Show that $1/(H - a) + 1/(H - b) = 1/a + 1/b$.

(e) Since 8 is the harmonic mean of 12 and 6, Philolaus, a Pythagorean of about 425 B.C., called the cube a "geometrical harmony." Explain this.

(f) Show that if a, b, c are in harmonic progression, so also are $a/(b + c)$, $b/(c + a)$, $c/(a + b)$.

(g) If a and c, $a < c$, are a pair of positive numbers, then any number b between a and c is, in some sense, **a mean** (or *average*) of a and c. The later Pythagoreans considered ten means b of a and c, defined as follows:

1. $(b - a)/(c - b) = a/a$	6. $(b - a)/(c - b) = c/b$
2. $(b - a)/(c - b) = a/b$	7. $(c - a)/(b - a) = c/a$
3. $(b - a)/(c - b) = a/c$	8. $(c - a)/(c - b) = c/a$
4. $(b - a)/(c - b) = c/a$	9. $(c - a)/(b - a) = b/a, \quad a < b$
5. $(b - a)/(c - b) = b/a$	10. $(c - a)/(c - b) = b/a, \quad a < b$

Assuming $0 < a < c$, show that in all ten cases $a < b < c$.

(h) Show that (1), (2), and (3) of (g) give the arithmetic, the geometric, and the harmonic means, respectively, of a and c.

3.5 Dissection Proofs of the Pythagorean Theorem

(a,b) Two areas, or 2 volumes, P and Q, are said to be **congruent by addition** if they can be dissected into corresponding pairs of congruent pieces. They are said to be **congruent by subtraction** if corresponding pairs of congruent pieces can be added to P and Q to give 2 new figures that are congruent by addition. There are many proofs of the Pythagorean theorem that achieve their end by showing that the square on the hypotenuse of the right triangle is congruent either by addition or subtraction to the combined squares on the legs of the right triangle. The proof given in Section 3–4 is a congruency-by-subtraction proof. Give 2 congruency-by-addition proofs of the Pythagorean theorem suggested by Figures 27 and 28, the first given by Henry Perigal (dates unknown) in 1873[11] and the second by H. E. Dudeney (1857–1930) in 1917.

(c) Give a congruency-by-subtraction proof of the Pythagorean theorem suggested by Figure 29, which is said to have been devised by Leonardo da Vinci (1452–1519).

It is interesting that any two equal polygonal areas are congruent by addition, and the dissection can always be carried out with

---

[11] This was a rediscovery, for the dissection was known to T̲âbit ibn Qorra (826–901).

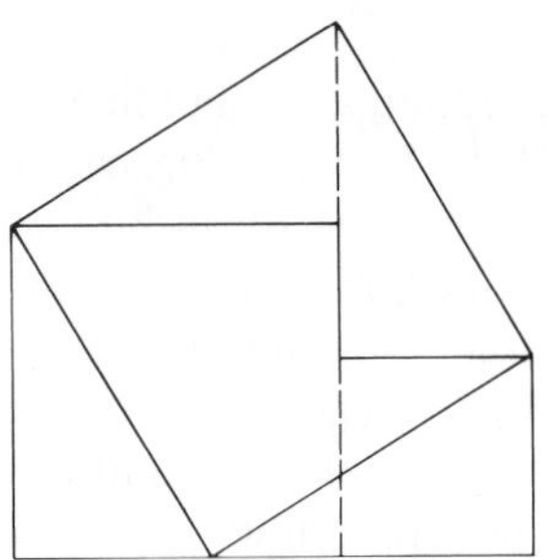

FIGURE 27

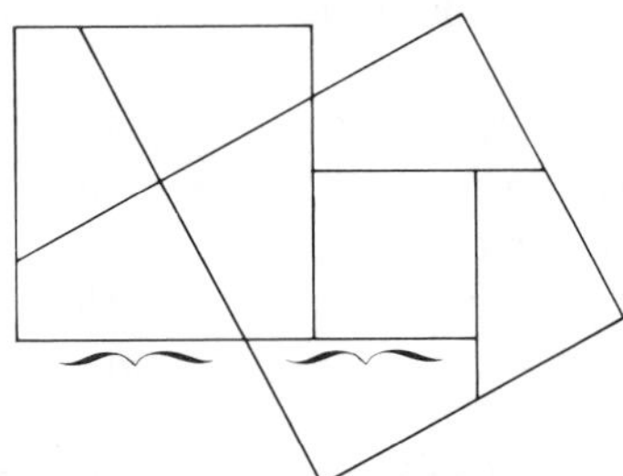

FIGURE 28

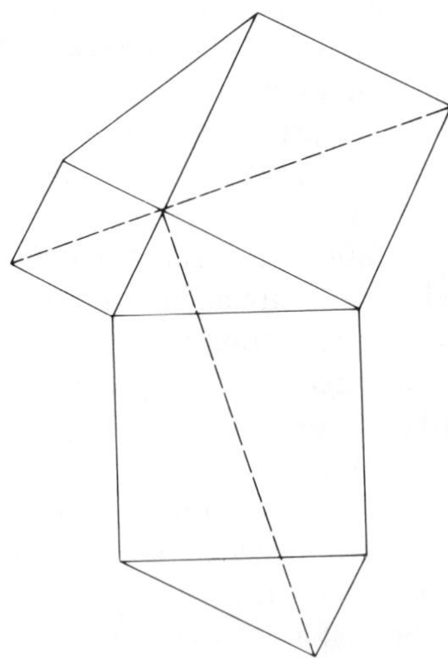

FIGURE 29

straightedge and compasses. In 1901, however, Max Dehn (1878–1952) showed that two equal polyhedral volumes are not necessarily congruent by either addition or subtraction. In particular, it is impossible to dissect a regular tetrahedron into polyhedral pieces that can be reassembled to form a cube. Dehn achieved these results in solving one of David Hilbert's (1862–1943) twenty-three Paris problems (see the final sentence of Section 15–8).

3.6 Pythagorean Triples

(a) What is the relation between the hypotenuse and the longer leg of the integral-sided right triangles given by the Pythagorean formula of Section 3-4?

(b) Find the Pythagorean triples given by the Pythagorean formula of Section 3–4 for which the hypotenuse does not exceed 100.

(c) Prove that no isosceles right triangle exists whose sides are integers.

(d) Prove that no Pythagorean triple exists in which 1 integer is a mean proportional between the other 2.

(e) Prove that (3,4,5) is the only Pythagorean triple containing three consecutive positive integers.

(f) Find the 16 *primitive* Pythagorean triples (a,b,c) for which b is even and $c < 100$. Now show that there are exactly 100 distinct Pythagorean triples (a,b,c) with $c < 100$.

(g) Show that if $(a,a + 1,c)$ is a Pythagorean triple, so is

$$(3a + 2c + 1, 3a + 2c + 2, 4a + 3c + 2).$$

It follows that, from a given Pythagorean triple whose legs are successive natural numbers, we can obtain another such Pythagorean triple with bigger sides.

(h) Starting with the Pythagorean triple (3,4,5), find 5 more Pythagorean triples whose legs are successive natural numbers and whose sides are progressively bigger.

(i) Prove that in each Pythagorean triple: (1) at least 1 is a multiple of 4, (2) at least one leg is a multiple of 3, (3) at least 1 side is a multiple of 5.

(j) Prove that for any natural number $n > 2$ there exists a Pythagorean triple with a leg equal to n.

(k) Prove that there are only a finite number of Pythagorean triples having a given leg a.

(l) Show that for any natural number n and for $k = 0, 1, 2, \ldots, n - 1$,

$$[2^{n+1}, 2^k(2^{2n-2k} - 1), 2^k(2^{2n-2k} + 1)]$$

are Pythagorean triples. It follows that for each natural number n there exist at least n different Pythagorean triples with the same leg $a = 2^{n+1}$. It can be shown, with more difficulty, that for each natural number n there exist at least n different *primitive* Pythagorean triples with a common leg.

(m) Let (a_k, b_k, c_k), $k = 1, 2, \ldots, n$, be n different *primitive* Pythagorean triples. Set

$$s_k = a_k + b_k + c_k \qquad \text{and} \qquad s = s_1 s_2 \ldots s_n.$$

Now set $a_k' = a_k s/s_k$, $b_k' = b_k s/s_k$, $c_k' = c_k s/s_k$ for $k = 1, 2, \ldots, n$. Show that (a_k', b_k', c_k') is a Pythagorean triple with

$$a_k' + b_k' + c_k' = s.$$

It now follows that for each natural number n there exist at least n noncongruent Pythagorean triples with the same perimeter.

3.7 Irrational Numbers

(a) Prove that the straight line through the points (0,0) and $(1, \sqrt{2})$ passes through no point, other than (0,0), of the coordinate lattice.

(b) Show how the coordinate lattice may be used for finding rational approximations of $\sqrt{2}$.

(c) If p is a prime number, show that $\sqrt{p}$ is irrational.

(d) Show that $\log_{10} 2$ is irrational.

(e) Generalize (d) by showing that $\log_a b$ is irrational if a and b are positive integers and 1 of them contains a prime factor not contained in the other.

(f) Draw a 60-30 right triangle; mark off the longer leg, from the 30° angle vertex, on the hypotenuse; draw a perpendicular to the hypotenuse from the dividing point. Using this figure, formulate a geometrical proof of the irrationality of $\sqrt{3}$.

(g) Prove that the sum (product) of a nonzero rational number and an irrational number is an irrational number.

3.8 Algebraic Identities

Indicate how each of the following algebraic identities might be established geometrically:

(a) $(a - b)^2 = a^2 - 2ab + b^2$

(b) $a(b + c) = ab + ac$

(c) $(a + b)(c + d) = ac + bc + ad + bd$

(d) $a^2 - b^2 = (a + b)(a - b)$

(e) The statement of Proposition 9 of Book II of Euclid's *Elements* is: *If a straight line is divided equally and also unequally, the sum of the squares on the two unequal parts is twice the sum of the squares on half the line and on the line between the points of section.* From this theorem, obtain the algebraic identity

$$(a + b)^2 + (a - b)^2 = 2(a^2 + b^2).$$

3.9 Geometric Algebra

Let the lengths of 3 given line segments be a, b, 1 ($a > b > 1$). With straightedge and compasses, construct line segments of lengths

(a) $a + b$ and $a - b$,
(b) ab,
(c) a/b,
(d) $\sqrt{a}$,
(e) a/n, n a positive integer,
(f) $\sqrt{ab}$,
(g) $a\sqrt{n}$, n a positive integer,
(h) $(a^3 + b^3)/(a^2 + b^2)$,
(i) $a[1 + \sqrt{2} + \sqrt{3}]^{1/2}$,
(j) $(abcd)^{1/4}$, where c and d are the lengths of 2 further given line segments,
(k) $x = (a^2 + b^2 - ab)^{1/2}$. If we form a triangle with sides a, b, x, what is the size of the angle between sides a and b?
(l) Show that $x = ab/(a^2 + b^2)^{1/2}$ is equal to the altitude of a right triangle with legs a and b.

3.10 Geometric Solution of Quadratic Equations

(a) Given a unit segment, solve the quadratic equation $x^2 - 7x + 12 = 0$ by the Pythagorean method.
(b) Given a unit segment, solve the quadratic equation $x^2 + 4x - 21 = 0$ by the Pythagorean method.
(c) With straightedge and compasses, divide a segment a into 2 parts such that the difference of their squares shall be equal to their product.
(d) Show that in (c) the longer segment is the mean proportional between the shorter segment and the whole line. The line segment is said to be divided in **extreme and mean ratio,** or in **golden section.**
(e) A quadratic equation $x^2 - gx + h = 0$ is given. On a rectangular Cartesian frame of reference, plot the points B:(0,1) and Q:(g,h). Draw the circle with BQ as a diameter and let it cut the x-axis in M and N. Show that the signed lengths of OM and ON represent the roots of the given quadratic equation. This geometrical solution of quadratic equations appeared in Leslie's *Elements of Geometry* with the remark: "The solution of this important problem now inserted in the text, was suggested to me by Mr. Thomas Carlyle, an ingenious young mathematician, and formerly my pupil."
(f) Solve the quadratic equations $x^2 - 7x + 12 = 0$ and $x^2 + 4x - 21 = 0$ by Carlyle's method.
(g) Again, the quadratic equation $x^2 - gx + h = 0$ is given. On a rectangular Cartesian frame of reference, plot the points (h/g,0) and ($4/g$,2), and let the join of these two points cut the unit circle of center (0,1) in points R and S. Project R and S from the point (0,2) onto points (r,0) and (s,0) on the x-axis. Show that r and s are the roots of the given

quadratic equation. This geometric solution of quadratic equations was given by the German geometer Karl Georg Christian von Staudt (1798–1867).

(h) Solve the quadratic equations $x^2 - 7x + 12 = 0$ and $x^2 + 4x - 21 = 0$ by Staudt's method.

(i) Verify the following geometrical solution of the quadratic equation $x^2 - gx + h = 0$, $h > 0$. First construct $\sqrt{h}$ as the mean proportional between 1 and h. Then on $AB = |g|$ as a diameter, construct a semicircle and draw the vertical half-chord $CD = \sqrt{h}$, where D is on AB. Then AD and DB, each taken with signs the same as g, are the roots of the quadratic equation. Solve, by this method, the quadratic equation $x^2 - 7x + 12 = 0$.

(j) Verify the following geometrical solution of the quadratic equation $x^2 - gx + h = 0$, $h < 0$. Draw a circle on $AB = |g|$ as a diameter and draw tangent $AC = \sqrt{-h}$. Draw the diametral secant CDE through C to cut the circle in D and E. Then CD and CE, taken with opposite signs and with that of CE the same as that of g, represent the roots of the quadratic equation. Solve, by this method, the quadratic equation $x^2 + 4x - 21 = 0$.

3.11 Transformation of Areas

(a) Draw an irregular hexagon and then construct, with straightedge and compasses, a square having the same area.

(b) With straightedge and compasses, divide a quadrilateral $ABCD$ into 3 equivalent parts by straight lines drawn through vertex A.

(c) Bisect a trapezoid by a line drawn from a point P in the smaller base.

(d) Transform triangle ABC so that the angle A is not altered, but the side opposite the angle A becomes parallel to a given line MN.

(e) Transform a given triangle into an isosceles triangle having a given vertex angle.

3.12 Regular Solids

(a) Show that there can be no more than 5 regular polyhedra.

(b) Find the volume and surface of a regular octahedron of edge e.

(c) For each of the 5 regular polyhedra, enumerate the number of vertices v, edges e, and faces f, and then evaluate the quantity $v - e + f$. One of the most interesting theorems relating to any convex (or more generally any *simply connected*) polyhedron, is that $v - e + f = 2$. This may have been known to Archimedes (ca. 225 B.C.), and was very nearly stated by Descartes about 1635. Since Euler later independently announced it in 1752, the result is often referred to as the Euler-Descartes formula.

(d) A **cuboctahedron** is a solid whose edges are obtained by joining together the midpoints of adjacent edges of a cube. Enumerate v, e, and f for a cuboctahedron.

(e) Consider a solid cube with regular pyramids built on a pair of opposite faces as bases. Now let a hole with square cross section, and with its axis on the line joining the vertices of the pyramids, be cut from the solid. Evaluate $v - e + f$ for this ring-shaped solid.[12]

3.13 Some Problems Concerning the Regular Solids

(a) In Section 3–9, the definition of regularity of a polyhedron involves 3 properties: regular faces, congruent faces, and congruent polyhedral angles. Many textbooks on solid geometry do not give all 3 of the defining properties. Show, by counterexamples, that all 3 properties are necessary.

(b) From the 3 defining properties listed in (a), one can deduce the regularity of the polyhedral angles. Do this, and then show that the 3 defining properties can be replaced by only 2: regular faces and regular polyhedral angles.

(c) The uninitiated will almost always intuitively believe that, when a regular dodecahedron (a solid having 12 faces) and a regular icosahedron (a solid having 20 faces) are inscribed in the same sphere, the icosahedron has the greater volume. Show that the reverse is actually the case, and also show that when a cube (a solid having 6 faces) and a regular octahedron (a solid having 8 faces) are inscribed in the same sphere, the cube has the larger volume.

(d) Show that a regular dodecahedron and a regular icosahedron inscribed in the same sphere have a common inscribed sphere.

(e) In Section 3–9, we noted that Kepler intuitively assumed that, of the 5 regular solids, for a given surface area, the icosahedron encloses the largest volume. Is this so?

(f) A regular dodecahedron, a regular icosahedron, and a cube are inscribed in the same sphere. Prove that the volume of the dodecahedron is to the volume of the icosahedron as the length of an edge of the cube is to the length of an edge of the icosahedron.

3.14 Golden Section

A point is said to divide a line segment in **extreme and mean ratio,** or in **golden section,** when the longer of the two segments formed is the mean proportional between the shorter segment and the whole line. The ratio of the shorter segment to the longer segment is called the **golden ratio.** The Pythagoreans showed considerable interest in the golden section and the golden ratio.

(a) Show that the golden ratio is $(\sqrt{5} - 1)/2$.

(b) The symbol of the Pythagorean brotherhood was the **pentagram,** or 5-pointed-star, formed by the 5 diagonals of a regular pentagon. Prove that each of the 5 sides of a pentagram divides into golden section the 2 sides of the pentagram that it intersects.

---

[12] Construction patterns for 100 different solids can be found in Miles C. Hartley, *Patterns of Polyhedrons*. Rev. Ed., Ann Arbor, Mich.: Edwards Brothers, 1957.

(c) Let point G divide line segment AB in golden section, where AG is the longer segment. On AB, mark off $AH = GB$. Show that H divides AG in golden section.

(d) Construct, with straightedge and compasses, a regular pentagon, given a side of the pentagon.

(e) Construct, with straightedge and compasses, a regular pentagon, given a diagonal of the pentagon.

(f) Inscribe a regular pentagon in a given circle, using straightedge and compasses alone.

3.15 Constructions of $\sqrt{n}$ by Theodorus

(a) Theodorus of Cyrene (born ca. 470 B.C.) constructed $\sqrt{n}$ as half the leg of a right triangle whose hypotenuse is $n + 1$ and other leg is $n - 1$. Justify this construction.

(b) It has been suggested that Theodorus also obtained $\sqrt{n}$ $(2 \leq n \leq 17)$ by constructing a spiral-like figure made up of a sequence of right triangles having a common vertex, where the first triangle in the sequence is the isosceles right triangle of leg 1, and where in each succeeding right triangle one leg is the hypotenuse of the previous triangle in the sequence and the other leg (opposite the common vertex) has length 1. Show that the hypotenuse of the nth triangle in the sequence has length $\sqrt{n + 1}$.

(c) Show how the construction process of (b) might explain why Theodorus cut off his consideration of $\sqrt{n}$ with $n = 17$.

3.16 An Interesting Relation

Prove geometrically that

$$1^3 + 2^3 + \ldots + n^3 = (1 + 2 + \ldots + n)^2.$$

Essay Topics

3/1 Possible reasons for the Greek introduction of deduction into mathematics.

3/2 Stories of Thales' prowess in engineering and astronomy, and their credibility.

3/3 Pythagorean number mysticism.

3/4 The case for Pythagoreanism, as evidenced by modern physical formulas.

3/5 Pythagoras justified, insofar as mathematics is concerned.

3/6 How the discovery of incommensurable magnitudes produced a crisis in the development of mathematics.

3/7 The golden ratio in art and architecture.

3/8 Simple examples of applied geometry for an elementary geometry class.

3/9 Early history of the regular solids, with patterns for their construction.

3/10 The debt of Greek mathematics to ancient Mesopotamia and Egypt.

3/11 Reasons for treating logistic and arithmetic as unrelated subjects.

3/12 Advantages and disadvantages of the Greek method of treating arithmetic from a geometric standpoint.

Bibliography

AABOE, ASGER *Episodes from the Early History of Mathematics.* New Mathematical Library. no. 13. New York: Random House and L. W. Singer, 1964.

ALLMAN, G. J. *Greek Geometry from Thales to Euclid.* Dublin: University Press, 1889. Reprinted by Bell & Howell, Cleveland, Ohio.

BELL, E. T. *The Magic of Numbers.* New York: McGraw-Hill, 1946.

BUNT, L. N. H.; P. S. JONES; and J. D. BEDIENT *The Historical Roots of Elementary Mathematics.* Englewood Cliffs, N.J.: Prentice-Hall, 1976.

COOLIDGE, J. L. *A History of Geometrical Methods.* New York: Oxford University Press, 1940.

COURANT, RICHARD, and H. E. ROBBINS *What is Mathematics?* New York: Oxford University Press, 1941.

DANTZIG, T. *Number: The Language of Science.* 3rd ed. New York: Macmillan, 1939.

——— *The Bequest of the Greeks.* New York: Charles Scribner's, 1955.

FRIEDRICHS, O. *From Pythagoras to Einstein.* New Mathematical Library, no. 16. New York: Random House, 1965.

GOW, JAMES *A Short History of Greek Mathematics.* New York: Hafner, 1923. Reprinted by Bell & Howell, Cleveland, Ohio.

HEATH, T. L. *History of Greek Mathematics,* vol. 1. New York: Oxford University Press, 1921. Reprinted by Dover, New York, 1981.

——— *A Manual of Greek Mathematics.* New York: Oxford University Press, 1931.

——— *The Thirteen Books of Euclid's Elements.* 2d ed., 3 vols. New York: Cambridge University Press, 1926. Reprinted by Dover, New York.

HERZ-FISCHLER, ROGER *A Mathematical History of Division in Extreme and Mean Ratio.* Waterloo, Canada: Wilfrid Laurier University Press, 1987.

KLEIN, JACOB *Greek Mathematical Thought and the Origin of Algebra.* Translated by Eva Brann. Cambridge, Mass.: M.I.T. Press, 1968.

LESLIE, SIR JOHN *Elements of Geometry, Geometrical Analysis, and Plane Trigonometry.* Edinburgh: Oliphant, 1809. Reprint available from Bell & Howell, Cleveland, Ohio.

LOOMIS, E. S. *The Pythagorean Proposition.* 2d ed. Ann Arbor, Mich.: Edwards Brothers, 1940. Reprinted by the National Council of Teachers of Mathematics, Washington, D.C., 1968.

MAZIARZ, EDWARD, and THOMAS GREENWOOD *Greek Mathematical Philosophy.* Cambridge, Mass.: M.I.T. Press, 1968.

MESCHKOWSKI, HERBERT *Ways of Thought of Great Mathematicians.* Translated by John Dyer-Bennet. San Francisco: Holden-Day, 1964.

MUIR, JANE *Of Men and Numbers.* New York: Dodd, Mead, 1961.

PROCLUS *A Commentary on the First Book of Euclid's Elements.* Translated by G. R. Morrow. Princeton, N.J.: Princeton University Press, 1970.

SHANKS, DANIEL *Solved and Unsolved Problems in Number Theory,* vol 1. Washington, D.C.: Spartan Books, 1962.

SIERPINSKI, WACLAW *Pythagorean Triangles.* Scripta Mathematica Studies Number Nine. Translated by Ambikeshwar Sharma. New York: Yeshiva University, 1962.

SZABO, ARPAD *The Beginnings of Greek Mathematics.* Dordrecht, Holland: D. Reidel, 1978.

THOMAS, IVOR, ed. *Selections Illustrating the History of Greek Mathematics. 2 vols. Cambridge, Mass.: Harvard University Press, 1939–1941.*

TURNBULL, H. W. *The Great Mathematicians.* New York: New York University Press, 1961.

VAN DER WAERDEN, B. L. *Science Awakening.* Translated by Arnold Dresden. New York: Oxford University Press, 1961. Paperback edition. New York: John Wiley, 1963.

WENNINGER, M. J. *Polyhedron Models for the Classroom.* National Council of Teachers of Mathematics, Washington, D.C., 1966.

——— *Polyhedron models.* New York: Cambridge University Press, 1970.

Chapter 4

DUPLICATION, TRISECTION, AND QUADRATURE

4–1 The Period from Thales to Euclid

The first three centuries of Greek mathematics, commencing with the initial efforts at demonstrative geometry by Thales about 600 B.C. and culminating with the remarkable *Elements* of Euclid about 300 B.C., constitute a period of extraordinary achievement. In the last chapter, we considered some of the Pythagorean contributions to this achievement. Besides the Ionian school founded by Thales at Miletus and the early Pythagorean school at Crotona, a number of mathematical centers arose and flourished at places and for periods that were largely governed by Greek political history.

It was about 1200 B.C. that the primitive Dorian tribes moved southward into the Greek peninsula, leaving their northern mountain fastnesses for more favorable territory. Their chief tribe, the Spartans, subsequently developed the city of Sparta. Many of the former inhabitants of the invaded region fled to Asia Minor and the Ionian islands of the Aegean Sea, where in time they established Greek trading colonies. It was in these colonies, in the sixth century B.C., that the Ionian school was founded, Greek philosophy blossomed, and demonstrative geometry was born.

Meanwhile, Persia had become a great military empire and, following the inevitable expansionist program induced by a slave-based economy, conquered the Ionian cities and the Greek colonies of Asia Minor in 546 B.C. As a result, a number of Greek philosophers, like Pythagoras and Xenophanes, abandoned their native land and moved to the prospering Greek colonies in southern Italy. Schools of philosophy and mathematics developed at Crotona, under Pythagoras, and at Elea, under Xenophanes, Zeno, and Parmenides.

The yoke of oppression rested uneasily on the conquered Ionian cities, and in 499 B.C. a revolt was fomented. Athens, which was becoming a center of Western civilization with political progress toward democracy, aided the revolution by sending armies. Although the revolt was crushed, the incensed King Darius of Persia decided to punish Athens. In 492 B.C., he organized a huge army and navy to attack the mainland of Greece, but his fleet was destroyed in a storm, and his land forces suffered expeditionary difficulties. Two years later, the Persian armies penetrated Attica, where they were decisively defeated by the Athenians at Marathon. Athens assumed the mantle of Greek leadership.

In 480 B.C. Xerxes, son of Darius, attempted another land and sea invasion of Greece. The Athenians met the Persian fleet in the great naval battle of

Salamis and won, and although the Greek land forces under Spartan leadership were defeated and wiped out at Thermopylae, the Greeks overcame the Persians the following year at Plataea and forced the invaders out of Greece. The hegemony of Athens was consolidated, and the following half century of peace was a brilliant period in Athenian history. This city of Pericles and Socrates became the center of democratic and intellectual development. Mathematicians were attracted from all parts of the Greek world. Anaxagoras, the last eminent member of the Ionian school, settled there. Many of the dispersed Pythagoreans found their way to Athens, and Zeno and Parmenides, of the Eleatic school, went to Athens to teach. Hippocrates,[1] from the Ionian island of Chios, visited Athens and is reputed by ancient writers to have published the first connected geometry there.

Peace came to an end in 431 B.C. with the start of the Peloponnesian War between Athens and Sparta. This proved to be a long, drawn-out conflict. Athens, at first successful, later suffered a devastating plague that killed off a fourth of its population; finally, in 404 B.C., Athens had to accept humiliating defeat. Sparta assumed political leadership, only to lose it, in 371 B.C., by defeat at the hands of a league of rebellious city-states. During these struggles, little progress was made in geometry at Athens, and once again development came from the more peaceful regions of Magna Graecia. The Pythagoreans of southern Italy had been allowed to return, purified of political association, and a new Pythagorean school at Tarentum arose under the influence of the gifted and much admired Archytas.

With the end of the Peloponnesian War, Athens, although reduced to a minor political power, regained her cultural leadership. Plato was born in or near Athens in 427 B.C., the year of the great plague. He studied philosophy under Socrates there, and then set out upon his extensive wanderings for wisdom. He studied mathematics under Theodorus of Cyrene on the African coast and became an intimate friend of the eminent Archytas. Upon his return to Athens around 387 B.C., he founded his famous Academy, an institution for the systematic pursuit of philosophical and scientific inquiry. He presided over his Academy for the rest of his life, dying in Athens in 347 B.C. at the venerable age of eighty. Almost all the important mathematical work of the fourth century B.C. was done by friends or pupils of Plato, making his Academy the link between the mathematics of the earlier Pythagoreans and that of the later, long-lived school of mathematics at Alexandria. Plato's influence on mathematics was not due to any mathematical discoveries he made, but rather to his enthusiastic conviction that the study of mathematics furnished the finest training for the mind and, hence, was essential for the cultivation of philosophers and those who should govern his ideal state. This explains the renowned motto over the door of his Academy: *Let no one unversed in geometry enter here*. Because of its logical element and the pure attitude of mind that he felt its study created,

[1] Not to be confused with Hippocrates of Cos, the famous Greek physician of antiquity.

mathematics seemed of utmost importance to Plato; for this reason, it occupied a valued place in the curriculum of the Academy. Some see in certain of Plato's dialogues what may be considered the first serious attempt at a philosophy of mathematics.

Eudoxus, who studied under both Archytas and Plato, founded a school at Cyzicus, in northern Asia Minor. Menaechmus, an associate of Plato and a pupil of Eudoxus, invented the conic sections. Dinostratus, brother of Menaechmus, was an able geometer and a pupil of Plato. Theaetetus, a man of unusual natural gifts, to whom we are probably indebted for much of the material of Euclid's tenth and thirteenth books, was another Athenian pupil of Theodorus. Mention should also be made of Aristotle who, although not a professed mathematician, was the systematizer of deductive logic and a writer on physical subjects; some parts of his *Analytica posteriora* show an unusual grasp of the mathematical method.

Some Greek Names of Classic Times
with accented syllable indicated

Anaxag'oras	Eudox'us	Phi'lon
An'tiphon	Euto'cius	Pla'to
Apollo'nius	Her'on	Polyc'rates
Archime'des	Hippar'chus	Pro'clus
Archy'tas	Hippa'sus	Ptol'emy
Aristae'us	Hip'pias	Pythag'oras
Aristar'chus	Hippoc'rates	Simpli'cius
Ar'istotle	Hypa'tia	Soc'rates
Co'non	Hyp'sicles	So'lon
Democ'ritus	Iam'blichus	Tha'les
Dinos'tratus	Menaech'mus	Theaete'tus
Di'ocles	Menela'us	Theodo'rus
Diophan'tus	Metrodor'us	Theodo'sius
Dosi'theus	Nicom'achus	The'on
Eratos'thenes	Nicome'des	Thymar'idas
Eu'clid	Pap'pus	Xenoc'rates
Eude'mus	Philola'us	Ze'no

PLATO
(David Smith Collection)

4–2 Lines of Mathematical Development

One can notice three important and distinct lines of development during the first 300 years of Greek mathematics. First, we have the development of the material that ultimately was organized into the *Elements*, ably begun by the Pythagoreans and then added to by Hippocrates, Eudoxus, Theodorus, Theaetetus, and others. We have already considered portions of this develop-

ARISTOTLE
(Brown Brothers)

ment and shall return to it in the next chapter. Second, there is the development of notions connected with infinitesimals and with limit and summation processes that did not attain final clarification until after the invention of the calculus in modern times. The paradoxes of Zeno, the method of exhaustion of Antiphon and Eudoxus, and the atomistic theory associated with the name of Democritus belong to this second line of development; these are discussed more logically in the early sections of Chapter 11 devoted to the origins of the calculus.

> Any student or instructor desirous of adhering strictly to chronological order can, at this point, turn to Sections 11–2 and 11–3.

The third line of development is that of higher geometry, or the geometry of curves other than the circle and straight line, and of surfaces other than the sphere and plane. Curiously enough, most of this higher geometry originated in continued attempts to solve three now famous construction problems. This chapter discusses these three famous problems.

4–3 The Three Famous Problems

The three famous problems are the following:

1. *The duplication of the cube,* or the problem of constructing the edge of a cube having twice the volume of a given cube.
2. *The trisection of an angle,* or the problem of dividing a given arbitrary angle into three equal parts.

3. *The quadrature of the circle,* or the problem of constructing a square having an area equal to that of a given circle.

The importance of these problems lies in the fact that they cannot be solved, except by approximation, with straightedge and compasses, although these tools successfully serve for the solution of so many other construction problems. The energetic search for solutions to these three problems profoundly influenced Greek geometry and led to many fruitful discoveries, such as that of the conic sections, many cubic and quartic curves, and several transcendental curves. A much later outgrowth was the development of portions of the theory of equations concerning domains of rationality, algebraic numbers, and group theory. The impossibility of the three constructions, under the self-imposed limitation that only the straightedge and compasses could be used, was not established until the nineteenth century, more than 2000 years after the problems were first conceived.

The great stimulation to the development and creation of new mathematics furnished by the continued efforts to solve the three famous problems of antiquity illustrates the heuristic value of attractive, unsolved problems in mathematics.

4–4 The Euclidean Tools

It is important to be clear as to just what we are permitted to do with the straightedge and compasses. *With the straightedge, we are permitted to draw a straight line of indefinite length through any two given distinct points. With the compasses, we are permitted to draw a circle with any given point as center and passing through any given second point.* The drawing of constructions with straightedge and compasses, viewed as a game played according to these two rules, has proved to be one of the most fascinating and absorbing games ever devised. One is surprised at the really intricate constructions that can be accomplished in this manner; accordingly, it is hard to believe that the seemingly simple construction problems presented in Section 4–3 cannot also be so accomplished.

Since the postulates of Euclid's *Elements* restrict the use of the straightedge and compasses in accordance with the above rules, these instruments, so used, have become known as **Euclidean tools**. Note that the straightedge is to be *unmarked.* We shall see that with a marked straightedge, it is possible to trisect a given angle. Also, we notice that the Euclidean compasses differ from our modern compasses, for with the modern compasses we are permitted to draw a circle having any point C as center and any segment AB as radius. In other words, we are permitted to transfer the distance AB to the center C, using the compasses as dividers. The Euclidean compasses, on the other hand, may be supposed to collapse if either leg is lifted from the paper. It might seem that the modern compasses are somewhat more powerful than the Euclidean, or collapsing, compasses. Curiously enough, the two are equivalent tools (see Problem Study 4.1).

4–5 Duplication of the Cube

There is evidence that the problem of duplicating a cube may have originated in the words of some mathematically unschooled ancient Greek poet (perhaps Euripides) who represented the mythical King Minos as dissatisfied with the size of a tomb erected to his son Glaucus. Minos ordered that the tomb be doubled in size. The poet then had Minos add, incorrectly, that this could be accomplished by doubling each dimension of the tomb. This faulty mathematics on the part of the poet led the geometers to take up the problem of finding how one can double a given solid while keeping the same shape. No progress seems to have been made on the problem until sometime later, when Hippocrates discovered his famous reduction, which we give below. Again, still later, it is told that the Delians were instructed by their oracle that, to get rid of a certain pestilence, they must double the size of Apollo's cubical altar. The problem reputedly was taken to Plato, who submitted it to the geometers. It is this latter story that led the duplication problem frequently to be referred to as the *Delian problem*. Whether the story is true or not, the problem was studied in Plato's Academy, and there are higher geometry solutions attributed to Eudoxus, Menaechmus, and even (though probably erroneously) to Plato himself.

The first real progress in the duplication problem was, no doubt, the reduction of the problem by Hippocrates (ca. 440 B.C.) to the construction of two mean proportionals between two given line segments of lengths s and $2s$. If we denote the two mean proportionals by x and y, then

$$s:x = x:y = y:2s.$$

From these proportions, we have $x^2 = sy$ and $y^2 = 2sx$. Eliminating y, we find that $x^3 = 2s^3$; thus x is the edge of a cube having twice the volume of the cube on edge s.

After Hippocrates made his reduction, subsequent attempts at duplicating the cube took the form of constructing two mean proportionals between two given line segments. One of the earliest, and certainly one of the most remarkable, higher geometry solutions in this form was given by Archytas (ca. 400 B.C.). His solution rests on finding a point of intersection of a right circular cylinder, a torus of zero inner diameter, and a right circular cone! The solution sheds some light on the unusual extent to which geometry must have been developed at this early date. The solution by Eudoxus (ca. 370 B.C.) is lost. Menaechmus (ca. 350 B.C.) gave two solutions of the problem and, as far as is known, invented the conic sections for the purpose. A later solution, using a mechanical contrivance, is credited to Eratosthenes (ca. 230 B.C.), and another of about the same time to Nicomedes. A still later solution was offered by Apollonius (ca. 225 B.C.). Diocles (ca. 180 B.C.) invented the cissoid curve to obtain the desired end. And, of course, many solutions using higher plane curves have been devised in more recent times.

A number of the solutions mentioned above may be found in the Problem Studies at the end of the chapter. To illustrate the spirit of the attempts, let us

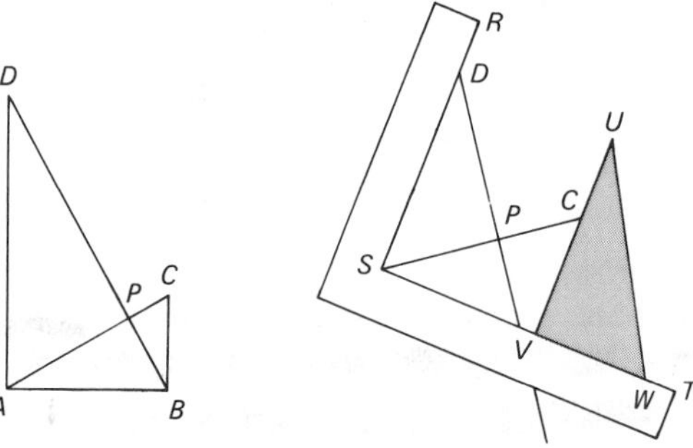

FIGURE 30

reproduce the one credited to Plato by Eutocius. Since the solution is by mechanical means, and since it is known that Plato objected to such methods, it is felt that the ascription to Plato is erroneous.

Consider two triangles (see first part of Figure 30), *CBA* and *DAB,* right angled at *B* and *A,* respectively, and lying on the same side of the common leg *AB*. Let the hypotenuses *AC* and *BD* of the triangles intersect perpendicularly in *P*. From the similar triangles *CPB, BPA, APD,* it follows that

$$PC:PB = PB:PA = PA:PD.$$

Thus, *PB* and *PA* are the two mean proportionals between *PC* and *PD*. It follows that the problem is solved if a figure can be constructed having $PD = 2(PC)$. The second part of Figure 30 shows how such a figure can be drawn by mechanical means. Draw two perpendicular lines intersecting in *P* and mark off *PC* and *PD* on them, with $PD = 2(PC)$. Now place a carpenter's square, with inner edge *RST,* on the figure so that *SR* passes through *D* and the vertex *S* of the right angle lies on *CP* produced. On *ST,* slide a right triangle *UVW,* with leg *VW* on *ST,* until leg *VU* passes through *C*. Now manipulate the apparatus[2] until *V* falls on *DP* produced.

4–6 Trisection of an Angle

Of the three famous problems of Greek antiquity, the trisection of an angle is pre-eminently the most popular among the mathematically uninitiated in America today. Every year the mathematics journals and the members of the mathematics teaching profession of the country receive many communications from "angle trisectors," and it is not unusual to read in a newspaper that someone has finally "solved" the elusive problem. The problem is certainly the simplest

[2] For an improved form of this apparatus see, for example, Richard Courant and H. E. Robbins, *What Is Mathematics?* p. 147.

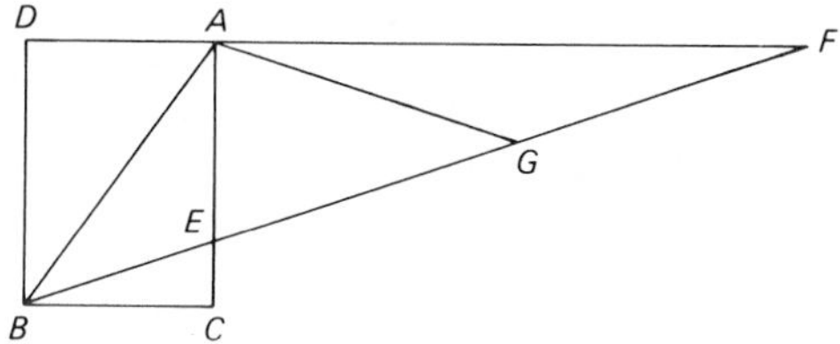

FIGURE 31

one of the three famous problems to comprehend, and since the bisection of an angle is so very easy, it is natural to wonder why trisection is not equally easy.

The multisection of a line segment with Euclidean tools is a simple matter, and it may be that the ancient Greeks were led to the trisection problem in an effort to solve the analogous problem of multisecting an angle. Or perhaps, more likely, the problem arose in efforts to construct a regular ninesided polygon, where the trisection of a 60° angle is required.

In dealing with the trisection problem, the Greeks seem first to have reduced it to what they called a **verging problem**. Any acute angle ABC (see Figure 31) may be taken as the angle between a diagonal BA and a side BC of a rectangle $BCAD$. Consider a line through B cutting CA in E and DA produced in F, and such that $EF = 2(BA)$. Let G be the midpoint of EF. Then

$$EG = GF = GA = BA,$$

whence

$$\measuredangle ABG = \measuredangle AGB = \measuredangle GAF + \measuredangle GFA = 2\measuredangle GFA = 2\measuredangle GBC,$$

and BEF trisects angle ABC. Thus, the problem is reduced to that of constructing a straight-line segment EF of given length $2(BA)$ between AC and the prolongation of DA so that FE verges toward B.

If, contrary to Euclidean assumptions, we permit ourselves to mark, on our straightedge, a segment $E'F' = 2(BA)$, and then to adjust the straightedge so that it passes through B and has the marked points E' and F' on AC and the prolongation of DA, the angle ABC will be trisected. This disallowed use of the straightedge may be referred to as an application of *the insertion principle*. For other applications of the principle, see Problem Study 4.6.

Various higher plane curves have been discovered that will solve the verging problem to which the trisection problem may be reduced. One of the oldest of these is the conchoid invented by Nicomedes (ca. 240 B.C.). Let c be a straight line and O any point not on c. On the prolongation of OP, where P is any point on c, mark off PQ equal to a given fixed length k. Then the locus of Q, as P moves along c, is (one branch of) the **conchoid** of c for the pole O and the constant k. It is not difficult to devise an apparatus that will draw conchoids,[3]

[3] See, for example, T. L. Heath, *A Manual of Greek Mathematics*, p. 150.

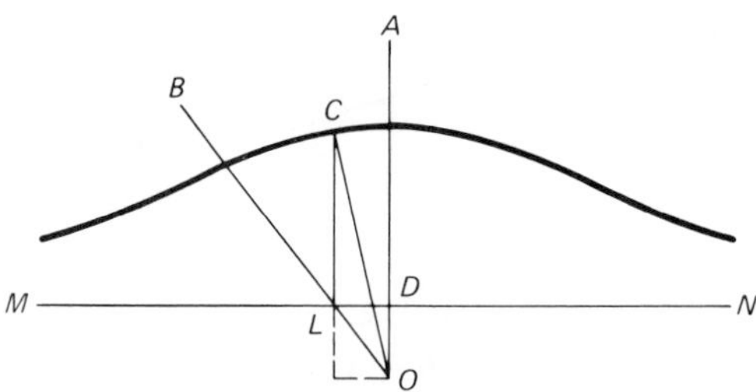

FIGURE 32

and with such an apparatus one may easily trisect angles. Thus, let AOB be any given acute angle. Draw a line MN perpendicular to OA, cutting OA and OB in D and L, as shown in Figure 32. Now draw the conchoid of MN for pole O and constant $2(OL)$. At L, draw the parallel to OA to cut the conchoid in C. Then OC trisects angle AOB.

A general angle may be trisected with the aid of a conic. The early Greeks were not familiar enough with the conics to accomplish this, and the earliest proof of this type was given by Pappus (ca. A.D. 300), using the focus and directrix property of conics. Two trisections using conics may be found in Problem Study 4.8.

There are transcendental (nonalgebraic) curves that will not only trisect a given angle but, more generally, multisect it into any number of equal parts. Among such curves are the *quadratrix*, invented by Hippias (ca. 425 B.C.) and the *spiral of Archimedes*. These two curves will also solve the problem of the quadrature of the circle. Applications of the quadratrix to both trisection and quadrature occur in Problem Study 4.10.

Over the years, many mechanical contrivances, linkage machines, and compound compasses, have been devised to solve the trisection problem.[4] An interesting and elementary implement of this kind is the so-called **tomahawk**. The inventor of the tomahawk is not known, but the instrument was described in a book in 1835. To construct a tomahawk, start with a line segment RU, trisected at S and T (see Figure 33). Draw a semicircle on SU as diameter, and draw SV perpendicular to RU. Complete the instrument as indicated in the accompanying figure. To trisect an angle ABC with the tomahawk, place the implement on the angle so that R falls on BA, SV passes through B, and the semicircle touches BC, at D, say. Then, since we may show that triangles RSB, TSB, TDB are all congruent, BS and BT trisect the given angle. The tomahawk may be constructed with straightedge and compasses on tracing paper and then adjusted on the given angle. By this subterfuge, we may trisect an angle with

[4] See R. C. Yates, *The Trisection Problem*.

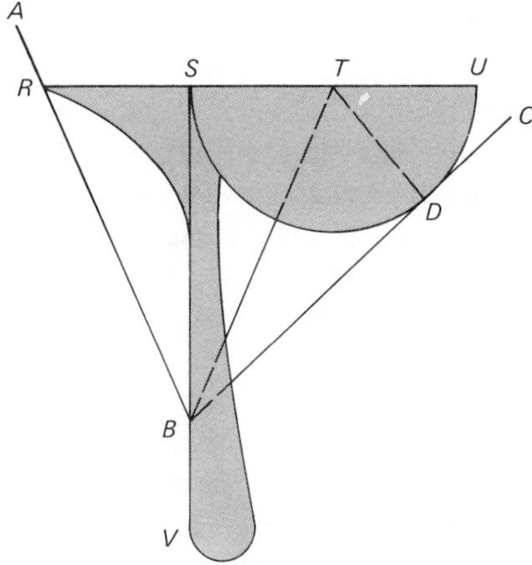

FIGURE 33

straightedge and compasses. (With two tomahawks, one may quintisect an angle.)

Although an arbitrary angle cannot be trisected exactly with Euclidean tools, there are constructions with these tools that give remarkably good approximate trisections. An excellent example is the construction given in 1525 by the famous etcher and painter, Albrecht Dürer. Take the given angle AOB as a central angle of a circle (see Figure 34). Let C be that trisection point of the chord AB that is nearer to B. At C, erect the perpendicular to AB to cut the circle in D. With B as center and BD as radius, draw an arc to cut AB in E. Let F be the trisection point of EC that is nearer to E. Again, with B as center and BF as radius, draw an arc to cut the circle in G. Then OG is an approximate trisecting line of angle AOB. It can be shown that the error in trisection increases with the size of the angle AOB, but is only about 1″ for angle $AOB = 60°$ and about 18″ for angle $AOB = 90°$.

Problem Study 4.9 describes an approximate trisection, using Euclidean tools, that may be made just as close to exact trisection as may be desired.

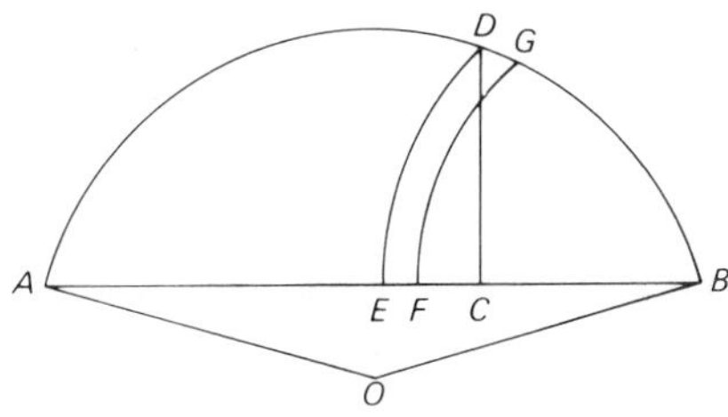

FIGURE 34

4–7 Quadrature of the Circle

Probably no other problem has exercised a greater or a longer attraction than that of constructing a square equal in area to a given circle. As far back as 1800 B.C. the ancient Egyptians "solved" the problem by taking the side of the square equal to 8/9 the diameter of the given circle. Since then, literally thousands of people have worked on the problem, and in spite of the present existence of a proof that the construction cannot be made with Euclidean tools,[5] not a year passes without its crop of "circle squarers."

The first Greek known to be connected with the problem is Anaxagoras (ca. 499–ca. 427 B.C.), but what his contribution was is not known. Hippocrates of Chios, who was a contemporary of Anaxagoras, succeeded in squaring certain special lunes, or moon-shaped figures bounded by two circular arcs, probably in the hope that his investigations might lead toward a solution of the quadrature problem. Some years later, Hippias of Elis (ca. 425 B.C.) invented the curve that became known as the *quadratrix*. This curve solves both the trisection and the quadrature problems, but traditions vary as to who first used it in the quadrature role. It may be that Hippias used it for trisecting angles, and that Dinostratus (ca. 350 B.C.), or some later geometer, realized its application to the quadrature problem. Some of the lunes of Hippocrates are considered in Problem Study 4.12; the quadratrix, in its dual role, is considered in Problem Study 4.10; and a few approximate quadratures are described in Problem Study 4.11.

A neat solution of the quadrature problem can be achieved with the **spiral of Archimedes,** and we are told that Archimedes (ca. 225 B.C.) actually used his spiral for this purpose. We may define the spiral, in dynamic terms, as the locus of a point P moving uniformly along a ray that, in turn, is uniformly rotating in a plane about its origin. If we take for the polar frame of reference the position OA of the rotating ray when P coincides with the origin O of the ray, we have that OP is proportional to angle AOP, and the polar equation of the spiral is $r = a\theta$, a being the constant of proportionality.

Let us draw the circle with center at O and radius equal to a. Then OP and the arc on this circle between the lines OA and OP are equal, since each is given by $a\theta$ (see Figure 35). It follows that if we take OP perpendicular to OA, then OP will have a length equal to one fourth the circumference of the circle. Since the area K of the circle is half the product of its radius and its circumference, we have

$$K = \left(\frac{a}{2}\right)(4OP) = (2a)(OP).$$

The side of the required square is thus the mean proportional between $2a$ and OP, or between the diameter of the circle and the length of that radius vector of the spiral that is perpendicular to OA.

---

[5] See, for example, Howard Eves, *A Survey of Geometry*, vol. 2, pp. 30–38.

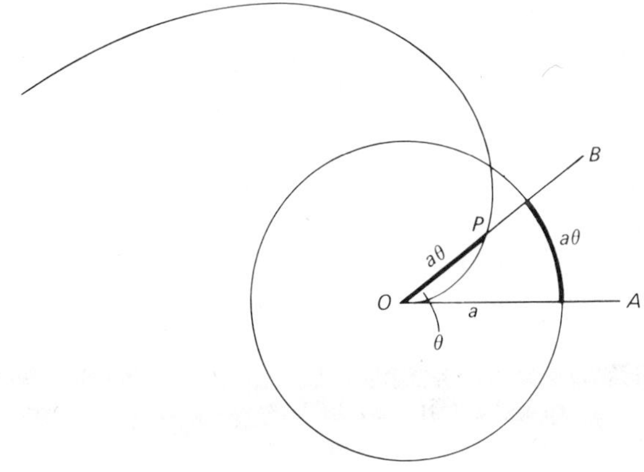

FIGURE 35

We may trisect (more generally, multisect) an angle AOB with the spiral of Archimedes. Let OB cut the spiral in P and trisect the segment OP by points P_1 and P_2. If the circles with O as center and OP_1 and OP_2 as radii cut the spiral in T_1 and T_2, then OT_1 and OT_2 trisect the angle AOB.

4–8 A Chronology of π[6]

Closely allied to the quadrature problem is the computation of π, the ratio of the circumference of a circle to its diameter. We have seen that in the ancient Orient the value of π was frequently taken as 3.[7] For the Egyptian quadrature of the circle given in the Rhind papyrus, we have $\pi = (4/3)^4 = 3.1604. \ldots$ The first scientific attempt to compute π, however, seems to be that of Archimedes, and we shall commence our chronology with his achievement.

ca. 240 B.C. To simplify matters, suppose we choose a circle with unit diameter. Now the (length of the) circumference of a circle lies between the perimeter of any inscribed regular polygon and that of any circumscribed regular polygon. Since it is a simple matter to compute the perimeters of the regular inscribed and circumscribed six-sided polygons, we easily obtain bounds for π. Now there are formulas (see Problem Study 4.13) that tell us how, from the perimeters of given regular inscribed and circumscribed polygons, we may obtain the perimeters of the regular inscribed and circumscribed polygons having twice the number of sides. By successive applications of this process, starting with the regular inscribed and circumscribed six-sided polygons, we can compute the perimeters of the regular inscribed and circumscribed poly-

---

[6] For a fuller chronology of π, containing over 120 entries, see H. C. Schepler, "The chronology of pi," *Mathematics Magazine* (January–February 1950): 165–70; (March–April 1950): 216–28; (May–June 1950): 279–83.

[7] See the Biblical references: I Kings 7:23; II Chron. 4:2.

gons of twelve, twenty-four, forty-eight, and ninety-six sides, in this way obtaining ever closer bounds for π. This is essentially what Archimedes did, finally obtaining the fact that π is between 223/71 and 22/7, or that, to two decimal places, π is given by 3.14. The work is found in Archimedes' *Measurement of a Circle,* a treatise containing only three propositions. The treatise, as it has come down to us, is not in its original form and may be only a fragment of a larger discussion. One inescapable conclusion, in view of the poor numeral system in use at the time, is that Archimedes was a very able computer. In the work are found some remarkable rational approximations to irrational square roots.

The above method of computing π by using regular inscribed and circumscribed polygons is known as the **classical method** of computing π.

ca. A.D. 150 The first notable value for π, after that of Archimedes, was given by Claudius Ptolemy of Alexandria in his famous *Syntaxis mathematica* (more popularly known by its Arabian title of the *Almagest*), the greatest ancient Greek work on astronomy. In this work, π is given, in sexagesimal notation, as 3 8′30″, which is 377/120, or 3.1416. Undoubtedly, this value was derived from the table of chords, which appears in the treatise. The table gives the lengths of the chords of a circle subtended by central angles of each degree and half degree. If the length of the chord of the 1° central angle is multiplied by 360, and the result divided by the length of the diameter of the circle, the above value for π is obtained.

ca. 480 The early Chinese worker in mechanics, Tsu Ch'ung-chih, gave the interesting rational approximation 355/113 = 3.1415929 . . . , which is correct to six decimal places. See Problem Study 4.11(c) for an application of this ratio to the quadrature problem.

ca. 530 The early Hindu mathematician Āryabhata gave 62,832/20,000 = 3.1416 as an approximate value for π. It is not known how this result was obtained. It may have come from some earlier Greek source or, perhaps, from calculating the perimeter of a regular inscribed polygon of 384 sides.

ca. 1150 The later Hindu mathematician, Bhāskara, gave several approximations for π. He gave 3927/1250 as an accurate value, 22/7 as an inaccurate value, and $\sqrt{10}$ for ordinary work. The first value may have been taken from Āryabhata. Another value, 754/240 = 3.1416, given by Bhāskara, is of uncertain origin; it is the same as that given by Ptolemy.

1429 Al-Kashi, astronomer royal to Ulugh Beg of Samarkand, computed π to sixteen decimal places by the classical method.

1579 The eminent French mathematician François Viète found π correct to nine decimal places by the classical method, using polygons having $6(2^{16})$ = 393,216 sides. He also discovered the equivalent of the interesting infinite product (see Problem Study 4.13)

$$\frac{2}{\pi} = \frac{\sqrt{2}}{2}\,\frac{\sqrt{(2+\sqrt{2})}}{2}\,\frac{\sqrt{\{2+\sqrt{(2+\sqrt{2})}\}}}{2}\cdots.$$

1585 Adriaen Anthoniszoon rediscovered the ancient Chinese ratio 355/113. This was apparently a lucky accident, since all he showed was that

$377/120 > \pi > 333/106$. He then averaged the numerators and the denominators to obtain the "exact" value of π. There is evidence that Valentin Otho, a pupil of the early table maker Rhaeticus, may have introduced this ratio for π into the western world at the slightly earlier date of 1573.

1593 Adriaen van Roomen, more commonly referred to as Adrianus Romanus, of the Netherlands, found π correct to fifteen decimal places by the classical method, using polygons having 2^{30} sides.

1610 Ludolph van Ceulen of the Netherlands computed π to thirty-five decimal places by the classical method, using polygons having 2^{62} sides. He spent a large part of his life on this task, and his achievement was considered so extraordinary that his widow had the number engraved on his tombstone (now lost) in St. Peter's churchyard in Leyden. To this day, the number is sometimes referred to as "the Ludolphine number."

1621 The Dutch physicist Willebrord Snell, best known for his discovery of the law of refraction, devised a trigonometric improvement of the classical method for computing π so that from each pair of bounds on π given by the classical method, he was able to obtain considerably closer bounds. By his method, he was able to get van Ceulen's thirty-five decimal places by using polygons having only 2^{30} sides. With such polygons, the classical method yields only fifteen places. For polygons of ninety-six sides, the classical method yields two decimal places, whereas Snell's improvement gives seven places. A correct proof of Snell's refinement was furnished in 1654 by the Dutch mathematician and physicist Christiaan Huygens.

1630 Grienberger, using Snell's refinement, computed π to thirty-nine decimal places. This was the last major attempt to compute π by the method of perimeters.

1650 The English mathematician John Wallis obtained the curious expression

$$\frac{\pi}{2} = \frac{2 \cdot 2 \cdot 4 \cdot 4 \cdot 6 \cdot 6 \cdot 8 \ldots}{1 \cdot 3 \cdot 3 \cdot 5 \cdot 5 \cdot 7 \cdot 7 \ldots}.$$

Lord Brouncker, the first president of the Royal Society, converted Wallis' result into the continued fraction

$$\frac{4}{\pi} = 1 + \cfrac{1^2}{2 + \cfrac{3^2}{2 + \cfrac{5^2}{2 + \ldots}}}.$$

Neither of these expressions, however, has served for an extensive calculation of π.

1671 The Scottish mathematician James Gregory obtained the infinite series

$$\arctan x = x - \frac{x^3}{3} + \frac{x^5}{5} - \frac{x^7}{7} + \ldots, \ (-1 \leqq x \leqq 1).$$

Not noted by Gregory is the fact that for $x = 1$, the series becomes

$$\frac{\pi}{4} = 1 - \frac{1}{3} + \frac{1}{5} - \frac{1}{7} + \cdots.$$

This very slowly converging series was known to Leibniz in 1674. Gregory attempted to prove that a Euclidean solution of the quadrature problem is impossible.

1699 Abraham Sharp found seventy-one correct decimal places by using Gregory's series with $x = \sqrt{1/3}$.

1706 John Machin obtained one hundred decimal places by using Gregory's series in connection with the relation (see Problem Study 4.13)

$$\frac{\pi}{4} = 4 \arctan\left(\frac{1}{5}\right) - \arctan\left(\frac{1}{239}\right).$$

1719 The French mathematician De Lagny obtained 112 correct places by using Gregory's series with $x = \sqrt{1/3}$.

1737 The symbol π was used by the early English mathematicians William Oughtred, Isaac Barrow, and David Gregory to designate the circumference, or periphery, of a circle. The first to use the symbol for the ratio of the circumference to the diameter was the English writer William Jones, in a publication in 1706. The symbol was not generally used in this sense, however, until Euler adopted it in 1737.

1754 Jean Étienne Montucla, an early French historian of mathematics, wrote a history of the quadrature problem.

1755 The French Academy of Sciences declined to examine any more solutions of the quadrature problem.

1767 Johann Heinrich Lambert showed that π is irrational.

1777 Comte de Buffon devised his famous **needle problem,** by which π may be approximated by probability methods. Suppose a number of parallel lines, distance a apart, are ruled on a horizontal plane, and suppose a homogeneous uniform rod of length $l < a$ is dropped at random onto the plane. Buffon showed that the probability[8] that the rod will fall across one of the lines in the plane is given by

$$p = \frac{2l}{\pi a}.$$

By actually performing this experiment a given large number of times and noting the number of successful cases, thus obtaining an empirical value for p,

[8] If a given event can happen in h ways and fail to happen in f ways, and if each of the $h + f$ ways is equally likely to occur, the *mathematical probability* p of the event happening is $p = h/(h + f)$.

we may use the above formula to compute an approximation for π. The best result obtained in this way was given by the Italian, Lazzerini, in 1901. From only 3408 tosses of the rod, he found π correct to six decimal places! His result is so much better than those obtained by other experimenters that it is sometimes regarded with suspicion. There are other probability methods for computing π. Thus, in 1904, R. Chartres reported an application of the known fact that if two positive integers are written down at random, the probability that they will be relatively prime is $6/\pi^2$.

1794 Adrien-Marie Legendre showed that π^2 is irrational.

1841 William Rutherford of England calculated π to 208 places, of which 152 were later found to be correct, by using Gregory's series in connection with the relation

$$\frac{\pi}{4} = 4 \arctan\left(\frac{1}{5}\right) - \arctan\left(\frac{1}{70}\right) + \arctan\left(\frac{1}{99}\right).$$

1844 Zacharias Dase, the lightning calculator, found π correct to 200 places using Gregory's series in connection with the relation

$$\frac{\pi}{4} = \arctan\left(\frac{1}{2}\right) + \arctan\left(\frac{1}{5}\right) + \arctan\left(\frac{1}{8}\right).$$

Dase, who was born in Hamburg in 1824, died at the early age of thirty-seven. He was perhaps the most extraordinary mental calculator who ever lived. Among his performances were the mental calculation of the product of two eight-digit numbers in fifty-four seconds, of two twenty-digit numbers in six minutes, of two forty-digit numbers in forty minutes, and of two one-hundred-digit numbers in eight hours and forty-five minutes. He mentally computed the square root of a hundred-digit number in fifty-two minutes. Dase used his powers more worthily when he constructed a seven-place table of natural logarithms and a factor table of all numbers between 7,000,000 and 10,000,000.

1853 Rutherford returned to the problem and obtained 400 correct decimal places.

1873 William Shanks of England, using Machin's formula, computed π to 707 places. For a long time, this remained the most fabulous piece of calculation ever performed.

1882 A number is said to be **algebraic** if it is a root of some polynomial having rational coefficients; otherwise, it is said to be **transcendental.** F. Lindemann showed that π is transcendental. This fact proves (see Section 14–2) that the quadrature problem cannot be solved by Euclidean tools.

1906 Among the curiosities connected with π are various mnemonics that have been devised for the purpose of remembering π to a large number of decimal places. The following, by A. C. Orr, appeared in the *Literary Digest*. One has merely to replace each word by the number of letters it contains to obtain π correct to thirty decimal places.

Now I, even I, would celebrate
In rhymes unapt, the great
Immortal Syracusan, rivaled nevermore,
Who in his wondrous lore,
Passed on before,
Left men his guidance
How to circles mensurate.

A few years later, in 1914, the following similar mnemonic appeared in the *Scientific American Supplement:* "See, I have a rhyme assisting my feeble brain, its tasks ofttimes resisting." Two other such mnemonics are: "How I want a drink, alcoholic of course, after the heavy lectures involving quantum mechanics," and "May I have a large container of coffee?"

1948 In 1946, D. F. Ferguson of England discovered errors, starting with the 528th place, in Shanks' value for π, and in January 1947 gave a corrected value to 710 places. In the same month, J. W. Wrench, Jr., of America, published an 808-place value of π, but Ferguson soon found an error in the 723rd place. In January 1948, Ferguson and Wrench jointly published the corrected and checked value of π to 808 places. Wrench used Machin's formula, whereas Ferguson used the formula

$$\frac{\pi}{4} = 3 \arctan\left(\frac{1}{4}\right) + \arctan\left(\frac{1}{20}\right) + \arctan\left(\frac{1}{1985}\right).$$

1949 The electronic computer, the ENIAC, at the Army Ballistic Research Laboratories in Aberdeen, Maryland, calculated π to 2037 decimal places.

1959 François Genuys, in Paris, computed π to 16,167 decimal places, using an IBM 704.

1961 Wrench and Daniel Shanks, of Washington, D.C., computed π to 100,265 decimal places, using an IBM 7090.

1965 The ENIAC, now obsolete, was dismembered and moved to the Smithsonian Institution as a museum piece.

1966 On February 22, M. Jean Guilloud and his co-workers at the Commissariat à l'Énergie Atomique in Paris attained an approximation to π extending to 250,000 decimal places on a STRETCH computer.

1967 Exactly one year later, the above workers found π to 500,000 places on a CDC 6600.

1973 Guilloud and his co-workers found π to 1,000,000 places on a CDC 7600.

1981 The two Japanese mathematicians Kazunori Miyoshi and Kazuhika Nakayama of the University of Tsukuba calculated π to 2,000,038 significant figures in 137.30 hours on a FACOM M-200 computer. They used the formula

$$\pi = 32 \arctan\left(\frac{1}{10}\right) - 4 \arctan\left(\frac{1}{239}\right) - 16 \arctan\left(\frac{1}{515}\right)$$

and checked their result with Machin's formula.

1986 In January of 1986, D. H. Bailey of the NASA Ames Research Center in California ran a Cray-2 supercomputer for 28 hours to get π to 29,360,000 digits. His code was based on an algorithm by J. M. and P. D. Borwein of Dalhousie University. Bailey checked his code against a slower algorithm, also developed by the Borweins, and verified the accuracy of his result. A little later, Yasumasa Kanada of the University of Tokyo, using an NEC SX-2 supercomputer and the Borweins' algorithm, computed π to 134,217,700 digits.

We have not placed in the above chronology of π any items from the vast literature supplied by sufferers of *morbus cyclometricus,* the circle-squaring disease. These contributions, often amusing and at times almost unbelievable, would require a publication all to themselves. To illustrate their tenor, consider the instance in 1892 when a writer announced in the *New York Tribune* the rediscovery of a long-lost secret that leads to 3.2 as the exact value of π. The lively discussion following this announcement won many advocates for the new value. Again, since its publication in 1931, a great many college and public libraries throughout the United States have received, from the obliging author, complimentary copies of a thick book devoted to the demonstration that $\pi = 3\frac{13}{81}$. And then there is a House Bill No. 246 of the Indiana State Legislature that attempted, in 1897, to determine the value of π by legislation. In Section I of the bill, we read: "Be it enacted by the General Assembly of the State of Indiana: It has been found that a circular area is to the square on a line equal to the quadrant of the circumference, as the area of an equilateral rectangle is to the square on one side. . . ." The bill passed the House but, because of some newspaper ridicule, was shelved by the Senate, in spite of the energetic backing of the State Superintendent of Public Instruction.[9]

There is more to the calculation of π to a large number of decimal places than just the challenge involved. Before 1767 (when π was proven irrational), one of the reasons was to see if the digits of π started to repeat, and, if so, to obtain π as an exact rational number, with perhaps a large denominator. In more recent times, a motivation is to secure statistical information concerning the "normalcy" of π. A real number is said to be **simply normal** if in its decimal expansion all ten digits occur with equal frequency; it is said to be **normal** if all blocks of digits of the same length occur with equal frequency.[10] It is not known if π (or even $\sqrt{2}$, for that matter) is normal or even simply normal. The calculations of π, starting with that on the ENIAC in 1949, were performed to secure statistical information on the matter. From counts on these extensive expansions of π, it would seem that the number is perhaps normal. The erroneous 707-place calculation of π made by Shanks in 1873 seemed to indicate that π was not even simply normal.

---

[9] See W. E. Edington, "House Bill No. 246 Indiana State Legislature, 1897," *Proceedings of the Indiana Academy of Science* 45 (1935): 206–10. Also see A. E. Hallerberg, "Indiana's squared circle," *Mathematics Magazine,* 50, no. 3 (May 1977): 136–140.

[10] The concept of normalcy of a number is due to Émile Borel (1871–1956), who showed that "almost all" numbers are normal.

There are other reasons for calculating π to a great number of decimal places. First of all, it is valuable in computer science because the designing of a program for such an extensive calculation leads to greater programming ability. Also, once a program has been successfully used on one computer, it can be used as a check to see if a new computer is operating properly.

In many situations, there is a need for a table of random numbers, such as in problems involving Markov chains, in applications of Monte Carlo methods to problems in mathematical physics, and in the drawing of random samples in statistics. The digits of π are not truly random, because each digit is uniquely determined. The digits of π may be sufficiently "jumbled," however, to serve practically as a random number table; tests (such as the "poker test") seem to indicate this.

In connection with the possible normalcy of π, it is interesting that the sequence 314159 of the first six digits of π appears six times in the first ten million digits of the decimal expansion of π, and the sequence 0123456789 does not appear at all.

The sequence 271828 of the first six digits of e (the base for natural logarithms) occurs eight times in the first ten million digits of the decimal expansion of e.

Problem Studies

4.1 Euclidean and Modern Compasses

A student reading Euclid's *Elements* for the first time might experience some surprise at the opening propositions of Book I. The first three propositions are the construction problems

1. To describe an equilateral triangle upon a given finite straight line.
2. From a given point to draw a straight line equal to a given straight line.
3. From the greater of two given straight lines to cut off a part equal to the less.

These three constructions are trivial with straightedge and *modern* compasses but require some ingenuity with straightedge and *Euclidean* compasses.

(a) Solve Proposition 1 of Book I with Euclidean tools.

(b) Solve Proposition 2 of Book I with Euclidean tools.

(c) Solve Proposition 3 of Book I with Euclidean tools.

(d) Show that Proposition 2 of Book I proves that the straightedge and *Euclidean* compasses are equivalent to the straightedge and *modern* compasses.

4.2 Duplication by Archytas and Menaechmus

(a) Archytas (ca. 400 B.C.), the Pythagorean philosopher, mathematician, general, and statesman, was one of the most respected and influential citizens of Tarentum (now Taranto), Italy. He is said to have been elected 7 times as general of the Tarentine forces, and he was noted for the concern he showed for the comfort and education of the children of

Tarentum. He tragically drowned in a shipwreck near Tarentum. Following is a description of his remarkable solution of the problem of inserting 2 mean proportionals between 2 given line segments.

Let a and b, $a > b$, be two given line segments. In a horizontal plane, draw a circle on $AD = a$ as diameter and construct chord $AB = b$. Let AB produced meet in point P, the tangent to the circle at D. Vertically erect the upper half of a right circular semicylinder on the semicircle ABD as base; generate a right circular cone by rotating AP about line AD; generate a torus of zero inner radius by rotating, about the element of the semicylinder through A, the vertical circle on AD as diameter. Denote by K the point common to the semicylinder, the cone, and the torus, and let I be the foot on the semicircle ABD of the element through K of the semicylinder. Prove that AK and AI are the two mean proportionals between a and b; that is, show that $AD:AK = AK:AI = AI:AB$.

(b) Menaechmus (ca. 350 B.C.) gave the following 2 solutions to the duplication problem. They utilize certain conic sections that, apparently, were invented by Menaechmus for the problem at hand.

1. Draw 2 parabolas having a common vertex, perpendicular axes, and such that the latus rectum of one is double that of the other. Denote by x the length of the perpendicular dropped from the other intersection of the two parabolas upon the axis of the smaller parabola. Then x is the edge of a cube having twice the volume of the cube that has the smaller latus rectum for edge. Prove this construction correct by using modern analytic geometry.
2. Draw a parabola of latus rectum s, then a rectangular hyperbola with transverse axis equal to $4s$ and having for asymptotes the axis of the parabola and the tangent to the parabola at its vertex. Let x be the length of the perpendicular dropped from the intersection of the 2 curves upon the axis of the parabola. Then $x^3 = 2s^3$. Prove this construction correct by using modern analytic geometry.

4.3 Duplication by Apollonius and Eratosthenes

Apollonius (ca. 225 B.C.) solved the duplication problem as follows. Draw a rectangle $OADB$, and then a circle concentric with the rectangle cutting OA and OB produced in A' and B' such that A', D, B' are collinear. Actually, it is impossible to construct this circle with Euclidean tools, but Apollonius gave a mechanical way of describing it.

(a) Show that BB' and AA' are 2 mean proportionals between OA and OB.

(b) If $OB = 2(OA)$, show that $(BB')^3 = 2(OA)^3$.

(c) Eratosthenes (ca. 230 B.C.) devised a mechanical "mean-finder" consisting of three equal rectangular frames, with a set of corresponding diagonals, capable of sliding in grooves so that the second frame can be slid under the first one, and the third frame under the second. Suppose the frames are slid, as indicated in Figure 36, so that points A', B', C' are collinear. Show that BB' and CC' are the 2 mean proportionals between AA' and DD'. A mean finder of this sort is easily made from a

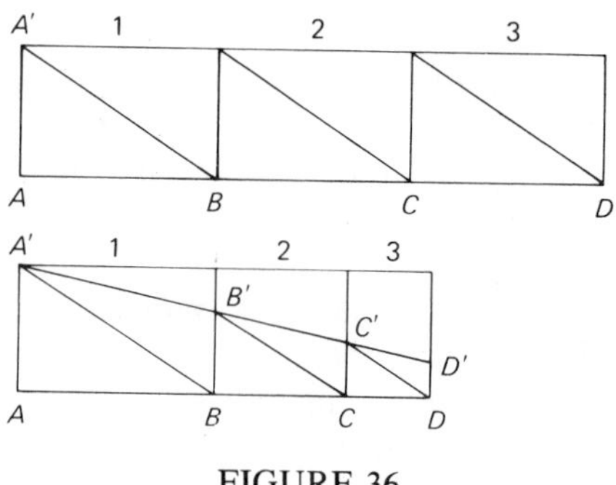

FIGURE 36

set of equal paper rectangles and can be generalized so as to insert n means between 2 given segments.[11]

4.4 The Cissoid of Diocles

Diocles (ca. 180 B.C.) invented the **cissoid** in order to solve the duplication problem. A general cissoid may be defined as follows: Let C_1 and C_2 be 2 given curves, and let O be a fixed point. Let P_1 and P_2 be the intersections of a variable line through O with the given curves. The locus of P on this line such that $OP = OP_2 - OP_1 = P_1P_2$ is called the *cissoid of* C_1 *and* C_2 *for the pole* O. If C_1 is a circle, C_2 a tangent to C_1 at point A, and O is the point on C_1 diametrically opposite A, then the cissoid of C_1 and C_2 for the pole O is the **cissoid of Diocles.**

(a) Taking O as origin and OA as the positive x-axis, show that the Cartesian equation of the cissoid of Diocles is $y^2 = x^3/(2a - x)$, where a is the radius of C_1. Show that the corresponding polar equation is $r = 2a \sin \theta \tan \theta$.

(b) On the positive y-axis, lay off $OD = n(OA)$. Draw DA to cut the cissoid in P. Let OP cut line C_2 in Q. Show that $(AQ)^3 = n(OA)^3$. When $n = 2$, we have a solution of the duplication problem.

(c) Newton has shown how the cissoid of Diocles may be generated by a carpenter's square. Let the outside edge of the square be ACB, AC being the shorter arm. Draw a line MN and mark a point R at distance AC from MN. Move the square so that A always lies on MN and BC always passes through R. Show that the midpoint P of AC describes a cissoid of Diocles.

(d) What is the cissoid of 2 concentric circles with respect to their common center? Of a pair of parallel lines with respect to any point not on either line?

(e) If C_1 and C_2 intersect in P, show that OP is a tangent at O to the cissoid of C_1 and C_2 for the pole O.

---

[11] For a more recent mechanical approach, see George E. Martin, "Duplicating the cube with a mira," *The Mathematics Teacher* (March 1979): 204–208.

4.5 Some Seventeenth-Century Duplications

Many eminent seventeenth-century mathematicians, like Huygens, Descartes, Grégoire de Saint-Vincent, and Newton, devised constructions for duplicating a cube. Following are two of these constructions.

(a) Grégoire de Saint-Vincent (1647) gave a construction for finding the two mean proportionals between two given line segments, based on the following theorem: *The hyperbola drawn through a vertex of a rectangle and having the two sides opposite this vertex for asymptotes meets the circumcircle of the rectangle in a point whose distances from the asymptotes are the mean proportionals between the adjacent sides of the rectangle.* Prove this theorem.

(b) Descartes (1659) pointed out that the curves

$$x^2 = ay, \qquad x^2 + y^2 = ay + bx$$

intersect in a point (x, y) such that x and y are the two mean proportionals between a and b. Show this.

4.6 Applications of the Insertion Principle

Let us be given 2 curves m and n, and a point O. Suppose we permit ourselves to mark, on a given straightedge, a segment MN, and then to adjust the straightedge so that it passes through O and cuts the curves m and n with M on m and N on n. The line drawn along the straightedge is then said to have been drawn by the **insertion principle.** Problems beyond the Euclidean tools can often be solved with these tools if we also permit ourselves to use the insertion principle. Establish the correctness of the following constructions, each of which uses the insertion principle.

(a) Let AB be a given segment. Draw angle $ABM = 90°$ and the angle $ABN = 120°$. Now draw ACD, cutting BM in C and BN in D such that $CD = AB$. Then $(AC)^3 = 2(AB)^3$. Essentially this construction was given in publications by Viète (1646) and Newton (1728).

(b) Let AOB be any central angle in a given circle. Through B, draw a line BCD, cutting the circle again in C, and AO produced in D, such that $CD = OA$, the radius of the circle. Then angle $ADB = \frac{1}{3}$ angle AOB. This solution of the trisection problem is implied by a theorem given by Archimedes (ca. 240 B.C.).

4.7 The Conchoid of Nicomedes

Little is known about Nicomedes (ca. 240 B.C.) beyond his invention of the **conchoid,** a curve with which one may solve both the trisection and the duplication problems. A general conchoid may be defined as follows: Let c be a given curve and O a fixed point. On the radius vector OP from O to a point P on c, mark off $PQ = \pm k$, where k is a constant. Then the locus of Q is called the *conchoid of c for pole O and constant k.* The complete curve consists of two branches, one corresponding to $PQ = +k$ and the other to $PQ = -k$. If c is a straight line and O is any point not on c, we get a **conchoid of Nicomedes.**

(a) Taking O as origin and the line through O parallel to the given line c as x-axis, show that the Cartesian equation of the conchoid of Nicomedes for constant k is $(y - a)^2(x^2 + y^2) = k^2y^2$, where a is the distance of O from c.

(b) Show how the conchoid of Nicomedes may be used to solve the duplication problem.

(c) A conchoid of a circle for a fixed point on the circle is called a **limaçon of Pascal** (erroneously named after Étienne Pascal (1588–1640), father of the famous Blaise Pascal, although the curve had already been given by Albrecht Dürer (1471–1528) in the early sixteenth century). If $k = a$, the radius of the given circle, we obtain a special limaçon known as the **trisectrix**. Establish the following construction for trisecting an angle with the trisectrix. Let AOB be any central angle in a circle with center O and radius OA. Draw the trisectrix for the circle with pole at A, and let BO produced cut the trisectrix in C. Then angle $ACB = \frac{1}{3}$ angle AOB.

(d) Show that the 2 branches of the conchoid of curve c for pole O and constant k constitute the cissoid of s and c for the pole O, where s is the circle with center O and radius k (see Problem Study 4.4).

4.8 Trisection by Conics

A general angle is easily trisected by the aid of conics. Establish the following constructions of this sort.

(a) Let the given angle be AOB. Draw the branch of an equilateral hyperbola having O as center and OA as an asymptote, cutting OB in P. With P as the center and $2(PO)$ as a radius, draw a circle cutting the hyperbola in R. Draw PM parallel to OA and RM perpendicular to OA, to intersect in M. Then angle $AOM = \frac{1}{3}$ angle AOB.

(b) Let angle AOB be taken as a central angle of a circle, and let OC be the bisector of angle AOB. Draw the branch of the hyperbola of eccentricity 2 having A for focus and OC for corresponding directrix, and let this branch cut arc AB in P. Then angle $AOP = \frac{1}{3}$ angle AOB. This construction was quoted by Pappus (ca. A.D. 300).

(c) A clever trisection of an arbitrary angle can be accomplished, not with a conic section, but with a right circular cone itself. Consider such a cone (made out of wood, for example) having its slant height equal to three times the radius of its base. On the circumference of the circular base of the cone, mark off arc AB of a central angle AOB equal to the angle we wish to trisect. Now wrap a sheet of paper around the cone and mark on the paper the positions of points A and B and the vertex V of the cone. Show that when the paper is flattened out, angle AVB is one-third of angle AOB. This novel procedure was described by Aubry in 1896.[12]

---

[12] For a more recent mechanical approach, see Johnny W. Lott and Iris Mack Dayoub, "What can be done with a mira?" *The Mathematics Teacher* (May 1977): 394–99.

4.9 Asymptotic Euclidean Constructions

A construction using Euclidean tools but requiring an infinite number of operations is called an **asymptotic Euclidean construction.** Establish the following two constructions of this type for solving the trisection and the quadrature problems.[13]

(a) Let OT_1 be the bisector of angle AOB, OT_2 that of angle AOT_1, OT_3 that of angle T_2OT_1, OT_4 that of angle T_3OT_2, OT_5 that of angle T_4OT_3, and so forth. Then $\lim_{i\to\infty} OT_i = OT$, one of the trisectors of angle AOB. (This construction was given by Fialkowski, 1860.)

(b) On the segment AB_1 produced, mark off $B_1B_2 = AB_1$, $B_2B_3 = 2(B_1B_2)$, $B_3B_4 = 2(B_2B_3)$, and so forth. With $B_1, B_2, B_3, \ldots$ as centers, draw the circles $B_1(A)$, $B_2(A)$, $B_3(A)$, ... Let M_1 be the midpoint of the semicircle on AB_2. Draw B_2M_1 to cut circle $B_2(A)$ in M_2, B_3M_2 to cut circle $B_3(A)$ in M_3, ... Let N_i be the projection of M_i on the common tangent of the circles at A. Then $\lim_{i\to\infty} AN_i =$ quadrant of circle $B_1(A)$.

4.10 The Quadratrix

Hippias (ca. 425 B.C.) invented a transcendental curve, called the **quadratrix,** by means of which one can multisect angles and square the circle. The quadratrix may be defined as follows: Let the radius OX of a circle rotate uniformly about the center O from OC to OA, at right angles to OC. At the same time, let a line MN parallel to OA move uniformly parallel to itself from CB to OA. The locus of the intersection P of OX and MN is the quadratrix.

(a) Taking $OA = 1$ and the positive x-axis along OA, show that the Cartesian equation of the quadratrix is $y = x \tan (\pi y/2)$.

(b) Show how an angle may be multisected with the quadratrix.

(c) Find the x-intercept of the quadratrix, and show how the curve may be used for squaring the circle.

4.11 Approximate Rectification

Many approximate constructions have been given for finding a line segment equal in length to the circumference of a given circle. An approximate quadrature of the circle is then easily obtained by constructing the square on the mean proportional between the radius of the circle and a segment equal in length to half the circumference of the circle.

(a) Show that the circumference of a circle is given approximately by 3 times the diameter of the circle increased by one-fifth the side of the inscribed square. This leads to what approximation for π?

(b) Let AOB be a diameter of the given circle. Find C on the tangent at B such that angle $COB = 30°$. Mark off CBD on the tangent equal to 3 times the radius of the circle. Then $2(AD)$ is approximately the circum-

---

[13] For an asymptotic Euclidean solution of the duplication problem; see T. L. Heath, *History of Greek Mathematics*, vol. 1, pp. 268–270.

ference of the circle. This leads to what approximation for π? This construction was given in 1685 by the Polish Jesuit Kochanski.

(c) Let $AB = 1$ be a diameter of the given circle. Draw $BC = \frac{7}{8}$, perpendicular to AB at B. Mark off $AD = AC$ on AB produced. Draw $DE = \frac{1}{2}$, perpendicular to AD at D, and let F be the foot of the perpendicular from D on AE. Draw EG parallel to FB to cut BD in G. Then GB is approximately the decimal part of π. Find the length of GB to seven decimal places. This construction was given in 1849 by de Gelder.

4.12 Lunes of Hippocrates

Hippocrates of Chios (ca. 440 B.C.) squared certain lunes, perhaps hoping that his investigations might throw some light on the quadrature problem. Following are 2 of Hippocrates' lune quadratures:[14]

(a) Let AOB be a quadrant of a circle. On AB as a diameter, draw a semicircle lying outside the quadrant. Show that the lune bounded by the quadrant and the semicircle has the same area as triangle AOB.

(b) Let $ABCD$ be half of a regular hexagon inscribed in a circle of diameter AD. Construct a lune by describing, exterior to the circle, a semicircle on AB as a diameter. Show that the area of the trapezoid $ABCD$ is equal to three times the area of the lune plus the area of the semicircle on AB.

4.13 Computation of π

(a) Prove that $\pi/4 = 4\tan^{-1}(\frac{1}{5}) - \tan^{-1}(\frac{1}{239})$. This is the formula utilized by Machin in 1706 to compute π to 100 decimal places.

(b) Establish Viète's formula given under the date 1579 in Section 4–8.

(c) Show that

$$\pi/6 = \sqrt{1/3}\{1 - 1/(3)(3) + 1/(3^2)(5) - 1/(3^3)(7) + \ldots\}.$$

(d) A common approximation in the Middle Ages for a square root was $\sqrt{n} = \sqrt{a^2 + b} = a + b/(2a + 1)$. By taking $n = 10 = 3^2 + 1$, show why it may be that $\sqrt{10}$ was so frequently used for π.

(e) Show that the theorem in House Bill No. 246, Indiana State Legislature, 1897 (see Section 4–8) makes the incorrect assumption that a circle and a square have equal areas if they have equal perimeters. This assumption leads to what value for π?

(f) If s_k denotes the side of a regular polygon of k sides inscribed in a circle of radius R, show that

$$s_{2n} = \{2R^2 - R(4R^2 - s_n^{\,2})^{1/2}\}^{1/2}.$$

(g) If S_k denotes the side of a regular polygon of k sides circumscribed about a circle of radius r, show that

[14] For an acount of modern developments leading to an analysis of quadrable lunes, see Tobias Dantzig, *The Bequest of the Greeks,* chap. 10.

$$S_{2n} = \frac{2rS_n}{2r + (4r^2 + S_n^2)^{1/2}}.$$

(h) If p_k and P_k denote, respectively, the perimeters of regular polygons of k sides inscribed in and circumscribed about the same circle, show that

$$P_{2n} = \frac{2p_nP_n}{p_n + P_n}, \qquad p_{2n} = (p_nP_{2n})^{1/2}.$$

(It was with these formulas that Archimedes, in his *Measurement of a Circle,* starting with p_6 and P_6, successively computed $P_{12}, p_{12}, P_{24}, p_{24}, P_{48}, p_{48}, P_{96}, p_{96}$.)

(i) If a_k and A_k denote, respectively, the areas of regular polygons of k sides inscribed in and circumscribed about the same circle, show that

$$a_{2n} = (a_nA_n)^{1/2}, \qquad A_{2n} = \frac{2a_{2n}A_n}{a_{2n} + A_n}.$$

4.14 The Snell Refinement

Let angle AOP (see Figure 37) be an acute central angle in a circle of unit radius. Produce diameter AOB to point S so that $BS = AO$. Draw SP to cut, in the point T, the tangent to the circle at A. Snell noticed that if angle AOP is sufficiently small, *the tangential segment AT is approximately equal in length to the arc AP.*

(a) Find the error in the Snell approximation when angle $AOP = 90°$.

(b) Designating angle AOP by θ and angle AST by ϕ, show that

$$AT = \frac{3 \sin \theta}{2 + \cos \theta} = 3 \tan \phi.$$

(c) Show that $\phi < \theta/3$, whence

$$\frac{\sin \theta}{2 + \cos \theta} < \tan\left(\frac{\theta}{3}\right).$$

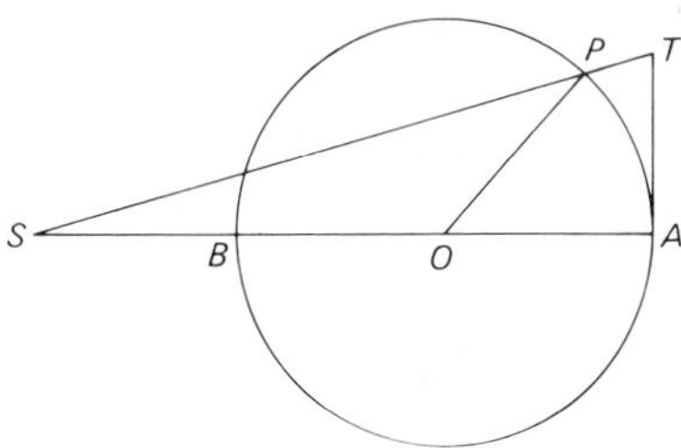

FIGURE 37

(d) Show how the Snell approximation may be used for approximately multisecting angles.
(e) Show how the Snell approximation may be used for approximately dividing a circumference into n equal parts.
(f) Show how the Snell approximation may be used for approximately squaring the circle.

4.15 Mnemonics for π

(a) How good is the following English mnemonic for π?

Sir, I bear a rhyme excelling
  In mystic force and magic spelling
Celestial sprites elucidate
  All my own striving can't relate.

(b) Show that the following French poem yields π correct to twenty-six decimal places:

Que j'aime à faire apprendre
Un nombre utile aux sages
Immortel Archimède artiste ingénieur
Qui de ton jugement peut priser la valeur
Pour moi ton problème
A les pareils avantages!

(c) How good for recalling the decimal expansion of π is the following pretty Spanish mnemonic?

Sol y Luna y Mundo proclaman al Eterne Autor del Cosmos.

(d) The most successful mnemonic given in the text [under A Chronology of π (1906)] yields 30 correct decimal places. No one has ever been able to make up a sentence mnemonic of this kind giving π to more than 31 correct decimal places. Why is this?
(e) The number π can be approximated by rational numbers. For example,

$$\begin{aligned} 22/7 &= 3.14|28, \\ 355/113 &= 3.141592|92, \\ 104348/33215 &= 3.141592653|92142, \\ 833719/265381 &= 3.14159265358|108, \end{aligned}$$

which, in turn, give π correct to 2, 6, 9, and 11 decimal places. Show that the following mnemonics may be used for recalling the last two fractions:

$$\frac{\text{calculator will get fair accuracy}}{\text{but not to } \pi \text{ exact}},$$

$$\frac{\text{dividing top lot through (a nightmare)}}{\text{by number below, you approach } \pi}.$$

It has been shown that for one-digit, two-digit, three-digit, five-digit, and six-digit denominators, the best rational approximations of π are, respectively, to 2, 3, 6, 10, and 11 correct decimal places. No improvement can be made over 6 correct decimal places using four-digit denominators.

Essay Topics

4/1 Plato's influence upon mathematics.
4/2 Aristotle's influence upon mathematics.
4/3 The importance of unsolved problems in mathematics.
4/4 Early steps in the history of the conic sections.
4/5 Euclidean constructions viewed as a game of geometric solitaire.
4/6 Modern versus Euclidean compasses.
4/7 The study of higher plane curves among the early Greeks.
4/8 Quadrable lunes.
4/9 Normal numbers.
4/10 Mnemonics in elementary mathematics.
4/11 Plato's educational concept of "transfer of training."
4/12 Pseudomaths.

Bibliography

ALLMAN, G. J. *Greek Geometry from Thales to Euclid*. Dublin: University Press, 1889.

BALL, W. W. R., and H. S. M. COXETER *Mathematical Recreations and Essays*. 11th ed. New York: Macmillan, 1939.

BECKMANN, PETR *A History of Pi*. Boulder, Colo.: Golem, 1970.

BOROFSKY, SAMUEL *Elementary Theory of Equations*. New York: Macmillan, 1950.

BRUMBAUGH, R. S. *Plato's Mathematical Imagination; The Mathematical Passages in the Dialogues; and their Interpretation*. Bloomington, Ind.: Indiana University Press, 1954.

BUNT, L. N. H.; P. S. JONES; and J. D. BEDIENT *The Historical Roots of Elementary Mathematics*. Englewood Cliffs, N.J.: Prentice-Hall, 1976.

COOLIDGE, J. L. *The Mathematics of Great Amateurs*. New York: Oxford University Press, 1949.

COURANT, RICHARD, and HERBERT ROBBINS *What Is Mathematics*? New York: Oxford University Press, 1941.

DANTZIG, TOBIAS *The Bequest of the Greeks*. New York: Charles Scribner's, 1955.

DE MORGAN, AUGUSTUS *A Budget of Paradoxes*. 2 vols., 2d ed., ed. D. E. Smith. Chicago: Open Court, 1915.

DICKSON, L. E. *New First Course in the Theory of Equations*. New York: John Wiley, 1939.

DUDLEY, UNDERWOOD *A Budget of Trisections*. New York: Springer-Verlag, 1987.

EVES, HOWARD *A Survey of Geometry*. 2 vols. Boston: Allyn and Bacon, 1963 and 1965. Vol. I revised in 1972.

EVES, HOWARD *Foundations and Fundamental Concepts of Mathematics*. 3rd ed. Boston: PWS-KENT Publishing Company, 1990.

FOWLER, D. H. *The Mathematics of Plato's Academy. A New Reconstruction*. Oxford: Clarendon Press, 1987.

GOW, JAMES *A Short History of Greek Mathematics*. New York: Hafner, 1923.

HEATH, T. L. *History of Greek Mathematics*, vol. 1. New York: Oxford University Press, 1921. Reprinted by Dover, New York, 1981.

——— *A Manual of Greek Mathematics*. New York: Oxford University Press, 1931. Reprinted by Dover, New York, 1963.

——— *Mathematics in Aristotle*. New York: Oxford University Press, 1949.

HOBSON, E. W. *"Squaring the Circle": A History of the Problem*. New York: Chelsea, 1953.

KNORR, WILBUR *The Ancient Tradition of Geometric Problems*. New York: Springer-Verlag, 1986.

LEE, H. D. P., ed. *Zeno of Elea*. New York: Cambridge University Press, 1935.

LOVITT, W. V. *Elementary Theory of Equations*. Englewood Cliffs, N.J.: Prentice-Hall, 1939.

PHIN, JOHN *The Seven Follies of Science*. 3rd ed. Princeton, N.J.: D. Van Nostrand, 1911.

THOMAS, IVOR, ed. *Selections Illustrating the History of Greek Mathematics*. 2 vols. Cambridge, Mass.: Harvard University Press, 1939 and 1941.

VAN DER WAERDEN, B. L. *Science Awakening*. Translated by Arnold Dresden. New York: Oxford University Press, 1961. Paperback edition. New York: John Wiley, 1963.

WEDBERG, ANDERS *Plato's Philosophy of Mathematics*. Stockholm: Almqvist & Wiksell, 1955.

WEISNER, LOUIS *Introduction to the Theory of Equations*. New York: Macmillan, 1938.

YATES, R. C. *The Trisection Problem*. Ann Arbor, Mich.: Edward Brothers, 1947. Reprinted by the National Council of Teachers of Mathematics, Washington, D.C., 1974.

Cultural Connection 

THE OIKOUMENE

The Persian Empire—550–330 B.C.;
Hellenistic Greece—336–31 B.C.;
The Roman Empire—31 B.C.–A.D. 476
(to accompany Chapters Five and Six)

Sometime during the latter half of the third century B.C., Eratosthenes (276–196? B.C.), mathematician, scientist, geographer, and curator of the great library in Alexandria, decided to make a new map of the world (see Section 6–3 for a reproduction of the map). It had been more than two centuries since the Greek historian Herodotus had drawn his map of the world, and many new places had been discovered in the interim. Eratosthenes knew that the explorer Pythias had made two trips into the Atlantic Ocean in approximately 300 B.C., visiting the British Isles, Scandinavia, Germany, and even a frigid, mysterious land where the sun never set. Pythias believed this frozen place to be the edge of the world, and called it *Ultima Thule;* it may have been Iceland. Hanno, a king of Carthage, had sailed south along the west coast of Africa in about 470 B.C., and Eratosthenes had reports of what he had seen. The librarian had accounts of Patrocles' excursion to the Caspian Sea as well. Daily, traders and merchants arrived in Alexandria's busy marketplace with tales of faraway lands. It was certainly time for a new map.

Resolutely, Eratosthenes spread out a fresh roll of papyrus and began to sketch. In the center, he placed Alexandria, the largest city in the world with 500,000 inhabitants. Laying astride several important trade routes, the city was the commercial and cultural center of what Eratosthenes and other Greek speakers called the *oikoumene,* or "inhabited world"—Greece, Egypt, and the Middle East. Alexandria's deep harbor was filled with ships from distant ports, guided there safely by its great *Pharos* lighthouse, one of the seven wonders of the ancient world. Alexandrian merchants ranged far overland, and in the city's marketplace, one could buy spices from India and Arabia, wood and ivory from Africa, cloth from Tyre, olives and wine from Greece, and salt and slaves from Rome. The palace of the king, Ptolemy IV Philopater (ruled 222–205 B.C.) had a prominent place in the city; next to it stood the University and its wondrous library, with 600,000 rolls of papyrus.

Of course, Eratosthenes well knew the story of Alexandria's founding. One hundred years before, in 338 B.C., as the city-states of Greece lay exhausted after nearly a century of fratricidal warfare, Philip II (382–336 B.C.) of Macedonia had united all Greece under his rule. When Philip died later that year, his new empire was claimed by his son, Alexander the Great (356–323 B.C.); two years later, in 334 B.C., Alexander led his armies on a daring invasion of the mighty Persian Empire, then the largest and most powerful nation in the world. Two hundred years before Alexander, in 550 B.C., Persia's first king, Cyrus the Great (died 529 B.C.), had conquered Babylonia, and its second king, Cambyses (died 522 B.C.), had annexed Egypt twenty-five years after that, creating the world's first truly polycultural empire. In 330 B.C., after six years of compaigning, Alexander's Macedonians captured the Persian capital, Persepolis, and the old empire fell. Two years prior to the seige of Persepolis, while in Egypt, Alexander had founded Alexandria as his western capital.

Surrounding Alexandria on his map, Eratosthenes drew in the three empires and several lesser states that had emerged in the *oikoumene* after Persia fell. After the conquest, Alexander had set about unifying Persia and Greece into a single cosmopolitan empire. He founded Greek colonies on Persian soil, created a Greek aristocracy in the Middle East and Egypt, and integrated Persian soldiers into the Macedonian army. Together, Greece, Egypt, and the Middle East became the *oikoumene,* considered by Greeks as the civilized world. The attempt at unity failed in 323 B.C., when Alexander died at the young age of thirty-three, and his empire was divided among his generals.

Egypt, with Alexandria its capital, was one of three principal states to emerge in the *oikoumene* after Alexander's death. It stretched southerly from the Mediterranean Sea along both banks of the Nile River, past the old Egyptian cities of Memphis and Thebes, past Syrene, and even past the city of Meroe, near modern Khartoum, at the juncture of the White Nile and the Blue Nile. Egypt was ruled by the Greek Ptolemaic dynasty, which was founded in 323 B.C. by Ptolemy I Soter (367?–283 B.C.), who turned Alexandria into a cosmopolitan commercial city dominated by a Greek aristocracy.

To the east of Egypt lay the Seleucid kingdom, the largest remnant of Alexander's empire. The Seleucid kingdom sprawled easterly from its capital, alabaster Antioch on the Mediterranean Sea, and comprised modern Palestine, Syria, Iraq, and Iran. Like Egypt, the Seleucid empire was ruled by a Greek upper class, and it boasted more than sixty Greek cities—colonies established by the government. The third major power of the *oikoumene* was Macedonia itself, which included most of the old Greek city-states.

Beyond the *oikoumene* were lands inhabited by people the Greeks considered barbarians. To the east lay mysterious, exotic India, where Alexander the Great had fought his last battles. To the west, in Italy and north Africa, two non-Greek city-states, Rome and Carthage, were building budding empires, although both were still republics. In southern Italy, a few old Greek colonies, most notably Syracuse, were still independent, although they were destined to be swallowed up shortly by Rome. Beyond these places lived savage hunters or illiterate farmers just emerged from the Stone Age. Farthest away was civilized China; merchants from the *oikoumene* carried on a thin trickle of trade with China, but China did not appear on Eratosthenes' map.

The *oikoumene* was dominated both politically and culturally by Greeks and has been called the Hellenistic (Greek-like) world by historians, and the period of time from Alexander the Great to the conquest of the city of Alexandria by the Romans (336–31 B.C.) has been called the *Hellenistic Age*. The Greeks who came to live in the new cities like Alexandria and Antioch superimposed a veneer of Greek culture on top of the older, Middle Eastern civilizations already there. They built cities and markets, academies and universities, museums and libraries. Now at the center of large political and economic empires, Greek scholars had access to information about new people, places, and things on an unprecedented scale, which spurred, among other things, the creation of the new science of geography by our friend Eratosthenes. Babylonian and Egyptian science were absorbed into Greek scholarship, invigorating both.

At the beginning of the Hellenistic Age, Greek science emerged as a separate discipline; it was no longer considered merely a subset of philosophy. Although Athenian scholars continued to concentrate on philosophy, history, and literature, thinkers at Alexandria emphasized science and mathematics. They were encouraged in their research by the Egyptian government. King Ptolemy II Philadelphus (308?–246? B.C.) endowed the University heavily, erecting a museum, a zoo, and an impressive series of academic buildings. Moreover, the kings accorded the scholars privacy and academic freedom and did not interfere with their studies.

Greek science reached its pinnacle at Alexandria during the first century and a half of the Hellenistic Age, between 300 and 150 B.C. After that came a long, slow decline, punctuated in 46 B.C., when much of the University at Alexandria, including the library, burned, and completed in A.D. 529, when the Academy at Athens closed its doors. The decline was caused by a combination of technological, political, economic, and social factors:

Technological Factors. The sciences of astronomy, biology, and geography had reached the point where they could not progress much further without telescopes, microscopes, and clocks. Theories and hypotheses needed to be tested, and the necessary equipment had yet to be invented.

Political Factors. In 149 B.C., Rome, a rising, aggressive power in the Mediterranean Sea region, completed its conquest of Carthage and turned its attention to the *oikoumene*. The Romans annexed Macedonia in 148 B.C., wealthy Pergamum fifteen years later, and powerful Pontus in 66 B.C. As it acquired conquered territory, Roman social and political life began to decay, resulting in a series of civil wars. One of these internecine conflicts pitted Julius Caesar (102?–44 B.C.) against Cneius Pompey (106–48 B.C.) and culminated in the latter's defeat. Pompey fled to Egypt, where Caesar followed him, only to find Pompey dead and himself trapped in Alexandria by the navy of Ptolemy XIII (died 44 B.C.). The wily Roman extricated himself by setting fire to the Egyptian fleet, but sparks from the blaze blew into the city. Much of Alexandria burned, including, to Caesar's dismay, the great library. Caesar made good his escape, after first fathering a child by the Egyptian queen, Cleopatra (69–30 B.C.), only to be assassinated by foes in Rome in 44 B.C., two years later. Another civil war ensued that was eventually won by Caesar's nephew, Augustus (63 B.C.–A.D. 14) in 31 B.C. After the war, Augustus declared himself dictator, gutted Rome's remaining republican institutions, and annexed Egypt as penalty for harboring one of his rivals. Rome continued to rule most of the *oikoumene* until the empire fell to barbarian invaders in A.D. 476.

Unlike the Egyptian kings, the Roman emperors, most of them professional soldiers, declined to use the public treasury to support scientific endeavor. The Roman Empire (31 B.C.–A.D. 476) was at its core a military dictatorship and, like most military regimes, was unsympathetic to independent scholarship. (See the description of Sparta in Cultural Connection III.) Imperial Rome was not totally lacking in intellectual achievement; it produced some good histories and a fine body of literature, for example, but it proved a relatively sterile environment for science.

Economic Factors. The Romans used slave labor to an almost unprecedented degree, especially after the founding of the Empire by Augustus in 31 B.C. More than half of the Empire's inhabitants were slaves. With slaves to do most of the backbreaking work, there was little perceived need for labor-saving devices, such as the pulleys and levers invented by Archimedes of Syracuse (287–212 B.C.); hence, scientists had little incentive to invent them.

Social Factors. Despite its initial successes, science interested Hellenistic and Roman scholars far less than philosophy, literature, and religion. The Hellenistic Age saw the development of Stoicism and Epicurianism as philosophical disciplines, and the Roman Empire witnessed the rise of Christianity (as well as several lesser religions and cults, such as Mithraism, that did not survive) and its establishment as the state religion by the emperor Constantine I (A.D. 288?–337) in A.D. 325. Religious leaders often opposed scientific inquiry, especially when scientific models appeared to challenge religious dogma. Despite the fact that the Christians themselves had been the victims of brutal repression before A.D. 325, an extremist minority in the Christian community found it difficult to tolerate scientists. The last scientist of Alexandria, Hypatia, was murdered savagely by Christian zealots in A.D. 415, and, in A.D. 529, Christian leaders in Greece persuaded the Byzantine emperor, Justinian I (A.D. 483–565), to close the Academy at Athens, ostensibly owing to the Academy's heretical activities.

Summary

Between 550 B.C. and A.D. 476, the western world was ruled by a series of great empires. The Persian Empire lasted until conquered by Alexander the Great in 330 B.C.; three Greek empires, Ptolemaic Egypt, the Seleucid kingdom, and Macedonia, shared control between 323 B.C. and 31 B.C.; and the Roman Empire dominated from 31 B.C. until A.D. 476. The Greek expansion into Asia and Africa after the fall of Persia transported Greek culture and science to new parts of the world. In Alexandria, Egypt, the Greek kings built and endowed a great university, and scholarship flourished for about 150 years, between 300 B.C. and 150 B.C. After that time, scientific endeavor began to wane, owing to a number of factors: lack of equipment, decline in government support after the Roman conquest of Egypt in 31 B.C., an increase in the use of slave labor, a competing interest in philosophy and religion, and opposition from some religious leaders. By A.D. 529, the last Greek school, the Academy at Athens, had closed, and the grand adventure of Greek science had ended. It would be nearly one thousand years before science would again flourish in the western world.

Chapter 5

EUCLID AND HIS ELEMENTS

5–1 Alexandria[1]

The period following the Peloponnesian War was one of political disunity among the Greek states; this disunity rendered them easy prey for the now-strong kingdom of Macedonia, which lay to the north. King Philip of Macedonia was gradually extending his power southward, and Demosthenes thundered his unheeded warnings. The Greeks rallied too late for a successful defense and, with the Athenian defeat at Chaeronea in 338 B.C., Greece became a part of the Macedonian empire.

Two years after the fall of the Greek states, ambitious Alexander the Great succeeded his father Philip and set out upon his unparalleled career of conquest, which added vast portions of the then-civilized world to the growing Macedonian domains. Wherever he led his victorious army, he created, at well-chosen places, a string of new cities. It was in this way, when Alexander entered Egypt, that the city of Alexandria was founded in 332 B.C.

It is said that the choice of the site, the drawing of the ground plan, and the process of colonization for Alexandria were directed by Alexander himself and that the actual building of the city was assigned to the eminent architect Dinocrates. From its inception, Alexandria showed every sign of fulfilling a remarkable future. In an incredibly short time, largely owing to its very fortunate location at a natural intersection of some important trade routes, it grew in wealth and became the most magnificent and cosmopolitan center of the world. By 300 B.C., it already had 500,000 inhabitants.

After Alexander the Great died in 323 B.C., his empire was partitioned among some of his military leaders, resulting in the eventual emergence of three empires, under separate rule, but nevertheless united by the bonds of the Hellenic civilization that had followed Alexander's conquests. Egypt fell to the lot of Ptolemy. It was not until about 306 B.C. that Ptolemy actually began his reign. He selected Alexandria as his capital and, to attract learned men to his city, immediately began the erection of the famed University of Alexandria. This was the first institution of its kind and, in its scope and setup, soon became much like the universities of today. Reputedly, it was highly endowed, and its attractive and elaborate plan contained lecture rooms, laboratories, gardens, museums, library facilities, and living quarters. The core of the institution was the great library, which for a long time was the largest repository of learned

[1] See R. E. Langer, "Alexandria—shrine of mathematics," *The American Mathematical Monthly* 48 (February 1941): 109–25.

works to be found anywhere in the world, boasting, within forty years of its founding, over 600,000 papyrus rolls. It was about 300 B.C. that the university opened its doors and Alexandria became, and remained for close to one thousand years, the intellectual metropolis of the Greek race.

For recognized scholars to staff the university, Ptolemy turned to Athens and invited the distinguished Demetrius Phalereus to take charge of the great library. Able and talented men were selected to develop the various fields of study. Euclid, who also may have come from Athens, was chosen to head the department of mathematics.

5–2 Euclid

Disappointingly little is known about the life and personality of Euclid except that he was a professor of mathematics at the University of Alexandria and apparently the founder of the illustrious and long-lived Alexandrian School of Mathematics. Even his dates and his birthplace are not known, but it seems probable that he received his mathematical training in the Platonic school at Athens. Many years later, when comparing Euclid with Apollonius, to the latter's discredit, Pappus praised Euclid for his modesty and consideration of others. Proclus augmented his *Eudemian Summary* with the frequently told story of Euclid's reply to Ptolemy's request for a short cut to geometric knowledge that "there is no royal road in geometry." But the same story has been told of Menaechmus when he was serving as instructor to Alexander the Great. Stobaeus told another story of a student studying geometry under Euclid who questioned what he would get from learning the subject, whereupon Euclid ordered a slave to give the fellow a penny, "since he must make gain from what he learns."

5–3 Euclid's "Elements"

Although Euclid was the author of at least ten works (fairly complete texts of five of these have come down to us), his reputation rests mainly on his *Elements*. It appears that this remarkable work immediately and completely superseded all previous *Elements;* in fact, no trace remains of the earlier efforts. As soon as the work appeared, it was accorded the highest respect, and from Euclid's successors on up to modern times, the mere citation of Euclid's book and proposition numbers was regarded as sufficient to identify a particular theorem or construction. No work, except the Bible, has been more widely used, edited, or studied, and probably no work has exercised a greater influence on scientific thinking. Over one thousand editions of Euclid's *Elements* have appeared since the first one printed in 1482; for more than two millennia, this work has dominated all teaching of geometry.

It is a misfortune that no copy of Euclid's *Elements* has been found that actually dates from the author's own time. Modern editions of the work are based upon a revision that was prepared by the Greek commentator Theon of

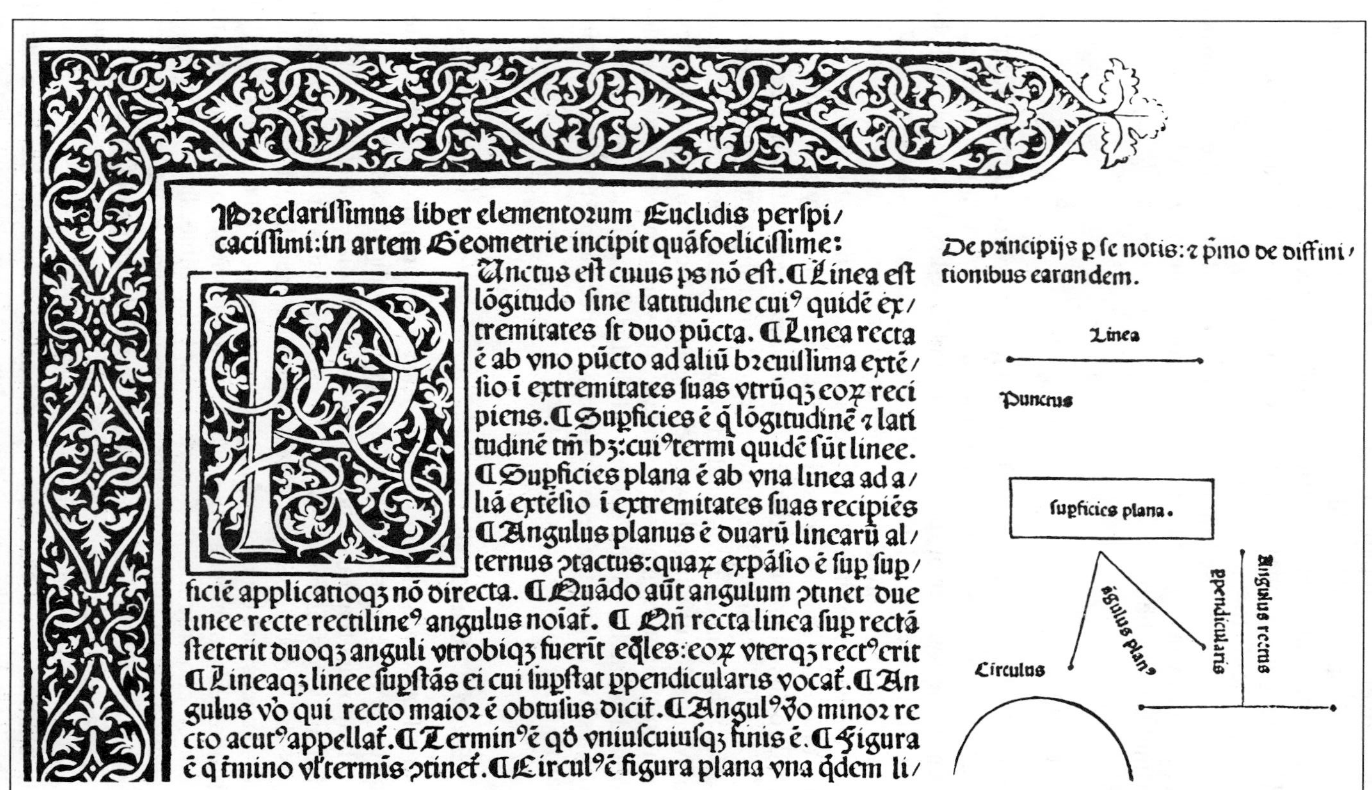
Preclarissimus liber elementorum Euclidis perspi-
cacissimi: in artem Geometrie incipit quã foelicissime:
Punctus est cuius ps nõ est. ¶ Linea est
lõgitudo sine latitudine cuiꝰ quidẽ extremitates st duo pũcta. ¶ Linea recta
ẽ ab vno pũcto ad aliũ breuissima extẽsio i extremitates suas vtrũqꝫ eoꝝ reci
piens. ¶ Supficies ẽ q̃ lõgitudinẽ ꝛ lati
tudinẽ tm̃ hꝫ: cuiꝰ termi quidẽ sũt linee.
¶ Supficies plana ẽ ab vna linea ad a-
liã extẽsio i extremitates suas recipiẽs
¶ Angulus planus ẽ duarũ linearũ al-
ternus ꝯtactus: quaꝝ expãsio ẽ sup sup-
ficiẽ applicatioqꝫ nõ directa. ¶ Quãdo aũt angulum ꝯtinet due
linee recte rectilineꝰ angulus noiaẽ. ¶ Qñ recta linea sup rectã
steterit duoqꝫ anguli vtrobiqꝫ fuerĩt eq̃les: eoꝝ vterqꝫ rectꝰ erit
¶ Lineaqꝫ linee supstãs ei cui supstat ppendicularis vocaẽ. ¶ An
gulus võ qui recto maior ẽ obtusus dicĩt. ¶ Angulꝰ võ minor re
cto acutꝰ appellaẽ. ¶ Terminꝰ ẽ qd vniuscuiusqꝫ finis ẽ. ¶ Figura
ẽ q̃ tmino vl termĩs ꝯtineẽ. ¶ Circulꝰ ẽ figura plana vna q̃dem li-

De principijs p se notis: ꝛ pmo de diffini-
tionibus earundem.

Part of a page of the first printed edition of Euclid's *Elements* made at Venice in 1482.

VIRESCIT VVLNERE VERITAS

Ptolomeus

Marinus

Strabo

Aratus

Polibius

Hipparchus

Geometria

Astronomia

Arithmetica

Musica

MERCVRIVS

THE ELEMENTS

[illegible]EOMETRIE

of the most aunci-
ent Philosopher
EVCLIDE
of Megara.

*Faithfully (now first) tran-
slated into the Englishe toung, by*
H. Billingsley, *Citizen of London.*
Whereunto are annexed certaine
[illegible]

With a very fruitfull Praface made by M. I. Dee,
*specifying the chiefe Mathematicall Sciẽces, what
they are, and wherunto commodious: where, also, are
disclosed certaine new Secrets Mathematicall
and Mechanicall, untill these our daies, greatly missed.*

Imprinted at London by *John Daye.*

Title page (reduced in size) of Billingsley's *Euclid* (1570).

Alexandria, who lived almost 700 years after the time of Euclid. Theon's revision was, until the early nineteenth century, the oldest edition of the *Elements* known to us. In 1808, however, when Napoleon ordered valuable manuscripts to be taken from Italian libraries and sent to Paris, F. Peyrard found, in the Vatican library, a tenth-century copy of an edition of Euclid's *Elements* that predates Theon's recension. A study of this older edition and a careful sifting of citations and remarks made by early commentators indicate that the introductory material of Euclid's original treatise undoubtedly underwent some editing in the subsequent revisions, but that the propositions and their proofs, except for minor additions and deletions, have remained essentially as Euclid wrote them.

The first complete Latin translations of the *Elements* were not made from the Greek but from the Arabic. In the eighth century, a number of Byzantine manuscripts of Greek works were translated by the Arabians, and in 1120 the English scholar, Adelard of Bath, made a Latin translation of the *Elements* from one of these older Arabian translations. Other Latin translations were made from the Arabic by Gherardo of Cremona (1114–1187) and, 150 years after Adelard, by Johannes Campanus. The first printed edition of the *Elements* was made at Venice in 1482 and contained Campanus' translation. This very rare book was beautifully executed and was the first mathematical book of any consequence to be printed. An important Latin translation from the Greek was made by Commandino in 1572. This translation served as a basis for many subsequent translations, including the very influential work by Robert Simson, from which, in turn, so many of the English editions were derived. The first complete English translation of the *Elements* was the monumental Billingsley translation issued in 1570.[2]

It is no reflection upon the brilliance of Euclid's work that there had been other *Elements* anterior to his own. According to the *Eudemian Summary,* Hippocrates of Chios made the first effort along this line, and the next attempt was made by Leon, who in age fell somewhere between Plato and Eudoxus. It is said that Leon's work contained a more careful selection of propositions than did that of Hippocrates, and that these propositions were more numerous and more serviceable. The textbook of Plato's Academy was written by Theudius of Magnesia and was praised as an admirable collection of elements. The geometry of Theudius seems to have been the immediate precursor of Euclid's work and was undoubtedly available to Euclid, especially if he studied in the Platonic School. Euclid was also acquainted with the important work of Theaetetus and Eudoxus. Thus, it is probable that Euclid's *Elements* is, for the most part, a highly successful compilation and systematic arrangement of works of earlier writers. No doubt Euclid had to supply a number of the proofs and to perfect many others, but the chief merit of his work lies in the skillful selection of the propositions and in their arrangement into a logical sequence, presumably following deductively from a small handful of initial assumptions.

5–4 Content of the "Elements"

Contrary to widespread impressions, Euclid's *Elements* is not devoted to geometry alone, but contains much number theory and elementary (geometric) algebra. The work is composed of thirteen books with a total of 465 proposi-

---

[2] See R. C. Archibald, "The first translation of Euclid's *Elements* into English and its source," *The American Mathematical Monthly* 57 (August–September 1950): 443–52, and W. F. Shenton, "The first English Euclid," *The American Mathematical Monthly* 35 (December 1928): 505–12.

tions. American high-school plane and solid geometry texts contain much of the material found in Books I, III, IV, VI, XI, and XII.

Book I commences with the necessary preliminary definitions, postulates, and axioms; we shall return to these in Section 5–7. The forty-eight propositions of Book I fall into three groups. The first twenty-six deal mainly with properties of triangles and include the three congruence theorems. Propositions I 27 through I 32 establish the theory of parallels and prove that the sum of the angles of a triangle is equal to two right angles. The remaining propositions of the book deal with parallelograms, triangles, and squares, with special reference to area relations. Proposition I 47 is the Pythagorean theorem, with a proof universally credited to Euclid himself, and the final proposition, I 48, is the converse of the Pythagorean theorem. The material of this book was developed by the early Pythagoreans.

It is worthwhile to comment further on a few propositions of Book I. The first three propositions are construction problems that show how, along with a straightedge, Euclidean compasses can transfer a line segment from a given position to any other desired position (see Problem Study 4.1). It follows that one can often shorten a construction by treating the Euclidean compasses as modern compasses.

Proposition I 4 establishes the congruence of two triangles having two sides and the included angle of one equal to two sides and the included angle of the other. The proof is by superposition, wherein it is shown that one triangle can be applied to the other by placing the given angle of the one triangle upon that of the other triangle, so that corresponding equal sides also coincide. Mathematicians later raised objections to proofs by superposition (see Section 15–1).

Proposition I 5, which proves that the base angles of an isosceles triangle are equal, is of interest because it is said that many beginners found the proof so confusing that they abandoned further study of geometry. The proposition has been dubbed the *pons asinorum*, or "bridge of fools," because of the fancied resemblance of the figure of the proposition to a simple trestle bridge too steep for some novices to pass over. Euclid's proof involves drawing some preliminary construction lines and is illustrated in the figure on our reproduction of a page of Isaac Barrow's *Euclid*. In this figure, the equal sides *AB* and *AC* of the given isosceles triangle *BAC* are produced the same lengths to *D* and *F*, and *CD* and *BF* are drawn. It then follows (by Proposition I 4) that triangles *AFB* and *ADC* are congruent, making $BF = DC$ and angle BDC = angle *CFB*. It now follows (again by Proposition I 4) that triangles *BDC* and *CFB* are congruent, guaranteeing the equality of angles *DBC* and *FCB* and, hence, angles *ABC* and *ACB*. Actually, the proof can be considerably shortened, as was later noted by Pappus (ca. A.D. 300), by applying Proposition I 4 directly to triangles *ABC* and *ACB*, wherein *AB* in the one equals *AC* in the other, *AC* in the one equals *AB* in the other, and angle *BAC* in the one equals angle *CAB* in the other.

Proposition I 6 establishes the converse of Proposition I 5. In this instance, we are given that in triangle *BAC*, angle *ABC* = angle *ACB*, and we wish to show that $BA = CA$. Euclid proceeds by *reductio ad absurdum* and

Liber I. 11

lum, sub æqualibus rectis lineis contentum, & basim BC *basi* EF *æqualem habebunt; eritque triangulum* BAC *triangulo* EDF *æquale, ac reliqui anguli* B, C *reliquis angulis* E, F *æquales erunt, uterque utrique, sub quibus æqualia latera subtenduntur.*

Si punctum D puncto A applicetur, & recta DE rectæ AB superponatur, cadet punctum E in B, quia DE [a] = AB. Item recta DF cadet in AC, quia ang. A [a] = D. Quinetiam punctum E puncto C coincidet, quia AC [a] = DF. Ergò rectæ EF, BC, cùm eosdem habeant terminos, [b] congruent, & proinde æquales sunt. Quare triangula BAC, EDF; & anguli B, E; itémq; anguli C, F etiam congruunt, & æquantur. Quod erat Demonstrandum.

a hyp.
b 14. ax.

PROP. V.

Isoscelium triangulorum ABC *qui ad basim sunt anguli* ABC, ACB *inter se sunt æquales. Et productis æqualibus rectis lineis* AB, AC *qui sub base sunt anguli* CBD, BCE *inter se æquales erunt.*

[a] Accipe AF = AD, & junge CD, ac BF.

Quoniam in triangulis ACD, ABF, sunt AB [c] = AC, & AF [d] = AD, angulúsq; A communis, [e] erit ang. ABF = ACD; & ang. AFB [e] = ADC, & bas. BF [e] = DC; item FC [f] = DB. ergò in triangulis BFC, BDC [g] erit ang. FCB = DBC. Q. E. D. Item ideo ang. FBC = DCB. atqui ang. ABF [h] = ACD. ergò ang. ABC [k] = ACB. Q. E. D.

a 3. 1.
b 1. post.
c hyp.
d constr.
e 4. 1.
f 3 ax.
g 4. 1.
h pr.
k 3. ax.

Corollarium.

Hinc, Omne triangulum æquilaterum est quoq; æquiangulum.

PROP.

Euclid's proof of Proposition I 5 (The base angles of an isosceles triangle are equal) as given in Isaac Barrow's *Euclid*.

assures, for example, that $BA > CA$. Then $BM = CA$ can be laid off on BA. By Proposition I 4, triangles CBM and BCA are congruent, but this is absurd, because the former triangle is a proper part of the latter one. It follows that $BA \ngtr CA$. Similarly, $CA \ngtr BA$, and it follows that $BA = CA$. This is the first place in the *Elements* where *reductio ad absurdum,* or the indirect method, is used. The method is frequently employed by Euclid thereafter.

Propositions I 9 through I 12 are construction problems, the first two giving the familiar constructions for the bisector of a given angle and the midpoint of a given line segment. One purpose of such construction problems is to serve as existence proofs; for instance, perhaps the best way to prove the existence of the bisector of a given angle is actually to construct that bisector.

Proposition I 47 is the Pythagorean theorem. Euclid's figure for this proposition and a précis of his beautiful proof can be found in Problem Study 5.3(b).

Book II, a short book of only fourteen propositions, deals with the transformation of areas and the geometric algebra of the Pythagorean school. It is in this book that we find the geometrical equivalents of a number of algebraic identities. In Section 3–6, for example, we have shown, how Propositions II 4, II 5, and II 6 establish the respective identities

$$(a + b)^2 = a^2 + 2ab + b^2,$$

$$(a + b)(a - b) = a^2 - b^2,$$

$$4ab + (a - b)^2 = (a + b)^2.$$

Of special interest are Propositions II 12 and II 13. These propositions, stated together and in more modern language, say: *In an obtuse-angled (acute-angled) triangle, the square of the side opposite the obtuse (acute) angle is equal to the sum of the squares of the other two sides increased (decreased) by twice the product of one of these sides and the projection of the other on it.* Thus, these two propositions establish the generalization of the Pythagorean theorem that we today refer to as the "law of cosines."

In spite of a long-held belief, there is at present a lively debate among some historians of mathematics as to whether the propositions of Book II are really intended to be a geometric form of algebra.

Book III, which consists of thirty-nine propositions, contains many of the familiar theorems about circles, chords, secants, tangents, and the measurement of associated angles that we find in our high-school geometry texts. Book IV, with only sixteen propositions, discusses the construction, with straightedge and compasses, of regular polygons of three, four, five, six, and fifteen sides, and the inscription of these polygons within a given circle and their circumscription about a given circle. Since little of the geometry of the circle given in Books III and IV is found in Pythagorean work, the material of these books was probably furnished by the early Sophists and the researchers on the three famous problems discussed in Chapter 4.

Book V is a masterly exposition of Eudoxus' theory of proportion. It was this theory, applicable to incommensurable as well as to commensurable magnitudes, that resolved the "logical scandal" created by the Pythagorean discovery of irrational numbers. The Eudoxian definition of proportion, or equality of two ratios, is remarkable and worth repeating here. *Magnitudes are said to be in the same ratio, the first to the second and the third to the fourth, when, if any equimultiples whatever be taken of the first and third, and any equimultiples whatever of the second and fourth, the former equimultiples alike exceed, are alike equal to, or are alike less than the latter equimultiples taken in corresponding order.* In other words, if A, B, C, D are any four unsigned magnitudes, A and B being of the same kind (both line segments, or angles, or areas, or volumes) and C and D being of the same kind, then the ratio of A to B is equal to that of C to D when, for arbitrary positive integers m and n, $mA \gtreqless nB$ according as $mC \gtreqless nD$. The Eudoxian theory of proportion provided a foundation, later developed by Dedekind and Weierstrass, for the real number system of mathematical analysis.

Book VI applies the Eudoxian theory of proportion to plane geometry. Here we find the fundamental theorems on similar triangles; constructions giving third, fourth, and mean proportionals; the geometric solution of quadratic equations that we considered in Chapter 3; the proposition that the internal bisector of an angle of a triangle divides the opposite side into segments proportional to the other two sides; a generalization of the Pythagorean theorem in which, instead of squares, three similar and similarly described figures are drawn on the three sides of a right triangle; and many other theorems. There probably is no theorem in this book that was not known to the early Pythagoreans, but the pre-Eudoxian proofs of many of them were at fault, since they were based upon the incomplete theory of proportion.

Books VII, VIII, and IX, which contain a total of 102 propositions, deal with elementary number theory. Book VII commences with the process, referred to today as the **Euclidean algorithm,** for finding the greatest common integral divisor of two or more integers and uses it as a test for two integers to be relatively prime (see Problem Study 5.1). We also find an exposition of the numerical, or Pythagorean, theory of proportion. Many basic number properties are established in this book.

Book VIII concerns itself largely with continued proportions and related geometric progressions. If we have the continued proportion $a{:}b = b{:}c = c{:}d$, then a, b, c, d form a geometric progression.

A number of significant theorems are found in Book IX. Proposition IX 14 is equivalent to the important **fundamental theorem of arithmetic**—namely, that *any integer greater than 1 can be expressed as a product of primes in one and, except for the order of the factors, only one way*. Proposition IX 35 gives a geometric derivation of the formula for the sum of the first n terms of a geometric progression, and the last proposition, IX 36, establishes the remarkable formula for perfect numbers that was stated in Section 3–3.

Euclid's proof of IX 20 (*the number of prime numbers is infinite*) has been universally regarded by mathematicians as a model of mathematical elegance. The proof employs the indirect method,[3] or *reductio ad absurdum,* and runs

---

[3] It is easy to formulate the proof so that the indirect method is avoided.

essentially as follows. Suppose there are only a finite number of prime numbers, which we shall denote by $a, b, \ldots, k$. Set $P = (a)(b) \ldots (k)$. Then $P + 1$ is either prime or composite. But since $a, b, \ldots, k$ are all the primes, $P + 1$, which is greater than each of $a, b, \ldots, k$, cannot be a prime. If $P + 1$ is composite, however, it must be divisible by some prime p. But p must be a member of the set $a, b, \ldots, k$ of all primes, which means that p is a divisor of P. Consequently, p cannot divide $P + 1$, since $p > 1$. Thus, our initial hypothesis that the number of primes is finite is untenable, and the theorem is established.

Book X deals with irrationals—that is, with line segments that are incommensurable with respect to some given line segment. Many scholars regard this book as perhaps the most remarkable book in the *Elements*. Much of the subject matter of this book is attributed to Theaetetus, but the extraordinary completeness, elaborate classification, and finish are usually credited to Euclid. It taxes one's credulity to realize that the results of this book were arrived at by abstract reasoning, unassisted by any convenient algebraic notation. The opening proposition (X 1) is the basis of the method of exhaustion later employed in Book XII—namely, that *if from any magnitude there be subtracted a part not less than its half, from the remainder another part not less than its half, and so on, there will at length remain a magnitude less than any assigned magnitude of the same kind*. In this book, we also find formulas yielding Pythagorean triples of numbers, formulas that the ancient Babylonians may have known over one thousand years earlier (see Section 2–6).

The remaining three books, XI, XII, and XIII, concern themselves with solid geometry and cover much of the material, with the exception of that on spheres, commonly found in high-school texts. The definitions, the theorems about lines and planes in space, and theorems concerning parallelepipeds are found in Book XI. The method of exhaustion plays an important role in the treatment of volumes in Book XII and will be reconsidered in some detail in Chapter 11. In Book XIII, constructions are developed for inscribing the five regular polyhedra in a sphere.

The frequently stated remark that Euclid's *Elements* was really intended to serve merely as a drawn-out account of the five regular polyhedra appears to be a lopsided evaluation. More likely, it was written as a beginning text in general mathematics. Euclid also wrote texts on higher mathematics.

Finally, a word concerning the meaning of the term "elements." Proclus has told us that the early Greeks defined the "elements" of a deductive study as the leading, or key, theorems that are of wide and general use in the subject. Their function has been compared to that of the letters of the alphabet in relation to language; as a matter of fact, letters are called by the same name in Greek. Aristotle, in his *Metaphysics,* speaks of elements in the same sense when he says, "Among geometrical propositions we call those 'elements' the proofs of which are contained in the proofs of all or most of such propositions." The selection of the theorems to be taken as the elements of the subject requires the exercise of considerable judgment, and it is in this respect, among others, that Euclid's *Elements* was so superior to all earlier efforts.

It follows that another frequently stated remark (that Euclid's *Elements* was meant to contain essentially all the plane and solid geometry known in his time) is patently false. Euclid knew a great deal more geometry than appears in his *Elements*.

5–5 The Theory of Proportion

It is interesting to note the differences between the Pythagorean, the Eudoxian, and the modern textbook proofs of a simple proposition involving proportions. Let us select Proposition VI 1, which states that *the areas of triangles having the same altitude are to one another as their bases*. We shall permit ourselves to use Proposition I 38, which says that *triangles having equal bases and equal altitudes have equal areas,* and a consequence of I 38 to the effect that *of any two triangles having the same altitude, that one has the greater area which has the greater base*.

Let the triangles be ABC and ADE, the bases BC and DE lying on the same straight line MN, as in Figure 38. The Pythagoreans, before the discovery of irrational numbers, tacitly assumed that any two line segments are commensurable; thus, BC and DE were assumed to have some common unit of measure, going, say, p times into BC and q times into DE. Mark off these points of division on BC and DE and connect them with vertex A. Then triangles ABC and ADE are divided, respectively, into p and q smaller triangles, all having, by I 38, the same area. It follows that $\triangle ABC{:}\triangle ADE = p{:}q = BC{:}DE$, and the proposition is established. With the later discovery that two line segments need not be commensurable, this proof, along with others, became inadequate, and the very disturbing ''logical scandal'' came into existence.

Eudoxus' theory of proportion cleverly resolved the ''scandal,'' as we shall now illustrate by reproving VI 1 in the manner found in the *Elements*. On CB produced, mark off, successively from B, $m - 1$ segments equal to CB and connect the points of division, $B_2, B_3, \ldots, B_m$, with vertex A, as shown in Figure 39. Similarly, on DE produced, mark off, successively from E, $n - 1$ segments equal to DE and connect the points of division, $E_2, E_3, \ldots, E_n$, with vertex A. Then $B_mC = m(BC)$, $\triangle AB_mC = m(\triangle ABC)$, $DE_n = n(DE)$, $\triangle ADE_n = n(\triangle ADE)$. Also, by I 38 and its corollary, $\triangle AB_mC \gtreqless \triangle ADE_n$ according as $B_mC \gtreqless DE_n$; that is, $m(\triangle ABC) \gtreqless n(\triangle ADE)$ according as $m(BC) \gtreqless n(DE)$, whence, by the Eudoxian definition of proportion, $\triangle ABC{:}\triangle ADE = BC{:}DE$, and the proposition is established. No mention was made of commensurable

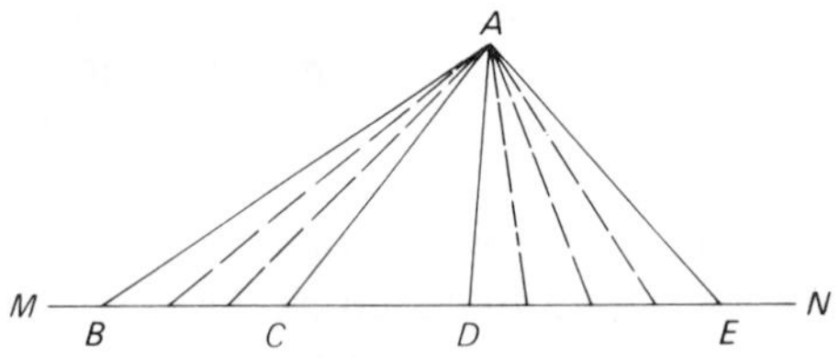

FIGURE 38

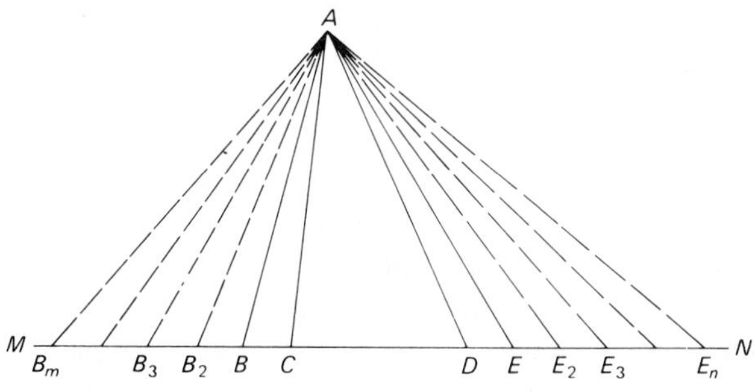

FIGURE 39

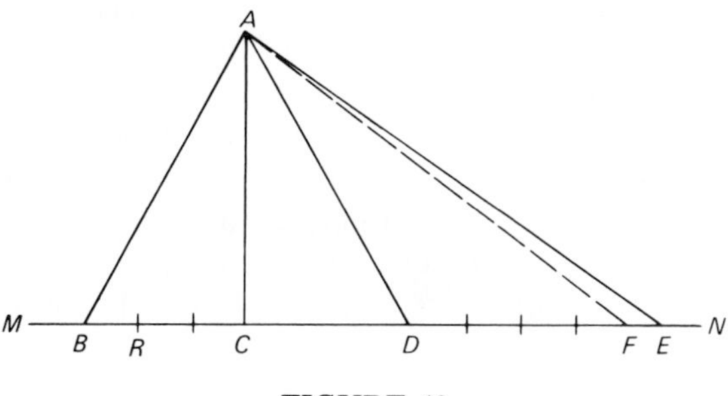

FIGURE 40

and incommensurable quantities, since the Eudoxian definition applies equally to both situations.

Until well into the twentieth century, many high-school textbooks advocated a proof of this theorem involving two cases, according as *BC* and *DE* are or are not commensurable. The commensurable case was handled as in the Pythagorean solution above, and simple limit notions were used to deal with the incommensurable case. Thus, suppose *BC* and *DE* are incommensurable. Divide *BC* into n equal parts, *BR* being one of the parts (see Figure 40). On *DE*, mark off a succession of segments equal to *BR*, finally arriving at a point *F* on *DE* such that $FE < BR$. By the commensurable case, already established, $\triangle ABC{:}\triangle ADF = BC{:}DF$. Now let $n \to \infty$. Then $DF \to DE$ and $\triangle ADF \to \triangle ADE$; hence, in the limit, $\triangle ABC{:}\triangle ADE = BC{:}DE$. This approach uses the fact that any irrational number may be regarded as the limit of a sequence of rational numbers, an approach that was rigorously developed in modern times by Georg Cantor (1845–1918).

5–6 Regular Polygons

We have noted that Euclid, in Book IV of his *Elements,* discusses the construction, done with straightedge and compasses, of regular polygons of three, four, five, six, and fifteen sides. By successive angle, or arc, bisections, we may then with Euclidean tools construct regular polygons having 2^n, $3(2^n)$, $5(2^n)$, or $15(2^n)$ sides. Not until almost the nineteenth century was it known that any other regular polygons could be constructed with these limited tools. In 1796, the eminent German mathematician Carl Friedrich Gauss initiated the theory that showed that a regular polygon having a *prime* number of sides can be constructed with Euclidean tools if and only if that number is of the form $f(n) = 2^{2^n} + 1$. For $n = 0, 1, 2, 3, 4$, we find $f(n) = 3, 5, 17, 257, 65{,}537$, all prime numbers. Thus, unknown to the Greeks, regular polygons of 17, 257, and 65,537 sides can be constructed with straightedge and compasses. For no other value of n than those listed above is it known that $f(n)$ is a prime number.

Many Euclidean constructions of the regular polygon of seventeen sides have been given. In 1832, Richelot published an investigation of the regular polygon of 257 sides, and a Professor Hermes of Lingen gave up ten years of his life to the problem of constructing a regular polygon of 65,537 sides. It has been said that it was Gauss' discovery, at the age of nineteen, that a regular polygon of seventeen sides can be constructed with straightedge and compasses that decided him to devote his life to mathematics. His pride in this discovery is evidenced by his request that a regular polygon of seventeen sides be engraved on his tombstone. Although this request was never fulfilled, such a polygon is found on the base of a monument to Gauss erected at his birthplace in Brunswick.

5–7 Formal Aspect of the "Elements"

Important as are the contents of the *Elements,* perhaps still more important is the formal manner in which those contents are presented. In fact, Euclid's *Elements* has become the prototype of modern mathematical form.

Certainly one of the greatest achievements of the early Greek mathematicians was the creation of the postulational form of thinking. In order to establish a statement in a deductive system, one must show that the statement is a necessary logical consequence of some previously established statements. These, in their turn, must be established from some still more previously established statements, and so on. Since the chain cannot be continued backward indefinitely, one must, at the start, accept some finite body of statements without proof or else commit the unpardonable sin of circularity, by deducing statement A from statement B and then later B from A. These initially assumed statements are called the **postulates,** or **axioms,** of the discourse, and all other statements of the discourse must be logically implied by them. When the statements of a discourse are so arranged, the discourse is said to be presented in postulational form.

So great was the impression made by the formal aspect of Euclid's *Elements* on following generations that the work became a model for rigorous mathematical demonstration. In spite of a considerable abandonment of the Euclidean form during the seventeenth and eighteenth centuries, the postulational method has today penetrated into almost every field of mathematics, and many mathematicians adhere to the thesis that, not only is mathematical thinking postulational thinking but, conversely, postulational thinking is mathematical thinking. A relatively modern outcome has been the creation of a field of study called **axiomatics,** devoted to examining the general properties of sets of postulates and of postulational thinking. We shall return to this in Section 15–2.

Most of the early Greek mathematicians and philosophers made a distinction between "postulates" and "axioms." At least three distinctions were advocated by various parties.

1. An axiom is a self-evident assumed statement about something, and a postulate is a self-evident assumed construction of something; thus, axioms and postulates bear a relation to one another much like that which exists between theorems and construction problems.
2. An axiom is an assumption common to all sciences, whereas a postulate is an assumption peculiar to the particular science being studied.
3. An axiom is an assumption of something that is both obvious and acceptable to the learner; a postulate is an assumption of something that is neither necessarily obvious nor necessarily acceptable to the learner. (This last is essentially the Aristotelian distinction.) In modern mathematics, no distinction is made, nor is the quality of being self-evident or obvious considered. There were some early Greeks who approached this viewpoint.

It is not certain precisely what statements Euclid assumed for his postulates and axioms, nor, for that matter, exactly how many he had, for changes and additions were made by subsequent editors. There is fair evidence, however, that he adhered to the second distinction and that he probably assumed the equivalents of the following ten statements, five "axioms," or common notions, and five geometric "postulates":

A1 *Things that are equal to the same thing are also equal to one another.*
A2 *If equals be added to equals, the wholes are equal.*
A3 *If equals be subtracted from equals, the remainders are equal.*
A4 *Things that coincide with one another are equal to one another.*
A5 *The whole is greater than the part.*

P1 *It is possible to draw a straight line from any point to any other point.*
P2 *It is possible to produce a finite straight line indefinitely in that straight line.*
P3 *It is possible to describe a circle with any point as center and with a radius equal to any finite straight line drawn from the center.*
P4 *All right angles are equal to one another.*
P5 *If a straight line intersects two straight lines so as to make the interior angles on one side of it together less than two right angles, these straight lines will intersect, if indefinitely produced, on the side on which are the angles which are together less than two right angles.*

Postulates P1 and P2 establish the existence of a line determined by two points; postulate P3 establishes the existence of a circle, given its center and radius. Because of this (as mentioned earlier in Section 4–4), the unmarked straightedge and collapsing compasses became the only tools allowed in Euclidean geometry for construction problems.

The *Elements* purports to derive all its 465 propositions from these ten statements! The development is the **synthetic** one of proceeding from the known and simpler to the unknown and more complex. Without a doubt, the reverse process, called **analysis,**[4] of reducing the unknown and more complex to the known, played a part in the discovery of the proofs of many of the theorems, but it plays no part in the exposition of the subject.

5–8 Euclid's Other Works

Euclid wrote several treatises besides the *Elements,* some of which have survived to the present day. One of the latter, called the *Data,* is concerned with the material of the first six books of the *Elements.* A **datum** may be defined as a set of parts or relations of a figure such that if all but any one are given, then that remaining one is determined. Thus, the parts A, a, R of a triangle, where A is one angle, a the opposite side, and R the circumradius, constitute a datum; given any two of these parts, the third is thereby determined. This is clear either geometrically or from the relation $a = 2R \sin A$. It is apparent that a collection of data of this sort could be useful in the analysis that precedes the discovery of a construction or a proof, and this is undoubtedly the purpose of the work.

Another work in geometry by Euclid, which has come down to us through an Arabian translation, is the book *On Divisions.* Here we find construction problems requiring the division of a figure by a restricted straight line so that the parts will have areas in a prescribed ratio. An example is the problem of dividing a given triangle into two equal areas by a line drawn through a given point within the triangle. Other examples occur in Problem Study 3.11(b) and (c).

Other geometrical works of Euclid that are now lost to us and are known only from subsequent commentaries are the *Pseudaria,* or book of geometrical fallacies; *Porisms,* about which there has been considerable speculation[5]; *Conics,* a treatise in four books that was later completed and then added to by Apollonius; and *Surface Loci,* about which nothing certain is known.

Euclid's other works concern applied mathematics, and two of these are extant: the *Phaenomena,* dealing with the spherical geometry required for

---

[4] The words *analysis* and *analytic* are used in several senses in mathematics. Thus we have *analytic* geometry, the large branch of mathematics called *analysis, analytic* functions, and so on.

[5] A **porism** is taken today to be a proposition stating a condition that renders a certain problem solvable, and then the problem has infinitely many solutions. For example, if r and R are the radii of two circles and d is the distance between their centers, the problem of inscribing a triangle in the circle of radius R, which will be circumscribed about the circle of radius r, is solvable if and only if $R^2 - d^2 = 2Rr$, and then there are infinitely many triangles of the desired sort. We do not know precisely Euclid's meaning of the term.

observational astronomy, and the *Optics,* an elementary treatise on perspective. Euclid is supposed also to have written a work on the *Elements of Music.*

Problem Studies

5.1 The Euclidean Algorithm

The **Euclidean algorithm,** or process, for finding the greatest common integral divisor (g.c.d.) of two positive integers is so named because it is found at the start of Book VII of Euclid's *Elements,* although the process no doubt was known considerably earlier. This algorithm is at the foundation of several developments in modern mathematics. Stated in the form of a rule, the process is this: *Divide the larger of the two positive integers by the smaller one, then divide the divisor by the remainder. Continue this process, of dividing the last divisor by the last remainder, until the division is exact. The final divisor is the sought g.c.d. of the two original positive integers.*

(a) Find, by the Euclidean algorithm, the g.c.d. of 5913 and 7592.

(b) Find, by the Euclidean algorithm, the g.c.d. of 1827, 2523, and 3248.

(c) Prove that the Euclidean algorithm does lead to the g.c.d.

(d) Let h be the g.c.d. of the positive integers a and b. Show that there exist integers p and q (not necessarily positive) such that $pa + qb = h$.

(e) Find p and q for the integers of (a).

(f) Prove that a and b are relatively prime if and only if there exist integers p and q such that $pa + qb = 1$.

5.2 Applications of the Euclidean Algorithm

(a) Prove, using Problem Study 5.1(f), that if p is a prime and divides the product uv, then either p divides u or p divides v.

(b) Prove from (a) the fundamental theorem of arithmetic: *Every integer greater than 1 can be uniquely factored into a product of primes.*

(c) Find integers a, b, c such that $65/273 = a/3 + b/7 + c/13$.

5.3 The Pythagorean Theorem

(a) Euclid's elegant proof of the Pythagorean theorem depends upon the diagram of Figure 41, sometimes referred to as the Franciscan's cowl or as the bride's chair. A précis of the proof runs as follows: $(AC)^2 = 2\triangle JAB = 2\triangle CAD = ADKL$. Similarly, $(BC)^2 = BEKL$, and so on. Fill in the details of this proof.

(b) Show how Figure 42 suggests a dynamical proof of the Pythagorean theorem, which might be given on movie film, wherein the square on the hypotenuse is continuously transformed into the sum of the squares on the legs of the right triangle.

(c) A few of our country's presidents have been tenuously connected with mathematics. George Washington was a noted surveyor, Thomas Jefferson did much to encourage the teaching of higher mathematics in the

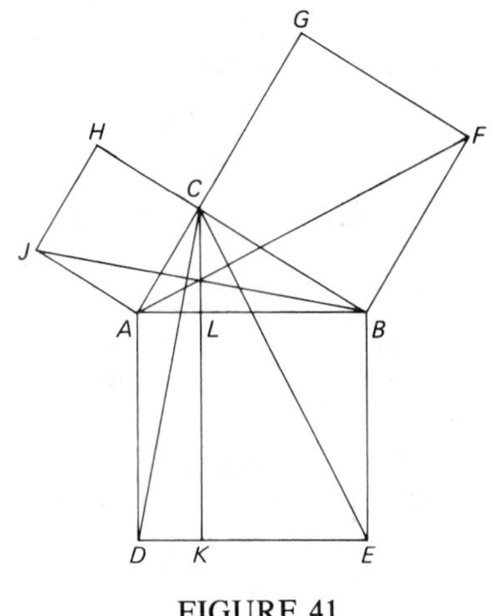

FIGURE 41

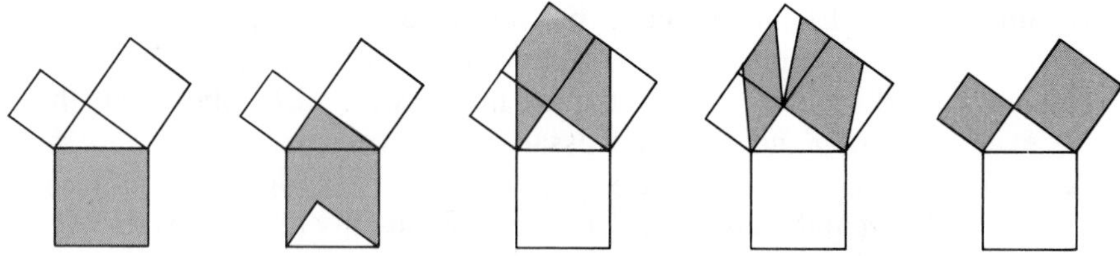

FIGURE 42

United States, and Abraham Lincoln is credited with learning logic by studying Euclid's *Elements*. More creative was James Abram Garfield (1831–1881), the country's twentieth president, who in his student days developed a keen interest and fair ability in elementary mathematics. It was in 1876, while he was a member of the House of Representatives and five years before he became President of the United States, that he independently discovered a very pretty proof of the Pythagorean theorem. He hit upon the proof in a mathematics discussion with some other members of Congress, and the proof was subsequently printed in the *New England Journal of Education*. The proof depends upon calculating the area of the trapezoid of Figure 43 in two different ways—first by the formula for the area of a trapezoid, and then as the sum of the three right triangles into which the trapezoid can be dissected. Carry out this proof in detail.

(d) State and prove the converse of the Pythagorean theorem.

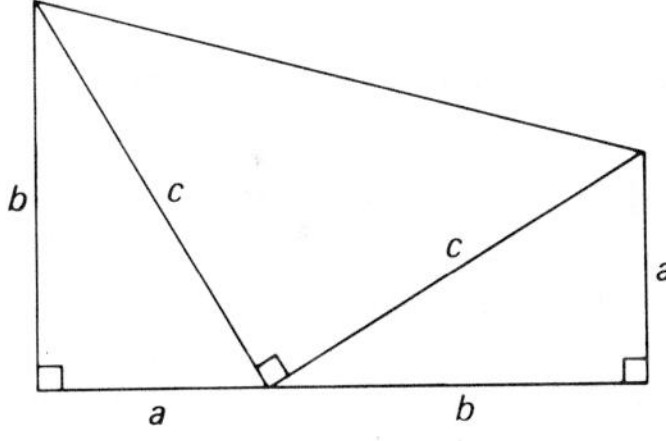

FIGURE 43

5.4 Euclid's Book II

(a) Following is Euclid's Proposition II 1: *If there be two straight lines, and one of them be cut into any number of segments whatever, the rectangle contained by the two straight lines is equal to the rectangles contained by the uncut straight line and each of the segments.* This is the geometrical counterpart of what familiar law of algebra?

(b) Show that Propositions II 12 and II 13 are essentially the law of cosines.

(c) Show how the Pythagorean theorem may be considered as a special case of the law of cosines.

5.5 Applications of the Fundamental Theorem of Arithmetic

The fundamental theorem of arithmetic says that, for any given positive integer a, there are unique nonnegative integers $a_1, a_2, a_3, \ldots$, only a finite number of which are different from zero, such that

$$a = 2^{a_1}3^{a_2}5^{a_3} \ldots,$$

where 2, 3, 5, . . . are the consecutive primes. This suggests a useful notation. We shall write

$$a = (a_1, a_2, \ldots, a_n),$$

where a_n is the last nonzero exponent. Thus, we have 12 = (2,1), 14 = (1,0,0,1), 27 = (0,3), and 360 = (3,2,1).

Prove the following theorems:

(a) $ab = (a_1 + b_1, a_2 + b_2, \ldots)$.

(b) b is a divisor of a if and only if $b_i \leqq a_i$ for each i.

(c) The number of divisors of a is $(a_1 + 1)(a_2 + 1) \ldots (a_n + 1)$.

(d) A necessary and sufficient condition for a number n to be a perfect square is that the number of divisors of n be odd.

(e) Set g_i equal to the smaller of a_i and b_i if $a_i \neq b_i$ and equal to either a_i or b_i if $a_i = b_i$. Then $g = (g_1, g_2, \ldots)$ is the g.c.d. of a and b.

(f) If a and b are relatively prime and b divides ac, then b divides c.
(g) If a and b are relatively prime and if a divides c and b divides c, then ab divides c.
(h) Show that $\sqrt{2}$ and $\sqrt{3}$ are irrational.

5.6 The Eudoxian Theory of Proportion

(a) Prove, by the Eudoxian method and by the early twentieth-century textbook method, Proposition VI 33: *Central angles in the same or equal circles are to each other as their intercepted arcs.*
(b) Prove, by the Pythagorean method and then complete by the use of limits, Proposition VI 2: *A line parallel to one side of a triangle divides the other two sides proportionally.*
(c) Prove Proposition VI 2 by using Proposition VI 1 (see Section 5–5).

5.7 Regular Polygons

(a) Suppose $n = rs$, where n, r, s are positive integers. Show that if a regular n-gon is constructible with Euclidean tools, then so also are a regular r-gon and a regular s-gon.
(b) Show that with Euclidean tools, it is impossible to construct a regular 27-gon.
(c) Suppose r and s are relatively prime positive integers and that a regular r-gon and a regular s-gon are constructible with Euclidean tools. Show that a regular rs-gon is also so constructible.
(d) Of the regular polygons having less than 20 sides, one can with Euclidean tools construct those having 3, 4, 5, 6, 8, 10, 12, 15, 16, and 17 sides. Actually construct these polygons, with the exception of the regular 17-gon.
(e) Construct a regular 17-gon by the following method (H. W. Richmond, "To construct a regular polygon of seventeen sides," *Mathematische Annalen* 67 (1909): 459).

Let OA and OB be two perpendicular radii of a given circle with center O. Find C on OB such that $OC = OB/4$. Now find D on OA such that angle OCD = (angle OCA)/4. Next find E on AO produced such that angle $DCE = 45°$. Draw the circle on AE as diameter, cutting OB in F, and then draw the circle $D(F)$, cutting OA and AO produced in G_4 and G_6. Erect perpendiculars to OA at G_4 and G_6, cutting the given circle in P_4 and P_6. These last points are the fourth and sixth vertices of the regular 17-gon whose first vertex is A.
(f) Establish Proposition XIII 10: *A side of a regular pentagon, of a regular hexagon, and of a regular decagon inscribed in the same circle constitute the sides of a right triangle.*
(g) Show that the smaller acute angle in a right triangle with legs 3 and 16 is very closely half the central angle subtended by one side of a regular 17-gon. Using this fact, give an approximate Euclidean construction of a regular 17-gon.

5.8 The Angle-Sum of a Triangle

Assuming the equality of alternate interior angles formed by a transversal cutting a pair of parallel lines, prove the following:

(a) The sum of the angles of a triangle is equal to a straight angle.

(b) The sum of the interior angles of a convex polygon of n sides is equal to $n - 2$ straight angles.

5.9 A Deductive Sequence Concerning Areas

Assuming the area of a rectangle is given by the product of its dimensions, establish the following chain of theorems:

(a) The area of a parallelogram is equal to the product of its base and altitude.

(b) The area of a triangle is equal to half the product of any side and the altitude on that side.

(c) The area of a right triangle is equal to half the product of its 2 legs.

(d) The area of a triangle is equal to half the product of its perimeter and the radius of its inscribed circle.

(e) The area of a trapezoid is equal to the product of its altitude and half the sum of its bases.

(f) The area of a regular polygon is equal to half the product of its perimeter and its apothem.

(g) The area of a circle is equal to half the product of its circumference and its radius.

5.10 A Deductive Sequence Concerning Angles

Assume (1) a central angle of a circle is measured by its intercepted arc; (2) the sum of the angles of a triangle is equal to a straight angle; (3) the base angles of an isosceles triangle are equal; (4) a tangent to a circle is perpendicular to the radius drawn to the point of contact. Establish the following chain of theorems:

(a) An exterior angle of a triangle is equal to the sum of the 2 remote interior angles.

(b) An inscribed angle in a circle is measured by one-half its intercepted arc.

(c) An angle inscribed in a semicircle is a right angle.

(d) An angle formed by 2 intersecting chords in a circle is measured by one-half the sum of the 2 intercepted arcs.

(e) An angle formed by 2 intersecting secants of a circle is measured by one-half the difference of the 2 intercepted arcs.

(f) An angle formed by a tangent to a circle and a chord through the point of contact is measured by one-half the intercepted arc.

(g) An angle formed by a tangent and an intersecting secant of a circle is measured by one-half the difference of the 2 intercepted arcs.

(h) An angle formed by 2 intersecting tangents to a circle is measured by one-half the difference of the 2 intercepted arcs.

5.11 Elements

(a) If you were to choose 2 of the following theorems for "elements" in a course in plane geometry, which would you choose?

1. The 3 altitudes of a triangle, produced if necessary, meet in a point.
2. The sum of the 3 angles of a triangle is equal to 2 right angles.
3. An angle inscribed in a circle is measured by one-half its intercepted arc.
4. The tangents drawn from any point on the common chord produced of 2 intersecting circles are equal in length.

(b) A geometry teacher is planning to present the topic of parallelograms to his or her class. After defining *parallelogram,* what theorems about parallelograms should the teacher offer as the "elements" of the subject?

(c) Preparatory to teaching the topic of similar figures, a geometry teacher gives a lesson or 2 on the theory of proportion. What theorems should the teacher select for the "elements" of the treatment, and in what order should she arrange them?

5.12 Data

Let A, B, C denote the angles of a triangle; a, b, c the opposite sides; h_a, h_b, h_c the altitudes on these sides; m_a, m_b, m_c the medians to these sides; t_a, t_b, t_c the angle bisectors drawn to these sides; R and r the circumradius and inradius; b_a and c_a the projections of sides b and c on side a; and r_a the radius of the circle touching side a and sides b and c produced. Show that each of the following constitutes a datum for a triangle.

(a) A, B, C (b) a/b, b/c, c/a
(c) b, A, h_c (d) $b + c$, A, $h_b + h_c$
(e) $b - c$, A, $h_c - h_b$ (f) h_a, t_a, $B - C$
(g) h_a, m_a, $b_a - c_a$ (h) R, $B - C$, $b_a - c_a$
(i) R, $r_a - r$, a (j) h_a, r, r_a

5.13 Constructions Employing Data

A datum may be useful in solving a construction problem if any one part of the datum can be constructed from the other parts. Construct a triangle given (for notation see Problem Study 5.12):

(a) a, A, $h_b + h_c$
(b) $a - b$, $h_b + h_c$, A
(c) R, r, h_a

5.14 Divisions

(a) Complete the details of the following solution (essentially found in Euclid's work *On Divisions*) of the problem of constructing a straight line GH passing through a given point D within triangle ABC, cutting sides BA and BC in G and H, respectively, such that triangles GBH and ABC have the same area (see Figure 44).

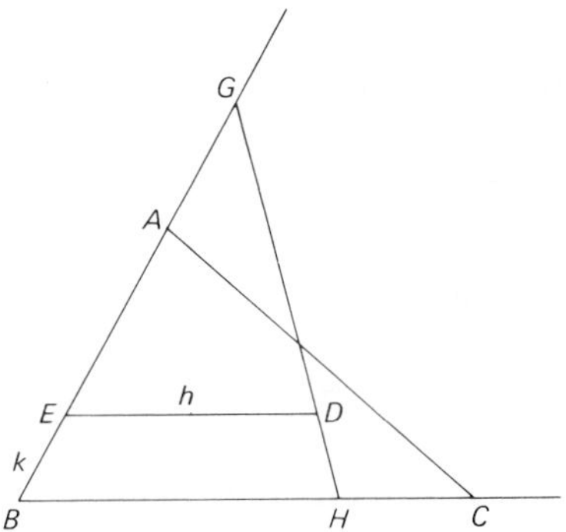

FIGURE 44

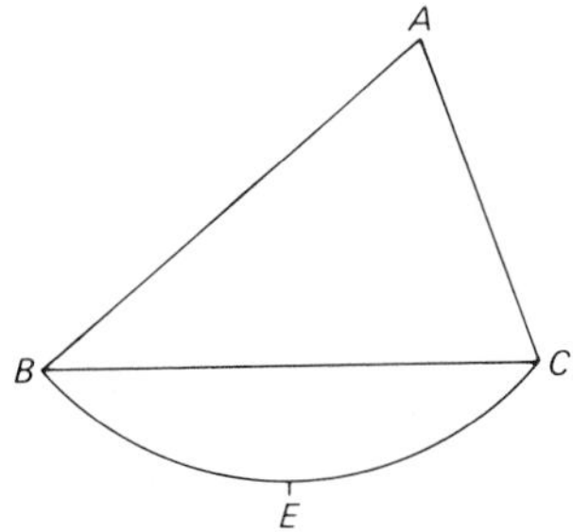

FIGURE 45

Draw DE parallel to CB to cut AB in E. Denote the lengths of DE and EB by h and k, respectively, and that of GB by x. Then $x(BH) = ac$. But $BH/h = x(x - k)$. Eliminating BH, we obtain $x^2 - mx + mk = 0$, where $m = ac/h$, and so on.

(b) Solve the following problem, which is Proposition 28 in Euclid's work *On Divisions:* In Figure 45, bisect the area $ABEC$ by a straight line drawn through the midpoint E of the circular arc BC.

(c) In Euclid's work *On Divisions,* the problem occurs of bisecting the area of a given trapezoid by a line parallel to the bases of the trapezoid. Solve this problem with straightedge and compasses.

Essay Topics

5/1 Origin of the axiomatic method, both evolutionary and revolutionary accounts.

5/2 Aristotle and Proclus on the axiomatic method.

5/3 Material axiomatics versus formal axiomatics.

5/4 The life, works, and influence of Euclid.

5/5 Euclid's sources for his *Elements*.
5/6 Algebra in Euclid's *Elements*.
5/7 Number theory in Euclid's *Elements*.
5/8 Applications of the Eudoxian theory of proportion to plane geometry.
5/9 Is there a royal road in geometry?
5/10 The most famous single utterance in the history of mathematics (Euclid's parallel postulate).
5/11 James Abram Garfield (1831–1881) and mathematics.
5/12 Sir Henry Billingsley.
5/13 Planar generalizations of the Pythagorean theorem.
5/14 De Gua's theorem.

Bibliography

AABOE, ASGER *Episodes from the Early History of Mathematics*. New Mathematical Library, no. 13. New York: Random House and L. W. Singer, 1964.

ARCHIBALD, R. C. *Euclid's Book on Division of Figures*. New York: Cambridge University Press, 1915.

BELL, E. T. *The Magic of Numbers*. New York: McGraw-Hill, 1946.

BUNT, L. N. H.; P. S. JONES; and J. D. BEDIENT *The Historical Roots of Elementary Mathematics*. Englewood Cliffs, N.J.: Prentice-Hall, 1976.

COHEN, M. R., and I. E. DRABKIN *A Source Book in Greek Science*. New York: McGraw-Hill, 1948. Reprinted by Harvard University Press, Cambridge, Mass., 1958.

COOLIDGE, J. L. *A History of Geometrical Methods*. New York: Oxford University Press, 1940.

DANTZIG, TOBIAS *The Bequest of the Greeks*. New York: Charles Scribner, 1955.

DAVIS, H. T. *Alexandria, the Golden City*. 2 vols. Evanston, Ill.: Principia Press of Illinois, 1957.

DUNNINGTON, G. W. *Carl Friedrich Gauss: Titan of Science*. New York: Hafner, 1955.

EVES, HOWARD *Foundations and Fundamental Concepts of Mathematics*. 3rd ed. Boston: PWS-KENT Publishing Company, 1990.

FORDER, H. G. *The Foundations of Euclidean Geometry*. New York: Cambridge University Press, 1927.

FRANKLAND, W. B. *The Story of Euclid*. London: Hodder and Stoughton, 1901.

——— *The First Book of Euclid's Elements with a Commentary Based Principally upon that of Proclus Diadochus*. New York: Cambridge University Press, 1905.

GOW, JAMES *A Short History of Greek Mathematics*. New York: Hafner, 1923. Reprinted by Chelsea, New York.

HEATH, T. L. *History of Greek Mathematics*, vol. 1. New York: Oxford University Press, 1921. Reprinted by Dover, New York, 1981.

——— *A Manual of Greek Mathematics*. New York: Oxford University Press, 1931. Reprinted by Dover, New York, 1963.

——— *The Thirteen Books of Euclid's Elements*. 2d ed., 3 vols. New York: Cambridge University Press, 1926. Reprinted by Dover, New York, 1956.

JAMES, GLENN, ed. *The Tree of Mathematics*. Pacoima, Calif.: The Digest Press, 1957.

KNORR, WILBUR *The Evolution of the Euclidean Elements*. Dordrecht, Holland: D. Reidel, 1985.

PROCLUS *A Commentary on the First Book of Euclid's Elements*. Translated by G. R. Morrow. Princeton, N.J.: Princeton University Press, 1970.

SARTON, GEORGE *Ancient Science and Modern Civilization*. Lincoln, Neb.: The University of Nebraska Press, 1954.

SMITH, D. E. *A Source Book in Mathematics*. New York: McGraw-Hill, 1929.

THOMAS, IVOR, ed. *Selections Illustrating the History of Greek Mathematics*. 2 vols. Cambridge, Mass.: Harvard University Press, 1939–1941.

THOMAS-STANFORD, CHARLES *Early Editions of Euclid's Elements*. London: Bibliographical Society, 1926.

VAN DER WAERDEN, B. L. *Science Awakening*. Translated by Arnold Dresden. New York: Oxford University Press, 1961. Paperback edition. New York: John Wiley, 1963.

Chapter 6

GREEK MATHEMATICS AFTER EUCLID

6–1 Historical Setting

The city of Alexandria enjoyed many advantages, not the least of which was long-lasting peace with the rest of the world. During the reign of the Ptolemies, which lasted for almost 300 years, the city, although on occasion beset with internal power struggles, remained free from external strife. This was ended by a short period of conflict when Egypt became a part of the Roman Empire, after which the Pax Romana settled over the land. It is no wonder that Alexandria became a haven for scholars and that for well over a half millenium so much of ancient scholastic attainment emanated from that city. Almost every mathematician of antiquity to be discussed in this chapter was either a professor or a student at the University of Alexandria.

The closing period of ancient times was dominated by Rome. In 212 B.C., Syracuse yielded to a Roman siege; in 146 B.C., Carthage fell before the power of imperial Rome, and in the same year, the last of the Greek cities, Corinth, also fell, and Greece became a province of the Roman Empire. Mesopotamia was not conquered until 65 B.C., and Egypt remained under the Ptolemies until 30 B.C. Greek civilization diffused through Roman life, and Christianity began to spread, especially among the slaves and the poor. The Roman administrators collected heavy taxes, but otherwise did not interfere with the underlying economic organization of the eastern colonies.

Constantine the Great was the first Roman emperor to embrace Christianity, and he pronounced it the official religion. In A.D. 330, Constantine moved his capital from Rome to Byzantium, which he renamed Constantinople. In A.D. 395, the Roman Empire was divided into the Eastern and the Western Empires, with Greece as a part of the eastern division.

The economic structure of both empires was essentially based on agriculture, with a spreading use of slave labor. The eventual decline of the slave market, with its disastrous effect on Roman economy, found science reduced to a mediocre level. The Alexandrian school gradually faded, along with the breakup of ancient society. Creative thinking gave way to compilation and commentarization. Hectic days followed the fight of Christianity against paganism, and finally, in A.D. 641, Alexandria was taken by the Arabs.

6–2 Archimedes

One of the greatest mathematicians of all time, and certainly the greatest of antiquity, was Archimedes, a native of the Greek city of Syracuse on the island of Sicily. He was born about 287 B.C. and died during the Roman pillage of Syracuse in 212 B.C. He was the son of an astronomer and was in high favor with (perhaps even related to) King Hieron of Syracuse. There is a report that he spent time in Egypt, in all likelihood at the University of Alexandria, for he numbered among his friends Conon, Dositheus, and Eratosthenes; the first two were successors to Euclid, the last was a librarian at the University. Many of Archimedes' mathematical discoveries were communicated to these men.

Roman historians have related many picturesque stories about Archimedes. Among these are the descriptions of the ingenious contrivances devised by Archimedes to aid the defense of Syracuse against the siege directed by the Roman general Marcellus. There were catapults with adjustable ranges, movable projecting poles for dropping heavy weights on enemy ships that approached too near the city walls, and great grappling cranes that hoisted enemy ships from the water. The story that he used large burning glasses to set the enemy's vessels afire is of later origin, but it could be true. There also is the story of how he lent credence to his statement, "Give me a place to stand on and I will move the earth," by effortlessly and singlehandedly moving with a compound-pulley arrangement a heavily weighted ship that had, with difficulty, been drawn up by a large contingent of laborers.

Apparently, Archimedes was capable of strong mental concentration, and tales are told of his obliviousness to surroundings when engrossed by a problem. The frequently told story of King Hieron's crown and the suspected goldsmith is typical. It seems that King Hieron had a goldsmith fashion him a

ARCHIMEDES
(Culver Service)

crown from a given weight of gold. Fearing that the goldsmith may have replaced some of the gold by hidden silver, and not wanting to cut the crown apart to find out, the king referred the matter to Archimedes, who, when in the public baths one day, hit upon a solution by discovering the first law of hydrostatics—that, when immersed in a fluid, a body is buoyed up by a force equal to the weight of the displaced fluid. In his excitement, forgetting to clothe himself, he rose from his bath and ran home through the streets shouting, "Eureka, eureka!" ("I have found it, I have found it!"). He placed the crown on one pan of a balance and an equal weight of gold on the other, and then set the whole thing under water. The pan containing the crown rose, showing that the crown contained some spurious material less dense than gold.

Archimedes worked much of his geometry from figures drawn in the ashes of the hearth or in the after-bathing oil smeared on his body. In fact, it is said that he met his end during the sack of Syracuse, while preoccupied with a diagram drawn on a sand tray. According to one version, he ordered a pillaging Roman soldier to stand clear of his diagram, whereupon the incensed looter ran a spear through the old man.

Because of Archimedes' defense machines, Syracuse resisted the Roman siege for close to three years. The city's defenses were finally broken only when, during a celebration within the city, the overconfident Syracusans relaxed their watches. Marcellus had built up an immense respect for his ingenious adversary, and when he finally managed to breach the city walls, he gave strict orders that no harm must come to the illustrious mathematician. Marcellus' affliction was very great upon hearing of Archimedes' death, and with all due honor and veneration, he buried the famous scholar in the city cemetery. Archimedes, justly proud of one of his great geometrical discoveries (to be described later) had expressed a desire that a figure showing a sphere and a circumscribed right circular cylinder be engraved upon his tombstone. Marcellus saw to it that Archimedes' request was carried out.

Many years later, in 75 B.C., when Cicero was serving as Roman quaestor in Sicily, he inquired as to the whereabouts of Archimedes' tomb. To his surprise, the Syracusans knew nothing of it. With considerable effort, Cicero examined all the monuments in the cemetery, of which there were a great many. Finally, he noticed a small column, standing out a little above overgrown briars and shrubs, with the figure of a sphere and circumscribed cylinder upon it; thus, the long-neglected and forgotten tomb of the greatest of all Syracusans was found. Cicero had men with scythes clear away the brush, and he left orders that the surrounding grounds be thenceforth preserved. How long this respect was kept up we do not know, for again the tomb completely vanished. Then, in 1965, while excavating for the foundations of a hotel in Syracuse, what is thought to be the long-vanished tomb was unexpectedly found once more.

Referring to the death of Archimedes, Sir William Rowan Hamilton once remarked, "Who would not rather have the fame of Archimedes than that of his conqueror Marcellus?". In the same vein, Alfred North Whitehead commented, "No Roman ever died in contemplation over a geometrical diagram." The twentieth-century English mathematician G. H. Hardy said, "Archimedes will be remembered when Aeschylus is forgotten, because languages die and mathematical ideas do not." Voltaire had similarly remarked, "There was more imagination in the head of Archimedes than in that of Homer."

The works of Archimedes are masterpieces of mathematical exposition and, to a remarkable extent, resemble modern journal articles. They are written with a high finish and an economy of presentation and exhibit great originality, computational skill, and rigor in demonstration. About ten treatises have come down to us, and there are various traces of lost works. Probably the most remarkable contribution made to mathematics in these works is the early development of some of the methods of the integral calculus. We shall return to this in a later chapter.

Three of Archimedes' extant works are devoted to plane geometry. They are *Measurement of a Circle, Quadrature of the Parabola,* and *On Spirals.* It was in the first of these that Archimedes inaugurated the classical method of computing π, which we have already described in Section 4–8. In the second work, which contains twenty-four propositions, it is shown that the area of a parabolic segment is four-thirds that of the inscribed triangle having the same base and having its opposite vertex at the point where the tangent is parallel to the base. The summation of a convergent geometric series is involved. The third work contains twenty-eight propositions devoted to properties of the curve today known as the spiral of Archimedes, which has $r = k\theta$ for a polar equation. In particular, the area enclosed by the curve and two radii vectors is found essentially as it would be today in a calculus exercise. There are allusions to many lost works on plane geometry by Archimedes, and there is reason to believe that some of the theorems of these works have been preserved in the *Liber assumptorum,* a collection that has reached us through the Arabians (see Problem Study 6.4) The Arabian scholar al-Biruni claims that Archimedes was the discoverer of the celebrated formula,

$$K = \sqrt{s(s - a)(s - b)(s - c)},$$

for the area of a triangle in terms of its three sides. This formula had hitherto been attributed to Heron of Alexandria.

Two of Archimedes' extant works are devoted to geometry of three dimensions—namely, *On the Sphere and Cylinder* and *On Conoids and Spheroids.* In the first of these, written in two books and containing fifty-three propositions, appear theorems giving the areas of a sphere and of a zone of one base and the volumes of a sphere and of a segment of one base (see Problem Study 6.2). It is shown, for example, that the surface area of a sphere is exactly two-thirds the total surface area of the circumscribed right circular cylinder, and the volume of the sphere is exactly two-thirds the volume of the same cylinder. In Book II of *On the Sphere and Cylinder,* the problem appears of dividing a sphere by a plane into two segments whose volumes shall be in a given ratio. This problem leads to a cubic equation whose solution is not given in the text as it has come down to us, but as it was found by Eutocius in an Archimedean fragment. There is a discussion concerning the conditions under which the cubic may have a real and positive root. Similar considerations do not appear again in European mathematics for over a thousand years. The treatise ends with the two interesting theorems: (1) *If V, V' and S, S' are the volumes of the segments and the areas of the zones into which a sphere is cut by a nondiametral plane, V and S pertaining to the greater piece, then*

$$S^{3/2}:S'^{3/2} < V:V' < S^2:S'^2.$$

(2) *Of all spherical segments of one base having equal zonal areas, the hemisphere has the greatest volume.* The treatise *On Conoids and Spheroids* contains thirty-two propositions that are concerned chiefly with an investigation of the volumes of quadrics of revolution. Pappus has ascribed to Archimedes thirteen semiregular polyhedra, but, unfortunately, Archimedes' own account of them is lost.[1]

Archimedes wrote two related essays on arithmetic, one of which is lost. The extant paper, entitled *The Sand Reckoner,* is addressed to Gelon, son of King Hieron, and applies an arithmetic system for the representation of large numbers to the finding of an upper limit to the number of grains of sand that would fill a sphere with center at the earth and radius reaching to the sun. It is here, among related remarks pertaining to astronomy, that we learn that Aristarchus (ca. 310–230 B.C.) had put forward the Copernican theory of the solar system. In addition to the two arithmetic essays, there is the so-called *Cattle Problem,* which, from a salutation, appears to have been communicated by Archimedes to Eratosthenes. It is a difficult indeterminate problem involving eight integral unknowns connected by seven linear equations and subjected to the two additional conditions that the sum of a certain pair of the unknowns be a perfect square, while the sum of another certain pair be a triangular number. Without the two additional conditions, the smallest values of the unknowns are numbers in the millions, and with the two additional conditions, one of the unknowns must be a number of more than 206,500 digits!

There are two extant treatises by Archimedes on applied mathematics: *On Plane Equilibriums* and *On Floating Bodies*. The first of these is in two books containing twenty-five propositions. Here, following a postulational treatment, are found the elementary properties of centroids and the determination of the centroids of a variety of plane areas, culminating with that of a parabolic segment and of an area bounded by a parabola and two parallel chords. The work *On Floating Bodies* is also in two books, containing nineteen propositions, and is the first application of mathematics to hydrostatics. The treatise, resting on two postulates, first develops those familiar laws of hydrostatics that are encountered in an elementary physics course. It then considers some rather difficult problems, concluding with a remarkable investigation of the positions of rest and stability of a right segment of a paraboloid of revolution floating in a

---

[1] Construction patterns for the Archimedean solids can be found in Miles C. Hartley, *Patterns of Polyhedra*. Rev. ed.

fluid. Archimedes wrote other, now lost, treatises on mathematical physics; Pappus mentions a work *On Levers,* and Theon of Alexandria quotes a theorem from another purported work on the properties of mirrors. It may be that there was originally a larger work by Archimedes of which the two books of *On Plane Equilibriums* formed only a part. It was not until the sixteenth-century work of Simon Stevin that the science of statics and the theory of hydrostatics were appreciably advanced beyond the points reached by Archimedes.

One of the most thrilling discoveries of modern times in the history of mathematics was that by Heiberg, in Constantinople, as late as 1906, of Archimedes' long-lost treatise entitled *Method.* This work is in the form of a letter addressed to Eratosthenes and is important because of the information it furnishes concerning a "method" that Archimedes used in discovering many of his theorems. Although the "method" can today be made rigorous by the modern integration process, Archimedes used the "method" only heuristically to discover results that he then rigorously established by the method of exhaustion. Since the "method" is so closely connected with the ideas of the integral calculus, we reserve a treatment of it until Chapter 11, which is devoted specially to the origin and development of the calculus.

Any student or instructor desirous of adhering strictly to chronological order can, at this point, turn to Section 11–4.

Archimedes has also been credited with a lost work *On the Calendar* and another lost work *On Sphere Making.* In the latter, Archimedes described a planetarium that he constructed to show the motions of the sun, the moon, and the five known planets of his day. The mechanism probably was operated by water. Cicero actually saw the mechanism and gave a description of it. The *Loculus Archimedius,* a teasing puzzle composed of fourteen assorted polygonal pieces to be assembled into a square, in all likelihood was not designed by Archimedes and probably received its name merely as a way of expressing that the puzzle is clever and difficult.

Archimedes' best known mechanical invention is the water-screw, devised by him for irrigating fields, draining marshes, and emptying water from holds of ships. The mechanism is still used in Egypt today.

6–3 Eratosthenes

Eratosthenes was a native of Cyrene, on the south coast of the Mediterranean Sea, and was only a few years younger than Archimedes. He spent many years of his early life in Athens and, at about the age of forty, was invited by Ptolemy III of Egypt to come to Alexandria as tutor to his son and to serve as chief librarian at the University there. It is told that in old age, about 194 B.C., he

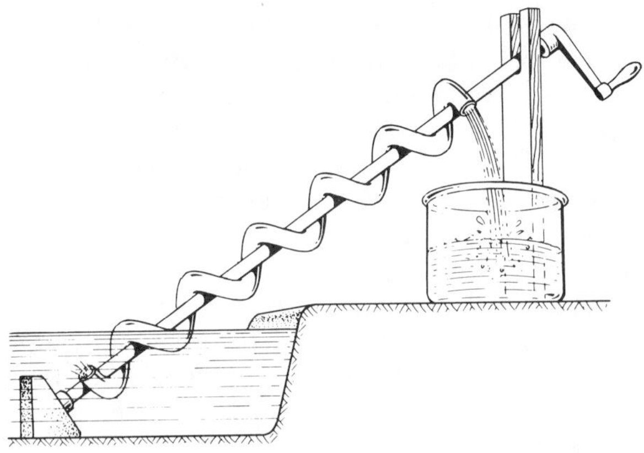

Drawing of an Archimedean water-screw.

became almost blind from ophthalmia and committed suicide by voluntary starvation.

Eratosthenes was singularly gifted in all the branches of knowledge of his time. He was distinguished as a mathematician, an astronomer, a geographer, an historian, a philosopher, a poet, and an athlete. It is said that the students at the University of Alexandria used to call him *Pentathlus,* the champion in five athletic sports. He was also called *Beta,* and some speculation has been offered as to the possible origin of this nickname. Some believe that it was because his broad and brilliant knowledge caused him to be looked upon as a second Plato. A less kind explanation is that, although he was gifted in many fields, he always failed to top his contemporaries in any one branch; in other words, he was always second best. Each of these explanations weakens somewhat when it is learned that a certain astronomer Apollonius (very likely Apollonius of Perga) was called *Epsilon.* Because of this, the historian James Gow has suggested that perhaps Beta and Epsilon arose simply from the Greek numbers (2 and 5) of certain offices or lecture rooms at the University particularly associated with the two men. On the other hand, Ptolemy Hephaestio claimed that Apollonius was called Epsilon because he studied the moon, of which the letter ε was a symbol.

Various of Eratosthenes' works are mentioned by later writers. We have already seen, in Problem Study 4.3(c), his mechanical solution of the duplication problem. His most scientific achievement, the measurement of the earth, is considered in Problem Study 6.1(c).

In arithmetic, Eratosthenes is noted for a device known as the **sieve,** which is used for finding all the prime numbers less than a given number n. One writes down, in order and starting with 3, all the odd numbers less than n. The composite numbers in the sequence are then sifted out by crossing off, from 3, every third number, then from the next remaining number, 5, every fifth number, then from the next remaining number, 7, every seventh number, from the next remaining number, 11, every eleventh number, and so on. In the process

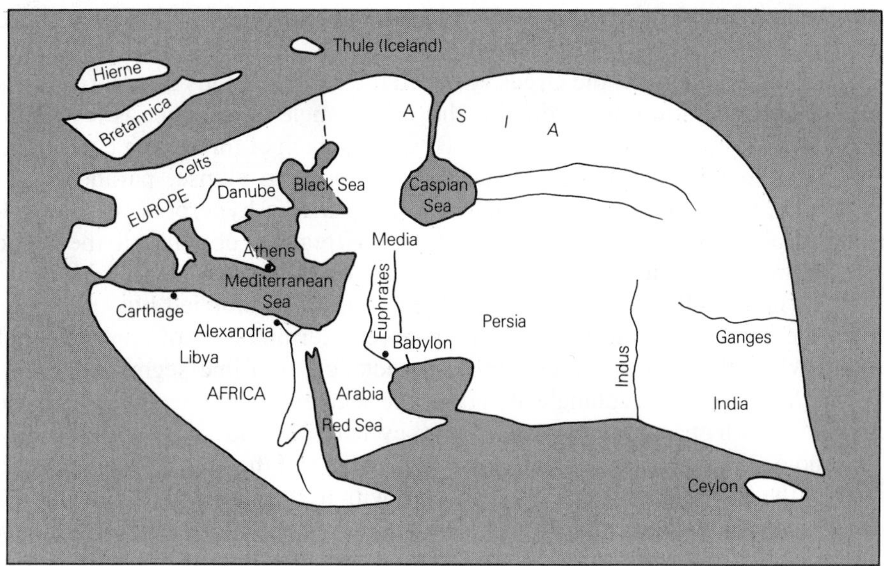

Eratosthenes' map of the world.

some numbers will be crossed off more than once. All the remaining numbers, along with the number 2, constitute the list of primes less than n.

6–4 Apollonius

Euclid, Archimedes, and Apollonius are the three mathematical giants of the third century B.C. Apollonius, who was younger than Archimedes by about twenty-five years, was born about 262 B.C. in Perga, in southern Asia Minor. The little that is known about the life of Apollonius is briefly told. As a young man he went to Alexandria, studied under the successors of Euclid, and remained there for a long time. Later, he visited Pergamum, in western Asia Minor, where there was a recently founded university and library patterned after that at Alexandria. He returned to Alexandria and died there sometime around 190 B.C.

Although Apollonius was an astronomer of note and although he wrote on a variety of mathematical subjects, his chief bid to fame rests on his extraordinary *Conic Sections,* a work that earned him the name, among his contemporaries, of "The Great Geometer." Apollonius' *Conic Sections,* in eight books and containing about 400 propositions, is a thorough investigation of these curves, and completely superseded the earlier works on the subject by Menaechmus, Aristaeus, and Euclid. Only the first seven of the eight books have come down to us, the first four in Greek and the following three from a ninth-century Arabian translation. The first four books, of which I, II, and III are presumably founded on Euclid's previous work, deal with the general elementary theory of conics, whereas the later books are devoted to more specialized investigations.

Prior to Apollonius, the Greeks derived the conic sections from three types of cones of revolution, according as the vertex angle of the cone is less than, equal to, or greater than a right angle. By cutting each of three such cones with a plane perpendicular to an element of the cone, an ellipse, parabola, and hyperbola, respectively, result. Only one branch of a hyperbola was considered. Apollonius, however, in Book I of his treatise, obtains all the conic sections in the now-familiar way from one right or oblique circular *double* cone.

The names **ellipse, parabola,** and **hyperbola** were supplied by Apollonius and were borrowed from the early Pythagorean terminology of application of areas. When the Pythagoreans applied a rectangle to a line segment (that is, placed the base of the rectangle along the line segment, with one end of the base coinciding with one end of the segment), they said they had a case of "ellipsis," "parabole," or "hyperbole"[2] according as the base of the applied rectangle fell short of the line segment, exactly coincided with it, or exceeded it. Now let AB (see Figure 46) be the principal axis of a conic, P any point on the conic, and Q the foot of the perpendicular from P on AB. At A, which is a vertex of the conic, draw a perpendicular to AB and mark off on it a distance AR equal to what we now call the *latus rectum,* or *parameter p,* of the conic. Apply to the segment AR a rectangle having AQ for one side and area equal to $(PQ)^2$. According as the application falls short of, coincides with, or exceeds the segment AR, Apollonius calls the conic an *ellipse,* a *parabola,* or a *hyperbola.* In other words, if we consider the curve referred to a Cartesian coordinate system having its x and y axes along AB and AR, respectively, and if we designate the coordinates of P by x and y, then the curve is an ellipse, parabola, or hyperbola according as $y^2 \lesseqgtr px$. Actually, in the cases of the ellipse and hyperbola,

$$y^2 = px \mp \frac{px^2}{d},$$

where d is the length of the diameter through vertex A. Apollonius derives the bulk of the geometry of the conic sections from the geometrical equivalents of these Cartesian equations. Facts like these cause some to defend the thesis that analytic geometry was an invention of the Greeks.

Book II of Apollonius' treatise on *Conic Sections* deals with properties of asymptotes and conjugate hyperbolas, and the drawing of tangents. Book III contains an assortment of theorems, including some area theorems like: *If the tangents at any two points A and B of a conic intersect in C and also intersect the diameters through B and A in D and E, then triangles CBD and CAE are equal in area.* One also finds the harmonic properties of poles and polars (a

[2] It is perhaps worth noting that we have corresponding names in English for the three figures of speech: ellipsis, parabole, hyperbole.

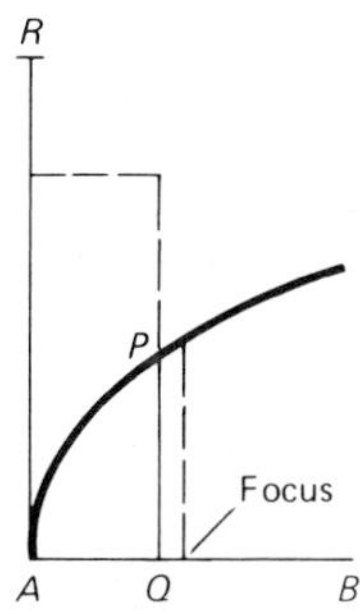

FIGURE 46

subject familiar to those who have had an elementary course in projective geometry), and theorems concerning the product of the segments of intersecting chords. As an example of the latter, there is the theorem (often today referred to as *Newton's theorem*): *If two chords PQ and MN, parallel to two given directions, intersect in O, then* $(PO)(OQ)/(MO)(ON)$ *is a constant independent of the position of O.* The well-known focal properties of the central conics occur toward the end of Book III. In the entire treatise, there is no mention of the focus-directrix property of the conics nor, for that matter, of the focus of the parabola. This is curious because, according to Pappus, Euclid was aware of these properties. The ancient Greeks had no specific name for "focus;" this term was introduced later by Johann Kepler (1571–1630). Book IV of the treatise proves the converses of some of those propositions of Book III concerning harmonic properties of poles and polars. There also are some theorems about pairs of intersecting conics. Book V is the most remarkable and original of the extant books. It treats normals as maximum and minimum line segments drawn from a point to the curve and deals with the construction and enumeration of normals from a given point. The subject is pushed to the point where one can write down the Cartesian equations of the evolutes (envelopes of normals) of the three conics! Book VI contains theorems and construction problems concerning equal and similar conics; thus, it is shown how, in a given right cone, to find a section equal to a given conic. Book VII contains a number of theorems involving conjugate diameters, such as the one about the constancy of the area of the parallelogram formed by the tangents to a central conic at the extremities of a pair of such diameters.

Conic Sections is a great treatise but, because of the extent and elaborateness of the exposition and the portentousness of the statements of many complicated propositions, is rather trying to read. Even from the above brief sketch of contents, we see that the treatise is considerably more complete than the usual present-day college course in the subject.

Pappus has given brief indications of the contents of six other works of Apollonius. These are *On Proportional Sections* (181 propositions), *On Spatial Section* (124 propositions), *On Determinate Section* (83 propositions), *Tangencies* (124 propositions), *Vergings* (125 propositions), and *Plane Loci* (147 prop-

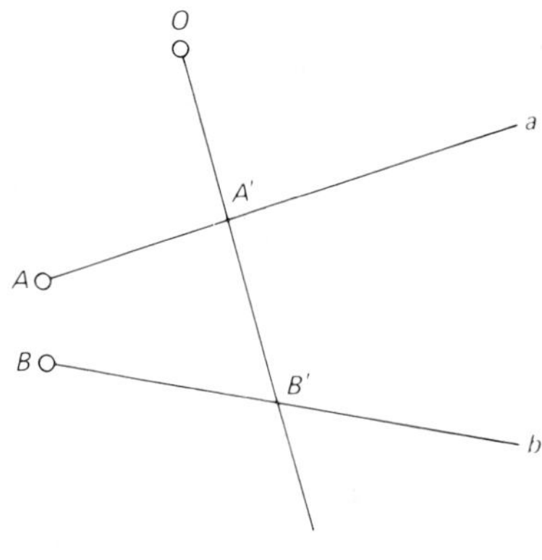

FIGURE 47

ositions). Only the first of these has survived, and this is in Arabic. It deals with the following general problem (see Figure 47): Given two lines a and b with the fixed points A on a and B on b, to draw through a given point O a line $OA'B'$, cutting a in A' and b in B' so that $AA'/BB' = k$, a given constant. The exhaustiveness of the treatment is indicated by the fact that Apollonius considers seventy-seven separate cases. The second work dealt with a similar problem, only here we wish to have $(AA')(BB') = k$. The third work concerned itself with the problem: Given four points A, B, C, D on a line, to find a point P on the line such that we have $(AP)(CP)/(BP)(DP) = k$. The work on *Tangencies* dealt with the problem of constructing a circle tangent to three given circles, where the given circles are permitted to degenerate independently into straight lines or points. This problem, now known as the **problem of Apollonius,** has attracted many mathematicians, among them Viète, Euler, and Newton. One of the first solutions applying the new Cartesian geometry was given by Descartes' pupil Princess Elizabeth, daughter of Frederick V of Bohemia. Probably the most elegant solution is that furnished by the French artillery officer and professor of mathematics Joseph-Diez Gergonne (1771–1859). The general problem in *Vergings* was that of inserting a line segment between two given loci, such that the line of the segment shall pass through a given point.

The last work, *Plane Loci,* contained, among many others, the two theorems:

1. *If A and B are fixed points and k a given constant, then the locus of a point P, such that $AP/BP = k$, is either a circle (if $k \neq 1$) or a straight line (if $k = 1$).*
2. *If $A, B, \ldots$ are fixed points and $a, b, \ldots, k$ are given constants, then the locus of a point P, such that $a(AP)^2 + b(BP)^2 + \ldots = k$, is a circle.*

The circle of the first theorem is known in modern college geometry texts as a **circle of Apollonius.**

Attempts have been made to restore all six of the above works, the first two works by Edmund Halley in 1706, the third by Robert Simson in 1749, the fourth by Viète in 1600, the fifth by Ghetaldi in 1607 and 1613, Alexander Anderson in 1612, and Samuel Horsley in 1770, and the last by Fermat in 1637 and, more completely, by Simson in 1746. In addition to these six works, a number of other lost works by Apollonius are referred to by ancient writers.

6–5 Hipparchus, Menelaus, Ptolemy, and Greek Trigonometry

The origins of trigonometry are obscure. There are some problems in the Rhind papyrus that involve the cotangent of the dihedral angles at the base of a pyramid, and, as we have seen in Section 2–6, the Babylonian cuneiform tablet Plimpton 322 essentially contains a remarkable table of secants. It may be that modern investigations into the mathematics of ancient Mesopotamia will reveal an appreciable development of practical trigonometry. The Babylonian astronomers of the fourth and fifth centuries B.C. had accumulated a considerable mass of observational data, and it is now known that much of this passed on to the Greeks. It was this early astronomy that gave birth to spherical trigonometry.

Probably the most eminent astronomer of antiquity was Hipparchus, who flourished about 140 B.C. Although there is an observation of the vernal equinox recorded by Hipparchus at Alexandria in 146 B.C., his most important observations were made at the famous observatory of the commercial center of Rhodes. Hipparchus was an extremely careful observer and is credited, in astronomy, with such feats as the determination of the length of the mean lunar month to within 1″ of the present accepted value, an accurate calculation of the inclination of the ecliptic, and the discovery and estimation of the annual precession of the equinoxes. He is said also to have computed the lunar parallax, to have determined the perigee and mean motion of the moon, and to have catalogued 850 fixed stars. It was Hipparchus, or perhaps Hipsicles (ca. 180 B.C.), who introduced into Greece the division of a circle into 360°, and he is known to have advocated location of positions on the earth by latitude and longitude. Knowledge of these achievements is secondhand, for almost nothing of Hipparchus' writings has reached us.

More important for us, though, than Hipparchus' achievements in astronomy is the part he played in the development of trigonometry. The fourth-century commentator, Theon of Alexandria, has credited to Hipparchus a twelve-book treatise dealing with the construction of a **table of chords.** A subsequent table, given by Claudius Ptolemy and believed to have been adopted from Hipparchus' treatise, gives the lengths of the chords of all central angles of a given circle by half-degree intervals from 1/2° to 180°. The radius of the circle is divided into sixty equal parts, and the chord lengths then expressed sexagesimally in terms of one of these parts as a unit. Thus, using the symbol crd α to represent the length of the chord of a central angle α, one finds recordings like

$$\text{crd } 36^\circ = 37^p4'55'',$$

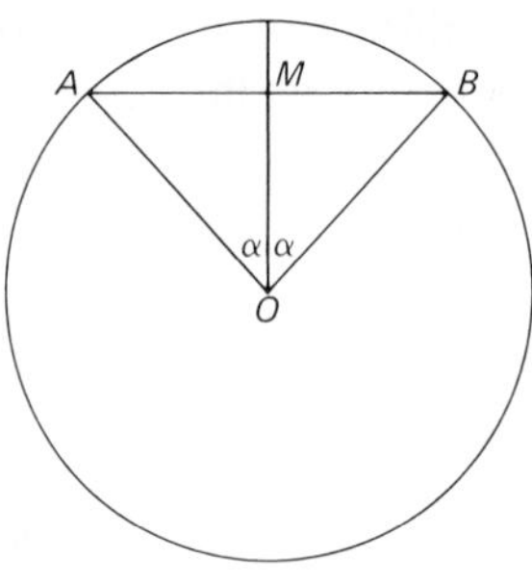

FIGURE 48

meaning, of course, that the chord of a central angle of 36° is equal to 37/60 (or thirty-seven small parts) of the radius, plus 4/60 of one of these small parts, plus 55/3600 more of one of the small parts. It is seen from Figure 48 that a table of chords is equivalent to a table of trigonometric sines, for

$$\sin \alpha = \frac{AM}{OA} = \frac{AB}{\text{diameter of circle}} = \frac{\text{crd } 2\alpha}{120}.$$

Thus, Ptolemy's table of chords gives, essentially, the sines of angles by 15′ intervals, from 0° to 90°. The mode of calculating these chord lengths, elegantly explained by Ptolemy, in all likelihood was known to Hipparchus. Evidence shows that Hipparchus made systematic use of his tables and was aware of the equivalents of several formulas now used in the solution of spherical right triangles.

Theon has also mentioned a six-book treatise on chords in a circle written by Menelaus of Alexandria, a contemporary of Plutarch (ca. A.D. 100). This work, along with a variety of others by Menelaus, is lost to us. Fortunately, however, Menelaus' three-book treatise *Sphaerica* has been preserved in the Arabic. This work throws considerable light on the Greek development of trigonometry. In Book I, there appears for the first time the definition of a *spherical triangle*. The book is devoted to establishing for spherical triangles many of the propositions Euclid established for plane triangles, such as the usual congruence theorems, theorems about isosceles triangles, and so on. In addition, it establishes the congruence of two spherical triangles having the angles of one equal to the angles of the other (for which there is no analogue in the plane) and the fact that the sum of the angles of a spherical triangle is greater than two right angles. Symmetrical spherical triangles are regarded as congruent. Book II contains theorems of interest in astronomy. In Book III is developed the spherical trigonometry of the times, largely deduced from the spherical case of the powerful proposition known to students of college geometry as **Menelaus' theorem:** *If a transversal intersects the sides BC, CA, AB of a triangle ABC in the points L, M, N, respectively, then*

$$\left(\frac{AN}{NB}\right)\left(\frac{BL}{LC}\right)\left(\frac{CM}{MA}\right) = -1.$$

In the spherical analogue, we have a great circle transversal intersecting the sides *BC, CA, AB* of a spherical triangle *ABC* in the points *L, M, N,* respectively. The corresponding conclusion is then equivalent to

$$\left(\frac{\sin \widehat{AN}}{\sin \widehat{NB}}\right)\left(\frac{\sin \widehat{BL}}{\sin \widehat{LC}}\right)\left(\frac{\sin \widehat{CM}}{\sin \widehat{MA}}\right) = -1.$$

The plane case is assumed by Menelaus as well known and is used by him to establish the spherical case. A great deal of spherical trigonometry can be deduced from this theorem by taking special triangles and special transversals. The converses of both the plane and spherical cases of the theorem are also true.

The definitive Greek work on astronomy was written by Claudius Ptolemy of Alexandria about A.D. 150. This very influential treatise, called the *Syntaxis mathematica,* or "Mathematical Collection," was based on the writings of Hipparchus and is noted for its remarkable compactness and elegance. To distinguish it from other lesser works on astronomy, later commentators assigned to it the superlative *magiste,* or "greatest." Still later, the Arabian translators prefixed the Arabian article *al,* and the work has ever since been known as the *Almagest.* The treatise is in thirteen books. Book I contains, among some preliminary astronomical material, the table of chords referred to above, along with a succinct explanation of its derivation from the fertile geometrical proposition now known as **Ptolemy's theorem:** *In a cyclic quadrilateral, the product of the diagonals is equal to the sum of the products of the two pairs of opposite sides* (see Problem Study 6.9). Book II considers phenomena depending on the sphericity of the earth. Books III, IV, and V develop the geocentric system of astronomy by epicycles. In Book IV, a solution appears of the **three-point problem** of surveying: To determine the point from which pairs of three given points are seen under given angles. This problem has had a long history and is sometimes referred to as the "Problem of Snell" (1617) or the "Problem of Pothenot" (1692). In Book VI, which gives the theory of eclipses, is found the four-place value of π alluded to in Section 4–8. Books VII and VIII are devoted to a catalogue of 1028 fixed stars. The remaining books are devoted to the planets. The *Almagest* remained the standard work on astronomy until the time of Copernicus and Kepler.

Ptolemy wrote on map projections (see Problem Study 6.10), optics, and music. He also attempted a derivation of Euclid's fifth (or parallel) postulate from the other axioms and postulates of the *Elements* in a vain effort to remove the postulate from Euclid's list of initial assumptions.

6–6 Heron

Another worker in applied mathematics belonging to this period was Heron of Alexandria. There has been much dispute as to the exact time that he lived, which has been variously estimated from 150 B.C. to A.D. 250. More recently, he has been placed in the second half of the first century A.D. His works on mathematical and physical subjects are so numerous and varied that it is customary to describe him as an encyclopedic writer in these fields. There are reasons to suppose he was an Egyptian with Greek training. At any rate, his writings, which so often aim at practical utility rather than theoretical completeness, show a curious blend of the Greek and the Oriental. He did much to furnish a scientific foundation for engineering and land surveying. Fourteen or so treatises by Heron, some evidently considerably edited, have come down to us, and there are references to additional lost works.

Heron's works may be divided into two classes, the geometric and the mechanical. The geometric works deal largely with problems on mensuration, and the mechanical ones with descriptions of ingenious mechanical devices.

The most important of Heron's geometrical works is his *Metrica,* written in three books and discovered in Constantinople by R. Schöne, as recently as 1896. Book I deals with the area mensuration of squares, rectangles, triangles, trapezoids, various other specialized quadrilaterals, the regular polygons from the equilateral triangle to the regular dodecagon, circles and their segments, ellipses, parabolic segments, and the surfaces of cylinders, cones, spheres, and spherical zones. It is in this book that we find Heron's clever derivation of the famous formula for the area of a triangle in terms of its three sides [see Problem Study 6.11(d)]. Of particular interest, also in this book, is Heron's method of approximating the square root of a nonsquare integer. It is a process frequently used today by computers—namely, if $n = ab$, then $\sqrt{n}$ is approximated by $(a + b)/2$, the approximation improving with the closeness of a to b. The method permits successive approximations. Thus, if a_1 is a first approximation to $\sqrt{n}$, then

$$a_2 = \frac{a_1 + \dfrac{n}{a_1}}{2}$$

is a better approximation, and

$$a_3 = \frac{a_2 + \dfrac{n}{a_2}}{2}$$

is still better, and so on. Book II of the *Metrica* concerns itself with the volume mensuration of cones, cylinders, parallelepipeds, prisms, pyramids, frustums of cones and pyramids, spheres, spherical segments, tori (anchorrings), the five regular solids, and some prismatoids [see Problem Study 6.11(g)]. Book III deals with the division of certain areas and volumes into parts having given ratios to one another. We have seen such problems in Problem Study 3.11(b) and (c).

In Heron's *Pneumatica* appear descriptions of about one hundred machines and toys, such as a siphon, a fire engine, a device for opening temple doors by a fire on the altar, and a wind organ. His work *Dioptra* concerns itself with the description and engineering applications of an ancient form of theodolite, or surveyor's transit. In *Catoptrica,* one finds the elementary properties of mirrors and problems concerning the construction of mirrors to satisfy certain requirements, such as for a person to see the back of his head or to appear upside down, and so on. Heron's works on mechanics show a fine grasp of the important basic principles of the subject.

6–7 Ancient Greek Algebra

In 1842, G. H. F. Nesselmann conveniently characterized three stages in the historical development of algebraic notation. First, we have **rhetorical algebra,** in which the solution of a problem is written, without abbreviations or symbols, as a pure prose argument. Then comes **syncopated algebra,** in which abbreviations are adopted for some of the more frequently recurring quantities and operations. Finally, as the last stage, we have **symbolic algebra,** in which solutions largely appear in a mathematical shorthand made up of symbols having little apparent connection with the entities they represent. It is fairly accurate to say that all algebra prior to the time of Diophantus (who will be considered in Section 6–8) was rhetorical. One of Diophantus' outstanding contributions to mathematics was the syncopation of Greek algebra. Rhetorical algebra, however, persisted quite generally in the rest of the world, with the exception of India, for many hundreds of years. Specifically, in Western Europe, most algebra remained rhetorical until the fifteenth century. Symbolic algebra made its first appearance in Western Europe in the sixteenth century, but did not become prevalent until the middle of the seventeenth century. It is not often realized that much of the symbolism of our elementary algebra textbooks is less than 400 years old.

One of our best sources of ancient Greek algebra problems is a collection known as the *Palatine,* or *Greek, Anthology*. This is a group of forty-six number problems, in epigrammatic form, assembled about A.D. 500 by the grammarian Metrodorus. Although some of the problems may have originated with the author, there is every reason to believe that many of them are considerably more ancient. The problems, apparently intended for mental recreation, are of a type alluded to by Plato and closely resemble some of the problems in the Rhind papyrus. Half of them lead to simple linear equations in one unknown, a dozen more to easy simultaneous equations in two unknowns, one to three equations in three unknowns, and one to four equations in four unknowns. There are also two cases of indeterminate equations of the first degree. A number of the problems are very much like many found in present-day elementary algebra textbooks. Some examples from the *Greek Anthology* are given in Problem Studies 6.13 and 6.14. Although these problems are easily solved with our modern algebraic symbolism, it must be conceded that a rhetorical solution would require close mental attention. It has been remarked that many of these problems can be readily solved by geometrical algebra, but it is believed that they were actually solved arithmetically, perhaps by applying the *rule of false position* (see Section 2–8). Just when Greek algebra changed from a geometrical form to an arithmetic one is not known, but this probably occurred as early as the time of Euclid.

6–8 Diophantus

Of tremendous importance to the development of algebra and of great influence on later European number theorists was Diophantus of Alexandria. Diophantus was another mathematician, like Heron, of uncertain date and nationality. Although there is some tenuous evidence that he may have been a contemporary, or near contemporary, of Heron, most historians tend to place him in the third century of our era. Beyond the fact that he flourished at Alexandria, nothing certain is known about him, although there is an epigram in the *Greek Anthology* that purports to give some details of his life [see Problem Study 6.15(a)].

Diophantus wrote three works: *Arithmetica,* his most important one, of which the first six of thirteen books are extant; *On Polygonal Numbers,* of which only a fragment is extant; and *Porisms,* which is lost. The *Arithmetica* had many commentators, but it was Regiomontanus who in 1463 called for a Latin translation of the extant Greek text. A very meritorious translation, with commentary, was made in 1575 by Xylander (the Greek name assumed by Wilhelm Holzmann, a professor at the University of Heidelberg). This was used in turn by the Frenchman Bachet de Méziriac, who in 1621 published the first edition of the Greek text, along with a Latin translation and notes. A second, carelessly printed, edition was brought out in 1670, which is historically important because it contained Fermat's famous marginal notes, which stimulated such extensive number theory research. French, German, and English translations appeared later.

The *Arithmetica* is an analytic treatment of algebraic number theory and marks the author as a genius in this field. The extant portion of the work is devoted to the solution of about 130 problems, of considerable variety, leading to equations of the first and second degree. One very special cubic is solved. The first book concerns itself with determinate equations in one unknown, and the remaining books with indeterminate equations of the second, and sometimes higher, degree in two and three unknowns. Striking is the lack of general methods and the repeated application of ingenious devices designed for the needs of each individual problem. Diophantus recognized only positive rational answers and was, in most cases, satisfied with only one answer to a problem.

There are some penetrating number theorems stated in the *Arithmetica;* thus, we find, without proof but with an allusion to the *Porisms,* that *the difference of two rational cubes is also the sum of two rational cubes*—a matter that was later investigated by Viète, Bachet, and Fermat. There are many propositions concerning the representation of numbers as the sum of two, three, or four squares, a field of investigation later completed by Fermat, Euler, and Lagrange. Perhaps it might be interesting to list a few of the problems found in the *Arithmetica;* they are all alluring, and some of them are challenging. It must be borne in mind that by "number" is meant "positive rational number."

Problem 28,[3] Book II: Find two square numbers such that their product added to either gives a square number. (Diophantus' answer: $(\frac{3}{4})^2$, $(\frac{7}{24})^2$.)

Problem 6, Book III: Find three numbers such that their sum is a square and the sum of any pair is a square. (Diophantus' answer: 80, 320, 41.)

Problem 7, Book III: Find three numbers in arithmetic progression such that the sum of any pair is a square. (Diophantus' answer: $120\frac{1}{2}$, $840\frac{1}{2}$, $1560\frac{1}{2}$.)

Problem 13, Book III: Find three numbers such that the product of any two added to the third is a square. [See Problem Study 6.16(d).]

Problem 15, Book III: Find three numbers such that the product of any two added to the sum of these two is a square. [See Problem Study 6.16(d).]

Problem 10, Book IV: Find two numbers such that their sum is equal to the sum of their cubes. (Diophantus' answer: $\frac{5}{7}$, $\frac{8}{7}$.)

Problem 21, Book IV: Find three numbers in geometric progression such that the difference of any two is a square number. (Diophantus' answer: $\frac{81}{7}$, $\frac{144}{7}$, $\frac{256}{7}$.)

Problem 1, Book VI: Find a Pythagorean triangle in which the hypotenuse minus each of the legs is a cube. (Diophantus' answer: 40, 96, 104).

Problem 16, Book VI: Find a Pythagorean triangle in which the length of the bisector of one of the acute angles is rational. [See Problem Study 6.15(c).]

Indeterminate algebraic problems in which one must find only the rational solutions have become known as **Diophantine problems.** In fact, modern usage of the terminology often implies the restriction of the solutions to integers. Diophantus did not originate problems of this sort, however. Also, he was not, as is sometimes stated, the first to work with indeterminate equations, or the first to solve quadratic equations nongeometrically. He may have been, however, the first to take steps towards an algebraic notation. These steps were in the nature of stenographic abbreviations.

Diophantus had abbreviations for the unknown, powers of the unknown up through the sixth, subtraction, equality, and reciprocals. Our word "arithmetic" comes from the Greek word *arithmetike,* a compound of the words *arithmos* for "number" and *techne* for "science." It has been rather convincingly pointed out by Heath that Diophantus' symbol for the unknown was probably derived by merging the first two Greek letters, α and ρ, of the word *arithmos*. This came, in time, to look like the Greek final sigma ς. Although there is doubt about this, the meaning of the notation for powers of the un-

[3] The numbering of the problems is that assigned to them in T. L. Heath's *Diophantus of Alexandria*. 2d ed.

known is quite clear; thus, "unknown squared" is denoted by Δ^{Y}, the first two letters of the Greek word *dunamis* (ΔΥΝΑΜΙΣ) for "power." Again, "unknown cubed" is denoted by K^{Y}, the first two letters of the Greek word *kubos* (ΚΥΒΟΣ) for "cube." Explanations are easily furnished for the succeeding powers of the unknown, $\Delta^{Y}\Delta$ (square-square), ΔK^{Y} (square-cube), and $K^{Y}K$ (cube-cube). Diophantus' symbol for "minus" looks like an inverted V with the angle bisector drawn in. This has been explained as a compound of Λ and Ι, letters in the Greek word *leipis* (ΛΕΙΨΙΣ) for "lacking." All negative terms in an expression are gathered together and preceded by the minus symbol. Addition is indicated by juxtaposition, and the coefficient of any power of the unknown is represented by the alphabetic Greek numeral (see Section 1–6) following the power symbol. If there is a constant term, then $\mathring{M}$, an abbreviation of the Greek word *monades* (ΜΟΝΑΔΕΣ), for "units," is used, with the appropriate number coefficient. Thus, $x^3 + 13x^2 + 5x$ and $x^3 - 5x^2 + 8x - 1$ would appear as

$$K^{Y}\alpha\Delta^{Y}\iota\gamma\varsigma\varepsilon \qquad \text{and} \qquad K^{Y}\alpha\varsigma\eta \pitchfork \Delta^{Y}\varepsilon\mathring{M}\alpha,$$

which can be read literally as

unknown cubed 1, unknown squared 13, unknown 5

and

(unknown cubed 1, unknown 8) minus (unknown squared 5, units 1).

It is thus that rhetorical algebra became syncopated algebra.

6–9 Pappus

The immediate successors to Euclid, Archimedes, and Apollonius prolonged the great Greek geometric tradition for a time, but then it began steadily to languish, and new developments were limited to astronomy, trigonometry, and algebra. Then, toward the end of the third century A.D., 500 years after Apollonius, there lived the enthusiastic and competent Pappus of Alexandria, who strove to rekindle fresh interest in the subject.

Pappus wrote commentaries on Euclid's *Elements* and *Data* and on Ptolemy's *Almagest* and *Planispherium,* but about all we know of these is through their influence on the writings of later commentators. Pappus' really great work is his *Mathematical Collection,* a combined commentary and guidebook of the existing geometrical works of his time, with numerous original propositions, improvements, extensions, and historical comments. Of the eight books, the first and part of the second are lost.

Judging from what remains, Book II of the *Mathematical Collection* dealt with a method developed by Apollonius for writing and working with large numbers. Book III is in four sections, the first two dealing with the theory of means [see, for example, Problem Study 6.17(a)], with attention given to the problem of inserting two mean proportionals between two given line segments, the third with some inequalities in a triangle, and the fourth with the inscription of the five regular polyhedra in a given sphere.

In Book IV is found Pappus' extension of the Pythagorean theorem [given in Problem Study 6.17(c)], the "ancient proposition" on the arbelos (stated at the end of Problem Study 6.4), the description, genesis, and some properties of the spiral of Archimedes, the conchoid of Nicomedes, and the quadratrix of Dinostratus, with applications to the three famous problems, and a discussion of a special spiral drawn on a sphere.

Book V is largely devoted to **isoperimetry,** or the comparison of the areas of figures having equal bounding perimeters, and of volumes of solids having equal bounding areas. This book also contains an interesting passage on bees and the maximum-minimum properties of the cells of their honey-combs. It is in this book that we find Pappus' reference, mentioned in Section 6–2, to the thirteen semiregular polyhedra of Archimedes. Book VI is on astronomy and deals with the treatises that were to be studied as an introduction to Ptolemy's *Almagest.*

Book VII is historically very important, for it gives an account of the works constituting *The Treasury of Analysis,* a collection that, after Euclid's *Elements,* purported to contain the material considered as essential equipment for the professional mathematician. The twelve treatises discussed are Euclid's *Data, Porisms,* and *Surface Loci;* Apollonius' *Conic Sections* and the six works considered toward the end of Section 6–4; Aristaeus' *Solid Loci,* and Eratosthenes' *On Means.* In this book, we find an anticipation of the centroid theorem of P. Guldin (see Problem Study 6.18). Also, a discussion is given of the famous "loci with respect to three or four lines": *If* p_1, p_2, p_3, p_4 *are the lengths of line segments drawn from a point P to four given lines, making given angles with these lines, and if* $p_1p_2 = kp_3^2$, *or* $p_1p_2 = kp_3p_4$, *where k is a constant, then the locus of P is a conic section.* This problem, solved by Apollonius, is historically important because in attempting to generalize it to *n* lines, Descartes was led in 1637 to formulate the method of coordinates; Pappus' contemporaries had unsuccessfully tried to generalize the problem. The linear case of the so-called **Stewart's theorem,** appearing in college geometry texts, is also found in this book—namely, *if A, B, C, D are any four points on a line, then*

$$(AD)^2(BC) + (BD)^2(CA) + (CD)^2(AB) + (BC)(CA)(AB) = 0,$$

where the segments involved are signed segments. Actually, Robert Simson anticipated Stewart in the discovery of the theorem for the more general case where *D* may be outside the line *ABC*. The **anharmonic,** or **cross, ratio** (AB,CD) of four collinear points *A, B, C, D* may be defined as $(AC/CB)/(AD/DB)$, that

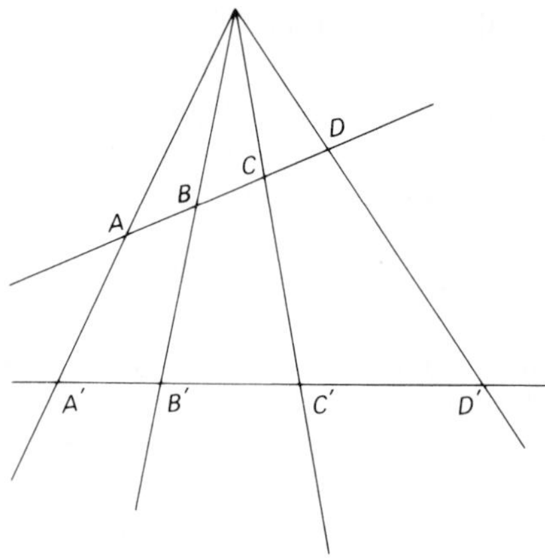

FIGURE 49

is, as the ratio of the ratios into which C and D divide the segment AB. In Book VII of the *Mathematical Collection,* Pappus proves that if four concurrent rays (see Figure 49) are cut by two transversals, giving the corresponding ranges A, B, C, D and A', B', C', D', then the two cross ratios (AB,CD) and $(A'B',C'D')$ are equal. In other words, the cross ratio of four collinear points is invariant under projection. This is a fundamental theorem of projective geometry. Book VII contains a solution of the problem: To inscribe in a given circle a triangle whose sides, produced if necessary, shall pass through three given collinear points. This has become known as the **Castillon-Cramer problem** because in the eighteenth century the problem was generalized by Cramer to the case where the three points need not be collinear, and a solution of this generalization was published by Castillon in 1776. Solutions were also given by Lagrange, Euler, Lhuilier, Fuss, and Lexell in 1780. A few years later, a gifted Italian lad of sixteen, named Giordano, generalized the problem to that of inscribing in a circle an n-gon whose sides shall pass through n given points, and he furnished an elegant solution. Poncelet extended the problem still further by replacing the circle with an arbitrary conic section. In Book VII also occurs the first recorded statement of the focus-directrix property of the three conic sections.

Book VIII, like Book VII, contains much that was probably original with Pappus. Here we find a solution of the problem of constructing a conic through five given points. An interesting proposition, probably due to Pappus and found in this book, is given in Problem Study 6.17(e).

Pappus' *Mathematical Collection* is a veritable mine of rich geometric nuggets. Comparisons, where possible, have shown the historical comments contained in the work to be trustworthy. We owe much of our knowledge of Greek geometry to this great treatise, which cites from or refers to the works of over thirty different mathematicians of antiquity. It may be called the requiem, or the swan song, of Greek geometry.

6–10 The Commentators

After Pappus, Greek mathematics ceased to be a living study, and we find merely its memory perpetuated by minor writers and commentators. Among these were Theon of Alexandria, his daughter Hypatia, Proclus, Simplicius, and Eutocius.

Theon lived in the turbulent closing period of the fourth century A.D. and was the author of a commentary, in eleven books, on Ptolemy's *Almagest*. Also, it will be recalled, the modern editions of Euclid's *Elements* are based upon Theon's revision of the original work.

Theon's daughter, Hypatia, was distinguished in mathematics, medicine, and philosophy, and wrote commentaries on Diophantus' *Arithmetica* and Apollonius' *Conic Sections*. She is the first woman mathematician to be mentioned in the history of mathematics. Her life and barbarous murder by a mob of fanatical Christians in March, 415, are reconstructed in Charles Kingsley's novel.[4]

Hypatia was taught by her father, who held an administrative post at the University of Alexandria. She traveled a number of years and then lectured on mathematics and philosophy, perhaps at the University or maybe in public in Alexandria. Her lectures attracted wide attendance and praise. Among her auditors was Synesius of Cyrene (later to become bishop of Ptolemais), who became one of her chief friends and admirers. Most of her writings are now lost, but a copy of her commentary on Diophantus was discovered in the Vatican library in the fifteenth century. She assisted her father in the latter's revision of Euclid's *Elements*. She never married, being, as she claimed, "wedded to truth."

As a leader of the neo-Platonic school of philosophy, Hypatia played a prominent role in the defense of paganism against Christianity. This aroused the ire of a new patriarch, Cyril of Alexandria, who, with frenzied zeal, opposed and oppressed all "heretics." The fact that Hypatia was a student of several religions particularly roused Cyril's wrath, and one day, as Hypatia was driving home, he had her dragged from her chariot, her hair pulled out, her flesh scraped from her bones with oyster shells, and the remnants of her body consigned to flame. In this way, the creative days of the famed University of Alexandria came to an end.

Historians of mathematics are indebted to the neo-Platonic philosopher and mathematician Proclus for his *Commentary on Euclid, Book I,* one of our principal sources of information on the early history of elementary geometry. Proclus had access to historical and critical works (or commentaries on such works) now lost to us, chief of which were Eudemus' *History of Geometry* in four books and Geminus' apparently comprehensive *Theory of the Mathematical Sciences*. Proclus' commentary on Plato's *Republic* also contains passages

[4] *Hypatia, or New Foes with an Old Face*. New York: E. P. Dutton, 1907.

of interest to the historian of mathematics. Proclus studied at Alexandria, became head of the Athenian school, and died in Athens in 485 when he was about seventy-five years old.

A debt is also owed to Simplicius, the commentator of Aristotle. He has given us accounts of Antiphon's attempt to square the circle, of the lunes of Hippocrates, and of a system of concentric spheres invented by Eudoxus to explain the apparent motions of the members of the solar system. He also wrote a commentary on the first book of Euclid's *Elements,* from which Arabian extracts were later made. Simplicius lived in the first half of the sixth century and studied at both Alexandria and Athens.

Probably contemporary with Simplicius was Eutocius, who wrote commentaries on Archimedes' *On the Sphere and Cylinder, Measurement of a Circle,* and *On Plane Equilibriums,* and on Apollonius' *Conic Sections.*

The Athenian school struggled on against growing opposition from the Christians until the latter finally, in A.D. 529, obtained a decree from Emperor Justinian that closed the doors of the school forever. Simplicius and some of the other philosophers and scientists fled to Persia, where they were well received by King Khosrau I, establishing there what could be called the Athenian academy of Persia. The seeds of Greek science then flourished centuries later under Muslim patronage.[5]

The school at Alexandria fared little better at the hands of the Christians than did the Athenian school, although it was, at least, partly in existence when Alexandria fell to the Arabs in 641. The Arabs then put the torch to what the Christians had left. The long and glorious history of Greek mathematics came to an end.

Problem Studies

6.1 Measurements by Aristarchus and Eratosthenes

Aristarchus of Samos (ca. 287 B.C.) applied mathematics to astronomy. Since he put forward the heliocentric hypothesis of the solar system, he has become known as the Copernicus of antiquity.

(a) Using crude instruments, Aristarchus observed that the angular distance between the moon, when at first quadrant, and the sun is $\frac{29}{30}$ of a right angle. On the basis of this measurement, he showed (without benefit of trigonometry) that the distance from the earth to the sun is between 18 and 20 times the distance from the earth to the moon. Verify this, using the result of Aristarchus' observation. (The angle concerned is actually about $89^\circ 50'$.)

[5] See George Sarton, *The History of Science,* vol. 1, p. 400.

(b) Aristarchus, in his tract *On Sizes and Distances of the Sun and Moon,* used the equivalent of the fact that

$$\frac{\sin a}{\sin b} < \frac{a}{b} < \frac{\tan a}{\tan b},$$

where $0 < b < a < \pi/2$. From a knowledge of the graphs of the functions $\sin x$ and $\tan x$, show that $(\sin x)/x$ *decreases* and $(\tan x)/x$ *increases* as x increases from 0 to $\pi/2$, and thus establish the above inequalities.

(c) Eratosthenes, in 240 B.C., made a famous measurement of the earth. He observed at Syene, at noon and at the summer solstice, that a vertical stick had no shadow, while at Alexandria (which he believed to be on the same meridian with Syene) the sun's rays were inclined $\frac{1}{50}$ of a complete circle to the vertical. He then calculated the circumference of the earth from the known distance of 5000 stades between Alexandria and Syene. Obtain Eratosthenes' result of 250,000 stades for the circumference of the earth. There is reason to suppose that an Eratosthenian stade was equal to about 559 feet. Assuming this, calculate from the above result the polar diameter of the earth in miles. (The actual polar diameter of the earth to the nearest mile is 7900 miles.)

6.2 On The Sphere and Cylinder

(a) Verify the following two results established by Archimedes in his work *On the Sphere and Cylinder:*

1. The volume of a sphere is $\frac{2}{3}$ that of the circumscribed cylinder.
2. The area of a sphere is $\frac{2}{3}$ of the total area of the circumscribed cylinder.

(b) Define *spherical zone* (of one and two bases), *spherical segment* (of one and two bases), and *spherical sector.*

(c) Assuming the theorem: *The area of a spherical zone is equal to the product of the circumference of a great circle and the altitude of the zone,* obtain the familiar formula for the area of a sphere, and establish the theorem: *The area of a spherical zone of one base is equal to that of a circle whose radius is the chord of the generating arc.*

(d) Assuming that the volume of a spherical sector is given by one-third the product of the area of its base and the radius of the sphere, obtain the following results:

1. The volume of a spherical segment of one base, cut from a sphere of radius R, having h as altitude and a as the radius of its base, is given by

$$V = \pi h^2\left(R - \frac{h}{3}\right) = \pi h\left(\frac{3a^2 + h^2}{6}\right).$$

2. The volume of a spherical segment of two bases, having h as altitude and a and b as the radii of its bases, is given by

$$V = \frac{\pi h(3a^2 + 3b^2 + h^2)}{6}.$$

3. The spherical segment of the second result is equivalent to the sum of a sphere of radius $h/2$ and two cylinders whose altitudes are each $h/2$ and whose radii are a and b, respectively.

(e) In Book II of *On the Sphere and Cylinder,* Archimedes considers the problem of cutting a given sphere by a plane so that the volumes of the two segments formed shall be in a given ratio. Show that, in modern notation, this leads to the cubic equation

$$n(R - x)^2(2R + x) = m(R + x)^2(2R - x),$$

where R is the radius of the sphere, x is the distance of the cutting plane from the center of the sphere, and $m/n < 1$ is the given ratio.

(f) Show how, with two parallel planes, to divide the surface area of a given sphere into three equal areas.

6.3 The Problem of the Crown

Proposition 7 of the first book of Archimedes' work, *On Floating Bodies,* is the famous law of hydrostatics: *A body immersed in a fluid is buoyed up by a force equal to the weight of the displaced fluid.*

(a) Let a crown of weight w pounds be made up of w_1 pounds of gold and w_2 pounds of silver. Suppose that w pounds of pure gold loses f_1 pounds when weighed in water, that w pounds of pure silver loses f_2 pounds when weighed in water, and that the crown loses f pounds when weighed in water. Show that

$$\frac{w_1}{w_2} = \frac{f_2 - f}{f - f_1}.$$

(b) Suppose the crown of (a) displaces a volume of v cubic inches when immersed in water, and that lumps of pure gold and pure silver that are of the same weight as the crown displace, respectively, v_1 and v_2 cubic inches when immersed in water. Show that

$$\frac{w_1}{w_2} = \frac{v_2 - v}{v - v_1}.$$

6.4 The Arbelos and the Salinon

The *Liber assumptorum,* or *Book of Lemmas,* which has been preserved in an Arabic version, contains some elegant geometrical theorems credited to Archimedes. Among them are some properties of the **arbelos** or "shoemaker's

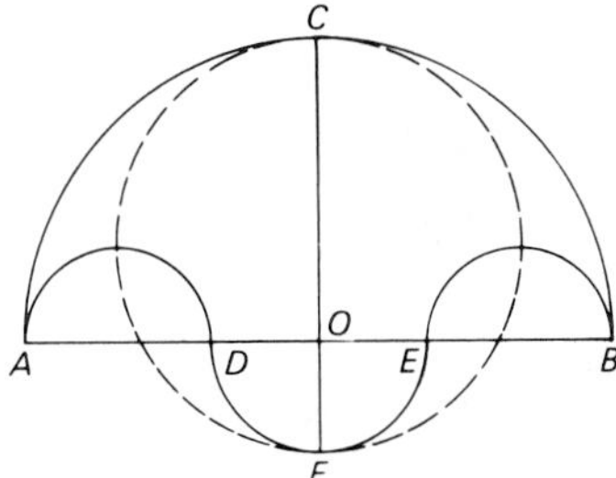

FIGURE 50

knife.'' Let A, C, B be three points on a straight line, C lying between A and B. Semicircles are drawn on the same side of the line and having AC, CB, AB as diameters. The arbelos is the figure bounded by these three semicircles. At C, erect a perpendicular to AB to cut the largest semicircle in G. Let the common external tangent to the two smaller semicircles touch these curves at T and W. Denote AC, CB, AB by $2r_1$, $2r_2$, $2r$. Establish the following elementary properties of the arbelos:

(a) GC and TW are equal and bisect each other.

(b) The area of the arbelos equals the area of the circle on GC as diameter.

(c) The lines GA and GB pass, respectively, through T and W.

The arbelos has many properties not so easily established. For example, it is alleged that Archimedes showed that the circles inscribed in the curvilinear triangles ACG and BCG are equal, the diameter of each being r_1r_2/r. The smallest circle that is tangent to and circumscribes these two circles is equal to the circle on GC, and therefore equal in area to the arbelos. Consider, in the arbelos, a chain of circles $c_1, c_2, \ldots$, all tangent to the semicircles on AB and AC, where c_1 is also tangent to the semicircle on BC, c_2 to c_1, and so on. Then, if r_n represents the radius of c_n, and h_n the distance of its center from ACB, we have $h_n = 2nr_n$. This last proposition is found in Book IV of Pappus' *Mathematical Collection* and is there referred to as an ''ancient proposition.''

(d) Proposition 14 of the *Liber assumptorum* concerns a figure called the **salinon** (''salt cellar''), which is pictured in Figure 50, wherein semicircles are described on the segments AB, AD, DE, and EB as diameters, with $AD = EB$. The proposition asserts that the total area of the salinon (which is bounded entirely by semicircular arcs) is equal to the area of the circle having for diameter the line FOC of symmetry of the figure. Prove this.

6.5 The Theorem of the Broken Chord

The Arabian scholar Abu'l Raihan al-Biruni (973–1048) has attributed to Archimedes the **theorem of the broken chord,** which asserts that if, as shown in Figure 51, AB and BC make up a broken chord in a circle, where $BC > AB$, and

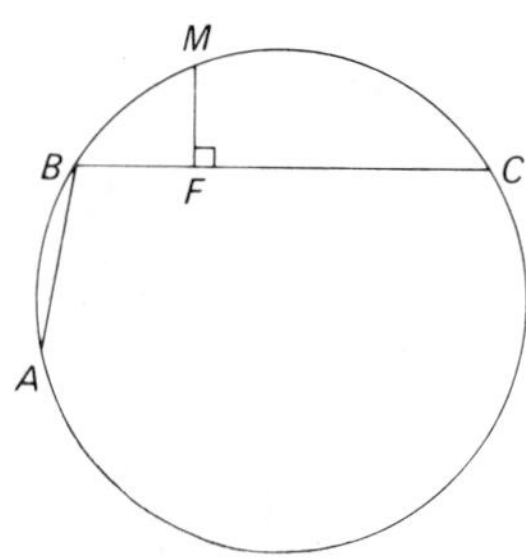

FIGURE 51

if M is the midpoint of arc ABC, the foot F of the perpendicular from M on BC is the midpoint of the broken chord ABC.

(a) Prove the theorem of the broken chord.

(b) Setting arc $MC = 2x$ and arc $BM = 2y$, successively show that $MC = 2 \sin x$, $BM = 2 \sin y$, $AB = 2 \sin(x - y)$, $FC = 2 \sin x \cos y$, $FB = 2 \sin y \cos x$. Now show that the theorem of the broken chord yields the identity

$$\sin(x - y) = \sin x \cos y - \sin y \cos x.$$

(c) Using the theorem of the broken chord, obtain the identity

$$\sin(x + y) = \sin x \cos y + \sin y \cos x.$$

6.6 The Focus-Directrix Property

(a) Although the Greeks defined the conic sections as sections of cones, it is customary in college courses in analytic geometry to define them by the focus-directrix property. Establish the following lemma (1) and then complete the simple proof in (2) that any section of a right circular cone possesses the focus-directrix property.

1. The lengths of any two line segments from a point to a plane are inversely proportional to the sines of the angles that the line segments make with the plane.
2. Denote the plane of the section of the right circular cone by p. Let a sphere touch the cone along a circle whose plane we shall call q, and also touch plane p at point F (see Figure 52). Let planes p and q intersect in line d. From P, any point on the conic section, drop a perpendicular PR on line d. Let the element of the cone through P cut plane q in point E. Finally, let α be the angle between planes p and q, and β the angle an element of the cone makes with plane q. Show that $PF/PR = PE/PR = (\sin \alpha)/(\sin \beta) = e$, a constant. Thus, F is a focus, d the corresponding directrix, and e the eccentricity of the conic section. [This simple and elegant approach was discovered

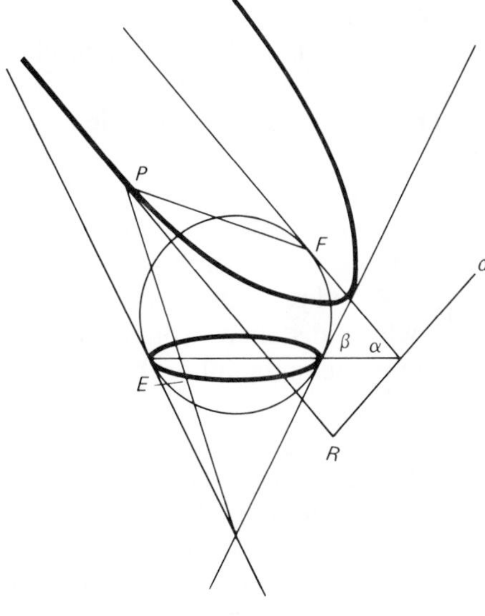

FIGURE 52

around the first quarter-mark of the nineteenth century by the two Belgian mathematicians Adolphe Quetelet (1796–1874) and Germinal Dandelin (1794–1847).]

(b) Show that if p cuts every element of one nappe of the cone, then $e < 1$; if p is parallel to 1 and only 1 element of the cone, then $e = 1$; if p cuts both nappes of the cone, then $e > 1$.

6.7 Tangencies

In his lost treatise on *Tangencies,* Apollonius considered the problem of drawing a circle tangent to three given circles A, B, C, where each of A, B, C may independently assume either of the degenerate forms of point or straight line. This problem has become known as the **problem of Apollonius.**

(a) Show that there are 10 cases of the problem of Apollonius, depending on whether each of A, B, C is a point, a line, or a circle. What is the number of solutions for each general case?

(b) Solve the problem where A, B, C are 2 points and a line.

(c) Reduce the problem where A, B, C are 2 lines and a point to the case of (b).

(d) The focus and directrix of a parabola p, and a line m are given. With Euclidean tools, find the points of intersection of p and m.

6.8 Problems from Apollonius

(a) Solve the following easy verging problem considered by Apollonius in his work *Vergings:* In a given circle, insert a chord of given length and verging to a given point.

A more difficult verging problem considered by Apollonius is the following: Given a rhombus with one side produced, insert a line segment of given length in the exterior angle so that it verges to the opposite vertex. Several solutions to this problem were furnished by Huygens (1629–1695).

(b) Establish, by analytic geometry, the two problems (1) and (2) stated in Section 6–4 in connection with Apollonius' work *Plane Loci.*

(c) Establish synthetically the first problem in (b) and also the following special case of the second problem in (b): The locus of a point, the sum of the squares of whose distances from 2 fixed points is constant, is a circle whose center is the midpoint of the segment joining the 2 points.

6.9 Ptolemy's Table of Chords

(a) Prove Ptolemy's theorem: *In a cyclic quadrilateral, the product of the diagonals is equal to the sum of the products of the pairs of opposite sides.*

(b) Derive, from Ptolemy's theorem, the following relations:

1. If a and b are the chords of 2 arcs of a circle of unit radius, then

$$s = \frac{a}{2}(4 - b^2)^{1/2} + \frac{b}{2}(4 - a^2)^{1/2}$$

is the chord of the sum of the 2 arcs.

2. If a and b, $a \geqq b$, are the chords of 2 arcs of a circle of unit radius, then

$$d = \frac{a}{2}(4 - b^2)^{1/2} - \frac{b}{2}(4 - a^2)^{1/2}$$

is the chord of the difference of the 2 arcs.

3. If t is the chord of an arc of a circle of unit radius, then

$$s = \{2 - (4 - t^2)^{1/2}\}^{1/2}$$

is the chord of half the arc.

In a circle of unit radius crd $60° = 1$, and one may show that crd $36°$ = larger segment of the radius when divided in golden section [see Problem Study 3.10(d)] $= 0.6180$. By (2), crd $24°$ = crd $(60° - 36°)$ $= 0.4158$. By (3), we may calculate the chords of $12°$, $6°$, $3°$, $90'$, $45'$, obtaining crd $90' = 0.0262$ and crd $45' = 0.0131$. By Problem Study 6.1(b), crd $60'$/crd $45' < 60/45 = 4/3$, or crd $1° < (4/3)(0.0131) = 0.0175$. Also, crd $90'$/crd $60' < 90/60 = 3/2$, or crd $1° > (2/3)(0.0262) = 0.0175$. Therefore, crd $1° = 0.0175$. By (3), we may find crd $(1/2)°$. Now one can construct a table of chords for $(1/2)°$ intervals. This is the gist of Ptolemy's method of constructing his table of chords.

(c) Show that the relations of (1), (2), and (3) of (b) are equivalent to the trigonometrical formulas for $\sin(\alpha + \beta)$, $\sin(\alpha - \beta)$, and $\sin(\theta/2)$.

(d) Establish the following interesting results as consequences of Ptolemy's theorem: If P lies on the arc AB of the circumcircle of
1. an equilateral triangle ABC, then $PC = PA + PB$,
2. a square $ABCD$, then $(PA + PC)PC = (PB + PD)PD$,
3. a regular pentagon $ABCDE$, then $PC + PE = PA + PB + PD$,
4. a regular hexagon $ABCDEF$, then $PD + PE = PA + PB + PC + PF$.

6.10 Stereographic Projection

In his *Planisphaerium,* Ptolemy developed **stereographic projection** as a mapping by which the points on a sphere are represented on the plane of its equator by projection from the South Pole. Under this mapping (see Figure 53), into what do

(a) the circles of latitude go?
(b) the meridian circles?
(c) small circles, on the sphere, passing through the South Pole?

It can be shown that any circle on the sphere, not through the South Pole, maps into a circle on the plane. Very important is the property that stereographic projection is a conformal mapping—that is, a mapping that preserves angles between curves. Why is this property important in mapping a small part of the earth's surface onto a plane? (An interesting development of spherical trigonometry from plane trigonometry by stereographic projection is given in J. D. H. Donnay, *Spherical Trigonometry after the Cesàro Method.* New York: Interscience, 1945.)

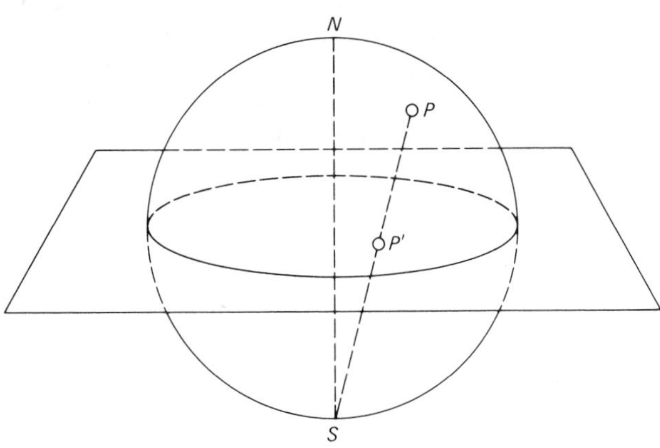

FIGURE 53

6.11 Problems from Heron

(a) A regular heptagon (7-sided polygon) cannot be constructed with Euclidean tools. In his work *Metrica,* Heron takes, for an approximate construction, the side of the heptagon equal to the apothem of a regular hexagon having the same circumcircle. How good an approximation is this?

(b) In *Catoptrica,* Heron proves, on the assumption that light travels by the shortest path, that the angles of incidence and reflection in a mirror are equal. Prove this.

(c) A man wishes to go from his house to the bank of a straight river for a pail of water, which he will then carry to his barn, on the same side of the river as his house. Find the point on the riverbank that will minimize the distance the man must travel.

(d) Complete the details of the following indication of Heron's derivation of the formula for the area Δ of a triangle ABC in terms of its sides a, b, c.

1. Let the incircle, with center I and radius r, touch the sides BC, CA, AB in D, E, F, as in Figure 54. On BC produced, take G such that $CG = AE$. Draw IH perpendicular to BI to cut BC in J and to meet the perpendicular to BC at C in H.
2. If $s = (a + b + c)/2$, then $\Delta = rs = (BG)(ID)$.
3. B, I, C, H are concyclic, whence $\measuredangle\, CHB$ is the supplement of $\measuredangle\, BIC$ and hence equal to $\measuredangle\, EIA$.
4. $BC/CG = BC/AE = CH/IE = CH/ID = CJ/JD$.
5. $BG/CG = CD/JD$.
6. $(BG)^2/(CG)(BG) = (CD)(BD)/(JD)(BD) = (CD)(BD)/(ID)^2$.
7. $\Delta = (BG)(ID) = \{(BG)(CG)(BD)(CD)\}^{1/2}$
$= \{s(s - a)(s - b)(s - c)\}^{1/2}$.

(e) Derive the formula of (d) by the following process: Let h be the altitude on side c, and let m be the projection of side b on side c. (1) Show that $m = (b^2 + c^2 - a^2)/2c$. (2) Substitute this value for m in $h = (b^2 - m^2)^{1/2}$. (3) Substitute this value for h in $\Delta = (ch)/2$.

(f) Approximate successively, by Heron's method, $\sqrt{3}$ and $\sqrt{720}$.

(g) A **prismatoid** is a polyhedron all of whose vertices lie in two parallel planes. The two faces in these parallel planes are called the **bases** of the prismatoid, the perpendicular distance between the two planes is called

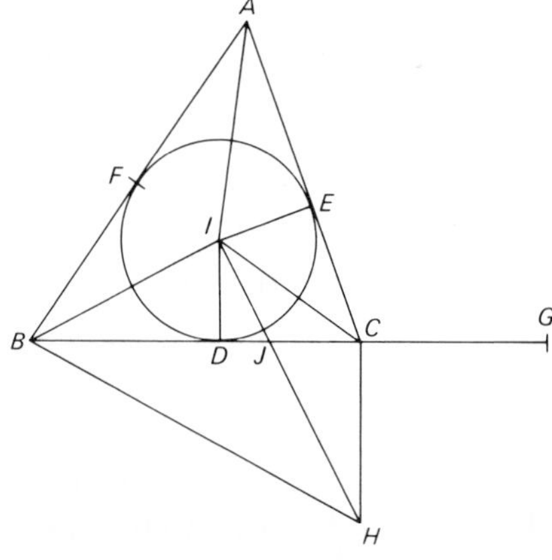

FIGURE 54

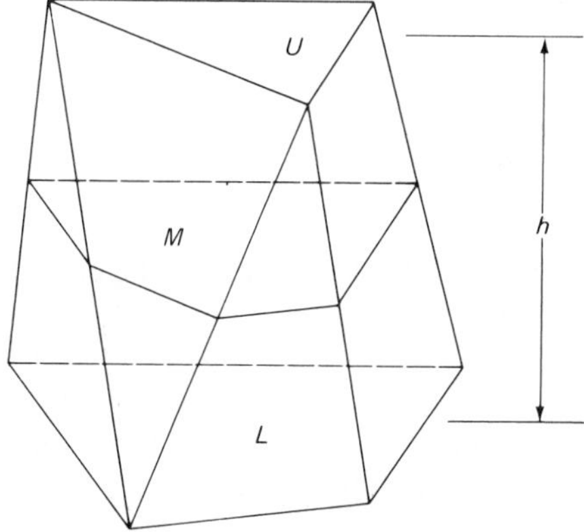

FIGURE 55

the **altitude** of the prismatoid, and the section parallel to the bases and midway between them is called the **midsection** of the prismatoid. Let us denote the volume of the prismatoid by V, the areas of the upper base, lower base, and midsection by U, L, M, and the altitude by h, as indicated in Figure 55. In books on solid geometry, it is shown that

$$V = \frac{h(U + L + 4M)}{6}.$$

In Book II of the *Metrica,* Heron gives, as the volume of a prismatoid having similarly oriented rectangular bases with corresponding pairs of dimensions a, b and c, d,

$$V = h\left[\frac{(a + c)(b + d)}{4} + \frac{(a - c)(b - d)}{12}\right].$$

Show that this result is equivalent to that given by the prismatoid formula above.

(h) Show that the "greatest Egyptian pyramid" [see Problem Study 2.13(a)] is a special case of the prismatoid formula of (g).

6.12 Simultaneous Equations

(a) Thymaridas, a lesser mathematician of the fourth century B.C., gave the following rule for solving a certain set of n simultaneous linear equations connecting n unknowns. The rule became so well known that it went by the title of the **bloom of Thymaridas:** *If the sum of n quantities be given, and also the sum of every pair which contains a particular one of them, then this particular quantity is equal to* $1/(n - 2)$ *of the difference between the sums of these pairs and the first given sum.* Prove this rule.

(b) In some problems given in the Heronian collections appear the formulas

$$a, b = \frac{(r + s) \pm \{(r + s)^2 - 8rs\}^{1/2}}{2},$$

for the legs a and b of a right triangle of perimeter $2s$ and inradius r. Obtain these formulas.

6.13 Problems from the "Greek Anthology"

(a) How many apples are needed if 4 persons out of 6 receive $\frac{1}{3}$, $\frac{1}{8}$, $\frac{1}{4}$, and $\frac{1}{5}$, respectively, of the total number, while the fifth receives 10 apples, and 1 apple remains for the sixth person?

(b) Demochares has lived a fourth of his life as a boy, a fifth as a youth, a third as a man, and has spent 13 years in his dotage. How old is he?

(c) After staining the holy chaplet of fair-eyed Justice that I might see thee, all-subduing gold, grow so much, I have nothing, for I gave forty talents under evil auspices to my friends in vain, while, O ye varied mischances of men, I see my enemies in possession of the half, the third, and the eighth of my fortune. (How many talents did the unfortunate man once possess?)

(d) The 3 Graces were carrying baskets of apples, and in each was the same number. The 9 Muses met them and asked each for apples and they gave the same number to each Muse and the 9 and the 3 each had the same number. Tell me how many they gave and how they all had the same number. (This problem is indeterminate. Find the smallest permissible solution.)

6.14 Type Problems from the "Greek Anthology"

Certain standard types of problems that are found in present-day elementary algebra texts date back to ancient times. Consider, for example, the following "work" problem, "cistern" problem, and "mixture" problem found in the *Greek Anthology*.

(a) Brickmaker, I am in a hurry to erect this house. Today is cloudless, and I do not require many more bricks, for I have all I want but 300. Thou alone in 1 day couldst make as many, but thy son left off working when he had finished 200, and thy son-in-law when he had made 250. Working all together, in how many days can you make these?

(b) I am a brazen lion; my spouts are my 2 eyes, my mouth, and the flat of my right foot. My right eye fills a jar in 2 days (1 day = 12 hours), my left eye in 3, and my foot in 4. My mouth is capable of filling it in 6 hours. Tell me how long all 4 together will take to fill it.

(c) Make a crown of gold, copper, tin, and iron weighing 60 minae: gold and copper shall be $\frac{2}{3}$ of it; gold and tin $\frac{3}{4}$ of it; and gold and iron $\frac{3}{5}$ of it; find the weights of gold, copper, tin, and iron required. [This is a numerical illustration of the *bloom of Thymaridas*. See Problem Study 6.12(a).]

6.15 Diophantus

(a) About all we know of Diophantus' personal life is contained in the following summary of an epitaph given in the *Greek Anthology:* "Diophantus passed $\frac{1}{6}$ of his life in childhood, $\frac{1}{12}$ in youth, and $\frac{1}{7}$ more as a bachelor. Five years after his marriage was born a son who died 4 years before his father, at $\frac{1}{2}$ his father's [final] age." How old was Diophantus when he died?

(b) Solve the following problem, which appears in Diophantus' *Arithmetica* (Problem 17, Book I): Find 4 numbers, the sum of every arrangement 3 at a time being given; say, 22, 24, 27, and 20.

(c) Solve the following problem, also found in the *Arithmetica* (Problem 16, Book VI): In the right triangle ABC, right angled at C, AD bisects angle A. Find the set of smallest integers for AB, AD, AC, BD, DC such that $DC:CA:AD = 3:4:5$.

(d) Augustus De Morgan, who lived in the nineteenth century, proposed the conundrum: "I was x years old in the year x^2." When was he born?

6.16 Some Number Theory in the "Arithmetica"

(a) Establish the identities

$$(a^2 + b^2)(c^2 + d^2) = (ac \pm bd)^2 + (ad \mp bc)^2$$

and use them to express $481 = (13)(37)$ as the sum of 2 squares in 2 different ways.

These identities were given later, in 1202, by Fibonacci in his *Liber abaci*. They show that the product of 2 numbers, each expressible as the sum of 2 squares, is also expressible as the sum of 2 squares. It can be shown that these identities include the addition formulas for the sine and cosine. The identities later became the germ of the Gaussian theory of arithmetical quadratic forms and of certain developments in modern algebra.

(b) Express $1105 = (5)(13)(17)$ as the sum of 2 squares in 4 different ways.

In the following 2 problems, "number" means "positive rational number."

(c) If m and n are numbers differing by 1, and if x, y, a are numbers such that $x + a = m^2$, $y + a = n^2$, show that $xy + a$ is a square number.

(d) If m is any number and $x = m^2$, $y = (m + 1)^2$, $z = 2(x + y + 1)$, show that the 6 numbers $xy + x + y$, $yz + y + z$, $zx + z + x$, $xy + z$, $yz + x$, $zx + y$ are all square numbers.

6.17 Problems from Pappus

(a) In Book III of Pappus' *Mathematical Collection,* we find the following interesting geometrical representation of some means. Take B on segment AC, B not being the midpoint O of AC. Erect the perpendicular to AC at B to cut the semicircle on AC in D, and let F be the foot of the perpendicular from B on OD. Show that OD, BD, FD represent the

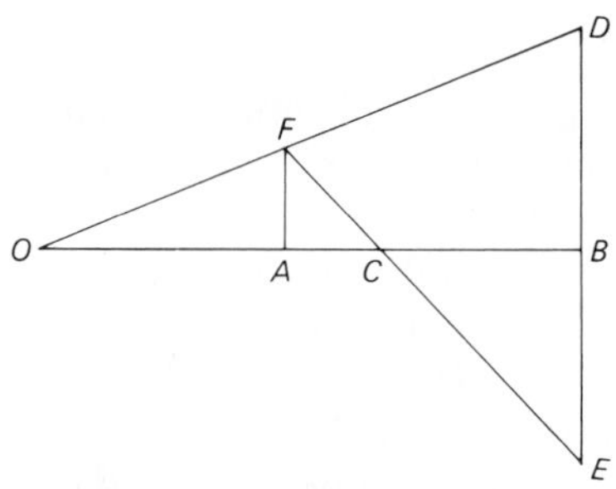

FIGURE 56

arithmetic mean, the geometric mean, and the harmonic mean of the segments AB and BC, and show that, if $AB \neq BC$,

$$\text{arithmetic mean} > \text{geometric mean} > \text{harmonic mean}.$$

(b) In Book III of *Mathematical Collection,* Pappus gives the following neat construction for the harmonic mean of the 2 given segments OA and OB in Figure 56. On the perpendicular to OB at B, mark off $BD = BE$, and let the perpendicular to OB at A cut OD in F. Draw FE to cut OB in C. Then OC is the sought harmonic mean. Prove this.

(c) Prove the following extension of the Pythagorean theorem given by Pappus in Book IV of *Mathematical Collection. Let ABC* (see Figure 57) *be any triangle, and ABDE, ACFG any parallelograms described externally on AB and AC. Let DE and FG meet in H and draw BL and CM equal and parallel to HA. Then*

$$\square\, BCML = \square\, ABDE + \square\, ACFG.$$

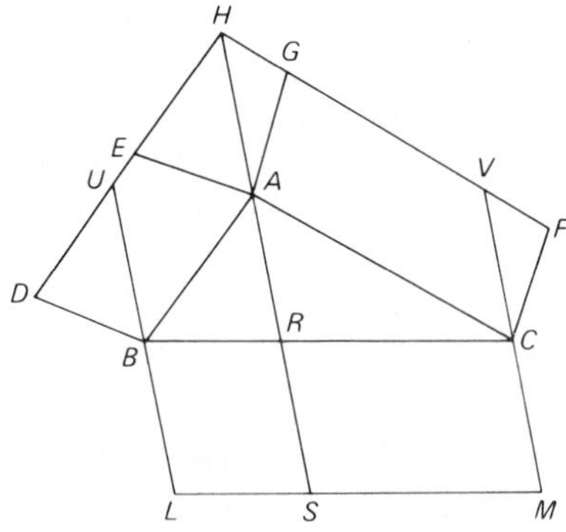

FIGURE 57

(d) Generalize the theorem of (c) to 3-dimensional space, replacing the triangle by a tetrahedron, and the parallelograms on the sides of the triangle by triangular prisms on the faces of the tetrahedron.

(e) In Book VIII of *Mathematical Collection,* Pappus establishes the following theorem: *If D, E, F are points on the sides BC, CA, AB of triangle ABC, such that* $BD/DC = CE/EA = AF/FB$, then triangles *DEF and ABC have a common centroid.* Prove this either synthetically or analytically.

6.18 The Centroid Theorems

In Book VII of *Mathematical Collection,* Pappus anticipated one of the centroid theorems sometimes credited to P. Guldin (1577–1642). These theorems may be stated as follows:

1. *If a planar arc be revolved about an axis in its plane, but not cutting the arc, the area of the surface of revolution so formed is equal to the product of the length of the arc and the length of the path traced by the centroid of the arc.*
2. *If a planar region be revolved about an axis in its plane, but not intersecting the region, the volume of the solid of revolution so formed is equal to the product of the area of the region and the length of the path traced by the centroid of the region.*

Using these theorems, find:

(a) The volume and surface area of the torus formed by revolving a circle of radius r about an axis, in the plane of the circle, at distance $R > r$ from the center of the circle.

(b) The centroid of a semicircular arc.

(c) The centroid of a semicircular area.

(It was the second of the above theorems that was anticipated by Pappus, making this the most general theorem involving the calculus to be found in antiquity.)

6.19 The Trammel Construction of an Ellipse

The following theorem has been ascribed to Proclus: *If a line segment of fixed length moves with its endpoints on 2 intersecting lines, then a fixed point on the segment, or on the segment produced, will describe an ellipse.*

(a) Choose a pair of rectangular axes Ox and Oy as the 2 lines in Proclus' theorem, and let AB be the segment of fixed length. Choose P on AB (produced if necessary) and denote AP by a and BP by b. Show that, as A moves on the y-axis and B moves on the x-axis, P describes the ellipse

$$\frac{x^2}{a^2} + \frac{y^2}{b^2} = 1.$$

(b) Design a simple mechanism (an **ellipsograph**) based on the result of (a) for describing an ellipse with given semiaxes a and b.

6.20 The Theorem of Menelaus

A point lying on a side line of a triangle, but not coinciding with a vertex of the triangle, is called a **menelaus point** of the triangle for this side. Prove the following chain of theorems, wherein all segments and angles are directed (or sensed) segments and angles:

(a) **Menelaus' theorem:** A necessary and sufficient condition for 3 menelaus points D, E, F for the sides BC, CA, AB of a triangle ABC to be collinear is that

$$\left(\frac{BD}{DC}\right)\left(\frac{CE}{EA}\right)\left(\frac{AF}{FB}\right) = -1.$$

(b) If vertex O of a triangle BOC is joined to a point D (other than B or C) on line BC, then

$$\frac{BD}{DC} = \frac{OB \sin BOD}{OC \sin DOC}.$$

(c) Let D, E, F be menelaus points for the sides BC, CA, AB of a triangle ABC, and let O be a point in space not in the plane of triangle ABC. Then the points D, E, F are collinear if and only if

$$\left(\frac{\sin BOD}{\sin DOC}\right)\left(\frac{\sin COE}{\sin EOA}\right)\left(\frac{\sin AOF}{\sin FOB}\right) = -1.$$

(d) Let D', E', F' be 3 menelaus points for the sides $B'C'$, $C'A'$, $A'B'$ of a spherical triangle $A'B'C'$. Then D', E', F' lie on a great circle of the sphere if and only if

$$\left(\frac{\sin \widehat{B'D'}}{\sin \widehat{D'C'}}\right)\left(\frac{\sin \widehat{C'E'}}{\sin \widehat{E'A'}}\right)\left(\frac{\sin \widehat{A'F'}}{\sin \widehat{F'B'}}\right) = -1.$$

(This is the spherical case of the Menelaus theorem that was used by Menelaus in his *Sphaerica*.)

6.21 More on Means

If a and b are 2 real numbers, the following means of a and b have been found useful:

1. arithmetic: $A = (a + b)/2$
2. geometric: $G = \sqrt{ab}$
3. harmonic: $H = 2ab/(a + b)$
4. heronian: $h = a + \sqrt{ab} + b$
5. contraharmonic: $c = (a^2 + b^2)/(a + b)$
6. root-mean-square: $r = \sqrt{(a^2 + b^2)/2}$
7. centroidal: $g = 2(a^2 + ab + b^2)/3(a + b)$

(a) If $a \neq b$, show that

$$c > r > g > A > h > G > H.$$

(b) If a^2, b^2, c^2 are in arithmetic progression, then $b + c, c + a, a + b$ are in harmonic progression.

(c) If a, b, c are in harmonic progression, so also are $a/(b + c)$, $b/(c + a)$, $c/(a + b)$.

(d) If between a and b there are inserted 2 arithmetic means A_1 and A_2, 2 geometric means G_1 and G_2, and 2 harmonic means H_1 and H_2, then $G_1G_2{:}H_1H_2 = A_1 + A_2{:}H_1 + H_2$.

(e) Let a and b, $a > b$, denote the lengths of the lower base and the upper base of a trapezoid. Then any line segment parallel to the bases and intercepted by the sides of the trapezoid is some *mean* of the bases a and b. Show that:

1. The arithmetic mean bisects the sides of the trapezoid.
2. The geometric mean divides the trapezoid into 2 similar trapezoids.
3. The harmonic mean passes through the intersection of the diagonals.
4. The heronian mean is $\frac{1}{3}$ the way from the arithmetic mean to the geometric mean.
5. The contraharmonic mean is as far below the arithmetic mean as the harmonic mean is above the arithmetic mean.
6. The root-mean-square bisects the area of the trapezoid.
7. The centroidal mean passes through the centroid of area of the trapezoid.

(f) Draw a trapezoid with bases a and b and construct the segments of (e). Now geometrically verify the inequalities of (a).

(g) The number $(a + wb)/(1 + w)$, $w > 0$, is called the **weighted mean,** for weight w, of a and b. Show that the following means of a and b have the indicated weights:

1. arithmetic: $w = 1$
2. geometric: $w = \sqrt{a/b}$
3. harmonic: $w = a/b$
4. heronian: $w = -(\sqrt{ab} + b - 2a)/(\sqrt{ab} + a - 2b)$
5. contraharmonic: $w = b/a$
6. root-mean-square: $w = -(\sqrt{a^2 + b^2} - a\sqrt{2})/(\sqrt{a^2 + b^2} - b\sqrt{2})$
7. centroidal: $w = -(a^2 + ab - 2b^2)/(b^2 + ab - 2a^2)$

(h) Let PT and PS be tangents drawn to a given circle from an external point P, and let TS cut the diametral secant PBA in C. Show that PC is the harmonic mean of PA and PB.

(i) Let CD and CE be the internal and the external bisector of angle C of a triangle ABC. Show that AB is the harmonic mean of AD and AE.

(j) Let s be the side of a square inscribed in a triangle and having one side lying along the base of the triangle. Show that s is half the harmonic mean of the base of the triangle and the altitude of the triangle on the base.

(k) Let s be the side of a square inscribed within a right triangle and having 1 angle coinciding with the right angle of the triangle. Show that s is half the harmonic mean of the legs of the triangle.

(l) Let ABC be a triangle having an angle of 120° at B, and let BT be the bisector of angle B. Show that BT is half the harmonic mean of BA and BC.

(m) Let s, a, b be chords of $\frac{1}{7}$, $\frac{2}{7}$, and $\frac{3}{7}$ of the circumference of a circle. Show that s is half the harmonic mean of a and b.

(n) A car travels at the rate of r_1 miles per hour from A to B, and then returns at the rate of r_2 miles per hour from B to A. Show that the average rate for the round trip is the harmonic mean of r_1 and r_2.

(o) A common precautionary procedure used with an equal-arm balance, when it is suspected that the arms are not quite equal, is known as **double weighing.** Here the unknown is first placed in the left-hand pan and balanced by a weight w_1, then the unknown is placed in the right-hand pan and balanced by a weight w_2. Show that the weight of the unknown is the geometric mean of w_1 and w_2.

(p) Show that the centroidal mean of a and b is equal to the heronian mean of a^2 and b^2 divided by the arithmetic mean of a and b.

(q) Show that $g = (H + 2c)/3 = (2A + c)/3$.

Essay Topics

6/1 Why is Archimedes considered the greatest mathematician of antiquity?

6/2 The Archimedean solids, with construction patterns.

6/3 The case for Archimedes as the inventor of the integral calculus.

6/4 The case for Menaechmus and Apollonius as inventors of analytic geometry.

6/5 The works of Eratosthenes.

6/6 The mathematical contributions of the Greek astronomers.

6/7 Heron's influence on the development of applied mathematics.

6/8 The first woman mathematician.

6/9 The Alexandrian school of mathematics.

6/10 Means.

6/11 Mathematics in the Roman civilization.

6/12 The "Greek Anthology."

6/13 Maps of the world according to Hecataeus, Eratosthenes, and Ptolemy.

Bibliography

AABOE, ASGER *Episodes from the Early History of Mathematics.* New Mathematical Library, no. 13. New York: Random House and L. W. Singer, 1964.

APOLLONIUS OF PERGA *Conics*. 3 vols. Translated by R. Catesby Taliaferro. (Classics of the St. John's program). Annapolis, Md.: R. C. Taliaferro, 1939.

BARON, MARGARET *Origins of the Infinitesimal Calculus*. New York: Pergamon, 1969. Reprinted by Dover, New York, 1987.

BUNT, L. N. H.; P. S. JONES; and J. D. BEDIENT *The Historical Roots of Elementary Mathematics*. Englewood Cliffs, N.J.: Prentice-Hall, 1976.

CLAGETT, MARSHALL *Archimedes in the Middle Ages*. 2 vols. Madison, Wis.: University of Wisconsin Press, 1964.

——— *Greek Science in Antiquity*. New York: Abelard Schuman, 1955. Paperback edition. New York: Collier Books, 1963.

COHEN, M. R., and I. E. DRABKIN *A Source Book in Greek Science*. New York: McGraw-Hill, 1948. Reprinted by Harvard University Press, Cambridge, Mass., 1958.

COOLIDGE, J. L. *History of the Conic Sections and Quadric Surfaces*. New York: Oxford University Press, 1945.

——— *History of Geometric Methods*. New York: Oxford University Press. Paperback edition. New York: Dover, 1963.

DANTZIG, TOBIAS *The Bequest of the Greeks*. New York: Charles Scribner's, 1955.

DAVIS, HAROLD T. *Alexandria, the Golden City*. 2 vols. Evanston: Principia Press of Illinois, 1957.

DIJKSTERHUIS, E. J. *Archimedes*. New York: Humanities Press, 1957.

EVES, HOWARD *A Survey of Geometry*, vol. 1. Boston: Allyn and Bacon, 1972.

GOW, JAMES *A Short History of Greek Mathematics*. New York: Hafner, 1923. Reprinted by Chelsea, New York.

HARTLEY, MILES C. *Patterns of Polyhedra*. Rev. ed. Ann Arbor, Mich.: Edwards Brothers, 1957.

HEATH, T. L. *Apollonius of Perga, Treatise on Conic Sections*. New York: Barnes and Noble, 1961.

——— *Aristarchus of Sama*. New York: Oxford University Press, 1913. Reprinted by Dover, New York, 1981.

——— *Diophantus of Alexandria*. Rev. ed. New York: Cambridge University Press, 1910. Reprinted by Dover, New York, 1964.

——— *History of Greek Mathematics* vol. 2. New York: Oxford University Press, 1921. Reprinted by Dover, New York, 1981.

——— *A Manual of Greek Mathematics*. New York: Oxford University Press, 1931.

——— *The Works of Archimedes*. New York: Cambridge University Press, 1897. Reprinted by Dover, New York.

JOHNSON, R. A. *Modern Geometry, an Elementary Treatise on the Geometry of the Triangle and the Circle*. Boston: Houghton Mifflin Company, 1929. Reprinted by Dover, New York.

MESCHKOWSKI, HERBERT *Ways of Thought of Great Mathematicians*. Translated by John Dyer-Bennett. San Francisco: Holden-Day, 1964.

ORE, OYSTEIN *Number Theory and Its History*. New York: McGraw-Hill, 1948.

PETERS, C. H. F., and E. B. KNOBEL *Ptolemy's Catalogue of Stars; A Revision of the Almagest*. Washington, D.C.: Carnegie Institution, 1915.

SARTON, GEORGE *Ancient Science and Modern Civilization*. Lincoln, Neb.: The University of Nebraska Press, 1954.

STAHL, W. H. *Ptolemy's Geography: A Select Bibliography*. Lincoln, Neb.: University of Nebraska Press, 1954.

——— *Roman Science*. Madison, Wis.: University of Wisconsin Press, 1962.

THOMAS, IVOR *Selections Illustrating the History of Greek Mathematics*. 2 vols. Cambridge, Mass.: Harvard University Press, 1939–41.

VAN DER WAERDEN, B. L. *Science Awakening*. Translated by Arnold Dresden. New York: Oxford University Press, 1961. Paperback edition. New York: John Wiley, 1963.

Cultural Connection

THE ASIAN EMPIRES

China Before A.D. 1260;
India Before A.D. 1206;
The Rise of Islam—A.D. 622 to 1258
(to accompany Chapter Seven)

In Cultural Connections III and IV, we examined the early growth and development of cultural, social, political, and economic life in the Mediterranean Sea basin. At the same time that the Greeks and Romans were forging many of the basic institutions of western society, eastern civilizations were being born as well: in China on the high plains surrounding the Yellow River valley, and in India in the shade of banyan trees below the towering peaks of the Himalayas. In the seventh century A.D., with the rise of Islam, the Arabs diverged from the rest of the western world and charted their own cultural path. It is to these three civilizations—China, India, and Arabia—that we now turn.

China

Chinese history may be divided into four general periods: Ancient China (ca. 2000–600 B.C.), Classical China (ca. 600 B.C.–A.D. 221), Imperial China (A.D. 221–1911), and Modern China (A.D. 1911 to the present). According to legend, in ancient times the dusty plains of northern China through which the Yellow River ran (a region the Chinese called the *Middle Kingdom* because they believed it to be located in the center of the world) were united under the semimythical Hsia (before 1500 B.C.), Shang (1500–1027 B.C.), and Chou (1027–256 B.C.) Dynasties. Whether or not these early dynasties ever really succeeded in establishing strong, centralized governments is unclear, but by 600 B.C., when China's Classical Age began, the authority of the Chou monarchs was only nominal, and real power lay in the hands of numerous petty lords who ruled small city-states and who engaged one another in countless wars, taxed their subjects mercilessly, and were generally unconcerned with the plight of the poor.

In response to the social chaos of early Classical China, the philosopher Confucius (551–479 B.C.) advocated a political and social restructuring. Confucius taught a combination of the golden rule, respect for authority, concern for the poor, humility, and the need for ethical governments. Although a few of his disciples rose to positions of authority, Confucius himself was largely ignored during his lifetime and was unable to convince the aristocracy to mend its ways. At about the same time, another Chinese philosopher, Lao-Tzu, is supposed to have devised Taoism, although in fact Lao-Tzu may never have existed. Taoism more probably was the invention of Chuang Tzu (399–295 B.C.) and other philosophers. Taoism declared that there is a natural order or harmony to the universe and urged simplicity, peace, and benevolent government. Later, the concept of *yin-yang* was associated with Taoism. This concept held that in all things there is a dialectical struggle between opposites and that a resolution can be achieved only through accommodating these opposites. Both Taoism and Confucianism were, in many respects, reactions against the misrule of the petty princes and the misery of their subjects.

In 221 B.C., the city-state of Chin, under the rule of a capable general and statesman the Chinese called simply Chin Shih-huang-ti (literally, "the first emperor of the Chin Dynasty"), united the warring city-states into a monolithic empire. Fifteen years later, the Chin Dynasty was supplanted by the Han Dynasty, which created an empire that would endure, with one minor interruption, until A.D. 221, when China's Classical Age ended. The Hans greatly expanded China, pushing south into the hilly rain forests of southern China and northern Vietnam, west into the deserts of central Asia, and northeast into Manchuria and Korea. The Han emperors were impressed by the ideas of Confucius about statecraft: three centuries after the sage's death, his philosophies attained the status of a state religion, somewhat akin to Christianity's eventual acceptance by the Roman emperors in the west a few hundred years later. In approximately A.D. 60, another philosophy, Buddhism, arrived in China over the Himalaya Mountains from India. To the Chinese, Buddhism resembled Taoism, and the two philosophies tended to merge. Buddhism remained a minority sect in China until approximately A.D. 800, when it acquired widespread appeal among the peasants. Confucianism retained its popularity with the upper classes. We will explore Buddhist philosophy when we look at Indian history and culture.

The Han Empire dissolved in A.D. 221, and for over 350 years China was again divided into warring factions, until A.D. 618, when Emperor Li Yüan united all China under the Tang Dynasty. The Tang emperors, like the emperors of the Sung and Yüan Dynasties that followed them, were patrons of art and literature, and their reigns mark China's imperial period, or Golden Age. Under these three dynasties, China attained its greatest size and influence, and trade was opened between east and west.

It is interesting to compare the Chinese Empire with Rome. Both were powerful, large, and long lived; however, Chinese unity persisted far longer than that of the Romans. The Roman Empire lasted for only about 500 years (31 B.C.–A.D. 476); the Chinese Empire, excluding the 397-year interlude between the Han and Tang Dynasties, remained in existence for more than 1500 years, until the Chinese Revolution in A.D. 1911. The Roman emperors were largely a collection of military dictators, often illiterate, whose short reigns usually ended in a bloody coup; the Chinese kings were more characteristically absolute monarchs with long reigns. The Romans eschewed scholarship; Chinese emperors, such as Li Yüan and Kublai Khan (A.D. 1216–1294), provided government support for the arts. Classical and Imperial China produced a rich culture and a solid intellectual base. Nevertheless, Chinese scholars were often more interested in philosophy, art, and literature than in science; consequently, Chinese mathematics and science lagged behind other disciplines.

European contact with China was very limited until after A.D. 1260, when three Italian merchants—Mateo, Niccolo, and Marco Polo—visited the fabulous court of the Yüan emperor Kublai Khan. Communication became more regular, and by 1600, European traders and Christian missionaries visited China regularly. At this point, the history of Chinese science and mathematics merged with that of Europe. We will return to China briefly in Cultural Connection X: The Atom and the Spinning Wheel.

India

Politically, India was China's opposite. The original Chinese, farmers of the Yellow River valley, went on to create great empires that dominated most of eastern Asia. The original Indians were wiped out by nomadic invaders in approximately 1500 B.C. For most of its history, China was united into a single empire. India was rarely united, but was instead divided into numerous small principalities. The Chinese were usually able to turn aside invaders. India was beset by conquering armies numerous times, as Aryans, Persians, Greeks, Arabs, and Englishmen marched through its forests and over its plains. When China did finally fall to its foes (the Mongols), the invaders were quickly assimilated into Chinese society. (Small wonder that the Chinese symbol for "strength" is the same as that for "water"—it takes the shape of its container but always eventually wears it away.) India's most successful invaders—the Aryans, Arabs, Turks, and British—established themselves as ruling aristocracies that remained apart from other peoples of India. Although China usually knew internal peace, wars were constant in India. Nevertheless, despite such a seemingly hostile environment for scholarship, Indians developed a full, rich culture that persevered for centuries.

Between 3000 and 1500 B.C., a city-dwelling, agricultural people lived in India in the valley of the Indus River on the edge of the Thar Desert. What happened to these people is a mystery. The only evidence we have that they ever existed are the archeological remains of some of their cities, the largest of which have been unearthed at Mohenjo Daro and at Harappa. They may have perished on their own, unable to sustain their urban cultures in a hostile environment. More probably they were destroyed by the Aryans, a nomadic, cattle-herding people who moved into India from central Asia about 1500 B.C.

By 500 B.C., the Aryans were firmly established in India, even though other peoples, such as the Tamils of southern India, lived there as well. There was little political unity among the Aryans, and they were split into a number of small, quarrelling kingdoms. Between 1500 and 500 B.C., the Aryans evolved Hinduism, a combination of religion, philosophy, and social structure that formed the cornerstone of their civilization. A complex set of beliefs and laws,

Hinduism was largely based on three principal ideas: the worship of a large pantheon of gods behind whom lies a single unity, the idea of transmigration (that is, that a person's soul is eternal and will be born again in different forms), and the caste system that rigidly divided Indian society into four distinct social classes—the *Brahmana* (priest class), *Kshatriya* (warrior class), *Vaisya* (merchant and artisan class), and *Sudra* (peasant class).

Hinduism grew increasingly formalistic, and by 500 B.C. various reform movements emerged, the most famous of which, Buddhism, was proposed by the wandering ascetic Gautama Buddha (563–483 B.C.). In a sermon delivered in the city of Banaras, Buddha condemned both excessive self-indulgence and excessive self-mortification, each of which he believed led inevitably to pain and suffering. Instead, Buddha advocated a "middle path" of moderation, knowledge, and tranquility. Such a path, Buddha told his audience, led to *nirvana,* which broke the endless series of reincarnations that doomed the soul to everlasting pain. Buddhism stressed the basic unity of the universe, an idea not dissimilar to Chinese Taoism; like Taoism and Confucianism, it may have been, in part, a response to the chaos and turmoil of the time. Buddhism flourished in India for about a thousand years, especially among the poor, until approximately A.D. 500, when it began to decline. By that time, however, Buddhism had spread to China, Japan, and southeast Asia, where it took firm root. Hinduism remains today the most prevalent religion in India.

In 320 B.C., Chandragupta Maurya (reigned ca. 320–ca. 296 B.C.), king of a small state in northern India, established suzerainity over his fellow princes and founded the Mauryan Empire, which, under his grandson Aśoka (272–232 B.C.), included most of India. By approximately 185 B.C., however, the Mauryan Empire had disintegrated, and India was again divided into several warring kingdoms, although one, the Andhra Empire, did control much of south-central India between approximately 180 B.C. and A.D. 200. Despite the lack of political unity, the period after the fall of the Mauryan Empire boasted a rich cultural life and a flowering of Indian literature, art, science, and philosophy. In A.D. 320, much of India was again united, under Chandragupta I (reigned A.D. 320–340?) and the Gupta Empire. This new empire endured until A.D. 470, a period that is considered India's classical age and that featured a renaissance of Sanskrit literature, art, and medicine.

For centuries, numerous invaders had forced their way into India from the west, through passes in the Hindu Kush Mountains. In repeated waves came the Aryans, Persians, Hellenistic Greeks, Sakas, Parthians, Kushans, Ephthalites, and Arabs. The latter introduced Islam to India in the eighth century A.D. and conquered parts of western India in the eighth, ninth, and tenth centuries. In 1206, the Arab general Kutb ud-Din-Aibak founded the Moslem Sultanate of Delhi in northern India, which remained the preeminent Indian kingdom until 1526, when the Turkish adventurer Babur (1483–1530), also a Moslem, forged the larger Mogul Empire. Under both the Sultans of Delhi and the Mogul emperors, India was a nation of Hindus ruled by an upper class of Moslems (although Islam did become the principal religion in those parts of India that now comprise the nations of Pakistan and Bangladesh). After 1206, Indian science and mathematics merged into those of Arabia.

The Rise of Islam

Before A.D. 622, when Mohammed made his famous *Hegira* from Mecca to Medina, Arabia was a disunited country peopled by nomadic herders and tribes of fierce warriors. Mostly desert, with only a little arable land, Arabia lay at the periphery of Middle Eastern, Egyptian, and Greek civilization. It was considered too poor and rough for conquest and had never been incorporated into any of the great western empires—Persia, the Hellenistic Greek empires, or Rome.

Beginning in 622, the Arab people suddenly found a vitality hitherto lacking. The warring tribes quickly united and, with missionary fervor, rode out of the searing desert to forge a great empire that, at its height, stretched from the Atlantic Ocean to India and included most of what the Hellenistic Greeks had called the *oikoumene*—the heartland of western civilization. The force that powered the expansion of Arabia was the new religion of Islam.

Islam was founded by the prophet Mohammed (ca. A.D. 570–632), a merchant from the town of Mecca on the Red Sea. At the age of forty, Mohammed had visions that he believed were messages from God and that he recorded in a book called the *Koran*. These messages revealed that there was only a single God (Allah), that Mohammed was His greatest prophet (although the Hebrew prophets and Jesus Christ were also considered prophets by Moslems), and that those who joined the new religion were members of a common brotherhood. The *Koran* instructed devout Moslems to care for the poor and to offer hospitality to strangers, two traditional Arab values. It also called upon believers to convert others to the faith, peacefully if possible, sometimes by force when necessary, although Jews and Christians, as members of related faiths, were supposedly exempt from forcible conversion. Holy wars (*jehads*) could be proclaimed against enemies of the religion. The faithful, the *Koran* told, would be rewarded in the afterlife. A combination of traditional values, optimism, and united brotherhood, the message of Islam was immensely popular among the Arabs. By the time Mohammed died in 632, most of Arabia had been converted.

Soon after Mohammed's death, the now-united Arabs expanded into neighboring countries, filling the power vacuum left by the fall of the Roman Empire. Palestine and Syria were incorporated into an Arabian Empire in 640, the valley of the Tigris and Euphrates Rivers in 641, and Egypt in 642. From these places, Arab horsemen thundered across north Africa and Iran, and by 715 the Arabian Empire included Spain and parts of western India and commanded most of the Mediterranean Sea basin. Islam as a religion spread even further than the Empire. Arab merchants carried the new faith to black Africa, central Asia, and Indonesia.

Mohammed's death also occasioned a power struggle for control of both the religion and the Empire. Civil war broke out in 656. The war was eventually won by the Ommeyad Dynasty, but it caused a split in the religion between the Sunnites (mostly Arabs), who supported the Ommeyads, and the Shiites (mostly Iranians), who did not. The Ommeyads ruled the Empire from the city of Damascus in Syria until 750, when they were overthrown by the Abbasid Dynasty, which moved the capital to Baghdad (now the capital of Iraq), near the ancient city of Babylon. The Arabian Empire under the Abbasids lasted five hundred years, into the thirteenth century. It never controlled Spain, and Egypt became independent in the tenth century. Most of the Arabian Empire fell to the Turks in 1258.

The Arabs greatest contributions to civilization were the religion of Islam and the versatile Arabic language, in which the *Koran* was written; however, the Arabs were also adept at assimilating the best elements they found in other cultures. They carefully preserved much of Greek science and excelled not only at mathematics, but at astronomy and medicine as well. Artistically, they eschewed human sculpture, which they considered idolatry, and concentrated on architecture and fine decorative work. A graceful Arabic literature emerged. Some of the best Moslem scientists, artists, and poets, however, were not Arabs, but Iranians (Persians) and Spaniards. Mecca, in Arabia, was Islam's religious capital, but the cultural, economic, and political center of the Arabian Empire was fabulous, golden Baghdad, with its rich blend of Arab and Iranian cultures.

With the rise of Islam, the western world was divided into two cultural regions: the Moslem southeast (north Africa, Egypt, the Middle East, and Iran) and Christian Europe. Of the two, the Arab world was the more culturally, artistically, and scientifically advanced between 622 and 1300, a time when Europe struggled through its so-called Dark Ages. We will look at Europe's cultural hiatus and subsequent resurgence in Cultural Connection VI.

Japan and Southeast Asia

Before returning to Europe, we should acknowledge that Asian civilization was not confined to China, India, and Arabia. People moved from the Asian mainland to Japan during the Stone Age, and there was a hunter/gatherer culture there as early as 4500 B.C. By the beginning of the fourth century A.D., Japan was united into a single kingdom, and Buddhism had reached there by the tenth century. Japan remained a strong, centralized kingdom well into the seventh century, when power began to drift into the hands of a court-based aristocracy. In the twelfth century, the aristocracy fell from power, and Japan entered its Feudal Age, in which a nominal emperor presided over a land divided into numerous baronies, and political and military power were held by a central military headquarters called the Shogunate. Southeast Asia, too, developed impressive civilizations that were influenced by China and India. Hinduism, Buddhism, and Islam all were carried into southeast Asia by missionaries. By 1600, Buddhism had taken firm root on the mainland (Thailand, Cambodia, and Vietnam), and Islam predominated on the offshore islands (Indonesia). The maritime Malay peoples, who lived in what is now Malaya and parts of Indonesia, roamed across the Indian Ocean to Africa (where they settled Madagascar) and into the Pacific Ocean.

Chapter 7

CHINESE, HINDU, AND ARABIAN MATHEMATICS

CHINA[1]

7–1 Sources and Periods

Although the civilizations in ancient China along the Yangtze and Hwang Ho rivers are probably not as old as the Egyptian civilization along the Nile and the Babylonian civilization between the Tigris and Euphrates, very little of a primary nature has come down to us from those early Chinese civilizations. This is because the peoples of the time in all likelihood recorded many of their findings on bamboo, which did not last through time. As a further complication, there is the infamous burning of the books ordered by the egotistical Emperor Shï Huang-ti in 213 B.C. Although, in spite of severe threats and reprisals, the emperor's edict was most certainly not completely carried out, and although many books burned were later restored from memory, we are now in doubt as to the genuineness of many items claimed to be older than that unfortunate date. It follows that much of our knowledge of very early Chinese mathematics rests on hearsay and on later renditions of original texts.

Until quite recently, English-speaking scholars unfamiliar with the Chinese language were severely handicapped and had to rely primarily on one book, *The Development of Mathematics in China and Japan,* which was published in 1913 by the Japanese mathematician Yoshio Mikami, and on a few scattered papers written by Europeans during the nineteenth century. With the publication in 1959 of the highly scholarly third volume of J. Needham's *Science and Civilization in China,* the situation was considerably improved. There are some accounts of Chinese mathematics in German, and recently (1987), Shen Kangshi of Hangzhou University published, in Chinese, an excellent introduction to the history of Chinese mathematics; it is hoped that this latter work will appear in English translation.

It is perhaps wise first to sketch, briefly, the principal periods of Chinese history prior to 1644. We start with the Shang period, which arose about 1500 B.C. The Shang government, which is the first dynasty in the recorded history of China, and which held sway over an area whose boundaries varied with the

[1] The material of the following sections on China has been largely adapted from D. J. Struik, "On ancient Chinese mathematics," *The Mathematics Teacher* 56 (1963): 424–432, and from scholarly notes kindly furnished by Ouyang Jiang and Zhang Liangjin of P.R. China.

fortunes of war, collapsed in 1027 B.C. and was succeeded by the feudal Chou period, regarded by the Chinese as their classical age. The Chou rule culminated in 256 B.C., after which occurred the short-lived Chin dynasty, which lasted from 221 to 206 B.C. The Chin was replaced by a powerful unified empire under the Han (206 B.C.–A.D. 221), which was followed by a post-Han period of division extending to about A.D. 600. It was in the post-Han period that Buddhism became well established in China. Then came the newly unified China under the Tang (618–906), during which printing was invented, the Five Dynasties of the Independent States (906–960), the Sung (960–1279), the Yüan (1279–1368), and the Ming (1368–1644). The last three all ruled over a united China. European influence in mathematics, as in other areas, began under the Ming with the arrival of the Jesuit missionaries.

Marco Polo (1254?–1324?) visited China from 1275 to 1292, and the "barbarian" Kublai Khan (1216–1294) consolidated China under the Yüan dynasty upon completion of his conquest of the country in 1279.

Pronunciation of Chinese Names

The English transliteration of Chinese ideograms is not uniform. We follow the plan adopted by the Royal Geographical Society of London and the United States Board of Geographic Names.

a as in *father*.

e, é as in *men*. The accent indicates that it does not form part of a dipthong.

i as in *pique*. When followed by *n* or a vowel it is short as in *pin*.

ï following a consonant indicates that the vowel is intoned with the consonant.

eï like *ey* in *they*.

u like in *oo* in *boot*. When preceding *n, a,* or *o* it is short.

ü like the French *u*. When preceding *n, a,* or *é* it is short.

óu is a diphthong with the two vowels distinctly intoned.

ui like *ooi* contracted into a diphthong.

Initial **k, p, t, hs, ts, tz** are not so hard as in English. When pronounced as hard as possible they are followed by (').

ch like *ch* in *church*.

y like *y* in *you*.

7–2 From the Shang to the Tang

An account of the history of the mathematics of ancient China starts in the Shang period, with some inscriptions on bone and tortoise shell that reveal a decimal numeral system closely akin to the traditional Chinese-Japanese multiplicative system described in Section 1–5. Even at this very early time, then, we find in China the seeds of a decimal positional numeral system. By the time of the Han, or perhaps earlier, the rod numeral system, which employed arrangements of bamboo sticks as described in Problem Study 1.4(c), and in which blank spaces appeared for zeros, was established. This instance of a decimal positional system was the most advanced system of numeration in the world at the time, and it played a very important role in the character of early Chinese mathematics, which was centered on calculation. The elementary arithmetic operations were carried out with bamboo sticks on counting boards. The familiar Chinese abacus, the **suan pan,** which consists of movable beads on parallel rods or wires, is a descendent of this early form of calculating. It is not known just when the suan pan was introduced; the earliest extant mention of the suan pan is found in a work of 1436, but the instrument could be much older.

One of the oldest Chinese works involving mathematics, the *I-king,* or *Book of Permutations,* also dates back to Shang days, for it is claimed to have been written by Wön-wang (1182–1135 B.C.). In it appears the *Liang I,* or "two principles" (the male *yang,* —, and the female *ying,* - -). From these are formed the *Pa-kua,* or eight figures

— - - — - - — - - — - -
— — - - - - — — - - - -
— — — — - - - - - - - -

These eight symbols had various attributes assigned to them, and they came to be used in divination. One cannot help but see in the *Pa-kua* an adumbration of a binary system of numeration. For if we take — as one and - - as zero, the successive trigrams shown above, beginning at the right, would represent the numbers 0, 1, 2, 3, 4, 5, 6, 7. Also in the *I-king* is found the oldest known example of a magic square (see Problem Study 7.3).

The most important of all ancient Chinese mathematical texts is the *K'ui-ch'ang Suan-shu,* or *Arithmetic in Nine Sections,* dating from the Han period but very likely containing material much older than the Han. It is the epitome of ancient Chinese mathematical knowledge, and it established the characteristic of the ancient mathematics of China as calculation oriented, with theory and practice connected in a sequence of applied problems. The work, which is rich in content, is a collection of 246 problems on agriculture, business procedures, engineering, surveying, solution of equations, and properties of right triangles. Rules of solution are given, but there are no proofs in the Greek sense. In Problem 36 of Section I, the area of a circular segment of base b and sagitta (height) s is given by the empirical formula $s(b + s)/2$. This may have been arrived at as indicated in Figure 58, where, when the secant lines are drawn so as to make the area of the isosceles triangle appear by eye to be equal to that of the circular segment, these lines seem to cut the base line prolonged a distance of $s/2$ in each direction. For a semicircle, the empirical formula leads to the value 3 for π. There are also problems in the text leading to simultaneous systems of linear equations, which are solved by what we today call the matrix method. A sample of the problems of the work can be found in Problem Studies 7.1 and 7.2.

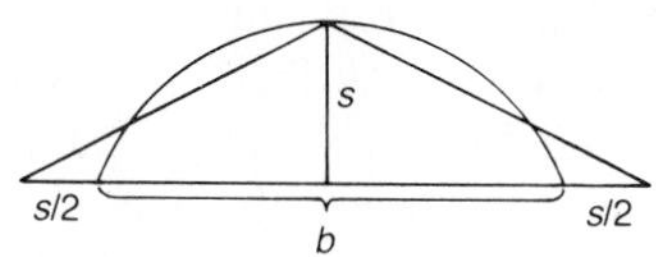

FIGURE 58

Following is a brief indication of the contents of the individual nine sections of the *K'ui-ch'ang Suan-shu:*

1. Surveying matters, with correct rules for the areas of a triangle, trapezoid, and circle, and the circle approximations $(\frac{3}{4})d^2$ and $(\frac{1}{12})c^2$, where π is taken as 3.
2. Percentage and proportion.
3. Partnership and the rule of three.
4. Finding sides of figures and including square and cube roots.
5. Volumes.
6. Motion problems and alligation.
7. The rule of false position.
8. Simultaneous linear equations and matrix procedures.
9. Pythagorean right triangles.

Another famous classic, perhaps even older than the *Arithmetic in Nine Sections,* is the *Chóu-peï,* which is only partly mathematical. Its chief interest to us is its discussion, based on the diagram of Figure 59 (but with no proof), of the Pythagorean theorem.

An interesting event occurred in January 1984, when there was unearthed in tombs of the Han dynasty an arithmetic book written on bamboo slips. The work was transcribed about the second century B.C. and is a collection of more

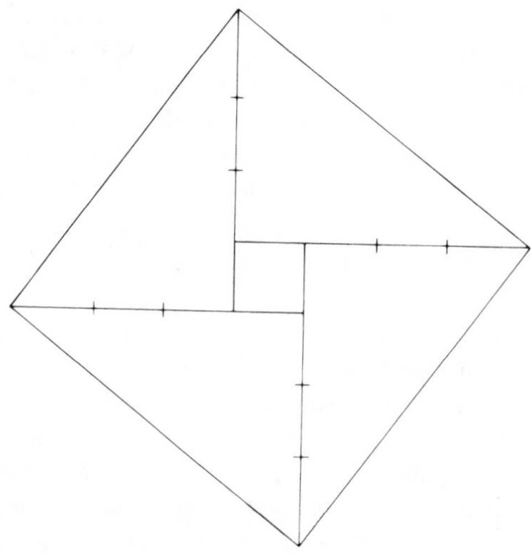

FIGURE 59

than ninety problems involving the four fundamental arithmetic operations on both integers and fractions, proportion, area, and volume. It is now the earliest Chinese mathematical work actually in existence.

Following the Han period lived the mathematician Sun-tzï, who wrote a book containing much material similar to that of the *Arithmetic in Nine Sections*. It is in this work that we encounter the first Chinese problem in indeterminate analysis: "There are things of an unknown number which when divided by 3 leave 2, by 5 leave 3, and by 7 leave 2. What is the (smallest) number?" Here we find the beginnings of the famous Chinese Remainder Theorem of elementary number theory.

In the post-Han period we also find a number of mathematicians devoting attention to the computation of π, the ratio of the circumference to the diameter of a circle. A general of the third century, named Wang Fan, has been credited with the rational approximation $\frac{142}{45}$ for π, yielding $\pi = 3.155$. A contemporary of Wang Fan, named Liu Hui, wrote a short commentary on the *Arithmetic in Nine Sections* called the *Sea Island Mathematical Manual*. In this, we find some new material on mensuration, among which is the relation

$$3.1410 < \pi < 3.1427.$$

About two centuries later, Tsu Ch'ung-chih (430–501) and his son, whose joint book is now lost, found

$$3.1415926 < \pi < 3.1415927$$

and the remarkable rational approximation $\frac{355}{113}$, which yields π correct to six decimal places. This rational approximation was not rediscovered in Europe until 1585 (see Section 4–8). The precision of π achieved by the Tsus seems not to have been surpassed until 1425 (?), when the astronomer Jamshid Al-Kashi (died ca. 1436) of Samarkand found π correct to sixteen decimal places. Western mathematicians did not surpass the Tsus' approximation until around 1600.

7–3 From the Tang through the Ming

During the Tang dynasty, a collection of the most important mathematical books available was assembled for official use in the imperial examinations. Printing originated in the eighth century, but the first mathematical work to be printed of which we are aware did not appear until 1084. In a work of about 625, written by one Wang Hs'iao-t'ung, appears the first cubic equation in Chinese mathematics more complicated than the $x^3 = a$ of the *Arithmetic in Nine Sections*.

An important printed edition of the *Arithmetic in Nine Sections* appeared during the Sung dynasty in 1115. The latter part of the Sung dynasty through the early part of the Yüan dynasty marks the greatest period in ancient Chinese mathematics. Many important mathematicians flourished and many worthy mathematical books appeared. Among the mathematicians were Ch'in Kiu-shao (whose book is dated 1247), Li Yeh (with books dated 1248 and 1259), Yang Hui (with books dated 1261 and 1275), and, the greatest of all, Chu Shï-kié (whose books are dated 1299 and 1303).

Ch'in took up indeterminate equations where Sun Tzï had left off. He also was the first Chinese to give a separate symbol, a circle, for zero. He was one of the mathematicians who generalized the method of extracting square roots (as given in the *Arithmetic in Nine Sections*) to equations of higher degree, leading to the numerical method of solving algebraic equations we today refer to as Horner's method, since it was independently found by the English school-master William George Horner (1786–1837) and published by him in 1819. He was completely unaware of the fact that he had rediscovered an ancient Chinese computational scheme. Li Yeh is of special interest because he introduced a notation for negative numbers by placing a diagonal stroke through the right-hand digit when the number is written in the Chinese scientific, or rod, system. Thus,—10724 appears as

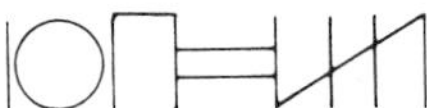

Yang Hui, whose books are a sort of extension of the *Arithmetic in Nine Sections,* worked deftly with decimal fractions by essentially our present methods. Yang Hui has also given us the earliest extant presentation of the so-called *Pascal arithmetic triangle* (see Section 9–9), which is again found in a later book written by Chu Shï-kié in 1303. Chu speaks of the triangle as already ancient in his time. It would appear, then, that the binomial theorem was known in China for a long time. Chu's books give the most accomplished presentation of Chinese arithmetic-algebraic methods that has come down to us. He employs familiar matrix methods of today, and his method of elimination and substitution has been compared with that of J. J. Sylvester (1814–1897).

The post-Sung period continued to produce mathematicians, who often served as astronomers, but little that is fundamentally new appeared in their mathematics. In the earlier Tang period, one can detect a Hindu influence; in the later Yüan period, one can find Arabic traces. There is very little in ancient Chinese mathematics that is directly traceable to Western (Greek or Latin) mathematics. It was only with the mathematics of the Ming era, after the Jesuit missionaries had penetrated China, that Western influence is noticeable.

7–4 Concluding Remarks

After the decline of classical Greek mathematics, the mathematics of China became one of the most prosperous in the world. While Western Europe was undergoing its period of Dark Ages, Chinese mathematics was thriving, and many of its achievements long predated the same achievements made later in Europe during and after the Renaissance. To mention a few of these achieve-

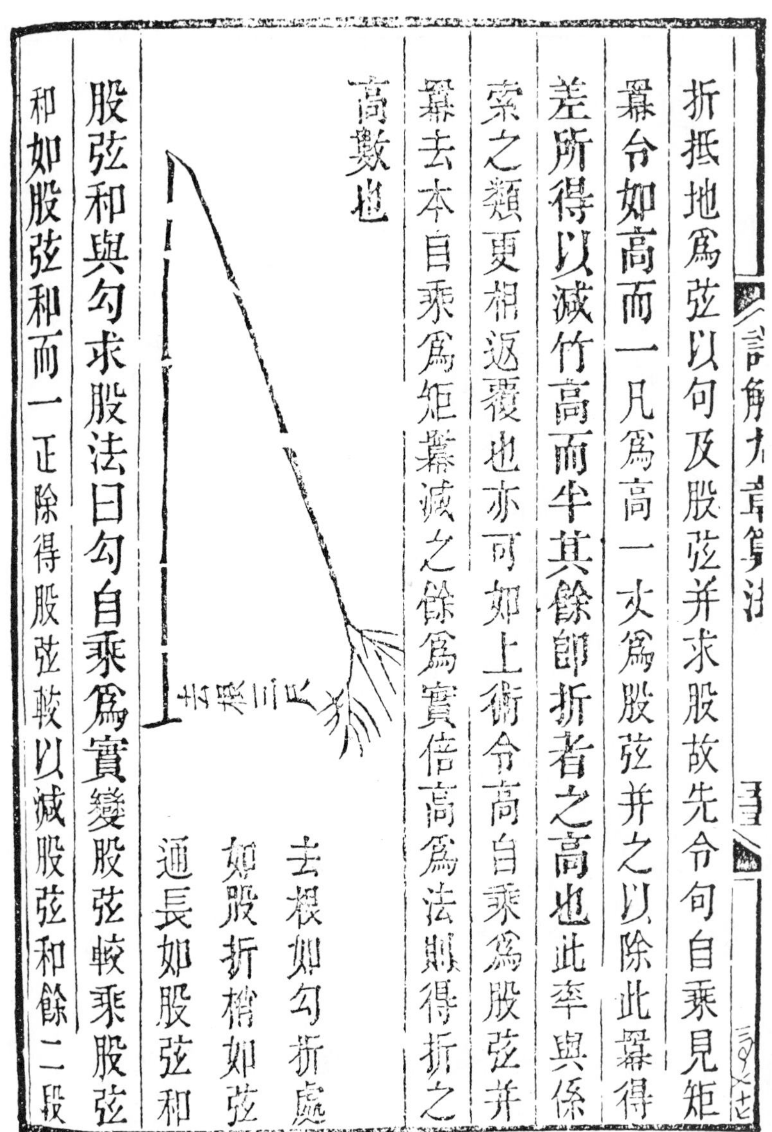

折抵地爲弦以句及股弦并求股故先令句自乘見矩冪令如高而一凡爲高一丈爲股弦并之以除此冪得差所得以減竹高而半其餘即折者之高也此率與係索之類更相返覆也亦可如上術令高自乘爲股弦并冪去本自乘爲矩冪減之餘爲實倍高爲法則得折之高數也

去根如勾折處如股折梢如弦通長如股弦和

股弦和與勾求股法曰勾自乘爲實變股弦較乘股弦和如股弦和而一正除得股弦較以減股弦和餘二段

The problem of the broken bamboo, from a work of Yang Hui (1261).

ments, we note that China was the first (1) to create a positional decimal numeral system, (2) to acknowledge negative numbers, (3) to obtain precise values of π, (4) to arrive at Horner's method for the numerical solution of algebraic equations, (5) to present the Pascal arithmetic triangle, (6) to be aware of the binomial theorem, (7) to employ matrix methods in the solution of systems of linear equations, (8) to solve systems of simultaneous congruences by the so-called Chinese Remainder Theorem, (9) to develop decimal fractions,

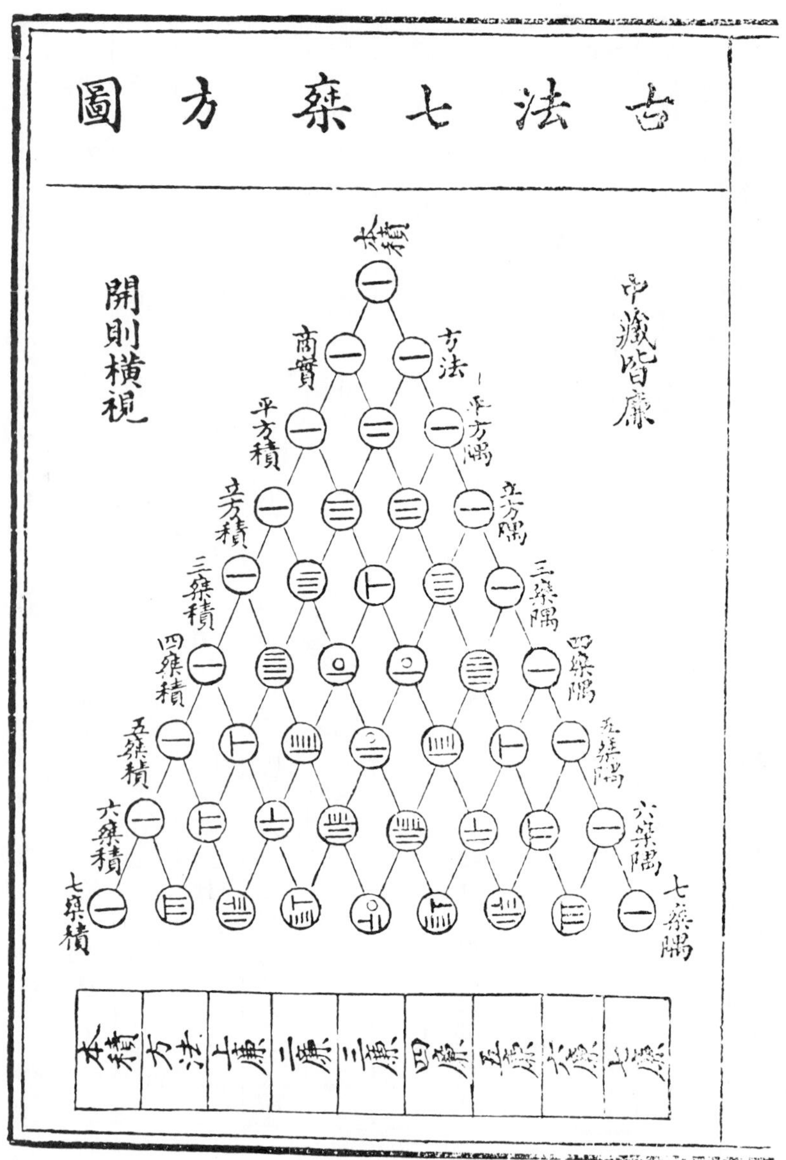

Pascal's arithmetic triangle as depicted in 1303 by Chu Shï-kié.

(10) to develop the rule of three, (11) to apply the rule of double false position, (12) to develop arithmetic series of higher order and their application to interpolation, and (13) to develop descriptive geometry.[2]

[2] This occurred in Nian Xi-yao's *Shi Xue,* or *Perspective Drawing,* which was published in 1729 and revised in 1735. Gaspard Monge's *Descriptive Geometry* did not appear until 1799.

Many of the Chinese findings in mathematics ultimately made their way to Europe via India and Arabia. On the other hand, it was not until the Jesuit missionaries entered China in the Ming period that western influence in mathematics was felt in China. An Italian Matteo Ricci (1552–1610), with the aid of Hsü Kuang-ching (1562–1634), in 1601–07 translated the first six books of Euclid's *Elements* into Chinese; this played a significant role in the subsequent development of mathematics in China.

Correspondents Ouyang Jiang and Zhang Liangjin have listed some twenty-six Chinese treatises on mathematics (some very comprehensive and encyclopedic) written prior to the nineteenth century.

INDIA

7–5 General Survey

Because of the lack of authentic records, very little is known of the development of ancient Hindu[3] mathematics. The earliest history is preserved in the 5000-year-old ruins of a city at Mohenjo Daro, located northeast of present-day Karachi in Pakistan. Evidence of wide streets, brick dwellings and apartment houses with tiled bathrooms, covered city drains, and community swimming pools indicates a civilization as advanced as that found anywhere else in the ancient Orient. These early peoples had systems of writing, counting, weighing, and measuring, and they dug canals for irrigation. All this required basic mathematics and engineering. It is not known what became of these peoples.

It was about 4000 years ago that wandering bands crossed the Himalaya passes into India from the great plains of central Asia. These people were called *Aryans,* from a Sanskrit word meaning "noblemen" or "owners of land." Many of these remained; others wandered into Europe and formed the root of the Indo-European stock. The influence of the Aryans gradually extended over all India. During their first thousand years, they perfected both written and spoken Sanskrit. They are also responsible for the introduction of the caste system. In the sixth century B.C., the Persian armies under Darius entered India but made no permanent conquests. To this period belong two great early Indians, the grammarian Panini and the religious teacher Buddha. This probably is also the approximate time of the *Śulvasūtras* ("the rules of the cord"), religious writings of interest in the history of mathematics because they embody geometric rules for the construction of altars by rope stretching and show an acquaintance with Pythagorean triples.

After the temporary conquest of northwest India by Alexander the Great in 326 B.C., the Maurya Empire was established and in time spread over all India and parts of central Asia. The most famous Maurya ruler was King Aśoka

[3] Because of confusion between western Indians and eastern Indians, writers frequently use the terms "Hindu" and "(eastern) Indian" interchangeably. Although this interchangeability is not strictly correct, it is common and convenient where there is no misunderstanding.

(272–232 B.C.), some of whose great stone pillars, erected in every important city in India of his day, still stand. These pillars are of interest to us because, as stated in Section 1–9, some of them contain the earliest preserved specimens of our present number symbols.

Pronunciation of Hindu Names

Some of the difficulty experienced in pronouncing Hindu names can be circumvented by observing the following accepted equivalents.

a like *u* in *but,* **ā** as in *father*.

e as in *they*.

i as in *pin,* **ī** as in *pique*.

o as in *so*.

u as in *put,* *ū* as in *rule*.

c like *ch* in *church*.

ś like English *sh*.

If the penult (next to last syllable) is long, it is accented; if it is short, the antipenult (third syllable from end) is accented.

After Aśoka, India underwent a series of invasions, that were finally followed by the Gupta dynasty of native Indian emperors. The Gupta period proved to be the golden age of the Sanskrit renaissance, and India became a center of learning, art, and medicine. Rich cities grew up and universities were

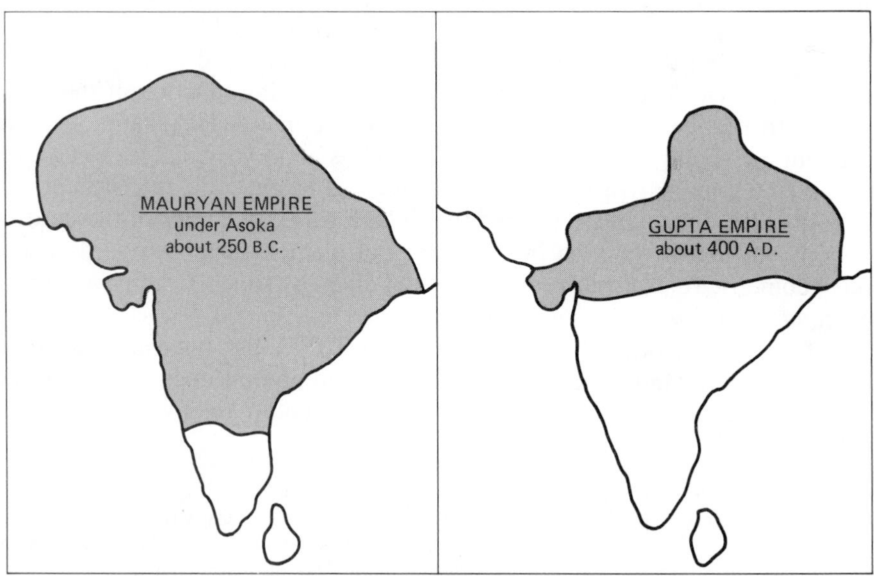

founded. The first important astronomical work, the anonymous *Sūrya Siddhānta* ("knowledge of the sun"), dates from this period, probably about the beginning of the fifth century. Hindu mathematics from here on became subservient to astronomy rather than religion. The sixth-century work *Pañca Siddhāntikā,* of the astronomer Varāhamihira of Ujjain and based on the earlier *Sūrya Siddhānta,* contains a good summary of early Hindu trigonometry and a table of sines apparently derived from Ptolemy's table of chords.

The degree of influence of Greek, Babylonian, and Chinese mathematics on Hindu mathematics, and vice versa, is still an unsettled matter, but there is ample evidence that influence in both directions was appreciable. One of the pronounced benefits of the Pax Romana was the diffusion of knowledge between East and West, and from a very early date, India exchanged diplomats with both the West and the Far East.

From about A.D. 450 until near the end of the 1400s, India was again subjected to numerous foreign invasions. First came the Huns, then the Arabs in the eighth century, and the Persians in the eleventh. During this period, there

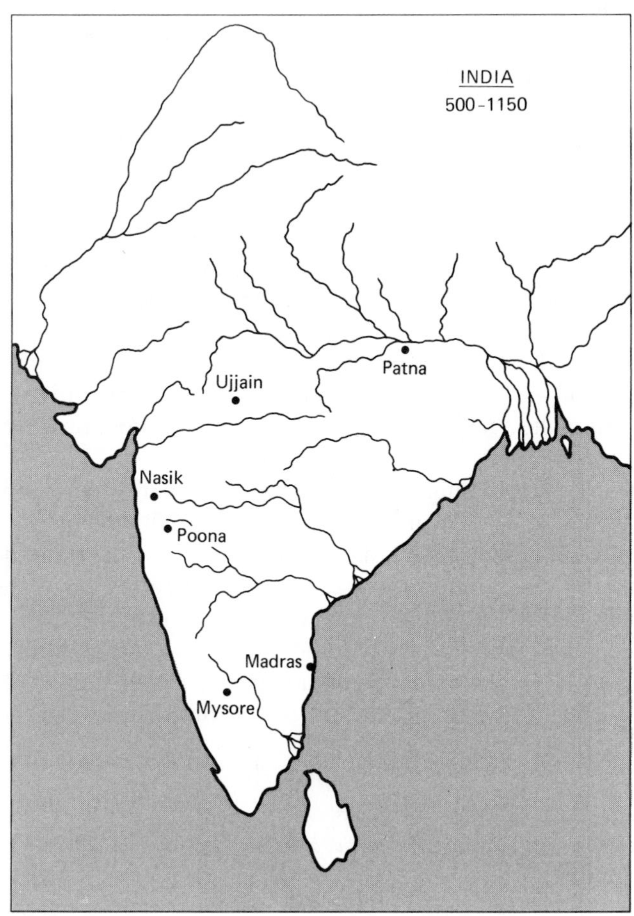

were several Hindu mathematicians of prominence, including the two Āryabhatas and Brahmagupta, Mahāvīra, and Bhāskara. The elder Āryabhata flourished in the sixth century and was born near present-day Patna on the Ganges. He wrote a work on astronomy entitled *Āryabhatiya,* of which the third chapter is devoted to mathematics. There is some confusion between the two Āryabhatas, and it may be that their work is not correctly differentiated. Brahmagupta was the most prominent Hindu mathematician of the seventh century. He lived and worked in the astronomical center of Ujjain, in central India. In 628, he wrote his *Brahma-sphuta-sidd'hānta* ("the revised system of Brahma"), a work on astronomy of twenty-one chapters, of which Chapters 12 and 18 deal with mathematics. Mahāvīra, who flourished about 850, was from Mysore in southern India and wrote on elementary mathematics. Bhāskara lived in Varāhamihira's and Brahmagupta's city of Ujjain. His work, *Siddhānta Śiromani* ("diadem of an astronomical system"), was written in 1150 and shows little advancement over the work of Brahmagupta of more than 500 years earlier. The important mathematical parts of Bhāskara's work are the *Lilāvati* ("the beautiful") and *Vijaganita* ("seed arithmetic"),[4] which deal with arithmetic and algebra, respectively. The mathematical parts of Brahmagupta's and Bhāskara's works were translated into English in 1817 by H. T. Colebrooke. The *Sūrya Siddhānta* was translated by E. Burgess in 1860, and Mahāvīra's work was published in 1912 by M. Rangācārya.

Hindu mathematics after Bhāskara made only spotty progress until modern times. In 1907, the Indian Mathematical Society was founded, and two years later the *Journal of the Indian Mathematical Society* started in Madras. The Indian statistics journal, *Sankhyā,* began publication in 1933.

Perhaps the most spectacular Indian mathematician of modern times was the impoverished clerk and untrained genius Srinivasa Ramanujan (1887–1920), who possessed amazing ability to see quickly and deeply into intricate number relations. He was "discovered" in 1913 by the eminent British number theorist, G. H. Hardy (1877–1947), whose efforts brought Ramanujan in the following year to England to study at Cambridge University. A most remarkable mathematical association resulted between the two men.

It is perhaps worth telling a couple of true anecdotes illustrating Ramanujan's amazing ability. Professor Hardy once visited Ramanujan in a hospital and incidentally remarked that he had arrived in a taxi with the dull number 1729. Without any hesitation, Ramanjan replied that 1729 was, on the contrary, very interesting, as it is the smallest integer expressible in two different ways as the sum of two cubes: $1^3 + 12^3 = 1729 = 9^3 + 10^3$. On another occasion, with no calculator other than his brain, Ramanujan remarked that $e^{\pi\sqrt{163}}$ is "very nearly" an integer: it is actually an integer followed by twelve zeros before another digit appears.

---

[4] It is not certain that the *Lilāvati* and *Vijaganita* are parts of the *Siddhānta Śiromani;* they may be separate works.

The publication in the 1920s of Ramanujan's notebooks, and the subsequent work done on them, has disclosed many facets of the man's unusual genius.

Texts on the history of mathematics show some contradictions and confusion when dealing with the Hindus. This is probably due, in no small measure, to the obscure and, at times, nearly unintelligible writing of the Hindu authors. The history of Hindu mathematics still awaits a more reliable and scholarly treatment.

7–6 Number Computing

In Section 1–9, we briefly considered the little that is known concerning the part played by the Hindus in the development of our present positional numeral system. We shall now give some account of Hindu methods of computing with this system. The key to an understanding of the algorithms that were elaborated lies in a realization of the writing materials that were at the disposal of the calculators. According to the German historian H. Hankel, they generally wrote either upon a small blackboard with a cane pen dipped in a thin, white paint that could easily be rubbed off or with a stick upon a white tablet less than a foot square and coated with a sprinkling of red flour. In either case the writing space was small and legibility demanded fairly large figures, but erasures and corrections were very easily effected. Accordingly, the calculation processes were schemed to conserve the writing space by erasing a digit as soon as it had served its purpose.

Early Hindu addition was perhaps done from left to right, instead of from right to left, as we prefer to do it today. As an example, consider the addition of 345 and 488. These would probably be written, one under the other, a little below the top of the computing tablet, as shown in the accompanying illustration. The computer would say 3 + 4 = 7, and write the 7 at the head of the left column. Next, 4 + 8 = 12, which changes the 7 to an 8, followed by a 2. The 7 is accordingly rubbed off and 82 written down. In our illustration, we have instead crossed out the 7 and written the 8 above it. Then 5 + 8 = 13, which changes the 2 to a 3, followed by another 3. Again things are corrected with a quick rub of the finger, and the final answer, 833, appears at the top of the tablet.

8	3	
~~7~~	~~2~~	3
3	4	5
4	8	8

Now the 345 and 488 can be rubbed off, and we have the rest of the tablet clear for further work.

In an undated commentary of Bhāskara's *Lilāvati,* we find another method, by which 345 and 488 would be added thus:

sum of units	5 + 8 = 13
sum of tens	4 + 8 = 12·
sum of hundreds	3 + 4 = 7··
sum of sums	= 833

Several methods were used for multiplication. The written work for the simple multiplication of, say, 569 by 5 might appear as follows, again working from left to right. On the tablet, a little below the top, write 569 followed, on the same line, by the multiplier 5. Then, since $5 \times 5 = 25$, 25 is written above the 569, as shown in the accompanying illustration. Next, $5 \times 6 = 30$, which changes the 5 in 25 to an 8 followed by a 0. A quick erasure fixes this. In the illustration we have instead crossed out the 5 and written the 8 above it. Then $5 \times 9 = 45$, which changes the 0 to a 4 followed by a 5. The final product, 2845, now appears at the top of the computing tablet.

  8 4
2 ~~5~~ ~~0~~ 5
  5 6 9   5

A more complicated multiplication, like 135×12, say, might be accomplished by first finding, as above, $135 \times 4 = 540$, then $540 \times 3 = 1620$, or by adding $135 \times 10 = 1350$ and $135 \times 2 = 270$ to get 1620. According to Hankel, it also might be accomplished as follows. A little below the top of the tablet, write the multiplicand 135 and the multiplier 12, so that the units digit in the multiplicand falls beneath the extreme left digit in the multiplier. Now $135 \times 1 = 135$, which is written at the top of the tablet. Next, by erasing, shift the multiplicand 135 one place to the right, and multiply by the 2 of the 12. In doing this, we find $2 \times 1 = 2$, which changes the 3 in our partial product to a 5. Then $2 \times 3 = 6$, which changes the two 5s in our new partial product to 61. Finally, $2 \times 5 = 10$, which changes the final 1 in our partial product to 2 followed by a 0. The finished product, 1620, now appears at the top of the tablet.

  6 2
  ~~5~~ ~~1~~
1 ~~3~~ ~~5~~ 0
    1 2
~~1~~ ~~3~~ ~~5~~
  1 3 5

Another method of multiplication, known to the Arabians and probably obtained from the Hindus that closely resembles our present process is indicated in the accompanying illustration, where we again find the product of 135 and 12. The lattice diagram is actually drawn, and the additions are performed diagonally. Note, because of the way each cell is divided in two by a diagonal, no carrying over is required in the multiplication.

The Arabians, who later borrowed some of the Hindu processes, were unable to improve on them and accordingly adapted them to "paper" work, where erasures were not easily effected, by crossing off undesired digits and

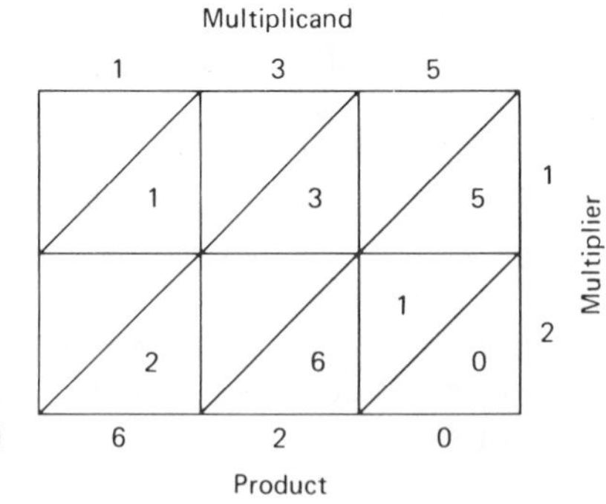

writing the new ones above or below the old ones, as we have done in the illustrations above.

The development of algorithms for our elementary arithmetic operations started in India, perhaps about the tenth or eleventh century, were adopted by the Arabians, and later carried to Western Europe, where they were modified into their present forms. This work received considerable attention from the fifteenth-century European writers on arithmetic.

7–7 Arithmetic and Algebra

The Hindus were gifted arithmeticians and made significant contributions to algebra.

Many of the arithmetical problems were solved by *false position*. Another favorite method of solution was that of **inversion,** where one works backward from a given piece of information. Consider, for example, the following problem, which appears in Bhāskara's *Lilāvati:* "Beautiful maiden with beaming eyes, tell me, as thou understandst the right method of inversion, which is the number which multiplied by 3, then increased by $\frac{3}{4}$ of the product, then divided by 7, diminished by $\frac{1}{3}$ of the quotient, multiplied by itself, diminished by 52, by the extraction of the square root, addition of 8, and division by 10 gives the number 2?" By the method of inversion we start with the number 2 and work backward. Thus, $[(2)(10) - 8]^2 + 52 = 196$, $\sqrt{196} = 14$, $(14)(\frac{3}{2})(7)(\frac{4}{7})/3 = 28$, the answer. Note that where the problem instructed us to divide by 10, we multiply by 10; where we were told to add 8, we subtract 8; where we were told to extract a square root, we take the square, and so forth. It is the replacement of each operation by its inverse that accounts for the name *inversion*. It is, of course, just what we would do if we were to solve the problem by modern methods. Thus, if we let x represent the sought number, we have

$$\frac{\sqrt{\left[\frac{(\frac{2}{3})(\frac{7}{4})(3x)}{7}\right]^2 - 52} + 8}{10} = 2.$$

To solve this we *multiply* both sides by 10, then *subtract* 8 from each side, then *square* both sides, and so forth. This problem also illustrates the Hindu practice of clothing arithmetical problems in poetic garb. This was because school texts were written in verse and because the problems were frequently used for social amusement.

The Hindus summed arithmetic and geometric progressions and solved commercial problems in simple and compound interest, discount, and partnership. They also solved *mixture* and *cistern* problems, similar to those found in modern texts. Several specimens of Hindu arithmetical problems may be found in Problem Studies 7.4, 7.5, and 7.6.

Much of our knowledge of Hindu arithmetic stems from Bhāskara's *Lilāvati*. A romantic story is told about this work. According to the tale, the stars foretold dire misfortune if Bhāskara's only daughter Lilāvati should marry other than at a certain hour on a certain propitious day. On that day, as the anxious bride was watching the sinking water level of the hour cup, a pearl fell unknowingly from her headdress and, stopping the hole in the cup, arrested the outflow of water, and so the lucky moment passed unnoticed. To console the unhappy girl, Bhāskara gave her name to his book.

The Hindus syncopated their algebra. Like Diophantus, addition was usually indicated by juxtaposition. Subtraction was indicated by placing a dot over the subtrahend, multiplication by writing *bha* (the first syllable of the word *bhavita,* "the product") after the factors, division by writing the divisor beneath the dividend, and square root by writing *ka* (from the word *karana,* "irrational") before the quantity. Brahmagupta indicated the unknown by *yā* (from *yāvattāvat,* "so much as"). Known integers were prefixed by *rū* (from *rūpa,* "the absolute number"). Additional unknowns were indicated by the initial syllables of words for different colors. Thus, a second unknown might be denoted by *kā* (from *kālaka,* "black"), and $8xy + \sqrt{10} - 7$ might appear as

$$\textit{yā kā}\ 8\ \textit{bha ka}\ 10\ \textit{rū}\ \dot{7}.$$

The Hindus admitted negative and irrational numbers, and recognized that a quadratic (having real answers) has two formal roots. They unified the algebraic solution of quadratic equations by the familiar method of completing the square. This method is today often referred to as the **Hindu method.** Bhāskara gave the two remarkable identities

$$\sqrt{a \pm \sqrt{b}} = \sqrt{(a + \sqrt{a^2 - b})/2} \pm \sqrt{(a - \sqrt{a^2 - b})/2},$$

which are sometimes employed in our algebra texts for finding the square root of a binomial surd. These identities are also found in Book X of Euclid's *Elements,* but they are presented in an involved language that is difficult to comprehend.

The Hindus showed remarkable ability in indeterminate analysis and were perhaps the first to devise general methods in this branch of mathematics. Unlike Diophantus, who sought *any one rational* solution to an indeterminate equation, the Hindus endeavored to find *all possible integral* solutions. Āryabhata and Brahmagupta found the integral solutions of the linear indeterminate equation $ax + by = c$, where a, b, c are integers. The indeterminate quadratic equation $xy = ax + by + c$ was solved by a method later reinvented by Euler. The work of Brahmagupta and Bhāskara on the so-called Pell equation,[5] $y^2 = ax^2 + 1$, where a is a nonsquare integer, is highly regarded by some. They showed how, from one solution x, y, where $xy \neq 0$, infinitely many others could be found. The complete theory of the Pell equation was finally worked

[5] This is an example of a misnomer that has stuck. The error of assignment is due to Euler, who mistakenly assumed that the Englishman John Pell (1611–1685) gave a method of solution of the equation that had really been given by Pell's countryman Lord Brouncker (ca. 1620–1684).

out by Lagrange in 1766–1769. The Hindu work on indeterminate equations reached western Europe too late to exert any beneficial influence.

7–8 Geometry and Trigonometry

The Hindus were not proficient in geometry. Rigid demonstrations were unusual, and postulational developments were nonexistent. Their geometry was largely empirical and generally connected with mensuration.

The ancient *Sulvasūtras* show that the early Hindus applied geometry to the construction of altars and in doing so made use of the Pythagorean relation. The rules furnished instructions for finding a square equal to the sum or difference of two given squares and of a square equal to a given rectangle. Solutions of the circle-squaring problem appear that are equivalent to taking $d = (2 + \sqrt{2})s/3$ and $s = 13d/15$, where d is the diameter of the circle and s the side of the equal square. There also appears the expression

$$\sqrt{2} = 1 + \frac{1}{3} + \frac{1}{(3)(4)} - \frac{1}{(3)(4)(34)},$$

which is interesting in that all the fractions are unit fractions and the expression is correct to five decimal places.

Both Brahmagupta and Mahāvīra not only gave Heron's formula for the area of a triangle in terms of the three sides, but also the remarkable extension,[6]

$$K = [(s - a)(s - b)(s - c)(s - d)]^{1/2},$$

for the area of a cyclic quadrilateral having sides a, b, c, d and semiperimeter s. It seems that later commentators failed to realize the limitation on the quadrilateral. The formula for the general case is

$$K^2 = (s - a)(s - b)(s - c)(s - d) - abcd \cos^2 \left(\frac{A + C}{2}\right),$$

where A and C are a pair of opposite vertex angles of the quadrilateral.

Most remarkable in Hindu geometry, and unique in their excellence, are Brahmagupta's theorems that the diagonals m and n of a cyclic quadrilateral having consecutive sides a, b, c, d are given by

$$m^2 = \frac{(ab + cd)(ac + bd)}{ad + bc},$$

$$n^2 = \frac{(ac + bd)(ad + bc)}{ab + cd},$$

---

[6] For a derivation of this formula see, for example, E. W. Hobson, *A Treatise on Plane Trigonometry*. 4th ed., p. 204, or R. A. Johnson, *Modern Geometry*, p. 81.

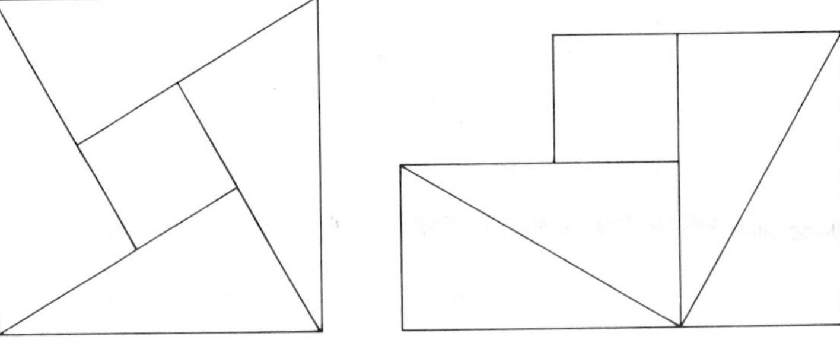

FIGURE 60

and that if a, b, c, A, B, C are positive integers, such that $a^2 + b^2 = c^2$ and $A^2 + B^2 = C^2$, then the cyclic quadrilateral having consecutive sides of aC, cB, bC, cA (called a **Brahmagupta trapezium**) has rational area and diagonals, and the diagonals are perpendicular to each other (see Problem Studies 7.9 and 7.10). Brahmagupta knew Ptolemy's theorem on the cyclic quadrilateral.

Many inaccuracies appear in Hindu mensuration formulas. Thus, Āryabhata gives the volume of a pyramid as *half* the product of the base and altitude, and the volume of a sphere as $\pi^{3/2}r^3$. The Hindus gave some accurate values for π, but also frequently used $\pi = 3$ and $\pi = \sqrt{10}$.

Most students of high-school geometry have seen Bhāskara's dissection proof of the Pythagorean theorem, in which the square on the hypotenuse is cut up, as indicated in Figure 60, into four triangles, each congruent to the given triangle, plus a square with side equal to the difference of the legs of the given triangle. The pieces are easily rearranged to give the sum of the squares on the two legs. Bhāskara drew the figure and offered no further explanation than the word "Behold!" A little algebra, however, supplies a proof; for if c is the hypotenuse and a and b are the legs of the triangle,

$$c^2 = 4\left(\frac{ab}{2}\right) + (b - a)^2 = a^2 + b^2.$$

This dissection proof is found much earlier in China. Bhāskara also gave a second demonstration of the Pythagorean theorem by drawing the altitude on the hypotenuse. From similar right triangles in Figure 61, we have

$$\frac{c}{b} = \frac{b}{m}, \qquad \frac{c}{a} = \frac{a}{n},$$

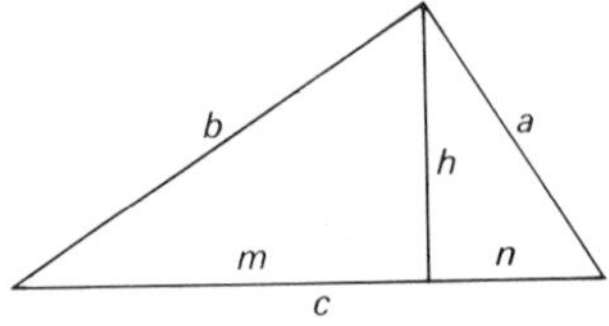

FIGURE 61

or

$$cm = b^2, \qquad cn = a^2.$$

Adding, we get

$$a^2 + b^2 = c(m + n) = c^2.$$

This proof was rediscovered by John Wallis in the seventeenth century.

The Hindus, like the Greeks, regarded trigonometry as a tool for their astronomy. They used our familiar degree, minute, and second divisions and constructed tables of sines. (That is, they constructed tables of half chords, and not tables of chords as the Greeks had done.) The Hindus employed the equivalents of sines, cosines, and versed sines (versin $A = 1 - \cos A$). They computed the sines of halves of angles by the relation versin $2A = 2 \sin^2 A$. In their astronomy, they solved plane and spherical triangles. The astronomy itself is of poor quality and shows an ineptness in observing, collecting, and collating facts, and inducing laws. Their trigonometry may be described as arithmetic rather than geometric.

7–9 Contrast Between Greek and Hindu Mathematics

There are many differences between Greek and Hindu mathematics. In the first place, the Hindus who worked in mathematics regarded themselves primarily as astronomers; thus, Hindu mathematics remained largely a handmaiden to astronomy. With the Greeks, mathematics attained an independent existence and was studied for its own sake. Also, as a result of the caste system, mathematics in India was cultivated almost entirely by the priests; in Greece, mathematics was open to any one who cared to study the subject. Again, the Hindus were accomplished computers but mediocre geometers; the Greeks excelled in geometry but cared little for computational work. Even Hindu trigonometry, which was meritorious, was arithmetic in nature; Greek trigonometry was geometric in character. The Hindus wrote in verse and often clothed their works in obscure and mystic language; the Greeks strove for clarity and logicality in presentation. Hindu mathematics is largely empirical, with proofs or derivations seldom offered; an outstanding characteristic of Greek mathematics is its insistence on rigorous demonstration. Hindu mathematics is of very uneven quality, good and poor mathematics often appearing side by side; the Greeks seemed to have an instinct that led them to distinguish good from poor quality and to preserve the former while abandoning the latter. As the Muslim writer al-Biruni put it in his well-known book *India,* in contrast to the uniformly high quality of Greek mathematics, Hindu mathematics was "a mixture of pearl shells and sour dates . . . of costly crystal and common pebbles."

Some of the contrast between Greek and Hindu mathematics is perpetuated today in the differences between many of our elementary geometry and algebra textbooks, because the former are deductive and the latter are often collections of rules.

ARABIA

7–10 The Rise of Moslem Culture

The rise and decline of the Arabian empire is one of the most spectacular episodes in history. Within the decade following Mohammed's flight from Mecca to Medina in A.D. 622, the scattered and disunited tribes of the Arabian peninsula were consolidated by a strong religious fervor into a powerful nation. Within a century, force of arms under the green and gold banner of Islam had extended the rule and influence of the Moslem star and crescent over a territory reaching from India, through Persia, Mesopotamia, and northern Africa, clear into Spain. Opposing contenders for the caliphate caused an east-west split in the empire in 755, resulting in one caliph reigning in Baghdad and another in Córdoba. Until about the year 1000, the eastern empire enjoyed spiritual supremacy. At that time, however, much of the eastern territory became overrun by the ruthless Seljuk Turks. Between 1100 and 1300, the Christian Crusades were launched to dislodge the Moslems from the Holy Land. In 1258, Baghdad was taken by the Mongols, the eastern caliph fell from power, and the Arabian empire began to decline. In 1492, Spain overthrew the last of its Moorish rulers, and the Arabs lost their European foothold.

Pronunciation of Arabian Names

The following equivalents will help in pronouncing Arabian names.

a as in *ask,* **â** as in *father.*

e as in *bed.*

i as in *pin,* **î** as in *pique.*

o as in *obey.*

u as in *put,* **û** as in *rule.*

d like *th* in *that,* **t** *like th* in *thin.*

h and **kh** like *ch* in the German *nach.*

q̇ like *c* or *k* in *cook.*

The accent is on the last syllable containing a long vowel or a vowel followed by two consonants. Otherwise the accent falls on the first syllable.

Of considerable importance for the preservation of much of world culture was the manner in which the Arabs seized upon Greek and Hindu erudition. The Baghdad caliphs governed well, and many became patrons of learning and invited distinguished scholars to their courts. Numerous Hindu and Greek works in astronomy, medicine, and mathematics were industriously translated into the Arabic tongue and thus were saved until later European scholars were able to retranslate them into Latin and other languages. But for the work of the Arabian scholars, much of Greek and Hindu science would have been irretrievably lost over the long period of the Dark Ages.

During the reign of the caliph al-Mansûr, Brahmagupta's works were brought to Baghdad (ca. 766) and, under royal patronage, translated into Arabic. It has been said that this was the means by which the Hindu numerals were brought into Arabic mathematics. The next caliph was Harun al-Rashid (Aaron the Just), who reigned from 786 to 808 and is known to us in connection with *The Arabian Nights*. Under his patronage, several Greek classics in science were translated into Arabic, among them part of Euclid's *Elements*. There was also a further influx of Hindu learning into Baghdad during his reign. Harun al-Rashid's son, al-Mâmûn, who reigned from 809 to 833, also was a patron of learning and was himself an astronomer. He built an observatory at Baghdad and undertook the measurement of the earth's meridian. The difficult task of obtaining satisfactory translations of Greek classics continued under his orders; the *Almagest* was put into Arabic and the translation of the *Elements* completed. Greek manuscripts were secured, as a condition in a peace treaty, from the emperor of the Byzantine Empire and were then translated by Syrian Christian scholars invited to al-Mâmûn's court. Many scholars wrote on mathematics and astronomy during this reign, the most famous being Mohammed ibn Mûsâ al-Khowârizmî (Mohammed, the son of Moses of Khwarezm). He wrote a treatise on algebra and a book on the Hindu numerals, both of which later exerted tremendous influence in Europe when translated into Latin in the twelfth century. A somewhat later scholar was T̲âbit ibn Qorra (826–901), famed as a physician, philosopher, linguist, and mathematician. He produced the first really satisfying Arabic translation of the *Elements*. His translations of Apollonius, Archimedes, Ptolemy, and Theodosius are said to rank among the best made. Especially important are his versions of Books V, VI, and VII of the *Conics* of Apollonius, as only through his versions have these books come to us. He also wrote on astronomy, the conics, elementary algebra, magic squares, and amicable numbers (see Problem Study 7.11).

Probably the most celebrated Moslem mathematician of the tenth century was Abû'l-Wefâ (940–998), born in the Persian mountain region of Khorâsân. He is known for his translation of Diophantus, his introduction of the *tangent* function into trigonometry, and his computation of a table of sines and tangents for 15′ intervals. To do this, he perfected Ptolemy's method, obtaining sin 30′ with nine exact decimal places. He wrote on a number of mathematical topics. Abû Kâmil and al-Karkhî, who wrote in the tenth and eleventh centuries, should be mentioned for their work in algebra. The former wrote a commentary on Al-Khowârizmî's algebra, which was later drawn upon by the European mathematician Fibonacci (1202). Al-Karkhî, who was a disciple of Diophantus, produced a work called the *Fakhrî,* one of the most scholarly of the Moslem works on algebra. But perhaps the deepest and most original algebraic contribution was the geometrical solution of cubic equations by Omar Khayyam (ca.

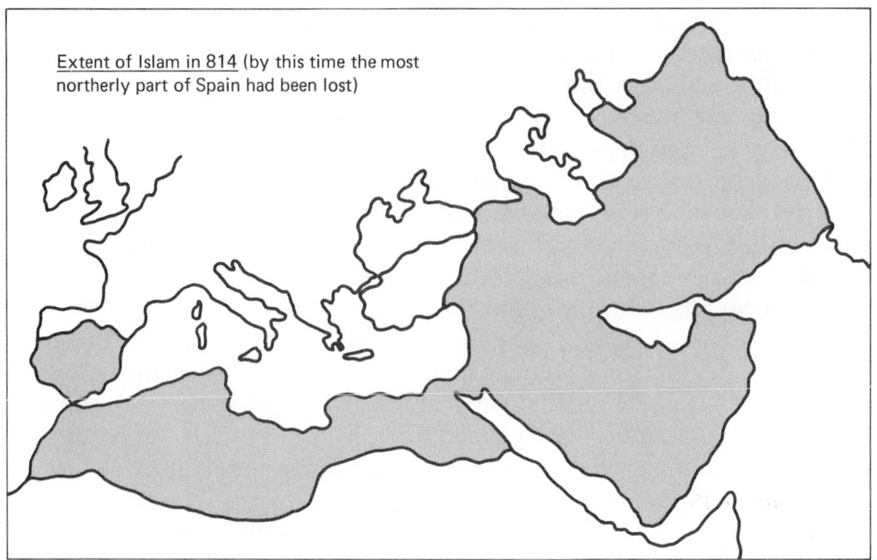
Extent of Islam in 814 (by this time the most
northerly part of Spain had been lost)

1100), another native of Khorâsân, known to the western world as the author of the exquisite *Rubaiyat*. Khayyam is also noted for his very accurate proposed calendar reform.

A considerably later writer was Nasîr ed-dîn (ca. 1250), also of Khorâsân. He wrote the first work on plane and spherical trigonometry considered independently of astronomy. Saccheri (1667–1733) started his work on non-Euclidean geometry through a knowledge of Nasîr ed-dîn's writings on Euclid's parallel postulate. His was the only attempt to prove this postulate in the period from the ancient Greeks up to the Renaissance. These writings were translated into Latin by John Wallis in the seventeenth century and used by him in his geometrical lectures at Oxford. Finally, there was Ulugh Beg, a fifteenth-century Persian astronomer of royal blood, who compiled remarkable tables of sines and tangents for $1'$ intervals correct to eight or more decimal places. In his court at Samarkand was Al-Kashi, mentioned in Section 7–2 for his accurate approximation of π. Al-Kashi did important work with decimal fractions and was the first Arabic author we know of who dealt with the binomial theorem in the "Pascal triangle" form.

7–11 Arithmetic and Algebra

Before Mohammed, the Arabians wrote out all numbers in words. The subsequent extensive administration of conquered lands was partly responsible for the introduction of a short symbolism. Sometimes local numeral systems were adopted, and at one time it was rather common practice to use a ciphered numeral system, like the Ionic Greek, employing the twenty-eight Arabic letters. This notation was, in turn, superseded by the Hindu notation, which was first adopted by merchants and writers on arithmetic. Strangely enough, the Hindu numerals are excluded from some of the later arithmetics of the eastern empire. Thus, Abû'l-Wefâ and al-Karkhî, of the tenth and eleventh centuries, wrote arithmetics in which all numbers are again written out in words. These later Arabian writers departed from Hindu teachings and became influenced by Greek methods. No trace of the use of an abacus has been discovered among the early Arabs.

The first Arabic arithmetic known to us is that of al-Khowârizmî; it was followed by a host of other Arabic arithmetics by later authors. These arithmetics generally explained the rules for computing, modeled after the Hindu algorithms. They also gave the process known as *casting out 9s,* used for checking arithmetical computations, and the rules of *false position* and *double false position,* by which certain algebra problems can be solved nonalgebraically (see Problem Studies 7.12 and 7.14). Square and cube roots, fractions, and the *rule of three* were also frequently explained.

The **rule of three,** which probably originated in ancient China, reached Arabia through India, and was actually called by this name by Brahmagupta and Bhāskara. For centuries, the rule was very highly regarded by merchants. It was mechanically stated without justification, and its connection with proportion was not recognized until the end of the fourteenth century. Here is how Brahmagupta stated the rule: *In the rule of three, Argument, Fruit, and Requisition are the names of the terms. The first and last terms must be similar. Requisition multiplied by Fruit, and divided by Argument, is the Produce.* For clarification, consider the following problem given by Bhāskara: If two and a half palas of saffron are purchased for three sevenths of a niska, how many palas will be purchased for nine niskas? Here $\frac{3}{7}$ and 9, which are of the same denomination, are the Argument and the Requisition, and $\frac{5}{2}$ is the Fruit. The answer, or Produce, is then given by $(9)(\frac{5}{2})/(\frac{3}{7}) = 52\frac{1}{2}$. Today we would regard the problem as a simple application of proportion,

$$x:9 = \tfrac{5}{2}:\tfrac{3}{7}.$$

Much space was devoted to the *rule of three* by the early European writers on arithmetic, the mechanical nature of the rule being observable in the doggerel verse and the schematic diagrams often used to explain it.

Al-Khowârizmî's algebra shows little originality. The four elementary operations are explained and linear and quadratic equations are solved, the latter both arithmetically and geometrically. The work contains some geometric mensuration and some problems on inheritance.

The Moslem mathematicians made their best contributions in the field of geometric algebra, the peak being reached in Omar Khayyam's geometric solution of cubic equations. Here cubic equations are systematically classified and a root obtained as the abscissa of a point of intersection of a circle and a rectangular hyperbola, or of two rectangular hyperbolas (see Problem Study 7.15). Khayyam rejected negative roots and frequently failed to discover all the positive ones. Cubic equations arose from the consideration of such problems as the construction of a regular heptagon and the Archimedean problem of cutting a sphere into two segments having a prescribed ratio. Abû'l-Wefâ gave geometric solutions to some special quartic equations.

Some of the Moslem mathematicians showed interest in indeterminate analysis; thus, a proof (probably defective and now lost) was given to the theorem that it is impossible to find two positive integers the sum of whose cubes is the cube of a third integer. This is a special case of Fermat's famous *last "theorem,"* to which we will return in Chapter 10. Mention has already been made of Tâbit ibn Qorra's rule for finding amicable numbers. This is said to be the first piece of original mathematical work done by an Arabian. Al-Kharkhî was the first Arabian writer to give and prove theorems furnishing the sums of the squares and cubes of the first n natural numbers.

Arabian algebra, except for that of the later western Arabs, was rhetorical.

7–12 Geometry and Trigonometry

The important role played by the Arabs in geometry was more one of preservation than one of discovery. The world owes them a large debt for their persevering efforts to translate satisfactorily the great Greek classics.

There was a nice geometric study done by Abû'l-Wefâ, in which he showed how to locate the vertices of the regular polyhedra on their circumscribed spheres, using compasses of fixed opening. We have already mentioned Omar Khayyam's geometric solution of cubic equations and Nasîr ed-dîn's influential work on the parallel postulate. Nasîr ed-dîn published, with comments and "corrections," part of an earlier work by Khayyam entitled *Discussion of the Difficulties in Euclid.* In this part of the earlier work, we find what was apparently the first consideration of the three alternatives later named, by Saccheri, the hypotheses of the acute, the obtuse, and the right angle (see Section 13–6). Nasîr ed-dîn is also credited with an original proof of the Pythagorean theorem. The proof is essentially the one we have suggested, in the notes to Problem Study 6.17(c), for Pappus' extension of the Pythagorean theorem.

The name al-Haiṯam, or, more popularly, Alhazen (ca. 965–1039), has been preserved in mathematics in connection with the so-called **problem of Alhazen:** To draw, from two given points in the plane of a given circle, lines that intersect on the circle and make equal angles with the circle at that point. The problem leads to a quartic equation that was solved in Greek fashion by an intersecting hyperbola and circle. Alhazen was born in Basra in southern Iraq and was perhaps the greatest of the Moslem physicists. The above problem arose in connection with his *Optics,* a treatise that later had great influence in Europe.

A pathetic story is told about Alhazen. He unfortunately once boasted that he could construct a machine that would control and regulate the annual inundation of the Nile River. He was accordingly summoned to Cairo by Caliph Hakim to explain and perhaps demonstrate his idea. Aware of the utter impracticality of his scheme, and fearing the anger of the Caliph, Alhazen feigned

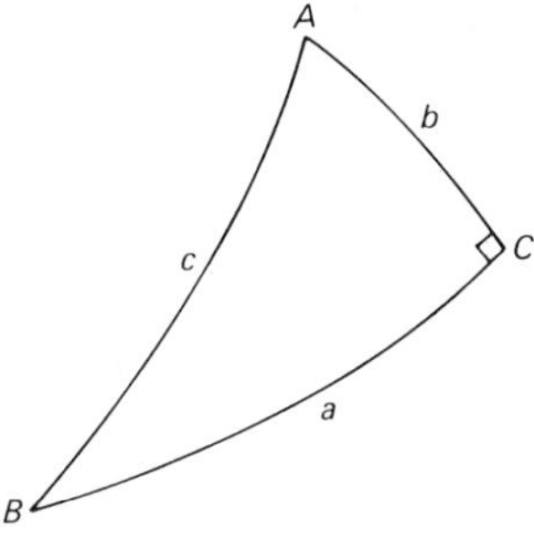

FIGURE 62

madness, for the insane were specially protected in those times. With great care, Alhazen had to keep up the hoax until Hakim's death in 1021.

Like the Hindus, the Arabian mathematicians generally regarded themselves primarily as astronomers and thus showed considerable interest in trigonometry. We have already mentioned some of the Moslem accomplishments in the construction of trigonometric tables. They may also be credited with using all six of the trigonometric functions and with improving upon the derivation of the formulas of spherical trigonometry. The law of cosines for an oblique spherical triangle,

$$\cos a = \cos b \cos c + \sin b \sin c \cos A,$$

was given by al-Battânî (Latinized as Albategnius, ca. 920). The formula

$$\cos B = \cos b \sin A,$$

for a spherical triangle ABC with a right angle at C (see Figure 62), is sometimes called **Geber's theorem,** after the western Moslem astronomer Jabir ibn Aflah (frequently called Geber, ca. 1130) who flourished at Seville.

7–13 Some Etymology

Many names and words used today may be traced back to the Arabian period; thus, anyone interested in observational astronomy probably is aware that a large number of star names, particularly those of the fainter stars, are Arabic. Well-known examples are Aldebaran, Vega, and Rigel, among the brighter stars, and Algol, Alcor, and Mizar, among the fainter ones. Many of the star names were originally expressions locating the stars in the constellations. These descriptive expressions when transcribed from Ptolemy's catalogue into the Arabic, later degenerated into single words, such as Betelgeuse (armpit of the Central One), Fomalhaut (mouth of the Fish), Deneb (tail of the Bird), Rigel (leg of the Giant), and so forth. In Section 6–5, we traced the derivation of *Almagest,* the Arabic name by which Ptolemy's great work is commonly known.

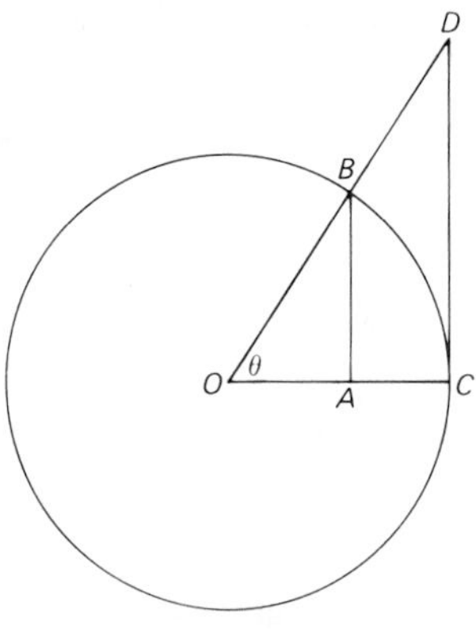

FIGURE 63

The origin of our word *algebra* from the title of al-Khowârizmî's treatise on the subject, *Hisâb al-jabr w'al-muqâ-balah,* is very interesting. This title has been literally translated as "science of the reunion and the opposition" or, more freely, as "science of transposition and cancellation."[7] The text, which is extant, became known in Europe through Latin translations, and made the word *al-jabr,* or *algebra,* synonymous with the science of equations. Since the middle of the nineteenth century, *algebra* has come, of course, to mean a great deal more.

The Arabic word *al-jabr,* used in a nonmathematical sense, found its way into Europe through the Moors of Spain. In Spain, an *algebrista* was a bonesetter (reuniter of broken bones), and it was usual for a barber of the times to call himself an *algebrista,* for bonesetting and bloodletting were sidelines of the medieval barber.

Al-Khowârizmî's book on the use of the Hindu numerals also introduced a word into the vocabulary of mathematics. This book is not extant in the original, but in 1857 a Latin translation was found that begins, "Spoken has Algoritmi, ..." Here the name *al-Khowârizmî* had become *Algoritmi,* from which, in turn, was derived our present word *algorithm,* meaning "the art of calculating in any particular way."

The meanings of the present names of the trigonometric functions, with the exception of *sine,* are clear from their geometrical interpretations when the angle is placed at the center of a circle of unit radius. Thus, in Figure 63, if the radius of the circle is one unit, the measures of tan θ and sec θ are given by the lengths of the *tangent* segment CD and the *secant* segment OD. Of course, *cotangent* merely means "complement's tangent," and so on. The functions tangent, cotangent, secant, and cosecant have been known by various other

[7] For a deeper analysis, see Solomon Gandz, "The origin of the term 'algebra'," *The American Mathematical Monthly* 33 (1926): 437–40.

names, but these particular names appeared as late as the end of the sixteenth century.

The origin of the word *sine* is curious. Āryabhata called it *ardhā-jyā* ("half chord") and also *jyā-ardhā* ("chord half"), and then abbreviated the term by simply using *jyā* ("chord"). From *jyā* the Arabs phonetically derived *jîba,* which, following Arabian practice of omitting vowels, was written as *jb*. Now *jîba,* aside from its technical significance, is a meaningless word in Arabic. Later writers who came across *jb* as an abbreviation for the meaningless *jîba* substituted *jaib* instead, which contains the same letters and is a good Arabic word meaning "cove" or "bay." Still later, Gherardo of Cremona (ca. 1150), when he made his translations from the Arabic, replaced the Arabian *jaib* by its Latin equivalent, *sinus,* from whence came our present word *sine*.

7–14 The Arabian Contribution

Estimates of the Arabian contribution to the development of mathematics are by no means in agreement. Some have assigned very high originality and genius to the Moslem writers, particularly in their work in algebra and trigonometry. Others see these writers as perhaps learned, but scarcely creative, and point out that their work is secondary, both in quantity and quality, to that of either the Greek or the modern writers. On the one hand, it must be admitted that they made at least small advances; on the other hand, it may be that their achievements, when viewed against the scientifically sterile backdrop of the rest of the world of the time, seem greater than they really were. There is still, in the balance in their favor, the outstanding fact that they served admirably as custodians of much of the world's intellectual possessions, which were transmitted to the later Europeans after the Dark Ages had passed.

Problem Studies

7.1 Some Problems From the "Arithmetic in Nine Sections"

Solve the following problems found in the *Arithmetic in Nine Sections*.

(a) Problem 11, Section IV. "Given a field of width, 1, $\frac{1}{2}$, $\frac{1}{3}$, $\frac{1}{4}$, $\frac{1}{5}$, $\frac{1}{6}$, $\frac{1}{7}$, $\frac{1}{8}$, $\frac{1}{9}$, $\frac{1}{10}$, $\frac{1}{11}$, and $\frac{1}{12}$ pu. It is known that the area of the field is 1 mu. What is the length of the field?" (A *pu* is a double pace; 1 *mu* = 240 square pu; the width of the field is $1 + \frac{1}{2} + \frac{1}{3} + \ldots + \frac{1}{12}$ pu.)

(b) Problem 14, Section IV. "Given a square field of 71,824 square pu. What is the side of the square?"

(c) Problem 16, Section I. "Given a field in the form of a segment of a circle, the base of which is $78\frac{1}{2}$ pu and the sagitta $13\frac{7}{9}$ pu. What is the area?" (Use the approximation formula $A = s(b + s)/2$.)

(d) Problem 1, Section VIII. "Three sheafs of good crop, 2 sheafs of mediocre crop, and 1 sheaf of bad crop are sold for 39 dou. Two sheafs of good, 3 of mediocre, and 1 of bad are sold for 34 dou. One sheaf of good, 2 of mediocre, and 3 of bad are sold for 26 dou. What is the price for a sheaf of good crop, mediocre crop, and bad crop?''

7.2 The Pythagorean Theorem

(a) Problem 11, Section IX, of the *Arithmetic in Nine Sections* reads: ''Given a door of which the height is larger than the width by 6 ch'ih 8 ts'un. The maximum distance between the vertices is 1 chang. What is the height and width of the door?'' (1 chang = 10 ch'ih, 1 ch'ih = 10 ts'un.)

(b) Solve the following problem, adapted from one in the *Arithmetic in Nine Sections:* ''There grows in the middle of a circular pond 10 feet in diameter a reed which projects one foot out of the water. When it is drawn down it just reaches the edge of the pond. How deep is the water?''

(c) Solve the *problem of the broken bamboo* (found in the *Arithmetic in Nine Sections* and later in a work of Yang Hui): ''There is a bamboo 10 feet high, the upper end of which being broken reaches the ground 3 feet from the stem. Find the height of the break.''

(d) Using a generalization of Figure 59, devise a proof of the Pythagorean theorem.

(e) Obtain a correct formula for the area of a segment of a circle in terms of the base b and sagitta s of the segment.

7.3 Magic Squares

No treatment, however brief, of ancient Chinese mathematics should omit mention of the so-called *lo-shu* magic square.

One of the oldest of the Chinese mathematical classics is the *I-king,* or *Book on Permutations*. In this appears a numerical diagram known as the lo-shu, later pictured as in Figure 64. The lo-shu is the oldest known example of a magic square, and myth claims that it was first seen by the Emperor Yu, in about 2200 B.C., decorating the back of a divine tortoise along a bank of the Yellow River. It is a square array of numerals indicated in Figure 64 by knots in strings—black knots for even numbers and white knots for odd numbers.

(a) An **nth order magic square** is a square array of n^2 distinct integers so arranged that the n numbers along any row, column, or main diagonal have the same sum, called the **magic constant** of the square. The magic square is said to be **normal** if the n^2 numbers are the first n^2 positive integers. Show that the magic constant of an nth-order normal magic square is $n(n^2 + 1)/2$.

(b) De la Loubère, when envoy of Louis XIV to Siam in 1687 through 1688, learned a simple method for finding a normal magic square of any odd order. Let us illustrate the method by constructing one of the fifth order. Draw a square and divide it into 25 cells (see Figure 65). Border the square with cells along the top and the right edge and shade the added cell in the top right corner. Write 1 in the middle top cell of the original square. The general rule is then to proceed diagonally upward

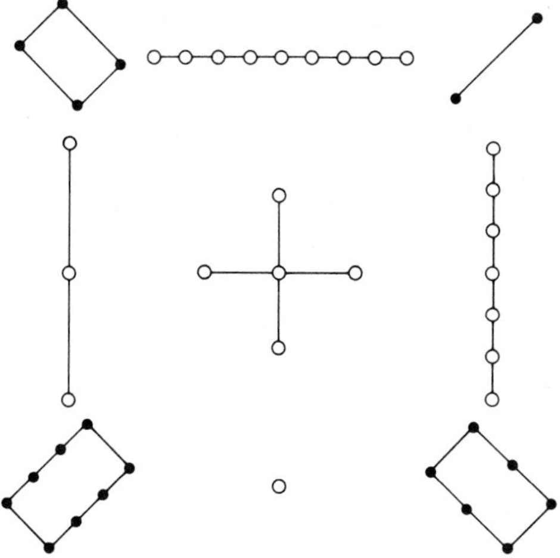

FIGURE 64

to the right with the successive integers. Exceptions to this general rule occur when such an operation takes us out of the original square or leads us into a cell already occupied. In the former situation, we get back into the original square by shifting clear across the square, either from top to bottom or from right to left, as the case may be, and continue with the general rule. In the second situation, we write the number in the cell immediately beneath the one last filled and then continue with the general rule. The shaded cell is to be regarded as occupied.

In our illustration, then, the general rule would place 2 diagonally upward from 1 in the fourth cell bordered along the top. We must therefore shift the 2 to the fourth cell in the bottom row of the original square. When we come to 4, it first falls in the third cell up bordered

	18	25	2	9	
17	24	1	8	15	17
23	5	7	14	16	23
4	6	13	20	22	4
10	12	19	21	3	10
11	18	25	2	9	

FIGURE 65

along the right edge. It must therefore be written clear across to the left in the third cell up in the first column of the original square. The general rule would place 6 in the cell already occupied by 1. It is accordingly written in the cell just below that occupied by the last written number, 5. And so on.

Construct a normal magic square of the seventh order.

(c) Show that the central cell of a normal magic square of the third order must be occupied by 5.

(d) Show that in a normal magic square of the third order 1 can never occur in a corner cell.

7.4 Some Early Hindu Problems

(a) Solve the following problem generalized from one given by Brahmagupta (ca. 630): "Two ascetics lived at the top of a cliff of height h, whose base was distant d from a neighboring village. One descended the cliff and walked to the village. The other, being a wizard, flew up a height x and then flew in a straight line to the village. The distance traversed by each was the same. Find x." In the original problem, $h = 100$ and $d = 200$.

(b) Solve the following version of the *problem of the broken bamboo* [see Problem Study 7.2(c)] given by Brahmagupta: "A bamboo 18 cubits high was broken by the wind. Its top touched the ground 6 cubits from the root. Tell the lengths of the segments of the bamboo."

(c) An anonymous arithmetic, known as the *Bakhshālī manuscript*,[8] was unearthed in 1881 at Bakhshali, in northwest India. It consists of 70 pages of birch bast. Its origin and date have been the subject of much conjecture, estimates of the date ranging from the third to the twelfth century A.D. Solve the following problem found in this manuscript: "A merchant pays duty on certain goods at three different places. At the first he gives $\frac{1}{3}$ of the goods, at the second $\frac{1}{4}$ [of the remainder], and at the third $\frac{1}{5}$ [of the remainder]. The total duty is 24. What was the original amount of goods?"

7.5 Problems from Mahāvīra

The nature of many of the Hindu arithmetical problems may be judged from the following, adapted from Mahāvīra (ca. 850). Solve these problems.

(a) A powerful, unvanquished, excellent black snake that is 80 angulas in length enters into a hole at the rate of $7\frac{1}{2}$ angulas in $\frac{5}{14}$ of a day; in the course of $\frac{1}{4}$ of a day, its tail grows $\frac{11}{4}$ of an angula. O ornament of arithmeticians, tell me by what time this serpent enters fully into the hole?

(b) Of a collection of mango fruits, the king took $\frac{1}{6}$, the queen $\frac{1}{5}$ of the remainder, and the three chief princes $\frac{1}{4}$, $\frac{1}{3}$, and $\frac{1}{2}$ of the successive

[8] See H. O. Midonick, *The Treasury of Mathematics*. New York: Philosophical Library, 1965, pp. 92–105.

remainders, and the youngest child took the remaining 3 mangoes. O you who are clever in miscellaneous problems on fractions, give out the measure of that collection of mangoes.

(c) The mixed price of 9 citrons and 7 fragrant wood apples is 107; again, the mixed price of 7 citrons and 9 fragrant wood apples is 101. O you arithmetician, tell me quickly the price of a citron and of a wood apple here, having distinctly separated those prices well.

(d) One fourth of a herd of camels was seen in the forest; twice the square root of that herd had gone to the mountain slopes; and 3 times 5 camels remained on the riverbank. What is the numerical measure of that herd of camels?

7.6 Problems from Bhāskara

Hindu arithmetical problems usually involved quadratics, the Pythagorean theorem, arithmetic progressions, and permutations. Consider the following problems adapted from Bhāskara (ca. 1150).

(a) The square root of half the number of bees in a swarm has flown out upon a jessamine bush, $\frac{8}{9}$ of the swarm has remained behind; a female bee flies about a male that is buzzing within a lotus flower into which he was allured in the night by its sweet odor, but is now imprisoned in it. Tell me, most enchanting lady, the number of bees.

(b) A snake's hole is at the foot of a pillar that is 15 cubits high, and a peacock is perched on its summit. Seeing a snake, at a distance of thrice the pillar's height, gliding toward his hole, he pounces obliquely upon him. Say quickly at how many cubits from the snake's hole do they meet, both proceeding an equal distance?

(c) In an expedition to seize his enemy's elephants, a king marched 2 yojanas the first day. Say, intelligent calculator, with what increasing rate of daily march did he proceed, since he reached his foe's city, a distance of 80 yojanas, in a week?

(d) How many are the variations in the form of the god Sambu (Siva) by the exchange of his 10 attributes held reciprocally in his several hands: namely, the rope, the elephant's hook, the serpent, the tabor, the skull, the trident, the bedstead, the dagger, the arrow, the bow: as those of Hari by the exchange of the mace, the discus, the lotus, and the conch?

(e) Arjuna, exasperated in combat, shot a quiver of arrows to slay Carna. With half his arrows he parried those of his antagonist; with 4 times the square root of the quiverful he killed his horse; with 6 arrows he slew Salya (Carna's charioteer); with 3 he demolished the umbrella, standard, and bow; and with 1 he cut off the head of the foe. How many were the arrows which Arjuna let fly?

7.7 Quadratic Surds

A numerical radical in which the radicand is rational but the radical itself is irrational is called a **surd.** A surd is called **quadratic, cubic,** and so on, according as its index is 2, 3, and so on.

(a) Show that a quadratic surd cannot be equal to the sum of a nonzero rational number and a quadratic surd.

(b) Show that if $a + \sqrt{b} = c + \sqrt{d}$, where $\sqrt{b}$ and $\sqrt{d}$ are surds and a and c are rational, then $a = c$ and $b = d$.

(c) Establish Bhāskara's identities given in Section 7–6, and use 1 of them to express $\sqrt{17 + \sqrt{240}}$ as the sum of 2 quadratic surds.

7.8 Indeterminate Equations of the First Degree

The Hindus solved the problem of finding all integral solutions of the linear indeterminate equation $ax + by = c$, where a, b, c are integers.

(a) If $ax + by = c$ has an integral solution, show that the greatest common divisor of a and b is a divisor of c. (This theorem says that there is no loss in generality if we consider a and b to be relatively prime.)

(b) If x_1 and y_1 constitute an integral solution of $ax + by = c$, where a and b are relatively prime, show that all integral solutions are given by $x = x_1 + mb$, $y = y_1 - ma$, where m is an arbitrary integer. [This theorem says that all integral solutions are known if just 1 integral solution can be found. A simple way of finding 1 integral solution is illustrated in the suggestions for Problem Study 7.8(c).]

(c) Solve $7x + 16y = 209$ for positive integral solutions.

(d) Solve $23x + 37y = 3000$ for positive integral solutions.

(e) In how many ways can the sum of 5 dollars be paid in dimes and quarters?

(f) Find the smallest permissible answer to the following indeterminate problem of Mahāvīra: "Into the bright and refreshing outskirts of a forest, which were full of numerous trees with their branches bent down with the weight of flowers and fruits, trees such as jambu trees, lime trees, plantains, areca palms, jack trees, date palms, hintala trees, palmyras, punnāga trees, and mango trees—outskirts, the various quarters whereof were filled with the many sounds of crowds of parrots and cuckoos found near springs containing lotuses with bees roaming about them—into such forest outskirts a number of weary travelers entered with joy. There were 63 numerically equal heaps of plantain fruits put together and combined with 7 more of those same fruits, and these were equally distributed among 23 travelers so as to have no remainder. You tell me now the numerical measure of a heap of plantains."

7.9 The Diagonals of a Cyclic Quadrilateral

Establish the following chain of theorems:

(a) The product of 2 sides of a triangle is equal to the product of the altitude on the third side and the diameter of the circumscribed circle.

(b) Let $ABCD$ be a cyclic quadrilateral of diameter δ. Denote the lengths of sides AB, BC, CD, DA, by a, b, c, d, the diagonals BD and AC by m and n, and the angle between either diagonal and the perpendicular upon the other by θ. Show that

$$m\delta \cos \theta = ab + cd, \qquad n\delta \cos \theta = ad + bc.$$

(c) Show, for the above quadrilateral, that

$$m^2 = \frac{(ac + bd)(ab + cd)}{ad + bc},$$

$$n^2 = \frac{(ac + bd)(ad + bc)}{ab + cd}.$$

(d) If, in the above quadrilateral, the diagonals are perpendicular to each other, then

$$\delta^2 = \frac{(ad + bc)(ab + cd)}{ac + bd}.$$

7.10 Brahmagupta's Quadrilaterals

(a) Brahmagupta gave the formula $K^2 = (s - a)(s - b)(s - c)(s - d)$ for the area K of a cyclic quadrilateral of sides a, b, c, d and semiperimeter s. Show that Heron's formula for the area of a triangle is a special case of this formula.

(b) Using Brahmagupta's formula of (a), show that the area of a quadrilateral possessing both an inscribed and a circumscribed circle is equal to the square root of the product of its sides.

(c) Show that a quadrilateral has perpendicular diagonals if and only if the sum of the squares of 1 pair of opposite sides is equal to the sum of the squares of the other pair of opposite sides.

(d) Brahmagupta showed that if $a^2 + b^2 = c^2$ and $A^2 + B^2 = C^2$, then any quadrilateral having aC, cB, bC, cA for consecutive sides has perpendicular diagonals. Prove this.

(e) Find the sides, diagonals, circumdiameter, and area of the Brahmagupta trapezium (see Section 7–7) determined by the 2 Pythagorean triples (3, 4, 5) and (5, 12, 13).

7.11 T̲âbit ibn Qorra, Al-Karkhî, and Nasîr ed-dîn

(a) T̲âbit ibn Qorra (826–901) invented the following rule for finding amicable numbers: *If* $p = 3 \cdot 2^n - 1$, $q = 3 \cdot 2^{n-1} - 1$, $r = 9 \cdot 2^{2n-1} - 1$ *are 3 odd primes then* $2^n pq$ and $2^n r$ *are a pair of amicable numbers*. Verify this for $n = 2$ and $n = 4$ (see Section 3–3).

(b) Establish the following generalization of the Pythagorean theorem, given by T̲âbit ibn Qorra: If triangle ABC is *any* triangle, and if B' and C' are points on BC, such that $\measuredangle AB'B = \measuredangle AC'C = \measuredangle A$, then $(AB)^2 + (AC)^2 = BC(BB' + CC')$.

Show that when angle A is a right angle this theorem becomes the Pythagorean theorem.

(c) The Arabians claimed that Archimedes wrote a work *On the Heptagon in a Circle*. Such a work by Archimedes has not come down to us, but the claim acquired more substance when the following remarkable theorem, handed down to us by Tâbit ibn Qorra, became known: If C and D are points on a segment AB, such that $(AD)(CD) = (DB)^2$, $(CB)(DB) = (AC)^2$, and if H is found, such that $CH = AC$, $DH = DB$, then HB is a side of a regular heptagon inscribed in the circumcircle of triangle AHB; furthermore, if HC and HD produced intersect the circle in F and E, respectively, then A, F, E are 3 consecutive vertices of the regular heptagon. Establish this theorem.

(d) Al-Karkhî (ca. 1020) wrote a work on algebra called the *Fakhrî,* named after his patron Fakhr al-Mulk, the grand vizier of Baghdad at the time. Problem 1 of Section 5 of the *Fakhrî* requests one to find 2 rational numbers such that the sum of their cubes is the square of a rational number. In other words, find rational numbers x, y, z such that

$$x^3 + y^3 = z^2.$$

Al-Karkhî essentially takes

$$x = \frac{n^2}{1 + m^3}, \qquad y = mx, \qquad z = nx,$$

where m and n are arbitrary rational numbers. Verify this, and find x, y, z for $m = 2$ and $n = 3$.

(e) Prove the following easy theorem, credited to Nasîr ed-dîn: The sum of 2 odd squares cannot be a square.

7.12 Casting Out 9s

(a) Show that when the sum of the digits of a natural number is divided by 9, one obtains the same remainder as when the number itself is divided by 9.

The act of obtaining the remainder when a given natural number is divided by an integer n is known as **casting out n's.** The above theorem shows that it is particularly easy to cast out 9s.

(b) Let us call the remainder obtained when a given natural number is divided by 9, the excess for that number. Prove the following 2 theorems:

1. *The excess for a sum is equal to the excess for the sum of the excesses of the addends.*
2. *The excess for the product of 2 numbers is equal to the excess for the product of the excesses of the 2 numbers.*

These two theorems furnish the basis for checking addition and multiplication by casting out 9s.

(c) Add and then multiply 478 and 993, and check by casting out 9s.

(d) Show that if the order of the digits of a natural number are permuted in any way to form a new number, then the difference between the old and the new numbers is divisible by 9.

This furnishes the basis for the **bookkeeper's check:** If the sums of the debit and credit entries in double entry bookkeeping do not balance, and the difference between the 2 sums is divisible by 9, then it is quite likely that the error is due to a transposition in digits made when transcribing a debit or a credit into the book.

(e) Explain the following number trick: Someone is asked to think of a number; form a new number by reversing the order of the digits; subtract the smaller from the larger number; multiply the difference by any number whatever; scratch out any nonzero digit in the product; and announce what is left. The conjurer finds the scratched-out digit by calculating the excess for the announced result and then subtracting this excess from 9.

(f) Generalize the theorem of (a) for an arbitrary base b.

7.13 Casting Out 11s

(a) Prove the following 3 theorems concerning casting out 11s:

1. Let s_1 be the sum of the digits in the odd places of any natural number n, and let s_2 be the sum of the digits in the even places. Then the excess of 11s for n is equal to the excess of 11s for the difference $s_1 - s_2$, where if $s_1 < s_2$, we increase s_1 by adding a multiple of 11.
2. To find the excess of 11s for any natural number, subtract the left-hand digit from its neighbor; subtract this difference from the next digit to the right, and so on; if at any time the subtrahend is greater than the minuend, add 11 to the minuend.
3. In casting out 11s, you may discard any pair of like consecutive digits.

(b) Find the excess of 11s for 180,927 and for 810,297, using the theorem of (a) 1. Find the excess of 11s for the same two numbers using the theorem of (a) 2. Find the excess of 11s for 148,337.

(c) Prove the following four theorems:

1. *The excess of 11s for a sum is equal to the excess for the sum of the excesses for the addends.*
2. *The excess of 11s for the minuend is equal to the excess for the sum of the excesses for the difference and subtrahend.*
3. *The excess of 11s for the product of 2 numbers is equal to the excess for the product of the excesses for the 2 numbers.*
4. *The excess of 11s for the dividend is equal to the excess for the product of the excesses for the divisor and quotient increased by the excess for the remainder.*

(d) Check the addition $104 + 454 + 1096 + 2195 + 3566 + 4090 = 11{,}505$ by casting out 11s.

(e) Check the subtraction $23{,}028 - 8476 = 14{,}552$ by casting out 11s.

(f) Check the multiplication $(8205)(536) = 4{,}397{,}880$ by casting out 11s.

(g) Check the division $62{,}540/207 = 302 + 26/207$ by casting out 11s.

7.14 Double False Position

(a) One of the oldest methods for approximating the real roots of an equation is the rule known as *regula duorum falsorum,* often called the **rule of double false position.** This method seems to have originated in China, from whence it spread to India and Arabia. In brief, and in modern form, the method is this: Let x_1 and x_2 be 2 numbers lying close to and on each side of a root x of the equation $f(x) = 0$. Then the intersection with the x-axis of the chord joining the points $(x_1, f(x_1))$, $(x_2, f(x_2))$ gives an approximation x_3 to the sought root (see Figure 66). Show that

$$x_3 = \frac{x_2 f(x_1) - x_1 f(x_2)}{f(x_1) - f(x_2)}.$$

The process can now be applied with the appropriate pair x_1, x_3 or x_3, x_2.

(b) Compute, by double false position, to three decimal places, the root of $x^3 - 36x + 72 = 0$ that lies between 2 and 3.

(c) Compute, by double false position, to three decimal places, the root of $x - \tan x = 0$ that lies between 4.4 and 4.5.

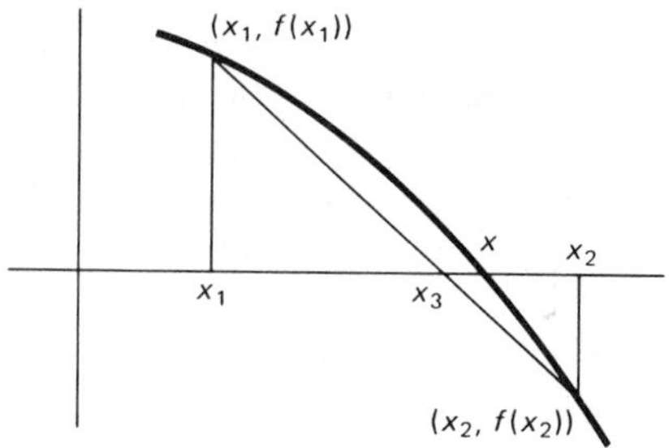

FIGURE 66

7.15 Khayyam's Solution of Cubics

(a) Given line segments of lengths a, b, n, construct a line segment of length $m = a^3/bn$.

(b) Omar Khayyam was the first to handle every type of cubic that possesses a positive root. Complete the details in the following sketch of Khayyam's geometrical solution of the cubic

$$x^3 + b^2x + a^3 = cx^2,$$

where a, b, c, x are thought of as lengths of line segments. Khayyam stated this type of cubic rhetorically as "a cube, some sides, and some numbers are equal to some squares."

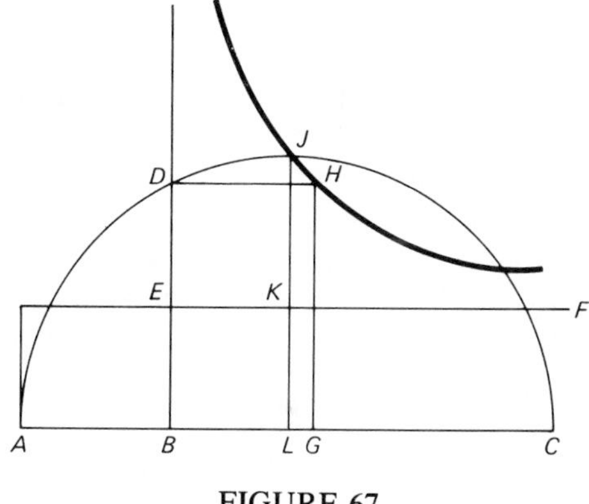

FIGURE 67

In Figure 67, construct $AB = a^3/b^2$ [by (a)] and $BC = c$. Draw a semicircle on AC as diameter and let the perpendicular to AC at B cut it in D. On BD, mark off $BE = b$ and through E draw EF parallel to AC. Find G on BC, such that $(BG)(ED) = (BE)(AB)$, and complete the rectangle $DBGH$. Through H, draw the rectangular hyperbola having EF and ED for asymptotes, and let it cut the semicircle in J. Let the parallel to DE through J cut EF in K and BC in L. Show, successively, that

1. $(EK)(KJ) = (BG)(ED) = (BE)(AB)$
2. $(BL)(LJ) = (BE)(AL)$
3. $(LJ)^2 = (AL)(LC)$
4. $(BE)^2/(BL)^2 = (LJ)^2/(AL)^2 = LC/AL$
5. $(BE)^2(AL) = (BL)^2(LC)$
6. $b^2(BL + a^3/b^2) = (BL)^2(c - BL)$
7. $(BL)^3 + b^2(BL) + a^3 = c(BL)^2$

Thus, BL is a root of the given cubic equation.

(c) Find geometrically, by Omar Khayyam's method, the positive roots of the cubic equation $x^3 + 2x + 8 = 5x^2$. Extend the method slightly to find the negative root.

7.16 A Geometric Solution of Cubics

(a) Show that the incomplete cubic equation

$$ax^3 + bx + c = 0$$

can be solved geometrically for its real roots on a rectangular Cartesian coordinate framework on which the cubic curve $y = x^3$ has already been drawn, by merely drawing the line $ay + bx + c = 0$.

(b) Solve, by the method of (a), the cubic equation $x^3 + 6x - 15 = 0$.

(c) Solve the cubic equation $4x^3 - 39x + 35 = 0$ geometrically.

(d) Show that any complete cubic equation

$$ax^3 + bx^2 + cx + d = 0$$

can be reduced to the incomplete form in the variable z by the substitution $x = z - b/3a$.

(e) Now solve the cubic equation $x^3 + 9x^2 + 20x + 12 = 0$ geometrically.

It is interesting that any complex imaginary roots possessed by either an incomplete or a complete cubic equation can also be found geometrically. (See, for example, Arthur Schultze, *Graphic Algebra.* New York: Macmillan Company, 1922, Sections 58, 59, 65.)

7.17 Geometrical Constructions on a Sphere

The Arabians were interested in constructions on a spherical surface. Consider the following problems, to be solved with Euclidean tools and appropriate plane constructions.

(a) Given a material sphere, find its diameter.

(b) On a given material sphere, locate the vertices of an inscribed cube.

(c) On a given material sphere, locate the vertices of an inscribed regular tetrahedron.

Essay Topics

7/1 The burning of the books in China in 213 B.C.

7/2 Chinese mathematical works prior to 1200.

7/3 *The Sea Island Mathematical Manual.*

7/4 Marco Polo's visit to China.

7/5 Matteo Ricci (1552–1610).

7/6 The rule of double false position.

7/7 The influence of Chinese and Hindu mathematics on European mathematics.

7/8 Hindu mathematical works prior to A.D. 1200.

7/9 The two Āryabhatas.

7/10 Mahāvīra and his work.

7/11 Srinivasa Ramanujan (1887–1920).

7/12 The Baghdad school.

7/13 Al-Khowârizmî's *Al-jabr.*

7/14 Abû'l-Wefâ (940–998).

7/15 The mathematical contributions of Omar Khayyam.

7/16 Al-Kashi's contributions to mathematics.

7/17 Greek mathematical works that would have been lost but for the Arabs.

7/18 Causes of decay of Moslem mathematics.

7/19 History of early Japanese mathematics.

7/20 The transmission of mathematical knowledge following the conquests of the Macedonians, the Moslems, and the Romans.

Bibliography

BERGGREN, J. L. *Episodes in the Mathematics of Medieval Islam*. New York: Springer-Verlag, 1986.

CAJORI, FLORIAN *A History of Mathematical Notations*. 2 vols. Chicago: Open Court, 1928–29.

CLARK, W. E., ed. *The Aryabhatiya of Aryabhata*. Chicago: Open Court, 1930.

COOLIDGE, J. L. *A History of Geometrical Methods*. New York: Oxford University Press, 1940.

——— *The Mathematics of Great Amateurs*. New York: Oxford University Press, 1949.

DATTA, B. *The Science of the Sulba: A Study in Early Hindu Geometry*. Calcutta: University of Calcutta, 1932.

———, and A. N. SINGH *History of Hindu Mathematics*. Bombay: Asia Publishing House, 1962.

HARDY, G. H. *A Mathematician's Apology*. Foreword by C. P. Snow. Cambridge: The University Press, 1967.

HEATH, T. L. *A Manual of Greek Mathematics*. New York: Oxford University Press, 1931.

HILL, G. F. *The Development of Arabic Numerals in Europe*. New York: Oxford University Press, 1915.

HOBSON, E. W. *A Treatise on Plane Trigonometry*. 4th ed. New York: Macmillan, 1902. Reprinted by Dover, New York.

JOHNSON, R. A. *Modern Geometry*. Boston: Houghton Mifflin Company, 1929. Reprinted by Dover, New York.

KAKHEL, ABDUL-KADER *Al-Kashi on Root Extraction*. Lebanon: 1960.

KASIR, D. S., ed. *The Algebra of Omar Khayyam*. New York: Columbia Teachers College, 1931.

KARPINSKI, L. C., ed. *Robert of Chester's Latin Translation of the Algebra of al-Khowarizmi*. New York: Macmillan, 1915.

——— *The History of Arithmetic*. New York: Russell & Russell, 1965.

KRAITCHIK, MAURICE *Mathematical Recreations*. New York: W. W. Norton, 1942.

KÛSHYÂR IBN LABBÂN *Principles of Hindu Reckoning*. Translated by Martin Levey and Marvin Petruck. Madison, Wis.: The University of Wisconsin Press, 1965.

LAMB, HAROLD *Omar Khayyam, A Life*. New York: Doubleday, 1936.

LARSEN, H. D. *Arithmetic for Colleges*. New York: Macmillan, 1950.

LEVEY, MARTIN *The Algebra of Abú Kâmil*. Madison, Wis.: The University of Wisconsin Press, 1966.

LI YAN and DU SHIRAN *Chinese Mathematics: A Concise History*. Translated by J. N. Crossley and A. W.-C. Lun. Oxford: Clarendon Press, 1987.

LOOMIS, E. S. *The Pythagorean Proposition*. 2d ed. Ann Arbor, Mich.: Edwards Brothers, 1940. Reprinted by the National Council of Teachers of Mathematics, Washington, D.C., 1968.

MACFALL, HALDANE *The Three Students*. New York: Alfred A. Knopf, 1926.

MIKAMI, YOSHIO *The Development of Mathematics in China and Japan*. New York: Hafner, 1913. Reprinted by Chelsea, New York, 1961.

NEEDHAM, J., with the collaboration of WANG LING *Science and Civilization in China,* vol. 3. New York: Cambridge University Press, 1959.

ORE, OYSTEIN *Number Theory and Its History*. New York: McGraw-Hill, 1948.

SAYILL, AYDIN *Logical Necessities in Mixed Equations by 'Abd al Hamid ibn Turk and the Algebra of His Time*. Ankara: 1962.

SMITH, D. E., and L. C. KARPINSKI *The Hindu-Arabic Numerals*. Boston: Ginn, 1911.

SMITH, D. E., and YOSHIO MIKAMI *A History of Japanese Mathematics*. Chicago: Open Court, 1914.

STORY, W. E. *Omar Khayyam as a Mathematician*. (Read at a meeting of the Omar Khayyam Club of America, April 6, 1918). Needham, Mass.: Rosemary Press, 1919.

WINTER, H. J. J. *Eastern Science*. London: John Murray, 1952.

WOLFE, H. E. *Introduction to Non-Euclidean Geometry*. New York: Holt, Rinehart and Winston, 1945.

Cultural Connection 

SERFS, LORDS, AND POPES

The European Middle Ages—A.D. 476–1492
(to accompany Chapter Eight)

Beginning in the fifth century A.D., with the fall of Rome to "barbarian" invaders, Europe was transformed from an ancient civilization to a medieval one. As we noted in Cultural Connection IV: The Oikoumene, the ancient agricultural societies of the west were merged politically, socially, and economically as a consequence of the Persian conquest of Egypt in 525 B.C. To be sure, the melding had never been total; Egyptian culture remained distinct from Greek, just as the Romans differed from Arabs or Jews. Nevertheless, there was a very real unity to western civilization in the thousand years between the founding of Persia and the fall of Rome—a unity that manifested itself through such things as shared commercial networks, similar economic systems, related religions, and often a single political hegemony. People living then sensed this sameness and gave voice to it in geographic terms; the Greeks referred to Greece, Italy, Egypt, and the Middle East collectively as the *oikoumene,* or "inhabited/civilized world."

Not only had the ancient West in many respects formed a single civilization, it also had been an expanding one. Over the course of a millenium, successive empires had carried western civilization to new places. The Persian Empire brought Middle Eastern and Egyptian culture to what is now Iran; the Greeks colonized the Mediterranean Sea coasts of Cyprus, Libya, Italy, and France, and the Black Sea coasts of Turkey and Russia; the Romans extended western civilization into the rest of Italy and France, northwestern Africa, Spain, and England. By the beginning of the fifth century A.D., western civilization extended through a region that ranged from the icy North Sea to the searing sands of Egypt, and from Gibraltar to the Persian Gulf.

After the Roman Empire collapsed in A.D. 476, however, western civilization changed in many respects. The West became divided into two very distinct cultural areas: the Arab-Iranian world and Europe. (As the reader will recall, we discussed the rise of the religion of Islam and Arab-Iranian culture in Cultural Connection V: The Asian Empires.) Furthermore, a second, although less severe, fragmentation sundered Europe into a Germanic-Latin west and a Greco-Slavic east, a split still recognizable in the twentieth century. Also, the political and cultural center of Europe shifted slowly northward, from the Mediterranean Sea basin (Greece and Rome), to lands bordering the North and Baltic Seas: France, England, the Netherlands, Germany, Scandinavia, Poland, and Russia. The great empires of the ancient world eventually gave way to feudal baronies. Slaves and yeomen farmers alike were replaced by serfs. Scholars and inventors ceased to be interested by pure science and mathematics and turned their energies more and more to engineering and religion.

Why did the ancient western civilization come to an end? We can suggest several causes: the breakdown of the Roman political system, the cataclysmic invasion of Germanic and Slavic "barbarians" (who conquered much of the Roman Empire in the fifth century A.D. and established feudalism), and the increased importance of the Christian church after the collapse of Roman civil authority. Nevertheless, we must acknowledge that ancient civilization in the West did not perish suddenly in a blazing bier, but instead faded away over several centuries, and was even occasionally resurrected by its conquerors. The Greco-Roman culture did not vanish completely; instead, it blended with other cultures and societies to birth a new civilization that was a synthesis of the Greeks, Romans, Germans, Slavs, and other peoples.

The Breakdown of the Roman Political System

Through much of its history, the Roman Empire suffered from twin problems: it was big, and therefore difficult to govern, and it had a political system that produced mediocre leaders. With a few notable exceptions, most Roman emperors came to power through military *coups d'état* and ruled for but a few years, only to be overthrown by another general with a better army. Rebellions were frequent, and emperors were forced to set aside affairs of state in order to quell uprisings led by rivals. The Romans addressed this situation not by devising methods of assuring themselves better emperors, but by reducing the size of the territory an emperor was required to govern. In A.D. 305, Emperor Diocletian (A.D. 245–313) divided the Empire into two halves. Thereafter, a western emperor reigned at Rome, and an eastern emperor sat at a Greek city called Byzantium, later renamed Constantinople in honor of eastern Emperor Constantine I (A.D. 272–337). The eastern emperor was regarded as senior to the western and was theoretically accorded more political authority, a situation that further contributed to the decline of leadership in the west. When the western half of the Empire was invaded by "barbarians" in the fifth century A.D., the western emperors were unable to meet the challenge.

The "Barbarian" Invasions

Late in the fourth century A.D., northern and eastern Europe were invaded by Huns, a tribe of fierce warriors from central Asia. As Hun horsemen thundered across Europe, they drove before them the Germanic and Slavic hunters who lived in the northern and eastern forests. Goths and Alans from the Ukraine, Franks and Burgundians from east of the Rhine River in Germany, Vandals from Czechoslovakia's Carpathian Mountains, Slavs from the Pripet Marshes in central Russia—all came into Rome, organized in warrior bands made refugees by Hun lances.

Once they reached Roman territory, the refugees became conquerors. Visigoths fled the Ukraine in approximately A.D. 350 to escape the Huns, settled for a time in the Roman province of Moesia (now Romania), and in 376 beseiged Constantinople. When repulsed, they were led by their chief, Alaric (ca. A.D. 370–410), on a rampage through Greece and Italy that ended only with Alaric's death in 410. In approximately 406, Huns drove the Vandals from their central European homeland. The Vandals marched through the Roman province of Gaul in 407 and 408, leaving such a trough of destruction in their wake that the very name "Vandal" became synonymous with "looter." They entered Spain in 409, deposed the Roman governor, and established their own kingdom, which was later transferred to northern Africa in the 420s and 430s. Bands of Franks followed the Vandals into Gaul and settled there. Britain, cut off from the rest of the Empire, was invaded and occupied by Angles and Saxons. After the Visigoths finished pillaging Italy (they moved on to Spain to trouble the Vandals), Ostrogoths came there to live, later followed by Lombards.

When the Huns, led by their infamous king Attila (ca. A.D. 406–453), invaded Roman Gaul in 451, they were defeated by a combined Roman-Frankish army under the command of the Roman general Aetius (ca. A.D. 396–454) at Chalons. It was a hollow victory. Attila turned his attention to Italy, destroyed much of the countryside, and left only when he ran out of food. Attila's march through the Roman heartland followed that of the Visigoths by a scant forty-six years. The country was in ruins. In 476, with little effort, the Ostrogoths deposed the last emperor, and the Western Empire fell.

The Greco-Slavic East

Fifty years later, under Emperor Justinian I (A.D. 483–565), the Eastern Roman Empire launched a valiant, but doomed, counterattack. Between 530 and 550, Justinian's generals, Belisarius (ca. A.D. 505–565) and Narses (ca. A.D. 478–ca. 573), reconquered Italy and northern Africa from the Ostrogoths and Vandals. The easterners soon found themselves stunned by a series of invasions from Asia and eastern Europe, however, first from a reconstituted Persian (Parthian) Empire, then by the Slavic Bulgarians, and after 640 from the powerful new Arabian Empire. By 600, the Eastern Empire was forced to abandon Italy to the Lombards, and by 700 northern Africa was lost to the Arabs, who also annexed Egypt and Palestine. Stripped of most of its territory, the Eastern Roman Empire became essentially just a middle-sized Greek kingdom, although it remained independent until conquered by Turks in 1453.

Although politically the government at Byzantium (as the Eastern Empire came to be called) was, after 700, little more than a shadow of Roman imperial might, the predominantly Greek city of Constantinople remained an important commercial and cultural center, in many ways resembling a latter-day Alexandria. Trade was particularly lively with the Slavic peoples who lived in eastern Europe and who borrowed many elements of Greek culture from Byzantine merchants. The Russian alphabet, for example, is based on the Greek, and the Christian religion, in the form of the Greek Orthodox Church, spread throughout most of eastern Europe. Culturally, the Byzantine Greeks' greatest achievements were in theology and law. The legal code promulgated by Justinian is considered a masterpiece, as well as a hallmark in the evolution of European jurisprudence. Nevertheless, the Byzantine Greeks were generally poor scholars. Their histories were simply panegyrics to living emperors. They continued the Roman trend of exalting religion over science, and were unable to reconcile the two. It was Justinian himself who, under pressure from religious leaders, ordered the closing of the sole remaining school of ancient western science and philosophy, the Academy at Athens, in 529.

Western Europe in the Middle Ages

After the fall of the Western Roman Empire, political power in western Europe shifted north to Gaul (now France), where the Franks founded a sturdy kingdom. Originally several disparate tribes of Germanic hunters, the Franks were united by the warlord Clovis I (ca. A.D. 466–511) in A.D. 481. The Franks adopted the Christian religion and the agricultural economy of the native Gallic Celts and intermarried with them, forging a society that combined elements of Frankish, Latin, and Celtic cultures.

In the 770s, during a dispute between the Lombard king in Italy and the Catholic pope, the Franks intervened on the side of the latter, a development that led to the Frankish annexation of much of Italy. When the Frankish king Charlemagne (A.D. 742–814) restored Pope Leo III (died A.D. 816) to the Papal See in 800, the grateful pontiff crowned him emperor of a "new Rome," which was styled the Holy Roman Empire. Charlemagne also carried on long wars against other Germanic peoples, the Saxons, Avars, and Wends, and forcibly converted them to Christianity. He built a palace at Aachen (Aix-la-Chapelle), grand by medieval standards, although modest when compared with ancient Roman edifices. Although barely literate himself, Charlemagne patronized art and literature. As a cultural center, however, Aachen was overshadowed by Constantinople, and Charlemagne's attempts to rejuvenate Latin civilization under Frankish auspices did not survive him. After the great monarch's death, his empire was divided among his three sons, and the importance of the Frank kingdom declined considerably.

A second Holy Roman Empire coalesced in Germany about 150 years after Charlemagne's death, when a German king, Otto I (912–973), united most of central Europe under his rule. This second Holy Roman Empire, which did not include France, bore little resemblance to the great monolithic empires of ancient times. Less a centralized kingdom than a confederation of several dozen German principalities, the second Holy Roman Empire was the archetype of the medieval state. With the exception of Otto and a few others, its emperors were figureheads, elected by the various German lords or barons. Each of these lesser lords ruled over his own small barony; collectively, the barons held the real power in the Empire. Nevertheless, this Holy Roman Empire established Germany as the cultural and commercial center of medieval western Europe, and it lasted, at least in name, until broken up by Napoleon in 1806.

Medieval Europe developed a unique social structure called *feudalism.* Most people were poor peasant farmers, or serfs, who were legally bound to farm the estates of the lords and pay a portion of their crops as rent. In theory, the lords were vassals of a king or of the Holy Roman Emperor, although in fact few kings or emperors had much real power. There was a small urban middle class of merchants and artisans. Upward mobility was minimal, as entrance into the aristocracy was by birth only.

Although individual lords had considerable authority on their own lands, they had little control over their neighbors. Ambitious barons therefore sought to further their interests at the court of the king or emperor by forming coalitions and battling one another in interminable dynastic wars. Kings were dependent on the nobility to supply them with money and soldiers, and woe unto any monarch foolish enough to displease his lords. When one of England's kings, John (1167?–1216), sought to streamline the English judicial system (placing himself at its center, of course), the English barons joined forces to halt him. Following a battle at Runnymede (1215), the barons forced the chastened monarch to sign the *Magna Charta,* or Great Charter, guaranteeing the continuance of England's system of traditional, largely unwritten law (the Common Law), which even today forms the basis of both English and American jurisprudence.

This mighty nobility, however, was generally an isolated country aristocracy, not centered about a large urban court. Most barons and dukes were poorly schooled. The best were brave generals and wise administrators, but few were learned scholars. Because the kings and emperors had minimal powers, no large capital cities grew up around their modest courts. Commerce, too, was limited, and medieval Europe was devoid of metropoli. Because its largest cities (outside of Rome, which housed the administrative superstructure of the Catholic Church) were not much more than oversized country towns, urban civilization was generally absent in medieval western Europe.

Alongside this civil social structure, yet also apart from it, stood the Catholic Church, which was under the control of a pope officed in the city of Rome who was assisted by an integrated bureaucracy. Bishops oversaw Church affairs in the principal towns, such as Cologne, Mainz, Venice, and Tours. The Church dabbled in secular as well as religious affairs. So powerful was the pope that the city of Rome and the surrounding countryside was an independent kingdom under his rule. The Church owned substantial estates throughout western Europe, and some of its bishops participated in the election of Holy Roman Emperors who, significantly, were crowned by the pope until into the 1500s. Monasteries and convents, homes to men and women engaged in the work of the Church, dotted the countryside. The monasteries were the only true loci of scholarship in medieval western Europe, and the monks naturally preferred the study of religion and philosophy to science; like the Byzantines, they often considered the two studies incompatible. The Middle Ages produced several justly celebrated theologians, among them Saint Benedict (died ca. 547), who first proposed the communal, monastic life, with its emphasis on manual labor, simplicity, and the preservation of knowledge, and Saint Francis of Assisi (1182?–1226), who advocated gentleness, concern for the poor, and respect for animal life. There were, however, few mathematicians and scientists.

Medieval people did display skill at engineering. Masons and carpenters designed and built immense and elegant cathedrals, replete with intricate and beautiful stained glass windows and remarkable flying buttresses. Smiths worked out methods for constructing accurate clocks. Millers perfected the water wheel. Long canals were dug, stone bridges spanned even the widest rivers, and marshes were diked and drained. Medieval engineers, however, were not university-trained "pure" scientists; they were poorly schooled artisans and mechanics whose work was often ignored by educated scholars. Indeed, the marriage of pure science and technology would not come about until the twentieth century.

The Renaissance

In the 1300s and 1400s, almost a thousand years after the fall of Rome, medieval European civilization at last began to give way to modern civilization. Ironically, the path towards modernity began with a rekindling of interest in ancient art and science. Commerce with the Moslems and the Byzantine Greeks gave impetus to the growth of several Italian cities after 1300, among them Venice, Genoa, and Florence. The aristocracy there became intrigued not only with eastern produce, but with eastern scholarship as well. The Arabs and Byzantine Greeks had carefully preserved much of ancient Greek and Roman art and science, and now transmitted this knowledge to Italian merchants. Wealthy Italian aristocratic families, such as the Estes and the Borgias, sponsored artists and poets who immersed themselves in the works of the ancient Greek and Roman masters. These artists also often dabbled in ancient science. Among the Italian scholars of the Renaissance were Leonardo Fibonacci (ca. 1175–1250), Leonardo da Vinci (1452–1519), Michaelangelo (1475–1564), and Benvenuto Cellini (1500–1571). This rebirth of ancient western culture soon spread to northern Europe, where it led to a new interest in science and art, stimulating the work of Polish astronomer Nicholas Copernicus (1473–1543) and his Danish successor, Tycho Brahe (1546–1601).

Unfortunately, Renaissance scholars were unable to reconcile easily their ideas about science with the Catholic Church's religious doctrines, and much of the scientific work of the period met stiff opposition from the Church. Afraid of prosecution for heresy, many Renaissance scholars were reluctant to publish their theories, especially in astronomy, a science particularly opposed by the Church. As modern Europe slowly developed out of medieval Europe, the

Catholic Church, once a force for change, found itself growing increasingly conservative. Not only did the Church disapprove of many of the discoveries of modern European scientists, it also came to stand athwart attempts by political reformers to replace feudalism with more democratic forms of government. Those stories will be told in Cultural Connection VII: Puritans and Sea Dogs, and in Cultural Connection VIII: The Revolt of the Middle Class.

Chapter 8

EUROPEAN MATHEMATICS, 500 TO 1600

8–1 The Dark Ages

The period starting with the fall of the Roman Empire, in the middle of the fifth century, and extending into the eleventh century is known as Europe's Dark Ages, for during this period, civilization in western Europe reached a very low ebb. Schooling became almost nonexistent, Greek learning all but disappeared, and many of the arts and crafts bequeathed by the ancient world were forgotten. Only the monks of the Catholic monasteries and a few cultured laymen preserved a slender thread of Greek and Latin learning. The period was marked by much physical violence and intense religious faith. The old social order gave way, and society became feudal and ecclesiastical.

The Romans had never taken to abstract mathematics; instead they contented themselves with practical aspects of the subject that were associated with commerce and civil engineering. With the fall of the Roman Empire and the subsequent cessation of much of east-west trade and the abandonment of state engineering projects, even these interests waned, and it is no exaggeration to say that very little in mathematics, beyond the development of the Christian calendar, was accomplished in the West during the whole of the half millennium covered by the Dark Ages.

Of the persons charitably credited with playing a role in the history of mathematics during the Dark Ages, we might mention the martyred Roman citizen Boethius, the British ecclesiastical scholars Bede and Alcuin, and the famous French scholar and churchman Gerbert, who became Pope Sylvester II.

The importance of Boethius (ca. 475–524) in the story of mathematics rests on the fact that his writings on geometry and arithmetic remained standard texts in the monastic schools for many centuries. These very meager works came to be considered as the height of mathematical achievement and, thus, well illustrate the poverty of the subject in Christian Europe during the Dark Ages. Boethius' *Geometry* consists of nothing but the statements of the propositions of Book I and a few selected propositions of Books III and IV of Euclid's *Elements,* along with some applications to elementary mensuration, and the *Arithmetic* is founded on the tiresome and half mystical, but once highly reputed, work of Nicomachus of four centuries earlier. (It is contended by some that at least part of the *Geometry* is spurious.) With these works and his writings on philosophy, Boethius became the founder of medieval scholasti-

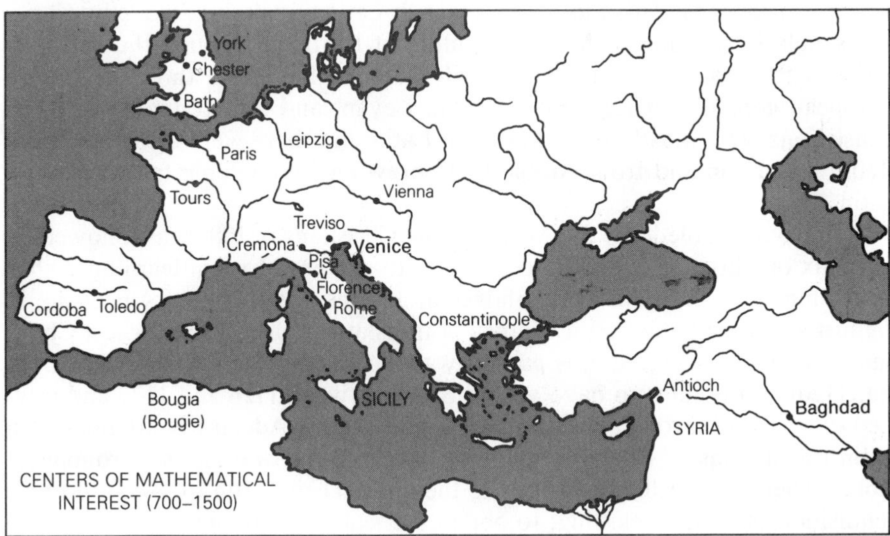

cism. His high ideals and inflexible integrity led him into political troubles and he suffered a cruel end, for which the Church declared him a martyr.

Bede (ca. 673–735), later qualified as Bede *the Venerable,* was born in Northumberland, England, and became one of the greatest of the medieval Church scholars. His numerous writings include some on mathematical subjects, chief of which are his treatises on the calendar and on finger reckoning. Alcuin (735–804), born in Yorkshire, was another English scholar. He was called to France to assist Charlemagne in his ambitious educational project. Alcuin wrote on a number of mathematical topics and is doubtfully credited with a collection of puzzle problems that influenced textbook writers for many centuries (see Problem Study 8.1).

Gerbert (ca. 950–1003) was born in Auvergne, France, and early showed unusual abilities. He was one of the first Christians to study in the Moslem schools of Spain, and there is evidence that he may have brought back the Hindu-Arabic numerals, without the zero, to Christian Europe. He is said to have constructed abaci, terrestrial and celestial globes, a clock, and perhaps an organ. Such accomplishments corroborated the suspicions of some of his contemporaries that he had traded his soul to the devil. Nevertheless, he steadily rose in the Church and was finally elected to the papacy as Sylvester II in 999. He was considered a profound scholar and wrote on astrology, arithmetic, and geometry [see Problem Study 8.1(f)], although his mathematical work is of no special value.

8–2 The Period of Transmission

About the time of Gerbert, the Greek classics in science and mathematics began to filter into western Europe. A period of transmission followed during which the ancient learning preserved by Moslem culture was passed on to the western Europeans. This took place through Latin translations made by Christian scholars traveling to Moslem centers of learning, through the relations between the Norman kingdom of Sicily and the East, and through western European commercial relations with the Levant and the Arabic world. The translations were mostly from Arabic to Latin, but there were also some from Hebrew to Latin and from Arabic to Hebrew and even some from Greek to Latin.

The loss of Toledo by the Moors to the Christians in 1085 was followed by an influx of Christian scholars to that city to acquire Moslem learning. Other Moorish centers in Spain were infiltrated, and the twelfth century became, in the history of mathematics, a century of translators. One of the earliest Christian scholars to engage in this pursuit was the English monk Adelard of Bath (ca. 1120), who seems to have visited Spain between 1126 and 1129 and traveled extensively through Greece, Syria, and Egypt. Adelard is credited with Latin translations of Euclid's *Elements* and of al-Khowârizmî's astronomical tables. There are thrilling allusions to the physical risks run by Adelard in his acquisition of Arabic learning; to obtain the jealously guarded knowledge, he disguised himself as a Mohammedan student. Another early translator was the Italian, Plato of Tivoli (ca. 1120), who translated the astronomy of al-Battânî, the *Spherics* of Theodosius, and various other works. A Jewish mathematician, Abraham bar Hiyya, known as Savasorda, is mentioned with Plato. He wrote in Hebrew a book titled *Practical Geometry,* which Plato, probably working with him, translated into Latin. It was through this work that the West first learned of the complete solution of the quadratic equation, and this had a great impact.

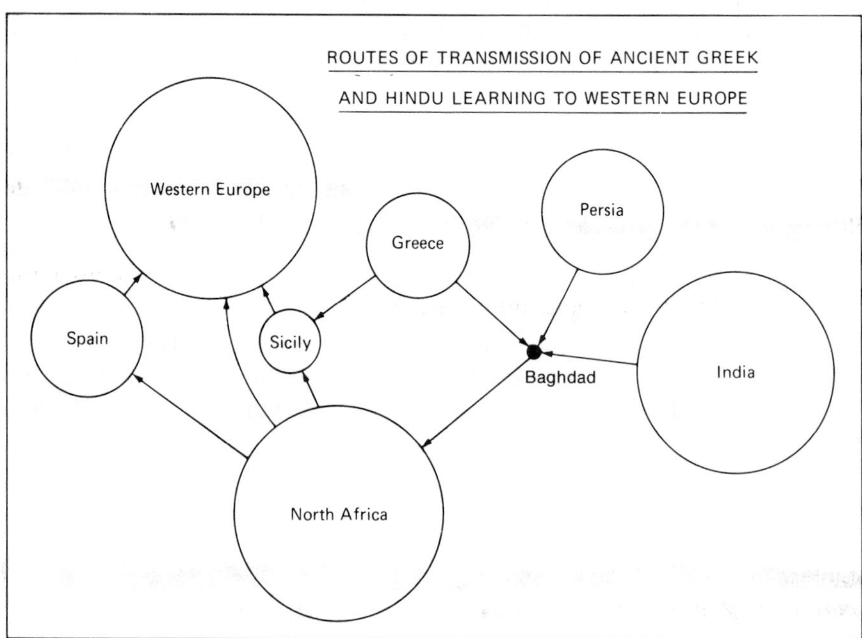

The most industrious translator of the period was Gherardo of Cremona (1114–1187), who translated into Latin over ninety Arabian works, among which were Ptolemy's *Almagest,* Euclid's *Elements,* and al-Khowârizmî's algebra. He certainly did not do all this himself, but also worked with members of the School of Translators founded by Archbishop don Raimundo shortly after the fall of Toledo. We have already, in Section 7–12, mentioned the part played by Gherardo of Cremona in the development of our word *sine*. Other noted translators of the twelfth century were John of Seville and Robert of Chester.

The location and political history of Sicily made that island a natural meeting ground of East and West. Sicily started as a Greek colony, became part of the Roman Empire, linked itself with Constantinople after the fall of Rome, was held by the Arabs for about fifty years in the ninth century, recaptured by the Greeks, and then taken over by the Normans. During the Norman regime, the Greek, Arabian, and Latin tongues were used side by side, and diplomats frequently traveled to Constantinople and Baghdad. Many Greek and Arabian manuscripts in science and mathematics were obtained and translated into Latin. This work was greatly encouraged by the two rulers and patrons of science Frederick II (1194–1250) and his son Manfred (ca. 1231–1266).

Among the first cities to establish mercantile relations with the Arabic world were the Italian commercial centers at Genoa, Pisa, Venice, Milan, and Florence. Italian merchants came in contact with much of Eastern civilization, picking up useful arithmetical and algebraical information. These merchants played an important part in the dissemination of the Hindu-Arabic numeral system.

In the period of transmission discussed above, Spain turned out to be the most important link between Islam and the Christian world.

8–3 Fibonacci and the Thirteenth Century

At the threshold of the thirteenth century appeared Leonardo Fibonacci ("Leonardo, son of Bonaccio," ca. 1175–1250), the most talented mathematician of the Middle Ages. Also known as Leonardo of Pisa (or Leonardo Pisano), Fibonacci was born in the commercial center of Pisa, where his father was connected with the mercantile business. Many of the large Italian businesses in those days maintained warehouses in various parts of the Mediterranean world. It was in this way, when his father was serving as a customs manager, that young Leonardo was brought up in Bougie on the north coast of Africa. The father's occupation early roused in the boy an interest in arithmetic, and subsequent extended trips to Egypt, Sicily, Greece, and Syria brought him in contact with Eastern and Arabic mathematical practices. Thoroughly convinced of the practical superiority of the Hindu-Arabic methods of calculation, Fibonacci, in 1202, shortly after his return home, published his famous work called the *Liber abaci*.

The *Liber abaci* is known to us through a second rendition that appeared in 1228. The work is devoted to arithmetic and elementary algebra and, although

LEONARDO FIBONACCI
(David Smith Collection)

essentially an independent investigation, shows the influence of the algebras of al-Khowârizmî and Abû Kâmil. The book profusely illustrates and strongly advocates the Hindu-Arabic notation and did much to aid the introduction of these numerals into Europe. The fifteen chapters of the work explain the reading and writing of the new numerals, methods of calculation with integers and fractions, computation of square and cube roots, and the solution of linear and quadratic equations both by false position and by algebraic processes. Negative and imaginary roots of equations are not recognized, and the algebra is rhetorical. Applications are given involving barter, partnership, alligation, and mensurational geometry. The work contains a large collection of problems that served

Incipit primum capitulum

Nouem figure indorum he sunt

9 8 7 6 5 4 3 2 1

Cym his itaque nouem figuris, et cum hoc signo 0, quod arabice zephirum appellatur, scribitur quilibet numerus, ut inferius demonstratur.

[These are the nine figures of the Indians

9 8 7 6 5 4 3 2 1

With these nine figures, and with the sign 0 which in Arabic is called zephirum, any number can be written, as will below be demonstrated.]

The opening sentence of Fibonacci's *Liber abaci* of 1202.
(Courtesy of West Virginia University Library.)

later authors as a storehouse for centuries. In Section 2–10, we mentioned one interesting problem from the collection that apparently evolved from a much older problem in the Rhind papyrus. Another problem, giving rise to the important **Fibonacci sequence** $(1, 1, 2, 3, 5, \ldots, x, y, x + y, \ldots)$, and some other problems from the *Liber abaci* may be found in Problem Studies 8.2, 8.3, and 8.4.

In 1220, Fibonacci's *Practica geometriae* appeared, a vast collection of material on geometry and trigonometry treated skillfully with Euclidean rigor and some originality. About 1225, Fibonacci wrote his *Liber quadratorum,* a brilliant and original work on indeterminate analysis, which has marked him as the outstanding mathematician in this field, between Diophantus and Fermat. These works were beyond the abilities of most of the contemporary scholars.

Fibonacci's talents came to the attention of the patron of learning, Emperor Frederick II, with the result that Fibonacci was invited to court to partake in a mathematical tournament. Three problems were set by John of Palermo, a member of the emperor's retinue. The first problem was to find a rational number x, such that $x^2 + 5$ and $x^2 - 5$ shall each be squares of rational numbers. Fibonacci gave the answer $x = 41/12$, which is correct, since $(41/12)^2 + 5 = (49/12)^2$ and $(41/12)^2 - 5 = (31/12)^2$. The solution appears in the *Liber quadratorum*. The second problem was to find a solution to the cubic equation $x^3 + 2x^2 + 10x = 20$. Fibonacci attempted a proof that no root of the equation can be expressed by means of irrationalities of the form $\sqrt{a + \sqrt{b}}$, or, in other words, that no root can be constructed with straightedge and compasses. He then obtained an approximate answer, which, expressed in decimal notation, is 1.3688081075, and is correct to nine places. The answer appears, without any accompanying discussion, in a work by Fibonacci entitled the *Flos* ("blossom" or "flower") and has excited some wonder. The third problem, also recorded in the *Flos,* is easier and may be found in Problem Study 8.4.

It is apparent that Fibonacci was an unusually capable mathematician, who had no rival during the whole nine hundred years of the Middle Ages. One of his better contemporaries was Jordanus Nemorarius, sometimes identified (but in all likelihood mistakenly) with the German monk Jordanus Saxus who, in 1222, was elected the second general of the rapidly expanding Dominican order. He wrote several works dealing with arithmetic, algebra, geometry, and statistics. These prolix works, some of which enjoyed considerable fame at one time, may now seem largely trivial, but his algebra was the first forward treatment composed in western Europe. He was perhaps the first one widely to use letters to represent general numbers, although his practice had little influence on subsequent writers. There is only one instance in which Fibonacci did this. In spite of the bleak picture often given of the mathematics of the thirteenth century, it was the early part of that century that saw the high point of medieval achievement in arithmetic, geometry, and algebra.

Perhaps mention should also be made of Sacrobosco (John of Holywood, or John of Halifax), Campanus, and Roger Bacon. The first taught mathematics in Paris and wrote a collection of arithmetical rules and a popular compilation of extracts from Ptolemy's *Almagest* and the works of Arabian astronomers.

Campanus' chief bid to fame is his Latin translation of Euclid's *Elements*, mentioned in Section 5–3. Roger Bacon, original genius that he was, had little ability in mathematics, but he was acquainted with many of the Greek works in geometry and astronomy, and, as his eulogies attest, fully appreciated the value of the subject.

The early part of the thirteenth century saw the rise of the universities at Paris, Oxford, Cambridge, Padua, and Naples. Universities later became potent factors in the development of mathematics, many mathematicians being associated with one or more such institutions.

8–4 The Fourteenth Century

The fourteenth century was relatively barren, mathematically. It was the century of the Black Death, which swept away more than a third of the population of Europe, and the Hundred Years' War, with its political and economic upheavals in northern Europe, got well under way.

The greatest mathematician of the period was Nicole Oresme, who was born in Normandy about 1323. He died in 1382 after a career that carried him from a college professorship to a bishopric. He wrote five mathematical works and translated some of Aristotle. In one of his tracts appears the first known use of fractional exponents (not, of course, in modern notation); in another tract, he locates points by coordinates, thus foreshadowing modern coordinate geometry. A century later, this last tract enjoyed several printings, and it may have influenced Renaissance mathematicians and even Descartes. In an unpublished manuscript, he also obtained the sum of the series

$$\frac{1}{2} + \frac{2}{4} + \frac{3}{8} + \frac{4}{16} + \frac{5}{32} + \cdots.$$

This makes him one of the forerunners of infinitesimal analysis.

Although European mathematics during the Middle Ages was essentially practical, speculative mathematics did not entirely die out. The meditations of scholastic philosophers led to subtle theorizing on motion, infinity, and the continuum, all of which are fundamental concepts in modern mathematics. The centuries of scholastic disputes and quibblings may, to some extent, account for the remarkable transformation from ancient to modern mathematical thinking, and might, as suggested by E. T. Bell, constitute a *submathematical analysis*. From this point of view, Thomas Aquinas (1226–1274), possessing perhaps the acutest mind of the thirteenth century, can well be considered as having played a part in the development of mathematics. Definitely more of the conventional mathematician was Thomas Bradwardine (1290–1349), who died as Archbishop of Canterbury. In addition to speculations on the basic concepts of the continuous and the discrete and on the infinitely large and the infinitely small, Bradwardine wrote four mathematical tracts on arithmetic and geometry.

8–5 The Fifteenth Century

The fifteenth century witnessed the start of the European Renaissance in art and learning. With the collapse of the Byzantine Empire, culminating in the fall of Constantinople to the Turks in 1453, refugees flowed into Italy, bringing with them treasures of Greek civilization. Many Greek classics, hitherto known only through the often inadequate Arabic translations, could now be studied from original sources. Also, about the middle of the century, printing was invented and revolutionized the book trade, enabling knowledge to be disseminated at an unprecedented rate. Toward the end of the century, America was discovered and soon the earth was circumnavigated.

Mathematical activity in the fifteenth century was largely centered in the Italian cities and in the central European cities of Nuremberg, Vienna, and Prague, and was concentrated on arithmetic, algebra, and trigonometry. Thus, mathematics flourished principally in the growing mercantile cities under the influence of trade, navigation, astronomy, and surveying.

Adhering to chronological order, we first mention Nicholas Cusa, who took his name from the city of Cues on the Mosel, where he was born in 1401. The son of a poor fisherman, he rose rapidly in the Church, finally becoming a cardinal. In 1448, he became governor of Rome. He was only incidentally a mathematician but did succeed in writing a few tracts on the subject. He is now remembered along these lines chiefly for his work on calendar reform and his attempts to square the circle and trisect the general angle (see Problem Study 8.6). He died in 1464.

A better mathematician was Georg von Peurbach (1423–1461), who numbered Nicholas Cusa as one of his teachers. After lecturing on mathematics in Italy, he settled in Vienna and made the university there the mathematical center of his generation. He wrote an arithmetic and some works on astronomy, and compiled a table of sines. Most of these works were not published until after his death. He also had started a Latin translation, from the Greek, of Ptolemy's *Almagest.*

The ablest and most influential mathematician of the century was Johann Müller (1436–1476), more generally known from the Latinized form of his birthplace of Königsberg ("king's mountain") as Regiomontanus. At a young age, he studied under Peurbach in Vienna and was later entrusted with the task of completing the latter's translation of the *Almagest.* He also translated, from the Greek, works of Apollonius, Heron, and Archimedes. His treatise *De triangulis omnimodis,* written about 1464 but posthumously published in 1533, is his greatest publication and was the first systematic European exposition of plane and spherical trigonometry considered independently of astronomy. Regiomontanus traveled considerably in Italy and Germany, finally settling in 1471 at Nuremberg, where he set up an observatory, established a printing press, and wrote some tracts on astronomy. He is said to have constructed a mechanical eagle that flapped its wings and was considered as one of the marvels of the age. In 1475, Regiomontanus was invited to Rome by Pope Sixtus IV to partake in the reformation of the calendar. Shortly after his arrival,

REGIOMONTANUS
(David Eugene Smith Collection, Rare Book and Manuscript Library, Columbia University)

at the age of forty, he suddenly died. Some mystery shrouds his death, for although most accounts claim he probably died of a pestilence, it was rumored that he was poisoned by an enemy.

Regiomontanus' *De triangulis omnimodis* is divided into five books, the first two devoted to plane trigonometry and the other three to spherical trigonometry. In it he shows much interest in the determination of a triangle satisfying three given conditions. On several occasions, he applies algebra, as in Propositions 12 and 23 of Book II: (II 12) Determine a triangle, given a side, the altitude on this side, and the ratio of the other two sides; (II 23) Determine a triangle, given the difference of two sides, the altitude on the third side, and the difference of the segments into which the altitude divides the third side. The algebra is rhetorical, an unknown part of the figure being found as a root of a quadratic equation. Although his methods were meant to be considered as general, he gives specific numerical values to the given parts. The only trigonometric functions employed in *De triangulis omnimodis* are the sine and cosine. Later, however, Regiomontanus computed a table of tangents. In another work, Regiomontanus applied algebra and trigonometry to the problem of constructing a cyclic quadrilateral, given the four sides.

The most brillant French mathematician of the fifteenth century was Nicolas Chuquet, who was born in Paris but lived and practiced medicine in Lyons. In 1484, he wrote an arithmetic known as *Triparty en la science des nombres,* which was not printed until the nineteenth century. The first of the three parts of this work concerns itself with computation with rational numbers, the second with irrational numbers, and the third with the theory of equations. Chuquet recognized positive and negative integral exponents and syncopated some of his algebra. His work was too advanced, for the time, to exert much influence on his contemporaries. He died around 1500. Some problems from Chuquet may be found in Problem Study 8.9.

In 1494 appeared the first printed edition of the *Summa de arithmetica, geometrica, proportioni et proportionalita,* usually referred to briefly as the *Sūma,* of the Franciscan friar Luca Pacioli (ca. 1445–1509). This work, freely compiled from many sources, purported to be a summary of the arithmetic, algebra, and geometry of the time. It contains little of importance not found in Fibonacci's *Liber abaci,* but does employ a superior notation.

The arithmetical portion of the *Sūma* begins with algorithms for the fundamental operations and for extracting square roots. The presentation is rather complete, containing, for example, no less than eight plans for the performance of a multiplication. Mercantile arithmetic is fully dealt with and illustrated by numerous problems; there is an important treatment of double-entry bookkeeping. The rule of false position is discussed and applied. In spite of many numerical mistakes, the arithmetical part of the work has become a standard authority on the practices of the time. The algebra in the *Sūma* goes through quadratic equations and contains many problems that lead to such equations. The algebra is syncopated by the use of abbreviations such as *p* (from *piu,* "more") for plus, *m* (from *meno,* "less") for minus, *co* (from *cosa,* "thing") for the unknown x, *ce* (from *censo*) for x^2, *cu* (from *cuba*) for x^3, and *cece* (from *censocenso*) for x^4. Equality is sometimes indicated by *ae* (from *aequalis*). Frequently appearing bars indicate abbreviations, as in *Sūma* for *Summa.* The work contains little of interest in geometry. As with Regiomontanus, algebra is employed in the solution of geometrical problems. After the *Sūma,* algebra, which had been neglected for two hundred years, experienced a strong growth in Italy and also progressed in Germany, England, and France.

Pacioli traveled extensively, taught in various places, and wrote a number of other works, not all of which were printed. In 1509, he published his *De divina proportione,* which contains woodcuts of the regular solids that were drawn by Leonardo da Vinci during the time he lived with Pacioli and received lessons in mathematics from Pacioli.

The first appearance in print of our present + and − signs is in an arithmetic published in Leipzig in 1489 by Johann Widman (born ca. 1460 in Bohemia). Here the signs are not used as symbols of operation but merely to indicate excess and deficiency. Quite likely the plus sign is a contraction of the Latin word *et,* which was frequently used to indicate addition, and it may be that the minus sign is contracted from the abbreviation $\overline{m}$ for minus. Other plausible explanations have been offered. The + and − signs were used as symbols of algebraic operation in 1514 by the Dutch mathematician Vander Hoecke, but they were probably so used earlier.[1]

8–6 The Early Arithmetics

With the interest in education that accompanied the Renaissance, and with the tremendous increase in commercial activity at the time, hosts of popular textbooks in arithmetic began to appear. Three hundred such books were printed in

[1] See J. W. L. Glaisher, "On the early history of the signs + and − and on the early German arithmeticians," *Messenger of Mathematics* 51 (1921–1922): 1–148.

Europe prior to the seventeenth century. These texts were largely of two types—those written in Latin by classical scholars who often were attached to the Church schools, and those written in the vernaculars by practical teachers interested in preparing boys for commercial careers. These teachers often also served as town surveyors, notaries, and gaugers, and included the influential Rechenmeisters supported by the Hanseatic League, a powerful protective union of commercial towns in the Teutonic countries.

The earliest printed arithmetic is the anonymous and now extremely rare *Treviso Arithmetic*, published in 1478 in the town of Treviso, which was located

Per fare de ducati grossi a oro.

1	fia	24	fa	24
2	fia	24	fa	48
3	fia	24	fa	72
4	fia	24	fa	96
5	fia	24	fa	120
6	fia	24	fa	144
7	fia	24	fa	168
8	fia	24	fa	192
9	fia	24	fa	216
0	fia	24	fa	0

Per fare de grossi a oro pizoli.

1	fia	32	fa	32
2	fia	32	fa	64
3	fia	32	fa	96
4	fia	32	fa	128
5	fia	32	fa	160
6	fia	32	fa	192
7	fia	32	fa	224
8	fia	32	fa	256
9	fia	32	fa	288
0	fia	32	fa	0

Per fare de quarti karatti.

1	fia	36	fa	36
2	fia	36	fa	72
3	fia	36	fa	108
4	fia	36	fa	144

A page (reduced in size) from the *Treviso Arithmetic* of 1478, showing the already well-formed appearance of the Hindu-Arabic numerals. (By permission of the Houghton Library, Harvard University.)

on the trade route linking Venice with the north. It is largely a commercial arithmetic devoted to explaining the writing of numbers, computation with them, and applications to partnership and barter. Like the earlier "algorithms" of the fourteenth century, it also contains some recreational questions. It is the first printed book in the Western world devoted to mathematics.

Far more influential in Italy than the *Treviso Arithmetic* was the commercial arithmetic written by Piero Borghi. This highly useful work was published in Venice in 1484 and ran through at least seventeen editions, the last appearing in 1557. In 1491, there appeared in Florence a less important arithmetic by Filippo Calandri, but interesting to us because it contains the first printed example of our modern process of long division and also the first illustrated problems published in Italy. We have already considered Pacioli's *Sūma,* published in 1494, a large portion of which is devoted to arithmetic. Much information regarding the Italian commercial customs of the time may be gleaned from the problems of this book.

Very influential in Germany was Widman's arithmetic, published in 1489 at Leipzig. Another important German arithmetic was that written by Jacob Köbel (1470–1533), a Rechenmeister of Heidelberg. The popularity of this arithmetic, published in 1514, is attested by the fact that it ran through at least twenty-two editions. But perhaps the most influential of the German commercial arithmetics was that of Adam Riese (ca. 1489–1559), published in 1522. So reputable was this work that, even today in Germany, the phrase *nach Adam Riese* is used to indicate a correct calculation.

A humorous anecdote is told about Adam Riese. It seems that one day Riese and a draftsman entered into a friendly contest to see which of them could, with straightedge and compasses, draw more right angles in one minute. The draftsman drew a straight line, and then proceeded, by the standard construction now taught in high school, to erect perpendiculars to the line. Adam Riese drew a semicircle on a straight line, and then in rapid order drew a large number of inscribed right angles. He easily won the contest.

England, too, produced some noted early arithmetics. The first published work in England devoted exclusively to mathematics was an arithmetic written by Cuthbert Tonstall (1474–1559). This book, founded on Pacioli's *Sūma,* was printed in 1552 and was written in Latin. During his eventful life, Tonstall filled a number of ecclesiastical and diplomatic posts. The regard of his contemporaries for his scholarship is indicated by the fact that the first printed edition of Euclid's *Elements* in Greek (1533) was dedicated to him. But the most influential English textbook writer of the sixteenth century was Robert Recorde (ca. 1510–1558). Recorde wrote in English, his works appearing as dialogues between master and student. He wrote at least five books, his first being an arithmetic fancifully entitled *The Grovnd of Artes* and published about 1542. This work enjoyed at least twenty-nine printings. Recorde studied at Oxford and then took a medical degree at Cambridge. He taught mathematics in private classes at both institutions while in residence there and after leaving Cambridge served as physician to Edward VI and Queen Mary. In later life, he became Comptroller of the Mines and Monies in Ireland. His last years were spent in prison, probably for some misdemeanor connected with his work in Ireland.

8–7 Beginnings of Algebraic Symbolism

Besides his arithmetic, mentioned in the last section, Robert Recorde wrote an astronomy, a geometry, an algebra, a book on medicine, and probably some other works now lost. The book on astronomy, printed in 1551, is called *The Castle of Knowledge* and was one of the first works to introduce the Copernican system to English readers. Recorde's geometry, *The Pathewaie to Knowledge,* was also printed in 1551 and contains an abridgment of Euclid's *Elements*. Of historical interest here is Recorde's algebra, *The Whetstone of Witte,* published in 1557, for it was in this book that our modern symbol for equality was used for the first time. Recorde justified his adoption of a pair of equal parallel line segments for the symbol of equality "bicause noe 2 thynges can be moare equalle."

Another of our modern algebraic symbols, the familiar radical sign (adopted perhaps because it resembles a small *r*, for *radix*), was introduced in 1525 by Christoff Rudolff in his book on algebra entitled *Die Coss*. This book was very influential in Germany, and an improved edition of the work was brought out by Michael Stifel (1486–1567) in 1553. Stifel has been described as the greatest German algebraist of the sixteenth century. His best known mathematical work is his *Arithmetica integra,* published in 1544. It is divided into three parts devoted, respectively, to rational numbers, irrational numbers, and algebra. In the first part, Stifel points out the advantages of associating an arithmetic progression with a geometric one, thus foreshadowing the invention of logarithms nearly a century later. He also gives, in this part, the binomial coefficients up to the seventeenth order. The second part of the book is essentially an algebraic presentation of Euclid's Book X, and the third part deals with equations. Negative roots of an equation are discarded, but the signs $+$, $-$, $\sqrt{}$ are used, and often the unknown is represented by a letter.

Stifel was one of the oddest personalities in the history of mathematics. He was originally a monk, was converted by Martin Luther, and became a fanatical reformer. His erratic mind led him to indulge in number mysticism. From an analysis of Biblical writings, he prophesied the end of the world on October 3, 1533 and was forced to take refuge in a prison after ruining the lives of many believing peasants who had abandoned work and property to accompany him to heaven. An extreme example of Stifel's mystical reasoning is his proof, by arithmography, that Pope Leo X was the "beast" mentioned in the *Book of Revelation*.[2] From LEO DECIMVS he retained the letters, L, D, C, I, M, V, since these have significance in the Roman numeral system. He then added *X*, for Leo *X* and because *Leo decimus* contains ten letters, and omitted the *M*, because it stands for *mysterium*. A rearrangement of the letters gave DCLXVI, or 666, the "number of the beast" in the *Book of Revelation*. This discovery gave Stifel such extreme comfort that he believed his interpretation must have resulted from an inspiration from God.

---

[2] "Let him that hath understanding count the number of the beast: for it is the number of a man; and his number is six hundred three score and six." See W. F. White, *A Scrap-Book of Elementary Mathematics*, pp. 180–82.

The Arte

as their workes doe extende) to distincte it onely into twoo partes. Whereof the firste is, *when one nomber is equalle unto one other*. And the seconde is, *when one nomber is compared as equalle unto .2. other nombers.*

Alwaies willyng you to remeber, that you reduce your nombers, to their leaste denominations, and smalleste formes, before you procede any farther.

And again, if your *equation* be soche, that the greateste denomination *Cosike*, be ioined to any parte of a compounde nomber, you shall tourne it so, that the nomber of the greateste signe alone, maie stande as equalle to the reste.

And this is all that neadeth to be taughte, concernyng this woorke.

Howbeit, for easie alteratiō of *equations*. I will propounde a fewe exāples, bicause the extraction of their rootes, maie the more aptly bee wroughte. And to auoide the tediouse repetition of these woordes: is equalle to: I will sette as I doe often in woorke use, a paire of paralleles, or Gemowe lines of one lengthe, thus: =====, bicause noe .2. thynges, can be moare equalle. And now marke these nombers.

1. 14.ze.—+—.15.ϱ ===== 71.ϱ.
2. 20.ze.———.18.ϱ ===== .102.ϱ.
3. 26.ȝ.—+—10ze ===== 9.ȝ——10ze—+—213.ϱ.
4. 19.ze—+—192.ϱ ===== 10ȝ—+—108ϱ——19ze.
5. 18.ze—+—24.ϱ. ===== 8.ȝ.—+—2.ze.
6. 34ȝ——12ze ===== 40ze—+—480ϱ——9.ȝ.

The page in Robert Recorde's *The Whetstone of Witte* (1557) wherein he introduced his symbol for equality.

Some years later, Napier, the inventor of logarithms, showed that 666 stands for the Pope of Rome, and his Jesuit contemporary, Father Bongus, declared that it stands for Martin Luther. Father Bongus' reasoning ran as follows. If from A to I represents 1 to 9, from K to S represents 10 to 90 (by tens), and T to Z represents 100 to 500 (by hundreds),[3] we obtain

M	A	R	T	I	N	L	V	T	E	R	A
30	1	80	100	9	40	20	200	100	5	80	1

which gives, as a sum, 666.

During World War I, arithmography was used to show that 666 must be interpreted as Kaiser Wilhelm, and later it was shown to represent Hitler. It has been shown that 666 spells Nero, when expressed in the letter symbols of the Aramaic language in which the *Book of Revelation* was originally written.

8–8 Cubic and Quartic Equations

Probably the most spectacular mathematical achievement of the sixteenth century was the discovery, by Italian mathematicians, of the algebraic solution of cubic and quartic equations. The story of this discovery, when told in its most colorful version, rivals any page ever written by Benvenuto Cellini. Briefly told, the facts seem to be these. About 1515, Scipione del Ferro (1465–1526), a professor of mathematics at the University of Bologna, solved algebraically the cubic equation $x^3 + mx = n$, probably basing his work on earlier Arabic sources. He did not publish his result but revealed the secret to his pupil Antonio Fior. Now about 1535, Nicolo Fontana of Brescia, commonly referred to as Tartaglia[4] (the stammerer) because of a childhood injury that affected his speech, claimed to have discovered an algebraic solution of the cubic equation $x^3 + px^2 = n$. Believing this claim was a bluff, Fior challenged Tartaglia to a public contest of solving cubic equations, whereupon the latter exerted himself and only a few days before the contest found an algebraic solution for cubics lacking a quadratic term. Entering the contest equipped to solve two types of cubic equations, whereas Fior could solve but one type, Tartaglia triumphed completely. Later Girolamo Cardano,[5] an unprincipled genius who taught mathematics and practiced medicine in Milan, upon giving a solemn pledge of secrecy, wheedled the key to the cubic from Tartaglia. In 1545, Cardano published his *Ars magna,* a great Latin treatise on algebra, at Neuremberg, Germany, and in it appeared Tartaglia's solution of the cubic. Tartaglia's vehement protests were met by Ludovico Ferrari, Cardano's most capable pupil, who argued that Cardano had received his information from del Ferro through a

[3] The Latin alphabet is like the English, except that it lacks j and w. Moreover, in the upper-case letters, a U appears as a V.

[4] The *g* is silent. The name also appears as Tartalea.

[5] The name also appears as Hieronymus Cardanus, Geronimo Cardano, and Jerome Cardan.

third party and accused Tartaglia of plagiarism from the same source. There ensued an acrimonious dispute from which Tartaglia was perhaps lucky to escape alive.

Since the actors in the above drama seem not always to have had the highest regard for truth, one finds a number of variations in the details of the plot.

The solution of the cubic equation $x^3 + mx = n$ given by Cardano in his *Ars magna* is essentially the following. Consider the identity

$$(a - b)^3 + 3ab(a - b) = a^3 - b^3.$$

If we choose a and b so that

$$3ab = m, \qquad a^3 - b^3 = n,$$

then x is given by $a - b$. Solving the last two equations simultaneously for a and b we find that

$$a = \sqrt[3]{(n/2) + \sqrt{(n/2)^2 + (m/3)^3}},$$

$$b = \sqrt[3]{-(n/2) + \sqrt{(n/2)^2 + (m/3)^3}},$$

and x is thus determined.

It was not long after the cubic had been solved that an algebraic solution was discovered for the general quartic (or biquadratic) equation. In 1540, the Italian mathematician Zuanne de Tonini da Coi proposed a problem to Cardano that led to a quartic equation (see Problem Study 8.15). Although Cardano was unable to solve the equation, his pupil Ferrari succeeded, and Cardano had the pleasure of publishing this solution also in his *Ars magna*.

Ferrari's method of solving quartics, summarized in modern notation, is as follows. A simple transformation [see Problem Study 8.14(a)] reduces the complete quartic to one of the form

$$x^4 + px^2 + qx + r = 0.$$

From this we obtain

$$x^4 + 2px^2 + p^2 = px^2 - qx - r + p^2$$

or

$$(x^2 + p)^2 = px^2 - qx + p^2 - r,$$

whence, for arbitrary y,

$$\begin{aligned}(x^2 + p + y)^2 &= px^2 - qx + p^2 - r + 2y(x^2 + p) + y^2\\ &= (p + 2y)x^2 - qx + (p^2 - r + 2py + y^2).\end{aligned}$$

Now let us choose y so that the right member of the above equation is a square. This is the case when[6]

$$4(p + 2y)(p^2 - r + 2py + y^2) - q^2 = 0.$$

But this is a cubic in y, and may be solved by previous methods. Such a value of y reduces the original problem to nothing but extraction of square roots.

Other algebraic solutions of the general cubic and quartic equations have been given. In the next section, we shall consider the methods devised by the sixteenth-century French mathematician François Viète. A solution of quartics given by Descartes in 1637 may be found in many of the standard college textbooks on the theory of equations [see Problem Study 10.4(e)].

Since the solution of the general quartic equation is made to depend on the solution of an associated cubic equation, Euler, about 1750, attempted similarly to reduce the solution of the general quintic equation to that of an associated quartic equation. He failed in this attempt, as did Lagrange about thirty years later. An Italian physician, Paola Ruffini (1765–1822), in 1803, 1805, and 1813 supplied an incomplete proof of what is now known to be a fact, that the roots of a general fifth, or higher, degree equation cannot be expressed by means of radicals in terms of the coefficients of the equation. This remarkable fact was independently and conclusively established later, in 1824, by the famous Norwegian mathematician Niels Henrik Abel (1802–1829). In 1858, Charles Hermite (1822–1901) gave a solution of the general quintic equation by means of elliptic functions. Hermite's success with the quintic equation later led to the fact that a root of the general equation of degree n can be represented in terms of the coefficients by means of Fuchsian functions. Modern developments in the theory of equations are very fascinating, but too advanced to be considered here, and involve such names as Bring, Jerrard, Tschirnhausen, Galois, Jordan, and many others.

Girolamo Cardano is one of the most extraordinary characters in the history of mathematics. He was born in Pavia in 1501 as the illegitimate son of a jurist and developed into a man of passionate contrasts. He commenced his turbulent professional life as a doctor, studying, teaching, and writing mathematics while practicing his profession. He once traveled as far as Scotland, and, upon his return to Italy, he successively held important chairs at the Universities of Pavia and Bologna. He was imprisoned for a time for heresy because he published a horoscope of Christ's life. Resigning his chair in Bologna he moved to Rome and became a distinguished astrologer, receiving a pension as astrologer to the papal court. He died in Rome in 1576, by his own hand, one story says, so as to fulfill his earlier astrological prediction of the date of his death. Many stories are told of his wickedness, as when in a fit of rage he cut off the ears of his younger son. Some of the stories could be exaggerations

---

[6] A necessary and sufficient condition for the quadratic $Ax^2 + Bx + C$ to be the square of a linear function is that the discriminant, $B^2 - 4AC$, vanish.

of his enemies, and it may be that he has been maligned. His autobiography, of course, supports this viewpoint.

One of the most gifted and versatile men of his time, Cardano wrote a number of works on arithmetic, astronomy, physics, medicine and other subjects. His greatest work is his *Ars magna,* the first great Latin treatise devoted solely to algebra. Here notice is taken of negative roots of an equation and some attention is paid to computations with imaginary numbers. There also occurs a crude method for obtaining an approximate value of a root of an equation of any degree. There is evidence that he was familiar with Descartes' rule of signs, explained in Problem Study 10.3. As an inveterate gambler, Cardano wrote a gambler's manual in which some interesting questions on probability are considered.

Tartaglia had a hard childhood. He was born about 1499 at Brescia to poor parents and was present at the taking of Brescia by the French in 1512. During the brutalities that accompanied this event, Tartaglia and his father (who was a postal messenger at Brescia) fled with many others into the cathedral for sanctuary, but the soldiers pursued and a massacre took place. The father was killed, and the boy, with a split skull and a severe saber cut that cleft his jaws and palate, was left for dead. When the boy's mother later reached the cathedral to look for her family, she found her son still alive and managed to carry him off. Lacking resources for medical assistance, she recalled that a wounded dog always licks the injured spot, and Tartaglia later attributed his recovery to this remedy. The injury to his palate caused a lifelong imperfection in his speech, from which he received his nickname of "the stammerer." His mother gathered together sufficient money to send him to school for fifteen days, and he made the best of the opportunity by stealing a copybook from which he

GIROLAMO CARDANO
(New York Public Library Collection)

subsequently taught himself how to read and write. It is said that lacking the means to buy paper, he was obliged to use the tombstones in the cemetery as slates. He later earned his livelihood teaching science and mathematics in various Italian cities. He died in Venice in 1557.

Tartaglia was a gifted mathematician. We have already reported his work on the cubic equation. He is also credited with being the first to apply mathematics to the science of artillery fire. He wrote what is generally considered the best Italian arithmetic of the sixteenth century, a two-volume treatise containing full discussion of the numerical operations and commercial customs of the time. He also published editions of Euclid and Archimedes.

In 1572, a few years before Cardano died, Rafael Bombelli published an algebra that was a noteworthy contribution to the solution of the cubic equation. It is shown in textbooks on the theory of equations that if $(n/2)^2 + (m/3)^3$ is negative, then the cubic equation $x^3 + mx = n$ has three real roots. But in this case, in the Cardano-Tartaglia formula, these roots are expressed by the difference of two cube roots of *complex imaginary numbers*. This seeming anomaly is known as the **irreducible case in cubics** and considerably bothered the early algebraists. Bombelli pointed out the reality of the apparently imaginary roots in the irreducible case. Bombelli also improved current algebraic notation. Consider, for instance, his use of a bracket symbol. Thus, the compound expression $\sqrt{7 + \sqrt{14}}$ would have been written by Pacioli as $RV\,7\,\bar{p}\,R\,14$, where RV, the *radix universalis,* indicates that the square root is to be taken of all that follows; Bombelli would have written this as $R\llcorner\!\_\ 7\ p\ R\ 14\ \_\!\lrcorner$. Bombelli distinguished square and cube roots by writing $R\ q$ and $R\ c$, and indicated $\sqrt{-11}$ by *di m R q* 11.

NICOLO TARTAGLIA
(David Smith Collection)

8–9 François Viète

The greatest French mathematician of the sixteenth century was François Viète, frequently called by his semi-Latin name of Vieta, a lawyer and member of parliament who devoted most of his leisure time to mathematics. He was born in 1540 at Fontenay and died in 1603 in Paris.

Some entertaining anecdotes are told about Viète. There is a story about an ambassador from the Low Countries who boasted to King Henry IV that France had no mathematician capable of solving a problem proposed in 1593 by his countryman Adrianus Romanus (1561–1615), which required the solution of a forty-fifth degree equation. Viète was summoned and shown the equation. Recognizing an underlying trigonometric connection, he was able, in a few minutes, to give two roots and later gave twenty-one more. The negative roots escaped him. In return, Viète challenged Romanus to solve the problem of Apollonius (see Section 6–4), but Romanus was unable to obtain a solution using Euclidean tools. When he was shown the proposer's elegant solution, he traveled to Fontenay to meet Viète, and a warm friendship developed. There is also a story of how Viète successfully deciphered a Spanish code containing several hundred characters, and for two years France profited thereby in its war against Spain. So certain was King Philip II that the code was undecipherable that he complained to the Pope that the French were employing magic against his country, "contrary to the practice of the Christian faith." It is said that when absorbed with mathematics, Viète would closet himself in his study for days.

Viète wrote a number of works on trigonometry, algebra, and geometry, chief of which are the *Canon mathematicus seu ad triangula* (1579), the *In*

FRANÇOIS VIÈTE
(Brown Brothers)

artem analyticam isagoge (1591), the *Supplementum geometriae* (1593), *De numerosa potestatum resolutione* (1600), and *De aequationum recognitione et emendatione* (published posthumously in 1615). These works, except the last, were printed and distributed at Viète's own expense.

The *Canon mathematicus seu ad triangula* contains some notable contributions to trigonometry. It is perhaps the first book in western Europe to develop, systematically, methods for solving plane and spherical triangles with the aid of all six trigonometric functions. Considerable attention is paid to analytical trigonometry (see Problem Study 8.17). Viète obtained expressions for $\cos n\theta$ as a function of $\cos \theta$ for $n = 1, 2, \ldots, 9$, and later suggested a trigonometric solution of the irreducible case in cubics.

Viète's most famous work is his *In artem,* which did much for the development of symbolic algebra. In this text, Viète introduced the practice of using the vowels to represent unknown quantities and the consonants to represent known ones. Our present custom of using the later letters of the alphabet for unknowns and the early letters for knowns was introduced by Descartes in 1637. Prior to Viète, it was common practice to use different letters or symbols for the various powers of a quantity. Viète used the same letter, properly qualified; thus, our x, x^2, x^3 were written by Viète as A, *A quadratum*, *A cubum*, and by later writers more briefly as A, $A\ q$, $A\ c$. Viète also qualified the coefficients of a polynomial equation so as to render the equation homogeneous, and he used our present + and − signs, but he had no symbol for equality. Thus, he would have written

$$5BA^2 - 2CA + A^3 = D$$

as

$$B\ 5 \text{ in } A \text{ quad} - C \text{ plano } 2 \text{ in } A + A \text{ cub aequatur } D \text{ solido.}$$

Note how the coefficients C and D are qualified so as to make each term of the equation three dimensional. Viète used the symbol = between two quantities, not to indicate the equality of the quantities, but rather the difference between them.

In *De numerosa,* Viète gives a systematic process, which was in general use until about 1680, for successively approximating to a root of an equation. The method becomes so laborious for equations of high degree that one seventeenth-century mathematician described it as "work unfit for a Christian." Applied to the quadratic equation

$$x^2 + mx = n$$

the method is as follows. Suppose x_1 is a known approximate value of a root of the equation, so that the sought root may be written as $x_1 + x_2$. Substitution in the given equation yields

$$(x_1 + x_2)^2 + m(x_1 + x_2) = n,$$

or

$$x_1^2 + 2x_1x_2 + x_2^2 + mx_1 + mx_2 = n.$$

Assuming x_2 so small that x_2^2 may be neglected, we obtain

$$x_2 = \frac{n - x_1^2 - mx_1}{2x_1 + m}.$$

Now from the improved approximation $x_1 + x_2$, we calculate in the same way a still better approximation $x_1 + x_2 + x_3$, and so on. Viète used this method to approximate a root of the sextic equation

$$x^6 + 6000x = 191{,}246{,}976.$$

Viète's posthumously published treatise contains much of interest in the theory of equations. In this treatise, we find the familiar transformations for either increasing or multiplying the roots of an equation by a constant. Viète was aware of the expressions for the coefficients of polynomials, up through the fifth degree, as symmetric functions of the roots, and he knew the transformation that rids the general polynomial of its next to the highest-degree term. In this treatise is found the following elegant solution of the cubic equation $x^3 + 3ax = 2b$, a form to which any cubic can be reduced. Setting

$$x = \frac{a}{y} - y,$$

the given equation becomes

$$y^6 + 2by^3 = a^3,$$

a quadratic in y^3. We thus find y^3, and then y, and then x. Viète's solution of the quartic is similar to Ferrari's. Consider the general depressed quartic

$$x^4 + ax^2 + bx = c,$$

which may be written as

$$x^4 = c - ax^2 - bx.$$

Adding $x^2y^2 + y^4/4$ to both sides yields

$$\left(x^2 + \frac{y^2}{2}\right)^2 = (y^2 - a)x^2 - bx + \left(\frac{y^4}{4} + c\right).$$

Now we choose y so that the right member is a perfect square. The condition for this is

$$y^6 - ay^4 + 4cy^2 = 4ac + b^2,$$

a cubic in y^2. Such a y may be found and the problem completed by extracting square roots.

Viète was an outstanding algebraist, so it is no surprise to learn that he applied algebra and trigonometry to his geometry. He contributed to the three famous problems of antiquity by showing that both the trisection and the duplication problems depend upon the solution of cubic equations. In Section 4–8, we have mentioned Viète's calculation of π and his interesting infinite product converging to $2/\pi$. In Section 6–4, we mentioned his attempted restoration of Apollonius' lost work on *Tangencies*.

In 1594, Viète acquired some unfortunate notoriety by conducting an angry controversy with Clavius on the Gregorian reform of the calendar. Viète's attitude in the matter was wholly unscientific.

8–10 Other Mathematicians of the Sixteenth Century

Our account of the mathematics of the sixteenth century would not be complete without at least a brief mention of some of the other contributors. Of these are the mathematicians, Clavius, Cataldi, and Stevin, and the mathematical astronomers Copernicus, Rhaeticus, and Pitiscus.

Christopher Clavius was born in Bamberg, Germany, in 1537 and died in Rome in 1612. He added little of his own to mathematics, but probably did more than any other German scholar of the century to promote a knowledge of the subject. He was a gifted teacher and wrote highly esteemed textbooks on arithmetic (1583) and algebra (1608). In 1574, he published an edition of Euclid's *Elements* that is valuable for its extensive scholia. He also wrote on trigonometry and astronomy and played an important part in the Gregorian reform of the calendar. As a Jesuit, he brought honor to his order.

Pietro Antonio Cataldi was born in Bologna in 1548, taught mathematics and astronomy in Florence, Perugia, and Bologna, and died in the city of his birth in 1626. He wrote a number of mathematical works, among which are an arithmetic, a treatise on perfect numbers, an edition of the first six books on the *Elements,* and a brief treatise on algebra. He is credited with taking the first steps in the theory of continued fractions.

The most influential mathematician of the Low Countries in the sixteenth century was Simon Stevin (1548–1620). He became quartermaster general for the Dutch army and directed many public works. In the history of mathematics, Stevin is best known as one of the earliest expositors of the theory of decimal fractions. In the history of physics, he is best known for his contribution to statics and hydrostatics. To the savants of his time, he was best known for his

CHRISTOPHER CLAVIUS
(David Smith Collection)

works on fortifications and military engineering. To the general populace of his time, he was best known for his invention of a carriage that was propelled by sails and that ran along the seashore carrying twenty-eight people, easily outstripping a galloping horse.

Astronomy has long contributed to mathematics; in fact, at one time, the name *mathematician* meant an astronomer. Prominent among the astronomers who stimulated mathematics was Nicolas Copernicus (1473–1543) of Poland.

NICOLAS COPERNICUS
(American Museum)

He was educated at the University of Cracow and studied law, medicine, and astronomy at Padua and Bologna. His theory of the universe was completed in 1530 but was not published until the year of his death in 1543. Copernicus' work necessitated the improvement of trigonometry, and Copernicus himself contributed a treatise on the subject.

The leading Teutonic mathematical astronomer of the sixteenth century, and a disciple of Copernicus, was Georg Joachim Rhaeticus (1514–1576). He spent twelve years with hired computers, forming two remarkable and still useful trigonometric tables. One was a ten-place table of all six of the trigonometric functions for every 10″ of arc; the other was a fifteen-place table for sines for every 10″ of arc, along with first, second, and third differences. Rhaeticus was the first to define the trigonometric functions as ratios of the sides of a right triangle. It was because of the importunities of Rhaeticus that Copernicus' great work was dramatically published just before the author died.

Rhaeticus' table of sines was edited and perfected in 1593 by Bartholomaus Pitiscus (1561–1613), a German clergyman with a preference for mathematics. His very satisfactory treatise on trigonometry was the first work on the subject to bear this title.

In summarizing the mathematical achievements of the sixteenth century, we can say that symbolic algebra was well started, computation with the Hindu-Arabic numerals became standardized, decimal fractions were developed, the cubic and quartic equations were solved and the theory of equations generally advanced, negative numbers were becoming accepted, trigonometry was perfected and systematized, and some excellent tables were computed. The stage was set for the remarkable strides of the next century.

It is interesting to note here that the first work on mathematics printed in the New World appeared in 1556 at Mexico City; it was a small commercial compendium by Juan Diez.

Problem Studies

8.1 Problems from the Dark Ages

Alcuin of York (ca. 775) may have been the compiler of the Latin collection entitled *Problems for the Quickening of the Mind.* Solve the following five problems from this collection.

(a) If 100 bushels of corn be distributed among 100 people in such a manner that each man receives 3 bushels, each woman 2, and each child $\frac{1}{2}$ of a bushel, how many men, women, and children were there?

(b) Thirty flasks—10 full, 10 half-empty, and 10 entirely empty—are to be divided among 3 sons so that flasks and contents should be shared equally. How may this be done?

(c) A dog chasing a rabbit, which has a start of 150 feet, jumps 9 feet every time the rabbit jumps 7. In how many leaps does the dog overtake the rabbit?

(d) A wolf, a goat, and a cabbage must be moved across a river in a boat holding only one besides the ferryman. How must he carry them across so that the goat shall not eat the cabbage, nor the wolf the goat?

(e) A dying man wills that if his wife, being with child, gives birth to a son, the son shall inherit $\frac{3}{4}$ and the widow $\frac{1}{4}$ of the property, but if a daughter is born, she shall inherit $\frac{7}{12}$ and the widow $\frac{5}{12}$ of the property. How is the property to be divided if both a son and a daughter are born? (This problem is of Roman origin. The solution given in Alcuin's collection is not acceptable.)

(f) In his *Geometry*, Gerbert solved the problem, considered very difficult at the time, of determining the legs of a right triangle whose hypotenuse and area are given. Solve this problem.

(g) Gerbert expressed the area of an equilateral triangle of side a as $(a/2)(a - a/7)$. Show that this is equivalent to taking $\sqrt{3} = 1.714$.

8.2 The Fibonacci Sequence

(a) Show that the following problem, found in the *Liber abaci*, gives rise to the *Fibonacci sequence:* $1, 1, 2, 3, 5, 8, \ldots, x, y, x + y, \ldots$.

How many pairs of rabbits can be produced from a single pair in a year if every month each pair begets a new pair, which, from the second month on, becomes productive?

(b) If u_n represents the nth term of the Fibonacci sequence, show that

1. $u_{n+1}\, u_{n-1} = u_n^2 + (-1)^n,\ n \geqq 2$.
2. $u_n = [(1 + \sqrt{5})^n - (1 - \sqrt{5})^n]/2^n\sqrt{5}$.
3. $\lim_{n\to\infty} (u_n/u_{n+1}) = (\sqrt{5} - 1)/2$.
4. u_n and u_{n+1} are relatively prime.

There is an extensive literature concerning the Fibonacci sequence. For some of the more esoteric applications to dissection puzzles, art, phyllotaxis, and the logarithmic spiral, see, for example, E. P. Northrop, *Riddles in Mathematics*.

8.3 Problems from the *Liber abaci*

Solve the following problems found in the *Liber abaci* (1202). The first was posed to Fibonacci by a magister in Constantinople; the second was designed to illustrate the rule of three; the third is an example of an inheritance problem that reappeared in later works by Chuquet and Euler.

(a) If A gets from B 7 denarii, then A's sum is fivefold B's; if B gets from A 5 denarii, then B's sum is sevenfold A's. How much has each?

(b) A certain king sent 30 men into his orchard to plant trees. If they could set out 1000 trees in 9 days, in how many days would 36 men set out 4400 trees?

(c) A man left to his oldest son 1 bezant and $\frac{1}{7}$ of what was left; then, from the remainder, to his next son he left 2 bezants and $\frac{1}{7}$ of what was left; then, from the new remainder, to his third son he left 3 bezants and $\frac{1}{7}$ of what was left. He continued this way, giving each son 1 bezant more than the previous son and $\frac{1}{7}$ of what remained. By this division, it developed that the last son received all that was left and all the sons shared equally. How many sons were there and how large was the man's estate?

8.4 Further Problems from Fibonacci

(a) Show that the squares of the numbers $a^2 - 2ab - b^2$, $a^2 + b^2$, $a^2 + 2ab - b^2$ are in arithmetic progression. If $a = 5$ and $b = 4$, the common difference is 720, and the first and third squares are $41^2 - 720 = 31^2$ and $41^2 + 720 = 49^2$. Dividing by 12^2, we obtain Fibonacci's solution to the first of the tournament problems—namely, find a rational number x such that $x^2 + 5$ and $x^2 - 5$ are each squares of rational numbers (see Section 8–3). The problem is insolvable if the 5 is replaced by 1, 2, 3, or 4. Fibonacci showed that if x and h are integers, such that $x^2 + h$ and $x^2 - h$ are perfect squares, then h must be divisible by 24. As examples we have $5^2 + 24 = 7^2$, $5^2 - 24 = 1^2$ and $10^2 + 96 = 14^2$, $10^2 - 96 = 2^2$.

(b) Find a solution to the following problem, which is the third of the tournament problems solved by Fibonacci: Three men possess a pile of money, their shares being $\frac{1}{2}$, $\frac{1}{3}$, $\frac{1}{6}$. Each man takes some money from the pile until nothing is left. The first man then returns $\frac{1}{2}$ of what he took, the second $\frac{1}{3}$, and the third $\frac{1}{6}$. When the total so returned is divided equally among the men, it is found that each then possesses what he is entitled to. How much money was in the original pile, and how much did each man take from the pile?

(c) Solve the following problem given by Fibonacci in the *Liber abaci*. This problem reappeared in a remarkable number of variations. It contains the essence of the idea of an annuity.

A man entered an orchard through 7 gates, and there took a certain number of apples. When he left the orchard, he gave the first guard half the apples that he had and 1 apple more. To the second guard, he gave half his remaining apples and 1 apple more. He did the same to each of the remaining five guards, and left the orchard with 1 apple. How many apples did he gather in the orchard?

8.5 Star-Polygons

A **regular star-polygon** is the figure formed by connecting with straight lines every ath point, starting with some given one, of the n points that divide a circumference into n equal parts, where a and n are relatively prime and $n > 2$. Such a star-polygon is represented by the symbol $\{n/a\}$, and is sometimes called a regular **n-gram.** When $a = 1$, we have a regular polygon. Star-polygons made their appearance in the ancient Pythagorean school, where the $\{\frac{5}{2}\}$ star-polygon, or pentagram, was used as a badge of recognition. Star-polygons also occur in the geometry of Boethius and the translations of Euclid from the Arabic by Adelard and Campanus. Bradwardine developed some of their geometric properties. They were considered also by Regiomontanus, Charles de Bouelles (1470–1533), and Johann Kepler (1571–1630).

(a) Construct, with the aid of a protractor, the star-polygons $\{\frac{5}{2}\}$, $\{\frac{7}{2}\}$, $\{\frac{7}{3}\}$, $\{\frac{8}{3}\}$, $\{\frac{9}{2}\}$, $\{\frac{9}{4}\}$, $\{\frac{10}{3}\}$.

(b) Let $\phi(n)$, called the **Euler ϕ function,** denote the number of numbers less than n and prime to it. Show that there are $[\phi(n)]/2$ regular n-grams.

(c) Show that if n is prime there are $(n - 1)/2$ regular n-grams.

(d) Show that the sum of the angles at the "points" of the regular $\{n/a\}$ star-polygon is given by $(n - 2a)180°$. (This result was given by Bradwardine.)

8.6 Jordanus and Cusa

(a) At the end of Book IV of his translation of Euclid's *Elements,* Campanus describes an angle trisection that is exactly the same as that given by Jordanus in his *De triangulis,* a geometric work in four books containing 72 standard propositions, along with some others on such topics as the centroid of a triangle, curved surfaces, and similar arcs. The trisection, which employes the insertion principle (see Problem Study 4.6), runs as follows: Let $\measuredangle AOB$, given as a central angle in a circle, be the angle we wish to trisect; through A, draw chord AD, cutting the diameter perpendicular to OB in E, such that $ED = OA$; then the line OF parallel to DA trisects $\measuredangle AOB$. Prove the correctness of this construction.

(b) In his *Tractatus de numeris datis,* Jordanus has problems in which a given number is to be divided in some stated fashion. Thus, one of the early problems in the work is: Separate a given number into 2 parts, such that the sum of the squares of the parts shall be another given number. Solve this problem when the 2 given numbers are 10 and 58, respectively.

(c) Cusa gave a number of ways of approximating the circumference of a given circle. His best attempt is the following: Let M be the center of an equilateral triangle ABC, and let D be the midpoint of side AB; let E be the midpoint of DB; then $(\frac{5}{4})ME$, Cusa claimed, is the radius of a circle having a circumference equal to the perimeter of the equilateral triangle. Now draw a right triangle with legs $RS = (\frac{5}{4})ME$ and $RT = (\frac{3}{2})AB$, and construct an angle α "of brass or wood" equal to angle RST. To rectify the circumference of a given circle, draw 2 pependicular diameters UOV and XOY; place the angle α with vertex at U and with 1 side along UOV; then the other side of the angle cuts XOY produced in Z, such that OZ is half the sought circumference of the circle. Show that Cusa's method approximates π by $(\frac{24}{35})\sqrt{21} = 3.142337...$

8.7 Dürer and Magic Squares of Doubly Even Order

In Albrecht Dürer's famous engraving *Melancholia* appears the fourth-order magic square pictured in Figure 68, wherein the date, 1514, in which the en-

16	3	2	13
5	10	11	8
9	6	7	12
4	15	14	1

FIGURE 68

Albrecht Dürer's *Melancholia*.
(The British Museum)

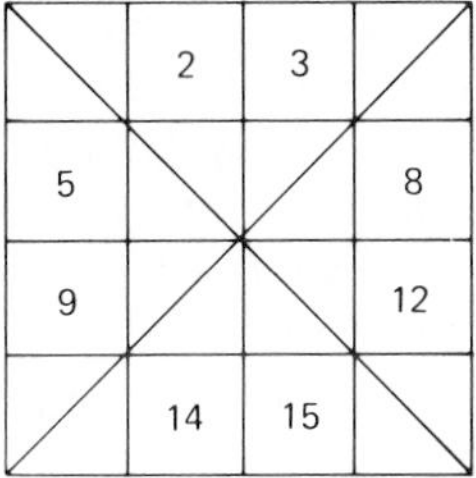

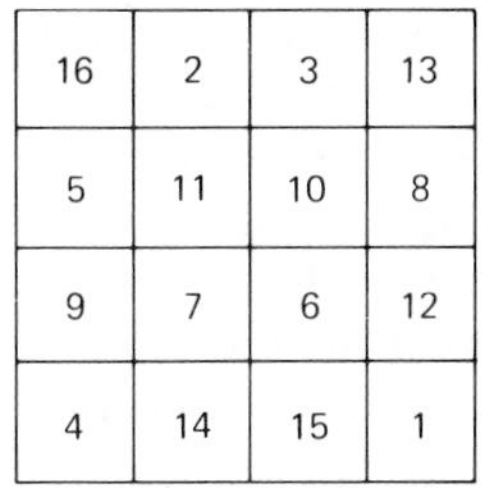

16	2	3	13
5	11	10	8
9	7	6	12
4	14	15	1

FIGURE 69

graving was made, appears in the 2 middle cells of the bottom row. In addition to the usual "magic" properties, show that:

(a) The sum of the squares of the numbers in the top 2 rows is equal to the sum of the squares of the numbers in the bottom 2 rows.

(b) The sum of the squares of the numbers in the first and third rows is equal to the sum of the squares of the numbers in the second and fourth rows.

(c) The sum of the numbers in the diagonals is equal to the sum of the numbers not in the diagonals.

(d) The sum of the squares of the numbers in the diagonals is equal to the sum of the squares of the numbers not in the diagonals.

(e) The sum of the cubes of the numbers in the diagonals is equal to the sum of the cubes of the numbers not in the diagonals.

There is an easy way to construct magic squares of doubly even order—that is, magic squares whose orders are a multiple of 4. Consider, first of all, a square of order 4, and visualize the diagonals as drawn (see Figure 69). Beginning in the upper left corner, count across the rows from left to right in descending succession, recording only the

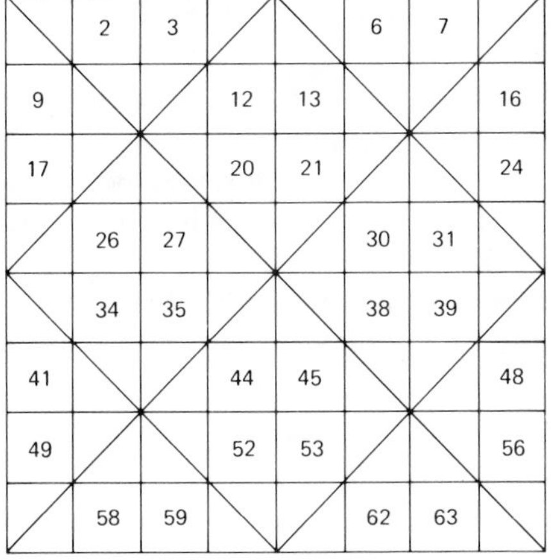

64	2	3	61	60	6	7	57
9	55	54	12	13	51	50	16
17	47	46	20	21	43	42	24
40	26	27	37	36	30	31	33
32	34	35	29	28	38	39	25
41	23	22	44	45	19	18	48
49	15	14	52	53	11	10	56
8	58	59	5	4	62	63	1

FIGURE 70

numbers in cells not cut by a diagonal. Now, beginning at the lower right corner, count across the rows from right to left in ascending succession, recording only the numbers in cells that are cut by a diagonal. The resulting magic square is little different from Dürer's square. The same rule applies to any magic square of order $4n$ if we visualize, as drawn in, the diagonals of all the n^2 principal 4×4 sub-blocks. Figure 70 shows the construction of an 8×8 magic square by this rule.

(f) Construct a magic square of order 12.

8.8 Problems from Regiomontanus

Solve the following 3 problems, the first 2 of which are found in Regiomontanus' *De triangulis omnimodis* (1464):

(a) Determine a triangle given the difference of 2 sides, the altitude on the third side, and the difference of the segments into which the altitude divides the third side. (The numerical values given by Regiomontanus are 3, 10, and 12.)

(b) Determine a triangle given a side, the altitude on this side, and the ratio of the other 2 sides. (The numerical values given by Regiomontanus are 20, 5, and $\frac{3}{5}$.)

(c) Construct a cyclic quadrilateral, given the 4 sides.

8.9 Problems from Chuquet

Solve the following problems adopted from Chuquet's *Triparty en la science des nombres* (1484):

(a) A merchant visited 3 fairs. At the first, he doubled his money and spent \$30; at the second he tripled his money and spent \$54; at the third, he quadrupled his money and spent \$72, and then had \$48 left. How much money had he at the start?

(b) A carpenter agrees to work under the conditions that he is to be paid \$5.50 every day he works, but must pay \$6.60 every day he does not work. At the end of 30 days, he finds he has paid out as much as he has received. How many days did he work?

(c) Two wine merchants enter Paris, one of them with 64 casks of wine, the other with 20. Since they have not enough money to pay the customs duties, the first pays 5 casks of wine and 40 francs, and the second pays 2 casks of wine and receives 40 francs in change. What is the price of each cask of wine and the duty on it?

(d) Chuquet gave the **regle des nombres moyens,** which says that if a, b, c, d are positive numbers, then $(a + b)/(c + d)$ lies between a/c and b/d. Prove this.

8.10 Problems from Pacioli

Solve the following 2 problems found in Pacioli's *Sūma* (1494). The second problem is an elaboration of the popular "frog-in-the-well problem" and has had many variants.

(a) The radius of the inscribed circle of a triangle is 4, and the segments into which one side is divided by the point of contact are 6 and 8. Determine the other two sides.

(b) A mouse is at the top of a poplar tree that is 60 feet high, and a cat is on the ground at its foot. The mouse descends $\frac{1}{2}$ of a foot each day, and at night it turns back $\frac{1}{6}$ of a foot. The cat climbs 1 foot a day and slips back $\frac{1}{4}$ of a foot each night. The tree grows $\frac{1}{4}$ of a foot between the cat and the mouse each day and shrinks $\frac{1}{8}$ of a foot every night. How long will it take the cat to reach the mouse?

8.11 Early Commercial Problems

Solve the following problems found in early European arithmetics.

(a) This problem, from Buteo's arithmetic of 1559, is based upon difficulties of the early Roman navigators.

Two ships that were 20,000 stadia apart weighed anchor to sail straight toward each other. It happened that the first one set sail at daybreak with the north wind blowing. Toward evening, when it had gone 1200 stadia, the north wind fell and the southwest wind rose. At this time, the other ship set sail and sailed 1400 stadia during the night. The first ship, however, was driven back 700 stadia by the contrary wind, but with the morning north wind it was driven ahead in the usual manner of outward sailing, while the other went back 600 stadia. Thus, alternately, night and day, the ships were carried along by a favorable wind and then driven back by an unfavorable one. I ask how many stadia the ships sailed in all and when they met.

(b) Here is a problem given by Tartaglia to illustrate the important matter of exchange.

If 100 lire of Modon money amounts to 115 lire in Venice, and if 180 lire in Venice comes to 150 in Corfu, and if 240 lire Corfu money is worth as much as 360 lire in Negroponte, what is the value in Modon coinage of 666 lire Negroponte money?

(c) The early arithmetics gave many problems involving custom duties. Following is a problem of this sort adapted from Clavius' arithmetic of 1583.

A merchant bought 50,000 pounds of pepper in Portugal for 10,000 scudi, paying a tax of 500 scudi. He carried it to Italy at a cost of 300 scudi and there paid another duty of 200 scudi. The transportation from the coast to Florence cost 100 scudi, and he was obliged to pay an impost of 100 scudi to that city. Lastly, the government demanded a tax from each merchant of 1000 scudi. Now he is perplexed to know what price to charge per pound so that, after all these expenses, he may make a profit of $\frac{1}{10}$ of a scudi a pound.

(d) In a practical manual for merchants written by the Florentine Ghaligai in 1521 occurs the following problem concerning profit and loss.

A man bought a number of bales of wool in London, each bale weighed 200 pounds, English measure, and each bale cost him 24 fl. He sent the wool to Florence and paid carriage duties, and other expenses, amounting to 10 fl. a bale. He wishes to sell the wool in Florence at such a price as to make 20 percent on his investment. How much should he charge a hundredweight if 100 London pounds are equivalent to 133 Florentine pounds?

(e) Interest problems were very common. Here is one from Fibonacci's *Liber abaci* of 1202.

A certain man puts 1 denarius at interest at such a rate that in 5 years he has 2 denarii, and in 5 years thereafter the money doubles. I ask how many denarii he would gain from this 1 denarius in 100 years.

(f) The following problem is from Humphrey Baker's *The Well Spring of Sciences* (1568) and concerns itself with partnership.

Two marchauntes haue companied together, the first hath layde in the first of Januarie, 640 li. The seconde can lay in nothing vntill the firste of April. I demaund how much he shall lay in, to the end that he may take halfe the gaynes. (Assume that the partnership is to last for 1 year from the date of the first man's investment.)

(g) Here is essentially an annuity problem from Tartaglia's *General trattato of* 1556. It should be borne in mind that this problem was proposed before the invention of logarithms.

A merchant gave a university 2814 ducats on the understanding that he was to be paid back 618 ducats a year for 9 years, at the end of which the 2814 ducats should be considered as paid. What compound interest was he getting on his money?

8.12 The Gelosia and Galley Algorithms

(a) The arithmetics of the fifteenth and sixteenth centuries contain descriptions of algorithms for the fundamental operations. Of the many schemes devised for performing a long multiplication, the so-called **gelosia,** or **"grating," method** was perhaps the most popular. The method, which is illustrated in Figure 71 by the multiplication of 9876 and 6789 to yield 67,048,164, is very old. It was probably first developed in India (see Section 7–5), for it appears in a commentary on the *Lilāvati* and in other Hindu works. From India, it made its way into Chinese, Arabian, and Persian works. It was long a favorite method among the Arabs, from whom it passed over to the Western Europeans. Because of its simplicity to apply, it could well be that the method would still be in use, but for the difficulty of printing, or even drawing, the needed net of lines. The pattern resembles the grating, or lattice, used in some windows. These were known as "gelosia," eventually becoming "jalousie" (meaning "blind," in French). Find the product of 80,342 and 7318 by the gelosia method.

(b) By far the most common algorithm for long division in use before 1600 was the so-called **galley,** or **scratch, method,** which in all likelihood was of Hindu origin. To clarify the method, consider the following steps in the division of 9413 by 37.

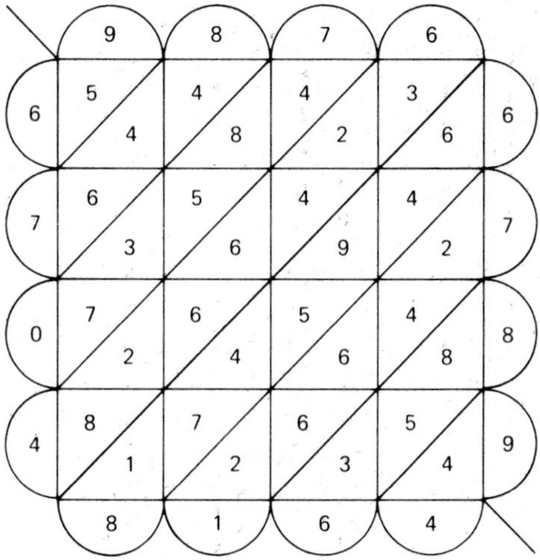

FIGURE 71

1. Write the divisor 37 below the dividend as shown. Obtain the first quotient digit, 2, in the usual manner, and write it to the right of the dividend.	9413 | 2 37
2. Think: 2 × 3 = 6, 9 − 6 = 3. Scratch 9 and 3 and write 3 above the 9. Think: 2 × 7 = 14, 34 − 14 = 20. Scratch 7, 3, 4 and write 2 above the 3 and 0 above the 4.	2 30 9413 | 2 37
3. Write the divisor 37 one place to the right, diagonally. The resultant dividend after Step 2 is 2013. Obtain the next quotient digit, 5. Think: 5 × 3 = 15, 20 − 15 = 5. Scratch 3, 2, 0 and write 5 above the 0. Think: 5 × 7 = 35, 51 − 35 = 16. Scratch 7, 5, 1 and write 1 above the 5 and 6 above the 1.	˙1 25 306 9413 | 25 377 3
4. Write the divisor 37 one more place to the right, diagonally. The resultant dividend after Step 3 is 163. Obtain the next quotient digit, 4. Think: 4 × 3 = 12, 16 − 12 = 4. Scratch 3, 1, 6 and write 4 above the 6. Think: 4 × 7 = 28, 43 − 28 = 15. Scratch 7, 4, 3 and write 1 above the 4 and 5 above the 3.	11 254 3065 9413 | 254 — 1/5 3777 33
5. The quotient is 254, with remainder 15.	

After a little practice, the galley method is found to be not nearly as difficult as it at first appears. Its popularity was due to the ease with which it can be used on a sand abacus, where the scratching is actually a simple erasing followed by a possible replacement. The name *galley* referred to a boat, which the outline of the finished problem was thought to resemble. The resemblance follows either by viewing the work from the bottom of the page, when the quotient appears as a bowsprit, or by viewing the work from the left side of the page, when the quotient appears as a mast. In this second viewpoint, the remainder was frequently written (as indicated above) like a flag on the top of the mast.

Divide 65,284 by 594, using the galley method. (This problem, solved in this way, appears in the *Treviso Arithmetic* of 1478.)

8.13 Gematria or Arithmography

Since many of the ancient numeral systems were alphabetical systems, it was natural to substitute the number values for the letters in a name. This led to a mystic pseudo-science known as **gematria,** or **arithmography,** which was very popular among the ancient Hebrews and others, and was revived during the Middle Ages.

(a) The word *amen* when written in Greek is $\alpha\mu\eta\nu$. On this basis, explain why, in certain Christian manuscripts, the number 99 appears at the end of a prayer.

(b) Using gematria, "prove," by English key, that of the 3 men, Roosevelt, Churchill, and Stalin, Roosevelt was the greatest political figure.

(c) "Beast" the following (all but the last in Roman, the last in Greek): LUDOVICUS (presumably Louis XIV), SILVESTER SECUNDUS (Gerbert, who reigned as Pope Sylvester II), PAULO V. VICE–DEO, VICARIUS FILII DEI, DOCTOR ET REX LATINUS, VICARIUS GENERALIS DEI IN TERRIS, DUX CLERI, GLADSTONE.

(d) Verify the following, found in De Morgan's *A Budget of Paradoxes:*

1. "A Mr. James Dunlop was popping at the Papists with a 666-rifled gun, when Dr. Chalmers quietly said, 'Why, Dunlop, you bear it yourself,' and handed him a paper on which the numerals in IACOBVS DVNLOPVS were added up."
2. "Mr. Davis Thom found a young gentleman of the name of St. Claire busy at the Beast number: he forthwith added the letters in $\sigma\tau$ $\kappa\lambda\alpha\iota\rho\varepsilon$ and found 666."

(e) John F. Bobalek submitted the following to English key: HOWARD W. EVES, A PROFESSOR OF MATHEMATICS AND DOCTOR OF PHILOSOPHY. Find Bobalek's frightening discovery.

8.14 Cubic Equations

(a) Show that the transformation $x = z - a_1/na_0$ converts the n-ic equation

$$a_0x^n + a_1x^{n-1} + a_2x^{n-2} + \ldots + a_n = 0$$

into an equation in z that lacks the $(n-1)$st degree term.

(b) By (a), the transformation $x = z - b/3a$ converts the cubic equation $ax^3 + bx^2 + cx + d = 0$ into one of the form $z^3 + 3Hz + G = 0$. Find H and G in terms of a, b, c, d.

(c) Derive the Cardano-Tartaglia formula,

$$x = \sqrt[3]{(n/2) + \sqrt{(n/2)^2 + (m/3)^3}} - \sqrt[3]{-(n/2) + \sqrt{(n/2)^2 + (m/3)^3}}$$

for solving the cubic equation $x^3 + mx = n$ (see Section 8–8).

(d) Solve $x^3 + 63x = 316$, for 1 root, by both the Cardano-Tartaglia formula and Viète's method.

(e) As an example of the irreducible case in cubics, solve $x^3 - 63x = 162$ by the Cardano-Tartaglia formula. Then show that $(-3 + 2\sqrt{-3})^3 = 81 + 30\sqrt{-3}$ and $(-3 - 2\sqrt{-3})^3 = 81 - 30\sqrt{-3}$, whence the root given by the formula is -6 in disguise.

8.15 Quartic Equations

(a) Cardano solved the particular quartic $13x^2 = x^4 + 2x^3 + 2x + 1$ by adding $3x^2$ to both sides. Do this and solve the equation for all 4 roots.

(b) Da Coi in 1540 proposed the following problem to Cardano: "Divide 10 into 3 parts, such that they shall be in continued proportion and that the product of the first 2 shall be 6." If the 3 parts be denoted by a, b, c, we have

$$a + b + c = 10, \qquad ac = b^2, \qquad ab = 6.$$

Show that when a and c are eliminated we obtain the quartic equation

$$b^4 + 6b^2 + 36 = 60b.$$

It was in trying to solve this quartic that Cardano's pupil Ferrari discovered his general method.

(c) Obtain, by both Ferrari's and Viète's methods, the cubic equations associated with the quartic of (b).

8.16 Sixteenth-Century Notation

(a) Write, in Bombelli's notation, the expression

$$\sqrt{\left[\sqrt[3]{\sqrt{68} + 2} - \sqrt[3]{\sqrt{68} - 2}\,\right]}.$$

(b) Write, in modern notation, the following expression that occurs in Bombelli's work:

R c⌊ 4 *p di m R q* 11 ⌋*p R c*⌊ 4 *m di m R q* 11 ⌋

(c) Write, in Viète's notation,

$$A^3 - 3BA^2 + 4CA = 2D.$$

8.17 Problems from Viète

(a) Establish the following identities given by Viète in his *Canon mathematicus seu ad triangula* (1579):

$$\sin \alpha = \sin (60° + \alpha) - \sin (60° - \alpha),$$

$$\csc \alpha + \cot \alpha = \cot \frac{\alpha}{2},$$

$$\csc \alpha - \cot \alpha = \tan \frac{\alpha}{2}.$$

(b) Express $\cos 5\theta$ as a function of $\cos \theta$.
(c) Starting with $x_1 = 200$, approximate, by Viète's method, a root of $x^2 + 7x = 60{,}750$.
(d) Find the x_2, of Viète's method of successive approximations, for the cubic equation $x^3 + px^2 + qx = r$ (see Section 8–9).
(e) Viète derived the formula

$$\sin x + \sin y = 2 \sin \frac{x + y}{2} \cos \frac{x - y}{2}$$

from the diagram of Figure 72, wherein the angles $x = DOA$ and $y = COD$ appear as central angles of a unit circle. Fill in the details of the following sketch of Viète's proof:

$$\sin x + \sin y = AB + CD = AE = AC \cos \frac{x - y}{2}$$

$$= 2 \sin \frac{x + y}{2} \cos \frac{x - y}{2}.$$

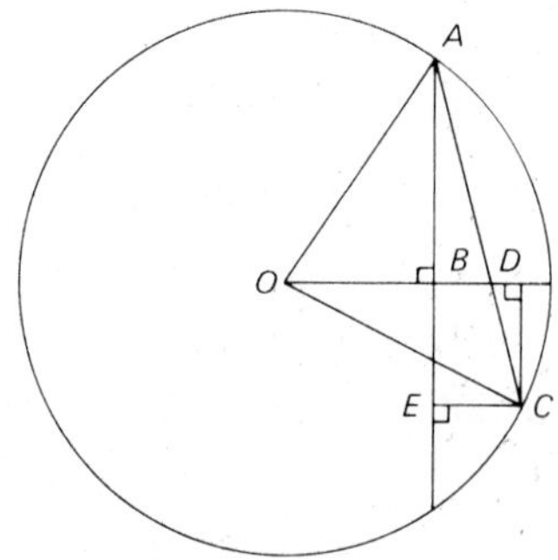

FIGURE 72

8.18 Problems from Clavius

Solve the following recreational problems in Clavius' algebra of 1608.

(a) In order to encourage his son in the study of arithmetic, a father agrees to pay his boy 8 cents for every problem correctly solved and to fine him 5 cents for each incorrect solution. At the end of 26 problems, neither owes anything to the other. How many problems did the boy solve correctly?

(b) If I were to give 7 cents to each of the beggars at my door, I would have 24 cents left. I lack 32 cents of being able to give them 9 cents apiece. How many beggars are there, and how much money have I?

(c) A servant is promised \$100 and a cloak as his wages for a year. After 7 months, he leaves this service and receives the cloak and \$20 as his due. How much is the cloak worth?

8.19 Some Geometry

(a) Books IV and VI of Bombelli's *Algebra* contain some geometry problems that are solved algebraically. In one problem, Bombelli asks for the side of a square inscribed in a triangle ABC, in which $AB = 13$, $BC = 14$, $CA = 15$, such that 1 side of the square lies along BC. Solve this problem.

(b) Johannes Werner (1468–1528) wrote a Latin work, in 22 books, on the *Elements of Conics*, printed at Nuremberg in 1522. In this work, Werner gives the following method of plotting points, with compass and straightedge, of a parabola having a given vertex V, axis VW, and latus rectum p. On WV produced, mark off $VA = p$. Draw any circle of radius greater than $p/2$, with center on AW, and passing through A. Let this circle cut AW in B and the perpendicular to AW at V in C and C'. Mark off on the perpendicular to AW at B the distances $BP = BP' = VC$. Then P and P' are points of the parabola. By drawing a sufficient number of circles, one can obtain as many points of the parabola as one pleases. Establish Werner's construction.

(c) Albrecht Dürer gave the following approximate construction of a regular nonagon inscribed in a given circle of center O. Draw the concentric circle of radius 3 times that of the given circle and let $AC'BA'CB'$ be a regular hexagon inscribed in this latter circle. With B' and C' as centers and with radii equal to OA, describe arcs connecting O and A and cutting the original circle in F and G. Then FG is very nearly equal to a side of the sought regular nonagon. It can be shown that angle FOG differs from 40° by less than 1°. Approximately inscribe a regular nonagon in a given circle by Dürer's method.

For Dürer's approximate trisection of an arbitrary angle, see the penultimate paragraph of Section 4–6.

(d) Campanus, at the end of Book IV of his translation of Euclid's *Elements*, gives the following method of trisecting a given angle. Let the given angle AOB be placed with its vertex at the center of a circle of any chosen radius $OA = OB$. From O draw a radius OC perpendicular to *OB*. Through *A*, place a straight line *AED* such that *ED* = *OA*. Finally, draw a radius *OF* parallel to *DEA*. Then angle *FOB* = $\frac{1}{3}$ (angle *AOB*). Establish the correctness of the construction, granting the use of the *insertion principle* (see Problem Study 4.6).

Essay Topics

8/1 Reasons for the low state of mathematics in Europe during most of the Middle Ages.

8/2 Mathematical recreations in the Middle Ages.

8/3 The number game Rithmomachia.

8/4 The effect of the loss of Toledo by the Moors in 1085 on European mathematics.

8/5 Gerbert and his influence on mathematics.

8/6 The transmission of ancient Greek and Hindu learning to Western Europe after the Dark Ages.

8/7 The ubiquitous Fibonacci sequence.

8/8 The patrons of science Frederick II and his son Manfred.

8/9 Important factors in the development of Renaissance mathematics.

8/10 Luca Pacioli (ca. 1445–1476).

8/11 Leonardo da Vinci and mathematics.

8/12 Regiomontanus (1436–1476).

8/13 Albrecht Dürer and mathematics.

8/14 Copernicus (1473–1543).

8/15 The importance of the solution of cubic equations in the development of imaginary numbers.

8/16 The life and works of Robert Recorde.

8/17 Matteo Ricci (1552–1610).

8/18 Viète as the first really modern mathematician.

8/19 The history of decimal fractions.

8/20 The leading printed mathematical works of the fifteenth century.

8/21 Reasons for the prominence of commercial arithmetics in the latter half of the fifteenth century.

8/22 The Rechenmeisters.

8/23 Gematria.

8/24 Algorithms for long multiplication.

8/25 Algorithms for long division.

Bibliography

ADAMCZEWSKI, JAN *Nicholas Copernicus and His Epoch*. Philadelphia: Copernican Society of America, 1973.

ARMITAGE, ANGUS *The World of Copernicus*. New York: The New Library (a Mentor Book), 1947.

CAJORI, FLORIAN *A History of Mathematical Notations*. 2 vols. Chicago: Open Court, 1928–1929.

CARDAN, JEROME *The Book of My Life*. Translated from the Latin by Jean Stoner. New York: Dover, 1963.

CARDANO, GIROLAMO *The Great Art, or the Rules of Algebra*. Translated by Richard Witmer. Cambridge, Mass.: M.I.T. Press, 1968.

CLAGETT, MARSHALL *The Science of Mechanics in the Middle Ages*. Madison, Wis.: The University of Wisconsin Press, 1959.

——— *Archimedes in the Middle Ages*. The Arabo-Latin Tradition, vol. 1. Madison, Wis.: The University of Wisconsin Press, 1964.

COOLIDGE, J. L. *The Mathematics of Great Amateurs*. New York: Oxford University Press, 1949.

CROSBY, H. L., JR. *Thomas of Bradwardine His Tractatus de Proportionibus: Its Significance for the Development of Mathematical Physics*. Madison, Wis.: The University of Wisconsin Press, 1955.

CUNNINGTON, SUSAN *The Story of Arithmetic, A Short History of Its Origin and Development*. London: Swan Sonnenschein, 1904.

DAVID, F. N. *Games, Gods and Gambling*. New York: Hafner, 1962.

DAY, M. S. *Scheubel as an Algebraist, Being a Study of Algebra in the Middle of the Sixteenth Century, Together with a Translation of and a Commentary upon an Unpublished Manuscript of Scheubel's Now in the Library of Columbia University*. New York: Teachers College, Columbia University, 1926.

DE MORGAN, AUGUSTUS *A Budget of Paradoxes*. 2 vols. New York: Dover, 1954.

FIERZ, MARKUS *Girolamo Cardano (1501–1576)*. Translated by Helga Niman. Boston: Birkhaüser, 1983.

GRANT, EDWARD, ed. *Nicole Oresme: De proportionibus proportionum and Ad pauca respicientes*. Madison, Wis.: The University of Wisconsin Press, 1966.

HAMBRIDGE, JAY *Dynamic Symmetry in Composition*. New Haven: Yale University Press, 1923.

——— *Practical Applications of Dynamic Symmetry*. Edited by Mary C. Hambridge. New Haven: Yale University Press, 1932.

HAY, CYNTHIA, ed. *Mathematics from Manuscript to Print*. Oxford: Oxford University Press, 1987.

HILL, G. F. *The Development of Arabic Numerals in Europe*. New York: Oxford University Press, 1915.

HOGGATT, V. E., JR. *Fibonacci and Lucas Numbers*. Boston: Houghton Mifflin, 1969.

HUGHES, BARNABAS *Regiomontanus on Triangles*. Madison, Wis.: The University of Wisconsin Press, 1964.

——— *De Numeris Datis*. University of California Press, 1983.

INFELD, LEOPOLD *Whom the Gods Love, The Story of Evariste Galois*. New York: McGraw-Hill, 1948.

JOHNSON, R. A. *Modern Geometry*. Boston: Houghton Mifflin Company, 1929.

KARPINSKI, L. C. *The History of Arithmetic*. New York: Russell & Russell, 1965.

KRAITCHIK, MAURICE *Mathematical Recreations*. New York: W. W. Norton, 1942.

MESCHKOWSKI, HERBERT *Ways of Thought of Great Mathematicians*. San Francisco: Holden-Day, 1964.

MOODY, E. A., and MARSHALL CLAGETT *The Medieval Science of Weights*. Madison, Wis.: The University of Wisconsin Press, 1952.

MORLEY, HENRY *Jerome Cardan, The Life of Girolamo Cardano of Milan, Physician*. 2 vols. London: Chapman & Hall, 1854.

NICOMACHUS OF GERASA *Introduction to Arithmetic*. Translated by M. L. D'Ooge, with *Studies in Greek Arithmetic,* by F. E. Robbins and L. C. Karpinski. Ann Arbor, Mich.: University of Michigan Press, 1938.

NORTHROP, E. P. *Riddles in Mathematics*. Princeton, N.J.: D. Van Nostrand, 1944.

ORE, OYSTEIN *Number Theory and Its History*. New York: McGraw-Hill, 1948.

——— *Cardano, The Gambling Scholar*. Princeton, N.J.: Princeton University Press, 1953.

——— *Niels Henrik Abel, Mathematician Extraordinary*. Minneapolis: University of Minnesota Press, 1957.

ORESME, NICOLE *An Abstract of Nicholas Oréme's Treatise on the Breadths of Forms*. Translated by C. G. Wallis. Annapolis, Md.: St. John's Book Store, 1941.

PISANO, LEONARDO *The Book of Squares*. Annotated translation by L. E. Sigler. Orlando, Fla.: Academic Press, 1987.

ROSE, PAULI *The Italian Renaissance of Mathematics*. Geneva: Libraire Droz, 1975.

SMITH, D. E. *Rara Arithmetica*. Boston: Ginn, 1908.

——— *A Source Book in Mathematics*. New York: McGraw-Hill, 1929.

———, and L. C. KARPINSKI *The Hindu-Arabic Numerals*. Boston: Ginn, 1911.

STIMSON, DOROTHY *The Gradual Acceptance of the Copernican Theory of the Universe*. Gloucester, Mass.: Peter Smith, 1972.

SULLIVAN, J. W. N. *The History of Mathematics in Europe, from the Fall of Greek Science to the Rise of the Conception of Mathematical Rigour*. New York: Oxford University Press, 1925.

SWETZ, F. J. *Capitalism and Arithmetic: The New Math of the 15th Century*. Chicago: Open Court, 1987.

TAYLOR, R. EMMETT *No Royal Road. Luca Pacioli and His Time*. Chapel Hill, N.C.: University of North Carolina Press, 1942.

WATERS, W. G. *Jerome Cardan, a Biographical Study*. London: Lawrence & Bullen, 1898.

WHITE, W. F. *A Scrap-Book of Elementary Mathematics*. Chicago: Open Court, 1927.

WILSON, CURTIS *William Heytesbury, Medieval Logic and the Rise of Mathematical Physics*. Madison, Wis.: The University of Wisconsin Press, 1956.

YATES, R. C. *The Trisection Problem*. Ann Arbor, Mich.: Edwards Brothers, 1947.

ZELLER, SISTER MARY CLAUDIA *The Development of Trigonometry from Regiomontanus to Pitiscus*. Ph.D. thesis. Ann Arbor, Mich.: University of Michigan, 1944.

Part 2

THE SEVENTEENTH CENTURY AND AFTER

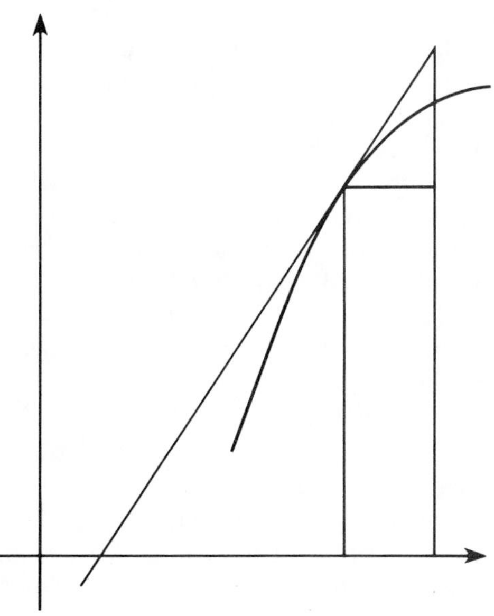

Cultural Connection

PURITANS AND SEADOGS

The Expansion of Europe—1492 to 1700
(to accompany Chapters Nine, Ten, and Eleven)

A starry, silver dusk settled over the great stone city of Tenochtitlan, capital of the Aztec Empire. It was a hot night, in the middle of June, 1520. An angry mob had gathered in a street outside a building used as a temporary headquarters by a small band of Spanish adventurers led by Hernán Cortés (1485–1547). A few days before, Cortés' troops, with the acquiesence of the Aztec emperor, Moctezuma (1480?–1520), had murdered two hundred Aztec nobles. Now, Moctezuma, in a desperate effort to calm the mob, emerged from the building onto a balcony. His furious subjects pelted him with stones, one of which struck his head as he staggered back inside. Three days later, Moctezuma died from his wounds, and Cortés' army frantically prepared to battle an enraged empire.

We can only speculate as to what Moctezuma thought as he gulped the warm, dry night air and prepared to address his subjects for a final time. We do not know what he planned to say, or whether he expected to be killed by or protected by Cortés, whom Moctezuma probably considered the reincarnation of the magical god, Quetzalcoatl. Almost certainly, perhaps the moment the stone struck his head, he must have wondered "why?"

It would have been a poignant question. In the summer of 1520, white strangers, Cortés' *conquistadores,* arrived on large boats from the eastern sea. They rode atop strange beasts, spoke an alien tongue, and sported mysterious weapons that spewed thunder and smoke and killed from great distances. They marched into Tenochtitlan, already having convinced many of the peasants in the countryside that they were gods; by the end of the summer, these strangers would conquer the most powerful empire in the Americas, tear down the great city of Tenochtitlan, and proclaim themselves rulers of all of Mexico. For centuries, the sandy land on Mexico's high plateau had been ruled by a series of native empires that governed with a mixture of splendor and savagery, their sprawling stone cities monuments to the vitality of their civilization. For Mexicans, Cortés' conquest marked the end of an era of native cultural brilliance that had brought irrigated agriculture, the domestication of animals, the emergence of variegated social classes, and the evolution of sophisticated governments to Central America. To the Europeans who conquered it, however, the fall of the Aztec Empire was but one of many chapters, albeit bloodier than most, in the story of the Age of Exploration, the expansion of European interests from the continent of Europe to every corner of the world.

The Age of Exploration began with commercial voyages. European merchants had begun to trade with Asia, through Moslem intermediaries, in the 1300s and 1400s, a commerce that, as we saw in Cultural Connection VI, sparked the Renaissance. Italian cities, such as Venice, Genoa, Florence, and Naples, were well-positioned geographically for this new commerce, in which eastern goods, like spices and cloth, were brought across the Mediterranean Sea by ship. Less well-positioned were European merchants in cities on the Atlantic Ocean. These merchants, often with official backing from their governments, sought out alternate trade routes. In the mid-1400s, shippers in Lisbon, Portugal, their voyages sponsored by the king's brother, Prince Henry the Navigator (1394–1460), commenced a search for a water route to India around Africa. This search culminated in Vasco da Gama's (ca. 1469–1524) successful expedition in 1497 to 1499. A chancier water route to Asia was postulated by the Italian Christopher Columbus (1451?–1506), who in 1492 to 1493, sponsored by the Spanish government, tried to reach India by sailing due west across the Atlantic Ocean. Instead of India, he reached America.

The race was on. Spanish, English, and French merchants, like Columbus, explored the Atlantic coast of North and South America, looking for a passageway to China and India. Dutch traders, who were more realistic, followed the Portugese around Africa. Russians poked into Siberia, hunting for an overland trail to China.

Whereas the first stage of the Age of Exploration was characterized by commerce, the next phase was marked by conquest and annexation. Spanish *conquistadores* like Hernán Cortés violently established hegemony over native empires—the Aztecs in Mexico (1520), the Incas in Peru (1530–1535), the Chibchas in Colombia (1536)—and over disunited tribal communities in the West Indies (1492–1511), Argentina (1530s), Chile (1540s), and the Philippines (1560s). The Portuguese erected fortresses on foreign soil in India (1510), the East Indies (1511), an island in the Persian Gulf (1515), and China (1557). By 1600, the Portuguese also controlled vast stretches of the African coast, several towns in India, the island of Timor in the East Indies, and most of Brazil. Between 1608 and 1703, the French built a string of forts in North America in the St. Lawrence and Mississippi River valleys, at the same time as the English were doing the same on the eastern seaboard of what would later become the United States. The Dutch constructed forts in Africa, India, the East Indies, and Taiwan between 1602 and 1700, including major enclaves at Batavia (1619), now Djakarta, on the island of Java, and at Cape Town (1652) at the southern tip of Africa. After 1462, Russia expanded eastward into Siberia, most of which was annexed by 1689.

Most of the lands conquered or otherwise annexed by European countries in the Age of Exploration were neither militarily powerful nor politically or technologically advanced. The Aztec and Inca Empires had just emerged from the first stages of their agricultural revolution and, although densely populated, lacked weaponry as advanced as Spanish cannon and firearms. Coastal Africa,

Siberia, and most of the Americas were populated by tribal peoples, had relatively low population densities (the more densely populated states of inland Africa remained independent until the nineteenth century), and were also without guns. More powerful non-European countries, such as the various Moslem states, China, and Japan, remained independent. European expansion during the Age of Exploration came largely at the cost of the weak.

The third stage of the Age of Exploration was colonization, the actual migration of Europeans to other continents. European colonies in America, Africa, and Asia were of several kinds. Some colonies were centered around the extraction of raw materials, like the Spanish silver-mining colonies in Peru and the British fishing colony of Newfoundland; others were commercial outposts, like the English and French fur-trading "factories" in Canada. Some colonies were military garrisons that guarded vital waterways, like the Dutch Cape Colony; others were agricultural colonies, such as English Virginia, Spanish Cuba, and Portuguese Brazil. Still other colonies served as havens for religious and political minorities, like the English settlements in Pennsylvania and New England. Most colonies served several of these functions. New England colonists, for example, caught cod and other fish, cut timber, traded for furs with the Indians, and farmed—all this in addition to their more celebrated goal of establishing model religious communities that they hoped would "be as a beacon unto the world." In most cases, native peoples were overwhelmed by the newcomers, either outnumbered or outgunned, and assumed the status of a lower class, which was the fate of blacks in Cape Colony and Indians in Peru. Natives either toiled in mines or as field hands, or were forcibly evicted and pushed further into the wilderness. Black Africans were often enslaved, either to work as menial laborers in the African colonies or to be shipped to colonies in the Americas, where there was a labor shortage. Every major European colonial power except Russia enslaved black Africans in large numbers, and the Spanish also enslaved Indians.

Europeans also fought with one another in the race for overseas riches. English raiders, called "sea dogs," preyed upon Spanish shipping. French settlements in Florida and Brazil were sacked by Spanish and Portuguese armies, and Frenchmen and Englishmen skirmished in the oak and pine forests of North America. Outposts often changed hands. The English were particularly adept at the game of flag changing; they acquired the Dutch outposts of Cape Town and New Amsterdam (New York), and several of the West Indies islands from Spain.

We have described the form and substance of European expansion overseas between 1492 and 1700. Now let us return to Moctezuma's "why?" Europeans had been aware of the general pattern of world geography for centuries. Eratosthenes, the third-century B.C. Alexandrian mathematician and geographer whom we met in Cultural Connection IV, knew that the world was spherical, was aware of its approximate circumference, and understood the general shapes of Europe, Africa, and Asia. Scandinavians had encountered North America in approximately A.D. 1000, and even planted unsuccessful settlements in Greenland and Newfoundland. Irish, English, French, and Basque fishermen almost certainly visited North America regularly before 1492, to take cod in the offshore waters; however, these activities never led to widespread commerce with overseas lands, nor to permanent conquest and colonization. Why, then, in the 1500s and 1600s, did European civilization so suddenly expand and change so dramatically the way of life of people living in America, Africa, and Asia?

We first must understand the Age of Exploration as a natural outgrowth of the European Renaissance of the 1300s and 1400s. Trade with the Moslem world stimulated a growing demand among wealthy Europeans for Asian consumer goods, such as spices and fine cloth. The demand quickly became too large to be met by the relatively small amount of such goods Europeans could obtain from their Moslem neighbors. As a consequence, European merchants began to look about for other suppliers.

The emergence of nation-states in Europe during the Late Middle Ages provided the necessary capital to finance the explorations, conquests, and colonies. As we saw in Cultural Connection VI, medieval Europe lacked large states with strong central governments. Instead, weak kings had little power, and authority was vested in a country nobility. By 1500, however, monarchs in France, Spain, England, and Portugal had wrested some power from their barons and had centralized political and economic control into themselves. Through taxes, these autocrats were able to acquire the huge amounts of capital required to finance overseas expeditions. Much as today, when only superpowers like the United States and the Soviet Union can afford to explore space, the Age of Exploration had to wait for the emergence of centralized nation-states in Europe.

Spain and Portugal took the lead in overseas expansion because the monarchs were more powerful in those countries, and because they were aggressive military powers that had already been expanding. Remember that most of the Iberian Peninsula, the region that comprises Spain and Portugal, had been annexed by the Arabian Empire at the beginning of the Middle Ages. Small Christian principalities remained independent in the northern part of the peninsula, however, and, over the course of several centuries in a long series of wars, these small states reclaimed much lost territory, eventually coalescing into the nation-states of Spain and Portugal. This process, called the *Reconquista,* was completed in 1492, when the last Moslem state on the peninsula, the Kingdom of Granada, fell to the armies of Ferdinand and Isabella. Put somewhat simplistically, the Spanish and Portuguese essentially just kept on going after that, expanding into Africa and America, propelled by a military tradition that was centuries old.

The 1400s also saw several technological advances crucial to European expansion. Navigation equipment and ship design were improved. More importantly, European weaponry, in the form of guns and cannon, had become far more sophisticated than anywhere else in the world, making large-scale conquest possible.

The Protestant Reformation of the 1500s also would seem to have had an impact on the expansion of Europe, although it is a difficult one to measure. The Reformation challenged the primacy of the Catholic Church in the western-European Christian community. Protestants called for Biblical fundamentalism, a restoration of faith, and local control of churches. Protestantism was strong only in northern Europe, however, and there was little unity among its several different sects, which ranged from the radical Puritans and other Calvinists, who believed only an elect few were destined for heaven, to the more moderate Lutherans and Dutch Reformed, to small egalitarian churches like the Quakers. The impact of the Reformation in the secular world was manifold: It stimulated the development of nation-states in England, the Netherlands, and Scandinavia by providing monarchs with an avenue to escape papal intervention in affairs of state and to seize and nationalize church lands. Still, highly Protestant Germany did not develop modern nation-states (Prussia was the first to emerge there) until the nineteenth century, and nation-states came into being in heavily Catholic Spain, Portugal, and France. The Reformation did weaken the influence of the Catholic Church in northern Europe, making its opposition to scientific inquiry less effective, and the Protestant churches were generally more receptive to science. Open debate of religious topics probably touched off discussions of secular matters as well, including science. Finally, antagonism between different Protestant sects in England led to the emigration of members of the more radical movements, Puritans and Quakers, to North America as colonists between 1620 and 1700.

Also, we must recognize that, once started, the expansion of Europe was a process that fed upon itself. The more gold and silver discovered in South and Central America, the more fortunes that were made from the sale of spices and silks imported from Asia, and the more good farmland found on new continents, the more did Europeans seek out additional riches. Two of the most lucrative commercial ventures of the Age of Exploration, the fur trade and the slave trade, developed as a consequence of European expansion. The first explorations had been conducted on a modest scale, but the movements of peoples and armies assumed vaster proportions with time. What began in 1520 as small trading posts on distant shores had become, by 1700, bustling colonial towns surrounded by hinterlands of farmers, trappers, and traders.

The Age of Exploration had a tremendous impact on Europe. There was a sudden influx of capital into the continent, especially into Spain (which quickly bypassed trade for the opportunity of looting gold and silver from America), although fortunes were also made in France, Portugal, England, and the Netherlands. European port cities, which on the Atlantic coast before 1500 had been little more than oversized towns, grew rapidly. Cadiz, Spain; Lisbon, Portugal; LaRochelle, France; Bristol, England; and Amsterdam, the Netherlands became important commercial centers with busy marketplaces. Also, much of the new wealth found its way into the capital cities, like London, Paris, and Madrid, to royal courts. As had been the case in ancient Alexandria, explorers brought back new information about new places, which constituted a veritable explosion of scientific data. Artists were influenced by Asian motifs. Artists, scientists, and philosophers, some with royal patronage, others hired by the growing new merchant middle class, found employment in the capitals and port towns. The Age of Exploration touched off a cultural and scientific revolution in Europe that was marked by an interest in new ideas and places, a flowering of the arts, and a perceived need for new technologies, especially in navigation. Europe stood at the dawn of the modern era.

Chapter 9

THE DAWN OF MODERN MATHEMATICS

9–1 The Seventeenth Century

The seventeenth century is outstandingly conspicuous in the history of mathematics. Early in the century, Napier revealed his invention of logarithms, Harriot and Oughtred contributed to the notation and codification of algebra, Galileo founded the science of dynamics, and Kepler announced his laws of planetary motion. Later in the century, Desargues and Pascal opened a new field of pure geometry, Descartes launched modern analytic geometry, Fermat laid the foundations of modern number theory, and Huygens made distinguished contributions to the theory of probability and other fields. Then, toward the end of the century, after a host of seventeenth-century mathematicians had prepared the way, the epoch-making creation of the calculus was made by Newton and Leibniz. We can see that many new and vast fields were opened up for mathematical investigation during the seventeenth century.

The great impetus given to mathematics in the seventeenth century was shared by all intellectual pursuits and was largely due, no doubt, to the political, economic, and social advances of the time. The century witnessed strong gains in the struggle for human rights, saw machines well advanced from the amusing toys of Heron's day to objects of increasing economic importance, and observed a growing spirit of intellectual internationalism and scientific skepticism. The more favorable political atmosphere of northern Europe and the general conquering of the cold and darkness of the long winter months by advances in heating and lighting probably largely account for the northward shift of mathematical activity in the seventeenth century from Italy to France and England.

It is only fair to note here two facts that will contribute to the somewhat unbalanced presentation of the history of the mathematics in the second part of this book. The first of these is that mathematical activity began to grow at so great a rate that henceforth many names must be omitted that might have been considered in a less productive period. The second fact is that, with the unfolding of the seventeenth century, an increasing amount of mathematical research occurred that cannot be appreciated by a general reader, for it has been rightfully claimed that the history of a subject cannot be properly understood without a knowledge of the subject itself.

In this chapter and the following one, we shall consider developments of the seventeenth century that can be appreciated without a knowledge of the calculus. Chapter 11 contains a sketch of the developments of the calculus from their beginnings in Greek antiquity up to the remarkable contributions made by Newton and Leibniz and their immediate precursors in the second half of the seventeenth century. The final chapters of the book describe the transition to the twentieth century; these last chapters must of necessity be very sketchy, for most of the mathematics of this period can be understood only by the expert.

9–2 Napier

Many of the fields in which numerical calculations are important, such as astronomy, navigation, trade, engineering, and war, made ever-increasing demands that these computations be performed more quickly and accurately. These increasing demands were met successively by four remarkable inventions: the Hindu-Arabic notation, decimal fractions, logarithms, and the modern computing machines. It is now time to consider the third of these great labor-saving devices, the invention of logarithms by John Napier in the early seventeenth century. The fourth invention will be considered later in Section 15–9.

John Napier (1550–1617), who was born when his father was only sixteen years of age, lived most of his life at the imposing family estate of Merchiston Castle, near Edinburgh, Scotland, and expended much of his energies in the political and religious controversies of his day. He was violently anti-Catholic and championed the causes of John Knox and James I. In 1593, he published a bitter and widely read attack on the Church of Rome entitled *A Plaine Discouery of the Whole Reuelation of Saint Iohn,* in which he endeavored to prove

JOHN NAPIER
(Culver Service)

that the Pope was the Antichrist and that the Creator proposed to end the world in the years between 1688 and 1700. The book ran through twenty-one editions, at least ten of them during the author's lifetime, and Napier sincerely believed that his reputation with posterity would rest upon this book.

Napier also wrote prophetically of various infernal war engines, accompanying his writings with plans and diagrams. He predicted the future would develop a piece of artillery that could "clear a field of four miles circumference of all living creatures exceeding a foot of height," that it would produce "devices for sayling under water," and that it would create a chariot with "a living mouth of mettle" that would "scatter destruction on all sides." In World War I, these were realized as the machine gun, the submarine, and the army tank, respectively.

It is no wonder that Napier's remarkable ingenuity and imagination led some to believe he was mentally unbalanced and others to regard him as a dealer in the black art. Many stories, probably unfounded, are told in support of these views. Once he announced that his coal-black rooster would identify for him which of his servants was stealing from him. The servants were sent one by one into a darkened room with instructions to pat the rooster on the back. Unknown to the servants, Napier had coated the bird's back with lampblack, and the guilty servant, fearing to touch the rooster, returned with clean hands. There was also the occasion when Napier became annoyed by his neighbor's pigeons eating his grain. He threatened to impound the birds if his neighbor did not restrict their flight. The neighbor, believing the capture of his pigeons to be virtually impossible, told Napier that he was welcome to the birds if he could catch them. The next day, the surprised neighbor observed his pigeons staggering on Napier's lawn with Napier calmly collecting them into a large sack. Napier had rendered the birds drunk by scattering some brandy-soaked peas about his lawn.

As relaxation from his political and religious polemics, Napier amused himself with the study of mathematics and science, with the result that four products of his genius are now recorded in the history of mathematics. These are (1) the invention of logarithms; (2) a clever mnemonic, known as the *rule of circular parts,* for reproducing the formulas used in solving right spherical triangles; (3) at least two trigonometric formulas of a group of four known as *Napier's analogies,* useful in the solution of oblique spherical triangles; and (4) the invention of a device, called *Napier's rods,* or *Napier's bones,* used for mechanically multiplying, dividing, and taking square roots of numbers. We turn now to the first, and most remarkable, of these four contributions; for a discussion of the other three, see Problem Studies 9.2 and 9.3.

9–3 Logarithms

As we know today, the power of logarithms as a computing device lies in the fact that by them multiplication and division are reduced to the simpler operations of addition and subtraction. A forerunner of this idea is apparent in the trigonometric identity

$$2 \cos A \cos B = \cos(A + B) + \cos(A - B),$$

which was well known in Napier's time. Here the product of the two numbers $2 \cos A$ and $\cos B$ is replaced by the sum of the two numbers $\cos(A + B)$ and $\cos(A - B)$. The formula is easily extended to convert the product of any two numbers to the sum of two other numbers. Suppose, for example, one desires the product of 437.64 and 27.327. From a table of cosines, find, using interpolation if necessary, angles A and B, where

$$\cos A = (0.43764)/2 = 0.21882 \qquad \text{and} \qquad \cos B = 0.27327.$$

Then, again using the cosine table, with interpolation if necessary, find $\cos(A + B)$ and $\cos(A - B)$, and add these two numbers. One now has the product of 0.43764 and 0.27327. Finally, by properly adjusting the decimal point in the answer, one obtains the sought product of 437.64 and 27.327. The problem of finding the product (437.54)(27.327) has been cleverly reduced to a simple problem in addition.

Allied to the preceding trigonometric identity are the three following identities:

$$2 \sin A \cos B = \sin(A + B) + \sin(A - B),$$
$$2 \cos A \sin B = \sin(A + B) - \sin(A - B),$$
$$2 \sin A \sin B = \cos(A - B) - \cos(A + B).$$

The four identities are sometimes called **Werner's formulas** because the German Johannes Werner (1468–1528) seems to have used them to simplify lengthy calculations arising in astronomy. The formulas became widely used by mathematicians and astronomers in the late sixteenth century as a method of converting products into sums and differences. The method became known as **prosthaphaeresis,** from a Greek word meaning "addition and subtraction." A long division may be similarly treated. Thus, again utilizing the first of Werner's formulas, we have

$$\frac{2 \cos A}{\csc B} = 2 \cos A \sin B = 2 \cos A \cos(90° - B)$$
$$= \cos[A + (90° - B)] + \cos[A - (90° - B)].$$

We know Napier was aware of the method of prosthaphaeresis, and it could be that he was influenced by it, since otherwise it is difficult to account for his initial restriction of logarithms to those of sines of angles. But Napier's approach to eliminating the bugbear of long multiplications and divisions differed considerably from prosthaphaeresis, and lay in the fact that if one associates with the terms of a geometric progression

$$b, b^2, b^3, b^4, \ldots, b^m, \ldots, b^n, \ldots$$

those of the arithmetic progression

$$1, 2, 3, 4, \ldots, m, \ldots, n, \ldots,$$

then to the *product* $b^m b^n = b^{m+n}$ of two terms of the former progression is associated the *sum* $m + n$ of the two corresponding terms of the latter progression. To keep the terms of the geometric progression sufficiently close together so that interpolation can be used to fill in the gaps between the terms in the preceding association, the number b must be chosen very close to 1. Napier accordingly chose $1 - 1/10^7 = 0.9999999$ for b. To avoid decimals, he multiplied each power by 10^7. Then, if

$$N = 10^7(1 - 1/10^7)^L,$$

he called L the "logarithm" of the number N. It follows that Napier's logarithm of 10^7 is 0 and that of $10^7(1 - 1/10^7) = 9999999$ is 1. If one should divide both N and L by 10^7, one would virtually obtain a system of logarithms to the base $1/e$, for

$$(1 - 1/10^7)^{10^7} \doteq \lim_{n\to\infty} (1 + 1/n)^n = 1/e.$$

Of course, one must keep in mind that Napier had no concept of a base for a system of logarithms.

Napier labored at least twenty years upon his theory, and finally explained the principles of his work in geometrical terms as follows. Consider a line segment AB and an infinite ray DE, as shown in Figure 73. Let points C and F start moving simultaneously from A and D, respectively, along these lines with the same initial rate. Suppose C moves with a velocity always numerically equal to the distance CB, and that F moves with a uniform velocity. Then Napier defined DF to be the logarithm of CB. That is, setting $DF = x$ and $CB = y$,

$$x = \text{Nap log } y.$$

In order to avoid the nuisance of fractions, Napier took the length of AB as 10^7, for the best tables of sines available to him extended to seven places. From Napier's definition and through the use of knowledge not available to Napier, it

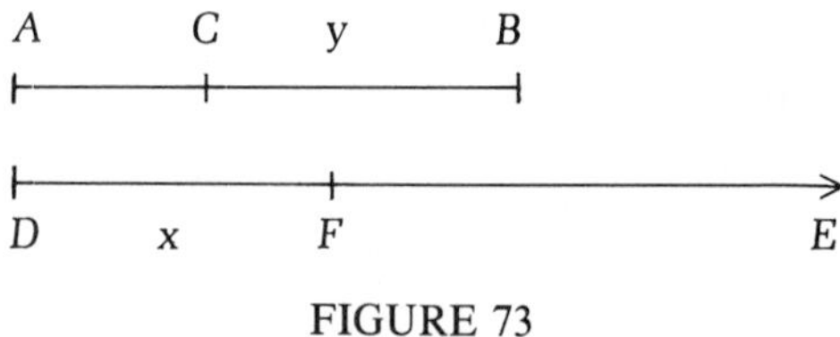

FIGURE 73

develops that[1]

$$\text{Nap log } y = 10^7 \log_{1/e}\left(\frac{y}{10^7}\right),$$

so that the frequently made statement that Napierian logarithms are natural logarithms is actually not true. One observes that the Napierian logarithm decreases as the number increases, contrary to what happens with natural logarithms.

It further develops that, over a succession of equal periods of time, y *decreases* in *geometric* progression while x *increases* in *arithmetic* progression. Thus, we have the fundamental principle of a system of logarithms, the association of a geometric and an arithmetic progression. It now follows, for example, that if $a/b = c/d$, then

$$\text{Nap log } a - \text{Nap log } b = \text{Nap log } c - \text{Nap log } d,$$

which is one of the many results established by Napier.

Napier published his discussion of logarithms in 1614 in a brochure entitled *Mirifici logarithmorum canonis descriptio* (A Description of the Wonderful Law of Logarithms). The work contains a table giving the logarithms of the sines of angles for successive minutes of arc. The *Descriptio* aroused immediate and widespread interest, and in the year following its publication, Henry Briggs (1561–1631), professor of geometry at Gresham College in London, and later professor at Oxford, traveled to Edinburgh to pay his respects to the great inventor of logarithms. It was upon this visit that both Napier and Briggs agreed that the tables would be more useful if they were altered so that the logarithm of 1 would be 0 and the logarithm of 10 would be an appropriate power of 10, thereby creating the so-called **Briggsian,** or **common,** logarithms of

---

[1] The result is easily shown with the aid of a little calculus. Thus we have $AC = 10^7 - y$, whence

$$\text{velocity of } C = -dy/dt = y.$$

That is, $dy/y = -dt$, or integrating, $\ln y = -t + C$. Evaluating the constant of integration by substituting $t = 0$, we find that $C = \ln 10^7$, whence

$$\ln y = -t + \ln 10^7.$$

Now

$$\text{velocity of } F = dx/dt = 10^7,$$

so that $x = 10^7 t$. Therefore

$$\begin{aligned}\text{Nap log } y = x = 10^7 t &= 10^7 (\ln 10^7 - \ln y)\\ &= 10^7 \ln (10^7/y) = 10^7 \log_{1/e} (y/10^7).\end{aligned}$$

today. Logarithms of this sort, which are essentially logarithms to the base 10, owe their superior utility in numerical computations to the fact that our number system also is based on 10. For a number system having some other base b, it would, of course, be most convenient for computational purposes to have tables of logarithms also to the base b.

Briggs devoted all his energies toward the construction of a table upon the new plan, and in 1624 published his *Arithmetica logarithmica,* which contained a fourteen-place table of common logarithms of the numbers from 1 to 20,000 and from 90,000 to 100,000. The gap from 20,000 to 90,000 was later filled in, with help, by Adriaen Vlacq (1600–1666), a Dutch bookseller and publisher. In 1620, Edmund Gunter (1581–1626), one of Briggs' colleagues, published a seven-place table of the common logarithms of the sines and tangents of angles for intervals of a minute of arc. It was Gunter who invented the words *cosine* and *cotangent;* he is known to engineers for his "Gunter's chain." Briggs and Vlacq published four fundamental tables of logarithms, which have only recently been superseded when, between 1924 and 1949, extensive twenty-place tables were calculated in England in partial celebration of the tercentenary of the discovery of logarithms.

The word **logarithm** means "ratio number" and was adopted by Napier after first using the expression *artificial number*. Briggs introduced the word *mantissa,* which is a late Latin term of Etruscan origin, originally meaning an "addition" or "makeweight," and which, in the sixteenth century, came to mean "appendix." The term *characteristic* was also suggested by Briggs and was used by Vlacq. It is curious that it was customary in early tables of common logarithms to print the characteristic as well as the mantissa; it was not until the eighteenth century that the present custom of printing only the mantissas was established.

Napier's wonderful invention was enthusiastically adopted throughout Europe. In astronomy, in particular, the time was overripe for such a discovery; as Laplace asserted, the invention of logarithms "by shortening the labors doubled the life of the astronomer." Bonaventura Cavalieri, about whom we shall have more to say in Chapter 11, did much to bring logarithms into vogue in Italy. A similar service was rendered by Johann Kepler in Germany and Edmund Wingate in France. Kepler will be considered more fully in Section 9–7; Wingate, who spent many years in France, became the most prominent seventeenth-century British textbook writer on elementary arithmetic.

Napier's only rival for priority in the invention of logarithms was the Swiss instrument maker Jobst Bürgi (1552–1632). Bürgi conceived and constructed a table of logarithms independently of Napier, publishing his results in 1620, six years after Napier had announced his discovery to the world. Although both men had conceived the idea of logarithms long before publishing, it is generally believed that Napier had the idea first. Whereas Napier's approach was geometrical, Bürgi's was algebraic. Nowadays, a logarithm is universally regarded as an exponent; thus, if $n = b^x$, we say x is the logarithm of n to the base b. From this definition, the laws of logarithms follow immediately from the laws of exponents. One of the anomalies in the history of mathematics is the fact that logarithms were discovered before exponents were in use.

In 1971, Nicaragua issued a series of postage stamps paying homage to the world's "ten most important mathematical formulas." Each stamp features a particular formula accompanied by an illustration and carries on its reverse side a brief statement in Spanish concerning the importance of the formula. One of the stamps is devoted to Napier's invention of logarithms. It must be pleasing to scientists and mathematicians to see their formulas so honored, for these formulas have certainly contributed far more to human development than the exploits of the kings and generals who are so often featured on postage stamps.[2]

For years, computation by logarithms has been taught in the late high-school or the early college mathematics courses; also, for years the logarithmic slide rule, hanging from the belt in a handsome leather case, was the badge of recognition of the engineering students of a university campus. Today, however, with the advent of the amazing and increasingly inexpensive little pocket calculators, no one in his right mind would use a table of logarithms or a slide rule for calculation purposes. The teaching of logarithms as a computing device is vanishing from the schools, the famous makers of precision slide rules are discontinuing their production, and noted handbooks of mathematical tables are considering the abandonment of the logarithm tables. The products of Napier's great invention have become museum pieces.

The logarithmic function, however, will never die, for the simple reason that logarithmic and exponential variations are a vital part of nature and of analysis. Accordingly, a study of the properties of the logarithmic function and of its inverse, the exponential function, will always remain an important part of mathematical instruction.

9–4 The Savilian and Lucasian Professorships

Because so many distinguished British mathematicians have held either a Savilian professorship at Oxford or a Lucasian professorship at Cambridge, a brief reference to these professorships is desirable.

Sir Henry Savile was at one time warden of Merton College at Oxford, later provost of Eton, and a lecturer on Euclid at Oxford. In 1619, he founded two professorial chairs at Oxford, one in geometry and one in astronomy. Henry Briggs was the first occupant of the Savilian chair of geometry at Oxford. The earliest professorship of mathematics established in Great Britain

---

[2] The other formulas featured on the stamps are the fundamental counting formula $1 + 1 = 2$, the Pythagorean relation $a^2 + b^2 = c^2$, the Archimedean law of the lever $w_1d_1 = w_2d_2$, Isaac Newton's universal law of gravitation, J. C. Maxwell's four famous equations of electricity and magnetism, Ludwig Boltzmann's gas equation, Konstantin Tsiolkovskii's rocket equation, Albert Einstein's famous mass-energy equation $E = mc^2$, and Louis de Broglie's revolutionary matter-wave equation.

was a chair in geometry founded by Sir Thomas Gresham in 1596 at Gresham College in London. Briggs also had the honor of being the first to occupy this chair. John Wallis, Edmund Halley, and Sir Christopher Wren are other seventeenth-century incumbents of Savilian professorships.

Henry Lucas, who represented Cambridge in parliament in 1639 to 1640, willed resources to the university for the founding in 1663 of the professorship that bears his name. Isaac Barrow was elected the first occupant of this chair in 1664 and six years later was succeeded by Isaac Newton.

9–5 Harriot and Oughtred

Thomas Harriot (1560–1621) was another mathematician who lived the longer part of his life in the sixteenth century but whose outstanding publication appeared in the seventeenth century. He is of special interest to Americans, because in 1585 he was sent by Sir Walter Raleigh as a surveyor with Sir Richard Grenville's expedition to the New World to map what was then called Virginia but is now North Carolina. As a mathematician, Harriot is usually considered the founder of the English school of algebraists. His great work in this field, the *Artis analyticae praxis,* was not published until ten years after his death and deals largely with the theory of equations. This work did much toward setting the present standards for a textbook on the subject. It includes a treatment of equations of the first, second, third, and fourth degrees; the formation of equations having given roots; the relations between the roots and the coefficients of an equation; the familiar transformations of an equation into another having roots bearing some specific relation to the roots of the original equation; and the numerical solution of equations. Much of this material is found, of course, in the works of Viète, but Harriot's is a more complete and better systematized treatment. Harriot followed Viète's plan of using vowels for unknowns and consonants for constants, but he adopted the lower-case rather than the upper-case letters. He improved on Viète's notation for powers by representing a^2 by aa, a^3 by aaa, and so forth. He was also the first to use the signs $>$ and $<$ for "is greater than" and "is less than," respectively, but these symbols were not immediately accepted by other writers.

Harriot has been erroneously credited with several other mathematical innovations and discoveries, such as a well-formed analytic geometry (before Descartes' publication of 1637), the statement that any polynomial of degree n has n roots, and "Descartes' rule of signs." Some of these errors of authorship seem due to insertions, made by later writers, among some of Harriot's preserved manuscripts. There are eight volumes of Harriot's manuscripts in the British Museum, but the part dealing with analytic geometry has been shown by D. E. Smith to be an interpolation by a later hand.

Harriot was also prominent as an astronomer, having discovered sunspots and having observed the satellites of Jupiter, independently of Galileo and at about the same time. He died in 1621 of a cancerous ulcer in his left nostril; the ulcer was brought on by inhalation of tobacco smoke, a practice taught to him

HARRIOT
(David Smith Collection)

by the local Indians when he was in America in 1586, thus rendering him as perhaps the first tobacco fatality to be recorded.

In the same year (1631) that Harriot's posthumous work on algebra appeared, there also appeared the first edition of William Oughtred's popular *Clavis mathematicae,* a work on arithmetic and algebra that did much toward spreading mathematical knowledge in England. William Oughtred (1574–1660)

WILLIAM OUGHTRED
(David Smith Collection)

was one of the most influential of the seventeenth-century English writers on mathematics. Although by profession a clergyman (of the parish of Bletchingdon), he gave free private lessons to pupils interested in mathematics. Among such pupils were John Wallis, Christopher Wren, and Seth Ward, later popularly famous, respectively, as a mathematician, an architect, and an astronomer.

Oughtred seems to have ignored the usual rules of good health and probably continued to ignore them throughout his long life. When he finally died, it is said that he did so in a transport of joy at receiving the news of the restoration of Charles II. To this, Augustus De Morgan once remarked, "It should be added, by way of excuse, that he was eighty-six years old."

In his writings, Oughtred placed emphasis on mathematical symbols, giving over 150 of them. Of these, only three have come down to present times: the cross ($\times$) for multiplication, the four dots ($::$) used in a proportion, and our frequently used symbol for difference between ($\sim$). The cross as a symbol for multiplication, however, was not readily adopted because, as Leibniz objected, it too closely resembles x. Although Harriot on occasion used the dot ($\cdot$) for multiplication, this symbol was not prominently used until Leibniz adopted it. Leibniz also used the cap symbol ($\cap$) for multiplication, a symbol that is used today to indicate intersection in the theory of sets. The Anglo-American symbol for division ($\div$) is also of seventeenth-century origin, having first appeared in print in 1659 in an algebra by the Swiss Johann Heinrich Rahn (1622–1676). The symbol became known in England some years later when this work was translated. This symbol for division has long been used in continental Europe to indicate subtraction. Our familiar signs in geometry—($\sim$) for similar and ($\simeq$) for congruent—are due to Leibniz.

Besides the *Clavis mathematicae,* Oughtred published *The Circles of Proportion* (1632) and *Trigonometrie* (1657). The second work is of some historical importance because of its early attempt to introduce abbreviations for the names of the trigonometric functions. The first work describes a circular slide rule. Oughtred, however, was not the first to describe in print a slide rule of the circular type, and an argument of priority of invention rests between him and Richard Delamain, one of his pupils. Oughtred does seem unquestionably to have invented, about 1622, the straight logarithmic slide rule. In 1620, Gunter constructed a logarithmic scale, or a line of numbers on which the distances are proportional to the logarithms of the numbers indicated (see Figure 74), and mechanically performed multiplications and divisions by adding and subtracting segments of this scale with the aid of a pair of dividers. The idea of carrying out these additions and subtractions by having two like logarithmic scales, one sliding along the other as shown in Figure 75, is due to Oughtred. Although Oughtred invented such a simple slide rule as early as 1622, he did not describe it in print until 1632. A runner for the slide rule was suggested by Isaac Newton in 1675, but was not actually constructed until nearly a century later. Several slide rules for special purposes, such as for commercial transactions, for measuring timber, and so forth, were devised in the seventeenth century. The

Notæ ſeu ſymbola quibus in ſequentibus utor:

Æquale = Simile *Sim.*

Majus ⊏. Proxime majus ⊏·.

Minus ⊐. Proxime minus ·⊐.

Non majus ⊏̸. Æquale vel minus ⊑.

Non minus ⊐̸. Æquale vel majus ⊒.

Proportio, ſive ratio æqualis ::

Major ratio ⊏:. Minor ratio ⊐:.

Continuè proportionales ∺.

Commenſurabilia ⊓.

Incommenſurabilia ⊓̸.

Commenſurabilia potentiâ ⊓̲.

Incommenſurabilia potentià ⊓̸̲.

Rationale, ῥητὸν, R, vel ℞.

Irrationale, ἄλογον, ℞̸.

Medium ſive mediale m℞

Linea ſecta ſecundum extremam & mediam rationem } ς

Major ejus portio σ

Minor ejus portio τ.

Z eſt A + E. ʒ eſt a + e.

X eſt A-E. ɛ eſt a-e

A 2 Z eſt

A page from Oughtred's *Clavis mathematicae* (1631), showing a number of his mathematical symbols.

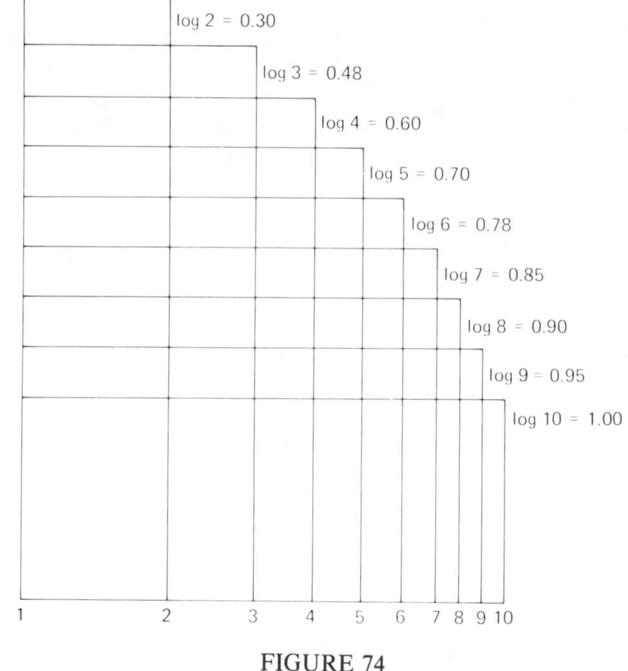

FIGURE 74

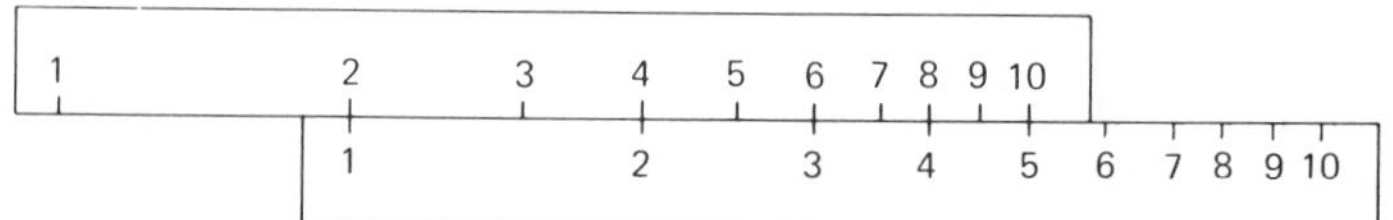

FIGURE 75

log log scale was invented in 1815, and it was in 1850 that the French army officer Amédée Mannheim (1831–1906) standardized the modern slide rule.

It is believed that Oughtred was the author of the remarkable anonymous sixteen-page Appendix to the 1618 English edition by Edward Wright of Napier's *Descriptio*. Here appears the first use of the cross for multiplication, the first invention of the radix method of calculating logarithms [see Problem Study 9.1(c)], and the first table of natural logarithms. Oughtred also wrote a work on **gauging** (the science of computing the capacities of casks and barrels), and he translated and edited a French work on mathematical recreations.

9–6 Galileo

Two outstanding astronomers contributed notably to mathematics in the early part of the seventeenth century: the Italian, Galileo Galilei, and the German, Johann Kepler.

Galileo, the son of an impoverished Florentine nobleman, was born in Pisa in 1564 on the day that Michelangelo died. At the age of seventeen, he was sent by his parents to the University of Pisa to study medicine. One day, while attending a service in the cathedral at Pisa, his mind was distracted by the great bronze lamp suspended from the high ceiling. The lamp had been drawn aside in order to light it more easily, and when released it oscillated to and fro with gradually decreasing amplitude. Using the beat of his pulse to keep time, he was surprised to find that the period of an oscillation of the lamp was independent of the size of the arc of oscillation.[3] Later, by experiments, he showed that the period of a swinging pendulum is also independent of the weight of the pendulum's bob and thus depends solely on the length of the pendulum. It is reported that Galileo's interest in science and mathematics was roused by this problem and then further stimulated by the chance attendance at a lecture on geometry at the university. The result was that he asked for, and secured, parental permission to abandon medicine and to devote himself to science and mathematics instead, fields in which he possessed strong natural talent.

When twenty-five, Galileo was appointed professor of mathematics at the University of Pisa, and while holding this appointment is said to have performed public experiments with falling bodies. According to the story, before a crowd of students, faculty, and priests, he dropped two pieces of metal, one ten times the weight of the other, from the top of the leaning tower of Pisa. The two pieces of metal struck the ground at practically the same moment, thus contradicting Aristotle, who said that a heavier body falls faster than a lighter one. Galileo arrived at the law that the distance a body falls is proportional to the square of the time of falling, in accordance with the familiar formula $s = gt^2/2$. Even the visual evidence of Galileo's experiments, however, did not shake the faith of the other professors at the university in the teaching of Aristotle. The authorities at the university were so shocked at Galileo's sacrilegious insolence in contradicting Aristotle that they made life unpleasant for him there, with the result that he resigned his professorship in 1591. The following year, he accepted a professorship at the University of Padua, where there was an atmosphere more friendly to scientific pursuits. Here, for nearly eighteen years, Galileo continued his experiments and his teaching and won widespread fame.

About 1607, an apprentice to the spectacle maker Hans Lippershey of Holland, while playing with some of his master's spectacle lenses, discovered that if he held two of the lenses at an appropriate distance apart, objects seen through the pair of lenses became enlarged. The apprentice brought his discovery to the attention of his master, who placed two lenses in a tube and displayed the device as a toy in his shop window. The toy was seen by a government official, who bought it and presented it to Prince Maurice of Nassau. As commander of the armed forces of the United Netherlands, Prince Maurice saw the possibilities of the toy as a spyglass for military use.

---

[3] This is only approximately true, the approximation being very close in the case of small amplitudes of oscillation.

GALILEO GALILEI
(David Smith Collection)

By 1609, the news of the invention of the spyglass reached Galileo, who soon made a spyglass greatly superior to the one made by Lippershey. Upon request, he demonstrated his instrument in Venice, where, from the top of the highest church in the city, Venetian senators were able to see the sails of an approaching ship a full two hours before they were visible by naked eye. Galileo presented his model to the Doge of Venice, who, like Prince Maurice, recognized the immense possibilities of the instrument in naval and military operations, and Galileo was given a sizably increased stipend.

Galileo went on and made four more telescopes, as his instruments were named (from the Greek *tele,* "far," *skopos,* "watching"), each more powerful than the last. With the fifth telescope, which had a power of thirty diameters, Galileo noticed, on the night of January 7, 1610, two small stars to the east of the planet Jupiter and one to the west. The following night, to his surprise, all three stars were to the west of the planet, and three nights later he found there was still another small star revolving about Jupiter. He had discovered Jupiter's four bright satellites and observed a striking confirmation of the Copernican theory of smaller bodies revolving about larger ones. With his telescope, Galileo observed sunspots, the mountains on the moon, the phases of Venus, and Saturn's rings. But these discoveries only aroused once more the bigoted opposition of many churchmen, who accepted the authority of Aristotle; Aristotle had asserted that the sun is without blemish and that the earth, and hence man, is the center of the universe. One churchman even accused Galileo of placing the four satellites of Jupiter inside his telescope.

Finally, in 1633, one year after his publication of a book that supported the Copernican theory, Galileo was summoned to appear before the Inquisition, and there, an ill and an old man, forced, under the threat of torture, to recant his scientific findings. His book was placed on the Index of prohibited works and remained there for two hundred years. Having perjured his conscience, the old scholar's life was broken. He was permitted to continue innocuous scientific work, but became blind and died in January, 1642, still under the supervision of the Inquisition and a virtual prisoner in his own home.[4]

There is a legend that, as Galileo rose to his feet after his forced recantation and denial of the earth's motion, he muttered softly under his breath to himself, "The earth *does* move all the same." Whatever the basis of this story, it has come to be a sort of proverb to the effect that truth shall prevail despite all attempts at suppression. And so it came to pass, for the year 1642, which saw the death of Galileo in captivity, also saw the birth of Isaac Newton.

To Galileo, we owe the modern spirit of science as a harmony between experiment and theory. He founded the mechanics of freely falling bodies and laid the foundation of dynamics in general, a foundation upon which Isaac Newton was able later to build the science. He was the first to realize the parabolic nature of the path of a projectile in a vacuum and speculated on laws involving momentum. He invented the first modern-type microscope and the once very popular sector compasses (see Problem Study 9.6). Historically interesting are statements made by Galileo showing that he grasped the idea of equivalence of infinite classes (see Problem Study 9.7), a fundamental point in Cantor's nineteenth-century theory of sets, which has been so influential in the development of modern analysis. These statements, and the bulk of Galileo's ideas in dynamics, can be found in his *Discorsi e dimonstrazioni matematiche intorno a due nuove scienze,* published in Leyden in 1638. Galileo has been quoted as saying: "In questions of sciences, the authority of a thousand is not worth the humble reasoning of a single individual."

It would seem that Galileo was jealous of his famous contemporary, Johann Kepler, for although Kepler had announced all three of his important laws of planetary motion by 1619, these laws were completely ignored by Galileo.

All his life, Galileo was a religious man and a devout Catholic. Accordingly, it distressed him to find the views to which he was irresistibly led by his observations and reasonings as a scientist condemned as contradicting the scriptures of the Church, of which he considered himself a loyal member. He therefore felt compelled to reason for himself the relation between science and scripture. Many scientists have, from time to time, found themselves in this position. It occurred, for example, in the middle of the nineteenth century, when difficulties were felt in reconciling Darwin's theory of evolution with the Biblical account of the creation of living things.

Galileo's conclusion was that the Bible is not, and never was intended to be, a textbook on astronomy, or biology, or any other science. In short, Galileo maintained, it was not intended as a book to teach us scientific truths that we

---

[4] In 1980, 347 years after being condemned by the Church for using telescopes to prove the earth revolves around the sun, the Vatican, under a call issued by Pope John Paul II, began to review Galileo's conviction of heresy.

can discover for ourselves. Rather, it was intended as a book to reveal spiritual truths that we could not have found out for ourselves. Now the conflict between science and scripture lies in the fact that these spiritual truths are expressed in the Bible in ways natural to the people to whom, and through whom, they were originally revealed. But this is clearly just an accident of time and should therefore be overlooked. A scientist should not be upset to find the Bible picturing the world in a way natural to the early Hebrews, and a churchman should not be upset to find a scientist picturing the world in a way contrary to the description in the Bible. The way in which the world is described is entirely incidental to the real aim of the Bible and in no way is inconsistent with the spiritual teachings of the Bible.

9–7 Kepler

Johann Kepler was born near Stuttgart in 1571 and educated at the University of Tübingen, with the original intention of becoming a Lutheran minister. His deep interest in astronomy led him to change his plans, and in 1594, when in his early twenties, he accepted a lectureship at the University of Grätz in Austria. In 1599, he became assistant to the famous but quarrelsome Danish-Swedish astronomer Tycho Brahe, who had moved to Prague as court astronomer to Kaiser Rudolph II. Shortly after, in 1601, Brahe suddenly died, and Kepler inherited both his master's position and his vast and very accurate collection of astronomical data on the motion of the planets.

It has often been said that almost any problem can be solved if one continuously worries over it and works at it a sufficiently long time. As Thomas Edison said of invention being one percent inspiration and ninety-nine percent perspi-

JOHANN KEPLER
(David Smith Collection)

ration, problem solving is one percent imagination and ninety-nine percent perseverance. Perhaps nowhere in the history of science is this more clearly demonstrated than in Kepler's incredible pertinacity in solving the problem of the motion of the planets about the sun. Thoroughly convinced of the Copernican theory that the planets revolve in orbits about the central sun, Kepler strenuously sought to determine the nature and position of those orbits and the manner in which the planets travel in their orbits. After many highly imaginative attempts, made when he had little data to aid in verification, Kepler inherited Tycho Brahe's enormous mass of very accurate observations on the motion of the planets. The problem then became this: to obtain a pattern of motion of the planets that would exactly jibe with Brahe's great set of observations. So dependable were Brahe's recordings, that any solution differing from Brahe's observed positions by even so little as a quarter of the moon's apparent diameter must be discarded as incorrect. Kepler needed, then, first to guess with his *imagination* some plausible solution, and then with painful *perseverance* to endure mountains of tedious calculation to confirm or reject his guess. He made hundreds of fruitless attempts and performed reams and reams of calculations, laboring with undiminished zeal and patience for twenty-one years. Finally, in 1609, he was able to formuate his first two laws, and then ten years later, in 1619, his third law, of planetary motion.

These laws of planetary motion are landmarks in the history of astronomy and mathematics, for in the effort to justify them Isaac Newton was led to create modern celestial mechanics. The three laws are:

I. *The planets move about the sun in elliptical orbits with the sun at one focus.*

II. *The radius vector joining a planet to the sun sweeps over equal areas in equal intervals of time.*

III. *The square of the time of one complete revolution of a planet about its orbit is proportional to the cube of the orbit's semimajor axis.*

The empirical discovery of these laws from Brahe's mass of data constitutes one of the most remarkable inductions ever made in science.

One never knows when a piece of pure mathematics may receive an unexpected application. As William Whewell once said, "If the Greeks had not cultivated the conic sections, Kepler could not have superseded Ptolemy." It is very interesting that 1800 years after the Greeks had developed the properties of the conics merely to satisfy their intellectual cravings, there should occur such an illuminating practical application of them. With justifiable pride, Kepler prefaced his *Harmony of the Worlds* of 1619 with the following outburst:

> I am writing a book for my contemporaries or—it does not matter—for posterity. It may be that my book will wait for a hundred years for a reader. Has not God waited for 6000 years for an observer?

Kepler was one of the precursors of the calculus. In order to compute the areas involved in his second law of planetary motion, he had to resort to a crude form of the integral calculus. He also, in his *Stereometria doliorum vinorum* (Solid Geometry of Wine Barrels, 1615), applied crude integration procedures to the finding of the volumes of ninety-three solids obtained by rotating segments of conic sections about an axis in their plane. Among these solids were the torus and two solids that he called **the apple** and **the lemon,** the last two solids being those obtained by revolving a major and a minor arc, respectively, of a circle about the arc's chord as an axis. Kepler became interested in this matter upon observing some of the poor methods in use by the wine gaugers of his time. It is quite possible that Cavalieri was influenced by this work of Kepler when he later carried the refinement of the infinitesimal calculus a stage further with his *method of indivisibles*. We shall return to a discussion of all this in Chapter 11.

Notable contributions were made by Kepler to the subject of polyhedra. He seems to have been the first to recognize an **antiprism** (obtained from a prism by rotating the top base in its own plane so as to make its vertices correspond to the sides of the lower base, and then joining, in zigzag fashion, the vertices of the two bases). He also discovered the cuboctahedron, rhombic dodecahedron, and rhombic triakontahedron.[5] The second of these polyhedra occurs in nature as a garnet crystal. Of the four possible regular star-polyhedra, two were discovered by Kepler and the other two in 1809 by Louis Poinsot (1777–1859), a pioneer worker in geometrical mechanics. The Kepler-Poinsot star-polyhedra are space analogues of the regular star-polygons in the plane (see Problem Study 8.5). Kepler also interested himself in the problem of filling the plane with regular polygons (not necessarily all alike) and filling space with regular polyhedra (see Problem Study 9.9).

Kepler solved the problem of determining the type of conic determined by a given vertex, the axis through this vertex, and an arbitrary tangent with its point of contact, and he introduced the word *focus* into the geometry of conics. He approximated the perimeter of an ellipse of semiaxes a and b by the use of the formula $\pi(a + b)$. He also laid down a so-called **principle of continuity,** which essentially postulates the existence at infinity, in a plane, of certain ideal points and an ideal line, having many of the properties of ordinary points and lines. He explained that a line can be considered as closed at infinity, that two parallel lines should be regarded as intersecting at infinity, and that a parabola may be regarded as a limiting case of either an ellipse or a hyperbola in which one of the foci has retreated to infinity. This concept was greatly extended in 1822 by the French geometer Poncelet when he made an effort to find in geometry a "real" justification for imaginaries that occur elsewhere in mathematics.

Kepler's work is often a blend of mystical and highly fanciful speculation, combined with a truly deep grasp of scientific truths. It is sad that his personal life was made almost unendurable by a succession of worldly misfortunes. An infection from smallpox when he was but four years old left his eyesight much

---

[5] Construction patterns for these solids can be found in Miles C. Hartley, *Patterns of Polyhedrons*. Rev. ed.

impaired. In addition to his general lifelong weakness, he spent a joyless youth; his marriage was a constant source of unhappiness; his favorite child died of smallpox; his wife went mad and died; he was expelled from his lectureship at the University of Grätz when that city fell to the Catholics; his mother was charged and imprisoned for witchcraft, and, for almost a year, he desperately tried to save her from the torture chamber; he himself very narrowly escaped condemnation of heterodoxy; and his stipend was always in arrears. One report says that his second marriage was even less fortunate than the first, although he took the precaution to analyze carefully the merits and demerits of eleven girls before choosing the wrong one. He was forced to augment his income by casting horoscopes, and in 1630 he died of a fever while on a journey to obtain some of his long overdue salary.

9–8 Desargues

In 1639, nine years after Kepler's death, there appeared in Paris a remarkably original but little-heeded treatise on the conic sections.[6] It was written by Gérard Desargues, an engineer, architect, and one-time French army officer, who was born in Lyons in 1591 and who died in the same city about 1662. The work was so generally neglected by other mathematicians that it was soon forgotten, and all copies of the publication disappeared. Two centuries later, when the French geometer Michel Chasles (1793–1880) wrote his still-valuable history of geometry, there was no means of estimating the value of Desargues' work. Six years later, however, in 1845, Chasles happened upon a manuscript copy of the treatise, made by Desargues' pupil, Philippe de la Hire (1640–1718); since that time, the work has been regarded as one of the classics in the early development of synthetic projective geometry.

Several reasons can be advanced to account for the initial neglect of Desargues' little volume. It was overshadowed by the more supple analytic geometry introduced by Descartes two years earlier. Geometers were generally expending their energies either developing this new powerful tool or trying to apply infinitesimals to geometry. Also, Desargues adopted an unfortunate and eccentric style of writing. He introduced some seventy new terms, many of a recondite botanical origin, of which only one, *involution,* has survived. Curiously enough, *involution* was preserved because it was the one piece of Desargues' technical jargon that was singled out for the sharpest criticism and ridicule by his reviewer.

Desargues wrote other books besides the one on conic sections, one of them being a treatise on how to teach children to sing well. But it is the little book on conic sections that marks him as the most original contributor to synthetic geometry in the seventeenth century. Starting with Kepler's doctrine

[6] *Brouillon projet d'une atteinte aux événements des rencontres d'un cone avec un plan.* (*Proposed Draft of an Attempt to Deal with the Events of the Meeting of a Cone with a Plane.*)

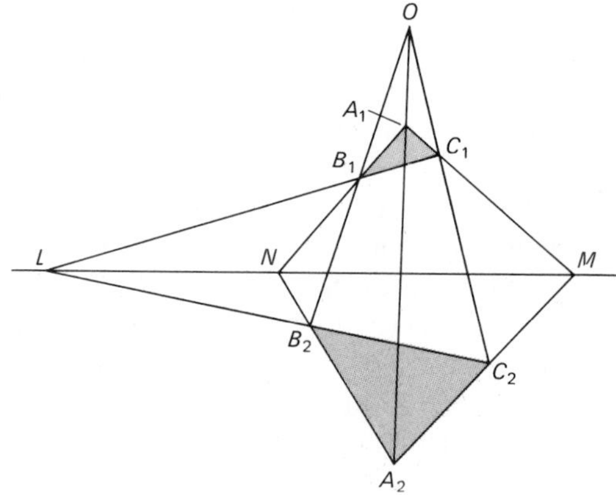

FIGURE 76

of continuity, the work develops many of the fundamental theorems on involution, harmonic ranges, homology, poles and polars, and perspective—topics familiar to those who have taken one of our present-day courses in projective geometry.[7] One interesting concept is that the notion of poles and polars may be extended to spheres and to certain other surfaces of the second degree. It is likely that Desargues was aware of only a few of the surfaces of second degree, many of these surfaces probably remaining unknown until their complete enumeration by Euler in 1748. Elsewhere we find Desargues' fundamental two-triangle theorem: *If two triangles, in the same plane or not, are so situated that lines joining pairs of corresponding vertices are concurrent, then the points of intersection of pairs of corresponding sides are collinear, and conversely* (see Figure 76).

Desargues, when he was in his thirties and living in Paris, made a considerable impression on his contemporaries through a series of gratuitous lectures. His work was appreciated by Descartes, and Blaise Pascal once credited Desargues as being the source of much of his inspiration. La Hire, with considerable labor, tried to show that all the theorems of Apollonius' *Conic Sections* can be derived from the circle by Desargues' method of central projection. In spite of all this, however, the new geometry took little hold in the seventeenth century, and the subject lay practically dormant until the early part of the nineteenth century, when enormous interest in the subject developed and great advances were made by such men as Gergonne, Poncelet, Brianchon, Dupin, Chasles, and Steiner. Whereas Desargues may have been motivated by the need of a theory of perspective for architects and draftsmen, these later writers developed the subject for its own intrinsic charm.

---

[7] That some of these concepts were known to the ancient Greeks has been pointed out in Chapter 6.

9–9 Pascal

One of the few contemporaries of Desargues who showed a real appreciation of his work was Blaise Pascal, a mathematical genius of high order. Pascal was born in the French province of Auvergne in 1623 and very early showed phenomenal ability in mathematics. Several stories of his youthful accomplishments have been told by his sister Gilberta, who became Madame Périer. Because of his delicate constitution, the boy was kept at home to ensure his not being overworked. His father decided that the youngster's education should be at first restricted to the study of languages and should not include any mathematics. The exclusion of mathematics from his studies aroused curiosity in the boy, and he inquired of his tutor as to the nature of geometry. The tutor informed him that it was the study of exact figures and the properties of their different parts. Stimulated by his tutor's description of the subject and by his father's injunction against it, he gave up his playtime and clandestinely, in a few weeks, discovered for himself many properties of geometric figures, in particular the fact that the sum of the angles of a triangle is equal to a straight angle. This latter was accomplished by some process of folding a paper triangle, perhaps by folding the vertices over to the center of the inscribed circle, as indicated in Figure 77, or by folding the vertices over to the foot of an altitude, as indicated in Figure 78. When his father came upon him one day during his geometric activities, he was so struck by the boy's ability that he gave his son a copy of Euclid's *Elements,* which the youngster read with avidity and quickly mastered.

At the age of fourteen, Pascal participated in the weekly gatherings of a group of French mathematicians from which the French Academy ultimately formed in 1666. When he was sixteen, he wrote an essay on conic sections that Descartes could not believe was the work of the boy, assuming it must be that of his father instead. At eighteen or nineteen, he invented the first calculating machine, which he devised to assist his father in the auditing of government accounts at Rouen. Pascal was to manufacture over fifty calculating machines, some of which are still preserved in the Conservatoire des Arts et Métiers at Paris. At twenty-one, he became interested in Torricelli's work on atmospheric

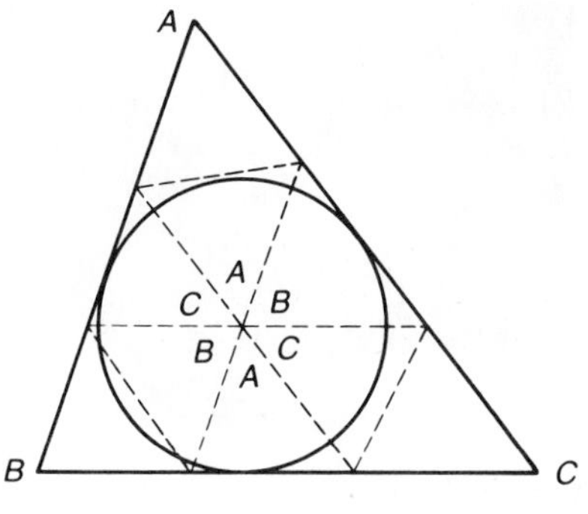

FIGURE 77

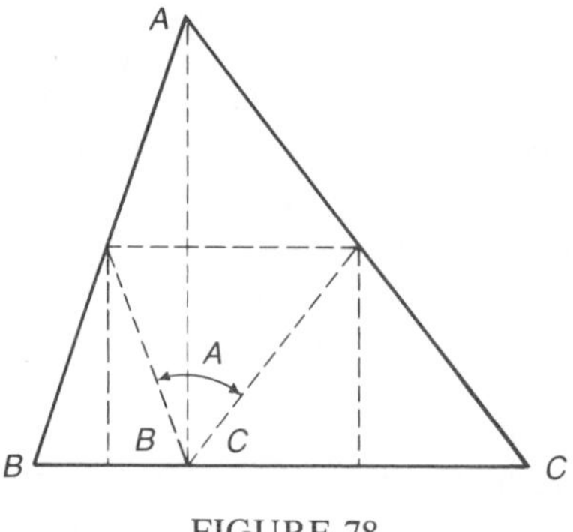

FIGURE 78

pressure and began to apply his unusual talents to physics, with the result that *Pascal's principle of hydrodynamics* is today known to every student of high school physics. A few years later, in 1648, he wrote a comprehensive, unpublished manuscript on conic sections.

This astonishing and precocious activity suddenly came to an end in 1650, when, suffering from frail health, Pascal decided to abandon his researches in mathematics and science and to devote himself to religious contemplation. Three years later, however, he returned briefly to mathematics. At this time, he wrote his *Traité du triangle arithmétique,* conducted several experiments on fluid pressure, and, in correspondence with Fermat, assisted in laying the foundations of the mathematical theory of probability. But late in 1654 he received what he regarded as a strong intimation that these renewed activities were not pleasing to God. The divine hint occurred when his runaway horses dashed

BLAISE PASCAL
(Brown Brothers)

over the parapet of the bridge at Neuilly, and he himself was saved only by the miraculous breaking of the traces. Fortified with a reference to the accident written on a small piece of parchment henceforth carried next to his heart, he dutifully went back to his religious meditations.

Only once again, in 1658, did Pascal return to mathematics. While suffering with toothache, some geometrical ideas occurred to him, and his teeth suddenly ceased to ache. Regarding this as a sign of divine will, he obediently applied himself assiduously for eight days toward developing his ideas, producing in this time a fairly full account of the geometry of the cycloid curve and solving some problems that subsequently, when issued as challenge problems, baffled other mathematicians. His famous *Provincial Letters* and his *Pensées,* which are read today as models of early French literature, were written toward the close of his brief life. He died in Paris in 1662. Desargues and Pascal died in the same year; Desargues was sixty-nine, but Pascal was only thirty-nine.

We might add here that Pascal's father, Étienne Pascal (1588–1640), was also an able mathematician; it is for the father that the *limaçon of Pascal* has been erroneously named [see Problem Study 4.7(c)].

Pascal has been described as the greatest "might have been" in the history of mathematics. With such unusual talents and such deep geometrical intuition, he might have produced, under more favorable conditions, a great deal more. But his health was such that much of his life was spent coping with physical discomfort, and from early manhood, he felt compelled to participate in the religious controversies of his time.

Pascal's manuscript on conic sections was founded on the work of Desargues and is now lost, but it was seen by Descartes and Leibniz. Here occurred Pascal's famous **mystic hexagram** theorem of projective geometry: *If a hexagon be inscribed in a conic, then the points of intersection of the three pairs of opposite sides are collinear, and conversely* (see Figure 79). He probably established the theorem, in Desargues' fashion, by first proving it true for a circle and then passing by projection to any conic section. Although the theorem is one of the richest in the whole of projective geometry (see Problem Study 9.12), we probably should take lightly the often-told tale that Pascal himself deduced over 400 corollaries from it. The manuscript was never pub-

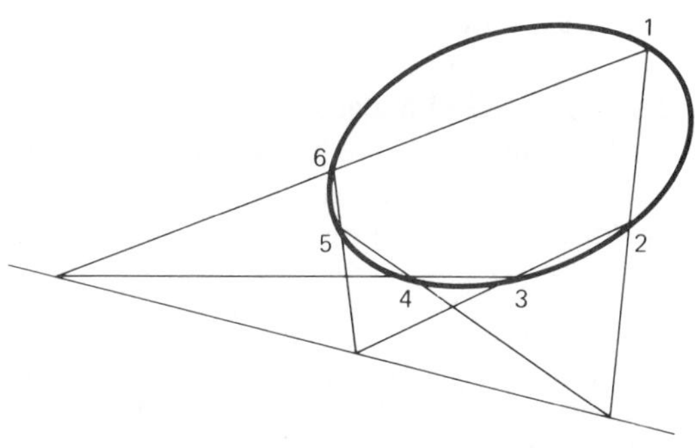

FIGURE 79

lished, and probably never completed, but in 1640 Pascal did print a one-page broadside entitled *Essay pour les coniques,* which announced some of his findings. Only two copies of this famous leaflet are known to be still in existence, one at Hanover among the papers of Leibniz, and the other in the Bibliothèque Nationale at Paris. Pascal's mystic hexagram theorem is involved in the second lemma of the leaflet.

Pascal's *Traité du triangle arithmétique* was written in 1653 but was not printed until 1665. He constructed his "arithmetic triangle" as indicated in Figure 80. Any element (in the second or a following row) is obtained as the sum of all those elements of the preceding row lying just above or to the left of the desired element. Thus, in the fourth row,

$$35 = 15 + 10 + 6 + 3 + 1.$$

The triangle, which may be of any order, is obtained by drawing a diagonal, as indicated in the figure. The student of college algebra will recognize that the numbers along such a diagonal are the successive coefficients in a binomial expansion. For example, the numbers along the fifth diagonal, namely 1, 4, 6, 4, 1, are the successive coefficients in the expansion of $(a + b)^4$. The finding of binomial coefficients was one of the uses to which Pascal put his triangle. He also used it, particularly in his discussions on probability, for finding the number of combinations of n things taken r at a time [see Problem Study 9.13(g)], which he correctly stated to be

$$\frac{n!}{r!(n - r)!},$$

where $n!$ is our present-day notation[8] for the product

$$n(n - 1)(n - 2) \ldots (3)(2)(1).$$

There are many relations involving the numbers of the arithmetic triangle, several of which were developed by Pascal (see Problem Study 9.13). Pascal was not the first to exhibit the arithmetic triangle, for such an array had been anticipated several centuries earlier by Chinese writers (see Section 7–3). Because Pascal was for so long (until 1935) the first known discoverer of the triangle in the western world, and because of his development and application of many of the triangle's properties, the array became known as **Pascal's triangle.** One of the earliest acceptable statements of the method of mathematical induction appears in Pascal's treatise on the triangle.

---

[8] The symbol $n!$, called *factorial n,* was introduced in 1808 by Christian Kramp (1760–1826) of Strasbourg, who chose this symbol so as to circumvent printing difficulties incurred by a previously used symbol. For convenience one defines $0! = 1$.

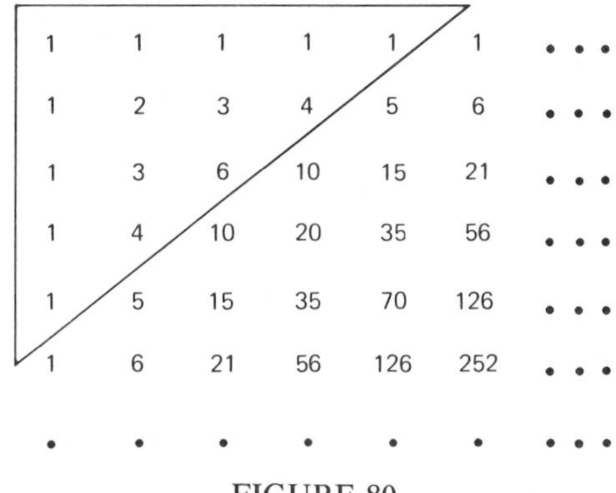

FIGURE 80

Although the Greek philosophers of antiquity discussed necessity and contingency at length, it is perhaps correct to say that there was no mathematical treatment of probability until the latter part of the fifteenth century and the early part of the sixteenth century, when some of the Italian mathematicians attempted to evaluate the chances in certain gambling games, like that of dice. Cardano, as was noted in Section 8–8, wrote a brief gambler's guidebook that involved some of the aspects of mathematical probability. But it is generally agreed that the one problem to which can be credited the origin of the science of probability is the so-called **problem of the points.** This problem requires the determination of the division of the stakes of an interrupted game of chance between two players of supposedly equal skills, knowing the scores of the players at the time of interruption and the number of points needed to win the game. Pacioli, in his *Sūma,* of 1494, was one of the first writers to introduce the problem of the points into a work on mathematics. The problem was also discussed by Cardano and Tartaglia, but a real advance was not made until the problem was proposed, in 1654, to Pascal by the Chevalier de Méré, an able and experienced gambler whose theoretical reasoning on the problem did not agree with his observations. Pascal became interested in the problem and communicated it to Fermat. There ensued a remarkable correspondence between the two men,[9] in which the problem was correctly but differently solved by each. Pascal solved the general case, obtaining many results through a use of the arithmetic triangle. In their correspondence, then, Pascal and Fermat laid the foundations of the science of probability.[10]

Pascal's last mathematical work was that on the **cycloid,** the curve traced by a point on the circumference of a circle as the circle rolls along a straight line (see Figure 81). This curve, which is very rich in mathematical and physical

---

[9] The correspondence appears in D. E. Smith, *A Source Book in Mathematics.*

[10] The methods of Pascal and Fermat for solving the problem of the points are described at the end of Section 10-3.

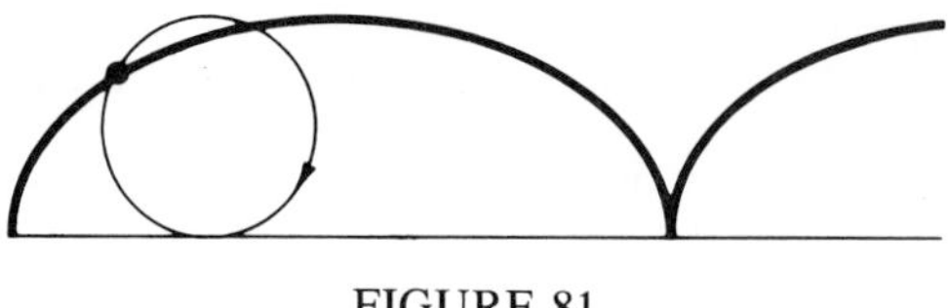

FIGURE 81

properties, played an important role in the early development of the methods of the calculus. Galileo was one of the first to call attention to the curve and once recommended that it be used for the arches of bridges. Shortly after, the area under one arch of the curve was found, and methods of drawing tangents to the curve were discovered. These discoveries led mathematicians to consider questions concerned with the surface and volume of revolution obtained by rotating a cycloidal arch about various lines. Such problems, as well as others concerned with the centroids of the figures formed, were solved by Pascal, and some of the results were issued by him as challenge problems to other mathematicians. Pascal's solutions were effected by the precalculus method of indivisibles and were equivalent to the evaluation of a number of definite integrals encountered in present-day calculus classes. The cycloid has so many attractive properties and has engendered so many quarrels that it has been called "the Helen of geometry" and "the apple of discord."

It is interesting that Pascal has been credited with the invention of the one-wheeled wheelbarrow as we know it today. Also, at the age of thirty-five, he conceived the omnibus—an idea that was soon put into practice at five sous a ride. Pascal sometimes wrote under the nom de plume of Lovis de Montalte, or its anagram, Amos Dettonville.

Problem Studies

9.1 Logarithms

(a) Using the familiar laws of exponents, establish the following useful properties of logarithms:
1. $\log_a mn = \log_a m + \log_a n$.
2. $\log_a (m/n) = \log_a m - \log_a n$.
3. $\log_a (m^r) = r \log_a m$.
4. $\log_a \sqrt[s]{m} = (\log_a m)/s$.

(b) Show that
1. $\log_b N = \log_a N/\log_a b$ (with this formula we may compute logarithms to a base b when we have available a table of logarithms to some base a).
2. $\log_N b = 1/\log_b N$.
3. $\log_N b = \log_{1/N} (1/b)$.

(c) By extracting the square root of 10, then the square root of the result thus obtained, and so on, the following table can be constructed:

$10^{1/2} = 3.16228$	$10^{1/256} = 1.00904$
$10^{1/4} = 1.77828$	$10^{1/512} = 1.00451$
$10^{1/8} = 1.33352$	$10^{1/1024} = 1.00225$
$10^{1/16} = 1.15478$	$10^{1/2048} = 1.00112$
$10^{1/32} = 1.07461$	$10^{1/4096} = 1.00056$
$10^{1/64} = 1.03663$	$10^{1/8192} = 1.00028$
$10^{1/128} = 1.01815$	. . . .

With this table, we may compute the common logarithm of any number between 1 and 10, and hence, by adjusting the characteristic, of any positive number whatever. Thus, let N be any number between 1 and 10. Divide N by the largest number in the table that does not exceed N. Suppose the divisor is $10^{1/p_1}$ and that the quotient is N_1. Then $N = 10^{1/p_1} N_1$. Treat N_1 in the same fashion, and continue the process, obtaining

$$N = 10^{1/p_1} 10^{1/p_2} \ldots 10^{1/p_n} N_n.$$

Let us stop when N_n differs from unity only in the sixth decimal place. Then, to 5 places,

$$N = 10^{1/p_1} 10^{1/p_2} \ldots 10^{1/p_n}$$

and

$$\log N = \frac{1}{p_1} + \frac{1}{p_2} + \cdots + \frac{1}{p_n}.$$

This procedure is known as the **radix method** of computing logarithms. Compute, in this manner, log 4.26 and log 5.00.

9.2 Napier and Spherical Trigonometry

(a) There are 10 formulas that are useful for solving right spherical triangles. There is no need to memorize these formulas, for it is easy to reproduce them by means of 2 rules devised by Napier. In Figure 82, a right spherical triangle is pictured, lettered in conventional manner. To the right of the triangle appears a circle divided into 5 parts, containing the same letters as the triangle, except C, arranged in the same order. The bars on c, B, A mean *the complement of* (thus $\overline{B}$ means $90° - B$). The angular quantities a, b, $\overline{c}$, $\overline{A}$, $\overline{B}$ are called the **circular parts.** In the circle, there are 2 circular parts contiguous to any given part, and 2 parts not contiguous to it. Let us call the given part the **middle part,** the 2 contiguous parts the **adjacent parts,** and the 2 noncontiguous parts the **opposite parts.** Napier's rules may be stated as follows:

1. The sine of any middle part is equal to the product of the cosines of the 2 opposite parts.
2. The sine of any middle part is equal to the product of the tangents of the 2 adjacent parts.

By applying each of these rules to each of the circular parts, obtain the 10 formulas for solving right spherical triangles.

(b) The formula connecting sides a, b, c of a right spherical triangle is called the **Pythagorean relation** for the triangle. Find the Pythagorean relation for a right spherical triangle.

(c) The following formulas are known as **Napier's analogies** (the word analogy being used in its archaic sense of "proportion"):

$$\frac{\sin \frac{1}{2}(A - B)}{\sin \frac{1}{2}(A + B)} = \frac{\tan \frac{1}{2}(a - b)}{\tan \frac{1}{2}c},$$

$$\frac{\cos \frac{1}{2}(A - B)}{\cos \frac{1}{2}(A + B)} = \frac{\tan \frac{1}{2}(a + b)}{\tan \frac{1}{2}c},$$

$$\frac{\sin \frac{1}{2}(a - b)}{\sin \frac{1}{2}(a + b)} = \frac{\tan \frac{1}{2}(A - B)}{\cot \frac{1}{2}C},$$

$$\frac{\cos \frac{1}{2}(a - b)}{\cos \frac{1}{2}(a + b)} = \frac{\tan \frac{1}{2}(A + B)}{\cot \frac{1}{2}C}.$$

These formulas, which are analogous to the law of tangents in plane trigonometry, may be used to solve oblique spherical triangles for which the given parts are 2 sides and the included angle, or 2 angles and the included side.

1. Find A, C, b for a spherical triangle in which $a = 125° 38'$, $c = 73° 24'$, $B = 102° 16'$.
2. Find A, B, c for a spherical triangle in which $a = 93° 8'$, $b = 46° 4'$, $C = 71° 6'$.

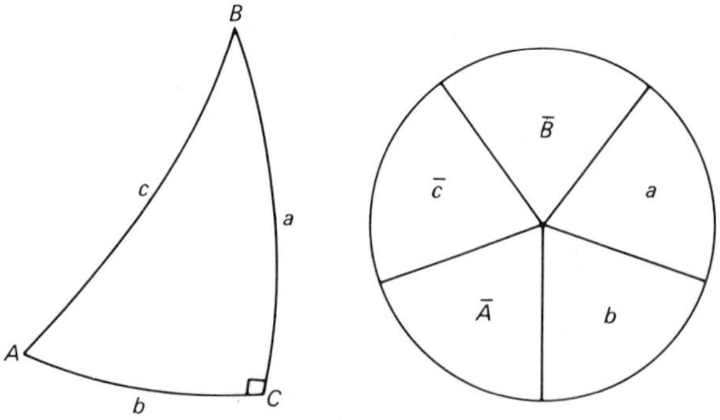

FIGURE 82

9.3 Napier's Rods

The difficulty that was so widely experienced in the multiplication of large numbers led to mechanical ways of carrying out the process. Famous in its time was Napier's invention, known as **Napier's rods,** or **Napier's bones,** which he described in his work *Rabdologiae,* which was published in 1617. In principle, the invention is the same as the Arabian lattice, or grating, method that we described in Section 7–5; in the invention, however, the process is carried out with the aid of rectangular strips of bone, metal, wood, or cardboard, prepared beforehand. For each of the 10 digits, one should have some strips, like the one shown to the left in Figure 83 for 6, bearing the various multiples of that digit. To illustrate the use of these strips in multiplication, let us select the example chosen by Napier in the *Rabdologiae,* the multiplication of 1615 by 365. Put strips headed 1, 6, 1, 5 side by side, as shown to the right in Figure 83. The results of multiplying 1615 by the 5, the 6, and the 3 of 365 are then easily read off as 8075, 9690, and 4845, some simple diagonal additions of 2 digits being necessary to obtain these results. The final product is then obtained by an addition, as illustrated in the figure.

(a) Make a set of Napier's rods and perform some multiplications.

(b) Explain how Napier's rods may be used to perform divisions.

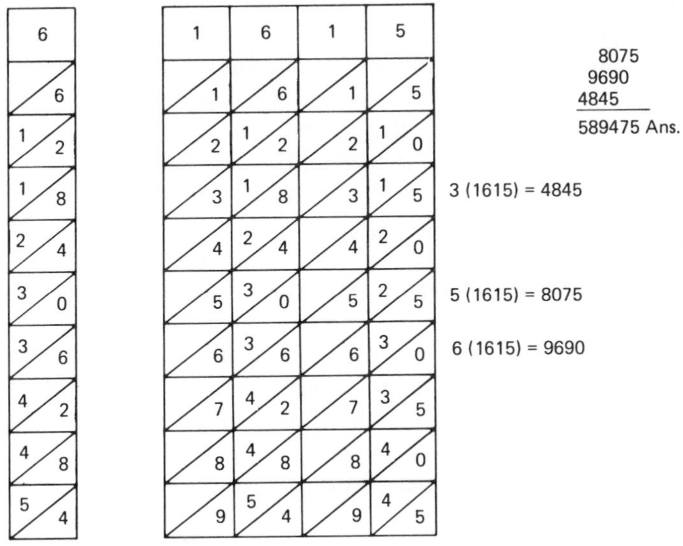

FIGURE 83

9.4 The Slide Rule

(a) Construct, with the aid of tables, a logarithmic scale, to be designated as the D scale, about 10 inches long. Use the scale, along with a pair of dividers, to perform some multiplications and divisions.

(b) Construct 2 logarithmic scales, to be called *C* and *D* scales, of the same size. By sliding *C* along *D*, perform some multiplications and divisions. [Refer to the laws of logarithms (Problem Study 9.1) for a suggestion.]

(c) Construct a logarithmic scale half as long as the preceding *D* scale, and designate, by *A*, 2 of these short scales placed end to end. Show how the *A* and *D* scales may be used for extracting square roots.

(d) How would one design a scale to be used with the *D* scale for extracting cube roots?

(e) Construct a scale just like the *C* and *D* scales, only running in the reverse direction, and call it the *CI* (*C inverted*) scale. Show how the *CI* and *D* scales may be used for performing multiplications. What is the advantage of the *CI* and *D* scales over the *C* and *D* scales for this purpose?

9.5 Freely Falling Bodies

Assuming that all bodies fall with the same constant acceleration g, Galileo showed that the distance d a body falls is proportional to the square of the time t of falling. Establish the following stages of Galileo's argument.

(a) If v is the velocity at the end of time t, then $v = gt$.

(b) If v and t refer to one falling body and V and T to a second falling body, then $v/V = t/T$, whence the right triangle having legs of numerical lengths v and t is similar to the right triangle having legs of numerical lengths V and T.

(c) Since the increase in velocity is uniform, the average velocity of fall is $v/2$, whence $d = vt/2 =$ area of right triangle with legs v and t.

(d) $d/D = t^2/T^2$. Show also that $d = gt^2/2$.

Galileo illustrated the truth of this final law by observing the times of descent of balls rolling down inclined planes.

9.6 Sector Compasses

About 1597, Galileo perfected the **sector compasses,** an instrument that enjoyed considerable popularity for more than two centuries. The instrument consists of 2 arms fastened together at one end by a pivot joint, as shown in Figure 84. On each arm, there is a simple scale radiating from the pivot and having the 0 of the scale at the pivot. In addition to these simple scales, other scales have often been used, some of which will be described below. Many problems can readily be solved using the simple scales of the compasses, the only theory required being that of similar triangles.

(a) Show how the sector compasses may be used to divide a given line segment into 5 equal parts.

(b) Show how the sector compasses may be used to change the scale of a drawing.

(c) Show how the sector compasses may be used to find the fourth proportional x to 3 given quantities a, b, c (that is, to find x where $a : b = c : x$), and thus applied to problems in foreign exchange.

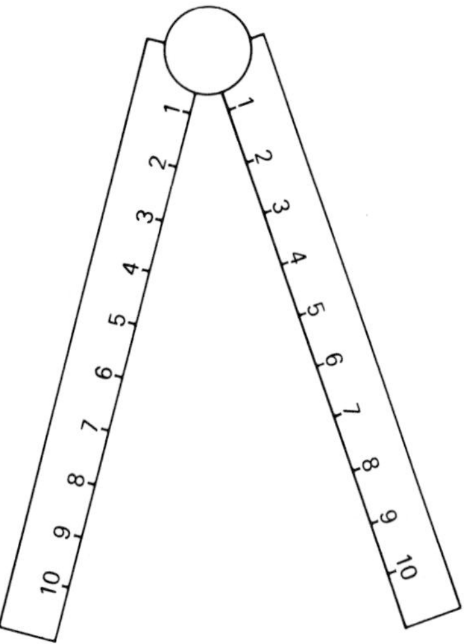

FIGURE 84

(d) Galileo illustrated the use of his sector compasses by finding the amount of money that should have been invested 5 years ago at 6 percent, compounded annually, to amount to 150 scudi today. Try to solve this problem with the sector compasses.

Among the additional scales frequently found on the arms of sector compasses was one (the *line of areas*) marked according to the squares of the numbers involved and used for finding squares and square roots of numbers. Another scale (the *line of volumes*) was marked according to the cubes of the numbers involved. Another gave the chords of arcs of specified numbers of degrees for a circle of unit radius and served engineers as a protractor. Still another (called the *line of metals*) contained the medieval symbols for gold, silver, iron, copper, and so forth, spaced according to the densities of these metals, and was used to solve such problems as finding the diameter of an iron sphere having its weight equal to that of a given copper sphere.

The sector compasses are neither as accurate nor as easy to manipulate as the slide rule.

9.7 Some Simple Paradoxes from Galileo's "Discorsi"

Explain the following 2 geometrical paradoxes considered by Galileo in his *Discorsi* of 1638.

(a) Suppose the large circle of Figure 85 has made one revolution in rolling along the straight line from A to B, so that AB is equal to the circumference of the large circle. Then the small circle, fixed to the large one, has also made 1 revolution, so that CD is equal to the circumference of the small circle. It follows that *the 2 circles have equal circumferences!*

FIGURE 85

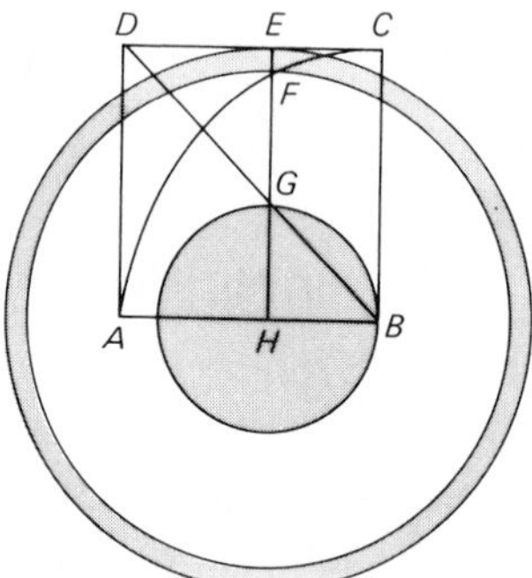

FIGURE 86

This paradox had been earlier described by Aristotle and is therefore sometimes referred to as **Aristotle's wheel.**

(b) Let $ABCD$ be a square and HE any line parallel to BC, cutting the diagonal BD in G, as shown in Figure 86. Let circle $B(C)$ cut HE in F, and draw the 3 circles $H(G)$, $H(F)$, $H(E)$. First show that the area of circle $H(G)$ is equal to the area of the ring between circles $H(F)$ and $H(E)$. Then let H approach B so that, in the limit, circle $H(G)$ becomes the point B, and the ring becomes the circumference $B(C)$. We now conclude that *the single point B* is equal to the whole circumference $B(C)$!

(c) Explain the remark in the *Discorsi* that "neither is the number of squares less than the totality of all numbers, nor the latter greater than the former."

9.8 Kepler's Laws

(a) Where is a planet in its orbit when its speed is greatest?

(b) Check, approximately, Kepler's third law using the following modern figures. (A. U. is an abbreviation for **astronomical unit,** the length of the semimajor axis of the earth's orbit.)

Planet	Time in Years	Semimajor Axis
Mercury	0.241	0.387 A.U.
Venus	0.615	0.723 A.U.
Earth	1.000	1.000 A.U.
Mars	1.881	1.524 A.U.
Jupiter	11.862	5.202 A.U.
Saturn	29.457	9.539 A.U.

(c) What would be the period of a planet having a semimajor axis of 100 A.U.?

(d) What would be the semimajor axis of a planet having a period of 125 years?

(e) Two hypothetical planets are moving about the sun in elliptical orbits having equal semimajor axes. The semiminor axis of one, however, is half that of the other. How do the periods of the planets compare?

(f) The moon revolves about the earth in 27.3 days in an elliptical orbit whose semimajor axis is 60 times the earth's radius. What would be the period of a hypothetical satellite revolving very close to the earth's surface?

9.9 Mosaics

A very interesting problem of mosaics is to fill the plane with congruent regular polygons. Let n be the number of sides of each polygon. Then the interior angle at each vertex of such a polygon is $(n - 2)180°/n$. Prove this statement.

(a) If we do not permit a vertex of 1 polygon to lie on a side of another, show that the number of polygons at each vertex is given by $2 + 4/(n - 2)$, and, hence, that we must have $n = 3$, 4, or 6. Construct illustrative mosaics.

(b) If we insist that a vertex of 1 polygon lie on a side of another, show that the number of polygons clustered at such a vertex is given by $1 + 2/(n - 2)$, whence we must have $n = 3$ or 4. Construct illustrative mosaics.

(c) Construct mosaics containing (1) 2 sizes of equilateral triangles, the larger having a side twice that of the smaller, such that sides of triangles of the same size do not overlap; (2) 2 sizes of squares, the larger having a side twice that of the smaller, such that sides of smaller squares do not overlap; (3) congruent equilateral triangles and congruent regular dodecagons; (4) congruent equilateral triangles and congruent regular hexagons; (5) congruent squares and congruent regular octagons.

(d) Suppose we have a mosaic composed of regular polygons of 3 different kinds of each vertex. If the 3 kinds of polygons have p, q, r sides, respectively, show that

$$\frac{1}{p} + \frac{1}{q} + \frac{1}{r} = \frac{1}{2}.$$

One integral solution of this equation is $p = 4$, $q = 6$, $r = 12$. Construct a mosaic of the type under consideration and composed of congruent squares, congruent regular hexagons, and congruent regular dodecagons.

9.10 Proving Theorems by Projection

(a) If l is a given line in a given plane π, and O is a given center of projection (not on π), show how to find a plane π' such that the projection of l onto π' will be the line at infinity on π'. (The operation of selecting a suitable center of projection O and plane of projection π' so that a given line on a given plane shall project into the line at infinity on π' is called the operation of "projecting a given line to infinity.")

(b) Show that under the projection of (a), the line at infinity in π will project into the intersection of π' with the plane through O parallel to π.

(c) Let UP, UQ, UR be three concurrent coplanar lines, cut by two lines OX and OY in P_1, Q_1, R_1 and P_2, Q_2, R_2, respectively (see Figure 87). Prove that the intersections of Q_1R_2 and Q_2R_1, R_1P_2 and R_2P_1, P_1Q_2 and P_2Q_1 are collinear.

(d) Prove that if $A_1B_1C_1$ and $A_2B_2C_2$ are two coplanar triangles such that B_1C_1 and B_2C_2 meet in L, C_1A_1 and C_2A_2 meet in M, A_1B_1 and A_2B_2 meet in N, where L, M, N are collinear, then A_1A_2, B_1B_2, C_1C_2 are concurrent. (This is the converse part of the statement of Desargues' two-triangle theorem as given in Section 9–8.)

(e) Show that by parallel projection (a projection where the center of projection is at infinity) an ellipse may always be projected into a circle.

(f) In 1678, the Italian Giovanni Ceva (1648–1734) published a work containing the following theorem (see Figure 88), which is now known by

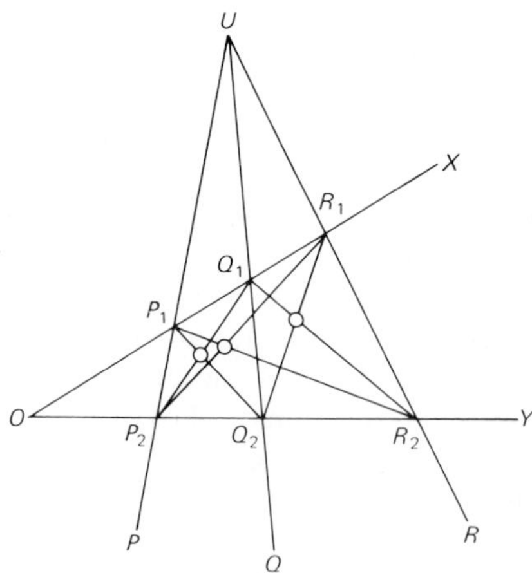

FIGURE 87

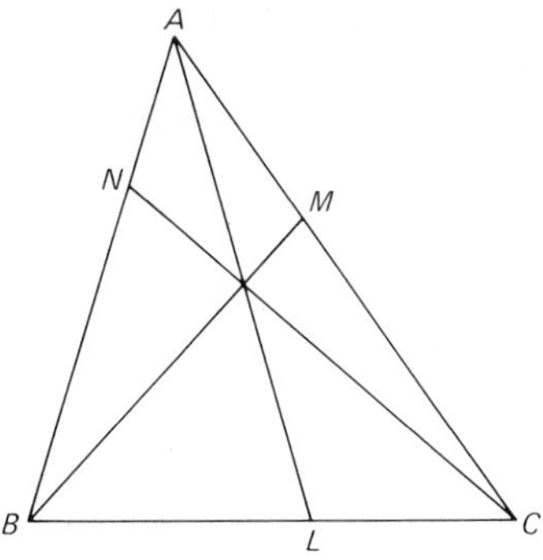

FIGURE 88

his name: *The 3 lines that join 3 points L, M, N on the sides BC, CA, AB of a triangle ABC to the opposite vertices are concurrent if and only if*

$$\left(\frac{AN}{NB}\right)\left(\frac{BL}{LC}\right)\left(\frac{CM}{MA}\right) = +1.$$

This is a companion theorem to Menelaus' theorem, which was stated in Section 6–5. Using Ceva's theorem, prove that the lines joining the vertices of a triangle to the opposite points of contact of the inscribed circle are concurrent. Then, by means of (e), prove that the lines joining the vertices of a triangle to the opposite points of contact of an inscribed ellipse are concurrent.

(g) La Hire invented the following interesting mapping of the plane onto itself (see Figure 89): Draw any 2 parallel lines a and b and select a point P in their plane. Through any second point M of the plane, draw a line cutting a in A and b in B. Then the map M' of M will be taken as the intersection with MP of the parallel to AP through B.

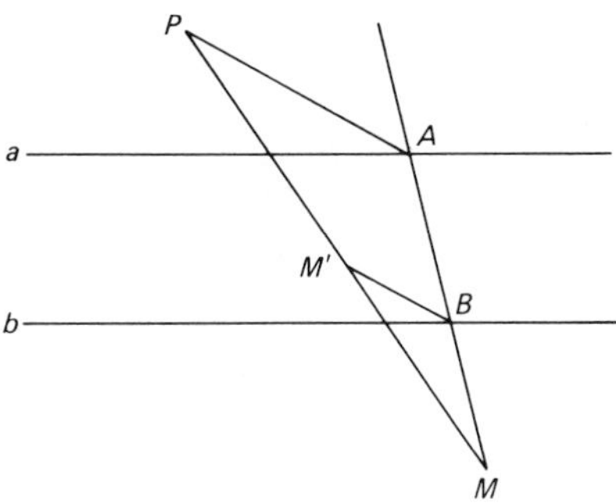

FIGURE 89

1. Show that M' is independent of the particular line MAB through M used in determining it.
2. Generalize La Hire's mapping to the situation where a and b need not be parallel.

9.11 Pascal's Youthful Empirical "Proof"

Fill in the details of the empirical "proofs" indicated by Figures 77 and 78.

9.12 Pascal's Theorem

The consequences of Pascal's mystic hexagram theorem are very numerous and attractive, and an almost unbelievable amount of research has been expended on the configuration. There are 60 possible ways of forming a hexagon from 6 points on a conic and, by Pascal's theorem, to each hexagon corresponds a *Pascal line*. These 60 Pascal lines pass 3 by 3 through 20 points, called *Steiner points,* which in turn lie 4 by 4 on 15 lines, called *Plücker lines*. The Pascal lines also concur 3 by 3 in another set of points, called *Kirkman points,* of which there are 60. Corresponding to each Steiner point, there are 3 Kirkman points such that all 4 lie upon a line, called a *Cayley line*. There are 20 of these Cayley lines, and they pass 4 by 4 through 15 points, called *Salmon points*. There are many further extensions and properties of the configuration, and the number of different proofs that have been supplied for the mystic hexagram theorem itself is now legion. In this Problem Study we shall consider a few of the many corollaries of the mystic hexagram theorem that can be obtained by making some of the 6 points coincide with one another. For simplicity, we shall number the points 1, 2, 3, 4, 5, 6. Then Pascal's theorem says that the intersections of the pairs of lines 12, 45; 23, 56; 34, 61 are collinear if and only if the 6 points lie on a conic.

(a) If a pentagon 12345 is inscribed in a conic, show that the pairs of lines 12, 45; 23, 51; 34 and the tangent at 1, intersect in 3 collinear points.

(b) Given 5 points, draw at any 1 of them the tangent to the conic determined by the 5 points.

(c) Given 4 points of a conic and the tangent at any 1 of them, construct further points on the conic.

(d) Show that the pairs of opposite sides of a quadrangle inscribed in a conic, together with the pairs of tangents at opposite vertices, intersect in 4 collinear points.

(e) Show that if a triangle is inscribed in a conic, then the tangents at the vertices intersect the opposite sides in 3 collinear points.

(f) Given 3 points on a conic and the tangents at 2 of them, construct the tangent at the third.

9.13 Pascal's Triangle

Establish the following relations, all of which were developed by Pascal, involving the numbers of the arithmetic triangle.

(a) Any element (not in the first row or the first column) of the arithmetic triangle is equal to the sum of the element just above it and the element just to the left of it.

(b) Any given element of the arithmetic triangle, decreased by 1, is equal to the sum of all the elements above the row and to the left of the column containing the given element.

(c) The mth element in the nth row is $(m + n - 2)!/(m - 1)!(n - 1)!$, where, by definition, $0! = 1$.

(d) The element in the mth row and nth column is equal to the element in the nth row and mth column.

(e) The sum of the elements along any diagonal is twice the sum of the elements along the preceding diagonal.

(f) The sum of the elements along the nth diagonal is 2^{n-1}.

(g) Let us be given a group of n objects. Any set of r of these objects, considered without regard to order, is called a *combination of the n objects taken r at a time,* or, more briefly, as an *r-combination* of the n objects. We shall use the symbol $C(n,r)$ to denote the number of such combinations. Thus, the 2-combinations of the 4 letters a, b, c, d are

$$ab,\ ac,\ ad,\ bc,\ bd,\ cd,$$

whence $C(4,2) = 6$. It is shown in textbooks on college algebra that

$$C(n,r) = \frac{n!}{r!(n - r)!}.$$

Show that $C(n,r)$ appears at the intersection of the $(n + 1)$st diagonal and the $(r + 1)$st column of the arithmetic triangle.

Essay Topics

9/1 Reasons for the upsurge of mathematics in the seventeenth century.

9/2 Napier as the science-fiction writer of his day.

9/3 The use of Napier's rods and Galileo's sector compasses.

9/4 The Nicaraguan science-formula stamps of 1971.

9/5 Reasons for the base e for logarithms and for radian measure for angles.

9/6 Harriot as the father of the modern theory of equations.

9/7 Harriot in America.

9/8 Oughtred's mathematical symbols.

9/9 Pernicious effects of the Inquisition.

9/10 Can science and religion be reconciled?

9/11 Kepler and the principle of continuity.

9/12 Art as a motivation for the development of projective geometry.

9/13 Pascal's triangle before Pascal.

9/14 A history of the cycloid curve.

Bibliography

BARLOW, C. W. C., and G. H. BRYAN *Elementary Mathematical Astronomy.* London: University Tutorial Press, 1923.

BELL, E. T. *Men of Mathematics.* New York: Simon and Schuster, 1937.

BISHOP, M. G. *Pascal, The Life of Genius.* New York: Reynal & Hitchcock, 1936.

BIXBY, WILLIAM *The Universe of Galileo and Newton.* New York: Harper & Row, 1964.

BRASCH, F. F., ed. *Johann Kepler, 1571–1630. A Tercentenary Commemoration of His Life and Work.* Baltimore: Williams & Wilkins, 1931.

CAJORI, FLORIAN *A History of the Logarithmic Slide Rule and Allied Instruments.* New York: McGraw-Hill, 1909.

——— *William Oughtred, A Great Seventeenth-Century Teacher of Mathematics.* Chicago: Open Court, 1916.

——— *A History of Mathematical Notation.* 2 vols. Chicago: Open Court, 1929.

CASPAR, MAX *Kepler.* Translated by C. Doris Hellman. New York: Abelard-Schuman, 1959.

COOLIDGE, J. L. *A History of Geometrical Methods.* New York: Oxford University Press, 1940.

——— *The Mathematics of Great Amateurs.* New York: Oxford University Press, 1949.

COXETER, H. S. M. *Regular Polytopes.* New York: Pitman, 1949.

DAVID, F. N. *Games, Gods and Gambling.* New York: Hafner, 1962.

DE SANTILLANA, GIORGIO *The Crime of Galileo.* Chicago: University of Chicago Press, 1955.

DRAKE, STILLMAN *Galileo at Work, His Scientific Biography.* Chicago: University of Chicago Press, 1978.

DREYER, J. L. E. *Tycho Brahe: A Picture of the Scientific Life and Work in the Sixteenth Century.* New York: Dover, 1963.

EDWARDS, A. W. F. *Pascal's Arithmetic Triangle.* Oxford: Oxford University Press, 1987.

FAHIE, J. J. *Galileo, His Life and Work.* London: John Murray, 1903.

FERMI, LAURA, and BERNADINI FERMI *Galileo and the Scientific Revolution.* New York: Basic Books, 1961.

GADE, J. A. *The Life and Times of Tycho Brahe.* Princeton, N.J.: Princeton University Press, 1970.

GALILEI, GALILEO *Dialogues Concerning Two New Sciences.* Translated by Henry Crew and Alfonso de Salvio. Introduction by Antonio Favaro. New York: Macmillan, 1914. Reprinted by Dover, New York.

——— *Discourses on the Two Chief Systems.* Edited by Stillman Drake. Berkeley, Calif.: University of California Press, 1953.

——— *Discourses on the Two Chief Systems.* Edited by Giorgio de Santillana. Chicago: University of Chicago Press, 1953.

GRAY, J. V., and J. J. FIELD *The Geometrical Work of Girard Desargues*. New York: Springer-Verlag, 1987.

HACKER, S. G. *Arithmetical View Points*. Pullman, Wash.: Mimeographed at Washington State College, 1948.

HARTLEY, MILES C. *Patterns of Polyhedrons*. Ann Arbor, Mich.: Edwards Brothers, 1957.

HOOPER, ALFRED *Makers of Mathematics*. New York: Random House, 1948.

IVINS, W. M., JR. *Art and Geometry*. Cambridge, Mass.: Harvard University Press, 1946.

KNOTT, C. G. *Napier Tercentenary Memorial Volume*. London: Longmans, Green, 1915.

KOESTLER, ARTHUR *The Watershed, a Life of Kepler*. New York: Doubleday, 1960.

KRAITCHIK, MAURICE *Mathematical Recreations*. New York: W. W. Norton, 1942.

McMULLIN, ERNAN, ed. *Galileo: Man of Science*. New York: Basic Books, 1967.

MESNARD, JEAN *Pascal, His Life and Works*. New York: Philosophical Library, 1952.

MORTIMER, ERNEST *Blaise Pascal: The Life and Work of a Realist*. New York: Harper and Brothers, 1959.

MUIR, JANE *Of Men and Numbers, The Story of the Great Mathematicians*. New York: Dodd, Mead, 1961.

NORTHROP, E. P. *Riddles in Mathematics*. Princeton, N.J.: D. Van Nostrand, 1944.

PEARCE, PETER, and SUSAN PEARCE *Polyhedra Primer*. New York: Van Nostrand, 1978.

RONAN, COLIN *Galileo*. New York: C.P. Putnam's Sons, 1974.

SMITH, D. E. *A Source Book in Mathematics*. New York: McGraw-Hill, 1929.

SULLIVAN, J. W. N. *The History of Mathematics in Europe, From the Fall of Greek Science to the Rise of the Conception of Mathematical Rigour*. New York: Oxford University Press, 1925.

TODHUNTER, ISAAC *A History of the Mathematical Theory of Probability, From the Time of Pascal to that of Laplace*. New York: Chelsea, 1949.

TURNBULL, H. W. *The Great Mathematicians*. New York: New York University Press, 1961.

Chapter 10

ANALYTIC GEOMETRY AND OTHER PRECALCULUS DEVELOPMENTS

10–1 Analytic Geometry

While Desargues and Pascal were opening the new field of projective geometry, Descartes and Fermat were conceiving ideas of modern analytic geometry. There is a fundamental distinction between the two studies, for the former is a *branch* of geometry whereas the latter is a *method* of geometry. Few academic experiences can be more thrilling to the student of elementary college mathematics than his or her introduction to this new and powerful method of attacking geometric problems. The essence of the idea, as applied to the plane, it will be recalled, is the establishment of a correspondence between points in the plane and ordered pairs of real numbers, thereby making possible a correspondence between curves in the plane and equations in two variables, so that for each curve in the plane there is a definite equation $f(x, y) = 0$, and for each such equation there is a definite curve, or set of points, in the plane. A correspondence is similarly established between the algebraic and analytic properties of the equation $f(x, y) = 0$ and the geometric properties of the associated curve. The task of proving a theorem in geometry is cleverly shifted to that of proving a corresponding theorem in algebra and analysis.

There are differences of opinion as to who invented analytic geometry, even as to what age should be credited with the invention, and the matter certainly cannot be settled without an agreement as to just what constitutes analytic geometry. We have seen that the ancient Greeks indulged in a good deal of geometric algebra, and it is well known that the idea of coordinates was used in the ancient world by the Egyptians and the Romans in surveying and by the Greeks in map making. Particularly strong in the favor of the Greeks is the fact that Apollonius derived the bulk of his geometry of the conic sections from the geometric equivalents of certain Cartesian equations of these curves, an idea that seems to have originated with Menaechmus. We also noted, in Section 8–4, that in the fourteenth century Nicole Oresme anticipated another aspect of analytic geometry when he represented certain laws by graphing the dependent variable *(latitudo)* against the independent one ((*longitudo*), as the latter variable was permitted to take on small increments. Proponents favoring Oresme as the inventor of analytic geometry see in his work such achievements as the first explicit introduction of the equation of a straight line and the extension of some of the notions of the subject to higher-dimensional spaces. A century after Oresme's text was written, it underwent several printings; in this way, it may have influenced later mathematicians.

The preceding views of analytic geometry seem to confuse the subject with one or more of its features. The real essence of the subject lies in the transference of a geometric investigation into a corresponding algebraic investigation. Before analytic geometry could assume this ability, it had to await the development of algebraic processes and symbolism. It would therefore seem more correct to agree with the majority of historians, who regard the decisive contributions made in the seventeenth century by the two French mathematicians René Descartes and Pierre de Fermat as the essential origin of the subject. Certainly not until after the stimulation given to the subject by these two men do we find analytic geometry in a form with which we are familiar.

10–2 Descartes

René Descartes was born near Tours in 1596. At the age of eight, he was sent to the Jesuit school at La Flèche. It was there that he developed (at first because of delicate health) his lifelong habit of lying in bed till late in the morning. These meditative hours of morning rest were later regarded by Descartes as his most productive periods. In 1612, Descartes left school and shortly after went to Paris, where, with Mersenne and Mydorge (see Section 10–6), he devoted some time to the study of mathematics. In 1617, he commenced several years of soldiering by joining the army of Prince Maurice of Orange. Upon quitting military life, he spent four or five years traveling through Germany, Denmark, Holland, Switzerland, and Italy. After resettling for a couple of years in Paris, where he continued his mathematical studies and his philosophical contemplations and where for a while he took up the construction of optical instruments, he decided to move to Holland, than at the height of its power. There he lived for twenty years, devoting his time to philosophy, mathematics, and science. In 1649, he reluctantly went to Sweden at the invitation of Queen Christina. A few months later, he contracted inflammation of the lungs and died in Stockholm early in 1650. The great philosopher-mathematician was entombed in Sweden, and efforts to have his remains transported to France failed. Then, seventeen years after Descartes' death, his bones, except for those of his right hand, were returned to France and reinterred in Paris. The bones of the right hand were secured, as a souvenir, by the French Treasurer-General who had arranged the transportation of the bones.

It was during his stay of twenty years in Holland that Descartes accomplished his writing. He spent the first four years writing *Le monde,* a physical account of the universe, but this was prudently abandoned and left incomplete when Descartes heard of Galileo's condemnation by the Church. He turned to the writing of a philosophical treatise on universal science under the title of *Discours de la méthode pour bien conduire sa raison et chercher la vérité dans les sciences* (A Discourse on the Method of Rightly Conducting the Reason and

Seeking Truth in the Sciences); this was accompanied by three appendices entitled *La dioptrique, Les météores,* and *La géométrie*. The *Discours,* with the appendices, was published in 1637; it is in the last of the three appendices that Descartes' contributions to analytic geometry appear. In 1641, Descartes published a work called *Meditationes,* which was devoted to a lengthy explanation of the philosophic views sketched in the *Discours*. In 1644, he issued his *Principia philosophiae,* which contains some inaccurate laws of nature and an inconsistent cosmological theory of vortices.

La géométrie, the famous third appendix of the *Discours,* occupies about one hundred pages of the complete work and is itself divided into three parts. It is the only mathematical writing published by Descartes. The first part contains an explanation of some of the principles of algebraic geometry and shows a real advance over the Greeks. To the Greeks, a variable corresponded to the length of some line segment, the product of two variables to the area of some rectangle, and the product of three variables to the volume of some rectangular parallelepiped. Beyond this the Greeks could not go. To Descartes, on the other hand, x^2 did not suggest an area, but rather the fourth term in the proportion $1:x = x:x^2$, and as such is representable by an appropriate line length that can easily be constructed when x is known. Using a unit segment, we can, in this way, represent any power of a variable, or the product of any number of variables, by a line length, and actually construct the line length with Euclidean tools when the values of the variables are assigned. With this arithmetization of geometry, Descartes, in the first part of *La géométrie,* marked off x on a given axis and then a length y at a fixed angle to this axis and endeavored to construct points whose x's and y's satisfy a given relation (see Figure 90). If we have the relation $y = x^2$, for example, then for each value of x we are able to construct the corresponding y as the fourth term of the above proportion. Descartes showed special interest in obtaining such relations for curves that are defined kinematically. As an application of his method, he discussed the problem: If $p_1, \ldots, p_m, p_{m+1}, \ldots, p_{m+n}$ are the lengths of $m + n$ line segments drawn from a point p to $m + n$ given lines, making given angles with these lines, and if

$$p_1p_2 \cdots p_m = kp_{m+1}p_{m+2} \cdots p_{m+n},$$

where k is a constant, find the locus of P. The ancient Greeks solved this problem for the cases where m and n do not exceed 2 (see Section 6–9), but the

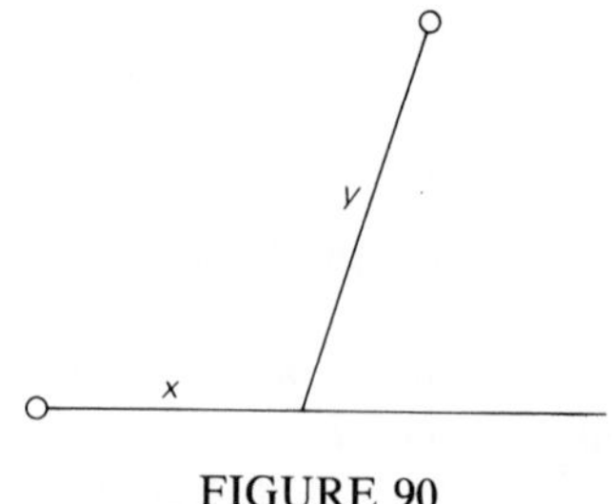

FIGURE 90

RENÉ DESCARTES
(David Smith Collection)

general problem had remained a baffling one. Descartes easily showed that higher cases of the problem lead to loci of degrees greater than two. In certain cases, he was actually able to construct with Euclidean tools points of the loci [see Problem Study 10.2(a)]. That Descartes' analytic geometry can cope with the general problem is a fine tribute to the power of the new method. It is said that it was Descartes' attempt to solve this problem that inspired his invention of analytic geometry.

The second part of *La géométrie* deals, among other things, with a now-obsolete classification of curves and with an interesting method of constructing tangents to curves. The method of drawing tangents is as follows (see Figure 91). Let the equation of the given curve be $f(x, y) = 0$, and let (x_1, y_1) be the coordinates of the point P of the curve at which we wish to construct a tangent.

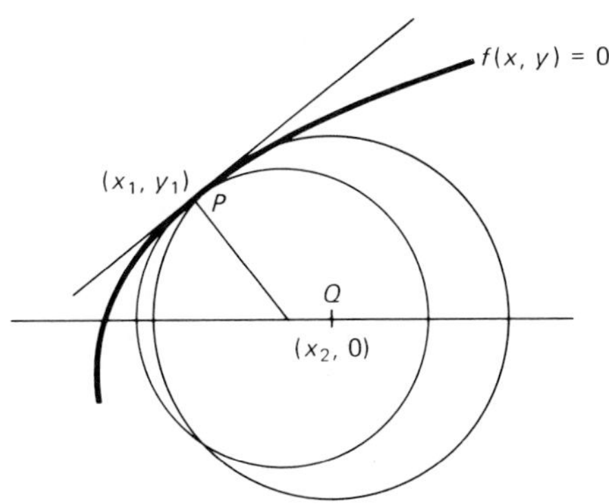

FIGURE 91

Let Q, having coordinates $(x_2, 0)$, be a point on the x-axis. Then the equation of the circle with Q as center and passing through P is

$$(x - x_2)^2 + y^2 = (x_1 - x_2)^2 + y_1^2.$$

If we eliminate y between this equation and the equation $f(x, y) = 0$, we obtain an equation in x leading to the abscissas of the points where the circle cuts the given curve. We now determine x_2 so that this equation in x will have a *pair* of roots equal to x_1. This condition fixes Q as the intersection of the x-axis and the normal to the curve at P, since the circle is now tangent to the given curve at P. Once this circle is drawn, we may easily construct the required tangent. As an example of the method, consider the construction of the tangent to the parabola $y^2 = 4x$ at the point $(1, 2)$. Here we have

$$(x - x_2)^2 + y^2 = (1 - x_2)^2 + 4.$$

The elimination of y gives

$$(x - x_2)^2 + 4x = (1 - x_2)^2 + 4,$$

or

$$x^2 + 2x(2 - x_2) + (2x_2 - 5) = 0.$$

The condition that this quadratic equation have two equal roots is that its discriminant vanish—that is, that

$$(2 - x_2)^2 - (2x_2 - 5) = 0,$$

or

$$x_2 = 3.$$

The circle with center $(3, 0)$ and passing through the point $(1, 2)$ of the curve may now be drawn and the required tangent finally constructed. This method of constructing tangents was applied by Descartes to a number of different curves, including one of the quartic ovals named after him.[1] Here we have a general process that tells us exactly what to do to solve our problem, but it must be confessed that in the more complicated cases, the required algebra may be quite forbidding. This is a well-recognized fault with elementary analytic geometry—we often know what to do but lack the technical ability to do it. There

[1] A **Cartesian oval** is the locus of a point whose distances, r_1 and r_2, from two fixed points satisfy the relation $r_1 + mr_2 = a$, where m and a are constants. The central conics will be recognized as special cases.

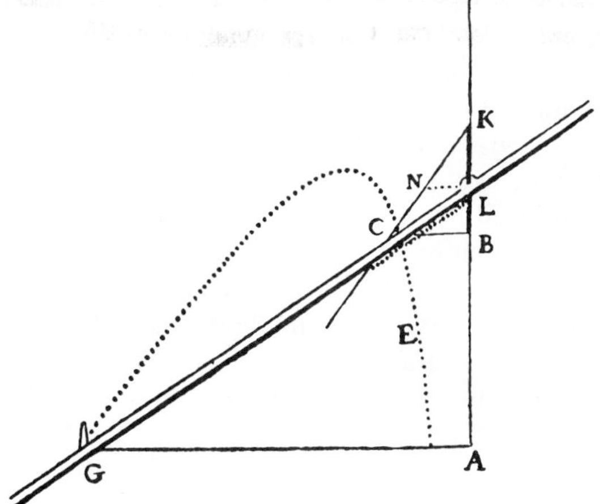

Aprés cela prenant vn point a diſcretion dans la courbe, comme C, ſur lequel ie ſuppoſe que l'inſtrument qui ſert a la deſcrire eſt appliqué, ie tire de ce point C la ligne C B parallele a G A, & pourceque C B & B A ſont deux quantités indeterminées & inconnuës, ie les nomme l'vne y & l'autre x. mais affin de trouuer le rapport de l'vne à l'autre; ie conſidere auſſy les quantités connuës qui determinent la deſcription de cete ligne courbe, comme G A que ie nomme a, K L que ie nomme b, & N L parallele a G A que ie nomme c. puis ie dis, comme N L eſt à L K, ou c à b, ainſi C B, ou y, eſt à B K, qui eſt par conſequent $\frac{b}{c}y$: & B L eſt $\frac{b}{c}y - b$, & A L eſt $x + \frac{b}{c}y - b$. de plus comme C B eſt à L B, ou y à $\frac{b}{c}y - b$, ainſi a, ou G A, eſt á L A, ou $x + \frac{b}{c}y - b$. de façon que mul-

S ſ tipliant

A page from Descartes' *La géométrie* (1637).

are, of course, much better methods than the above for finding tangents to curves.

The third part of *La géométrie* concerns itself with the solution of equations of degree greater than two. Use is made of what we now call **Descartes' rule of signs,** a rule for determining limits to the number of positive and the number of negative roots possessed by a polynomial (see Problem Study 10.3). In *La géométrie,* Descartes fixed the custom of employing the first letters of the alphabet to denote known quantities, and the last letters to denote unknown ones. He also introduced our present system of indices (such as a^3, a^4, and so forth), which is a great improvement over Viète's way of designating powers, and he realized that a letter might represent any quantity, positive or negative. Here we also find the first use of the **method of undetermined coefficients.** In the example of the last paragraph, then, we used the vanishing of the discriminant to determine the value of x_2 so that the quadratic equation

$$x^2 + 2x(2 - x_2) + (2x_2 - 5) = 0$$

should have both roots equal to 1. As an illustration of the method of undetermined coefficients, we might accomplish this by saying that we want

$$x^2 + 2(2 - x_2)x + (2x_2 - 5) \equiv (x - 1)^2 \equiv x^2 - 2x + 1,$$

whence we must have, by equating coefficients of like powers of x,

$$2(2 - x_2) = -2 \qquad \text{and} \qquad 2x_2 - 5 = 1.$$

Either of these leads to $x_2 = 3$.

La géométrie is not in any sense a systematic development of the method of analytics, and the reader must pretty much construct the method for himself from certain isolated statements. There are thirty-two figures in the text, but in none do we find the coordinate axes explicitly set forth. The work was written with intentional obscurity and as a result was too difficult to be widely read. In 1649, a Latin translation appeared with explanatory notes by F. de Beaune, edited with commentary by Frans van Schooten the Younger. This, and the revised 1659–1661 edition, had a wide circulation. One hundred years or more later, the subject achieved the familiar form found in our present-day college textbooks. The words *coordinates, abscissa,* and *ordinate,* as now technically used in analytic geometry, were contributed by Leibniz in 1692.

There are a couple of legends describing the initial flash that led Descartes to the contemplation of analytic geometry. According to one story, it came to him in a dream. On St. Martin's Eve, November 10, 1616, while encamped in the army's winter quarters on the banks of the Danube, Descartes experienced three singularly vivid and coherent dreams that, he claimed, changed the whole course of his life. The dreams, he said, clarified his purpose in life and determined his future endeavors by revealing to him "a marvelous science" and "a wonderful discovery." Descartes never explicitly disclosed just what were the marvelous science and the wonderful discovery, but some believe them to have been analytic geometry, or the application of algebra to geometry, and then the reduction of all science to geometry. It was eighteen years later that he expounded some of his ideas in his *Discours*.

Another story, perhaps on a par with the story of Isaac Newton and the falling apple, says that the initial flash of analytic geometry came to Descartes when watching a fly crawling about on the ceiling near a corner of his room. It struck him that the path of the fly on the ceiling could be described if only one knew the relation connecting the fly's distances from two adjacent walls. Even though this second story may be apocryphal, it has good pedagogic value.

Of the other two appendices to the *Discours*, one is devoted to optics and the other to an explanation of numerous meteorologic, or atmospheric, phenomena, including the rainbow.

Among other mathematical items credited to Descartes is the near discovery of the relation $v - e + f = 2$, connecting the numbers of vertices v, edges e, and faces f of a convex polyhedron (see Problem Study 3.12). He was the first to discuss the so-called **folium of Descartes,** a nodal cubic curve found today in many of our calculus texts, but he did not completely picture the curve. In correspondence he considered parabolas of higher order ($y^n = px$, $n > 2$) and gave a remarkably neat construction of a tangent to the cycloid.

10–3 Fermat

At the same time that Descartes was formulating the basis of modern analytic geometry, the subject was also occupying the attention of another French mathematical genius, Pierre de Fermat. Fermat's claim to priority rests on a letter written to Roberval in September, 1636, in which it is stated that the ideas of the writer were even then seven years old. The details of the work appear in the posthumously published paper *Isogoge ad locus planos et solidos*. Here we find the equation of a general straight line and of a circle, and a discussion of hyperbolas, ellipses, and parabolas. In a work on tangents and quadratures, completed before 1637, Fermat defined many new curves analytically. Where Descartes suggested a few new curves, generated by mechanical motion, Fermat proposed many new ones, defined by algebraic equations. The curves $x^m y^n = a$, $y^n = ax^m$, and $r^n = a\theta$ are still known as **hyperbolas, parabolas,** and **spirals of Fermat.** Fermat also proposed, in connection with work on quadratures, the cubic curve later called the **witch of Agnesi,** after the distinguished eighteenth-century mathematician, linguist, and philosopher Maria Gaetana Agnesi (see Section 12–7). Thus, where to a large extent Descartes began with a locus and then found its equation, Fermat started with the equation and then studied the locus. These are the two inverse aspects of the fundamental principle of analytic geometry. Fermat's work is written in Viète's notation and thus has an archaic look when compared with Descartes' more modern symbolism.

There is a seemingly reliable report that Fermat was born at Beaumont de Lomagne, near Toulouse, on August 17, 1601. It is known that he died at

Castres or Toulouse on January 12, 1665. His tombstone, originally in the church of the Augustines in Toulouse and later moved to the local museum, gives the preceding date of death and Fermat's age at death as fifty-seven years. Because of this conflicting data, Fermat's dates are usually listed as (1601?–1665). Indeed, for various reasons, Fermat's birth year, as given by different writers, ranges from 1590 to 1608.

Fermat was the son of a leather merchant and received his early education at home. At the age of thirty, he obtained the post of councilor for the local parliament at Toulouse and there discharged his duties with modesty and punctiliousness. Working as a humble and retiring lawyer, he devoted the bulk of his leisure time to the study of mathematics. Although he published very little during his lifetime, he was in scientific correspondence with many leading mathematicians of his day and, in this way, considerably influenced his contemporaries. He enriched so many branches of mathematics with so many important contributions that he has been called the greatest French mathematician of the seventeenth century.

Of Fermat's varied contributions to mathematics, the most outstanding is the founding of the modern theory of numbers. In this field, Fermat possessed extraordinary intuition and ability. It was probably the Latin translation of Diophantus' *Arithmetica,* made by Bachet de Méziriac in 1621, that first directed Fermat's attention to number theory. Many of Fermat's contributions to the field occur as marginal statements made in his copy of Bachet's work. In 1670, five years after his death, these notes were incorporated in a new, but unfortunately carelessly printed, edition of the *Arithmetica,* brought out by his son Clément-Samuel. Many of the unproved theorems announced by Fermat have later been shown to be correct. The following examples illustrate the tenor of Fermat's investigations.

PIERRE DE FERMAT
(David Smith Collection)

1. *If p is a prime and a is prime to p, then $a^{p-1} - 1$ is divisible by p.* For example, if $p = 5$ and $a = 2$, then $a^{p-1} - 1 = 15 = (5)(3)$. This theorem, known as the **little Fermat theorem,** was given by Fermat without proof in a letter to Frénicle de Bessy, dated October 18, 1640. The first published proof of it was given by Euler in 1736 (see Problem Study 10.5).
2. *Every odd prime can be expressed as the difference of two squares in one and only one way.* Fermat gave a simple proof of this. If p is an odd prime, then one easily verifies that

$$p = \left(\frac{p+1}{2}\right)^2 - \left(\frac{p-1}{2}\right)^2.$$

On the other hand, if $p = x^2 - y^2$, then $p = (x + y)(x - y)$. But since p is prime, its only factors are p and 1; hence, $x + y = p$ and $x - y = 1$, or $x = (p + 1)/2$ and $y = (p - 1)/2$.
3. *A prime of the form $4n + 1$ can be represented as the sum of two squares.* For example, $5 = 4 + 1$, $13 = 9 + 4$, $17 = 16 + 1$, $29 = 25 + 4$. This theorem was first stated by Fermat in a letter to Mersenne, dated December 25, 1640. The first published proof was given by Euler in 1754, who, moreover, succeeded in showing that the representation is unique.
4. *A prime of the form $4n + 1$ is only once the hypotenuse of an integral-sided right triangle; its square is twice; its cube is three times; and so forth.* For example, consider $5 = 4(1) + 1$. Now $5^2 = 3^2 + 4^2$; $25^2 = 15^2 + 20^2 = 7^2 + 24^2$; $125^2 = 75^2 + 100^2 = 35^2 + 120^2 = 44^2 + 117^2$.
5. *Every non-negative integer can be represented as the sum of four or fewer squares.* This difficult theorem was established by Lagrange in 1770.
6. *The area of an integral-sided right triangle cannot be a square number.* This also was established later by Lagrange.
7. *There is only one solution in integers of $x^2 + 2 = y^3$, and only two of $x^2 + 4 = y^3$.* This problem was issued as a challenge problem to English mathematicians. The solutions are $x = 5$, $y = 3$, for the first equation, and $x = 2$, $y = 2$ and $x = 11$, $y = 5$, for the second equation.
8. *There do not exist positive integers x, y, z such that $x^4 + y^4 = z^2$.*
9. *There do not exist positive integers x, y, z, n such that $x^n + y^n = z^n$, when $n > 2$.* This famous conjecture is known as **Fermat's last "theorem."** It was stated by Fermat in the margin of his copy of Bachet's translation of Diophantus, at the side of Problem 8 of Book II: "To divide a given square number into two squares." Fermat's marginal note reads, "To divide a cube into two cubes, a fourth power, or in general any power whatever into two powers of the same denomination above the second is impossible, and I have assuredly found an admirable proof of this, but the margin is too narrow to contain it." Whether Fermat really possessed a sound demonstration of this problem will probably forever remain an engima. Many of the most prominent mathematicians since his time have tried their skill on the problem, but the general conjecture still remains open. There is a proof

given elsewhere by Fermat for the case $n = 4$, and Euler supplied a proof (later perfected by others) for $n = 3$. About 1825, independent proofs for the case $n = 5$ were given by Legendre and Dirichlet; in 1839, Lamé proved the theorem for $n = 7$. Very significant advances in the study of the problem were made by the German mathematician E. Kummer (1810–1893). In 1843, Kummer submitted a purported proof to Dirichlet, who pointed out an error in the reasoning. Kummer then returned to the problem with renewed vigor, and a few years later, after developing an important allied subject in higher algebra called the *theory of ideals,* derived very general conditions for the insolubility of the Fermat relation. Almost all important subsequent progress on the problem has been based on Kummer's investigations. It is now known that Fermat's last "theorem" is certainly true for all $n < 125{,}000$,[2] and for many other special values of n. In 1908, the German mathematician Paul Wolfskehl bequeathed 100,000 marks to the Academy of Science at Göttingen as a prize for the first complete proof of the "theorem." The result was a deluge of alleged proofs by glory- and money-seeking laymen; ever since then, the problem has haunted amateurs, as does the trisection of an arbitrary angle and the squaring of the circle. Fermat's last "theorem" has the peculiar distinction of being the mathematical problem for which the greatest number of incorrect proofs have been published.

10. Fermat conjectured that $f(n) = 2^{2^n} + 1$ is prime for any non-negative integral n. The conjecture proved to be incorrect when Euler showed that $f(5)$ is a composite number. It is known that $f(n)$ is composite for $5 \leqslant n \leqslant 16$ and at least forty-seven other values of n, perhaps the largest being $n = 1945$. The prime factors of $f(5)$, $f(6)$, and $f(8)$ have been found, and one prime factor of $f(9)$ has been found.

In 1879, a paper was found in the library at Leyden, among the manuscripts of Christiaan Huygens, in which Fermat describes a general method by which he may have made many of his discoveries. The method is known as Fermat's **method of infinite descent** and is particularly useful in establishing negative results. In brief, the method is this. To prove that a certain relation connecting positive integers is impossible, assume, on the contrary, that the relation can be satisfied by some particular set of positive integers. From this assumption, show that the same relation then holds for another set of smaller positive integers. Then, by a reapplication, the relation must hold for another set of still smaller positive integers, and so on ad infinitum. Since the positive integers cannot be decreased in magnitude indefinitely, it follows that the assumption at the start is untenable and, therefore, that the original relation is impossible. Fermat used this method to establish result 8 above. To make the method clear, let us apply it by proving anew that $\sqrt{2}$ is irrational. Suppose $\sqrt{2} = a/b$, where a and b are positive integers. Now

$$\sqrt{2} + 1 = \frac{1}{\sqrt{2} - 1},$$

[2] This was accomplished in recent years with the aid of electronic computers.

whence

$$\frac{a}{b} + 1 = \frac{1}{\dfrac{a}{b} - 1} = \frac{b}{a - b},$$

and

$$\sqrt{2} = \frac{a}{b} = \frac{b}{a - b} - 1 = \frac{2b - a}{a - b} = \frac{a_1}{b_1}, \text{ say.}$$

But, since $1 < \sqrt{2} < 2$, after replacing $\sqrt{2}$ by a/b and then multiplying through by b, we have $b < a < 2b$. Now, since $a < 2b$, it follows that $0 < 2b - a = a_1$. And since $b < a$, it follows that $a_1 = 2b - a < a$. Thus, a_1 is a positive integer less than a. By a reapplication of our procedure, we find $\sqrt{2} = a_2/b_2$, where a_2 is a positive integer less than a_1. The process may be repeated indefinitely. Since the positive integers cannot be decreased in magnitude indefinitely, it follows that $\sqrt{2}$ cannot be rational.

We have already mentioned, in Section 9–9, the Pascal-Fermat correspondence that laid the foundations of the science of probability. It will be recalled that it was the so-called **problem of the points** that started the matter: "Determine the division of the stakes of an interrupted game of chance between two supposedly equally skilled players, knowing the scores of the players at the time of interruption and the number of points needed to win the game." Fermat discussed the case in which one player A needs two points to win, and the other player B needs three points. Here is Fermat's solution for this particular case. Since it is clear that four more trials will decide the game, let a indicate a trial where A wins and b a trial where B wins, and consider the sixteen permutations of the two letters a and b taken four at a time:

aaaa	*aaab*	*abba*	*bbab*
baaa	*bbaa*	*abab*	*babb*
abaa	*baba*	*aabb*	*abbb*
aaba	*baab*	*bbba*	*bbbb*

The cases in which a appears two or more times are favorable to A; there are eleven of them. The cases where b appears three or more times are favorable to B; there are five of them. Therefore, the stakes should be divided in the ratio 11 : 5. For the general case, where A needs m points to win and B needs n, one writes down the 2^{m+n-1} possible permutations of the two letters a and b taken $m + n - 1$ at a time. One then finds the number α of cases where a appears m or more times and the number β of cases where b appears n or more times. The stakes are then to be divided in the ratio $\alpha : \beta$.

Pascal solved the problem of the points by utilizing his "arithmetic triangle," described in Section 9–9. Letting $C(n, r)$ represent the number of combinations of n objects taken r at a time [see Problem Study 9.13(g)], one can easily show that the numbers along the fifth diagonal of the "arithmetic triangle" are, respectively,

$$C(4, 4) = 1,\ C(4, 3) = 4,\ C(4, 2) = 6,\ C(4, 1) = 4,\ C(4, 0) = 1.$$

Since, returning to the special problem of the points considered above, $C(4, 4)$ is the number of ways to obtain 4 a's, $C(4, 3)$ is the number of ways to obtain 3 a's, and so forth, it follows that the solution of the problem is given by

$$\begin{aligned}[C(4, 4) + C(4, 3) + C(4, 2)]:[C(4, 1) + C(4, 0)] &= (1 + 4 + 6):(4 + 1)\\ &= 11:5.\end{aligned}$$

In the general case, where A needs m points to win and B needs n, one chooses the $(m + n)$th diagonal of Pascal's arithmetic array. One then finds the sum α of the first n numbers of this diagonal and the sum β of the last m numbers. The stakes are then to be divided in the ratio $\alpha:\beta$.

Pascal and Fermat, in their historic correspondence of 1654, reflected upon other problems related to the problem of the points, such as the division of stakes when there are more than two players, or when there are two unevenly skilled players. It was this work by Pascal and Fermat that launched the mathematical theory of probability. In 1657, Christiaan Huygens (1629–1695) wrote the first formal treatise on probability, basing his work on the Pascal-Fermat correspondence. This was the best account of the subject until the posthumous appearance, in 1713, of the *Ars conjectandi* of Jakob Bernoulli (1654–1705), which contained a reprint of the earlier treatise by Huygens. After these pioneering efforts, we find the subject carried forward by such men as Abraham De Moivre (1667–1754), Daniel Bernoulli (1700–1782), Leonhard Euler (1707–1783), Joseph Louis Lagrange (1736–1813), Pierre-Simon Laplace (1749–1827), and a host of other contributors.

It is impressive, and somewhat surprising, that mathematicians have been able to develop a science (namely, the mathematical theory of probability) that establishes rational laws that can be applied to situations of pure chance. This science is far from being impractical, as evinced by experiments performed in great laboratories, by the existence of highly respected insurance companies, and by the logistics of big business and of war.

We shall return to Fermat in the next chapter (Section 11–7), where we will consider his use of infinitesimals in geometry—particularly his application of them to questions of maxima and minima—marking him an important forerunner of the differential calculus.

10–4 Roberval and Torricelli

We devote this section to Gilles Persone de Roberval and Evangelista Torricelli, a Frenchman and an Italian, respectively, who were contemporaries, were both accomplished geometers and physicists, possessed similar mathematical tastes and talents, and became embroiled with one another in priority disputes.

Gilles Persone, a contentious individual, was born in Roberval, near Beauvais, in 1602, and died in Paris in 1675. He assumed the seignioral name of de Roberval, to which he was not entitled. His extensive correspondence served as a medium for the intercommunication of mathematical ideas in those prejournal days. He became well known for his method of drawing tangents and his discoveries in the field of higher plane curves. He endeavored to consider a curve as generated by a point whose motion is compounded from two known motions. Then the resultant of the velocity vectors of the two known motions gives the tangent line to the curve. For example, in the case of a parabola, we may consider the two motions as away from the focus and away from the directrix. Since the distances of the moving point from the focus and the directrix are always equal to each other, the velocity vectors of the two motions must also be of equal magnitude. It follows that the tangent at a point of the parabola bisects the angle between the focal radius to the point and the perpendicular through the point to the directrix (see Figure 92).

The preceding idea of tangents was also held by Torricelli, and an argument of priority ensued. Roberval also claimed to be the inventor of Cavalieri's precalculus method of indivisibles (discussed in Section 11–6) and to have squared the cycloid before Torricelli. These matters of priority are difficult to settle, for Roberval was consistently tardy in disclosing his discoveries. This tardiness has been explained by the fact that for forty years, starting in 1634, Roberval held a professorial chair at the Collège Royale. This chair automatically became vacant every three years, to be filled by open competition in mathematical contests in which the questions were set by the incumbent. To retain his position, Roberval saved his discoveries to formulate contest questions that he would be able to answer but that his competitors would probably find troublesome. In any event, Roberval successfully employed the method of indivisibles to the finding of a number of areas, volumes, and centroids. In spite of his geometric successes, his chief interest lay in physics.

Evangelista Torricelli, a sensitive soul, was born in or near Faenza, Italy, in 1608, and died in Florence in 1647. He was, for a very brief time, a student of Galileo, during the latter's final year of life. Although he was forty-four years

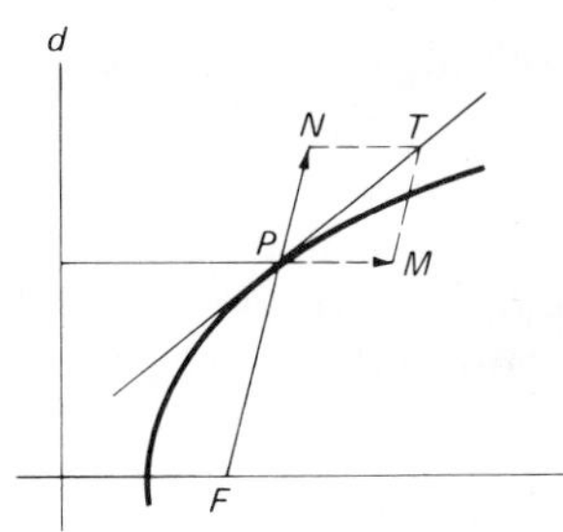

FIGURE 92

Galileo's junior, he survived the master by only five years, dying at the age of thirty-nine, as Pascal was to do fifteen years later. A perhaps overly romantic account claims that Torricelli died from dismay and chagrin at being accused of plagiarism by Roberval.

We have noted that Galileo valued the cycloid for the graceful form it would give to arches in architecture. In 1599, he also attempted to ascertain the area under one arch of the curve by balancing a cycloidal template against circular templates of the size of the generating circle. He incorrectly concluded that the area under an arch was very nearly, but not exactly, three times the area of the circle. The first published mathematical demonstration that the area is *exactly* three times that of the generating circle was furnished in 1644, by his pupil Torricelli, using early infinitesimal methods. Torricelli at the same time published a construction of the tangent to the cycloid at any given point on the curve. He made no reference to the fact that Roberval had earlier arrived at both the area and the tangent, so in 1646 the irritated Roberval wrote a letter accusing Torricelli of plagiarism. It is now clear that priority of discovery goes to Roberval, but priority of publication belongs to Torricelli, who probably independently rediscovered the two results.

For finding the tangent, both men employed the method of composition of motions, described before in connection with drawing a tangent to the parabola. In the case of the cycloid, a point P on the curve can be thought of as subject to two equal motions, one a translation and one a rotation. As the generating circle rolls along the horizontal base line AB (see Figure 93), point P is carried horizontally while at the same time rotating about O, the center of the circle. One therefore draws through P a horizontal vector PR, for the translation component, and a vector PS tangent to the generating circle, for the rotary component. Inasmuch as the two vectors have equal magnitudes, the required tangent to the cycloid lies along the bisector PT of the angle RPS formed by the two vectors.

Fermat proposed to Torricelli the problem of determining a point in the plane of a triangle such that the sum of its distances from the three vertices be a minimum. Torricelli's solution was published in 1659 by his pupil Viviani. This point, now known as the **isogonic center** of the triangle, was the first notable point of the triangle discovered since the period of ancient Greek mathematics. An elegantly simple analysis of the problem was later furnished by Jacob Steiner.[3] In 1640, Torricelli found the length of an arc of the logarithmic spiral. This curve had also been rectified two years earlier by Descartes and was the first curve after the circle to be rectified.

In 1641, Torricelli noted that an infinite area, when revolved about an axis in its plane, can sometimes yield a finite volume for the solid of revolution. For example, the area bounded by the hyperbola $xy = k^2$, the ordinate $x = b$ $(b > 0)$, and the x-axis is infinite, whereas the volume of the solid obtained by revolving

---

[3] See, for example, R. A. Johnson, *Modern Geometry*, pp. 218–25, and Richard Courant and H. E. Robbins, *What Is Mathematics?* pp. 354–61.

EVANGELISTA TORRICELLI
(David Smith Collection)

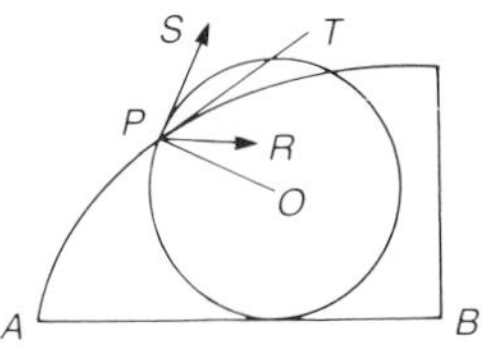

FIGURE 93

the area about the x-axis is finite. Torricelli, however, was not the first to note this seeming anomaly.

Torricelli is much better known for his contributions to physics, where he developed the theory of the barometer and worked on such questions as the value of the acceleration due to gravity, the theory of projectiles, and the motion of fluids.

10–5 Huygens

The great Dutch genius, Christiaan Huygens, lived an uneventful but remarkably productive life. He was born at The Hague in 1629 and studied at Leyden under Frans van Schooten the Younger. In 1651, when he was twenty-two, he published a paper pointing out fallacies committed by Saint-Vincent in his work on the quadrature of the circle. This was followed by a number of tracts dealing with the quadrature of the conics and with Snell's trigonometric improvement of the classical method of computing π (see Section 4–8). In 1654, he and his brother devised a new and better way of grinding and polishing lenses; consequently, Huygens was able to settle a number of questions in observational astronomy, such as the nature of Saturn's appendages. Huygen's work in astronomy led him, a couple of years later, to invent the pendulum clock, so that he might have more exact means of measuring time.

As remarked in Section 10–3, it was in 1657 that Huygens wrote the first formal treatise on probability, basing his work on the Pascal-Fermat correspondence. Many interesting and challenging problems were solved by Huygens, and he introduced the important concept of "mathematical expectation": If p denotes the probability that a person will win a certain sum s, then sp is called his **mathematical expectation.** Huygens showed, among other things, that if p is the probability of a person winning a sum a, and q that of winning a sum b, then he may expect to win the sum $ap + bq$.

Pascal, in his *Pensées,* or *Thoughts on Religion and Other Subjects,* published eight years after his death, made a specious application of the notion of mathematical expectation. He argued that, since the value of eternal happiness must be infinite, then, even if the probability that a religious life would ensure happiness were very small, still the expectancy (which is measured by the product of the two) must be sufficient to render it worthwhile to be religious.

In 1665, Huygens moved to Paris in order to benefit from a pension offered to him by Louis XIV. While there, in 1668, he communicated to the Royal Society of London a paper in which he demonstrated experimentally that the combined momentum of two bodies in a given direction is the same before and after a collision.

In 1673, in Paris, Huygens' greatest publication, *Horologium oscillatorium,* appeared. This work is in five parts, or chapters. The first part concerns

CHRISTIAAN HUYGENS
(David Smith Collection)

itself with the pendulum clock that the author had invented in 1656. The second part is devoted to a discussion of bodies falling freely in a vacuum, sliding on a smooth inclined plane, or sliding along a smooth curve. Shown here is the isochronous property of an inverted cycloid—that a heavy particle will reach the bottom of an inverted cycloidal arch in the same length of time, no matter from what point on the arch it begins its descent. The third part includes a treatment of evolutes and involutes. The **evolute** of a plane curve is the envelope of the normals to the curve, and any curve having a given curve for its evolute is called an **involute** of that given curve. As applications of his general theory, Huygens finds the evolute of a parabola and of a cycloid. In the former case, he obtains a semicubical parabola; in the latter, he obtains another cycloid of the same size. A treatment of the compound pendulum with a proof that the center of oscillation and the point of suspension are interchangeable is found in the fourth part of the *Horologium*. The last part of the work concerns itself with the theory of clocks. Here we find a description of the cycloidal pendulum (see Problem Study 10.7), in which the period of oscillation is the same no matter how great or how small the amplitude of the oscillation, something that is only approximately true of the period of oscillation of a simple pendulum. This last part closes with thirteen theorems related to centrifugal force in circular motion, proving, among other things, the now familiar fact that for uniform circular motion, the magnitude of the centrifugal force is directly proportional to the square of the linear speed and inversely proportional to the radius of the circle. In 1675, under Huygens' directions, the first watch regulated by a balance spring was made; it was presented to Louis XIV.

Huygens returned to Holland in 1681, constructed some lenses of very large focal lengths, and invented the achromatic eyepiece for telescopes. In 1689, he visited England and made the acquaintance of Isaac Newton, whose work he greatly admired. Shortly after his return to Holland in the following year, he published a treatise expounding the wave theory of light. On the basis of this theory, he was able to deduce geometrically the laws of reflection and refraction and to explain the phenomenon of double refraction. Newton, however, supported the emission theory of light, and his greater eminence caused contemporary scientists to favor that theory to the wave theory.

Huygens also wrote a number of minor tracts. He rectified the cissoid of Diocles; investigated the geometry of the catenary (the curve assumed by a perfectly flexible inextensible chain of uniform linear density, hanging from two supports not in the same vertical line); wrote on the logarithmic curve; gave, in modern form, for polynomials, Fermat's rule for maxima and minima; and made numerous applications of mathematics to physics.

Like many of the demonstrations given by Newton, Huygens' proofs are almost entirely accomplished, with great rigor, by the methods of Greek geometry. Reading his works, one would not realize that he was acquainted with the powerful new methods of analytic geometry and the calculus. Huygens died in the city of his birth in 1695.

10–6 Some Seventeenth-Century Mathematicians of France and Italy

The works of some lesser mathematicians of the seventeenth century should be mentioned here, even if only briefly. We devote the present and the following two sections to this purpose, treating the men by geographic areas.

An early noteworthy European Diophantist was the Frenchman Claude-Gaspar Bachet, Sieur de Méziriac (1581–1638), commonly referred to as Bachet de Méziriac. He was a broad scholar—a mathematician, philosopher, theologian, poet, and writer. His charming and classic *Problèmes plaisants et délectables,* which appeared in 1612 and again, enlarged, in 1624, contains many arithmetic tricks and questions that have reappeared in practically all subsequent collections of mathematical puzzles and recreations. In 1621, he published an edition of the Greek text of Diophantus' *Arithmetica,* along with a Latin translation and notes. It was in a copy of this work that Fermat made his famous marginal notes.

Another French number theorist, and a voluminous writer in many fields, was the Minimite friar Marin Mersenne (1588–1648). He maintained a constant correspondence with the greatest mathematicians of his day and served admirably, in those prejournal times, as a clearinghouse for mathematical ideas. He edited the works of many of the Greek mathematicians and wrote on a variety of subjects. He is especially known today in connection with the so-called **Mersenne primes,** or prime numbers of the form $2^p - 1$, which he discussed in a couple of places in his work *Cogitata physico-mathematica* of 1644. The connection between Mersenne primes and perfect numbers was pointed out in Section 3–3. The Mersenne prime for $p = 4253$ is the first known prime number to possess more than 1000 digits in its decimal expansion, and the Mersenne prime for $p = 216091$ was the largest prime known in 1986. With the remarkable improvement in modern computing machinery, it is probably pointless to continue recording further updatings of this sort.

Claude Mydorge (1585–1647), born in Paris and a Parisian by predilection, was a close friend of Descartes. He was a geometer and a physicist. He published some works on optics and a synthetic treatment of the conic sections in which he simplified many of Apollonius' prolix proofs. He left an interesting manuscript containing the statements and solutions of over a thousand geometric problems and edited the popular *Récréations mathématiques* of Leurechon.

We have, in Section 9–8, already said something of the work of the Frenchman Phillipe de la Hire (1640–1718). He has been described as a man of varied genius, having been a painter, an architect, an astronomer, and a mathematician. In addition to his work on conic sections described earlier, he wrote on graphical methods, various types of higher plane curves, and magic squares. He constructed maps of the earth by **globular projection,** where the center of projection is not a pole of the sphere, as in Ptolemy's stereographic projection (see Problem Study 6.10), but on the radius produced through a pole to a distance of $r \sin 45°$ outside the sphere.

Among the lesser Italian mathematicians to be mentioned here is Vicenzo

Viviani (1622–1703), another of Galileo's disciples, who interested himself in both physics and geometry. He was highly honored during his lifetime. Among his geometric accomplishments was determining the tangent to the cycloid; however, several had solved this problem previously. In 1692, he proposed the following problem, which attracted wide attention: A hemispheric dome has four equal windows of such size that the rest of the surface can be exactly squared; show how this is possible. Correct solutions were furnished by a number of eminent contemporary mathematicians. Viviani solved the trisection of an angle by using an equilateral hyperbola.

Mention should be made of the remarkable Italian-French Cassini family, several members of which contributed notably to astronomy and made skillful applications of mathematics in this field. The Cassini scientific dynasty started with Giovanni Domenico Cassini, who was born at Perinaldo, Italy, in 1625, and who died in Paris in 1712. The **Cassinian curve,** which is the locus of a point the product of whose distances from two fixed points is a constant, was studied by Giovanni Cassini in 1680 in connection with work on the relative motions of the earth and sun. In a family of confocal Cassinian curves is found the figure-eight-shaped lemniscate of Bernoulli, a fact not noted until the end of the eighteenth century. The Cassinian curves can be found as the intersections of a torus by planes parallel to the axis of the torus. Giovanni Cassini served as a professor of astronomy at Bologna, but in 1669 was invited by Louis XIV to come to Paris, where, in 1671, he became the first astronomer royal of France. Since he became naturalized in France and his second son, Jacques Cassini (1677–1756), was born there, this branch of the Cassini line ceased to be Italian. In 1712, Jacques succeeded his father as astronomer royal, and César-François Cassini, the son of Jacques, succeeded his father as astronomer royal in 1756, and was in turn succeeded by one of his sons, Jacques Dominique Cassini (1748–1845). All these men upheld the family tradition of making contributions to science.

10–7 Some Seventeenth-Century Mathematicians of Germany and the Low Countries

The auspicious progress made in mathematics by Germany during the sixteenth century did not continue in the seventeenth century. The Thirty Years' War (1618–1648) and the subsequent unrest in the Teutonic countries made that century inhospitable to intellectual progress. Kepler and Leibniz stand out as the only first-class German mathematicians of the period, and the only minor German mathematician whom we shall mention here is Ehrenfried Walther von Tschirnhausen (1651–1708). Tschirnhausen devoted much time to mathematics and physics, leaving his mark on the study of curves and the theory of equations. In 1682, he introduced and studied **catacaustic curves,** such a curve being the envelope of light rays, emitted from a point source, after reflection from a given curve. The special sinusoidal spiral, $a = r \cos^3 (\theta/3)$, is known as

Tschirnhausen's cubic. The general sinusoidal spiral, $r^n = a \cos n\theta$, where n is rational, was studied by Colin Maclaurin in 1718 (see Problem Study 10.8). In the theory of equations, Tschirnhausen is particularly known for a transformation that converts an nth degree polynomial equation in x into an nth degree polynomial equation in y in which the coefficients of y^{n-1} and y^{n-2} are both zero. Later, in 1834, G. B. Jerrard found a Tschirnhausen transformation that converts an nth degree polynomial equation in x into an nth degree polynomial equation in y in which the coefficients of y^{n-1}, y^{n-2}, y^{n-3} are all zero. This transformation, as applied to a quintic polynomial, had been given earlier, in 1786, by E. S. Bring and is of importance in the transcendental solution of the quintic equation by means of elliptic functions.

In spite of disturbed times, the geographic region now known as the Low Countries produced a number of lesser mathematicians in the seventeenth century. Willebrord Snell (1580 or 1581–1626) has already been mentioned in connection with his work on the mensuration of the circle. He was an infant prodigy, and it is said that by the age of twelve he had acquainted himself with the standard mathematical works of his time. The name **loxodrome,** for a path on a sphere that makes a constant angle with the meridians, is due to Snell, and he was an early investigator of the properties of polar spherical triangles. The latter were first discussed by Viète.

Albert Girard (1595–1632), who seems to have lived chiefly in Holland, also interested himself in spherical geometry and trigonometry. In 1626, he published a treatise on trigonometry that contains the earliest use of our abbreviations *sin, tan,* and *sec* for sine, tangent, and secant. He gave the expression for the area of a spherical triangle in terms of its spherical excess. Girard was also an algebraist of considerable power. He edited the works of Simon Stevin.

Grégoire de Saint-Vincent (1584–1667) was a prominent circle squarer of the seventeenth century. He applied precalculus methods to various quadrature problems.

Frans van Schooten the Younger (1615–1660 or 1661), a professor of mathematics who edited two Latin editions of Descartes' *La géométrie,* taught mathematics to Huygens, Hudde, and Sluze. He wrote on perspective and edited Viète's works. His father, Frans van Schooten the Elder, and his half brother, Petrus van Schooten, also were professors of mathematics.

Johann Hudde (1633–1704) was a burgomaster of Amsterdam. He wrote on maxima and minima and the theory of equations. In the latter subject, he gave an ingenious rule for finding multiple roots of a polynomial that is equivalent to our present method, in which we find the roots of the highest common factor of the polynomial and its derivative.

René François Walter de Sluze (1622–1685), a canon in the Church, wrote numerous tracts on mathematics. He discussed spirals, points of inflection, and the finding of geometric means. The family of curves, $y^n = k(a - x)^p x^m$, where the exponents are positive integers, are called **pearls of Sluze.**

We conclude with Nicolaus Mercator (ca. 1620–1687), who was born in Holstein, then a part of Denmark, but who spent most of his life in England. He edited Euclid's *Elements* and wrote on trigonometry, astronomy, the computation of logarithms, and cosmography. The series

$$\ln(1+x) = x - \frac{x^2}{2} + \frac{x^3}{3} - \frac{x^4}{4} + \cdots,$$

which was independently discovered by Saint-Vincent, is sometimes referred to as **Mercator's series.** It converges for $-1 < x \leqq 1$ and can be used very satisfactorily for computing logarithms (see Problem Study 10.11). The familiar map of a sphere known as **Mercator's projection,** in which loxodromes appear as straight lines, is not due to Nicolaus Mercator, but to Gerhardus Mercator (1512–1594).

10–8 Some Seventeenth-Century British Mathematicians

Great Britain had its share of lesser mathematicians in the seventeenth century. We have already mentioned William, Viscount Brouncker (1620–1684), elsewhere. He was one of the founders and the first president of the Royal Society of London and maintained relations with Wallis, Fermat, and other leading mathematicians. He wrote on the rectification of the parabola and the cycloid and had no qualms in using infinite series to express quantities that he could not determine otherwise. Thus, he proved that the area bounded by the rectangular hyperbola $xy = 1$, the x-axis, and the ordinates $x = 1$ and $x = 2$ is equal to

$$\frac{1}{(1)(2)} + \frac{1}{(3)(4)} + \frac{1}{(5)(6)} + \cdots$$

and to

$$1 - \frac{1}{2} + \frac{1}{3} - \frac{1}{4} + \cdots.$$

Brouncker was the first British writer to investigate and use properties of continued fractions. We have given, in Section 4–8, his interesting continued fraction development of $4/\pi$.

The Scottish mathematician James Gregory (1638–1675) has also been mentioned elsewhere (Section 4–8). He became successively, in 1668 and 1674, professor of mathematics at St. Andrews and at Edinburgh. He was equally interested in physics and published a work on optics in which he described the reflecting telescope now known by his name. In mathematics, he expanded in infinite series arc tan x, tan x, and arc sec x (1667) and was one of the first to distinguish between convergent and divergent series. He gave an ingenious but unsatisfactory proof that the Euclidean quadrature of the circle is impossible.

The series

$$\text{arc tan } x = x - \frac{x^3}{3} + \frac{x^5}{5} - \frac{x^7}{7} + \ldots,$$

which has played so great a part in calculations of π, is known by his name. He died at an early age, shortly after going blind from the eyestrain induced by his astronomical observations. It is interesting that his nephew, David Gregory (1661–1708), also served as professor of mathematics at Edinburgh, from 1684 to 1691, after which he was appointed Savilian professor of astronomy at Oxford. He, too, was interested in optics, writing on that subject as well as on geometry and the Newtonian theory.

It has been said that, but for London's Great Fire of 1666, Sir Christopher Wren (1632–1723) would have become popularly known as a mathematician instead of an architect. He was Savilian professor of astronomy and taught geometry at Oxford from 1661 to 1673. He was also a founder of the Royal Society and, for a time, president of the Society. He wrote on the laws of collision of bodies, on subjects connected with optics, the resistance of fluids, and other topics in mathematical physics and celestial mechanics. He is credited with the discovery, in 1669, of the two systems of rulings on a hyperboloid of one sheet. He independently showed, in 1658, that an arch of the cycloid is equal in length to eight times the radius of the generating circle. After the great fire, however, Wren took such a prominent part in rebuilding St. Paul's cathedral and some fifty or more other churches and public buildings that his fame as an architect came to overshadow his reputation as a mathematician. When Wren died, he was buried in St. Paul's with the fitting epitaph, *Si monumentum requiris, circumspice* (If you seek a monument, look about you).[4]

Mention should perhaps also be made of Robert Hooke (1635–1703) and Edmund Halley (1656–1742), although these men achieved fame in allied fields rather than in mathematics itself. For almost forty years, Hooke served as professor of geometry at Gresham College. He is known to every student of elementary physics by his law relating the stress and strain of a stretched elastic string. He invented the conical pendulum and attempted to find the law of force (later shown by Newton to be the inverse square law) under which the planets revolve about the sun. He and Huygens both designed watches regulated by a balance spring. Halley succeeded Wallis as Savilian professor of geometry and later became astronomer royal. He restored the lost Book VIII of Apollonius' *Conic Sections* by inference and edited various works of the ancient Greeks, translating some of these from the Arabic even though he did not know a single word of the language. He also compiled a set of mortality tables

[4] It is of interest to Americans that at the College of William and Mary in Williamsburg, Virginia, is the Wren Building, dating from 1695 and attributed in design to Sir Christopher Wren. This is the oldest academic building still in use in America. There is also a large portrait of Wren in a stained glass panel in the William and Mary Law School.

SIR CHRISTOPHER WREN
(David Smith Collection)

of the sort now basic in the life-insurance business. His major original contributions, however, were chiefly in astronomy, and of excellent quality. He was as kind and generous in his dealings with other scholars as Hooke was jealous and irritable. Much of his work was done in the eighteenth century.

Problem Studies

10.1 Geometric Algebra

(a) Given a unit segment and a segment of length x, construct with straightedge and compasses segments of lengths $x^2, x^3, x^4, \ldots$.

(b) Given a unit segment and segments of lengths x, y, z, construct segments of lengths xy and xyz.

(c) Given a unit segment, show that, if $f(x)$ and $g(x)$ are polynomials in x having coefficients represented by given line segments, we may construct a segment of length $y = f(x)/g(x)$ corresponding to any line segment chosen for x.

(d) Given a quadratic equation $x^2 - gx + h = 0$, $g > 0$, $h > 0$. On a line segment of length g as diameter, draw a semicircle C, and then draw a line parallel to the diameter of C at a distance $\sqrt{h}$ from it, cutting C in a point P. From P, drop a perpendicular upon the diameter of C, dividing the diameter into parts r and s. Show that r and s represent the roots of the given quadratic equation. Solve $x^2 - 7x + 12 = 0$ by this method.

(e) Given a quadratic equation $x^2 + gx - h = 0$, $g > 0$, $h > 0$. On a segment of length g as diameter, draw a circle C, and then draw a tangent to C, and mark off on it from the point of contact a length equal to $\sqrt{h}$. From the other extremity of this tangent segment, draw a secant passing through the center of C. Denoting the whole secant by r and its external segment by s, show that $-r$ and s represent the roots of the given quadratic equation. Solve $x^2 + 4x - 21 = 0$ by this method.

10.2 Descartes' "La Géométrie"

(a) Five lines are given, $L_1, \ldots, L_5$, arranged as in Figure 94. Let p_i denote the distance of a point P from line L_i. Taking L_5 and L_4 as x and y axes, find the equation of the locus of a point P moving such that

$$p_1p_2p_3 = ap_4p_5.$$

(The locus is a cubic that Newton called a **Cartesian parabola** and that has also sometimes been called a **trident;** it appears frequently in *La géométrie*.)

(b) Show that with Euclidean tools we may construct as many points as we wish on the locus in (a).

(c) Given any 4 lines L_1, L_2, L_3, L_4, let p_i denote the distance of a point P from line L_i. Show that the locus of P such that $p_1p_2 = kp_3p_4$ is a conic.

(d) Carry through Descartes' method of drawing a tangent at a general point (x_1, y_1) of the parabola $y^2 = 2mx$, and show that it leads to the fact that the subnormal (the projection upon the axis of the parabola of the segment of the normal lying between the curve and the axis) is of constant length, equal to half the latus rectum of the parabola.

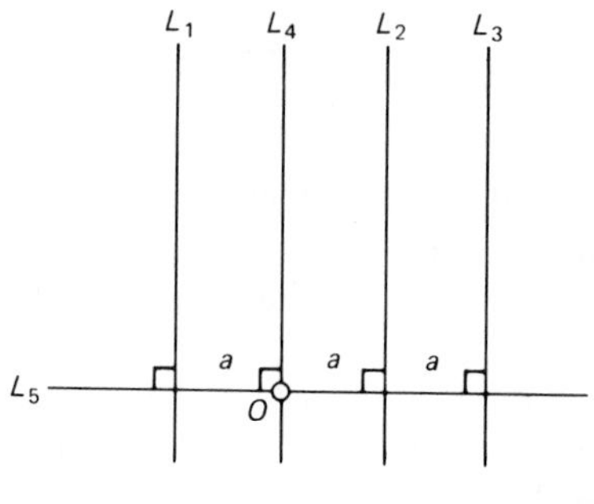

FIGURE 94

10.3 Descartes' Rule of Signs

(a) If $c_1, c_2, \ldots, c_m$ are any m nonzero real numbers, and if 2 consecutive terms of this sequence have opposite signs, we say that these 2 terms present a variation of sign. With this concept, we may state **Descartes' rule of signs,** a proof of which may be found in any textbook on the theory of equations, as follows: *Let* $f(x) = 0$ *be a polynomial equation* *with real coefficients and arranged in descending powers of x. The number of positive roots of the equation is either equal to the number of variations of signs presented by the coefficients of f(x), or less than this number of variations by a positive even number. The number of negative roots is either equal to the number of variations of signs presented by the coefficients of f(−x), or less than this number of variations by a positive even number. A root of multiplicity m is counted as m roots.* Investigate the nature of the roots of the following equations by means of Descartes' rule of signs:

1. $x^9 + 3x^8 - 5x^3 + 4x + 6 = 0$,
2. $2x^7 - 3x^4 - x^3 - 5 = 0$,
3. $3x^4 + 10x^2 + 5x - 4 = 0$.

(b) Show that $x^n - 1 = 0$ has exactly 2 real roots if n is even, and only 1 real root if n is odd.

(c) Show that $x^5 + x^2 + 1 = 0$ has 4 imaginary roots.

(d) Prove that if p and q are real and $q \neq 0$, the equation $x^3 + px + q = 0$ has 2 imaginary roots when p is positive.

(e) Prove that if the roots of a polynomial equation are all positive, the signs of the coefficients are alternately positive and negative.

10.4 Problems from Descartes

(a) Draw the graph of the folium of Descartes,

$$x^3 + y^3 = 3axy.$$

The line $x + y + a = 0$ is an asymptote.

(b) Find the corresponding polar equation of the folium of Descartes.

(c) Set $y = tx$ and obtain a parametric representation of the folium of Descartes in terms of t as parameter. Find the ranges for t leading to the loop, the lower arm, and the upper arm.

(d) Find the Cartesian equation of the folium of Descartes when the node is taken as the origin and the line of symmetry of the curve as the x-axis.

(e) Descartes' solution of a depressed quartic equation employs the method of undetermined coefficients. As an example consider the quartic equation

$$x^4 - 2x^2 + 8x - 3 = 0.$$

Set the left member of the equation equal to the product of 2 quadratic factors of the forms $x^2 + kx + h$ and $x^2 - kx + m$. Obtain 3 relations connecting k, h, m by equating *corresponding* coefficients on the 2 sides of the equation. Eliminate h and m from the 3 relations, obtaining a sextic equation in k that can be regarded as a cubic equation in k^2. Thus, the solution of the original quartic equation is reduced to the solution of an associated cubic equation. Knowing that 1 root of the cubic in k^2 is $k^2 = 4$, obtain the 4 roots of the original quartic equation.

10.5 Fermat's Theorems

About 1760, Euler proposed and solved the problem of determining the number of positive integers less than a given positive integer n and prime to n. This number is now usually denoted by $\phi(n)$ and is called **Euler's ϕ-function** of n (also sometimes called the **indicator,** or **totient,** of n). Thus, if $n = 42$, it is found that the 12 integers 1, 5, 11, 13, 17, 19, 23, 25, 29, 31, 37, and 41 are the only positive integers less than and prime to 42. Therefore $\phi(42) = 12$.

(a) Find $\phi(n)$ for $n = 2, 3, \ldots, 12$. A table giving the values $\phi(n)$ for all $n \leqq 10{,}000$ has been computed by J. W. L. Glaisher (1848–1928).

(b) If p is a prime show that $\phi(p) = p - 1$ and $\phi(p^a) = p^a(1 - 1/p)$.

(c) It can be shown that if $n = ab$, where a and b are relatively prime, then $\phi(n) = \phi(a)\phi(b)$. Using this fact, calculate $\phi(42)$ from the results of (a), and also show that if $n = p_1^{a_1}p_2^{a_2} \ldots p_r^{a_r}$, where $p_1, p_2, \ldots, p_r$ are primes, then

$$\phi(n) = n(1 - 1/p_1)(1 - 1/p_2) \ldots (1 - 1/p_r).$$

Use this last formula to calculate $\phi(360)$.

(d) Euler showed that if a is any positive integer relatively prime to n, then $a^{\phi(n)} - 1$ is divisible by n. Show that the little Fermat theorem is a special case of this.

(e) Show that to establish Fermat's last "theorem," it is sufficient to consider only prime exponents $p > 2$.

(f) Assuming Fermat's last "theorem," show that the curve $x^n + y^n = 1$, where n is a positive integer greater than 2, contains no points with rational coordinates except those points where the curve crosses a coordinate axis.

(g) Assuming item (6) of Section 10–3 (that the area of an integral-sided right triangle cannot be a square number), show that the equation $x^4 - y^4 = z^2$ has no solution in positive integers x, y, z, and then prove Fermat's last "theorem" for the case $n = 4$.

(h) Using Fermat's method of infinite descent, prove that $\sqrt{3}$ is irrational.

10.6 The Problem of the Points

Find the division of the stakes in a game of chance between 2 equally skilled players A and B where

(a) A needs 1 more point to win, and B needs 4 more points to win, using Fermat's method.

(b) A needs 3 more points to win, and B needs 4 more points to win, using Pascal's method.

10.7 Problems from Huygens

(a) A gambler is to win \$300 if a 6 is thrown with a single die. What is his mathematical expectation?

(b) Suppose a gambler is to win \$300 if he throws a 6 with a single die, but \$600 if he throws a 5. What is his mathematical expectation?

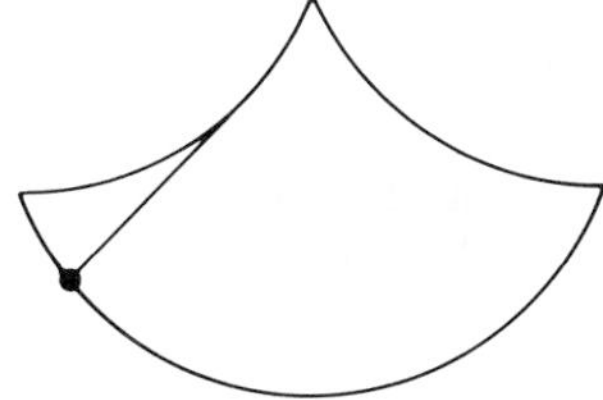

FIGURE 95

Following are some examples of probability problems solved by Huygens:

1. A and B cast alternately with a pair of ordinary dice. A wins if he throws 6 before B throws 7, and B wins if he throws 7 before A throws 6. If A begins, then his chance of winning is to B's chance of winning as 30:31.
2. A and B each take 12 counters and play with 3 dice as follows: if 11 is thrown, A gives a counter to B; if 14 is thrown, B gives a counter to A; and he wins the game who first obtains all the counters. Then A's chance is to B's as 244,140,625:282,429,536,481.
3. A and B play with 2 dice; if 7 is thrown, A wins; if 10 is thrown, B wins; if any other number is thrown, the game is drawn. Then A's chance of winning is to B's as 13:11.

(c) Using the isochronous property of the cycloid and the fact that the **evolute** of a cycloid is another cycloid of the same size, show that a pendulum constrained to swing between two successive arches of an inverted cycloid (see Figure 95) must oscillate with a constant period.

(d) A ball swings uniformly in a circle at the end of a string, making 1 revolution per minute. If the length of the string is doubled and the period of revolution halved, how does the centrifugal force compare with that of the first situation?

10.8 Higher Plane Curves

(a) Taking the foci of a Cassinian curve at the points $(-a,0)$ and $(a,0)$ on a rectangular Cartesian frame of reference and denoting the constant product of distances by k^2, find the Cartesian equation of the curve.

(b) Show that the corresponding polar equation of the curve is

$$r^4 - 2r^2a^2 \cos 2\theta + a^4 = k^4.$$

Note that if $k = a$, the curve becomes the **lemniscate of Bernoulli,**[5]

$$r^2 = 2a^2 \cos 2\theta.$$

---

[5] So named (from a Greek word meaning "ribbon") by Jakob Bernoulli (1654–1705) in 1694. Its principal properties were found in 1750 by the Italian Count G. C. Fagnano (1682–1766), who also showed that its rectification leads to elliptic integrals.

(c) Show that the lemniscate of Bernoulli is the cissoid (see Problem Study 4.4) of a circle of radius $a/2$, and itself, for a pole O distant $a\sqrt{2}/2$ units from its center.

(d) Plot carefully a rectangular hyperbola $xy = k^2$ and draw several members of the family of circles with centers on the hyperbola and passing through the origin. The envelope of this family of circles is a lemniscate of Bernoulli.

(e) Using the fact that the normal at a point of the lemniscate of Bernoulli in (b) makes an angle 2θ with the radius vector to the point, show how we may construct tangents to the lemniscate.

(f) Show that we have the following special cases of the sinusoidal spiral, $r^n = a \cos n\theta$, where n is a rational number.

n	Curve
-2	Rectangular hyperbola
-1	Straight line
$-\frac{1}{2}$	Parabola
$-\frac{1}{3}$	Tschirnhausen cubic
$\frac{1}{2}$	Cardioid
1	Circle
2	Lemniscate of Bernoulli

(g) An **epicycloid** is the path traced by a point on a circle rolling externally upon a fixed base circle. The catacaustic of a circle for a light source at infinity is an epicycloid of two cusps whose base circle is concentric with the given circle and whose radius is half the radius of the given circle. An epicycloid of two cusps is called a **nephroid.** The catacaustic of a circle for a light source on the circumference of the circle is an epicycloid of one cusp whose base circle is concentric with the given circle and whose radius is one-third the radius of the given circle. An epicycloid of one cusp is a **cardioid.** Jakob Bernoulli showed, in 1692, that the catacaustic of a cardioid, when the light source is at the cusp of the cardioid, is a nephroid. Catacaustics of a circle can be seen as the bright curves on the surface of coffee in a cup or upon the table inside a circular napkin ring. Observe some catacaustics of a circle using a cup of liquid and a movable light source.

10.9 Recreational Problems from Bachet

Following are some arithmetic recreations found in Bachet's *Problèmes plaisants et délectables*. They, and other problems from Bachet, can also be found in Ball-Coxeter, *Mathematical Recreations and Essays*.

(a) (1) Ask a person to choose secretly a number, and then to treble it. (2) Inquire if the product is even or odd. If it is even, ask him to take half of it; if it is odd, ask him to add 1 and then take half of it. (3) Tell him to multiply the result in (2) by 3 and to tell you how many times, say n, 9 will divide integrally into the product. (4) Then the number originally chosen was $2n$ or $2n + 1$, according as the result in step (1) was even or odd. Prove this.

(b) Ask a person to choose secretly a number less than 60, and to announce the remainders, say a, b, c, when the number is divided by 3, by 4, and by 5. Then the number originally chosen can be found as the remainder obtained by dividing $40a + 45b + 36c$ by 60. Prove this.

(c) Tell A to choose secretly any number, greater than 5, of counters, and B to take 3 times as many. Ask A to give 5 counters to B, and then ask B to transfer to A 3 times as many counters as A has left. You may now tell B that he has 20 counters. Explain why this is so and generalize to the case where the 3 and 5 are replaced by p and q.

(d) A secretly selects either a pair of numbers, 1 of which is odd and the other even, and the other number is given to B. Ask A to double his number, and B to triple his. Request the sum of the 2 products. If the sum is even, then A selected the odd number; otherwise, A selected the even number. Explain this.

(e) Ask someone to think of an hour, say m, and then to touch on a watch the number that marks some other hour, say n. If, beginning with the number touched, he taps successively in the counterclockwise direction the numbers on the watch, meanwhile mentally counting the taps as m, $m + 1$, and so on, until he reaches the number $n + 12$, then the last number tapped will be that of the hour he originally thought of. Prove this.

10.10 Some Geometry

(a) Show, by Roberval's method, that the tangent and normal at a point on a central conic bisect the angles between the 2 focal radii drawn to the point.

(b) A **spherical degree** is defined to be any spherical area that is equivalent to (1/720)th of the entire surface of the sphere. Show that the area of a lune whose angle is $n°$ is equal to $2n$ spherical degrees.

(c) Show that the area of a spherical triangle, in spherical degrees, is equal to the spherical excess of the triangle.

(d) Show that the area A of a spherical triangle of spherical excess E is given by

$$A = \frac{\pi r^2 E}{180°},$$

where r is the radius of the sphere.

(e) Find the area of a trirectangular triangle on a sphere whose diameter is 28 inches.

(f) Show (by the integral calculus) that the area bounded by the hyperbola $xy = 6$, the ordinate $x = 2$, and the x-axis is infinite. On the other hand, show that the volume obtained by revolving the area about the x-axis is finite.

This problem gives rise to the following *paint paradox*. Since the above area is infinite, an infinite amount of paint is needed to paint the area. Because the above volume is finite, however, it requires only a finite amount of paint to fill the volume. But the volume contains the concerned area within itself. Explain this paradox.

10.11 Computation of Logarithms by Series

The Mercator series

$$\ln(1 + x) = x - \frac{x^2}{2} + \frac{x^3}{3} - \frac{x^4}{4} + \cdots$$

converges for $-1 < x \leqq 1$. Replacing x by $-x$, it follows that the series

$$\ln(1 - x) = -x - \frac{x^2}{2} - \frac{x^3}{3} - \frac{x^4}{4} - \cdots$$

must converge for $-1 \leqq x < 1$. Since a series whose terms are the differences of the corresponding terms of 2 given series certainly converges for all values of x for which both of the given series converge, it follows that, for $-1 < x < 1$,

$$\ln\left(\frac{1 + x}{1 - x}\right) = \ln(1 + x) - \ln(1 - x)$$
$$= 2\left(x + \frac{x^3}{3} + \frac{x^5}{5} + \frac{x^7}{7} + \cdots\right).$$

If we set $x = 1/(2N + 1)$, we observe that $-1 < x < 1$ for all positive N, and $(1 + x)/(1 - x) = (N + 1)/N$. Substituting in the last equation, we find

$$\ln(N + 1) = \ln N + 2\left[\frac{1}{2N + 1} + \frac{1}{3(2N + 1)^3} + \frac{1}{5(2N + 1)^5} + \cdots\right],$$

the series converging, and rather rapidly, for all positive N.

(a) By setting $N = 1$, compute ln 2 to 4 decimal places.
(b) Compute ln 3 to 4 decimal places.
(c) Compute ln 4 to 4 decimal places.

Essay Topics

10/1 The transform-solve-invert technique.

10/2 Analytic geometry as the example *par excellence* of the transform-solve-invert technique.

10/3 Analytic geometry as a method of discovery.

10/4 Who invented analytic geometry?

10/5 The greatest French mathematician of the seventeenth century.
10/6 The five most important French mathematicians of the seventeenth century.
10/7 Fermat's method of infinite descent.
10/8 The origin of mathematical probability.
10/9 The anomaly of an infinite area generating a finite volume of revolution.
10/10 The Roberval-Torricelli method of drawing tangents.
10/11 Huygens' perfect seconds pendulum.
10/12 The isogonic center of a triangle.
10/13 Theorems of Ceva and Commandino.
10/14 The first two higher plane curves to receive practical application.
10/15 The remarkable Cassini family.
10/16 Modern electronic computers and new findings in number theory.

Bibliography

ADAMS, O. S. *A Study of Map Projections in General*. Washington: Coast and Geodetic Survey, Special Publication No. 60, Department of Commerce, 1919.

———, and C. H. DEETZ *Elements of Map Projection, With Applications to Map and Chart Construction*. Washington: Coast and Geodetic Survey, Special Publication No. 68, Department of Commerce, 1938.

ARCHIBALD, R. C. *Mathematical Table Makers*. New York: *Scripta Mathematica*, Yeshiva University, 1948.

AUBREY, JOHN *Brief Lives*. Edited by Richard Berber, B & N Imports, 1983.

BALL, W. W. R., and H. S. M. COXETER *Mathematical Recreations and Essays*. 12th ed. Toronto: University of Toronto Press, 1974. Reprinted by Dover, New York.

BELL, A. E. *Christian* [sic] *Huygens and the Development of Science in the Seventeenth Century*. London: Edward Arnold, 1948.

BELL, E. T. *Men of Mathematics*. New York: Simon and Schuster, 1937.

——— *The Last Problem*. Washington, D.C.: Math. Assoc. of Am., 1991

BOYER, C. B. *History of Analytic Geometry*. New York: *Scripta Mathematica*, Yeshiva University, 1956.

——— *The History of the Calculus and Its Conceptual Development*. New York: Dover, 1959.

CONKWRIGHT, N. B. *Introduction to the Theory of Equations*. Boston: Ginn, 1941.

COOLIDGE, J. L. *A History of Geometrical Methods*. New York: Oxford University Press, 1940.

——— *A History of the Conic Sections and Quadric Surfaces*. New York: Oxford University Press, 1947.

——— *The Mathematics of Great Amateurs*. New York: Oxford University Press, 1949.

COURANT, RICHARD, and HERBERT ROBBINS *What is Mathematics?* New York: Oxford University Press, 1941.

DAVID, F. N. *Games, Gods and Gambling*. New York: Hafner, 1962.

DESCARTES, RENÉ *The Geometry of René Descartes*. Translated by D. E. Smith and Marcia L. Latham. New York: Dover, 1954.

HACKER, S. G. *Arithmetical View Points*. Pullman, Wash.: Mimeographed at Washington State College, 1948.

HACKING, IAN *The Emergence of Probability*. London: Cambridge University Press, 1975.

HALDANE, ELIZABETH S. *Descartes: His Life and Times*. New York: E. P. Dutton, 1905.

JOHNSON, R. A. *Modern Geometry*. Boston: Houghton Mifflin, 1929. Reprinted by Dover, New York.

KRAITCHIK, MAURICE *Mathematical Recreations*. New York: W. W. Norton, 1942.

MAHONEY, M. S. *The Mathematical Career of Pierre de Fermat, 1601–1665*. Princeton, N.J.: Princeton University Press, 1972.

MAISTROV, L. E. *Probability: A Historical Sketch*. New York: Academic Press, 1974.

MERRIMAN, MANSFIELD *The Solution of Equations*. 4th ed. New York: John Wiley, 1906.

MILLER, G. A. *Historical Introduction to Mathematical Literature*. New York: Macmillan, 1916.

MUIR, JANE *Of Men and Numbers, The Story of the Great Mathematicians*. New York: Dodd, Mead, 1961.

ORE, OYSTEIN *Number Theory and Its History*. New York: McGraw-Hill, 1948.

SMITH, D. E. *History of Modern Mathematics*. 4th ed. New York: John Wiley, 1906.

——— *A Source Book in Mathematics*. New York: McGraw-Hill, 1929.

SULLIVAN, J. W. N. *The History of Mathematics in Europe, From the Fall of Greek Science to the Rise of the Conception of Mathematical Rigour*. New York: Oxford University Press, 1925.

SUMMERSON, JOHN *Sir Christopher Wren*. No. 9 in a series of *Brief Lives*. New York: Macmillan 1953.

TODHUNTER, ISAAC *A History of the Mathematical Theory of Probability from the Time of Pascal to that of Laplace*. New York: Chelsea, 1949.

TURNBULL, H. W. *The Great Mathematicians*. New York: New York University Press, 1961.

VEITCH, JOHN *The Method, Meditation and Philosophy of René Descartes*. New York: Tudor, 1901.

VROOMAN, JACK *René Descartes. A Biography*. New York: C.P. Putnam's Sons, 1970.

WILLIAMSON, BENJAMIN *An Elementary Treatise on the Differential Calculus*. London: Longmans, Green, 1899.

WINGER, R. M. *An Introduction to Projective Geometry*. Boston: D. C. Heath, 1923.

YATES, R. C. *A Handbook on Curves and Their Properties*. Ann Arbor, Mich.: J. W. Edwards, 1947.

Chapter 11

THE CALCULUS AND RELATED CONCEPTS

11–1 Introduction

We have seen that many new and extensive fields of mathematical investigation were opened up in the seventeenth century, making that era outstandingly productive in the development of mathematics. Unquestionably, the most remarkable mathematical achievement of the period was the invention of the calculus, toward the end of the century, by Isaac Newton and Gottfried Wilhelm Leibniz. With this invention, creative mathematics passed to an advanced level, and the history of elementary mathematics essentially terminated. This chapter is devoted to a brief account of the origins and development of the important concepts of the calculus, concepts that are so far reaching and that have exercised such an impact on the modern world that it is perhaps correct to say that without some knowledge of them a person today can scarcely claim to be well educated.

It is interesting that, contrary to the customary order of presentation found in our beginning college courses, where we start with differentiation and later consider integration, the ideas of the integral calculus developed historically before those of the differential calculus. The idea of integration first arose in its role of a summation process in connection with the finding of certain areas, volumes, and arc lengths. Some time later, differentiation was created in connection with problems on tangents to curves and with questions about maxima and minima of functions. And still later it was observed that integration and differentiation are related to each other as inverse operations.

Although the major part of our story lies in the seventeenth century, we must, for the beginning, go back to ancient Greece and the fifth century B.C.

11–2 Zeno's Paradoxes

Should we assume that a magnitude is infinitely divisible or that it is made up of a very large number of small indivisible atomic parts? The first assumption appears the more reasonable to most of us, but the utility of the second assumption in the making of discoveries causes it to lose some of its seeming absurdity. There is evidence that, in Greek antiquity, schools of mathematic reasoning developed that employed each of the above two assumptions.

Some of the logical difficulties encountered in either assumption were strikingly brought out in the fifth century B.C. by some paradoxes devised by the Eleatic philosopher Zeno (ca. 450 B.C.). These paradoxes, which have had a profound influence on mathematics, assert that motion is impossible whether we assume a magnitude to be infinitely divisible or to be made up of a large number of atomic parts. We illustrate the nature of the paradoxes by the following two.

The Dichotomy: If a straight line segment is infinitely divisible, then motion is impossible, for in order to traverse the line segment it is necessary first to reach the midpoint, and to do this one must first reach the one-quarter point, and to do this one must first reach the one-eighth point, and so on, ad infinitum. It follows that the motion can never even begin.

The Arrow: If time is made up of indivisible atomic instants, then a moving arrow is always at rest, for at any instant the arrow is in a fixed position. Since this is true of every instant, it follows that the arrow never moves.

Many explanations of Zeno's paradoxes have been given, and it is not difficult to show that they challenge the common intuitive beliefs that the sum of an infinite number of positive quantities is infinitely large, even if each quantity is extremely small ($\Sigma_{i=1}^{\infty} \varepsilon_i = \infty$), and that the sum of either a finite or an infinite number of quantities of dimension zero is zero ($n \times 0 = 0$ and $\infty \times 0 = 0$). Whatever might have been the intended motive of the paradoxes, their effect was to exclude infinitesimals from Greek demonstrative geometry.[1]

11–3 Eudoxus' Method of Exhaustion

The first problems occurring in the history of the calculus were concerned with the computation of areas, volumes, and lengths of arcs. In their treatment, one finds evidence of the two assumptions about the divisibility of magnitudes that we considered above.

One of the earliest important contributions to the problem of squaring the circle was that of Antiphon the Sophist (ca. 430 B.C.), a comtemporary of Socrates. Antiphon, we are told, advanced the idea that by successively doubling the number of sides of a regular polygon inscribed in a circle, the difference in area between the circle and the polygon would at last be exhausted. Because a square can be constructed equal in area to any given polygon, it will then be possible to construct a square equal to the circle. This argument met immediate criticism on the grounds that it violated the principle that magnitudes are divisible without limit, and that, accordingly, Antiphon's process could never use up the whole area of the circle. Nevertheless, Antiphon's bold pronouncement contained the germ of the famous Greek **method of exhaustion.**

The method of exhaustion is usually credited to Eudoxus (ca. 370 B.C.) and can perhaps be considered as the Platonic school's answer to the paradoxes of Zeno. The method assumes the infinite divisibility of magnitudes and

[1] For an excellent informative historical treatment of Zeno's paradoxes, see Florian Cajori, "History of Zeno's Arguments on Motion," *The American Mathematical Monthly* 22 (1915): 1–6, 39–47, 77–82, 109–115, 145–149, 179–186, 215–220, 253–258, 292–297.

has, as a basis, the proposition: *If from any magnitude there be subtracted a part not less than its half, from the remainder another part not less than its half, and so on, there will at length remain a magnitude less than any preassigned magnitude of the same kind.* Let us employ the method of exhaustion to prove that if A_1 and A_2 are the areas of two circles having diameters d_1 and d_2, then

$$A_1:A_2 = d_1^2:d_2^2.$$

We first show, with the aid of the basic proposition, that the difference in area between a circle and an inscribed regular polygon can be made as small as desired. Let AB, in Figure 96, be a side of a regular inscribed polygon, and let M be the midpoint of the arc AB. Since the area of triangle AMB is half that of the rectangle $ARSB$, and hence greater than half the area of the circular segment AMB, it follows that by doubling the number of sides of the inscribed regular polygon, we increase the area of the polygon by more than half the difference in area between the polygon and the circle. Consequently, by doubling the number of sides sufficiently often, we can make the difference in area between the polygon and the circle less than any assigned area, however small.

We now return to our theorem. Suppose that instead of equality we have

$$A_1:A_2 > d_1^2:d_2^2.$$

Then we can inscribe in the first circle a regular polygon whose area P_1 differs so little from A_1 that

$$P_1:A_2 > d_1^2:d_2^2.$$

Let P_2 be a regular polygon similar to P_1, but inscribed in the second circle. Then, from a known theorem about similar regular polygons,

$$P_1:P_2 = d_1^2:d_2^2.$$

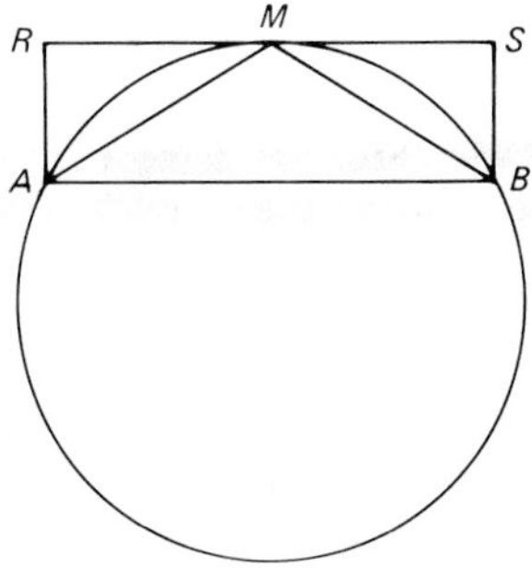

FIGURE 96

It follows that $P_1:A_2 > P_1:P_2$, or $P_2 > A_2$, an absurdity, because the area of a regular polygon cannot exceed the area of its circumcircle. In a similar way, we can show that we cannot have

$$A_1:A_2 < d_1^2:d_2^2.$$

Consequently, by this double *reductio ad absurdum* process, our theorem is established. Thus, if A is the area and d the diameter of a circle, $A = kd^2$, where k is a constant (actually $\pi/4$), which is the same for all circles.

Archimedes claimed that Democritus (ca. 410 B.C.) stated that the volume of a pyramid on any polygonal base is one-third that of a prism with the same base and altitude. Very little is known of Democritus, but he could hardly have given a rigorous demonstration of this theorem. Since a prism can be dissected into a sum of prisms all having triangular bases, and, in turn, a prism of this latter sort can be dissected into three triangular pyramids having, in pairs, equivalent bases and equal altitudes, it follows that the crux of Democritus' problem is to show that two pyramids of the same height and equivalent bases have equal volumes. A demonstration of that was later furnished by Eudoxus, using the method of exhaustion.

How, then, might Democritus have arrived at this last result? A clue is furnished by Plutarch, who quotes a dilemma encountered by Democritus on an occasion when he considered a cone as made up of infinitely many plane cross sections parallel to the base. If two "adjacent" sections are of the same size, the solid would be a cylinder and not a cone. On the other hand, if two "adjacent" sections are different in area, the surface of the solid would be broken into a series of small steps, which certainly is not the case. Here we have an assumption concerning the divisibility of magnitudes that is somewhat intermediate to the two assumptions already considered, for here we assume the volume of the cone is infinitely divisible (namely, into an infinite number of plane atomic sections), but that these sections are countable in the sense that, given one of them, there is one next to it. Now Democritus may have argued that if two pyramids with equivalent bases and equal heights are cut by planes parallel to the bases and dividing the heights in the same ratio, then the corresponding sections formed are equivalent. Therefore, the pyramids contain the same infinite number of equivalent plane sections and hence, must be equal in volume. This would be an early instance of Cavalieri's *method of indivisibles,* considered below in Section 11–6.

Of the ancients, it was Archimedes who made the most elegant applications of the method of exhaustion and who came the nearest to actual integration. As one of his earliest examples, consider his quadrature of a parabolic segment. Let C, D, E be points on the arc of the parabolic segment (see Figure 97) obtained by drawing LC, MD, NE parallel to the axis of the parabola through the midpoints L, M, N of AB, CA, CB. From the geometry of the parabola, Archimedes shows that

$$\Delta CDA + \Delta CEB = \frac{\Delta ACB}{4}.$$

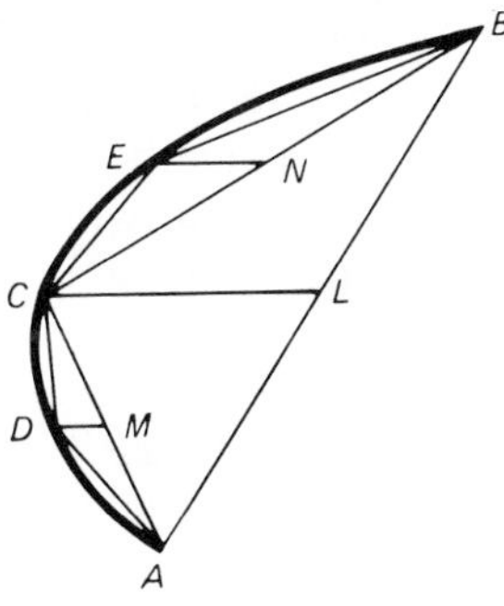

FIGURE 97

By repeated applications of this idea, it follows that the area of the parabolic segment is given by

$$\begin{aligned}\Delta ABC &+ \frac{\Delta ABC}{4} + \frac{\Delta ABC}{4^2} + \frac{\Delta ABC}{4^3} + \cdots \\ &= \Delta ABC\left(1 + \frac{1}{4} + \frac{1}{4^2} + \frac{1}{4^3} + \cdots\right) \\ &= \frac{4}{3}\,\Delta ABC.\end{aligned}$$

Here we have shortened the work by taking the limit of the sum of a geometric progression; Archimedes employs the double *reductio ad absurdum* apparatus of the method of exhaustion.

In his treatment of certain areas and volumes, Archimedes arrived at equivalents of a number of definite integrals found in our elementary calculus textbooks.

11–4 Archimedes' Method of Equilibrium

The method of exhaustion is a rigorous but sterile method. In other words, once a formula is known, the method of exhaustion may furnish an elegant tool for establishing it, but the method does not lend itself to the initial discovery of the result. The method of exhaustion is, in this respect, very much like the process of mathematical induction. How, then, did Archimedes discover the formulas that he so neatly established by the method of exhaustion?

This question was finally answered in 1906, with the discovery by Heiberg, in Constantinople, of a copy of Archimedes' long-lost treatise *Method,* addressed to Eratosthenes. The manuscript was found on a palimpsest (see Section 1–8); that is, it had been written in the tenth century on parchment, and then later, in the thirteenth century, washed off and the parchment reused for a religious text. Fortunately, most of the first text was able to be restored from beneath the later writing.

The fundamental idea of Archimedes' method is this. To find a required area or volume, cut it up into a very large number of thin parallel plane strips, or thin parallel layers, and (mentally) hang these pieces at one end of a given lever in such a way as to be in equilibrium with a figure whose content and centroid are known. Let us illustrate the method by using it to discover the formula for the volume of a sphere.

Let r be the radius of the sphere. Place the sphere with its polar diameter along a horizontal x-axis with the north pole N at the origin (see Figure 98). Construct the cylinder and the cone of revolution obtained by rotating the rectangle $NABS$ and the triangle NCS about the x-axis. Now cut from the three solids thin vertical slices (assuming that they are flat cylinders) at distance x from N and of thickness Δx. The volumes of these slices are, approximately,

$$\text{sphere: } \pi x(2r - x)\,\Delta x$$

$$\text{cylinder: } \pi r^2\,\Delta x$$

$$\text{cone: } \pi x^2\,\Delta x.$$

Let us hang at T the slices from the sphere and the cone, where $TN = 2r$. Their combined moment[2] about N is

$$[\pi x(2r - x)\,\Delta x + \pi x^2\,\Delta x]2r = 4\pi r^2 x\,\Delta x.$$

This, we observe, is four times the moment of the slice cut from the cylinder when that slice is left where it is. Adding a large number of these slices together, we find

$$2r[\text{volume of sphere} + \text{volume of cone}] = 4r[\text{volume of cylinder}],$$

or

$$2r\left[\text{volume of sphere} + \frac{8\pi r^3}{3}\right] = 8\pi r^4,$$

or

$$\text{volume of sphere} = \frac{4\pi r^3}{3}.$$

This, we are told in the *Method*, was Archimedes' way of discovering the formula for the volume of a sphere. His mathematical conscience would not permit him to accept such a method as a proof, however, and he accordingly

[2] The **moment** of a volume about a point is the product of the volume and the distance from the point to the centroid of the volume.

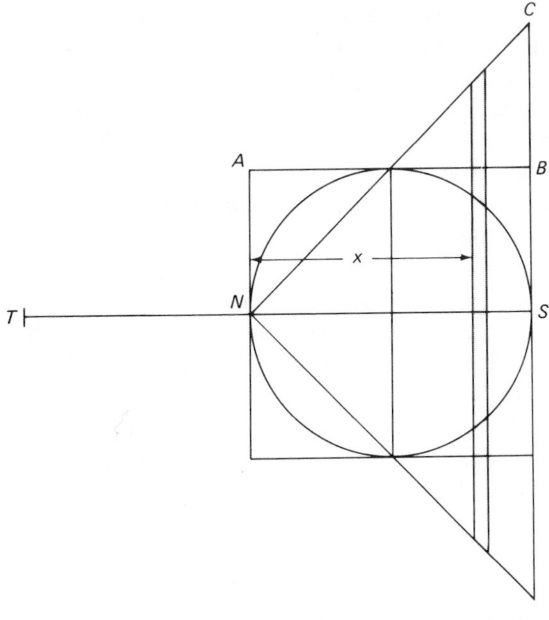

FIGURE 98

supplied a rigorous demonstration by means of the method of exhaustion. In the method of equilibrium, we see the fertility of the loosely founded idea of regarding a magnitude as composed of a large number of atomic pieces. Needless to say, with the modern method of limits, Archimedes' method of equilibrium can be made perfectly rigorous and becomes essentially the same as present-day integration.

11–5 The Beginnings of Integration in Western Europe

The theory of integration received very little stimulus after Archimedes' remarkable achievements until relatively modern times. It was about 1450 that Archimedes' works reached western Europe through a translation of a ninth-century copy of his manuscripts found at Constantinople. This translation was revised by Regiomontanus and was printed in 1540. A few years later, a second translation appeared. It was not until about the beginning of the seventeenth century, however, that we find Archimedes' ideas receiving further development.

Two early writers of modern times who used methods comparable to those of Archimedes were the Flemish engineer Simon Stevin (1548–1620) and the Italian mathematician Luca Valerio (ca. 1552–1618). Each of these men tried to avoid the double *reductio ad absurdum* of the method of exhaustion by making a direct passage to the limit, much as we did toward the end of Section 11–3 in our treatment of the area of a parabolic segment. Stevin used such a method in his work on hydrostatics, where he found the force due to fluid pressure against a vertical rectangular dam by dividing the dam into thin horizontal strips and then rotating these strips about their upper and lower edges until they became parallel to a horizontal plane. This is fundamentally the method we use today in our elementary textbooks on calculus.

Of the early modern Europeans who developed ideas of infinitesimals in connection with integration, particular mention must be made of Johann Kepler. We have already remarked (in Section 9–7) that Kepler had to resort to an integration procedure in order to compute the areas involved in his second law of planetary motion and also the volumes dealt with in his treatise on capacities of wine barrels. But Kepler, like others of the time, had little patience with the careful rigor of the method of exhaustion and, under the temptation to save time and trouble, freely adopted processes that Archimedes considered as merely heuristic. Thus, Kepler regarded the circumference of a circle as a regular polygon possessing an infinite number of sides. If each of these sides is taken as the base of a triangle whose vertex is at the center of the circle, then the area of the circle is divided into an infinite number of thin triangles, all having an altitude equal to the radius of the circle. Since the area of each thin triangle is equal to half the product of its base and altitude, it turns out that the area of a circle is equal to half the product of its circumference and radius. Similarly, the volume of a sphere was regarded as composed of an infinite number of thin pyramids having a common vertex at the center of the sphere. It follows that the volume of a sphere is one-third the product of its surface area and radius. Objectionable as such methods are from the standpoint of mathematical rigor, they produce correct results in a very simple manner. Even well into the twentieth century, such "atomic" methods were used quite regularly by physicists and engineers for setting up a mathematical problem, leaving the rigorous "limit" treatment to the professional mathematician.[3] Geometers frequently resorted to the convenient concept of "consecutive" points and "consecutive" curves and surfaces in a one-parameter family of such entities.[4]

11–6 Cavalieri's Method of Indivisibles

Bonaventura Cavalieri was born in Milan in 1598, became a Jesuat (not a Jesuit, as is frequently incorrectly stated] at the age of fifteen, studied under Galileo,

---

[3] "Thus, so far as first differentials are concerned, a small part of a curve may be treated as straight and a part of a surface near a point as plane; during a short time dt, a particle may be considered as moving with constant speed and a physical process as occurring at a constant rate." H. B. Phillips, *Differential Equations*, 3rd ed., p. 28.

[4] "In other words, *the characteristic of a surface* [of a one-parameter family of surfaces] *is the curve in which a consecutive surface intersects it*." E. P. Lane, *Metric Differential Geometry of Curves and Surfaces*, p. 81.

BONAVENTURA CAVALIERI
(David Smith Collection)

and served as a professor of mathematics at the University of Bologna from 1629 until his death in 1647 at the age of forty-nine. He was one of the most influential mathematicians of his time and wrote a number of works on mathematics, optics, and astronomy. He was largely responsible for the early introduction of logarithms into Italy. But his greatest contribution to mathematics was a treatise, *Geometria indivisibilibus,* published in its first form in 1635, devoted to the precalculus **method of indivisibles.** Although the method can be traced back to Democritus (ca. 410 B.C.) and Archimedes (ca. 287–212 B.C.), very likely it was Kepler's attempts to find certain areas and volumes that directly motivated Cavalieri.

Cavalieri's treatise is verbose and not clearly written, and it is difficult to know precisely what is to be understood by an "indivisible." It seems that an indivisible of a given planar piece is a chord of that piece, and an indivisible of a given solid is a plane section of that solid. A planar piece is considered as made up of an infinite set of parallel chords and a solid as made up of an infinite set of parallel plane sections. Now, Cavalieri argued, if we slide each member of the set of parallel chords of some given planar piece along its own axis, so that the end points of the chords still trace a continuous boundary, then the area of the new planar piece so formed is the same as that of the original planar piece, inasmuch as the two pieces are made up of the same chords. A similar sliding of the members of a set of parallel planar sections of a given solid will yield another solid having the same volume as the original one. (This last result can be strikingly illustrated by taking a vertical stack of cards and then pushing the sides of the stack into curved surfaces; the volume of the disarranged stack is the same as that of the original stack.) These results, slightly generalized, give the so-called *Cavalieri principles:*

1. *If two planar pieces are included between a pair of parallel lines, and if the lengths of the two segments cut by them on any line parallel to the including lines are always in a given ratio, then the areas of the two planar pieces are also in this ratio.*
2. *If two solids are included between a pair of parallel planes, and if the areas of the two sections cut by them on any plane parallel to the including planes are always in a given ratio, then the volumes of the two solids are also in this ratio.*

Cavalieri's principles constitute a valuable tool in the computation of areas and volumes, and their intuitive bases can easily be made rigorous with the modern integral calculus. Accepting these principles as intuitively apparent, one can solve many problems in mensuration that normally require the more advanced techniques of the calculus.

Let us illustrate the use of Cavalieri's principles, first employing the planar case to find the area of an ellipse of semiaxes a and b, and then the solid case to find the volume of a sphere of radius r.

Consider the ellipse and circle

$$\frac{x^2}{a^2} + \frac{y^2}{b^2} = 1,\ a > b, \text{ and } x^2 + y^2 = a^2,$$

plotted on the same rectangular coordinate frame of reference, as shown in Figure 99. Solving each of the equations above for y, we find, respectively,

$$y = \frac{b}{a}(a^2 - x^2)^{1/2},\ y = (a^2 - x^2)^{1/2}.$$

It follows that corresponding ordinates of the ellipse and the circle are in the ratio b/a. It then follows that corresponding vertical chords of the ellipse and the circle are also in this ratio and, by Cavalieri's first principle, so are the areas of the ellipse and the circle. We conclude that

$$\begin{aligned}\text{area of ellipse} &= \frac{b}{a}(\text{area of circle}) \\ &= \frac{b}{a}(\pi a^2) = \pi ab.\end{aligned}$$

This is basically the procedure Kepler employed in finding the area of an ellipse of semiaxes a and b.

Now let us find the familiar formula for the volume of a sphere of radius r. In Figure 100, we have a hemisphere of radius r on the left, and on the right a circular cylinder of radius r and altitude r with a cone removed whose base is the upper base of the cylinder and whose vertex is the center of the lower base of the cylinder. The hemisphere and the gouged-out cylinder are resting on a common plane. We now cut both solids by a plane parallel to the base plane and

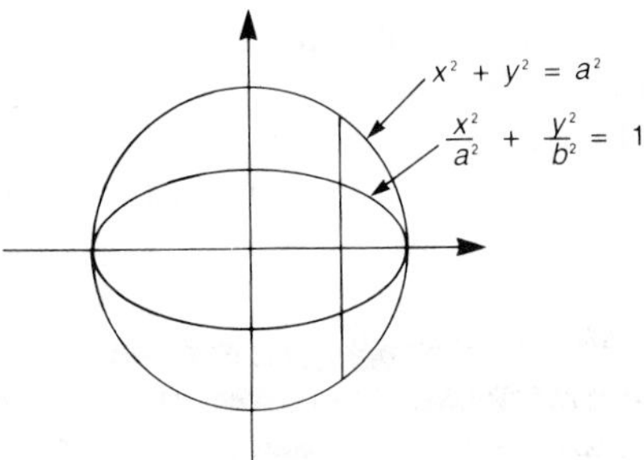

FIGURE 99

at distance h from it. This plane cuts the one solid in a circular section and the other in an annular, or ring-shaped, section. By elementary geometry, we easily show that each of the two sections has an area equal to $\pi(r^2 - h^2)$. It follows, by Cavalieri's principle, that the two solids have equal volumes. Therefore, the volume V of a sphere is given by

$$\begin{aligned} V &= 2(\text{volume of cylinder} - \text{volume of cone}) \\ &= 2\left(\pi r^3 - \frac{\pi r^3}{3}\right) = \frac{4\pi r^3}{3}. \end{aligned}$$

The assumption and then consistent use of Cavalieri's second principle can greatly simplify the derivation of many of the volume formulas encountered in a beginning treatment of solid geometry. This procedure has been adopted by a number of textbook writers and has been advocated on pedagogic grounds. In deriving the familiar formula for the volume of a tetrahedron ($V = Bh/3$), for example, the sticky part is first to show that any two tetrahedra having equivalent bases and equal altitudes on those bases have equal volumes. The inherent difficulty here is reflected in all treatments of solid geometry from Euclid's *Elements* on. With Cavalieri's second principle, however, the difficulty simply disappears.

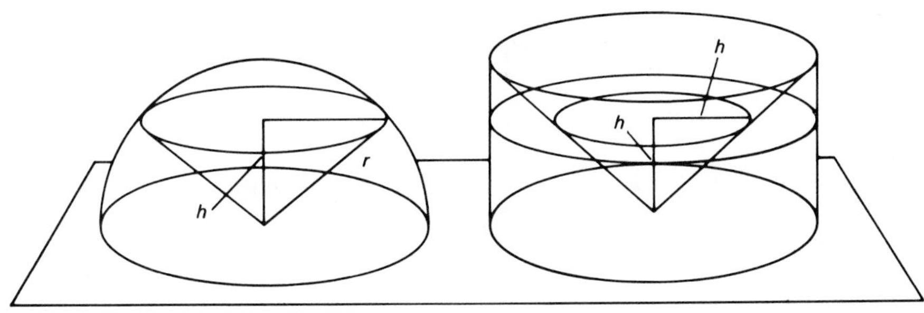

FIGURE 100

Cavalieri's hazy conception of indivisibles, as sort of atomic parts of a figure, led to much discussion and serious criticism by some students of the subject, particularly by the Swiss goldsmith and mathematician Paul Guldin (1577–1642). Cavalieri recast his treatment in the vain hope of meeting these objections. The French mathematician Roberval ably handled the method and claimed to be an independent inventor of the conception. The method of indivisibles, or some process very similar to it, was effectively used by Torricelli, Fermat, Pascal, Saint-Vincent, Barrow, and others. In the course of the work of these men, results were reached that are equivalent to the integration of expressions such as x^n, $\sin \theta$, $\sin^2 \theta$, and $\theta \sin \theta$.

11–7 The Beginning of Differentiation

Differentiation may be said to have originated in the problem of drawing tangents to curves and in finding maximum and minimum values of functions. Although such considerations go back to the ancient Greeks, it seems fair to assert that the first really marked anticipation of the method of differentiation stems from ideas set forth by Fermat in 1629.

Kepler had observed that the increment of a function becomes vanishingly small in the neighborhood of an ordinary maximum or minimum value. Fermat translated this fact into a process for determining such a maximum or minimum. The method will be considered in brief. If $f(x)$ has an ordinary maximum or minimum at x, and if e is very small, then the value of $f(x - e)$ is almost equal to that of $f(x)$. Therefore, we tentatively set $f(x - e) = f(x)$ and then make the equality correct by letting e assume the value zero. The roots of the resulting equation then give those values of x for which $f(x)$ is a maximum or a minimum.

Let us illustrate the above procedure by considering Fermat's first example—to divide a quantity into two parts such that their product is a maximum. Fermat used Viète's notation, where constants are designated by upper case consonants and variables by upper case vowels. Following this notation, let B be the given quantity and denote the desired parts by A and $B - A$. Forming

$$(A - E)[B - (A - E)]$$

and equating it to $A(B - A)$, we have

$$A(B - A) = (A - E)(B - A + E)$$

or

$$2AE - BE - E^2 = 0.$$

After dividing by E, one obtains

$$2A - B - E = 0.$$

Now setting $E = 0$, we obtain $2A = B$ and thus find the required division.

Although the logic of Fermat's exposition leaves much to be desired, it is seen that his method is equivalent to setting

$$\lim_{h \to 0} \frac{f(x + h) - f(x)}{h} = 0,$$

that is, to setting the derivative of $f(x)$ equal to zero. This is the customary method for finding ordinary maxima and minima of a function $f(x)$, and it is sometimes referred to in our elementary textbooks as **Fermat's method.** Fermat, however, did not know that the vanishing of the derivative of $f(x)$ is only a necessary, but not a sufficient, condition for an ordinary maximum or minimum. Also, Fermat's method does not distinguish between a maximum and a minimum value.

Fermat also devised a general procedure for finding the tangent at a point of a curve whose Cartesian equation is given. His idea is to find the **subtangent** for the point—that is, the segment on the x-axis between the foot of the ordinate drawn to the point of contact and the intersection of the tangent line with the x-axis. The method employs the idea of a tangent as the limiting position of a secant when two of its points of intersection with the curve tend to coincide. Using modern notation, the method is as follows. Let the equation of the curve (see Figure 101) be $f(x,y) = 0$, and let us seek the subtangent a of the curve for the point (x,y). By similar triangles, we easily find the coordinates of a near point on the tangent to be $[x + e, y(1 + e/a)]$. This point is tentatively treated as if it were also on the curve, giving us

$$f\left[x + e, y\left(1 + \frac{e}{a}\right)\right] = 0.$$

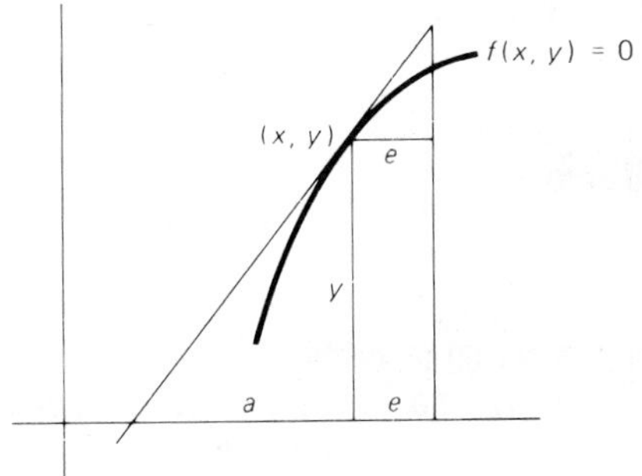

FIGURE 101

The equality is then made correct by letting e assume the value zero. We then solve the resulting equation for the subtangent a in terms of the coordinates x and y of the point of contact. This, of course, is equivalent to setting

$$a = -y \frac{\dfrac{\partial f}{\partial y}}{\dfrac{\partial f}{\partial x}},$$

a general formula that appeared later in the work of Sluze. Fermat, in this way, found tangents to the ellipse, cycloid, cissoid, conchoid, quadratrix, and folium of Descartes. Let us illustrate the method by finding the subtangent at a general point on the folium of Descartes:

$$x^3 + y^3 = nxy.$$

Here we have

$$(x + e)^3 + y^3 \left(1 + \frac{e}{a}\right)^3 - ny(x + e)\left(1 + \frac{e}{a}\right) = 0,$$

or

$$e\left(3x^2 + \frac{3y^3}{a} - \frac{nxy}{a} - ny\right) + e^2\left(3x + \frac{3y^3}{a^2} - \frac{ny}{a}\right) + e^3\left(1 + \frac{y^3}{a^3}\right) = 0.$$

Now, dividing by e and then setting $e = 0$, we find

$$a = -\frac{3y^3 - nxy}{3x^2 - ny}.$$

Fermat did pioneering work not only in connection with differentiation, but also, as intimated at the end of Section 11–6, in connection with integration. Fermat was a singularly brilliant and versatile mathematician.

11–8 Wallis and Barrow

Isaac Newton's immediate predecessors in England were John Wallis and Isaac Barrow.

John Wallis, who was born in 1616, was one of the ablest and most original mathematicians of his day. He was a voluminous and erudite writer in a number of fields and is said to have been one of the first to devise a system for teaching deaf mutes. He was a student of Oughtred, and in 1649 he was appointed Savilian professor of geometry at Oxford, a position he held for fifty-four years until his death in 1703. He introduced series systematically in analysis, and his

JOHN WALLIS
(Library of Congress)

work in this field did much to prepare the way for his great contemporary, Isaac Newton.

Wallis was one of the first to discuss conics as curves of second degree rather than as sections of a cone. In 1655 appeared his *Arithmetica infinitorum* (dedicated to Oughtred)—a book that, in spite of some logical blemishes, remained a standard treatise for many years. In this book, the methods of Descartes and Cavalieri are systematized and extended and a number of remarkable results are induced from particular cases. Thus, the formula that we would now write as

$$\int_0^1 x^m \, dx = \frac{1}{m+1},$$

where m is a positive integer, is claimed to hold even when m is fractional or negative but different from -1. Wallis was the first to explain with any completeness the significance of zero, negative, and fractional exponents, and he introduced our present symbol (∞) for infinity.

Wallis endeavored to determine π by finding an expression for the area, $\pi/4$, of a quadrant of the circle $x^2 + y^2 = 1$. This is equivalent to evaluating $\int_0^1 (1 - x^2)^{1/2} \, dx$, which Wallis was unable to do directly since he was not acquainted with the general binomial theorem. He accordingly evaluated $\int_0^1 (1 - x^2)^0 \, dx$, $\int_0^1 (1 - x^2)^1 \, dx$, $\int_0^1 (1 - x^2)^2 \, dx$, and so forth, obtaining the sequence $1, \frac{2}{3}, \frac{8}{15}, \frac{16}{35}, \ldots$. He then considered the problem of finding the law that for $n = 0, 1, 2, 3, \ldots$ would yield the preceding sequence. It was the interpolated value of this law for $n = \frac{1}{2}$ that Wallis was seeking. By a long and complicated process, he finally arrived at his infinite product expression for $\pi/2$, given in Section 4–8.

Mathematicians of his day frequently resorted to interpolation processes in order to calculate quantities that they could not evaluate directly.

Wallis accomplished other things in mathematics. He was the mathematician who came nearest to solving Pascal's challenge questions on the cycloid (see Section 9–9). It can be argued fairly that he obtained an equivalent of the formula

$$ds = \left[1 + \left(\frac{dy}{dx}\right)^2\right]^{1/2} dx$$

for the length of an element of arc of a curve. His *De algebra tractatus; historicus & practicus,* written in 1673 but published in English in 1685 and in Latin in 1693, is considered as the first serious attempt at a history of mathematics in England. It is in this work that we find the first recorded effort to give a graphic interpretation of the complex roots of a real quadratic equation. Wallis edited parts of the works of a number of the great Greek mathematicians and wrote on a wide variety of physical subjects. He was one of the founders of the Royal Society, and for years he assisted the government as a cryptologist.

Whereas Wallis' chief contributions to the development of the calculus lay in the theory of integration, Isaac Barrow's most important contributions were perhaps those connected with the theory of differentiation.

Isaac Barrow was born in London in 1630. A story is told that in his early school days he was so troublesome that his father was heard to pray that should God decide to take one of his children he could best spare Isaac. Barrow completed his education at Cambridge and won renown as one of the best Greek scholars of his day. He was a man of high academic caliber who

ISAAC BARROW
(David Smith Collection)

achieved recognition in mathematics, physics, astronomy, and theology. Entertaining stories are told of his physical strength, bravery, ready wit, and scrupulous conscientiousness. After serving for two years as professor of geometry at Gresham College, London, he became, in 1664, the first to occupy the Lucasian chair at Cambridge. In 1669, he resigned from his position at Cambridge to accept a call as chaplain to Charles II. The vacated Lucasian chair was then, at Barrow's suggestion, given to his young colleague Isaac Newton, whose remarkable abilities he was one of the first to recognize and acknowledge. He died in Cambridge in 1677.

Barrow's most important mathematical work is his *Lectiones opticae et geometricae,* which appeared in the year he resigned his chair at Cambridge. The preface of the treatise acknowledges indebtedness to Newton for some of the material of the book, probably the parts dealing with optics. It is in this book that we find a very near approach to the modern process of differentiation, utilizing the so-called *differential triangle* that we find in our present-day textbooks. Let it be required to find the tangent at a point P on the given curve represented in Figure 102. Let Q be a neighboring point on the curve. Then triangles PTM and PQR are very nearly similar to one another, and, Barrow argued, as the little triangle becomes indefinitely small, we have

$$\frac{RP}{QR} = \frac{MP}{TM}.$$

Let us set $QR = e$ and $RP = a$. Then if the coordinates of P are x and y, those of Q are $x - e$ and $y - a$. Substituting these values into the equation of the curve and neglecting squares and higher powers of both e and a, we find the ratio a/e.

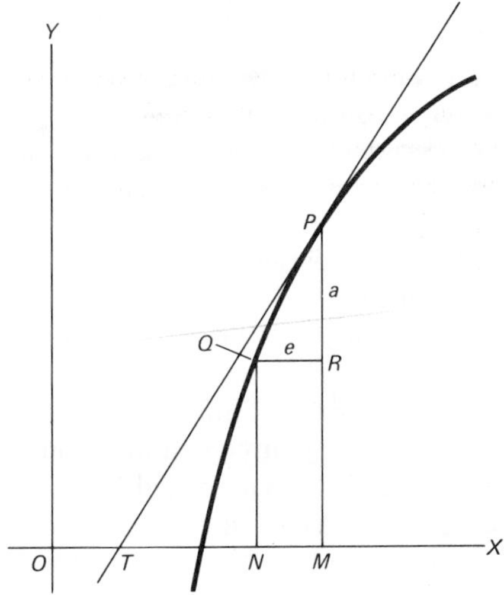

FIGURE 102

We then have

$$OT = OM - TM = OM - MP\left(\frac{QR}{RP}\right) = x - y\left(\frac{e}{a}\right),$$

and the tangent line is determined. Barrow applied this method of constructing tangents to the curves: (a) $x^2(x^2 + y^2) = r^2y^2$ (the **kappa curve**), (b) $x^3 + y^3 = r^3$ (a special **Lamé curve**), (c) $x^3 + y^3 = rxy$ (the **folium of Descartes,** but called **la galande** by Barrow), (d) $y = (r - x) \tan \pi x/2r$ (the **quadratrix**), (e) $y = r \tan \pi x/2r$ (a **tangent curve**). As an illustration, let us apply the method to curve (b). Here we have

$$(x - e)^3 + (y - a)^3 = r^3,$$

or

$$x^3 - 3x^2e + 3xe^2 - e^3 + y^3 - 3y^2a + 3ya^2 - a^3 = r^3.$$

Neglecting the square and higher powers of e and a, and using the fact that $x^3 + y^3 = r^3$, this reduces to

$$3x^2e + 3y^2a = 0,$$

from which we obtain

$$\frac{a}{e} = -\frac{x^2}{y^2}.$$

The ratio a/e is, of course, our modern dy/dx, and Barrow's questionable procedure can easily be made rigorous by the use of the theory of limits.

In spite of tenuous evidence pointing elsewhere, Barrow is generally credited as the first to realize in full generality that differentiation and integration are inverse operations. This important discovery is the so-called **fundamental theorem of the calculus** and appears to be stated and proved in Barrow's *Lectiones*.

Although Barrow devoted most of the latter part of his life to theology, he did, in 1675, publish an edition (with commentary) of the first four books of Apollonius' *Conic Sections* and of the extant works of Archimedes and Theodosius.

At this stage of the development of differential and integral calculus, many integrations had been performed; many cubatures, quadratures, and rectifications effected; a process of differentiation had been evolved and tangents to many curves constructed; the idea of limits had been conceived; and the fundamental theorem recognized. What more remained to be done? There still remained the creation of a general symbolism with a systematic set of formal analytical rules and also a consistent and rigorous redevelopment of the fundamentals of the subject. It is precisely the first of these, the creation of a suitable and workable *calculus,* that was furnished by Newton and Leibniz, who were working independently of each other. Thus, although Newton and Leibniz were preceded by a number of precursors of the calculus, it is generally conceded that they are the subject's essential inventors. The redevelopment of the fundamental concepts on an acceptably rigorous basis had to outwait the period of energetic application of the subject, and was the work of the great French analyst Augustin-Louis Cauchy (1789–1857) and his nineteenth-century successors. This story will be told in a later chapter.

11–9 Newton

Isaac Newton was born in Woolsthorpe hamlet on Christmas Day, 1642, the year in which Galileo died. His father, who died before Isaac was born, was a farmer, and it was at first planned that the son also should devote his life to farming. The youngster, however, showed great skill and delight in devising clever mechanical models and in conducting experiments. Thus, he made a toy gristmill that ground wheat to flour, with a mouse serving as motive power, and a wooden clock that worked by water. The result was that his schooling was extended. When eighteen years of age, he was allowed to enter Trinity College, Cambridge. It was not until this stage in his schooling that his attention came to be directed to mathematics, by a book on astrology picked up at the Stourbridge Fair. As a consequence, he first read Euclid's *Elements,* which he found too obvious, and then Déscartes' *La géométrie,* which he found somewhat difficult. He also read Oughtred's *Clavis,* works of Kepler and Viète, and the *Arithmetica infinitorum* by Wallis. From reading mathematics, he turned to

ISAAC NEWTON
(David Smith Collection)

creating it, early discovering the generalized binomial theorem and creating his method of fluxions, as he called what today is known as differential calculus. From late summer of 1665 until late summer of 1667, except for a brief temporary reopening from mid-March to mid-June of 1666, Cambridge University essentially closed down because of the rampant bubonic plague. It has been generally reported that it was in 1665, during the first year of this closing of the university and while living at home in Woolsthorpe, that Newton developed his calculus (to the point where he could find the tangent and radius of curvature at an arbitrary point of a curve), became interested in various physical questions, performed his first experiments in optics, and formulated the basic principles of his theory of gravitation. Recent research, however, has shown that this account is a myth, later promulgated by Newton himself to help assure him of primacy in the discovery of the calculus, and that these discoveries were actually not made until he was at Cambridge in 1666 during the university's brief, temporary reopening.

Newton returned to Cambridge in 1667 and for two years occupied himself with optical researches. In 1669, Barrow resigned the Lucasian professorship, to be succeeded by Newton, who began his eighteen years of university lecturing. His first lectures, which were on optics, were later communicated in a paper to the Royal Society and aroused considerable interest and discussion. His theory of colors and certain deductions from his optical experiments were vehemently attacked by some scientists. Newton found the ensuing argument so distasteful that he vowed never to publish anything on science again. His tremendous dislike of controversy, which seems to have bordered on the pathological, had an important bearing on the history of mathematics, for the result was that most all of his findings remained unpublished until many years after their discovery. This postponement of publication later led to the undignified dispute with Leibniz concerning priority of discovery of the calculus. It was owing to this controversy that the English mathematicians, backing Isaac Newton as their leader, cut themselves off from continental developments, and mathematical progress in England was retarded for practically one hundred years.

Newton continued his work in optics and, in 1675, communicated his work on the emission, or corpuscular, theory of light to the Royal Society. His reputation and his ingenious handling of the theory led to its general adoption, and it was not until many years later that the wave theory was shown to be a better hypothesis for research. Newton's university lectures from 1673 to 1683 were devoted to algebra and the theory of equations. It was in this period, in 1679, that he verified his law of gravitation[5] by using a new measurement of the earth's radius in conjunction with a study of the motion of the moon. He also established the compatibility of his law of gravitation with Kepler's laws of

[5] Any two particles in the universe attract one another with a force that is directly proportional to the product of their masses and inversely proportional to the square of the distance between them.

planetary motion, on the assumption that the sun and the planets may be regarded as heavy particles. But these important findings were not communicated to anyone until five years later, in 1684, when Halley visited Newton at Cambridge to discuss the law of force that causes the planets to move in elliptical orbits about the sun. With his interest in celestial mechanics rearoused in this way, Newton proceeded to work out many of the propositions later to become fundamental in the first book of his *Principia*. When Halley, sometime after, saw Newton's manuscript, he realized its tremendous importance and secured the author's promise to send the results to the Royal Society, which Newton did. At about the same time, he finally solved a problem that had been bothering him for some years—namely, that a spherical body whose density at any point depends only on its distance from the center of the sphere attracts an external particle as if its whole mass were concentrated at the center. This theorem completed his justification of Kepler's laws of planetary motion, for the slight departure of the sun and the planets from true sphericity is here negligible. Newton now worked in earnest on his theory and, by a gigantic intellectual effort, wrote the first book of the *Principia* by the summer of 1685. A year later, the second book was completed and a third begun. Jealous accusations by Hooke, and the resulting unpleasantness of the matter to Newton, almost led to the abandonment of the third book, but Halley finally persuaded Newton to finish the task. The complete treatise, entitled *Philosophiae naturalis principia mathematica,* was published, at Halley's expense, in the middle of 1687 and immediately made an enormous impression throughout Europe.

In 1689, Newton represented the university in parliament. In 1692, he suffered a curious illness that lasted about two years and that involved some form of mental derangement. Most of his later life was devoted to chemistry, alchemy, and theology. As a matter of fact, even during the earlier part of his life, he probably spent about as much time on these pursuits as he did on mathematics and natural philosophy. Although his creative work in mathematics practically ceased, he did not lose his remarkable powers, for he masterfully solved numerous challenge problems that were submitted to him and that were quite beyond the powers of the other mathematicians in England. In 1696, he was appointed Warden of the Mint; in 1699, he was promoted to Master of the Mint. In 1703, he was elected president of the Royal Society, a position to which he was annually reelected until his death; in 1705, he was knighted. The last part of his life was made unhappy by the unfortunate controversy with Leibniz. He died in 1727 when eighty-four years old after a lingering and painful illness, and was buried in Westminister Abbey.

All of Newton's important published works, except the *Principia,* appeared years after the author had discovered their contents, and almost all of them finally appeared only because of pressure from friends. The dates of these works, in order of publication, are as follows: *Principia,* 1687; *Opticks,* with two appendices on *Cubic Curves* and *Quadrature and Rectification of Curves by the Use of Infinite Series,* 1704; *Arithmetica universalis,* 1707; *Analysis per Series, Fluxiones, etc.,* and *Methodus differentialis,* 1711; *Lectiones opticae,* 1729; and *The Method of Fluxions and Infinite Series,* translated from Newton's Latin by J. Colson, 1736. One should also mention two important letters written in 1676 to H. Oldenburg, secretary of the Royal Society, in which Newton describes some of his mathematical methods.

It is in the letters to Oldenburg that Newton described his early induction of the generalized binomial theorem, which he enunciated in the form

$$(P + PQ)^{m/n} = P^{m/n} + \frac{m}{n} AQ + \frac{m-n}{2n} BQ + \frac{m-2n}{3n} CQ + \ldots,$$

where A represents the first term (namely, $P^{m/n}$), B represents the second term [namely, $(m/n)AQ$], C represents the third term, and so forth. The correctness, under proper restrictions, of the binomial expansion for all complex values of the exponent was established over 150 years later by the Norwegian mathematician N. H. Abel (1802–1829).

A more important mathematical discovery made by Newton at about the same time was his method of fluxions, the essentials of which he communicated to Barrow in 1669. His *Method of Fluxions* was written in 1671 but was not published until 1736. In this work, Newton considered a curve as generated by the continuous motion of a point. Under this conception, the abscissa and the ordinate of the generating point are, in general, changing quantities. A changing quantity is called a **fluent** (a flowing quantity), and its rate of change is called the **fluxion** of the fluent. If a fluent, such as the ordinate of the point generating a curve, be represented by y, then the fluxion of this fluent is represented by $\dot{y}$. In modern notation, we see that this is equivalent to dy/dt, where t represents time. In spite of this introduction of time into geometry, the idea of time can be evaded by supposing that some quantity, say the abscissa of the moving point, increases constantly. This constant rate of increase of some fluent is called the **principal fluxion,** and the fluxion of any other fluent can be compared with this principal fluxion. The fluxion of $\dot{y}$ is denoted by $\ddot{y}$, and so on for higher ordered fluxions. On the other hand, the fluent of y is denoted by the symbol y with a small square drawn about it, or sometimes by $\acute{y}$. Newton also introduces another concept, which he calls the **moment** of a fluent; it is the infinitely small amount by which a fluent such as x increases in an infinitely small interval of time o. Thus, the moment of the fluent x is given by the product $\dot{x}o$. Newton remarks that we may, in any problem, neglect all terms that are multiplied by the second or higher power of o and thus obtain an equation between the coordinates x and y of the generating point of a curve and their fluxions $\dot{x}$ and $\dot{y}$. As an example, he considers the cubic curve $x^3 - ax^2 + axy - y^3 = 0$. Replacing x by $x + \dot{x}o$ and y by $y + \dot{y}o$, we get

$$\begin{aligned} &x^3 + 3x^2(\dot{x}o) + 3x(\dot{x}o)^2 + (\dot{x}o)^3 \\ &\quad - ax^2 - 2ax(\dot{x}o) - a(\dot{x}o)^2 \\ &\quad + axy + ay(\dot{x}o) + a(\dot{x}o)(\dot{y}o) + ax(\dot{y}o) \\ &\quad - y^3 - 3y^2(\dot{y}o) - 3y(\dot{y}o)^2 - (\dot{y}o)^3 = 0. \end{aligned}$$

Now, using the fact that $x^3 - ax^2 + axy - y^3 = 0$, rejecting all terms containing the second or higher power of o, and then dividing by o, we find

$$3x^2\dot{x} - 2ax\dot{x} + ay\dot{x} + ax\dot{y} - 3y^2\dot{y} = 0.$$

Newton considers two types of problems. In the first type, we are given a relation connecting some fluents, and we are asked to find a relation connecting these fluents and their fluxions, which is what we did above; this is, of course, equivalent to differentiation. In the second type, we are given a relation connecting some fluents and their fluxions, and we are asked to find a relation connecting the fluents alone. This is the inverse problem and is equivalent to solving a differential equation. The idea of discarding terms containing the second and higher powers of o was later justified by Newton by the use of primitive limit notions. Newton made numerous and remarkable applications of his method of fluxions. He determined maxima and minima, tangents to curves, curvature of curves, points of inflection, and convexity and concavity of curves, and he applied his theory to numerous quadratures and to the rectification of curves. In the integration of some differential equations, he showed extraordinary ability. Also included in this work is a method (a modification of which is now known by Newton's name) for approximating the values of the real roots of either an algebraic or a transcendental numerical equation.

The *Arithmetica universalis* contains the substance of Newton's lectures of 1673 to 1683. In it are found many important results in the theory of equations, such as the fact that imaginary roots of a real polynomial must occur in conjugate pairs, rules for finding an upper bound to the roots of a real polynomial, his formulas expressing the sum of nth powers of the roots of a polynomial in terms of the coefficients of the polynomial, an extension of Descartes' rule of signs to give limits to the number of imaginary roots of a real polynomial, and many other things.

Cubic Curves, which appeared as an appendix to the work on *Opticks,* investigates the properties of cubic curves by analytic geometry. In his classification of cubic curves, Newton enumerates seventy-two out of the possible seventy-eight forms that a cubic may assume. Many of his theorems are stated without proof. The most attractive of these, as well as the most baffling, was his assertion that just as all conics can be obtained as central projections of a circle, so all cubics can be obtained as central projections of the curves

$$y^2 = ax^3 + bx^2 + cx + d.$$

This theorem remained a puzzle until a proof was discovered in 1731.

Of course, Newton's greatest work is his *Principia,* in which there appears for the first time a complete system of dynamics and a complete mathematical formulation of the principal terrestrial and celestial phenomena of motion. It proved to be the most influential and most admired work in the history of science. It is interesting that the theorems, although perhaps some may have been discovered by fluxional methods, are all masterfully established by classical Greek geometry aided, here and there, with some simple notions of limits. Until the development of the theory of relativity, all physics and astronomy rested on the assumption, made by Newton in this work, of a privileged frame of reference. In the *Principia* are found many results concerning higher plane curves and proofs of such attractive geometric theorems as the two following:

1. The locus of the centers of all conics tangent to the sides of a quadrilateral is the line **(Newton's line)** through the midpoints of its diagonals.
2. If a point P moving along a straight line is joined to two fixed points O and O', and if lines OQ and $O'Q$ make fixed angles with OP and $O'P$, then the locus of Q is a conic.

Newton was never beaten by any of the various challenge problems that circulated among the mathematicians of his time. In one of these, proposed by Leibniz, he solved the problem of finding the orthogonal trajectories of a family of curves.

Newton was a skilled experimentalist and a superb analyst. As a mathematician, he is ranked almost universally as the greatest the world has yet produced. His insight into physical problems and his ability to treat them mathematically has probably never been excelled. One can find many testimonials by competent judges as to his greatness, such as the noble tribute paid by Leibniz, who said, "Taking mathematics from the beginning of the world to the time when Newton lived, what he did was much the better half." And there is the remark by Lagrange to the effect that Newton was the greatest genius that ever lived, and the most fortunate, for we can find only once a system of the universe to be established. His accomplishments were poetically expressed by Alexander Pope in the lines,

> Nature and Nature's laws lay hid in night;
> God said, 'Let Newton be,' and all was light.

In contrast to these eulogies is Newton's own modest estimate of his work: "I do not know what I may appear to the world; but to myself I seem to have been only like a boy playing on the seashore, and diverting myself in now and then finding a smoother pebble or a prettier shell than ordinary, whilst the great ocean of truth lay all undiscovered before me." In generosity to his predecessors he once explained that if he had seen farther than other men, it was only because he had stood on the shoulders of giants.

It has been reported that Newton often spent eighteen or nineteen hours a day in writing, and that he possessed remarkable powers of concentration. Amusing tales, perhaps apocryphal, are told in testimony to his absent-mindedness when engaged in thought.

There is the story that, when giving a dinner to some friends, Newton left the table for a bottle of wine and, becoming mentally engaged, forgot his errand, went to his room, donned his surplice, and ended up in chapel.

On another occasion, Newton's friend Dr. Stukeley called on him for a chicken dinner. Newton was out, but the table was already laid with the cooked fowl in a dish under a cover. Forgetful of his dinner engagement, Newton overstayed his time, and Dr. Stukeley finally lifted the cover, removed and ate the chicken, and then replaced the bones in the covered dish. When Newton later appeared, he greeted his friend and sitting down he, too, lifted the cover, only to discover the remains. "Dear me," he said, "I had forgotten that we had already dined."

And then there was the occasion when, riding home one day from Grantham, Newton dismounted from his horse to walk the animal up Spittlegate Hill, just beyond the town. Unknown to Newton, the horse slipped away on the way up the hill, leaving only the empty bridle in his master's hands, a fact that Newton discovered only when, at the top of the hill, he endeavored to vault into the saddle.

11–10 Leibniz

Gottfried Wilhelm Leibniz, the great universal genius of the seventeenth century and Newton's rival in the invention of the calculus, was born in Leipzig in 1646. Having taught himself to read Latin and Greek when he was a mere child, he had, before he was twenty, mastered the ordinary textbook knowledge of mathematics, philosophy, theology, and law. At this young age, he began to develop the first ideas of his *characteristica generalis,* which involved a universal mathematics that later blossomed into the symbolic logic of George Boole (1815–1864) and, still later, in 1910, into the great *Principia mathematica* of Whitehead and Russell. When, ostensibly because of his youth, he was refused the degree of doctor of laws at the University of Leipzig, he moved to Nurem-

GOTTFRIED WILHELM LEIBNIZ
(David Smith Collection)

berg. There he wrote a brilliant essay on teaching law by the historical method and dedicated it to the Elector of Mainz. This led to his appointment by the Elector to a commission for the recodification of some statutes. The rest of Leibniz' life from this point on was spent in diplomatic service, first for the Elector of Mainz and then, from about 1676 until his death, for the estate of the Duke of Brunswick at Hanover.

In 1672, while in Paris on a diplomatic mission, Leibniz met Huygens, who was then residing there, and the young diplomat prevailed upon the scientist to give him lessons in mathematics. The following year, Leibniz was sent on a political mission to London, where he made the acquaintance of Oldenburg and others and where he exhibited a calculating machine to the Royal Society. Before he left Paris to take up his lucrative post as librarian for the Duke of Brunswick, Leibniz had already discovered the fundamental theorem of the calculus, developed much of his notation in this subject, and worked out a number of the elementary formulas of differentiation.

Leibniz' appointment in the Hanoverian service gave him leisure time to pursue his favorite studies, with the result that he left behind him a mountain of papers on all sorts of subjects. He was a particularly gifted linguist, winning some fame as a Sanskrit scholar, and his writings on philosophy have ranked him high in that field. He entertained various grand projects that came to nought, such as that of reuniting the Protestant and Catholic churches, and then later, just the two Protestant sects of his day. In 1682, he and Otto Mencke founded a journal called the *Acta eruditorum,* of which he became editor-in-chief. Most of his mathematical papers, which were largely written in the ten-year period from 1682 to 1692, appeared in this journal. The journal had a wide circulation in continental Europe. In 1700, Leibniz founded the Berlin Academy of Science and endeavored to create similar academies in Dresden, Vienna, and St. Petersburg.

The closing seven years of Leibniz' life were embittered by the controversy that others had brought upon him and Newton concerning whether he had discovered the calculus independently of Newton. In 1714, his employer became the first German King of England, and Leibniz was left, neglected, at Hanover. It is said that when he died two years later, in 1716, his funeral was attended only by his faithful secretary.

Leibniz' search for his *characteristica generalis* led to plans for a theory of mathematical logic and a symbolic method with formal rules that would obviate the necessity of thinking. Although this dream has only today reached a noticeable stage of realization, Leibniz had, in current terminology, stated the principal properties of logical addition, multiplication, and negation, had considered the null class and class inclusion, and had noted the similarity between some properties of the inclusion of classes and the implication of propositions (see Problem Study 11.10).

Leibniz invented his calculus sometime between 1673 and 1676. It was on October 29, 1675, that he first used the modern integral sign, as a long letter S derived from the first letter of the Latin word *summa* (sum), to indicate the sum of Cavalieri's indivisibles. A few weeks later, he was writing differentials and derivatives as we do today, as well as integrals like $\int y\,dy$ and $\int y\,dx$. His first published paper on differential calculus did not appear until 1684. In this paper, he introduces dx as an arbitrary finite interval and then defines dy by the proportion

$$dy{:}dx = y{:}\text{subtangent}.$$

Many of the elementary rules for differentiation, which a student learns early in a beginning course in the calculus, were derived by Leibniz. The rule for finding the nth derivative of the product of two functions (see Problem Study 11.6) is still referred to as **Leibniz' rule.**

Leibniz had a remarkable feeling for mathematical form and was very sensitive to the potentialities of a well-devised symbolism. His notation in the calculus proved to be very fortunate and is unquestionably more convenient and flexible than the fluxional notation of Newton. The English mathematicians, though, clung to the notation of their leader. As late as the nineteenth century, the Analytical Society, as it was named by one of its founders, Charles Babbage, was formed at Cambridge. It was formed for the purpose of advocating "the principles of pure *d*-ism as opposed to the *dot*-age of the university." It should be recalled that the rationalistic philosophy deism was in vogue among many of the intelligentsia of the time.

The theory of determinants is usually said to have originated with Leibniz, in 1693, when he considered these forms with reference to systems of simultaneous linear equations, although a similar consideration had been made ten years earlier in Japan by Seki Kōwa. The generalization of the binomial theorem into the multinomial theorem, which concerns itself with the expansion of

$$(a + b + \ldots + n)^r,$$

is due to Leibniz. He also did much to lay the foundation of the theory of envelopes, and he defined the osculating circle and showed its importance in the study of curves.

We shall not enter here into a discussion of the unfortunate Newton-Leibniz controversy. The universal opinion today is that each discovered the calculus independently of the other. Although Newton's discovery was made first, Leibniz was the earlier in publishing results. If Leibniz was not as penetrating a mathematician as Newton, he was perhaps a broader one, and although inferior to his English rival as an analyst and mathematical physicist, he probably had a keener mathematical imagination and a superior instinct for mathematical form. The controversy, which was brought upon the two principals by machinations of other parties, led to a long British neglect of European developments, much to the detriment of English mathematics.

For some time after Newton and Leibniz, the foundations of the calculus remained obscure and little heeded, for it was the remarkable applicability of

MARQUIS DE L'HOSPITAL
(David Smith Collection)

the subject that attracted the early researchers. By 1700, most of our undergraduate college calculus had been founded, along with sections of more advanced fields, such as the calculus of variations. The first textbook of the subject appeared in 1696, written by the Marquis de l'Hospital (1661–1704), when, under an odd agreement, he published the lectures of his teacher, Johann Bernoulli. In this book is found the so-called *l'Hospital's rule* for finding the limiting value of a fraction whose numerator and denominator tend simultaneously to zero.

Leibniz was an inveterate optimist. Not only did he hope to reunite the conflicting religious sects of his time into a single universal church, but he felt he might have a way of Christianizing all of China by what he believed to be the image of creation in the binary arithmetic. Since God may be represented by unity, and nothing by zero, he imagined that God created everything from nothing just as in the binary arithmetic all numbers are expressed by means of unity and zero. This idea so pleased Leibniz that he communicated it to the Jesuit Grimaldi, President of the Mathematical Board of China, with the hope that it might convert the reigning Chinese emperor (who was particularly attached to science), and thence all of China, to Christianity. As another instance of Leibniz' theological simulacrums, we have his remark that imaginary numbers are like the Holy Ghost of Christian scriptures—a sort of amphibian, midway between existence and nonexistence.

We conclude our account of Leibniz with a closing paean to his unique talent. There are two broad and antithetical domains of mathematical thought, the continuous and the discrete; Leibniz is the one man in the history of mathematics who possessed both of these qualities of thought to a superlative degree.

Problem Studies

11.1 The Method of Exhaustion

(a) Assuming the so-called **axiom of Archimedes:** *If we are given two magnitudes of the same kind, then we can find a multiple of the smaller that exceeds the larger,* establish the basic proposition of the method of exhaustion: *If from any magnitude there be subtracted a part not less than its half, from the remainder another part not less than its half, and so on, there will at length remain a magnitude less than any preassigned magnitude of the same kind.* (The axiom of Archimedes is implied in the fourth definition of Book V of Euclid's *Elements,* and the basic proposition of the method of exhaustion is found as Proposition 1 of Book X of the *Elements.*)

(b) Show, with the aid of the basic proposition of the method of exhaustion, that the difference in area between a circle and a circumscribed regular polygon can be made as small as desired.

11.2 The Method of Equilibrium

Figure 103 represents a parabolic segment having AC as chord. CF is tangent to the parabola at C, and AF is parallel to the axis of the parabola. OPM is also parallel to the axis of the parabola. K is the midpoint of FA and $HK = KC$. Take HC as a lever, or balance bar, with fulcrum at K. Place OP with its center at H, and leave OM where it is.

(a) Using the geometrical fact that $OM/OP = AC/AO$, show, by Archimedes' method of equilibrium, that the area of the parabolic segment is $\frac{1}{3}$ the area of triangle AFC.

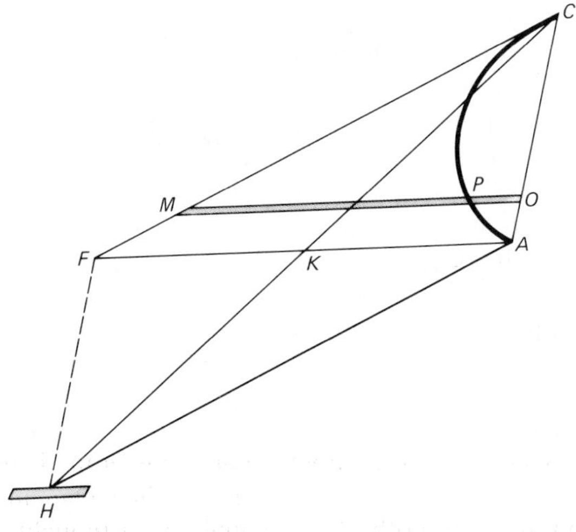

FIGURE 103

(b) Deduce, from (a), that the area of a parabolic segment is $\frac{2}{3}$ the area of the triangle bounded by the chord of the segment and the 2 tangents to the parabola at the endpoints of the chord.

11.3 Some Archimedean Problems

Archimedes devoted a number of tracts to solving volume and area problems. He established his results by the "method of exhaustion." By modern methods, solve the following Archimedean problems.

(a) Find the area of a spherical zone of height h and radius r.

(b) Find the centroid of a spherical segment.

(c) Find the volume of a **cylindrical wedge** or **hoof,** cut from a right circular cylinder by a plane passing through a diameter of the base of the cylinder.

(d) Find the volume common to 2 right circular cylinders of equal radii and having their axes intersecting perpendicularly.

11.4 The Method of Indivisibles

(a) (1) Show that any triangular prism can be dissected into 3 triangular pyramids having, in pairs, equivalent bases and equal altitudes. (2) Show, by Cavalieri's second principle, that 2 triangular pyramids having equivalent bases and equal altitudes have equal volumes. (3) Now show that the volume of a triangular pyramid is equal to $\frac{1}{3}$ the product of the area of the base of the pyramid and the altitude of the pyramid.

(b) Establish Cavalieri's principles by modern integration.

(c) Find, by Cavalieri's second principle, the volume of a *cylindrical wedge,* or *hoof* [see Problem Study 11.3(c)], in terms of the radius r of the associated cylinder and the altitude h of the hoof. (Divide the hoof into 2 equal parts by a plane p through the axis of the cylinder, and let A be the area of the resulting triangular cross section of the hoof. Construct a right prism having as its base a square of area A, the base lying in the plane p, and having an altitude equal to r. Cut from the prism a pyramid whose base is the base of the prism not lying in p and whose vertex is a point in the other base of the prism. This gouged-out prism may serve as a comparison solid for 1 of the halves of the hoof.)

(d) Find, by Cavalieri's second principle, the volume of the **spherical ring** obtained by removing from a solid sphere a cylindrical boring that is coaxial with the polar axis of the sphere. (Use for a comparison solid a sphere with diameter equal to the altitude h of the ring.)

(e) Show that all spherical rings of the same altitude have the same volume, irrespective of the radii of the spheres of the rings.

(f) Devise a *polyhedron* that can be used as a comparison solid for obtaining the volume of a sphere of radius r by means of Cavalieri's second principle. [Let AB and CD be 2 line segments in space such that (1) $AB = CD = 2r\sqrt{\pi}$, (2) AB and CD are each perpendicular to the line joining their midpoints, (3) the segment joining these midpoints has length $2r$, (4) AB is perpendicular to CD. The tetrahedron $ABCD$ may serve as the comparison polyhedron.]

(g) Find, by Cavalieri's second principle, the volume of a **torus,** or **anchor ring,** formed by revolving a circle of radius r about a line in the plane of the circle at distance $c \geqq r$ from the center of the circle. (Place the torus on a plane p perpendicular to the axis of the torus. Take for a comparison solid a right circular cylinder of radius r and altitude $2\pi c$, and place it lengthwise on the plane p.]

(h) Find, by Cavalieri's first principle, the area enclosed by the curve

$$b^2y^2 = (b + x)^2(a^2 - x^2),$$

where $b \geqq a > 0$.

(i) Show that there cannot exist a polygon that can be used as a comparison area for obtaining the area of a given circle by means of Cavalieri's first principle.

11.5 The Prismoidal Formula

A **prismatoid** is a polyhedron all of whose vertices lie in 2 parallel planes. The faces in these 2 parallel planes are called the **bases** of the prismatoid. If the 2 bases have the same number of sides, the prismatoid is called a **prismoid.** A **generalized prismoid** is any solid having 2 parallel base planes and having the areas of its sections parallel to the bases given by a quadratic function of their distances from 1 base.

(a) Show that the volumes of a prism, a wedge (a right triangular prism turned so as to rest on one of its lateral faces as a base), and a pyramid are given by the **prismoidal formula:**

$$V = \frac{h(U + 4M + L)}{6},$$

where h is the altitude, and U, L, and M are the areas of the upper and lower bases and midsection, respectively.

(b) Show that the volume of any convex prismatoid is given by the prismoidal formula.

(c) Show, by Cavalieri's principle, that the volume of any generalized prismoid is given by the prismoidal formula.

(d) Establish (c) by integral calculus.

(e) Show by integral calculus that the prismoidal formula gives the volume of any solid having two parallel base planes and having the areas of its sections parallel to the bases given by a *cubic* function of their distances from one base.

(f) Using the prismoidal formula, find the volumes of (1) a sphere, (2) an ellipsoid, (3) a cylindrical wedge, (4) the solid in Problem Study 11.3 (d).

11.6 Differentiation

(a) Find the slope of the tangent at the point (3,4) on the circle $x^2 + y^2 = 25$ by:

1. Fermat's method.
2. Barrow's method.
3. Newton's method of fluxions.
4. The modern method.

(b) If $y = uv$, where u and v are functions of x, show that the nth derivative of y with respect to x is given by

$$y^{(n)} = uv^{(n)} + nu'v^{(n-1)} + \frac{n(n-1)}{2!}u''v^{(n-2)} + \frac{n(n-1)(n-2)}{3!}u'''v^{(n-3)} + \ldots + u^{(n)}v.$$

This is known as **Leibniz' rule.**

11.7 The Binomial Theorem

(a) Show that Newton's enunciation of the binomial theorem as given in Section 11-9 is equivalent to the familiar expansion

$$(a + b)^r = a^r + ra^{r-1}b + \frac{r(r-1)}{2!}a^{r-2}b^2 + \frac{r(r-1)(r-2)}{3!}a^{r-3}b^3 + \ldots .$$

(b) Show by the binomial theorem that if $(a + ib)^k = p + iq$, where a, b, p, q are real, k is a positive integer, and $i = \sqrt{-1}$, then $(a - ib)^k = p - iq$.

(c) Show by using (b) that imaginary roots of a polynomial with real coefficients occur in conjugate pairs. (This result was given by Newton.)

11.8 An Upper Bound for the Roots of a Polynomial Equation

(a) By using the binomial theorem, or otherwise, show that if $f(x)$ is a polynomial of degree n, then

$$f(y + h) \equiv f(h) + f'(h)y + f''(h)\frac{y^2}{2!} + \ldots + f^{(n)}(h)\frac{y^n}{n!}.$$

(b) Show that any number that makes a real polynomial $f(x)$, and all of its derivatives $f'(x), f''(x), \ldots, f^{(n)}(x)$, positive, is an upper bound for the real roots of $f(x) = 0$. (This result was given by Newton.)

(c) Show that if for $x = a$ we have $f^{(n-k)}(x), f^{(n-k+1)}(x), \ldots, f^{(n)}(x)$ all positive, then these functions will also all be positive for any number $x > a$.

(d) Results (b) and (c) may be used to find a close upper bound for the real roots of a real polynomial equation. The general procedure is as follows: *Take the smallest integer that will make $f^{(n-1)}(x)$ positive. Substitute this integer in $f^{(n-2)}(x)$. If we obtain a negative result, increase the integer successively by units until an integer is found that makes this function positive. Now proceed with the new integer as before. Continue in this way until an integer is found that makes all of the functions $f(x), f'(x), \ldots, f^{(n-1)}(x)$ positive.* Find, by this procedure, an upper bound for the real roots of

$$x^4 - 3x^3 - 4x^2 - 2x + 9 = 0.$$

11.9 Approximate Solution of Equations

(a) Newton devised a method for approximating the values of the real roots of a numerical equation that applies equally well to either an algebraic or a transcendental equation. The modification of this method, now known as **Newton's method,** says: *If $f(x) = 0$ has only 1 root in the interval $[a, b]$, and if neither $f'(x)$ nor $f''(x)$ vanishes in this interval, and if x_0 be chosen as that 1 of the 2 numbers a and b for which $f(x_0)$ and $f''(x_0)$ have the same sign, then*

$$x_1 = x_0 - \frac{f(x_0)}{f'(x_0)}$$

is nearer to the root than is x_0. Establish this result.

(b) Solve by Newton's method the cubic $x^3 - 2x - 5 = 0$ for the root lying between 2 and 3.

(c) Solve by Newton's method the equation $x = \tan x$ for the root lying between 4.4 and 4.5.

(d) Find by Newton's method $\sqrt{12}$ correct to 3 decimal places.

(e) By means of the hyperbola $xy = k$, $k > 0$, show that if x_1 is an approximation to $\sqrt{k}$, then $x_2 = (x_1 + k/x_1)/2$ is a better approximation, and so on. (This is Heron's method of approximating a square root. See Section 6–6.)

(f) Obtain the procedure of (e) from Newton's method applied to $f(x) = x^2 - k$.

(g) By Newton's method applied to $f(x) = x^n - k$, n a positive integer, show that if x_1 is an approximation to $\sqrt[n]{k}$, then

$$x_2 = \frac{(n - 1)x_1 + \dfrac{k}{{x_1}^{n-1}}}{n}$$

is a better approximation, and so on.

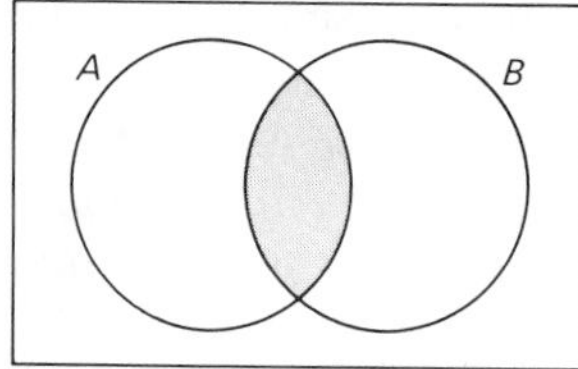

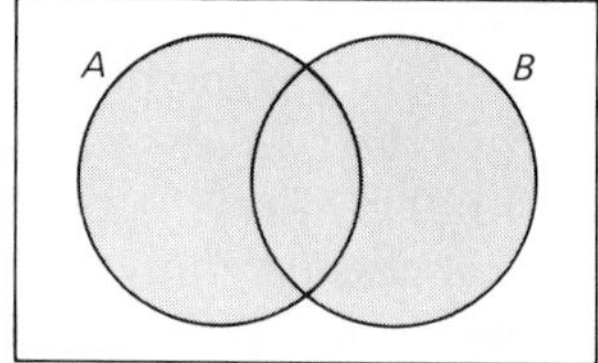

FIGURE 104

(h) In a text on the theory of equations, look up the so-called **Fourier's theorem,** which states a guarantee under which the Newton method is bound to succeed.

[In 1690, Joseph Raphson (1648–1715), a fellow of the Royal Society of London, published a tract, *Analysis aequationum universalis,* that describes essentially the Newton method for approximating the real roots of a numerical equation. For this reason, the method is today often referred to as the **Newton-Raphson method.** Newton had described his method, illustrating it on the cubic of (b), in his *Method of Fluxions,* which, though written in 1671, was not published until 1736. The earliest printed account of Newton's method appeared in Wallis' *Algebra* of 1685.][6]

11.10 Algebra of Classes

The concept of a "class of objects" is fundamental in logic. Leibniz developed some of the elementary algebra of classes. Using modern notation, if A and B are classes of objects, then $A \cap B$ (called the **intersection,** or **product,** of A and B) represents the class of all objects belonging to both A and B, and $A \cup B$ (called the **union,** or **sum,** of A and B) represents the class of all objects belonging to either A or B.

The algebra of classes can be illustrated graphically by means of so-called **Venn diagrams,**[7] where a class A is represented by a given region. Thus, if we represent classes A and B by the interiors of the 2 circles A and B, as indicated in Figure 104, the set $A \cap B$ is represented by the region common to these 2 circles, and the set $A \cup B$ is represented by the region made up of all the points in either 1 or the other of the 2 circles. If we represent all our classes inside a surrounding rectangle, then by A', called the **complement** of A, we mean the region inside the rectangle but outside the region that represents A.

(a) On a Venn diagram, shade each of the following regions: $A \cap (B' \cup C)$, $(A' \cap B) \cup (A \cap C')$, $(A \cup B') \cup C'$.

---

[6] See F. Cajori, "Historical Notes on the Newton-Raphson Method of Approximation," *The American Mathematical Monthly* 18 (1911): 29–33.

[7] Named after John Venn (1834–1923), an English logician, who employed the device in 1876 in a paper on Boole's logical system, and again in 1894 in his excellent book *Symbolic Logic*.

(b) By shading the appropriate regions on a Venn diagram, verify the following equations in the algebra of classes: $A \cap (B \cap C) = (A \cap B) \cap C$, $A \cap (B \cup C) = (A \cap B) \cup (A \cap C)$, $(A \cup B)' = A' \cap B'$.

(c) By shading the appropriate regions on a Venn diagram, determine which of the following equations are valid: $(A' \cup B)' = A \cap B'$, $A' \cup B' = (A \cup B)'$, $A \cup (B \cap C)' = (A \cup B') \cap C'$.

Essay Topics

11/1 The relation of Zeno's paradoxes to the calculus.

11/2 The Greek contribution to the development of the integral calculus.

11/3 Modern forerunners of Newton and Leibniz in the development of the calculus.

11/4 The use of Cavalieri's second principle in a beginning course in solid geometry.

11/5 The greatest mathematical discovery of the seventeenth century.

11/6 Leibniz' conception of the differential.

11/7 Barrow and the fundamental theorem of the calculus.

11/8 The Newton-Leibniz controversy.

11/9 The four greatest mathematical books of the seventeenth century.

11/10 The five most important British mathematicians of the seventeenth century.

11/11 Men who were prominent in both mathematics and physics in the seventeenth century.

11/12 The six leading mathematical countries of the seventeenth century, arranged in order of importance.

11/13 The Japanese Newton.

11/14 The history of continued fractions.

11/15 Determinants in seventeenth-century Japanese mathematics.

11/16 Roger Cotes (1682–1716).

11/17 The Royal Society.

Bibliography

ANTHONY, H. D. *Sir Isaac Newton.* New York: Abelard-Schuman, 1960.

BARON, M. E. *The Origins of the Infinitesimal Calculus.* New York: Dover Publications, 1987.

BELL, E. T. *Men of Mathematics.* New York: Simon and Schuster, 1937.

BOYER, C. B. *The History of the Calculus and Its Conceptual Development.* New York: Dover Publications, 1959.

BREWSTER, SIR DAVID *Life of Newton.* London: John Murray, 1831.

BRODETSKY, SELIG *Sir Isaac Newton: A Brief Account of His Life and Work.* London: Methuen & Co., 1927.

CAJORI, FLORIAN *A History of the Conceptions of Limits and Fluxions in Great Britain, From Newton to Woodhouse*. Chicago: Open Court, 1919.

——— *A History of Mathematical Notations*. 2 vols. Chicago: Open Court, 1928–1929.

——— *Sir Isaac Newton's Mathematical Principles of Natural Philosophy and His System of the World*. Revision of the translation of 1729 by Andrew Motte. Berkeley, Calif.: University of California Press, 1934.

CHILD, J. M. *The Early Mathematical Manuscripts of Leibniz*. Chicago: Open Court, 1920.

——— *The Geometrical Lectures of Isaac Barrow*. Chicago: Open Court, 1916.

CHRISTIANSON, G. E. *In the Presence of the Creator: Isaac Newton and His Times*. New York: The Free Press, 1984.

COOLIDGE, J. L. *Geometry of the Complex Domain*. New York: Oxford University Press, 1924.

——— *A History of Geometrical Methods*. New York: Oxford University Press, 1940.

——— *The Mathematics of Great Amateurs*. New York: Oxford University Press, 1949.

DE MORGAN, AUGUSTUS *Essays on the Life and Work of Newton*. Chicago: Open Court, 1914.

EDWARDS, C. H., Jr. *The Historical Development of the Calculus*. New York: Springer-Verlag, 1980.

GROWING, RONALD *Roger Cotes: Natural Philosopher*. Cambridge: Cambridge University Press, 1983.

HALL, A. R., and M. B. HALL, eds. *Unpublished Scientific Papers of Isaac Newton*. New York: Cambridge University Press, 1962.

——— *Philosophers at War: The Quarrel Between Newton and Leibniz*. New York: Cambridge University Press, 1980.

HEATH, T. L. *History of Greek Mathematics*. Vol. 2. New York: Oxford University Press, 1921. Reprinted by Dover Publications, New York, 1981.

——— *A Manual of Greek Mathematics*. New York: Oxford University Press, 1931.

——— *The Method of Archimedes Recently Discovered by Heiberg*. New York: Cambridge University Press, 1912. Contained in *The Works of Archimedes*. Reprinted by Dover Publications, New York.

——— *The Works of Archimedes*. New York: Cambridge University Press, 1897. Reprinted by Dover Publications, New York.

HERIVEL, JOHN *The Background to Newton's Principia*. Oxford: Oxford University Press, 1965.

HOFFMAN, J. E. *Leibniz in Paris, 1672–1676*. Cambridge: Cambridge University Press, 1964.

KERN, W. F., and J. R. BLAND *Solid Mensuration: With Proofs*. 2d ed. New York: John Wiley, 1938.

LANE, E. P. *Metric Differential Geometry of Curves and Surfaces*. Chicago: University of Chicago Press, 1940.

LEE, H. D. P., ed. *Zeno of Elea*. Cambridge: Cambridge University Press, 1936.

LOVITT, W. V. *Elementary Theory of Equations*. Englewood Cliffs, N.J.: Prentice-Hall, 1939.

MACFARLANE, ALEXANDER *Lectures on Ten British Mathematicians of the Nineteenth Century*. Mathematical Monographs, No. 17. New York: John Wiley, 1916.

MANHEIM, J. H. *The Genesis of Point Set Topology*. New York: Macmillan, 1964.

MANUEL, F. E. *A Portrait of Isaac Newton*. Cambridge, Mass.: Harvard University Press, 1968.

MELLONE, S. H. *The Dawn of Modern Thought—Descartes, Spinoza, and Newton*. London: Oxford University Press, 1930.

MERZ, JOHN *Leibniz*. New York: Hecker Press, 1948.

MESCHKOWSKI, HERBERT *Ways of Thought of Great Mathematicians*. San Francisco: Holden-Day, 1964.

MEYER, R. W. *Leibniz and the Seventeenth Century Revolution*. Translated by J. P. Stern. Cambridge: Bowes and Bowes, 1952.

MORE, L. T. *Isaac Newton, a Biography*. New York: Dover Publications, 1962.

MUIR, JANE *Of Men and Numbers, The Story of the Great Mathematicians*. New York: Dodd, Mead, 1961.

MUIR, THOMAS *The Theory of Determinants in the Historical Order of Development*. 4 vols. New York: Dover Publications, 1960.

NEUGEBAUER, OTTO *The Exact Sciences in Antiquity*. 2d ed. New York: Harper & Row, 1962.

NEWTON, SIR ISAAC *Mathematical Principles of Natural Philosophy*. Translated by Andrew Motte. Edited by Florian Cajori. Berkeley, Calif.: University of California Press, 1934.

——— *Mathematical Works*. 2 vols. Edited by D. T. Whiteside. New York: Johnson Reprint, 1964–1967.

——— *Mathematical Papers*. 7 vols. Edited by D. T. Whiteside. New York: Cambridge University Press, 1967.

PHILLIPS, H. B. *Differential Equations*. 3d ed. New York: John Wiley, 1934.

PRIESTLEY, W. M. *Calculus: An Historical Approach*. New York: Springer-Verlag, 1979.

ROSENTHAL, A. "The History of Calculus," *The American Mathematical Monthly* 58 (1951): 75–86.

ROYAL SOCIETY OF LONDON *Newton Tercentenary Celebrations, 15–19 July, 1946*. New York: Macmillan, 1947.

SABRA, A. I. *Theories of Light, from Descartes to Newton*. London: Oldburne Book Company, 1957.

SCOTT, J. T. *The Mathematical Work of John Wallis* (1616–1703). London: Taylor and Francis, 1938.

SMITH, D. E. *History of Modern Mathematics*. 4th ed. New York: John Wiley, 1906.

——— *A Source Book in Mathematics*. New York: McGraw-Hill, 1929.

SULLIVAN, J. W. N. *The History of Mathematics in Europe, From the Fall of Greek Science to the Rise of the Conception of Mathematical Rigour*. New York: Oxford University Press, 1925.

——— *Isaac Newton 1642–1727*. New York: Macmillan, 1938.

TAYLOR, E. G. R. *The Mathematical Practitioners of Tudor and Stuart England*. Cambridge: Cambridge University Press, 1954.

THOMSON, THOMAS *History of the Royal Society from Its Institution to the End of the 18th Century*. Ann Arbor, Mich.: University Microfilms, 1967.

TOEPLITZ, OTTO *The Calculus, a Genetic Approach*. Chicago: University of Chicago, Press, 1963.

TURNBULL, H. W. *Mathematical Discoveries of Newton*. Glasgow: Blackie & Sons, 1945.

——— *The Great Mathematicians*. New York: New York University Press, 1961.

———, ed. *Correspondence of Isaac Newton*. 7 vols. Cambridge: Cambridge University Press, 1959–1977.

WALKER, EVELYN *A Study of the Traité des Indivisibles of Gilles Persone de Roberval*. New York: Teachers College, Columbia University, 1932.

WELD, CHARLES *A History of the Royal Society*. Reprint ed. New Yok: Arno Press, 1975.

WESTFALL RICHARD *Never at Rest: A Biography of Isaac Newton*. Cambridge: Cambridge University Press, 1980.

Cultural Connection

THE REVOLT OF THE MIDDLE CLASS

The Eighteenth Century in Europe and America
(to accompany Chapter Twelve)

The eighteenth century was a time of turbulence and revolution in Europe and America. A newly emerged middle class, the *bourgeoisie,* overthrew the old aristocratic order in England, France, and America. Feudal political, social, and economic ideas, rooted in subsistence agriculture, were replaced by a philosophy of Classical Liberalism, a system that emphasized limited democracy, equality of opportunity, and the sanctity of private property, and consequently advanced the Industrial Revolution of the nineteenth century.

In 1690, in a book entitled *Two Treatises of Government,* the English philosopher John Locke (1632–1704) proposed the idea of Classical Liberalism as a social, political, and economic framework. Socially, Locke believed that all human beings, whether they be poor or well born, male or female, peasant or lord, were naturally equal or, as he put it, free "to order their possessions and persons, as they think fit, within the bounds of the law of nature." Locke especially argued for religious toleration. In politics, this gaunt Englishman espoused a revolutionary rhetoric. Leaders, he maintained, governed only by the consent of their subjects, who had a moral right to depose unjust rulers. Economically, Locke believed property should be owned privately, so long as it was managed usefully and on a small scale. "As much land as a man tills, plants, improves, cultivates, and can use the product of," he wrote, "so much is his property." Such were the tenets of eighteenth-century Liberalism: private ownership of modest amounts of property, government by the consent of the governed, and equality under the law.

Locke's ideas were echoed by others, like the French philosopher Jean Jacques Rousseau (1712–1778). "In the great family [of society]," wrote Rousseau in 1755, "all the members . . . are naturally equal." Unlike Locke, however, Rousseau did not mean that *all* people are equal; he believed that women were inferior to men, and he opposed equal rights for women. Other Liberal thinkers excluded black slaves and some religious minorities, like Jews, from their rubric of equality.

By 1776 Liberalism had become a revolutionary credo. Thomas Jefferson (1743–1826), providing the theoretical underpinning for the American Revolution, echoed Locke when he penned his renowned *Declaration of Indepen-*

dence: "We hold these truths to be self-evident, that all men are created equal, that they are endowed by their Creator with certain unalienable Rights, that among these are Life, Liberty and the pursuit of Happiness. That to secure these rights, Governments are instituted among Men, deriving their just powers from the consent of the governed. That whenever any Form of Government becomes destructive of these ends it is the Right of the People to alter or to abolish it . . . ."

Classical Liberalism embodied the perspectives of a newly emerged property-owning social class that we call the middle class and the French term the *bourgeoisie*. This class (consisting of well-to-do farmers, merchants, bankers, prosperous artisans, lawyers, doctors, civil servants, and others) had become increasingly prominent in Europe in the 1400s, 1500s, and 1600s. During the Middle Ages, about ninety percent of the European population had been peasant farmers, many of whom were non-landowning serfs. The peasants wielded little political or economic power, most of which was instead concentrated into a relatively tiny class, the aristocracy, that comprised only about two percent of the population. The remaining eight percent (artisans, merchants, lawyers, and beggars) lived in the cities.

The *bourgeoisie* began to grow in numbers and power when European cities began to grow larger during the late Middle Ages and during the Age of Exploration. Urban growth was fueled both by overpopulation in the countryside, which drove many rural people into the cities looking for work, and by the greatly expanded commerce of the Age of Exploration, which we discussed in Cultural Connection VII. As workers and capital poured into European cities like London, Paris, Frankfurt, Antwerp, Milan, and Seville, many merchants and artisans expanded their businesses. In a few cases, they became wealthy enough that they no longer had to work, but instead sold their businesses and invested their money in land or loaned it at interest to the aristocracy. Even in the Middle Ages, European cities had stood apart from the mainstream of feudal society, which was dominated by the hereditary aristocracy. Cities had their own governments, and urban residents enjoyed special privileges denied to rural peasants, such as freedom from forced labor on the aristocrats' estates. A saying of the Middle Ages proclaimed that "city air is free air." Such freedom, however, was relative. The urban poor who worked as laborers in the shops and small factories had little say in civic affairs, and city hall was usually dominated by the wealthier merchants and bankers.

As the *bourgeoisie* came into prominence between 1400 and 1700, its members grew resentful of the old aristocracy. Wealthy merchants and *rentiers* (a French word meaning, roughly, investment capitalists) had good education, understood affairs of state, and stood at the heart of both worldwide trading networks and government finance. Yet they were denied positions in the government reserved for aristocrats, were required to pay taxes from which aristocrats were exempt, and could not engage in certain businesses that the government granted to favored aristocrats as royal monopolies. Even more galling, the aristocrats, partially because of their jealousy of the new urban fortunes, treated the earnest *rentiers* like upstarts and ignored them socially. As the middle class grew, the aristocracy, anxious to maintain its position of control over European society, strove to exclude the *bourgeoisie* from positions of authority and power. The *bourgeoisie,* for its part, began to think about revolution.

Locke's Liberalism provided the *bourgeoisie* with a theoretical base for political and economic change. In the economic sphere, Scotsman Adam Smith (1723–1790), in his 1776 book *Wealth of Nations,* built upon Locke's ideas about the primacy of private property to advocate capitalism as an economic system. In England, Locke himself was involved in the Glorious Revolution of 1688, which limited the power of the monarchy, and his ideas in part served as justification for the change in English government from an absolute monarchy to a constitutional one. In the American Revolution (1776–1783), the French Revolution (1789–1799), and the Latin American revolutions (1800–1825), the old governments were condemned because aristocratic privilege flew in the face of ideas about equality and the right of people to control their governments was invoked.

Between 1688 and 1825, the bourgeois class seized power, either peacefully or violently, in England, France, and most parts of the Americas. The *bourgeoisie* did not fight these revolutions by itself. Like the middle class, the rural peasants and the urban poor were also unhappy with the old structure of European society. Despite the grumbling of the middle class, the peasants paid most of the taxes. Years of overpopulation and consequent overfarming had resulted in declining productivity and widespread rural poverty. Peasants had always resented the aristocracy, and, in the Middle Ages, had frequently rebelled. Still, in much of Europe, peasants also disliked the *bourgeoisie*. As they grew poorer, peasants sometimes borrowed money from the *rentiers* to see them through bad harvests and droughts. Rarely could a peasant repay such a loan, and farms were frequently forfeited to *rentiers* as collateral, forcing the peasant to become a tenant of either the *rentier* or the local aristocrat. Only in America, then, did the farmers, who were independent yeomen with comparatively larger farms and fewer debts, really join the *bourgeoisie* in revolution against the old order, and even here farmers often mistrusted the urban middle class. More important to the *bourgeoisie* as allies against the aristocracy were the urban poor. Threatened by spiraling food prices set artificially high by government monopolies and high agricultural taxes, the urban poor played an active role in revolution, especially in Paris, Boston, and Caracas. Nevertheless, with the exception of the United States, where many revolutionary leaders were wealthy farmers, leadership of the eighteenth-century revolts against the aristocracy was largely in the hands of the urban *bourgeoisie*.

Let's look briefly at this course of revolution.

England had been in political turmoil since 1603, when its great queen, Elizabeth I (reigned 1558–1603), died. Elizabeth was succeeded by a new dynasty, the Stuarts, who angered many Englishmen by their sympathy to Catholicism (most of the English were Protestants), their alien nationality (the Stuarts were Scots), and by their advocacy of the idea that kings had a divine and absolute right to rule as they saw fit (the Divine Right of Kings). The attempt of the Stuart kings to further their power led them into conflict with the London *bourgeoisie,* who resented royal meddling in their city's internal affairs. An association of Londoners, their economic allies, and Protestant Puritans led by Oliver Cromwell (1599–1658) challenged Stuart authority, and, in the English Civil War (1641–1649), deposed the Stuart king. Although the Stuarts regained the English throne in 1660, antagonisms engendered by the Civil War lingered and smouldered. When a Stuart king attempted forcibly to reconvert England to Catholicism in the 1680s, the army removed him in a bloodless coup known as the Glorious Revolution (1688). Parliament installed the more tractable Mary II (reigned 1689–1694) and her husband William III (reigned 1689–1702) as queen and king, provided they agreed to proclaim the English Bill of Rights (1689), which abolished the absolute monarchy and replaced it with a constitutional monarchy in which Parliament's House of Commons, controlled by the *bourgeoisie,* held real power. Before anywhere else, the *bourgeoisie* came to power in England.

The English Civil War and the Glorious Revolution had repercussions in England's American colonies, where the colonial *bourgeoisie* of merchants, well-to-do farmers, and land investors were to enjoy considerable local authority. This autonomy was eroded in the 1760s when, following a costly war with France, the home government enacted a series of taxes in the colonies, over the objections of the colonial assemblies. Disputes over taxes, civil liberties, trade restrictions, and the closing of the frontier by imperial edict sparked the American Revolution, which resulted in American independence. Unlike middle-class revolutions elsewhere, the American Revolution was waged by an alliance of urban *bourgeoisie,* wealthy planters, and yeoman farmers. A significant portion of the revolutionary leadership, like Thomas Jefferson and George Washington (1732–1799), were rural, but others, like merchant John Hancock (1737–1793) and lawyer John Adams (1735–1826), both of Boston, publisher Benjamin Franklin (1705–1790) of Philadelphia, and *bon vivant* Alexander Hamilton (1755–1804) of New York were urban *bourgeoisie.* Conflict between these two groups for power in the new republic continued well into the next century.

The middle class allied with the urban poor of Paris in 1789 to depose the French king, a movement set off by high bread prices in Paris, but also the result of decades of antagonism between the French *bourgeoisie* and the aristocracy. In France, the middle class lost control of the revolution in the 1790s to the Paris mob, which executed many *bourgeois* leaders. The *bourgeoisie* retaliated by supporting Napoleon (1769–1821), who seized power as dictator (later as emperor) in 1799. Napoleon enacted the kinds of economic reforms advocated by the *bourgeoisie,* although he squelched civil liberties.

Following his 1799 coup, Napoleon launched a campaign to conquer most of Europe and incorporate it into a French Empire. His annexation of Spain in 1800 weakened Spanish power in America, where bourgeoisie-led revolts broke out in the Spanish colonial cities of Mexico City, Caracas, and Buenos Aires. By 1825, republican governments under the control of the middle class had been established in most of Latin America. Further, Napoleon installed puppet governments, republican in form if not in fact, in conquered Germany, Italy, and Poland. Many people in those places hoped that this might lead to the establishment of true republics or constitutional monarchies—hopes that were never completely abandoned.

Between 1688 and 1825, therefore, middle-class revolutions were staged against the old aristocratic order throughout Europe and America. In England and the United States, the newly emerged *bourgeoisie* assumed political power. Even where revolts ultimately failed, as in France, the middle class was accorded a greater voice in affairs of state and finance. Throughout the West, by 1825 the *bourgeoisie* was well on its way toward supplanting the old medieval aristocracy as the new ruling class.

Napoleon's defeat in 1815 by the English/Prussian/Russian/Austrian alliance ended the French Empire and restored aristocratic rule in France, Germany, Italy, and Poland. These new aristocratic governments were shaky, however, and, in the nineteenth century, would face opposition from bourgeois republicans, nationalists, and socialists. Society in nineteenth-century Europe would face even more such political ferment; it would also encounter the Industrial Revolution. That, however, is a tale for Cultural Connection IX.

Chapter 12

THE EIGHTEENTH CENTURY AND THE EXPLOITATION OF THE CALCULUS

12–1 Introduction and Apology

The arithmetic, beginning algebra, geometry, and trigonometry ordinarily taught in the schools today, along with college algebra, analytic geometry, and beginning calculus usually taught during the freshman or sophomore year in college, constitute what is generally called "elementary mathematics." At this point of our book, then, we have virtually concluded the historic treatment of elementary mathematics in the form that we have it today. It is interesting to note, without carrying the generalization too far, that the sequence of mathematics courses taught in the classroom closely follows the evolutionary development of the subject.

It is rightfully claimed that one cannot properly study the history of a subject without a knowledge of the subject itself. It follows that for one who would study, with genuine understanding, what has happened in mathematics during the eighteenth, nineteenth, and twentieth centuries, an extensive study of advanced courses beyond the calculus is requisite. When the student possesses such a background, the excellent books, *The Development of Mathematics* by E. T. Bell, *A History of Mathematics* by C. B. Boyer, and *Mathematical Thought from Ancient to Modern Times* by Morris Kline, are recommended. Nevertheless, it seems advisable to add the present chapter, and the following three concluding chapters, in an attempt to provide some of the highlights of the mathematics of the eighteenth, nineteenth, and twentieth centuries that are within the purview of our intended reader, and that briefly show the more recent trend of mathematical development from the elementary basis. The field of elementary mathematics will then appear in its proper setting, as a prelude to the more remarkable achievements of modern times.

One cannot point out too strongly the increasing sketchiness and incompleteness of what follows. Moritz Cantor's great history of mathematics, which terminates with the end of the eighteenth century, consists of four large volumes averaging almost a thousand pages each. It has been conservatively estimated that if the history of mathematics of the nineteenth century should be written with the same detail it would require at least fourteen more such volumes! No one has yet hazarded an estimate of the number of such volumes needed for a similar treatment of the history of the mathematics of the twentieth century, which is by far the most active era of all. And, as indicated above, little of this additional material could be properly appreciated by the ordinary undergraduate; indeed, an understanding of much of the material would require the deep background of a mathematical expert.

The almost explosive growth of mathematical research in modern times is further illustrated by the fact that prior to 1700 there were only, by one count, seventeen periodicals containing mathematical articles. In the eighteenth century, there were 210 such periodicals, and in the nineteenth century, 950 of them. The number has increased enormously during the twentieth century, reaching, by one count, some 2600. Furthermore, it was not until the nineteenth century that journals appeared devoted either primarily or exclusively to mathematics. It has been remarked, probably quite properly, that the articles in these research journals constitute the true history of modern mathematics, and it must be confessed that very few of the present-day articles can be read by anyone but the specialist.

As still one further piece of statistics, pointing up the intense mathematical activity of the present century, one might mention that it has been estimated that more than fifty percent of all known mathematics was created during the past fifty years, and that fifty percent of all mathematicians who have ever lived are alive today.

The calculus, aided by analytic geometry, was the greatest mathematical tool discovered in the seventeenth century. It proved to be remarkably powerful and capable of attacking problems quite unassailable in earlier days. It was the wide and astonishing applicability of the discipline that attracted the bulk of the mathematical researchers of the day, with the result that papers were turned out in great profusion with little concern for the very unsatisfactory foundations of the subject. The processes employed were justified largely on the ground that they worked, and it was not until the eighteenth century had almost elapsed, after a number of absurdities and contradictions had crept into mathematics, that mathematicians felt it was essential that the basis of their work be logically examined and rigorously established. The painstaking effort to place analysis on a logically rigorous foundation was a natural reaction to the pell-mell employment of intuition and formalism of the previous century. The task proved to be a difficult one, its various ramifications occupying the better part of the next hundred years. A result of this careful work in the foundations of analysis was that it led to equally careful work in the foundations of all branches of mathematics and to the refinement of many important concepts. Thus, the function idea itself had to be clarified, and such notions as limit, continuity, differentiability, and integrability had to be very carefully and clearly defined. This task of refining the basic concepts of mathematics led, in turn, to intricate generalizations. Such concepts as space, dimension, convergence, and integrability, to name only a few, underwent remarkable generalization and abstraction. A good part of the mathematics of the twentieth century has been devoted to this sort of thing, until generalization and abstraction have now become striking features of present-day mathematics. But some of these developments brought about, in turn, a fresh batch of paradoxical situations.

The generalization to transfinite numbers, for example, and the abstract study of sets have widened and deepened many branches of mathematics; at the same time, however, they have revealed some very disturbing paradoxes that appear to lie in the innermost depths of mathematics. Here is where we seem to be today, and it may be that the final years of the twentieth century will witness the resolution of some of these critical problems.

In summarizing the last paragraph, we may say, with a fair element of truth, that the eighteenth century was largely spent in exploiting the new and powerful methods of the calculus, that the nineteenth century was largely devoted to the effort of establishing on a firm, logical foundation the enormous but shaky superstructure erected in the preceding century, that the twentieth century has, in large part, been spent in generalizing as far as possible the gains already made, and that at present many mathematicians are becoming concerned with even deeper foundational problems. This general picture is complicated by the various sociological factors that affect the development of any science. Such matters as the growth of life insurance, the construction of the large navies of the eighteenth century, the economic and technological problems brought about in the nineteenth century by the industrialization of western Europe and America, the twentieth-century world-wide war atmosphere, the advance of the remarkable modern electronic computers, and today's concentrated effort to conquer outer space have led to many practical developments in the field of mathematics. A division of mathematics into "pure" and "applied" has come about, research in the former being carried on to a great extent by those specialists who have become interested in the subject for its own sake, and in the latter, by those who remain attached to immediately practical uses.

We now, in the remainder of the book, fill in some of the details of the general picture just sketched.

12–2 The Bernoulli Family

The principal contributions to mathematics in the eighteenth century were made by members of the Bernoulli family, Abraham De Moivre, Brook Taylor, Colin Maclaurin, Leonhard Euler, Alexis Claude Clairaut, Jean-le-Rond d'Alembert, Johann Heinrich Lambert, Joseph Louis Lagrange, Pierre-Simon Laplace, Adrien-Marie Legendre, Gaspard Monge, and Lazare Carnot. It will be observed that the bulk of the mathematics of these men found its genesis and its goal in the applications of the calculus to the fields of mechanics and astronomy. It was not until well into the nineteenth century that mathematical research generally emancipated itself from this viewpoint. This section describes the remarkable Bernoulli family.

One of the most distinguished families in the history of mathematics and science is the Bernoulli family of Switzerland, which, from the late seventeenth century on, produced an unusual number of capable mathematicians and scientists. The family record starts with the two brothers, Jakob Bernoulli (1654–1705) and Johann Bernoulli (1667–1748), some of whose mathematical accomplishments have already been mentioned in this book. These two men gave up earlier vocational interests and became mathematicians when Leibniz' papers began to appear in the *Acta eruditorum*. They were among the first mathematicians to realize the surprising power of the calculus and to apply the tool to a great diversity of problems. From 1687 until his death, Jakob occupied the mathematics chair at Basel University. Johann, in 1697, became a professor at Gröningen University, and then, on Jakob's death in 1705, succeeded his brother in the chair at Basel University, to remain there for the rest of his life. The two brothers, often bitter rivals, maintained an almost constant exchange of ideas with Leibniz and with each other.

Among Jakob Bernoulli's contributions to mathematics are the early use of polar coordinates (see Section 14–5), the derivation in both rectangular and polar coordinates of the formula for the radius of curvature of a plane curve, the study of the catenary curve with extensions to strings of variable density and strings under the action of a central force, the study of a number of other higher plane curves, the discovery of the so-called **isochrone**—or curve along which a body will fall with uniform vertical velocity (it turned out to be a semicubical parabola with a vertical cusptangent), the determination of the form taken by an elastic rod fixed at one end and carrying a weight at the other, the form assumed by a flexible rectangular sheet having two opposite edges held horizontally fixed at the same height and loaded with a heavy liquid, and the shape of a rectangular sail filled with wind. He also proposed and discussed the problem of **isoperimetric figures** (planar closed paths of given species and fixed perimeter that include a maximum area), and was thus one of the first mathematicians to work in the calculus of variations. He was also (as was pointed out in Section 10–3) one of the early students of mathematical probabil-

JAKOB BERNOULLI
(David Smith Collection)

ity; his book in this field, the *Ars conjectandi,* was posthumously published in 1713. Several things in mathematics now bear Jakob Bernoulli's name. Among these are the *Bernoulli distribution* and *Bernoulli theorem* of statistics and probability theory; the *Bernoulli equation,* met by every student of a first course in differential equations; the *Bernoulli numbers* and *Bernoulli polynomials* of number-theory interest; and the *lemniscate of Bernoulli,* encountered in any first course in the calculus. In Jakob Bernoulli's solution to the problem of the isochrone curve, which was published in the *Acta eruditorum* in 1690, we meet for the first time the word *integral* in a calculus sense. Leibniz had called the integral calculus *calculus summatorius;* in 1696, Leibniz and Johann Bernoulli agreed to call it *calculus integralis.* Jakob Bernoulli was struck by the way the equiangular spiral reproduces itself under a variety of transformations and asked, in imitation of Archimedes, that such a spiral be engraved on his tombstone, along with the inscription "Eadem mutata resurgo" ("Though changed, I arise again the same.").

Johann Bernoulli was an even more prolific contributor to mathematics than was his brother Jakob. Though he was a jealous and cantankerous man, he was one of the most successful teachers of his time. He greatly enriched the calculus and was very influential in making the power of the new subject appreciated in continental Europe. As we have seen (in Section 11–10), it was his material that the Marquis de l'Hospital (1661–1704), under a curious financial agreement with Johann, assembled in 1696 into the first calculus textbook. In this way, the familiar method of evaluating the indeterminate form 0/0 became incorrectly known, in later calculus texts, as **l'Hospital's rule.** Johann Bernoulli wrote on a wide variety of topics, including optical phenomena connected with reflection and refraction, the determination of the orthogonal trajectories of families of curves, rectification of curves and quadrature of areas by series, analytical trigonometry, the exponential calculus, and other subjects. One of his more noted pieces of work is his contribution to the problem of the **brachistochrone**—the determination of the curve of quickest descent of a weighted particle moving between two given points in a gravitational field; the curve turned out to be an arc of an appropriate cycloid curve. This problem was also discussed by Jakob Bernoulli. The cycloid curve is also the solution to the problem of the **tautochrone**—the determination of the curve along which a weighted particle will arrive at a given point of the curve in the same time interval no matter from what initial point of the curve it starts. This latter problem, which was more generally discussed by Johann Bernoulli, Euler, and Lagrange, had earlier been solved by Huygens (1673) and Newton (1687), and applied by Huygens in the construction of pendulum clocks [see Problem Study 10.7(c)].

Johann Bernoulli had three sons, Nicolaus (1695–1726), Daniel (1700–1782), and Johann II (1710–1790), all of whom won renown as eighteenth-century mathematicians and scientists. Nicolaus, who showed great promise in the field of mathematics, was called to the St. Petersburg Academy, where he unfortunately died by drowning, only eight months later. He wrote on curves, differential equations, and probability. A problem in probability, which he pro-

JOHANN BERNOULLI
(David Smith Collection)

posed from St. Petersburg, later became known as the **Petersburg paradox.** The problem is: If A receives a penny when a head appears on the first toss of a coin, two pennies if a head does not appear until the second toss, four pennies if a head does not appear until the third toss, and so on, what is A's expectation? Mathematical theory shows that A's expectation is infinite, which seems a paradoxical result. The problem was investigated by Nicolaus' brother Daniel, who succeeded Nicolaus at St. Petersburg. Daniel returned to Basel seven years later. He was the most famous of Johann's three sons, and devoted most of his energies to probability, astronomy, physics, and hydrodynamics. In probability he devised the concept of *moral expectation,* and in his *Hydrodynamica,* of 1738, appears the principle of hydrodynamics that bears his name in all present-day elementary physics texts. He wrote on tides, established the kinetic theory of gases, studied the vibrating string, and pioneered in partial differential equations. Johann II, the youngest of the three sons, studied law but spent his later years as a professor of mathematics at the University of Basel. He was particularly interested in the mathematical theory of heat and light.

There was another eighteenth-century Nicolaus Bernoulli (1687–1759), a nephew of Jakob and Johann, who achieved some fame in mathematics. This Nicolaus held, for a time, the chair of mathematics at Padua once filled by Galileo. He wrote extensively on geometry and differential equations. Later in life, he taught logic and law.

Johann Bernoulli II had a son Johann III (1744–1807) who, like his father, studied law but then turned to mathematics. When barely nineteen years old, he was called as a professor of mathematics to the Berlin Academy. He wrote on astronomy, the doctrine of chance, recurring decimals, and indeterminate equations.

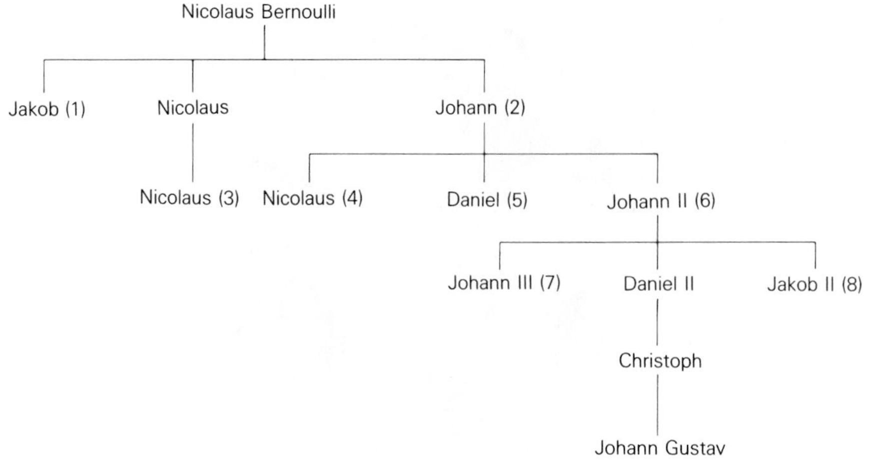

FIGURE 105

Lesser Bernoulli descendants are Daniel II (1751–1834) and Jakob II (1759–1789), two other sons of Johann II, Christoph (1782–1863), a son of Daniel II, and Johann Gustav (1811–1863), a son of Christoph.

Figure 105 shows the Bernoulli genealogical table.

12–3 De Moivre and Probability

In the eighteenth century, the pioneering ideas of Fermat, Pascal, and Huygens in probability theory were considerably elaborated, and the theory made rapid advances, with the result that the *Ars conjectandi* of Jakob Bernoulli was followed by further treatments of the subject. Important among those contributing to probability theory was Abraham De Moivre (1667–1754), a French Hugenot who moved to the more congenial political climate of London after the revocation of the Edict of Nantes in 1685. He earned his living in England by private tutoring, and he became an intimate friend of Isaac Newton.

De Moivre is particularly noted for his work *Annuities upon Lives,* which played an important role in the history of actuarial mathematics, his *Doctrine of Chances,* which contained much new material on the theory of probability, and his *Miscellanea analytica,* which contributed to recurrent series, probability, and analytic trigonometry. De Moivre is credited with the first treatment of the probability integral,

$$\int_0^\infty e^{-x^2}\,dx = \frac{\sqrt{\pi}}{2},$$

and of (essentially) the normal frequency curve

$$y = ce^{-hx^2}, \qquad c \text{ and } h \text{ constants},$$

so important in the study of statistics. The misnamed **Stirling's formula,** which says that for very large n

$$n! \approx (2\pi n)^{1/2} e^{-n} n^n,$$

is due to De Moivre and is highly useful for approximating factorials of large numbers. The familiar formula

$$(\cos x + i \sin x)^n = \cos nx + i \sin nx, \qquad i = \sqrt{-1},$$

known by De Moivre's name and found in every theory of equations textbook, was familiar to De Moivre for the case where n is a positive integer. This formula has become the keystone of analytic trigonometry.

An interesting fable is often told of De Moivre's death. According to the story, De Moivre noticed that each day he required a quarter of an hour more sleep than on the preceding day. When this arithmetic progression reached 24 hours, De Moivre passed away.

The insurance business made great strides in the eighteenth century, and a number of mathematicians were attracted to the underlying probability theory. As a consequence, interest developed in efforts to apply probability theory to new fields. Along these lines, Georges Louis Leclerc, Comte de Buffon (1707–1788), who was director of the Paris Jardin du Roi and noted for his delightful thirty-six volume work on natural history, gave in 1777 the first example of a geometrical probability, his famous "needle problem" for experimentally approximating the value of π (see Section 4–8 and Problem Study 12.13). Efforts also were made to apply probability theory to situations of human judgment, such as computing the chance a tribunal will arrive at a true verdict if to each of the jurymen a number can be assigned that measures the chance he will speak or understand the truth. This *probabilité des jugements,* with its overtones of the Enlightenment philosophy, was prominent in the work of Antoine-Nicolas Caritat, Marquis de Condorcet (1743–1794), who, though an advocate of the French Revolution, was one of the unfortunate victims among the intelligentsia of the excesses following the Revolution. One of Condorcet's conclusions was that capital punishment should be abolished because, however great the probability of the correctness of a single decision, there will be a large probability that in the course of many decisions, some innocent person will be wrongfully condemned.

12–4 Taylor and Maclaurin

Every student of the calculus is familiar with the name of the Englishman Brook Taylor (1685–1731) and the name of the Scotsman Colin Maclaurin (1698–1746), through the very useful Taylor's expansion and Maclaurin's expansion of a function. It was in 1715 that Taylor published (with no consideration of convergence) his well-known expansion theorem,

$$f(a + h) = f(a) + hf'(a) + \frac{h^2}{2!}f''(a) + \dots .$$

In 1717, Taylor applied his series to the solution of numerical equations as follows: Let a be an approximation to a root of $f(x) = 0$; set $f(a) = k$, $f'(a) = k'$, $f''(a) = k''$, and $x = a + h$; expand $0 = f(a + h)$ by the series; discard all powers of h above the second; substitute the values of k, k', k'', and then solve for h. By successive applications of this process, closer and closer approximations can be obtained. Some work done by Taylor in the theory of perspective has found a modern application in the mathematical treatment of photogrammetry, the science of surveying by means of photographs taken from an airplane.

Recognition of the full importance of Taylor's series awaited until 1755, when Euler applied them in his differential calculus, and still later, when Lagrange used the series with a remainder as the foundation of his theory of functions.

Taylor was educated at St. John's College of Cambridge University and early showed great promise in mathematics. He was admitted to the Royal Society and became its secretary, only to resign at the age of thirty-four so that he might devote his time to writing.

Maclaurin was one of the ablest mathematicians of the eighteenth century. The so-called Maclaurin expansion is nothing but the case where $a = 0$ in the Taylor expansion above and was actually explicitly given by Taylor and also by James Stirling (1692–1770) some years before Maclaurin used it, with acknowledgment, in his *Treatise of Fluxions* (two volumes, 1742). Maclaurin did very notable work in geometry, particularly in the study of higher plane curves, and he showed great power in applying classical geometry to physical problems.

BROOK TAYLOR
(David Smith Collection)

Among his many papers in applied mathematics is a prizewinning memoir on the mathematical theory of tides. In his *Treatise of Fluxions* appears his investigation of the mutual attraction of two ellipsoids of revolution.

Maclaurin probably knew as early as 1729 the rule for solving systems of simultaneous linear equations by determinants that today is called *Cramer's rule*. The rule first appeared in print in 1748 in Maclaurin's posthumous *Treatise of Algebra*. The Swiss mathematician Gabriel Cramer (1704–1752) independently published the rule in 1750 in his *Introduction à l'analyse des lignes courbes algébriques,* and it is probably his superior notation that led the general mathematical world to learn the rule from him rather than from Maclaurin.

Maclaurin was a mathematical prodigy. He matriculated at the University of Glasgow at the age of eleven. At fifteen, he took his master's degree and gave a remarkable public defense of his thesis on the power of gravity. At nineteen, he was elected to the chair of mathematics at the Marischal College in Aberdeen; at twenty-one, he published his first important work, *Geometria organica*. At twenty-seven, he became deputy, or assistant, to the professor of mathematics at the University of Edinburgh. There was some difficulty in obtaining a salary to cover his assistantship, and Newton offered to bear the cost personally so that the university could secure the services of so outstanding a young man. In time, Maclaurin succeeded the man he assisted. His treatise on fluxions appeared when he was forty-four, only four years before he died; this was the first logical and systematic exposition of Newton's method of fluxions and was written by Maclaurin as a reply to Bishop Berkeley's attack on the principles of the calculus (see Problem Study 14.24).

Having considered Taylor and Maclaurin, two men whose names are met by students of beginning calculus, we might mention here a third man whose name is similarly encountered. This third, and somewhat earlier, man is Michel

COLIN MACLAURIN
(David Smith Collection)

Rolle, who was born at Ambert, Auvergne, in 1652 and died in Paris in 1719. He was connected with the French war department and wrote on both geometry and algebra. He is known to all calculus students for the theorem of beginning calculus that bears his name and says that $f'(x) = 0$ has at least one real root lying between two successive real roots of $f(x) = 0$. It is from this theorem that textbooks usually derive the highly useful "theorem of mean value" of the calculus course. Few calculus students, however, know that Rolle was one of the most vocal critics of the calculus and that he strove to demonstrate that the subject gave erroneous results and was based upon unsound reasoning. He once characterized calculus as "a collection of ingenious fallacies". So vigorous were his quarrels with the calculus that on several occasions the Académie des Sciences felt obliged to intervene. Later in life, he moderated his attitude and came to see calculus as being useful.

12–5 Euler

The name Leonhard Euler has already been referred to many times in this book. Euler was born in Basel, Switzerland, in 1707. After an essay into the field of theology, Euler found his true vocation in mathematics. Here his father, a Calvinist pastor with an interest in mathematics, helped his son by teaching him the basics of the subject. The father had studied mathematics under Jakob Bernoulli, and it was arranged for his son to study under Johann Bernoulli.

In 1727, when Euler was only twenty years old, his two friends Daniel and Nicolaus Bernoulli, who were connected with the new St. Petersburg Academy formed by Peter the Great, secured a position for Euler at the Russian academy. Daniel left Russia soon after to occupy the chair of mathematics at Basel, and Euler became the Academy's chief mathematician.

LEONHARD EULER
(Library of Congress)

After gracing the St. Petersburg Academy for fourteen years, Euler accepted an invitation from Frederick the Great to go to Berlin to head the Prussian Academy. Euler remained at the Prussian Academy for twenty-five years, but his unsophisticated character did not harmonize with the more scintillating type admired by Frederick, and he suffered many years of petty unpleasantnesses. The Russians had held Euler in high respect, and even after he left for Prussia continued to advance him some salary.

The warmth of the Russian feeling toward him, as contrasted with the coolness of the court of Frederick the Great, led Euler in 1766 to accept an invitation from Catherine the Great to return to the St. Petersburg Academy. There he stayed for the remaining seventeen years of his life. He died very suddenly in 1783 when he was seventy-six years old. It is interesting that throughout his varied career, Euler never held a teaching post.

Euler was a voluminous writer on mathematics, indeed, far and away the most prolific writer in the history of the subject; his name is attached to every branch of the study. It is of interest that his amazing productivity was not in the least impaired when, shortly after his return to the St. Petersburg Academy, he had the misfortune to become totally blind. He had already, since 1735, been blind in his right eye, accounting for the poses assumed in his portraits. Blindness would seem to be an insurmountable barrier to a mathematician, but, like Beethoven's loss of hearing, Euler's loss of sight in no way impaired his amazing productivity. Aided by a phenomenal memory and an ability to concentrate even amidst loud disturbances, he continued his creative work by dictating to a secretary and by writing formulas in chalk on a large slate for his secretary to copy down. Euler published 530 books and papers during his lifetime, and at his death left enough manuscripts to enrich the *Proceedings of the St. Petersburg Academy* for another forty-seven years. A monumental edition of Euler's complete works, containing 886 books and papers, was initiated in 1909 by the Swiss Society of Natural Science and is planned to run to over one hundred large quarto volumes.

Euler's contributions to mathematics are too numerous to expound completely here, but we may note some of his contributions to the elementary field. First of all, we owe to Euler the conventionalization of the following notations:

$f(x)$ for functional notation,
e for the base of natural logarithms,
a, b, c for the sides of triangle ABC,
s for the semiperimeter of triangle ABC,
r for the inradius of triangle ABC,
R for the circumradius of triangle ABC,
Σ for the summation sign,
i for the imaginary unit, $\sqrt{-1}$.

To Euler is also due the very remarkable formula

$$e^{ix} = \cos x + i \sin x,$$

which, for $x = \pi$, becomes

$$e^{i\pi} + 1 = 0,$$

a relation connecting five of the most important numbers in mathematics. By purely formal processes, Euler arrived at an enormous number of curious relations, like

$$i^i = e^{-\pi/2},$$

and he succeeded in showing that any nonzero real number r has an infinite number of logarithms (for a given base), all imaginary if $r < 0$ and all imaginary but one if $r > 0$. In college geometry, we find the *Euler line* of a triangle (see Problem Study 14.1); in college courses in the theory of equations, the student sometimes encounters *Euler's method* for solving quartic equations; and in even the most elementary course in number theory, one meets *Euler's theorem* and the *Euler ϕ-function* (see Problem Study 10.5). The beta and gamma functions of advanced calculus are credited to Euler, although they were adumbrated by Wallis. Euler employed the idea of an integrating factor in the solution of differential equations, gave us our systematic method of solving linear differential equations with constant coefficients, and made the distinction between linear homogeneous and nonhomogeneous differential equations. The differential equation

$$x^n y^{(n)} + a_1 x^{n-1} y^{(n-1)} + \ldots + a_n y^{(0)} = f(x),$$

where exponents in parentheses indicate orders of differentiation, is today known as an **Euler differential equation.** Euler showed that the substitution $x = e^t$ reduces the equation to a linear differential equation with constant coefficients. The theorem, "If $f(x,y)$ is homogeneous of order n, then $xf_x + yf_y = nf$," is today known as **Euler's theorem on homogeneous functions.** Euler was one of the first to develop a theory of continued fractions. He contributed notably to the fields of differential geometry, the calculus of finite differences, and the calculus of variations, and he greatly enriched number theory. In one of his smaller papers occurs the relation

$$v - e + f = 2$$

connecting the number of vertices v, edges e, and faces f of any simple closed polyhedron. In another paper, he investigates **orbiform curves,** or curves that, like the circle, are convex ovals of constant width. Several of his papers are devoted to mathematical recreations, such as unicursal and multicursal graphs (inspired by the seven bridges of Königsberg), the re-entrant knight's path on a chess board, and Graeco-Latin squares. Of course, his chief field of publication was in areas of applied mathematics, particularly lunar theory, tides, the three-body problem of celestial mechanics, the attraction of ellipsoids, hydraulics, ship building, artillery, and a theory of music. The device of *Euler diagrams,* used to test the validity of deductive arguments, was given by Euler in one of his letters to Princess Phillipine von Schwedt, niece of Frederick the Great. During the Seven Years' War (1756–1763), the entire Berlin court sojourned in Magdeburg, and Euler tutored the Princess by letters written from his home in Berlin.

Euler was a masterful writer of textbooks, in which he presented his material with great clarity, detail, and completeness. Among these texts are his prestigious two-volume *Introductio in analysin infinitorum* of 1748, his exceedingly rich *Institutiones calculi differentialis* of 1755, and the allied three-volume *Institutiones calculi integralis* of 1768–74. These books, along with others on mechanics and algebra, served more than any other writings as models in style, scope, and notation for many of the college textbooks of today. Euler's texts enjoyed a marked and a long popularity and, to this day, make very interesting and profitable reading. One cannot but be surprised at Euler's enormous fertility of ideas, and it is no wonder that so many of the great mathematicians coming after him have admitted their indebtedness to him.

It is perhaps only fair to point out that some of Euler's works represent outstanding examples of eighteenth-century formalism, or the manipulation, without proper attention to matters of convergence and mathematical existence, of formulas involving infinite processes. He was incautious in his use of infinite series, often applying to them laws valid only for finite sums. Regarding power series as polynomials of infinite degree, he heedlessly extended to them well-known properties of finite polynomials. Frequently, by such careless approaches, he luckily obtained truly profound results (see Problem Study 12.6 for an example).

Euler's knowledge and interest were by no means confined to just mathematics and physics. He was an excellent scholar, with extensive knowledge of astronomy, medicine, botany, chemistry, theology, and oriental languages. He attentively read the eminent Roman writers, was well informed on both the civil and the literary history of all ages and nations, and showed a wide acquaintance with languages and with many branches of literature. Undoubtedly, he was greatly aided in these diverse fields by his uncommon memory.

Many glowing tributes have been paid to Euler, such as the following two made by the physicist and astronomer François Arago (1786–1853): "Euler could have been called, almost without metaphor, and certainly without hyperbole, analysis incarnate." "Euler calculated without any apparent effort, just as men breathe and as eagles sustain themselves in the air."

Euler had thirteen children. His first son, Johann Albrecht Euler (1734–1800), attained some fame in the field of physics.

12–6 Clairaut, d'Alembert, and Lambert

Alexis Claude Clairaut was born in Paris in 1713 and died there in 1765. He was a youthful mathematical prodigy, composing in his eleventh year a treatise on curves of the third order. This early paper, and a singularly elegant subsequent

ALEXIS CLAUDE CLAIRAUT
(David Smith Collection)

one on the differential geometry of twisted curves in space, won him a seat in the French Academy of Sciences at the illegal age of eighteen. In 1736, he accompanied Pierre Louis Moreau de Maupertuis (1698–1759) on an expedition to Lapland to measure the length of a degree of one of the earth's meridians. The expedition was undertaken to settle a dispute as to the shape of the earth. Newton and Huygens had concluded, from mathematical theory, that the earth is flattened at the poles. But about 1712, the Italian astronomer and mathematician Giovanni Domenico Cassini (1625–1712), and his French-born son Jacques Cassini (1677–1756), measured an arc of longitude extending from Dunkirk to Perpignan, and obtained a result that seemed to support the Cartesian contention that the earth is elongated at the poles. The measurement made in Lapland unquestionably confirmed the Newton-Huygens belief and earned Maupertuis the title of "earth flattener." In 1743, after his return to France, Clairaut published his definitive work, *Théorie de la figure de la Terre*. In 1752, he won a prize from the St. Petersburg Academy for his paper *Théorie de la Lune*, a mathematical study of lunar motion that cleared up some unanswered questions. He applied the process of differentiation to the differential equation

$$y = px + f(p),\ p = \frac{dy}{dx},$$

now known in elementary textbooks on differential equations as **Clairaut's equation,** and he found the singular solution, but this process had been used earlier by Brook Taylor. In 1759, he calculated, with an error of about a month, the 1759 return of Halley's comet.

Clairaut had a brother, three years his junior and known in the history of mathematics only as "le cadet Clairaut" (1716–1732), who tragically died of smallpox when only sixteen, but who at fourteen read a paper on geometry before the French Academy and at fifteen published a work on geometry. The father of the Clairaut children, Jean Baptiste Clairaut (died soon after 1765), was a teacher of mathematics, a correspondent of the Berlin Academy, and a writer on geometry; he had twenty children of whom only one survived him.

This is perhaps the place to mention another differential equation that is met today by any college student of a first course in differential equations and, at the same time, another celebrated mathematical family. The equation is the so-called **Riccati equation,**

$$y' = p(x)y^2 + q(x)y + r(x),$$

named after Giacomo Riccati (1676–1754), a man of independent wealth who studied at Padua when Nicolaus Bernoulli (the nephew of Jakob and Johann) was teaching there. In addition to an extensive study of the above equation, Giacomo Riccati wrote on physics, mensuration, and philosophy and did much to make Newton's work known in Italy. Special cases of the Riccati equation had been studied by Jakob Bernoulli and others, and it was Euler who first pointed out that if a particular solution $v = f(x)$ of the equation is known, then the substitution $y = v + 1/z$ converts the equation into a linear differential equation in z. Giacomo Riccati's second son, Vincenzo Riccati (1707–1775), became a Jesuit professor of mathematics and worked on differential equations, infinite series, quadratures, and hyperbolic functions. Giacomo's third son, Giordano Riccati (1709–1790), wrote on Newton's work, geometry, cubic equations, and physical problems. The fifth son, Francesco Riccati (1718–1791), wrote on applications of geometry to architecture.

Jean-le-Rond d'Alembert (1717–1783), like Alexis Clairaut was born in Paris and died in Paris. As a newborn, he was abandoned near the church of Saint Jean-le-Rond and was discovered there by a gendarme who had him hurriedly christened with the name of the place where he was found. Later, for reasons not known, the name d'Alembert was added.

A scientific rivalry, often unfriendly, existed between d'Alembert and Clairaut. At the age of twenty-four, d'Alembert was admitted to the French Academy. In 1743, he published his *Traite de dynamique,* based upon the great principle of kinetics that now bears his name. It says that the internal actions and reactions of a system of rigid bodies in motion are in equilibrium. In 1714, he applied his principle in a treatise on the equilibrium and motion of fluids, and, in 1746, in a treatise on the causes of winds. In each of these works, and also in one of 1747 devoted to vibrating strings, he was led to partial differential equations, and he became a pioneer in the study of such equations. The problem of vibrating strings led him to the partial differential equation

$$\frac{\partial^2 u}{\partial t^2} = \frac{\partial^2 u}{\partial x^2},$$

for which, in 1747, he gave the solution

$$u = f(x + t) + g(x - t),$$

where f and g are arbitrary functions. With the aid of his principle, he was able to obtain a complete solution to the baffling problem of the precession of the equinoxes. D'Alembert showed interest in the foundations of analysis. In 1754, he made the important suggestion that a sound theory of limits was needed to put analysis on a firm foundation, but most of his contemporaries paid little heed to his suggestion. D'Alembert worked so diligently in an effort to prove the fundamental theorem of algebra (that every polynomial equation $f(x) = 0$ having complex coefficients and of degree $n \geqq 1$ has at least one complex root) that the theorem is today known in France as **d'Alembert's theorem.** It was d'Alembert who gave the name *Riccati equation* to the differential equation considered above.

D'Alembert, like Euler, was broadly educated, with especial knowledge in law, medicine, mathematics, and science. Sharing many common interests, the two men corresponded with one another on a number of matters. D'Alembert was puzzled, as were other mathematicians of the time, as to the nature of logarithms of negative numbers—feeling that one must have $\log(-x) = \log(x)$, on the grounds that $(-x)^2 = (x)^2$, whence $\log(-x)^2 = \log(x)^2$, thence $2 \log(-x) = 2 \log(x)$, and finally $\log(-x) = \log(x)$. In 1747, Euler was able to write to d'Alembert explaining the correct status of logarithms of negative numbers. When, toward the close of Euler's residency in Berlin, Frederick the Great invited d'Alembert to head the Prussian Academy, d'Alembert declined, claiming that it would not be appropriate to place any contemporary in a position of academic superiority over the great Euler. D'Alembert was also invited by Catherine the Great to serve in Russia, but, in spite of the offer of a handsome stipend, he declined that invitation too. In 1754, d'Alembert became permanent secretary of the French Academy. During his later years, he worked on the great French *Encyclopédie,* which had been begun by Denis Diderot and himself. D'Alembert died in 1783, the same year in which Euler died.

A famous and oft-quoted remark made by d'Alembert (and well worth citing on occasion in an elementary algebra class) is: "Algebra is generous; she often gives more than is asked of her." He also once aptly remarked: "Geometrical truths are in a way asymptotic to physical truths; that is to say, the latter approach the former indefinitely near without ever reaching them exactly." Perhaps the most perceptive of d'Alembert's comments on mathematics is the following: "I have no doubt that if men lived separate from each other, and could in such a situation occupy themselves about anything but self-preservation, they would prefer the study of the exact sciences to the cultivation of the agreeable arts. It is chiefly on account of others that a man aims at excellence in the latter; it is on his own account that he devotes himself to the former. In a desert island, accordingly, I should think that a poet could scarcely be vain, whereas a mathematician might still enjoy the pride of discovery."

A little younger than Clairaut and d'Alembert was Johann Heinrich Lam-

JEAN-LE-ROND D'ALEMBERT
(Library of Congress)

bert (1728–1777), born in Mulhouse (Alsace), then part of Swiss territory. Lambert was a mathematician of high quality. As the son of a poor tailor, he was largely self-taught. He possessed a fine imagination, and he established his results with great attention to rigor. In fact, Lambert was the first to prove rigorously that the number π is irrational. He showed that if x is rational, but not zero, then $\tan x$ cannot be rational; since $\tan \pi/4 = 1$, it follows that $\pi/4$, or π, cannot be rational. We also owe to Lambert the first systematic development

JOHANN HEINRICH LAMBERT
(David Smith Collection)

of the theory of hyperbolic functions and, indeed, our present notation for these functions. Lambert was a many-sided scholar who made noteworthy contributions to the mathematics of numerous other topics, such as descriptive geometry, the determination of comet orbits, and the theory of projections employed in the making of maps (a much-used one of these projections is now named after him). At one time, he considered plans for a mathematical logic of the sort once outlined by Leibniz. In 1766, he wrote his posthumously published investigation of Euclid's parallel postulate entitled *Die Theorie der Parallellinien,* a work that places him among the forerunners of the discovery of non-Euclidean geometry (see Section 13–7).

For a short time, Lambert was an associate of Euler in the Prussian Academy. It has been said that when Frederick the Great once inquired of Lambert in which science he was most competent, Lambert curtly replied, "All." Lambert died in 1777, the year in which Carl Freidrich Gauss was born.

12–7 Agnesi and du Châtelet

Noted in mathematics and, indeed, in a number of other areas, is the gifted and erudite Maria Gaetana Agnesi (pronounced än yā′ zē). She was born in Milan in 1718, the first of her father's twenty-one children from three marriages. At an early age, she mastered Latin, Greek, Hebrew, French, Spanish, German, and a number of other foreign languages. When she was only nine years old, her Latin discourse defending higher education for women was published. During her childhood, her father, a professor of mathematics at the University of Bologna, hosted gatherings of the intelligentsia at which Maria would converse with learned professors on any topics of their choice in their native languages. Later, when she was twenty, there appeared her *Propositiones philosophicae,* a series of 190 essays that in addition to mathematics, dealt with logic, mechan-

MARIA GAETANA AGNESI
(David Smith Collection)

ics, hydromechanics, elasticity, gravitation, celestial mechanics, chemistry, botany, zoology, and mineralogy. These essays arose from discussions at her father's gatherings.

In 1748, at the age of thirty, Agnesi published a two-volume work entitled *Instituzioni Analitiche,* written by her primarily for the education of one of her younger brothers who showed interest and ability in mathematics. The work represents a course in elementary and advanced mathematics specially geared for young minds. The first volume deals with arithmetic, algebra, trigonometry, analytic geometry, and, principally, calculus, and is the first calculus text written primarily for young people. The second volume deals with infinite series

INSTITUZIONI
ANALITICHE
AD USO
DELLA GIOVENTU' ITALIANA
DI D.NA MARIA GAETANA
AGNESI
MILANESE
Dell' Accademia delle Scienze di Bologna.
TOMO I.

IN MILANO, MDCCXLVIII.

NELLA REGIA-DUCAL CORTE.
CON LICENZA DE' SUPERIORI.

Title page (reduced in size) of Volume 1 of Maria Gaetana Agnesi's *Instituzioni Analitiche* (1748).
(Courtesy of the Trustees of the Boston Public Library.)

and differential equations. The 1070 pages constitute a remarkable contribution to mathematical education. So that young people could read the work, she shunned the customary Latin and wrote in Italian. Later, in 1801, an English translation appeared, stemming from an earlier unpublished translation made by John Colson, who at one time occupied the Lucasian chair at Cambridge. The name of the English translation is *Analytical Institutions*.

In 1749, Pope Benedict XIV appointed Agnesi an honorary member of the University of Bologna, but (contrary to inaccurately told stories) she never lectured there.

Agnesi greatly disliked publicity and endeavored at different times to lead a secluded life. She finally succeeded when her father died in 1752, devoting the remainder of her life to charitable works and religious study. In 1771, she was appointed director of a beneficient institution in Milan, and it was there that she died in 1799. She had a younger sister, Maria Teresa Agnesi (1724–1780), who became an accomplished musician and composer.

During her lifetime, Maria Gaetano Agnesi achieved fame not only as a mathematician, linguist, and philosopher, but as a somnambulist. On several occasions, she proceeded, while in a somnambulistic state, to her study, lighted a lamp, and solved some problem that she had left incomplete when awake. In the morning, she would be surprised to find the solution carefully and completely worked out on paper on her desk.

Pierre de Fermat at one time interested himself in the cubic curve, which, in present-day notation, would be indicated by the Cartesian equation

$$y(x^2 + a^2) = a^3.$$

Fermat did not name the curve, but it was later studied by Guido Grandi (1672–1742), who named it *versoria*. This is a Latin word for a rope that guides a sail. It is not clear why Grandi assigned this name to the cubic curve. There is a similar obsolete Italian word, *versorio,* which means "free to move in every direction," and the doubly asymptotic nature of the cubic suggests that perhaps Grandi meant to associate this word with the curve. At any rate, when Agnesi wrote her *Instituzioni analitiche,* she confused Grandi's *versoria* or *versorio* with *versiera,* which, in Latin, means "devil's grandmother" or "female goblin." Later, when John Colson translated Agnesi's text into English, he rendered *versiera* as "witch." The curve has ever since in English been called the "witch of Agnesi," although in other languages it is generally more simply referred to as the "curve of Agnesi." The witch of Agnesi possesses a number of pretty properties, some of which can be found in Problem Study 12.11.

Contemporary with Agnesi was another woman mathematician, the Marquise du Châtelet (Gabrielle Émilie Tonnelier de Breteuil), who can be considered more an expositor than a creator of mathematics. She was born in Paris in 1706 and died there in 1749 at the young age of forty-three. She was a mathematician, physicist, and linguist, and a musician who performed skillfully on the clavicembalo (an early form of the piano). She was popularly known for her long entente cordiale with Voltaire. In 1740, she wrote *Institutions de physique,* a work diffused with the views of Leibniz. Her most important contribution to

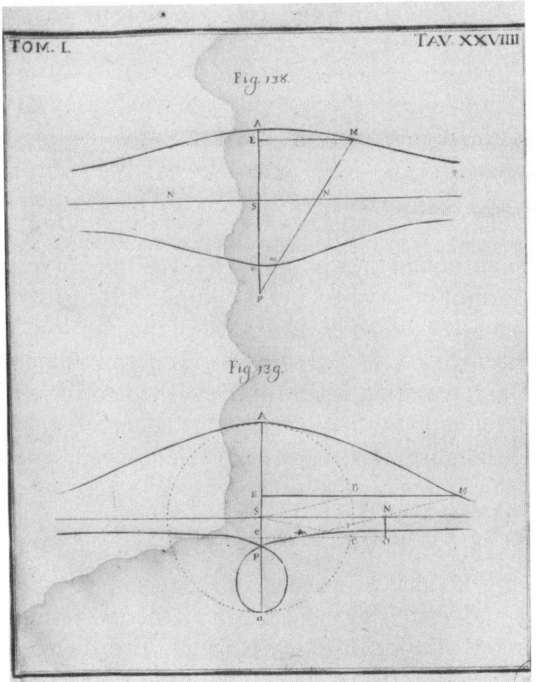

A sample page (reduced in size) from Volume 1 of Maria Gaetana Agnesi's *Instituzioni Analitiche* (1748), showing graphs of some of the curves she studied.
(Courtesy of the Trustees of the Boston Public Library)

mathematics was the first French translation of Newton's *Principia,* which was posthumously published in 1756 with a preface by Voltaire and under the direction of A. C. Clairaut. She also wrote a number of posthumously published treatises on philosophy and religion, and she did much to free French thought from subservience to Cartesianism.

MADAME DU CHÂTELET
(David Smith Collection)

12–8 Lagrange

The two greatest mathematicians of the eighteenth century were Euler and Joseph Louis Lagrange (1736–1813), and which of the two was the greater is a matter of debate that often reflects the differing mathematical sensitivities of the debaters. Lagrange was born in Turin, Italy, into a formerly prosperous family of French and Italian backgrounds; he was the youngest of eleven children and the only one to survive beyond infancy. He was educated in Turin and, as a young man, served as professor of mathematics at the military academy there. In 1766, when Euler left Berlin, Frederick the Great wrote to Lagrange that "the greatest king in Europe" wished to have at his court "the greatest mathematician of Europe." Lagrange accepted the invitation and for twenty years held the post vacated by Euler. A few years after leaving Berlin, in spite of the chaotic political situation in France, Lagrange accepted a professorship at the newly established École Normale, and then at the École Polytechnique. The first of these schools was short-lived, but the second one became famous in the history of mathematics, inasmuch as many of the great mathematicians of modern France were trained there and many held professorships there. Lagrange did much to develop the high degree of scholarship in mathematics that has become associated with the École Polytechnique.

Lagrange was revolted by the cruelties of the Terror that followed the French Revolution. When the great chemist Lavoisier went to the guillotine, Lagrange expressed his indignation at the stupidity of the execution: "It took the mob only a moment to remove his head; a century will not suffice to reproduce it."

Later in life, Lagrange was subject to great fits of loneliness and despondency. He was rescued from these, when he was fifty-six, by a young girl

JOSEPH LOUIS LAGRANGE
(Brown Brothers)

nearly forty years his junior. She was the daughter of his friend, the astronomer Lemonnier. She was so touched by Lagrange's unhappiness that she insisted on marrying him. Lagrange submitted, and the marriage turned out ideal. She proved to be a very devoted and competent companion, and succeeded in drawing her husband out and reawakening his desire to live. Of all his prizes in the world, Lagrange claimed, with honesty and simplicity, the one he most valued was his tender and devoted young wife.

Lagrange's work had a very deep influence on later mathematical research, for he was the earliest first-rank mathematician to recognize the thoroughly unsatisfactory state of the foundations of analysis and accordingly to attempt a rigorization of the calculus. The attempt, which was far from successful, was made in 1797 in his great publication *Théorie des fonctions analytiques contenant les principes du calcul différentiel*. The cardinal idea here was the representation of a function $f(x)$ by a Taylor's series. The derivatives $f'(x)$, $f''(x)$, . . . were then defined as the coefficients of h, $h^2/2!$, . . . in the Taylor expansion of $f(x + h)$ in terms of h. The notation $f'(x)$, $f''(x)$, . . . , very commonly used today, is due to Lagrange. Lagrange felt that his approach had side-stepped the use of limits, but, since he neglected to give sufficient attention to matters of convergence and divergence, and since these concepts depend upon the limit idea, his approach failed in its aim. Nevertheless, we have here the first "theory of functions of a real variable." Two other great works of Lagrange are his *Traité de résolution des équations numériques de tous degrés* (1767) and his monumental *Mécanique analytique* (1788); the former gives a method of approximating the real roots of an equation by means of continued fractions, the latter (which Sir William Rowan Hamilton described as a "scientific poem") contains the general equations of motion of a dynamical system known today as *Lagrange's equations*. His work in differential equations (for example, the method of variation of parameters), and particularly in partial differential equations, is very notable, and his contributions to the calculus of variations did much for the development of that subject. Lagrange had a penchant for number theory and wrote important papers in this field also, such as the first published proof of the theorem that every positive integer can be expressed as the sum of not more than four squares. Some of his early work on the theory of equations later led Galois to his theory of groups. In fact, the important theorem of group theory that states that the order of a subgroup of a finite group G is a factor of the order of G is called **Lagrange's theorem.** Lagrange has been mentioned a number of times in earlier parts of our book.

Whereas Euler wrote with a profusion of detail and a free employment of intuition, Lagrange wrote concisely and with attempted rigor. A formal manipulator in mathematics often experiences the discomforting feeling that his pencil surpasses him in intelligence; this was a feeling that Euler confessed he often could not lose. Lagrange seemed to have a greater mathematical conscience; he was "modern" in style and can be characterized as the first true analyst. All great musicians can be classified as either accomplished performers or composers, and a few have been both. Similarly, all great mathematicians can be classified as either expert formal operators or expert creators of theory, and a few have been both. Euler was primarily a great formal operator, Lagrange a great theorist, and Gauss was pre-eminently accomplished as both. Thus, Euler was like a Heifetz, Lagrange a Beethoven, and Gauss a Johann Sebastian Bach.

Lagrange once remarked that a mathematician has not thoroughly understood a piece of his own work until he has so clarified it that he can effectively explain it to the first man he meets in the street. Though this ideal often appears impossible, time frequently renders it attainable. Newton's law of universal gravitation, which at first was incomprehensible to even highly educated persons, has today become common knowledge. Einstein's relativistic theory of gravitation is now undergoing a similar transmutation.

Napoleon Bonaparte, who hobnobbed with a number of the great mathematicians of France, summed up his estimate of Lagrange by saying, "Lagrange is the lofty pyramid of the mathematical sciences."

12–9 Laplace and Legendre

Laplace and Legendre were contemporaries of Lagrange, although they published their principal works in the nineteenth century. Pierre-Simon Laplace was born of poor parents in 1749. His mathematical ability early won him good teaching posts; as a political opportunist, he ingratiated himself with whichever party happened to be in power during the uncertain days of the French Revolution. His most outstanding work was done in the fields of celestial mechanics, probability, differential equations, and geodesy. He published two monumental works, *Traité de mécanique céleste* (five volumes, 1799–1825) and *Théorie analytique des probabilités* (1812), each of which was preceded by an extensive nontechnical exposition. The five-volume *Traité de mécanique céleste,* which

PIERRE-SIMON LAPLACE
(Brown Brothers)

earned him the title of "the Newton of France," embraced all previous discoveries in this field along with Laplace's own contributions, and marked the author as the unrivaled master in the subject. It may be of interest to repeat a couple of anecdotes often told in connection with this work. When Napoleon teasingly remarked that God was not mentioned in his treatise, Laplace replied, "Sire, I did not need that hypothesis." The American astronomer, Nathaniel Bowditch, when he translated Laplace's treatise into English, remarked, "I never come across one of Laplace's 'Thus it plainly appears' without feeling sure that I have hours of hard work before me to fill up the chasm and find out and show how it plainly appears." Laplace's name is connected with the *nebular hypothesis* of cosmogony and with the so-called *Laplace equation* of potential theory (though neither of these contributions originated with Laplace), with the so-called *Laplace transform* that later became the key to the operational calculus of Heaviside, and with the *Laplace expansion* of a determinant. Laplace died in 1827, exactly one hundred years after the death of Isaac Newton. According to one report, his last words were: "What we know is slight; what we don't know is immense."

The following story about Laplace is of interest and offers a valuable suggestion to one applying for a position. When Laplace arrived as a young man in Paris seeking a professorship of mathematics, he submitted his recommendations by prominent people to d'Alembert, but he was not received. Returning to his lodgings, Laplace wrote d'Alembert a brilliant letter on the general principles of mechanics. This opened the door, and d'Alembert replied: "Sir, you notice that I paid little attention to your recommendations. You don't need any; you have introduced yourself better." A few days later, Laplace was appointed professor of mathematics at the Military School of Paris.

Lagrange and Laplace have often been contrasted with one another. First of all, there is a marked contrast in their styles, summed up as follows by W. W. Rouse Ball. "Lagrange is perfect both in form and matter, he is careful to explain his procedure, and though his arguments are general they are easy to follow. Laplace on the other hand explains nothing, is indifferent to style, and, if satisfied that his results are correct, is content to leave them either with no proof or with a faulty one." There is also a marked contrast in the viewpoints of mathematics held by the two men. For Laplace, mathematics was merely a kit of tools used to explain nature. To Lagrange, mathematics was a sublime art and was its own excuse for being.

Laplace was very generous to beginners in mathematical research. He called these beginners his stepchildren, and there are several instances in which he withheld publication of a discovery to allow a beginner the opportunity to publish first. Sadly, such generosity is rare in mathematics.

We close our brief account of Laplace with two quotations due to him. "All the effects of nature are only mathematical consequences of a small number of immutable laws." "In the final analysis, the theory of probability is only common sense expressed in numbers."

Adrien-Marie Legendre (1752–1833) is known in the history of elementary mathematics principally for his very popular *Éléments de géométrie,* in which he attempted a pedagogical improvement of Euclid's *Elements* by considerably rearranging and simplifying many of the propositions. This work was very favorably received in America and became the prototype of the geometry textbooks in this country. In fact, the first English translation of Legendre's geometry was made in 1819 by John Farrar of Harvard University. Three years later another English translation was made, by the famous Scottish littérateur Thomas Carlyle, who early in life was a teacher of mathematics. Carlyle's translation, as later revised by Charles Davies, and later still by J. H. Van Amringe, ran through thirty-three American editions. In later editions of his geometry, Legendre attempted to prove the parallel postulate (see Section 13–7). Legendre's chief work in higher mathematics centered about number theory, elliptic functions, the method of least squares, and integrals; this work is too advanced to be discussed here. He was also an assiduous computer of mathematical tables. Legendre's name is today connected with the second-order differential equation

$$(1 - x^2)y'' - 2xy' + n(n + 1)y = 0,$$

which is of considerable importance in applied mathematics. Functions satisfying this differential equation are called **Legendre functions** (of order n). When n is a nonnegative integer, the equation has polynomial solutions of special interest called **Legendre polynomials.** Legendre's name is also associated with the symbol $(c|p)$ of number theory. The **Legendre symbol** $(c|p)$ is equal to ± 1 according as the integer c, which is prime to p, is or is not a quadratic residue of the odd prime p. [For example, $(6|19) = 1$ since the congruence $x^2 \equiv 6 \pmod{19}$ has a solution, and $(39|47) = -1$ since the congruence $x^2 \equiv 39 \pmod{47}$ has no solution.]

ADRIEN-MARIE LEGENDRE
(David Smith Collection)

In addition to his *Éléments de géométrie,* which appeared in 1794, Legendre published a two-volume 859-page work, *Essai sur la théorie des nombres* (1797–1798), which was the first treatise devoted exclusively to number theory. He later wrote a three-volume treatise, *Exercises du calcul intégral* (1811–1819), which, for comprehensiveness and authoritativeness, rivaled the similar work of Euler. Legendre later expanded parts of this work into another three-volume treatise, *Traité des fonctions elliptiques et des intégrals eulériennes* (1825–1832). Here Legendre introduced the term *Eulerian integrals* for the beta and gamma functions. In geodesy, Legendre achieved considerable fame for his triangulation of France.

12–10 Monge and Carnot

The last two outstanding mathematicians whom we shall consider in this chapter are the two geometers Gaspard Monge (1746–1818) and Lazare Carnot (1753–1823). Unlike the three L's (Lagrange, Laplace, and Legendre), who remained aloof from the French Revolution, Monge and Carnot supported it and played active roles in revolutionary matters.

Monge was educated at the college of the Oratorians in Beaunne, the town of his birth, and at their college in Lyons, where at the age of sixteen he became an instructor in physics. A skillfully constructed large-scale map of his home town led to his acceptance at the military school in Mézières as a draftsman. Asked to work out from supplied data the gun emplacements of a proposed fortress, Monge circumvented the long and tedious arithmetic procedure of the time by a rapid geometric one. His method, which was one of cleverly representing three-dimensional objects by appropriate projections on the two-dimen-

GASPARD MONGE
(New York Public Library Collection)

sional plane, was adopted by the military and classified as top-secret. It later became widely taught as **descriptive geometry.** In 1768, Monge became a professor of mathematics, and in 1771, a professor of physics, at Mézières. In 1780, he was appointed to a chair of hydraulics at the Lyceum in Paris.

Monge served as Minister of Marine and engaged in the manufacture of arms and gunpowder for the army. He was the principal force, under the Directory, in the founding of the École Polytechnique in 1795, and was a professor of mathematics there. He gained the close friendship and admiration of Napoleon and accompanied the latter, along with the mathematician Joseph Fourier (1768–1830), on the ill-fated Egyptian expedition of 1798. Upon returning to France, Monge resumed his position at the École Polytechnique, where he proved to be a singularly gifted teacher. His lectures there inspired a large following of able geometers, among whom were Charles Dupin (1784–1873) and Jean Victor Poncelet (1788–1867), the former a contributor to the field of differential geometry, and the latter to that of projective geometry.

In addition to creating descriptive geometry, Monge is considered as the father of differential geometry. His work entitled *Application de l'analyse à la géométrie* ran through five editions and was one of the most important of the early treatments of the differential geometry of surfaces. It is here that Monge introduced, among other things, the concept of lines of curvature of a surface in three-space. Monge's contributions to differential geometry are devoted principally to the *extrinsic geometry* of surfaces (see Section 14–7).

It was with Monge's lectures at the École Polytechnique that solid analytic geometry came into its own. The material of these lectures was written up by Monge and Jean-Nicolas-Pierre Hachette (1769–1834) in 1802 in an extensive memoir on *Application d'algèbre à la géométrie* that appeared in the *Journal de l'École Polytechnique*. The opening theorem of the work is the well known eighteenth-century generalization of the Pythagorean theorem: *The sum of the squares of the orthogonal projections of a planar area upon three mutually perpendicular planes is equal to the square of the planar area.* Farther in the work we find much of the material of present-day texts devoted to solid analytic geometry, such as the formulas for the translation and rotation of axes, the customary treatment of lines and planes in space, and the determination of the principal planes of a conicoid. It is shown that the plane through a given point (x', y', z') orthogonal to the intersection of two given planes

$$ax + by + cz + d = 0 \text{ and } ex + fy + gz + h = 0$$

is given by

$$A(x - x') + B(y - y') + C(z - z') = 0,$$

where

$$A = bg - fc, B = ce - ga, C = af - eb.$$

Formulas are given for the distance between a point and a line in space and for the shortest distance between two skew lines in space. Among new results given by Monge we find.

1. *The six planes through the midpoints of the edges of a tetrahedron and perpendicular to the respectively opposite edges are concurrent in a point that is the reflection of the circumcenter of the tetrahedron in the centroid of the tetrahedron.* (This point is now called the **Monge point** of the tetrahedron.)
2. *The locus of the vertex of the trirectangular angle whose faces are tangent to a given central conicoid is a sphere concentric with the conicoid.* (This sphere is now called the **Monge sphere,** or **director sphere,** of the conicoid. The analogous locus in two dimensions is today called the **Monge circle** of the associated central conic, though this locus had been found a century earlier by La Hire using synthetic methods.)

Later, in 1809, Monge gave several proofs of the fact that the lines joining the midpoints of opposite edges of a tetrahedron concur at the centroid of the tetrahedron.

Monge had two brothers who also were professors of mathematics.

Lazare Nicolas Marguerite Carnot (1753–1823), following the custom of many of the sons of well-to-do French families, prepared himself for the army and was thus led to the military school of Mézières, where he studied under Monge, becoming a captain in the engineers in 1783. In 1784, he wrote his first mathematical work, on mechanics, which contains the earliest proof that kinetic energy is lost in the collision of imperfectly elastic bodies. With the advent of the French Revolution, he threw himself into politics and embraced the Revolution with enthusiasm and dedication. He succeeded to a number of important political posts and, in 1793, voted for the execution of Louis XVI as a traitor. Also in 1793, when a united Europe launched a million-man army against France, Carnot undertook the seemingly impossible task of organizing fourteen armies to successfully oppose the enemy, winning for himself the name "the Organizer of Victory." In 1796, he opposed Napoleon's coup d'état, and had to flee to Geneva, where he wrote a semiphilosophical work on the metaphysics of the calculus. His two important contributions to geometry, *Géométrie de position* and *Essai sur la théorie des transversals,* were published in 1803 and 1806. As an "irreconcilable enemy of kings" he offered, in 1814, after the Russian campaign, to fight for France but not for the empire. With the restoration, he was exiled, dying in Madgeburg in straitened circumstances in 1823.

It is in Carnot's *Géométrie de position* that sensed magnitudes were first systematically employed in synthetic geometry. By means of sensed magnitudes, several separate statements or relations can often be combined into a single inclusive statement or relation, and a single proof can frequently be formulated that would otherwise require the treatment of a number of different cases (see Problem Study 12.17). The idea of sensed magnitudes was further exploited by Augustus Ferdinand Möbius (1790–1868) in his *Der barycentische Calcul* of 1827.

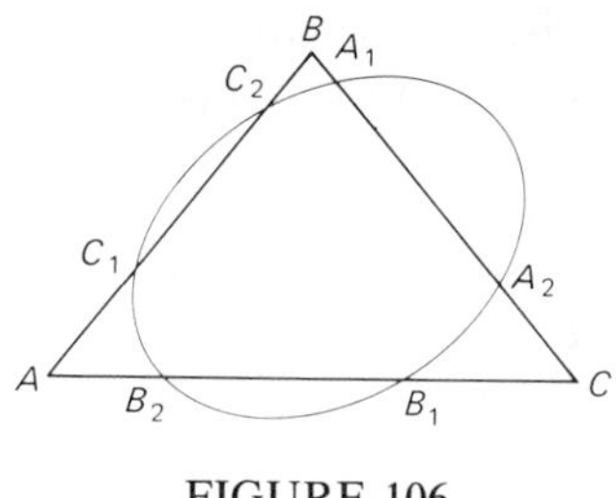

FIGURE 106

The theorem of Menelaus (see Section 6-5) is basic in Carnot's *Essai sur la théorie des transversals*. Here Carnot extends the theorem of Menelaus to the situation where the transversal in that theorem is replaced by an arbitrary algebraic curve of degree n. As an illustration, we have, for the case $n = 2$ (see Figure 106): *If the sides BC, CA, AB of a triangle ABC cut a conic in the* (real or imaginary) *points* A_1 *and* A_2, B_1 *and* B_2, C_1, *and* C_2, *respectively, then*

$$(AC_1)(AC_2)(BA_1)(BA_2)(CB_1)(CB_2) = (AB_1)(AB_2)(BC_1)(BC_2)(CA_1)(CA_2),$$

where all segments are sensed segments. The theorem can be further generalized by replacing the triangle by an arbitrary polygon.

Carnot also found the volume of a tetrahedron in terms of its six edges, and he obtained a formula (containing 130 terms) expressing any one of the ten segments joining five random points in terms of the other nine.

Carnot had a son, Hippolyte, who became minister of public instruction in 1848; another son, Sadi, who became a celebrated physicist; a grandson, also

LAZARE CARNOT
(David Smith Collection)

named Sadi and son of Hippolyte, who became the fourth president of the third French Republic; and a second grandson, Adolphe, the son of Hippolyte, who became an eminent chemist.

Monge and Carnot were both ardent Revolutionists, but surely Carnot was the intellectually more honest and consistent of the two. Both voted for the death of Louis XVI, but Carnot, although he was willing to serve under Napoleon as a soldier and as an administrator, was the only Tribune with the courage and conviction to vote against naming Napoleon emperor, and he went into exile for his stand. Monge, on the other hand, slavishy supported his idol all the way from the early idealistic and revolutionary corporal to the selfish and despotic emperor, and it was Monge who willingly accepted the detestable task of determining which art treasures should be brought back from Italy to Paris as war booty.

12–11 The Metric System

Measurements of length, area, volume, and weight play an important part in the practical applications of mathematics. Basic among the units of these measurements is that of length, for given a unit of length, units for the other quantities can easily be devised. One of the important accomplishments of the eighteenth century was the construction of the metric system, designed to replace the world's vast welter of chaotic and unscientific systems of weights and measures by one that is orderly, uniform, scientific, exact, and simple.

The development of our present metric system was not the first attempt to initiate a scientific system of measurement. In 1670, the French mathematician and vicar of the church of St. Paul at Lyons, Abbé Gabriel Mouton, suggested one minute of the earth's circumference as a unit of length, and he divided and multiplied this unit decimally, assigning appropriate Latin terminology to the various divisions and multiples. About the same time, Sir Christopher Wren, in England, proposed taking the length of a pendulum beating half seconds as the unit of length; this would have approximated one-half of the length commonly assigned to the ancient cubit (the distance from a man's elbow to the tip of his extended middle finger). In 1671, the French astronomer Jean Picard, and in 1673, the Dutch physicist Christiaan Huygens, advocated the length of a seconds pendulum at sea level at 45° of latitude; this would have been only about 6 millimeters shorter than the present-day meter. In 1747, La Condamine suggested the seconds pendulum at the equator. In 1775, Messier very carefully determined the length of a seconds pendulum at 45° of latitude and unsuccessfully endeavored to have this adopted as the standard unit.

Motivated by the widespread agitation for a new system of measures, in 1789 the French Académie des Sciences appointed a committee to work out an acceptable plan. In the following year, Sir John Miller proposed, in the House of Commons, a uniform system of measurement for Great Britain. About the same time, Thomas Jefferson proposed a uniform system for the United States, suggesting the length of a seconds pendulum at 38° latitude, this being the mean latitude for the United States of his day.

In its work, the committee of the Académie des Sciences agreed on a decimal system and considered two alternatives for the unit of length of the system. One was the length of a seconds pendulum. Since the pendulum equation is $T = 2\pi\sqrt{L/g}$, this would make the standard length, or meter, g/π^2. Inasmuch as g varies with both latitude and elevation, and in view of the accuracy with which Legendre and others had measured the length of a terrestrial meridian, the committee finally agreed to take the meter to be the ten-millionth part of the meridional distance from the North Pole to the equator. In 1793, due to political forces, the Académie des Sciences was suppressed, but the Committee on Weights and Measures was retained, purged of some of its members, such as Lavoisier, and enlarged by others, in time including Lagrange, Laplace, Legendre, and Monge. By 1799, the work of the Committee was completed, and our present-day metric system became a reality.

It was in June, 1799, that the Republic of France adopted the metric system of weights and measures; in 1837, its use was made compulsory. Today the system has been adopted by all the civilized nations of the world except the United States, which has been making preparations to join. Of course, the metric system has long been employed in the United States for scientific and other purposes.

An International Bureau of Weights and Measures has been set up on a piece of international territory at Sèvres, France, close to Paris. It was established by representatives from all over the world, and there the international standards for the kilogram and the meter are preserved. The standard for the kilogram is made of a special alloy of platinum and iridium, and an exact duplicate of this standard is kept by each of the represented nations. The one in the United States was received by President Benjamin Harrison on January 2, 1890, and is housed at the Bureau of Standards in Washington, D.C. Prior to 1960, the standard for the meter was a platinum-iridium bar, but today the standard meter is more accurately defined as 1,650,763.73 wavelengths of the orange-red light from the isotope krypton-86, measured in a vacuum.

12–12 Summary

We conclude our brief survey of eighteenth-century mathematics by noting that, though the century witnessed considerable further development in such subjects as trigonometry, analytic geometry, calculus, theory of numbers, theory of equations, probability, differential equations, and analytic mechanics, it also witnessed the creation of a number of new fields, such as actuarial science, the calculus of variations, higher functions, partial differential equations, descriptive geometry, and differential geometry. Much of the mathematical research of the century found its source and inspiration in mechanics and astronomy. But in d'Alembert's concern over the shaky basis of analysis, Lambert's work on the parallel postulate, Lagrange's effort to rigorize the calculus, and the philosophical thoughts of Carnot, we have intimations of the coming liberation of geometry and algebra and the deep concern over the foundations of mathematics that took place in the nineteenth century. Moreover, mathematicians with specialization in restricted fields, like Monge in geometry, began to emerge. Also, on June 22, 1799, following the French Revolution, the Republic of France adopted the metric system of weights and measures.

Another important event that took place in the eighteenth century was the serious entrance by women into the fields of mathematics and the exact sciences. Such pursuits by women had been socially frowned upon, and opportunities for them were virtually nonexistent. The first significant break was made by the Marquise du Châtelet (1706–1749) and Maria Gaetana Agnesi (1718–1799), who made worthy impressions in the area of mathematics. In our next chapter, we shall see how further liberation occurred in the early nineteenth century, with the work of Sophie Germain (1776–1831) and Mary Fairfax Somerville (1780–1872).

Problem Studies

12.1 Bernoulli Numbers

The formulas

$$1 + 2 + 3 + \cdots + (k - 1) = \frac{k^2}{2} - \frac{k}{2},$$

$$1^2 + 2^2 + 3^2 + \cdots + (k - 1)^2 = \frac{k^3}{3} - \frac{k^2}{2} + \frac{k}{6},$$

$$1^3 + 2^3 + 3^3 + \cdots + (k - 1)^3 = \frac{k^4}{4} - \frac{k^3}{2} + \frac{k^2}{4},$$

which express the sums

$$S_n(k) \equiv 1^n + 2^n + 3^n + \ldots + (k - 1)^n$$

for $n = 1, 2, 3$ as polynomials in k, have been known since remote times. Jakob Bernoulli became interested in the coefficients $B_1, B_2, B_3, \ldots$ when $S_n(k)$ is expressed as a polynomial in k of a form

$$S_n(k) = \frac{k^{n+1}}{n + 1} - \frac{k^n}{2} + B_1C(n,1)\frac{k^{n-1}}{2} - B_2C(n,3)\frac{k^{n-3}}{4} + \cdots,$$

where $C(n,r) = n(n - 1) \ldots (n - r + 1)/r!$. These coefficients, which are now known as **Bernoulli numbers,** play an important role in analysis and possess some remarkable arithmetic properties.

(a) If $n = 2r + 1$, it can be shown that

$$B_1C(n,2) - B_2C(n,4) + B_3C(n,6) - \cdots + (-1)^{r-1}B_rC(n,2r) = r - 1/2.$$

Using this formula, compute B_1 through B_5.

(b) A prime p is said to be **regular** if it divides none of the numerators of B_1, $B_2, \ldots, B_{(p-3)/2}$ when these numbers are written in their lowest terms. Otherwise p is said to be **irregular.** Knowing that

$$B_{16} = \frac{7709321041217}{510},$$

show that 37 is irregular.

In 1850, E. Kummer proved that Fermat's last "theorem" is true for every exponent that is a regular prime, and the only irregular primes below 100 are 37, 59, 67.

(c) K. C. G. von Staudt established the remarkable theorem: $B_r = G + (-1)^r(1/a + 1/b + 1/c + \ldots)$, where G is an integer and $a, b, c, \ldots$ are all the primes p such that $2r/(p-1)$ is an integer. Verify Staudt's theorem for $B_4 = \frac{1}{30}$ and $B_8 = \frac{3617}{510}$.

12.2 De Moivre's Formula

(a) Establish De Moivre's formula:

$$(\cos x + i \sin x)^n = \cos nx + i \sin nx,$$

where $i = \sqrt{-1}$ and n is a positive integer.

(b) Using the formula of (a), express $\cos 4x$ and $\sin 4x$ in terms of $\sin x$ and $\cos x$.

(c) Using De Moivre's formula, show that $(-1 - i)^{15} = -128 + 128i$.

(d) Prove that $i^n = \cos(n\pi/2) + i \sin(n\pi/2)$.

(e) Using De Moivre's formula, find the eight eighth roots of 1.

12.3 Distributions

(a) Six coins were simultaneously tossed 1000 times. Of these 1000 tosses, there were 9 in which no heads appeared, 99 in which 1 head appeared, 241 in which 2 heads appeared, 313 in which 3 heads appeared, 233 in which 4 heads appeared, 95 in which 5 heads appeared, and 10 in which 6 heads appeared. Display this frequency distribution by drawing a frequency curve.

(b) Plot the normal frequency curve $y = 10e^{-x^2}$.

(c) Calculate the arithmetic mean of the collection of heads per toss in the experiment of (a).

(d) The **median** of a collection of numerical values is the middle term after the values have been arranged in ascending or descending order of magnitude. What is the median of the collection of heads per toss in the experiment of (a)?

(e) If, in a collection of numerical values, 1 number occurs more often than any other, it is called the **mode** of the collection. What is the mode of the collection of heads per toss in the experiment of (a)?

(f) Consider the situation where a millionaire joins the population of a small community of low-income people. What is the effect on the mean income, the median income, and the modal income of the community?
(g) Is a shoe merchant most interested in the arithmetic mean, the median, or the mode of the shoe sizes of the people in his community?
(h) What can one say about the arithmetic mean, the median, and the mode of a normal frequency distribution?
(i) Approximate 1000! by Stirling's formula.

12.4 Formal Manipulation of Series

(a) Develop the Maclaurin expansions for $\sin z$, $\cos z$, and e^z.
(b) Show that the Maclaurin expansion for $\cos z$ can be obtained by differentiating, term by term, the Maclaurin expansion for $\sin z$.
(c) Show formally, using the expansions of (a), that

$$\cos x + i \sin x = e^{ix}.$$

(d) Using the Maclaurin expansion for $\sin z$, show that

$$\lim_{z \to 0} \left(\frac{\sin z}{z}\right) = 1.$$

(e) Using the Taylor's expansions about $x = a$ for $f(x)$ and $g(x)$ show, when $f(a) = f'(a) = \cdots = f^{(k)}(a) = 0$, $g(a) = g'(a) = \cdots = g^{(k)}(a) = 0$, $g^{(k+1)}(a) \neq 0$, that

$$\lim_{x \to a} \frac{f(x)}{g(x)} = \frac{f^{(k+1)}(a)}{g^{(k+1)}(a)}.$$

12.5 A Conjecture and a Paradox

(a) Euler conjectured that for $n > 2$ at least n nth powers are required to provide a sum that is itself an nth power. In 1966, L. J. Lander and T. R. Parkin, using high-speed computers, discovered that

$$27^5 + 84^5 + 110^5 + 133^5 = 144^5.$$

Check the truth of this counterexample.
(b) Explain the following paradox that bothered mathematicians of Euler's time: Since $(-x)^2 = (x)^2$, we have $\log(-x)^2 = \log(x)^2$, whence $2 \log(-x) = 2 \log(x)$, and thence $\log(-x) = \log(x)$.

12.6 Euler and an Infinite Series

(a) Oldenburg, in a letter to Leibniz in 1673, asked for the sum of the infinite series

$$1/1^2 + 1/2^2 + 1/3^2 + 1/4^2 + \cdots.$$

Leibniz was unable to find an answer, and, in 1689, Jakob Bernoulli admitted that he could not find an answer. Carry out the details of the following formal procedure used by Euler for solving this problem.

Start with the Maclaurin series

$$\sin z = z - z^3/3! + z^5/5! - z^7/7! + \cdots.$$

Then $\sin z = 0$ can (after dividing through by z) be considered as the infinite polynomial

$$1 - z^2/3! + z^4/5! - z^6/7! + \cdots = 0,$$

or, replacing z^2 by w, as the equation

$$1 - w/3! + w^2/5! - w^3/7! + \cdots = 0.$$

By the theory of equations, the sum of the reciprocals of the roots of this equation is the negative of the coefficient of the linear term—namely, $\frac{1}{6}$. Since the roots of the polynomial in z are $\pi, 2\pi, 3\pi, \ldots$, it follows that the roots of the polynomial in w are $\pi^2, (2\pi)^2, (3\pi)^2, \ldots$. Therefore

$$\tfrac{1}{6} = 1/\pi^2 + 1/(2\pi)^2 + 1/(3\pi)^2 + \cdots,$$

or

$$\pi^2/6 = 1/1^2 + 1/2^2 + 1/3^2 + \cdots.$$

(b) Apply Euler's procedure of (a) to the Maclaurin expansion of $\cos z$ to find

$$\pi^2/8 = 1/1^2 + 1/3^2 + 1/5^2 + \cdots.$$

(c) Using (a) and (b), formally show that

$$\pi^2/12 = 1/1^2 - 1/2^2 + 1/3^2 - 1/4^2 + \cdots.$$

In his *Introductio* of 1748, Euler gave the sum of

$$1/1^n + 1/2^n + 1/3^n + \cdots$$

for even values of n from $n = 2$ through $n = 26$. The cases in which n is odd are still intractable today, and it is not even known if the sum of the reciprocals of the cubes of the positive integers is a rational multiple of π^3. Euler arrived at many results, now known to be true, by employing a carefree application to infinite polynomials (power series) rules valid for finite polynomials.

12.7 Orbiform Curves

An **orbiform curve,** or **curve of constant width,** is a planar convex oval characterized by the property that the distance between 2 parallel tangents to the curve is constant.

(a) Show that the **Reuleaux triangle,** defined by 3 circular arcs with centers at the vertices of an equilateral triangle and with radii equal to a side of the triangle, is an orbiform curve. (Drills based on the shape of a Reuleaux triangle have been devised for drilling *square* holes.)

(b) Show how, starting from any triangle, one can construct an orbiform curve composed of 6 circular arcs.

(c) Starting from a pentagon whose diagonals are all equal, construct an orbiform curve composed of 5 circular arcs.

(d) Show how, starting from any convex pentagon, one can construct an orbiform curve composed of 10 circular arcs.

(e) Construct an orbiform curve containing no circular arcs.

(f) A point P on an orbiform curve is said to be an **ordinary point** if the curve has a continuously turning tangent at point P. Opposite extremities of a maximum chord of an orbiform curve are known as **opposite points** of the curve. Try to establish the following theorems about orbiform curves.

1. No part of an orbiform curve is straight.
2. If P_1 and P_2 are a pair of ordinary opposite points of an orbiform curve, then P_1P_2 is normal to the curve at P_1 and P_2.
3. If r_1 and r_2 are the radii of curvature at a pair of ordinary opposite points P_1 and P_2 of an orbiform curve of constant width d, then $r_1 + r_2 = d$.
4. **Barbier's theorem:** The circumference of an orbiform curve of constant width d is πd.

(g) Show that if a Reuleaux triangle is rotated about an axis of symmetry, one obtains a solid of constant width. (Much less is known about solids of constant width than about curves of constant width. Though there is no direct analog of Barbier's theorem, Minkowski has pointed out that the shadows, formed by orthogonal projection, of a solid of constant width are of constant circumference.)

12.8 Unicursal and Multicursal Graphs

In 1736, Euler resolved a question then under discussion as to whether it was possible to take a walk in the town of Königsberg in such a way that every bridge in the town would be crossed once and only once and the walker return to his starting point. The town was located close to the mouth of the Pregel River, had 7 bridges, and included an island, as pictured in Figure 107. Euler reduced the problem to that of tracing the associated graph of Figure 108 in such a way that each line of the graph is traced once and only once, and the tracing point ends up at its starting point

In considering the general problem, the following definitions are useful. A **node** is a point of a graph from which lines radiate. A **branch** is a line of a graph

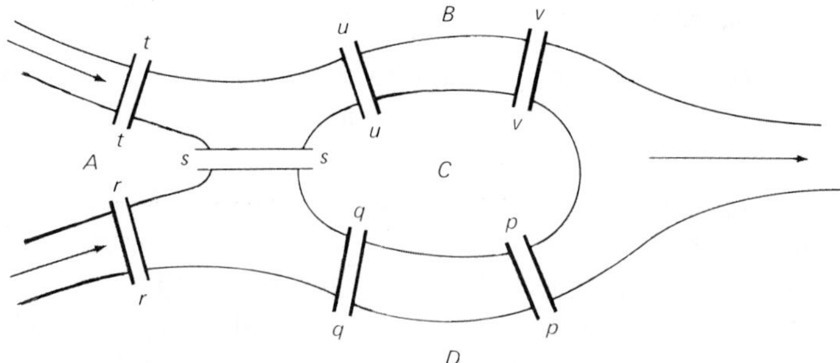

FIGURE 107

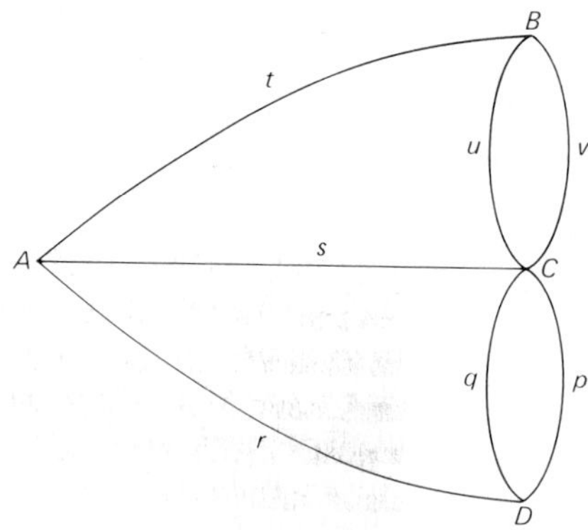

FIGURE 108

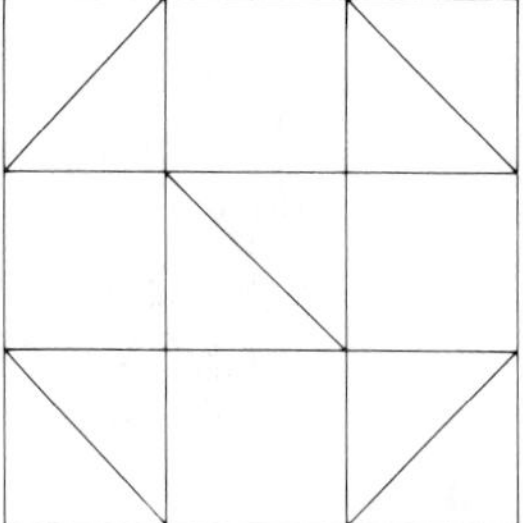
FIGURE 109

connecting 2 consecutive nodes. The **order** of a node is the number of branches radiating from it. A node is said to be **even** or **odd** according as its order is even or odd. A **route** consists of a number of branches that can be traced consecutively without traversing any branch twice. A graph that can be traced in 1 route is said to be **unicursal;** otherwise, it is said to be **multicursal.** About these concepts, Euler succeeded in establishing the following propositions:

1. *In any graph the number of odd nodes is even.*
2. *A graph with no odd nodes can be traversed unicursally in a reentrant route that terminates at its starting point.*
3. *A graph with exactly 2 odd nodes can be traversed unicursally by starting at one of the odd nodes and then terminating at the other.*
4. *A graph with more than 2 odd nodes is multicursal.*

(a) Using Euler's theorems, answer the Königsberg bridge question in the negative.

(b) Show that the graph of Figure 109 is unicursal, whereas that of Figure 110 is multicursal.

(c) Figure 111 represents a house with rooms and doors as marked. Is it possible to walk in succession through each door once and only once?

(d) Try to prove Euler's theorems stated above.

(e) Try to prove Listing's corollary to Euler's fourth theorem: *A graph with exactly 2n odd nodes can be traversed completely in n separate routes.* Verify this corollary for the graph in Figure 110.

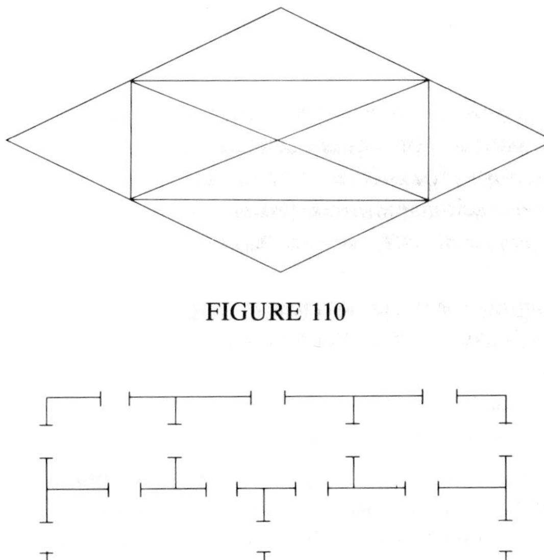

FIGURE 110

FIGURE 111

12.9 Some Differential Equations

(a) The differential equation

$$y^{n-1}(dy/dx) + a(x)y^n = f(x)$$

is known as **Bernoulli's equation.** Show that the transformation $v = y^n$ converts Bernoulli's equation into a linear differential equation.

(b) The differential equation

$$y = px + f(p),$$

where $p = dy/dx$, is known as **Clairaut's equation.** Show that the solution of Clairaut's equation is

$$y = cx + f(c).$$

(c) The differential equation

$$x^n y^{(n)} + a_1 x^{n-1} y^{(n-1)} + \cdots + a_n y^{(0)} = f(x),$$

where the exponents in parentheses indicate orders of differentiation, is known as **Euler's equation.** Show that the substitution $x = e^t$ reduces the Euler equation to a linear differential equation with constant coefficients.

(d) The differential equation

$$dy/dx = p(x)y^2 + q(x)y + r(x)$$

is known as **Riccati's equation.** Show that if $v = f(x)$ is a particular solution of the equation, then the substitution $y = v + 1/z$ converts the equation into a linear differential equation in z.

12.10 Hyperbolic Functions

(a) The **hyperbolic sine** and **hyperbolic cosine** functions may be defined by

$$\sinh u = \frac{e^u - e^{-u}}{2}, \qquad \cosh u = \frac{e^u + e^{-u}}{2},$$

and then the **hyperbolic tangent, hyperbolic cotangent, hyperbolic secant,** and **hyperbolic cosecant** functions by $\tanh u = \sinh u/\cosh u$, $\coth u = 1/\tanh u$, $\operatorname{sech} u = 1/\cosh u$, $\operatorname{csch} u = 1/\sinh u$. Show that

1. $\cosh^2 u - \sinh^2 u = 1$.
2. $\tanh u = (e^u - e^{-u})/(e^u + e^{-u})$.
3. $\coth^2 u - \operatorname{csch}^2 u = 1$.
4. $\tanh^2 u + \operatorname{sech}^2 u = 1$.
5. $\operatorname{csch}^2 u - \operatorname{sech}^2 u = \operatorname{csch}^2 u \operatorname{sech}^2 u$.
6. $\sinh(u \pm v) = \sinh u \cosh v \pm \cosh u \sinh v$.
7. $\cosh(u \pm v) = \cosh u \cosh v \pm \sinh u \sinh v$.
8. $d(\cosh u)/du = \sinh u$, $d(\sinh u)/du = \cosh u$.

(b) Consider a unit circle $x^2 + y^2 = 1$ and a unit equilateral hyperbola $x^2 - y^2 = 1$, as pictured in Figure 112. Represent the sector area $OPAP'$ by u. Show that, for the circle, $x = \cos u$, $y = \sin u$, and for the hyperbola, $x = \cosh u$, $y = \sinh u$, where (x,y) are the coordinates of P.

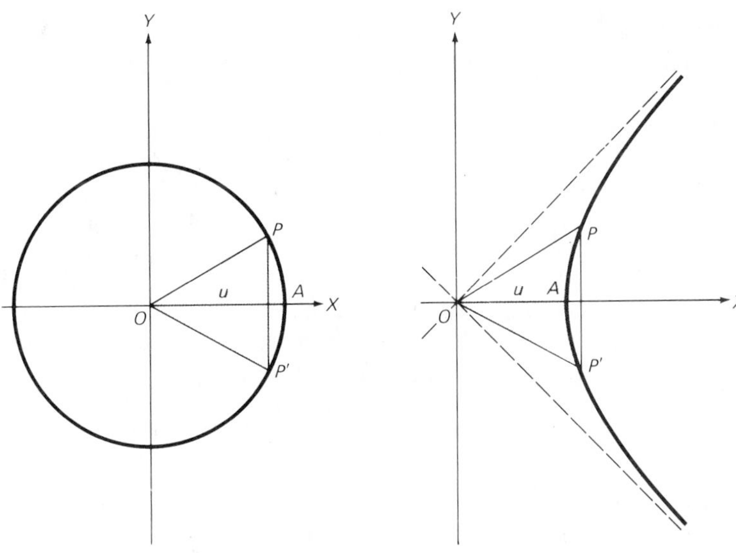

FIGURE 112

12.11 The Witch of Agnesi

The **witch of Agnesi** may be neatly described as the locus of a point P as follows. Consider a circle of radius a and having a diameter OK lying along the positive y-axis, where O is the origin of coordinates. Let a variable secant OA through O cut the circle again in Q and cut the tangent at K to the circle in A. The curve is then the locus of the point P of intersection of the lines QP and AP, parallel and perpendicular, respectively, to the x-axis.

(a) Show that the equation of the above witch is $y(x^2 + a^2) = a^3$.

(b) Show that the above witch is symmetric in the y-axis and has the x-axis as an asymptote.

(c) Show that the area between the witch and its asymptote is πa^2—that is, exactly four times the area of the associated circle.

(d) Show that the centroid of the area in (c) lies at the point $(0,a/4)$—that is, one-fourth the way from O to K.

(e) Show that the volume generated by rotating the witch about its asymptote is $\pi^2 a^3/2$.

(f) Show that the points of inflection of the witch occur where OQ makes angles of 60° with the asymptote.

An associated curve called the **pseudo-witch** is obtained by doubling the ordinates of the above witch. This curve was studied by James Gregory in 1658 and was used by Leibniz in 1674 in deriving his famous expression

$$\frac{\pi}{4} = 1 - \frac{1}{3} + \frac{1}{5} - \frac{1}{7} + \cdots.$$

12.12 Lagrange and Analytic Geometry

Lagrange gave (essentially) the formulas

$$A = (1/2)\begin{vmatrix} x_1 & y_1 & 1 \\ x_2 & y_2 & 1 \\ x_3 & y_3 & 1 \end{vmatrix} \text{ and } V = (1/6)\begin{vmatrix} x_1 & y_1 & z_1 & 1 \\ x_2 & y_2 & z_2 & 1 \\ x_3 & y_3 & z_3 & 1 \\ x_4 & y_4 & z_4 & 1 \end{vmatrix}$$

for the area A of the triangle whose vertices are the points (x_1,y_1), (x_2,y_2), (x_3,y_3) and the volume V of the tetrahedron whose vertices are the points (x_1,y_1,z_1), (x_2,y_2,z_2), (x_3,y_3,z_3), (x_4,y_4,z_4). He also gave the formula

$$D = \frac{ap + bq + cr - d}{\sqrt{a^2 + b^2 + c^2}}$$

for the distance D of the point (p,q,r) from the plane $ax + by + cz = d$.

(a) Establish the formula for the area of a triangle.

(b) Establish the formula for the distance of a point from a plane.

12.13 Buffon's Needle Problem

A problem proposed and solved by Comte de Buffon in 1777 is as follows: Let a homogeneous uniform needle of length l be tossed at random onto a horizontal plane ruled with parallel lines spaced at a distance $a > l$ apart. What is the probability that the needle will intersect 1 of the lines?

Let us assume that "at random" here means that all points of the center of the needle and all orientations of the needle are equally probable and that these 2 variables are independent. Let x denote the distance of the center of the needle from the nearest of the parallel lines, and let ϕ denote the needle's orientation referred to the direction of the parallel lines.

(a) Show, from Figure 113(1), that the needle will intersect a line if and only if $x < (1/2)l \sin \phi$.

(b) In a plane with rectangular Cartesian coordinates x and ϕ, consider [see Figure 113(2)] the rectangle OA, whose interior points satisfy the inequalities

$$0 < x < a/2, \qquad 0 < \phi < \pi.$$

To each point in this rectangle corresponds 1 and only 1 position (x) and orientation (ϕ) of the needle; to each point in the shaded area of Figure 113(2) corresponds 1 and only 1 position (x) and orientation (ϕ) of the needle such that the needle intersects 1 of the parallel lines. Show that the probability we seek is the ratio of the shaded area to the total area of the rectangle OA.

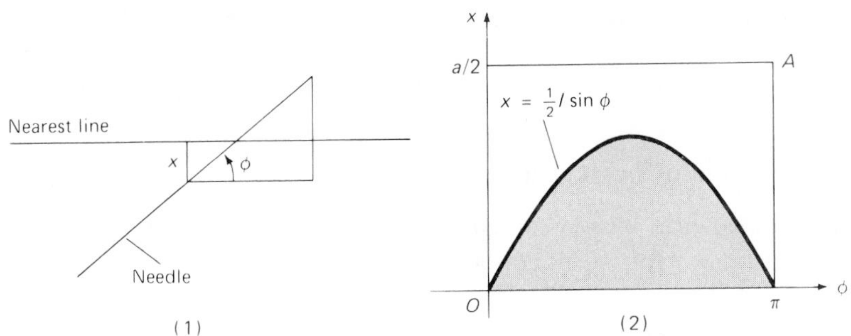

FIGURE 113

(c) Now show that the desired probability is given by

$$p = \frac{\frac{l}{2}\int_0^{\pi} \sin \phi \, d\phi}{\frac{\pi a}{2}} = \frac{2l}{\pi a}.$$

(d) Laplace, in his *Théorie analytique des probabilités* of 1812, extended Buffon's result by showing that if we have 2 orthogonal sets of equidistant parallel lines, the distance being a for one set and b for the other, then the probability p that a randomly tossed needle of length $l < a,b$ will fall on one of the lines is

$$p = \frac{2l(a + b) - l^2}{\pi ab}.$$

Obtain Buffon's result by letting $b \to \infty$ in Laplace's result.

Under the date 1777 in the chronology of π given in Section 4–8, we pointed out how experimenters have approximated π by using Buffon's result.

12.14 Random Chord in a Circle

This problem study illustrates the difficulty frequently encountered in deciding, in a geometrical probability problem, what set of equally likely cases is most desirable. Consider the following problem: What is the probability that a random chord drawn in a given circle is longer than a side of an inscribed equilateral triangle?

(a) Choose *any* point A on the given circle and draw the random chord through A. Assuming that all chords through A are equally likely, show that the sought probability is $\frac{1}{3}$.

(b) Choose *any* direction d and draw the random chord parallel to d. Assuming that all chords parallel to d are equally likely, show that the sought probability is $\frac{1}{2}$.

(c) Choose *any* point inside the given circle for the midpoint of the random chord, and draw the chord. Assuming that all points inside the given circle are equally likely as midpoints, show that the sought probability is $\frac{1}{4}$.

12.15 The Method of Least Squares

As a simple case of a basic problem in the method of least squares, suppose observations have led to $n > 2$ approximate linear equations

$$a_i x + b_i y + c_i = 0,\ i = 1, 2, \ldots, n,$$

satisfied by the two variables x and y. Then, by arguments based on the theory of probability, it has been shown that the "best" values to adopt for x and y are those given by the simultaneous solution of the two equations

$$(\Sigma a_i^2)x + (\Sigma a_i b_i)y + \Sigma a_i c_i = 0,$$

$$(\Sigma b_i a_i)x + (\Sigma b_i^2)y + \Sigma b_i c_i = 0.$$

(a) Using the method of least squares, find the "best" values of x and y satisfying the system

$$x - y + 1 = 0,$$

$$3x - 2y - 2 = 0,$$

$$2x + 3y - 2 = 0,$$

$$2x - y = 0.$$

(b) In determining the coefficient c of linear expansion of a certain bar of metal, the length of the bar was measured at different temperatures, yielding the following table:

Temperature (degrees centigrade)	Observed Length (millimeters)
20	1000.22
40	1000.65
50	1000.90
60	1001.05

Letting L_0 denote the length of the bar at 0°C and L the length at any temperature T, we have

$$L_0 + T_C = L.$$

Find, by the method of least squares, the "best" value of c furnished by the given measurements.

(c) Show that if, in the formulas introduced at the start of this Problem Study, we should take $n = 2$, then the "best" values for x and y are given by the simultaneous solution of the 2 equations

$$a_1x + b_1y + c_1 = 0, \qquad a_2x + b_2y + c_2 = 0.$$

12.16 Some Mongean Geometry

The interested student is invited to try to establish, either synthetically or analytically, the following theorems:

(a) The sum of the squares of the orthogonal projections of a planar area upon 3 mutually perpendicular planes is equal to the square of the planar area.

(b) Monge's theorem on the tetrahedron, as stated in Section 12–9.

(c) **Mannheim's theorem.** The 4 planes determined by the 4 altitudes of a tetrahedron and the orthocenters of the corresponding faces concur in the Monge point of the tetrahedron.

(d) The Monge point of a tetrahedron is equidistant from any altitude of the tetrahedron and the perpendicular to the corresponding face at its orthocenter.

(e) The center of the sphere determined by the midpoints of the medians of a tetrahedron lies on the Euler line of the tetrahedron. (The line containing the circumcenter, the centroid, and the Monge point has become known as the **Euler line** of the tetrahedron.)

(f) The Monge point and centroid of a tetrahedron coincide if and only if the tetrahedron is isosceles. (A tetrahedron is **isosceles** when and only when each edge of the tetrahedron is equal to its opposite edge.)

(g) The 5 lines joining each of 5 given cospherical points with the Monge point of the tetrahedron determined by the remaining 4 given points are concurrent.

12.17 Sensed Magnitudes

Carnot introduced the systematic use of sensed magnitudes in his *Géométrie de position* of 1803. Under this concept, we choose on each line, for example, 1 direction as the positive direction and the other as the negative direction. A segment AB on the line is then considered positive or negative according as the direction along the line from point A to point B is the positive or the negative direction of the line. Using sensed line segments, we then have $AB = -BA$ and $AB + BA = 0$. Establish the following theorems wherein all segments are sensed segments.

(a) For any 3 collinear points A, B, C, $AB + BC + CA = 0$.

(b) Let O be any point on the line of the segment AB. Then $AB = OB - OA$.

(c) **Euler's theorem** (1747). If A, B, C, D are any 4 collinear points, then $(AD)(BC) + (BD)(CA) + (CD)(AB) = 0$.

(d) If A, B, P are collinear and M is the midpoint of AB, then $PM = (PA + PB)/2$.

(e) If O, A, B, C are collinear and $OA + OB + OC = 0$, and if P is any point on the line AB, then $PA + PB + PC = 3\ PO$.
(f) If on the same line we have $OA + OB + OC = 0$ and $O'A' + O'B' + O'C' = 0$, then $AA' + BB' + CC' = 3\ OO'$.
(g) If A, B, C, are collinear and P, Q, R are the midpoints of BC, CA, AB, respectively, then the midpoints of CR and PQ coincide.
(h) If 2 lines through a point P cut a circle in points A and B and in points C and D, respectively, then $(PA)(PB) = (PC)(PD)$.

12.18 Carnot's Theorem

(a) State Carnot's theorem (see Section 12–9) for a triangle intersected by an n-ic (an algebraic curve of degree n).
(b) State the generalization of Carnot's theorem wherein the triangle is replaced by an arbitrary polygon.
(c) Through any point O, not on a given n-ic, 2 lines are drawn in fixed directions cutting the n-ic in $P_1, P_2, \ldots, P_n$ and $Q_1, Q_2, \ldots, Q_n$, respectively. Show, by employing an oblique Cartesian coordinate system with axes parallel to the 2 given directions, that

$$(OP_1)(OP_2) \ldots (OP_n)/(OQ_1)(OQ_2) \ldots (OQ_n)$$

is independent of the position of O.
(d) Using (c), prove the generalization of Carnot's theorem given in (b).

Essay Topics

12/1 Famous mathematical families.
12/2 Inscriptions found on mathematicians' tombstones.
12/3 L'Hospital and his rule.
12/4 How the logarithmic (or equiangular) spiral reproduces itself.
12/5 Bishop George Berkeley (1685–1753).
12/6 Colin Maclaurin (1698–1746).
12/7 The little-known William Whiston (1667–1752).
12/8 James Stirling (1692–1770) and his formula.
12/9 The Euler-Diderot anecdote.
12/10 Euler diagrams versus Venn diagrams.
12/11 Euler as a writer of great textbooks.
12/12 Who was the most eminent mathematician of the eighteenth century?
12/13 Napoleon Bonaparte and mathematics.
12/14 The story of d'Alembert's christening.
12/15 The St. Petersburg and Berlin Academies.
12/16 Legendre's influence on the teaching of geometry in America.
12/17 Thomas Carlyle and mathematics.

12/18 Guido Grandi (1671–1742) and his *rosaces* curves.
12/19 Nicholas Saunderson (1682–1739), the blind mathematician of Cambridge.
12/20 Pierre Louis Moreaux de Maupertuis (1698–1759), the "earth flattener."
12/21 Gabriel Cramer (1704–1752).
12/22 Thomas Simpson (1716–1761), the weaver mathematician.
12/23 John Wilson (1741–1793) and his single mathematical accomplishment.
12/24 Jean Étienne Montucla (1725–1799), an early historian of mathematics.
12/25 Alexandre Theophile Vandermonde (1735–1796).
12/26 Jean-Baptiste Joseph Delambre (1749–1822).
12/27 Sylvestre François Lacroix (1765–1845).
12/28 Three outstanding French mathematicians who supported the Revolution.
12/29 Weights and measures before the metric system.
12/30 Definitions of *are, stere, liter, gram, carat.*
12/31 Pierre Méchain's tragic error.
12/32 History of the dollar mark.

Bibliography

BALL, W. W. R., and H. S. M. COXETER *Mathematical Recreations and Essays.* 12th ed. Toronto: University of Toronto Press, 1974. Reprinted by Dover, New York.

BELL, E. T. *Men of Mathematics.* New York: Simon and Schuster, 1937.

BOYER, C. B. *The History of the Calculus and Its Conceptual Development.* New York: Dover, 1959.

BURLINGAME, A. E. *Condorcet, the Torch Bearer of the French Revolution.* Boston: Stratford, 1930.

CADWELL, J. N. *Topics in Recreational Mathematics.* New York: Cambridge University Press, 1966.

COOLIDGE, J. L. *The Mathematics of Great Amateurs.* New York: Oxford University Press, 1949.

DICKSON, L. E. *History of the Theory of Numbers.* 3 vols. New York: Chelsea, 1952.

DUGAS, RENÉ *A History of Mechanics.* New York: Central Books, 1955.

EULER, LEONHARD *Elements of Algebra.* Translated by John Hewlett. New York: Springer-Verlag, 1984.

EVES, HOWARD *A Survey of Geometry,* vol. 2. Boston: Allyn and Bacon, 1965.

GILLESPIE, C. C. *Lazare Carnot, Savant.* Princeton, N.J.: Princeton University Press, 1971.

GRIMSLEY, RONALD *Jean d'Alembert (1717–83).* Oxford: Clarendon Press, 1963.

HOFFMAN, J. E. *Classical Mathematics.* New York: Philosophical Library, 1959.

LAGRANGE, J. L. *Lectures on Elementary Mathematics*. 2d ed. Translated by T. J. McCormack. Chicago: Open Court, 1901.

LAPLACE, P. S. *The System of the World*. 2 vols. Translated by H. H. Harte. London: Longmans, Green, 1830.

——— *A Philosophical Essay on Probabilities*. Translated by F. Truscot and F. Emory. New York: Dover, 1951.

LEGENDRE, A. M. *Elements of Geometry and Trigonometry*. Translated by Davis Brewster. Revised by Charles Davies. New York: A. S. Barnes, 1851.

MACLAURIN, COLIN *A Treatise of Fluxions*. 2 vols. Edinburgh, 1742.

MAISTROV, L. E. *Probability Theory, A Historical Sketch*. Translated by Samuel Kotz. New York: Academic Press, 1974.

MUIR, JANE *Of Men and Numbers, the Story of the Great Mathematicians*. New York: Dodd, 1961.

NORTHROP, E. P. *Riddles in Mathematics, A Book of Paradoxes*. New York: Van Nostrand Reinhold, 1944.

ORE, OYSTEIN *Number Theory and Its History*. New York: McGraw-Hill, 1948.

ROEVER, W. H. *The Mongean Method of Descriptive Geometry*. New York: Macmillan, 1933.

TAYLOR, E. G. R. *The Mathematical Practitioners of Tudor and Stuart England*. New York: Cambridge University Press, 1954.

——— *The Mathematical Practitioners of Hanoverian England*. New York: Cambridge University Press, 1966.

TODHUNTER, ISAAC *A History of the Progress of the Calculus of Variations during the Nineteenth Century*. London, 1861.

——— *History of the Mathematical Theory of Attraction and the Figure of the Earth*. London, 1873.

——— *A History of the Mathematical Theory of Probability from the Time of Pascal to that of Laplace*. New York: Chelsea, 1949.

TURNBULL, H. W. *Bi-centenary of the Death of Colin Maclaurin*. Aberdeen: Aberdeen University Press, 1951.

——— *The Great Mathematicians*. New York: New York University Press, 1961.

TWEEDIE, CHARLES *James Stirling. Sketch of His Life and Works*. Oxford: The Clarendon Press, 1922.

WATSON, S. J. *Carnot*. London: Bodley Head, 1954.

WEIL, ANDRE *Number Theory: An Approach through History from Hammurabi to Legendre*. Boston: Birkhauser, 1984.

Cultural Connection 

THE INDUSTRIAL REVOLUTION

The Nineteenth Century
(to accompany Chapters Thirteen
and Fourteen)

There have been two great global revolutions in world history (global in the sense that they profoundly altered human culture and society across the entire world)—the Agricultural Revolution of the third millenium B.C. and the Industrial Revolution of the nineteenth century A.D.

In Cultural Connection II, we explored the Agricultural Revolution that began around 3000 B.C. in Egypt, China, and the Middle East. Before that, people had lived as hunters and gatherers, scattered in small bands across vast, grassy savannas, constantly moving from place to place in search of food. Preagricultural people could not read or write, and scientific knowledge was minimal. After 3000 B.C., people started to become sedentary farmers. They invented writing, machines, and complex political systems. Human civilization was irrevocably changed, and humankind lived primarily as agriculturalists for almost 5000 years. To be sure, not everyone was a farmer; there were also soldiers, artisans, poets, kings, merchants, scientists, and philosophers. Yet, most people cultivated the soil for a living, and agriculture remained the principal focus of human endeavor.

The Industrial Revolution of the nineteenth century changed the world. It signaled a thorough-going reorganization of human civilization. No longer would farmers constitute the majority of the population; no longer would agriculture remain the bedrock of human economies. Instead, the age of plants, draught animals, and fields was supplanted by the epoch of the machine. Industrial workers became the largest segment of the labor force, and industry assumed the status of economic bulwark.

The Industrial Revolution brought far-reaching social changes in its wake—among them industrial capitalism; increased urbanization; the factory system; mammoth corporations; the emergence of a new social class, the proletariat, or industrial workers; global imperialism on an unprecedented scale; impressive technological breakthroughs; a more mechanistic world view; and, in an attempt to reclaim some of the old, preindustrial values, romanticism. Let's briefly examine the Industrial Revolution in terms of its causes, the process whereby it came about, and its effects on human civilization.

The Causes of the Industrial Revolution

According to economic historians, in order for industrialization to occur naturally, several factors must be present: an appropriate technology, accumulated investment capital, markets for industrial products, a large pool of workers, an efficient means of transporting raw materials and finished goods in bulk, and a social climate conducive to entrepreneurial activity. Some of these things had existed in Europe for several centuries. Historian Jean Gimpel, in his book *The Medieval Machine* (1976), notes that European technology was primed for industrialization as early as 1300. The development of centralized nation-states in Europe after 1400 provided potential national markets, although in some countries, like France, a bewildering array of internal customs and taxes on transported merchandise hindered the formation of national markets. Investment capital accumulated among wealthy urban elites after 1500. Canals, relatively efficient and inexpensive means of transportation, existed in Europe even in the Middle Ages. Missing were two things: a large available work force, and *bourgeois* control of European national economies.

The rapid growth of European cities after 1500, with concentrated populations of urban poor, provided a potential work force of substantial size after 1750. The small factories that had existed in such cities since the Middle Ages now had plenty of cheap labor, and many factory owners expanded their businesses. Workers were paid very low wages and worked as many as eighty hours each week. Women and children were hired, as they could do the same work in a factory as men, but were paid much less. Working conditions were dismal and serious accidents were common. Workers lived in unsanitary, dirty ghettoes, often in large families crowded into tiny, unheated tenements. German philosopher Friedrich Engels (1820–1895) was shocked after his visit to workers' homes in the English city of Manchester. He described a scene of abject poverty—rat infested buildings, open sewers in the streets, dank workplaces, all hidden in back alleys, out of the sight of the middle-class residents of the city.

Besides a ready work force, industrialization required an entrepreneurial class with access to investment capital, entrenched in positions of authority. At the close of the eighteenth century, the *bourgeoisie* (a French word meaning, roughly, urban middle class) emerged as just such a class. The old feudal aristocracy had been hostile to commercial enterprise. Their landed estates were not easily converted into cash for investment, and the aristocracy profited from the heavy taxes, internal customs, and government monopolies that discouraged industrialization. After the aristocracy was overthrown in the eighteenth century (discussed in Cultural Connection VIII), the *bourgeoisie* in Europe and America assumed control of the financial and (sometimes) political institutions in their countries. Once in power, the *bourgeoisie* established a political and economic climate favorable to industrial capitalism.

The Course of the Industrial Revolution

Like the Agricultural Revolution, the Industrial Revolution was a historical process that took place over the course of many years and that has not yet happened in all corners of the globe. Not surprisingly, the Industrial Revolution began in England, in about 1750. This was not because England was more "advanced" culturally or technologically than other countries. Instead, the process started in the island kingdom because, by accident of history, the *bourgeoisie* came to power in England before anywhere else, in the so-called Glorious Revolution. From its origins in England, the Industrial Revolution diffused into other parts of Europe and America. By 1900, an industrialized "core region" comprised parts of England, Scotland, France, Belgium, the Netherlands, and Germany. In addition, sections of Italy, the United States, and Japan were becoming industrialized.

The Effects of the Industrial Revolution

Mentioned above were some of the effects of the Industrial Revolution: the factory system, giant industrial corporations, new socio-economic classes, the creation of new fortunes and the crumbling of old ones, even more rapid urbanization, and a more mechanistic world view. There were more—in fact, we could devote many pages just to listing all of the repercussions of industrialization. Some are obvious, like new technologies; others less so, such as global imperialism and romanticism. Let us discuss a few of the more important effects of the Industrial Revolution.

Global Imperialism. The nineteenth-century industrial nations quickly ran short of raw materials for their factories. The textile industries in Belgium and England found it necessary to import cotton from the United States and India. Iron and steel foundries sought ore and coal abroad. Tin, rubber, and other vital materials were either rare or nonexistent in Europe. As a result of such shortages, factory owners in industrial countries pressured their governments into establishing colonies in other parts of the world richer in natural resources. The Dutch turned the East Indies into a rubber plantation. England established India as a cotton colony, and metal ores were mined in England's several African colonies. The industrial nations were also interested in hewing out foreign markets for surplus goods. When countries such as China resisted trade, European, American, and Japanese armies arrived to supervise forced commerce, the most heinous of which was the opium trade. Furthermore, international trade necessitated that the industrial countries establish coaling stations around the globe to refuel their merchant ships. During the nineteenth century, England, France, Belgium, the Netherlands, Italy, Germany, the United States, and Japan all either enlarged existing empires or established new ones, mostly in Africa, Asia, and the Pacific islands. In Africa, only Ethiopia remained independent; in east Asia, only China, Thailand, and industrializing Japan. All of the Pacific islands were colonized.

Increased Industrial Production. Industrial production increased astronomically during the nineteenth century. For example, there were just 30,000 tons of iron made in England in 1770; two million tons were smelted in 1850. Although England mined ten million tons of coal in 1800, it produced fifty million tons in 1850. Similar rapid increases in industrial production occurred in other industrial countries.

The Factory System and Social Change. The factory system became the most common method of producing goods. It was efficient and could turn out products in great quantity. At the same time, however, it impoverished many workers and led to widespread discontent, manifested in drives to win voting rights for industrial workers, the establishment of labor unions, and socialism. In 1848, Engels and Karl Marx (1818–1883), in the *Communist Manifesto,* advocated the eventual elimination of industrial capitalism, believing that any system responsible for such misery was immoral.

Technological Progress. The Industrial Revolution created a need for new technologies, a need met by nineteenth century inventors. The textile industry, for example, saw the invention of the flying shuttle (1733), the spinning jenny (1764), the water frame (1771), the steam loom (1789), and the cotton gin (1793), all before 1800. During the nineteenth century, the steam engine was perfected and the gasoline engine invented, railroad transportation developed, steel production improved, and steamships became common. The airplane was invented in 1903, just after the close of the century. There was little, however, to link technology to pure science. Most technological breakthroughs were the accomplishments, not of scientists, but of artisans and tinkerers. Not until the twentieth century would pure science and technology merge.

Romanticism. The Age of the Machine did not please everybody. Concurrent with the beginning of the Industrial Revolution in the mid-eighteenth century was the Romantic Movement among poets, artists, and other scholars, which idealized the past. Romantics considered the Middle Ages to have been an age of dashing knights and fair maidens, and spun stories about Robin Hood and King Arthur. Among the more influential Romantics were Scots novelist Sir Walter Scott (1771–1823), the French writers René de Châteaubriand (1768–1848) and Victor Hugo (1802–1885), the German author Johann Wolfgang von Goethe (1749–1832), and the English poets John Keats (1795–1821), Percy Bysshe Shelly (1792–1822), and William Wordsworth (1770–1850). In contrast to the emotionless machine, Romantic literature was gloriously sentimental, and its music dark, brooding, and forceful.

Summary

The Industrial Revolution that birthed modern society began in the 1700s in England. During the nineteenth century, it spread to the European mainland and America. As great factories and sprawling cities were built, the structure of society changed radically. Among these changes, rapid technological progress touched off an era of unprecedented scientific inquiry, especially in mechanics and chemistry. Although at first most discoveries were made by artisans and tinkerers, by the twentieth century the needs of industry demanded university-educated mathematicians and scientists. Not everyone was pleased with the Industrial Revolution. Socialists, although not opposed to industrialization, decried the unequal division of wealth that characterized the nineteenth century. Romantics, for their part, advocated a return to past ideals.

Chapter 13

THE EARLY NINETEENTH CENTURY AND THE LIBERATION OF GEOMETRY AND ALGEBRA

13–1 The Prince of Mathematicians

A man of awesome mathematical stature and talent, Carl Friedrich Gauss straddled the eighteenth and nineteenth centuries like a mathematical Colossus of Rhodes. He is universally regarded as the greatest mathematician of the nineteenth century and, along with Archimedes and Isaac Newton, as one of the three greatest mathematicians of all time.

Carl was born in Brunswick, Germany, in 1777. His father was a hard-working laborer with stubborn and unappreciative views of education. His mother, however, though uneducated herself, encouraged the boy in his studies and maintained a lifelong pride in her son's achievements.

Carl was one of those remarkable infant prodigies who appear from time to time. It is said that at the age of three he detected an arithmetic error in his father's bookkeeping. There is a story that when Carl was ten years old and in the public schools, his teacher, to keep the class occupied, set the pupils to adding the numbers 1 through 100. Almost immediately, Carl placed his slate on the annoyed teacher's desk. When all the slates were finally turned in, the amazed teacher found that Carl alone had the correct answer, 5050, but with no accompanying calculation. Carl had mentally summed the arithmetic progression $1 + 2 + 3 + \ldots + 98 + 99 + 100$ by noting that $100 + 1 = 101$, $99 + 2 = 101$, $98 + 3 = 101$, and so on for fifty such pairs, whence the answer is 50×101, or 5050. Later in life, Gauss used to claim jocularly that he could figure before he could talk.

Gauss' precocity came to the attention of the Duke of Brunswick, who, as a kindly and understanding patron, saw the boy enter the college in Brunswick at the age of fifteen, and then Göttingen University at the age of eighteen. Vacillating between becoming a philologist or a mathematician (although he had already devised the method of least squares a decade before Legendre independently published it), his mind was dramatically made up in favor of mathematics on March 30, 1796, when he was still a month short of his nineteenth birthday. The event was his surprising contribution to the theory of the Euclidean construction of regular polygons and, in particular, the discovery that a regular polygon of seventeen sides can be so constructed. We have already told this story in Section 5–6.

On the same day that Gauss made his discovery concerning regular polygons, he commenced his famous mathematical diary, to which he confided in cryptic fashion many of his greatest mathematical achievements. Because Gauss, like Newton, was both slow and reluctant to publish, this diary, which was not found until 1898, has settled a number of disputes on priority. The diary contains 146 brief entries, the last dated July 9, 1814. As an illustration of the cryptic nature of the entries in the diary consider that for July 10, 1796, which reads

$$\text{ΕΥΡΗΚΑ!} \qquad \text{num} = \Delta + \Delta + \Delta,$$

and records Gauss' discovery of a proof of the fact that every positive integer is the sum of three triangular numbers. All the entries of the diary except two have, for the most part, been deciphered. The entry for March 19, 1797, shows that Gauss had already at that time discovered the double periodicity of certain elliptic functions (he was not yet twenty years old), and a later entry shows that he had recognized the double periodicity for the general case. This discovery alone, had Gauss published it, would have earned him mathematical fame. But Gauss never published it!

In his doctoral dissertation, at the University of Helmstädt and written at the age of twenty, Gauss gave the first wholly satisfactory proof of the *fundamental theorem of algebra* (that a polynomial equation with complex coefficients and of degree $n > 0$ has at least one complex root). Unsuccessful attempts to prove this theorem had been made by Newton, Euler, d'Alembert, and Lagrange. The idea behind Gauss' proof is the replacement of z in a general polynomial equation $f(z) = 0$ by $x + iy$. Then the separation of the real and imaginary parts of the resulting equation yields two real equations $g(x,y) = 0$ and $h(x,y) = 0$ in the real variables x and y. Gauss showed that the Cartesian graphs of $g(x,y) = 0$ and $h(x,y) = 0$ always have at least one real point of intersection (a,b). It follows that $f(z) = 0$ has the complex root $a + ib$. The proof involves geometrical considerations. Almost twenty years later, in 1816, Gauss published two new proofs, and still later, in 1850, a fourth proof, in an effort to find an entirely algebraic proof.[1]

Gauss' greatest single publication is his *Disquisitiones arithmeticae*, a work of fundamental importance in the modern theory of numbers. Gauss' findings on the construction of regular polygons appear in this work, as does his facile notation for congruence (see Problem Study 13.2), and a proof of the beautiful quadratic reciprocity law, which says, in terms of the Legendre symbol defined toward the end of Section 12–9, that if $p = 2P + 1$ and $q = 2Q + 1$ are unequal odd primes, then

$$(p|q)(q|p) = (-1)^{PQ}.$$

[1] For an English translation of the second proof, see David Eugene Smith, *A Source Book in Mathematics* (1958), pp. 292–306. Today it is believed that a proof of the fundamental theorem of algebra must involve topological considerations.

Gauss made notable contributions to astronomy, geodesy, and electricity. In 1801, he calculated, by a new procedure and from meager data, the orbit of the then recently discovered planetoid Ceres and, in the following year, that of the planetoid Pallas. In 1807, he became professor of mathematics and director of the observatory at Göttingen, a post that he held until his death. In 1821, he carried out a triangulation of Hanover, measured a meridional arc, and invented the heliotrope (or heliograph). In 1831, he commenced collaboration with his colleague Wilhelm Weber (1804–1891) in basic research in electricity and magnetism; in 1833, the two scientists devised the electromagnetic telegraph.

In 1812, in a paper on hypergeometric series, Gauss made the first systematic investigation of the convergence of a series. Gauss' masterpiece on surface theory, his *Disquisitiones generales circa superficies curvas,* appeared in 1827, and inaugurated the study of the intrinsic geometry of surfaces in space (see Section 14–7). His anticipation of non-Euclidean geometry will be discussed in Section 13–7.

Famous is Gauss' assertion that "mathematics is the queen of the sciences, and the theory of numbers is the queen of mathematics." Gauss has been described as "the mathematical giant who from his lofty heights embraces in one view the stars and the abysses." In his scientific writing, Gauss was a perfectionist. Claiming that a cathedral is not a cathedral until the last piece of scaffolding is removed, he strove to make each of his works complete, concise, polished, and convincing, with every trace of the analysis by which he reached his results removed. He accordingly adopted as his seal a tree bearing only a few fruits and carrying the motto: *Pauca sed matura* (*Few, but ripe*). Gauss chose for his second motto the following lines from *King Lear:*

CARL FRIEDRICH GAUSS
(Library of Congress)

Thou, nature, art my goddess; to thy laws
My services are bound.

Gauss believed that mathematics, for inspiration, must touch the real world. As Wordsworth put it, "Wisdom oft is nearer when we stoop than when we soar."

Gauss died in his home at the Göttingen Observatory on February 23, 1855, and right after, the King of Hanover ordered that a commemorative medal be prepared in honor of Gauss. This seventy-millimeter medal was in time (1877) completed by the well-known sculptor and medalist, Friedrich Brehmer, of Hanover. On it appears the inscription:

Georgius V. rex Hannoverge
Mathematicorum principi
(George V. King of Hanover
to the Prince of Mathematicians)

Ever since, Gauss has been known as "the Prince of Mathematicians."

Caspar Wessel (1745–1818), Jean Robert Argand (1768–1822), and Gauss were the earliest authors to note the now familiar association of the complex numbers with the real points of a plane.[2] Wessel and Argand were not professional mathematicians; Wessel was a surveyor, born in Josrud, Norway, and Argand was a bookkeeper, born in Geneva, Switzerland.

There is little doubt that priority for the idea is to be credited to Wessel, from a paper he presented to the Royal Danish Academy of Sciences in 1797 and published in that Academy's *Transactions* in 1799. Argand's contribution lies in a paper published in 1806 and later, in 1814, reported on in Gergonne's *Annales de Mathématiques*. But Wessel's paper lay buried from the general mathematical world until it was unearthed by an antiquary some ninety-eight years after it was written. It was then republished on the hundredth anniversary of its first appearance. This delay in the general recognition of Wessel's accomplishment is why the plane of complex numbers came to be called the **Argand plane,** rather than the **Wessel plane.**

Gauss' contribution is found in a memoir presented to the Royal Society of Göttingen in 1831, and later reproduced in his *Collected Works*. Gauss pointed out that the basic idea of the representation can be found in his doctoral dissertation of 1799. The claim seems to be well taken and explains why the plane of complex numbers is frequently referred to as the **Gauss plane.**

The simple idea of considering the real and imaginary parts of a complex number $a + bi$ as the rectangular coordinates of a point in a plane made mathematicians feel much more at ease with imaginary numbers, for these numbers could now actually be visualized, in the sense that to each complex number there corresponds a point in the plane, and vice versa. Seeing is believing, and former ideas about the nonexistence or fictitiousness of imaginary numbers were generally abandoned.

---

[2] The idea, however, is latent in a suggestion made by John Wallis (1616–1703) as early as 1673, that pure imaginary numbers should be represented by a line perpendicular to the axis of real numbers. See F. Cajori, "Historical notes on the graphic representation of imaginaries before the time of Wessel," *The American Mathematical Monthly* 19 (1912): 167.

Om

Directionens analytiske Betegning,

et Forsøg,

anvendt fornemmelig

til

plane og sphæriske Polygoners Opløsning.

Af

Caspar Wessel,

Landmaaler.

Kiobenhavn 1798.

Trykt hos Johan Rudolph Thiele.

Title page of Caspar Wessel's paper "Om Directionens analytiske Betregning" ("On the Analytical Representation of Vectors") presented at the Royal Danish Academy of Sciences in 1797 and published in that Academy's *Transactions* in 1799. The article presented for the first time the association of the complex numbers with the real points of a plane.

(Courtesy of the Department of Rare Books and Special Collections, The University of Michigan Library.)

13–2 Germain and Somerville

We now briefly consider two other mathematicians who, like Gauss, were born in the last quarter of the eighteenth century but did their important work in the early nineteenth century. These two mathematicians, Sophie Germain and Mary Fairfax Somerville, each in her own way helped to further the emancipation of women in mathematics.

Sophie Germain was born in Paris in 1776 and developed a deep interest in mathematics. As a woman, she was barred from enrolling at the École Polytechnique. Nevertheless, she procured the lecture notes of various professors there, and, from written comments submitted under the male pseudonym M. Leblanc, won high praise from Lagrange. In 1816, she was awarded a prize by the French Academy for a paper on the mathematics of elasticity. In the mid-1820s, she proved that for each odd prime $p < 100$, the Fermat equation $x^p + y^p = z^p$ has no solution in integers not divisible by p. In 1831, she introduced into differential geometry the fruitful notion of the *mean curvature* of a surface at a point of the surface (see Section 14–7). Though a much abler mathematician, she is frequently called the Hypatia of the nineteenth century.

Germain corresponded with Gauss under her psuedonym of M. Leblanc and was exuberantly commended and complimented by Gauss. It was not until some time later that Gauss learned M. Leblanc was a woman. It is sad that Gauss and Germain never met, and equally sad that Germain died (in 1831) before the University of Göttingen was able to award her the honorary doctorate recommended by Gauss.

It is said that Sophie Germain resolved to study mathematics when, during the violent days following the fall of the Bastille, she read with fascination the life and death of Archimedes during the similarly violent days following the siege of Syracuse. In her memoir on elasticity, she remarked: "Algebra is but written geometry and geometry is but figured algebra."

SOPHIE GERMAIN
(David Eugene Smith Collection, Rare Book and Manuscript Library, Columbia University)

Mary Fairfax Somerville (1780–1872) was a remarkable self-taught Scotswoman who, on her own, had studied Laplace's *Traité de mécanique céleste* and was induced by The Society for the Diffusion of Useful Knowledge to write a popular exposition of the great work. Though she was almost fifty years old and lacked formal training, she produced such a meritorious exposition (completed in 1830 and titled *The Mechanisms of the Heavens*) that it ran through many printings and became for close to a century required reading for students of mathematics and astronomy in the British universities. The work contains full mathematical explanations and diagrams to render Laplace's difficult work comprehensible. This necessary mathematical background was later (in 1832) issued separately under the title *A Preliminary Dissertation on the Mechanisms of the Heavens*.

To point up the absurd difficulties suffered by women during the nineteenth century, there is the story that as a young woman Mary Somerville wanted a copy of Euclid's *Elements* but had to get a male friend to purchase it at a bookstore, Euclid being considered improper reading for a young lady. She married at 24 to a man with little interest in intellectual pursuits by women. Fortunately for mathematics, her husband died after three years of marriage, leaving her a substantial sum of money that afforded her the opportunity to buy mathematics books. She married again, but this time to a man sympathetic to her intellectual activities.

Somerville was eventually granted a governmental pension, and the Royal Society of London placed her bust in its Great Hall. The astronomer John Couch Adams said that it was a reference in Somerville's *The Mechanisms of the Heavens* that caused him to consider looking for a new planet (Neptune) to explain the observed perturbations of Uranus. Somerville continued to work until the day she died, at the age of 92. Somerville College, one of the five women's colleges at Oxford, is named after her.

MARY FAIRFAX SOMERVILLE
(David Eugene Smith Collection, Rare Book and Manuscript Library, Columbia University)

13–3 Fourier and Poisson

As we enter the nineteenth century, the number of competent and productive mathematicians becomes almost legion, and we are forced to select for discussion only a few of the brighter stars in the dazzling mathematical firmament. Two of these stars, if not of the first magnitude then certainly of the second, are Jean Baptiste Joseph Fourier and Siméon Denis Poisson. Close contemporaries, these men were both born in France, both worked in applied mathematics, and both held teaching posts at the École Polytechnique.

Fourier was born in Auxerre in 1768 and died in Paris in 1830. The son of a tailor, he was orphaned at the age of eight and educated in a military school conducted by the Benedictines, where he was given a lectureship in mathematics. He assisted in the promotion of the French Revolution and was rewarded by a chair at the École Polytechnique. He resigned from this position so that he, along with Monge, could accompany Napoleon on the Egyptian expedition. In 1798, he was appointed governor of Lower Egypt. Following the British victories and the capitulation of the French in 1801, Fourier returned to France and was made prefect of Grenoble. It was while at Grenoble that he started his experiments on heat.

In 1807, Fourier presented a paper before the French Academy of Sciences that initiated a new and highly fruitful chapter in the history of mathematics. The paper dealt with the practical problem of the flow of heat in metallic rods, plates, and solid bodies. In the course of the presentation of the paper, Fourier made the startling claim that any function, defined in a finite closed interval by an arbitrarily drawn graph, can be resolved into a sum of sine and cosine functions. To be more explicit, he claimed that *any* function whatever, no matter how capriciously it is defined in the interval $(-\pi,\pi)$, can be represented in that interval by

$$\frac{a_0}{2} + \sum_{n=1}^{\infty} (a_n \cos nx + b_n \sin nx),$$

where the a's and the b's are suitable real numbers. Such a series is known as a **trigonometric series,** and was not new to the mathematicians of the time. Indeed, a number of more or less well behaved functions had been shown to be representable by such a series. But Fourier claimed that *any* function defined in $(-\pi,\pi)$ can be so represented. The savants at the Academy were very skeptical of Fourier's claim, and the paper, which was judged by Lagrange, Laplace, and Legendre, was rejected. However, to encourage Fourier to develop his ideas more carefully, the French Academy made the problem of heat propagation the subject of a grand prize to be awarded in 1812. Fourier submitted a revised paper in 1811, which was judged by a group containing, among others, the former three judges, and the paper won the prize, though it was criticized for lack of rigor and so was not recommended for publication in the Academy's *Mémoires*.

JOSEPH FOURIER
(David Smith Collection)

Resentful, Fourier continued his researches on heat, and, in 1822, after a move to Paris in 1816, he published one of the great classics of mathematics, his *Théorie analytique de la chaleur* (The Analytical Theory of Heat). Two years after the publication of his great work, Fourier became secretary of the French Academy, and, in that capacity, was able to have his 1811 paper published in its original form in the Academy's *Mémoires*.

Although it has been shown that Fourier's claim that *any* function can be represented by a trigonometric series (or **Fourier series,** as they are commonly called today) is too extravagant, the class of functions so representable is very broad indeed. The Fourier series have proved to be highly valuable in such fields of study as acoustics, optics, electrodynamics, thermodynamics, and many others, and they play a cardinal role in harmonic analysis, beam and bridge problems, and in the solution of differential equations. In fact, it was the Fourier series that motivated the modern method in mathematical physics involving the integration of partial differential equations subject to boundary conditions. In Section 15–3, we shall see the important role played by the Fourier series in the evolution of the function concept.

In an unpublished work that was edited and posthumously published in 1831, we find, among other original matters, Fourier's work on the position of the roots of a polynomial equation (considered in present-day texts on the theory of equations). This subject had interested him off and on since 1789. Fourier's contemporary, Sadi Carnot (1796–1832), son of the eminent geometer discussed in Section 12–10, also interested himself in the mathematical theory of heat, which initiated the modern theory of thermodynamics.

Lord Kelvin (William Thomson, 1824–1907) claimed that his whole career in mathematical physics was influenced by Fourier's work on heat, and Clerk

Maxwell (1831–1879) pronounced Fourier's treatise "a great mathematical poem."

An amusing story is told about Fourier and his interest in heat. It seems that from his experience in Egypt, and maybe his work on heat, he became convinced that desert heat is the ideal condition for good health. He accordingly clothed himself in many layers of garments and lived in rooms of unbearably high temperature. It has been said by some that this obsession with heat hastened his death, by heart disease, so that he died, thoroughly cooked, in his sixty-third year.

Perhaps Fourier's most quoted sentence (it appeared in his early work on the mathematical theory of heat) is: "The deep study of nature is the most fruitful source of mathematical discovery."

Poisson was born in Pithiviers in 1781 and died in Paris in 1840. He was educated by his father, a private soldier who on retirement received a small administrative post in his village and, when the French Revolution broke out, assumed the governing of the place. Relatives wished to press the young Poisson, much against his own wishes, into medicine. The education was undertaken by an uncle, who started the boy off with pricking veins in cabbage leaves with a lancet. When he had perfected himself in this, he was graduated to putting on blisters. But in almost the first case in which he did this by himself, his patient died within a few hours. Although the doctors assured him that "The event was a very common one," he vowed to have nothing more to do with the profession.

Strong mathematical interests led Poisson in 1798 to enter the École Polytechnique to study the subject, where his abilities impressed Lagrange and Laplace. Upon graduation, he was made a lecturer at the École Polytechnique.

SIMÉON POISSON
(David Smith Collection)

The rest of his life was spent in various government posts and professorships. Somewhat of a socialist, he remained a staunch republican until 1815, when he joined the legitimists.

Poisson's mathematical publications were numerous, numbering between 300 and 400. His chief treatises are his two-volume *Traité de mécanique*, published in 1811 and 1833, his *Théorie nouvelle de l'action capillaire* of 1831, his *Théorie mathématique de la chaleur* of 1835, and his *Recherches sur la probabilité des jugements* of 1837. In his papers, he considered such matters as the mathematical theory of electricity and magnetism, physical astronomy, the attraction of ellipsoids, definite integrals, series, and the theory of elasticity. The student encounters *Poisson's brackets* (in differential equations), *Poisson's constant* (in electricity), the *Poisson ratio* (in elasticity), *Poisson's integral* and *Poisson's equation* (in potential theory), and *Poisson's law* (in probability theory).

A droll story links Poisson to one of his professional interests. When a boy, he was put in the care of a nurse. One day, when his father came to see him, the nurse had gone out and left the youngster suspended by his straps to a nail in the wall—to protect the boy, the nurse said, from the disease and dirt of the floor. Poisson said that his gymnastic efforts when thus suspended caused him to swing back and forth, and it was in this way that he early became familiar with the pendulum, the study of which occupied much of his later life.

Poisson once remarked: "Life is good for only two things, discovering mathematics and teaching mathematics." He excelled in both pursuits.

13–4 Bolzano

Bernhard Bolzano was born in 1781 in Prague, Czechoslovakia, where he died in 1848. He became a priest, but was unfrocked for heresy and dismissed from his teaching post as professor of religion at the University of Prague. He had a leaning toward logic and mathematics, especially analysis, and can be considered a forerunner of the "arithmetization of analysis" (see Section 14–9). In fact, by 1817 he was so fully aware of the need for rigorization in analysis that Felix Klein later referred to him as "The Father of Arithmetization".

Unfortunately, Bolzano's mathematical work was largely ignored by his contemporaries, and many of his results awaited later rediscovery. In 1843, for example, he produced a function continuous in an interval, but surprisingly, having no derivative at any point of the interval. His function didn't become known, and it is Weierstrass, about forty years later, who is usually credited for the first example of this kind. There is in analysis a famous theorem that bears both of these mathematicians' names, the *Bolzano-Weierstrass theorem*, which says that every bounded infinite set of points contains at least one accumulation point. The theorem was first proved by Weierstrass in his Berlin lectures of the 1860s and is basic in the foundations of set theory. The highly useful **intermediate-value theorem** of the calculus is often referred to as **Bolzano's theorem.** This theorem says that if $f(x)$ is a real-valued continuous function in an open interval R, and takes on the values α and β at a and b in R, then f takes on any value γ lying between α and β at at least one point c in R between a and b.

Bolzano discussed a number of examples analogous to Galileo's paradox concerning the one-to-one correspondence between the positive integers and their squares (see Problem Study 9.7(c)), and he seems to have recognized that the infinity of the set of all real numbers is of a type different from the infinity of the set of all integers. In a posthumous work of 1850, *Paradoxien des Unendlichen* (*Paradoxes of the Infinite*), Bolzano exhibited many important properties of infinite sets.

There is an amusing story told about Bolzano when he once was suffering from an illness manifested by bodily aches and chills. To take his mind off his troubles, he picked up Euclid's *Elements* and for the first time read the masterly exposition of the Eudoxian doctrine of ratio and proportion set out in Book V. Lo and behold, his pains vanished. It has been said that after that, when anyone became similarly discomforted, Bolzano would recommend that the ill one read Euclid's Book V.

13–5 Cauchy

With Lagrange and Gauss, the nineteenth-century rigorization of analysis got under way. This work was considerably furthered and strengthened by the great French mathematician Augustin-Louis Cauchy, the most outstanding analyst of the first half of the nineteenth century.

Cauchy was born in Paris in 1789 and received his early education from his father. Later, at the École Centrale du Panthéon, he excelled in ancient classical studies. In 1805, he entered the École Polytechnique and won the admira-

AUGUSTIN-LOUIS CAUCHY
(David Smith Collection)

tion of Lagrange and Laplace. Two years later, he enrolled at the École des Ponts et Chaussées, where he prepared himself to be a civil engineer. Under the persuasion of Lagrange and Laplace, he decided to give up civil engineering in favor of pure science and accepted a teaching post at the École Polytechnique.

Cauchy wrote extensively and profoundly in both pure and applied mathematics, and he can probably be ranked next to Euler in volume of output. His collected works contain, in addition to several books, 789 papers, some of which are very extensive works, and fill twenty-four large quarto volumes. This work is of uneven quality; consequently, Cauchy (quite unlike the case of Gauss) has been criticized for overproduction and over-hasty composition. A story is told in connection with Cauchy's prodigious productivity. In 1835, the Academy of Sciences began publishing its *Comptes rendus*. So rapidly did Cauchy supply this journal with articles that the Academy became alarmed over the mounting printing bill and accordingly passed a rule, still in force today, limiting all published papers to a maximum length of four pages. Cauchy had to seek other outlets for his longer papers, some of which exceeded a hundred pages.

Cauchy's numerous contributions to advanced mathematics include researches in convergence and divergence of infinite series, real and complex function theory, differential equations, determinants, probability, and mathematical physics. His name is met by the student of calculus in the so-called *Cauchy root test* and *Cauchy ratio test* for convergence or divergence of a series of positive terms, and in the *Cauchy product* of two given series. Even in a first course in complex function theory, one encounters the *Cauchy inequality, Cauchy's integral formula, Cauchy's integral theorem,* and the basic *Cauchy-Riemann differential equations.*

Much of the treatment in our present-day college calculus texts, such as the basic concepts of limit and continuity, is due to Cauchy. Cauchy defined the derivative, with respect to x, of $y = f(x)$ as the limit, when $\Delta x \to 0$, of the difference quotient

$$\frac{\Delta y}{\Delta x} = \frac{f(x + \Delta x) - f(x)}{\Delta x}$$

Although he was well aware of the operational facility of differentials, he relegated them to a secondary role. If dx is a finite quantity, he defined dy of $y = f(x)$ simply to be $f'(x)dx$. Whereas during the eighteenth century integration was generally treated as the inverse of differentiation, Cauchy preferred to define the definite integral as the limit of the sum of an infinitely increasing set of vanishingly small parts, much as we do today. The relation between an integral and an antiderivative was then established by the theorem of mean value.

Cauchy's contributions to determinant theory, starting with a large eighty-four page memoir in 1812, mark him as the most prolific contributor in this field. It was in his 1812 paper that Cauchy gave the first proof of the important and useful theorem that if A and B are both $n \times n$ matrices, then $|AB| = |A||B|$. Incidentally, it was Cauchy who, in 1840, introduced the word "characteristic" into matrix theory, by calling the equation $|A - \lambda I| = 0$ the **characteristic equation** of matrix A.

Cauchy's work exhibits great attention to rigor, and as such was largely responsible for inspiring other mathematicians to attempt the banishment of blind formal manipulation and of intuitive proofs from analysis.

Cauchy was an ardent partisan of the Bourbons and, after the revolution of 1830, was forced to give up his professorship at the École Polytechnique and was excluded from public employment for eighteen years. Part of this time he spent in exile in Turin and Prague, and part in teaching in some church schools in Paris. In 1848, he was allowed to return to a professorship at the École Polytechnique without having to take the oath of allegiance to the new government. In religion, he was bigoted; he spent much of his time trying to convert others to his particular belief. Throughout his life, he was an indefatigable worker, and it is regrettable that he possessed a narrow conceit and often ignored the meritorious efforts of younger men. Nevertheless, on the other side of the coin, it should be pointed out that in 1843 Cauchy published, in the form of an open letter, a magnificent defense of freedom of conscience and thought. This letter helped to bring home to the government the stupidity of academic repression, and when Louis Philippe was ousted, one of the first acts of the succeeding Provisional Government was to abolish the detestable oath of allegiance.

Cauchy died suddenly on May 23, 1857, when he was sixty-eight years old. He had gone to the country to rest and to cure a bronchial trouble, only to be smitten by a fatal fever. Just before his death, he was talking with the Archbishop of Paris. His last words, addressed to the Archbishop, were: "Men pass away, but their deeds abide."

13–6 Abel and Galois

It is natural, for one reason or another, to associate certain men in the history of mathematics in pairs. Such was the case with Harriot and Oughtred (two contemporary English algebraists), Wallis and Barrow (two immediate predecessors of Isaac Newton in the field of the calculus), Taylor and Maclaurin (two contemporary British mathematicians chiefly known for their contributions to infinite series expansions), Monge and Carnot (two contemporary French geometers), and Fourier and Poisson (two contemporary researchers in mathematical physics). Another such pair was Niels Henrik Abel and Évariste Galois. These two men, though contemporaries, are not related by nationality or similar mathematical interest; each, like a streaking meteor in the mathematical heavens, flashed to an early brilliance and then was suddenly and pathetically extinguished by premature death, leaving remarkable material for future mathematicians to work upon. Abel died of tuberculosis and malnutrition in his twenty-sixth year, and Galois died in a foolish duel in his twenty-first year; neither man was properly appreciated during his lifetime for his genius.

Abel, born in 1802 at Findö in Norway, was the son of a country minister. When a student in Christiania, he thought he had discovered how to solve the general quintic equation algebraically, but soon corrected himself in a famous pamphlet published in 1824. In this early paper, Abel established the impossibility of solving the general quintic equation by means of radicals, thus finally laying to rest a difficult problem that had puzzled mathematicians from Bombelli to Viète (see Section 8–8). As a result of this paper, Abel obtained a small stipend that permitted him to travel in Germany, Italy, and France. During these travels, he wrote a number of papers in various areas of mathematics, such as on the convergence of infinite series, on the so-called *Abelian integrals*, and on elliptic functions.

Abel's researches on elliptic functions arose in exciting and friendly competition with Jacobi. The older Legendre, who had done pioneer work on elliptic functions, was deeply impressed with Abel's discoveries. Luckily Abel secured an outlet for his papers in the newly founded *Journal für die reine und angewandte Mathematik* (more popularly known as *Crelle's Journal*); in fact, the first volume of the journal (1826) contained no less than five of Abel's papers, and the second volume (1827) contained Abel's work that gave birth to the theory of doubly periodic functions.

Every student of analysis encounters *Abel's integral equation* and *Abel's theorem* on the sum of integrals of algebraic functions that leads to *Abelian functions*. In infinite series work, there is *Abel's convergence test* and *Abel's theorem* on power series. In abstract algebra, commutative groups are today called *Abelian groups*.

Plagued by poverty all his life and suffering from a pulmonary condition, Abel was unable to obtain a teaching position. He died tragically at Froland in

NIELS HENRIK ABEL
(David Smith Collection)

Norway in 1829. Two days after his death, a delayed letter was delivered in which Abel was belatedly offered a teaching post at the University of Berlin.

Although Abel received little recognition from his government when alive, he now appears on some of his country's smaller-denominational postage stamps.[3] But the mathematicians, in their characteristic manner, have erected far more lasting monuments to Abel, for today Abel's name is perpetuated in an abundance of theorems and theories. Of Abel, Hermite once said, "He has left mathematicians something to keep them busy for five hundred years." Abel's close friend, Mathias Keilhau, conceived the idea of erecting a more conventional monument to Abel, at the site of his friend's last resting place. The visitor of today pilgrimaging to Froland Church can see there Kielhau's monument to his friend.

When asked his formula for so rapidly forging ahead to the first ranks of his discipline, Abel replied, "By studying the masters and not their pupils."

Évariste Galois had an even shorter and more tragic life than did Abel. Born near Paris in 1811 as the son of a small-town mayor, he began to exhibit an extraordinary mathematical talent shortly after his fifteenth birthday. He tried twice to enter the École Polytechnique, but both times was refused admission because of his inability to meet the formal requirements of his examiners, who completely failed to recognize his genius. Then came another blow; his father, feeling himself persecuted by the clerics, committed suicide. Persevering, Galois finally entered the École Normale in 1829 to prepare himself to teach, but, drawn by democratic sympathy into the turmoils of the Revolution of 1830, he was expelled from school and spent several months in prison. Shortly after his release, in 1832, when not yet twenty-one years old, he was manipulated into a pistol duel over a love affair and was slain.

Galois mastered the mathematical textbooks of his time with the ease of reading novels, went on to the important papers of Legendre, Jacobi, and Abel, and then turned to creating mathematics of his own. In his seventeenth year, he reached results of great importance, but two memoirs that he sent to the French Academy were mislaid and lost, adding to his frustration. A short paper of his on equations was published in 1830 and gave results apparently based on a very general theory. The night before his duel, fully realizing he would in all probability be killed, he wrote his scientific testament in the form of a letter to one of his friends. This testament referred to some of his unpublished discoveries, the

---

[3] Among other mathematicians appearing on postage stamps are Archimedes, Aristotle, Farkas Bolyai, János Bolyai, Boscovich, Brahe, Buffon, L. N. M. Carnot, N. L. S. Carnot, Ch'ang Hong, Ch'unh Chih, Chaplygin, Copernicus, Cristescu, Cusanus, d'Alembert, da Vinci, Descartes, de Witt, Dürer, Einstein, Euler, Galileo, Gauss, Gerbert, Hamilton, Helmholtz, Hipparchus, Huygens, Kepler, Kovalevsky, Krylov, Lagrange. Laplace, Leibniz, Liapunov, Lobachevsky, Lorentz, Mercator, Monge, Nasir-ed din, Newton, Ostrogradsky, Pascal, Poincaré, Popov, Pythagoras, Ramanujan, Riese, Stevin, Teixeira, Titeica, and Torricelli. Russia and France have been the most generous in representing mathematicians on postage stamps; England has only recently done so, and the United States only twice. Da Vinci, Galileo, Copernicus, and Einstein have each been represented by four or more different countries.

ÉVARISTE GALOIS
(David Smith Collection)

later unraveling of which required the talents of some great mathematicians, and that turned out to contain the theory of groups and the so-called Galois theory of equations. The Galois theory of equations, based upon concepts of group theory, supplies criteria for the possibility of solving a geometrical construction with Euclidean tools and for the possibility of solving an algebraic equation by radicals.

Several of Galois' memoirs and manuscripts found among his papers after his death were published by Joseph Liouville (1809–1882) in 1846 in his *Journal de mathématique*. Full appreciation of Galois' accomplishments, however, had to await until 1870, when Camille Jordan (1838–1922) expounded them in his *Traité des substitutions,* and still later when Felix Klein (1849–1925) and Sophus Lie (1842–1899) brilliantly applied them to geometry.[4]

Galois essentially created the study of groups; he was the first (in 1830) to use the word "group" in its technical sense. Researches in the theory of groups were then carried on by Augustin-Louis Cauchy (1789–1857) and his successors under the particular guise of substitution groups. With the subsequent magnificent work of Arthur Cayley (1821–1895), Ludwig Sylow (1832–1918), Sophus Lie, Georg Frobenius (1848–1917), Felix Klein, Henri Poincaré (1854–1912), Otto Hölder (1859–1937), and others, the study of groups assumed its independent abstract form and developed at a rapid pace. The notion of group came to play a great codification role in geometry (see Section 14–8), and in

[4] For a discussion of myths about Galois and his work, see Tony Rothman, "Genius and biographers: The fictionalization of Évariste Galois," *The American Mathematical Monthly* 89 (1982): 84–106.

algebra it served as an atomic structure of cohesion that became an important factor in the twentieth-century rise of abstract algebra. The theory of groups is still, well into the last half of the twentieth century, a very active field of mathematical research.

13–7 Jacobi and Dirichlet

The French Revolution, with its ideological break from the past and its many sweeping changes, created highly favorable conditions for the growth of mathematics. Thus, in the nineteenth century, mathematics underwent a great forward surge, first in France and then later, as the motivating forces spread over northern Europe, in Germany, and still later in Britain. The new mathematics began to free itself from its ties to mechanics and astronomy, and a purer outlook evolved. Two notable German mathematicians who played an early part in the shift of the center of mathematical activity from France to Germany were Carl Gustav Jacob Jacobi (1804–1851) and Peter Gustav Lejeune Dirichlet (1805–1859).

Jacobi was born of Jewish parents in Potsdam in 1804 and was educated at the University of Berlin, where he obtained his doctorate in 1825. Two years later, he was appointed Extraordinary Professor of Mathematics at Königsberg, and two years after that was promoted to Ordinary Professor of Mathematics there. In 1842, under a pension from the Prussian government, he relinquished his chair at Königsberg and moved to Berlin, where he resided until his early death in 1851.

Rarely is an outstanding researcher in mathematics also an outstanding teacher of mathematics. Jacobi was one of the exceptions and was unquestion-

CARL GUSTAV JACOBI
(David Smith Collection)

ably the greatest university mathematics teacher of his generation, stimulating and influencing an unprecedented number of able students. His most celebrated researches in mathematics are those concerning elliptic functions. He and Abel independently and simultaneously established the theory of these functions, and Jacobi introduced what is essentially our present-day notation for them. Jacobi, next to Cauchy, was perhaps the most prolific contributor to determinant theory. It was with him that the word *determinant* received final acceptance. He early used the functional determinant that Sylvester later called the *Jacobian,* and that is encountered by all students of function theory. He also contributed to the theory of numbers, the theory of both ordinary and partial differential equations, the calculus of variations, the three-body problem, and other dynamical problems.

Most students feel that before doing research they should first master what has already been accomplished. To offset this notion, and to stimulate early interest in independent work, Jacobi would deliver the parable: "Your father would never have married, and you would not be born, if he had insisted on knowing *all* the girls in the world before marrying *one*." In defending pure research against applied research, he remarked, "The real end of science is the honor of the human mind." In imitation of Plato, who said, "God ever geometrizes," Jacobi said, "God ever arithmetizes."

Jacobi was always generous in his statements about his great contemporaries in the field of mathematics. Of one of Abel's masterpieces he said, "It is above my praise as it is above my own work."

Dirichlet was born at Düren in 1805, and successively held professorships at Breslau and Berlin. At Gauss' death in 1855 he was appointed Gauss' successor at Göttingen, a fitting honor for so talented a mathematician who was a former student of Gauss and a lifelong admirer of his mentor. While at Göttingen, he had hoped to finish Gauss' incomplete works, but his early death in 1859 prevented this.

Fluent in both German and French, Dirichlet served admirably as a liaison between the mathematics and the mathematicians of the two nationalities. Perhaps his most celebrated mathematical accomplishment was his penetrating analysis of the convergence of Fourier series, an undertaking that led him to generalize the function concept (see Section 15–3). He did much to facilitate the comprehension of some of Gauss' more abstruse methods, and he himself contributed notably to number theory; his beautiful *Vorlesungen über Zahlentheorie* still constitutes one of the most lucid introductions to Gauss' number theory investigations. We are indebted to him for applying infinitesimal methods in this branch of mathematics. Dirichlet was a close friend, expositor, and admirer, of Jacobi. His name is met by college mathematics majors in connection with *Dirichlet's series,* the *Dirichlet function,* and the *Dirichlet principle*.

A touching story is told of Dirichlet and his great teacher, Gauss. On July 16, 1849, exactly fifty years after the awarding to Gauss of his doctorate, Gauss enjoyed the celebration at Göttingen of his golden jubilee. As part of the "show," Gauss, at one point of the proceedings, was to light his pipe with a piece of the original manuscript of his *Disquisitiones arithmeticae*. Dirichlet,

LE JEUNE DIRICHLET
(David Smith Collection)

who was present at the celebration, was appalled at what seemed to him a sacrilege. At the last moment, he boldly rescued the paper from Gauss' hands and treasured the memento the rest of his life; it was found by his editors among his papers after he died.

Dirichlet has been described as possessing a noble, sincere, human, and modest disposition, but, unlike Jacobi, he seemed unable to communicate with young minds. When a schoolmate expressed envy because Dirichlet's son could always receive help from his gifted father, the son gave this lamentable but memorable reply: "Oh! My father doesn't know the little things anymore." Dirichlet's waggish nephew, Sebastian Hensel, wrote in his memoirs that the mathematics instruction he received in his sixth and seventh years at the gymnasium from his uncle was the most dreadful experience of his life.

Dirichlet was very lax in maintaining family correspondence. When his first child arrived, he failed to write of the event to his father-in-law, who was living in London at the time. The father-in-law, when he finally found out, commented that he thought Dirichlet "should have at least been able to write $2 + 1 = 3$." This witty father-in-law was none other than Abraham Mendelssohn, a son of the philosopher Moses Mendelssohn, and father of the composer Felix Mendelssohn.

Dirichlet's brain, and also that of Gauss, are preserved in the department of physiology at Göttingen University.

13–8 Non-Euclidean Geometry

Two very remarkable and revolutionary mathematical developments occurred in the first half of the nineteenth century. The first one was the discovery, about 1829, of a self-consistent geometry different from the customary geometry of

Euclid; the second one was the discovery, in 1843, of an algebra different from the familiar algebra of the real number system. We now turn to a consideration of these two developments, first discussing the one in the field of geometry.

There is evidence that a logical development of the theory of parallels gave the early Greeks considerable trouble. Euclid met the difficulties by defining parallel lines as coplanar straight lines that do not meet one another however far they may be produced in either direction, and by adopting as an assumption his now famous parallel postulate. This postulate (see Section 5–7 for its statement) lacks the terseness and simple comprehensibility of the others and in no sense possesses the characteristic of being "self-evident." Actually, it is the converse of Proposition I 17, and to the early Greeks, it seemed more like a proposition than a postulate. Moreover, Euclid made no use of the parallel postulate until he reached Proposition I 29. It was natural to wonder if the postulate was really needed at all and to think that perhaps it could be derived as a theorem from the remaining nine "axioms" and "postulates," or, at least, that it could be replaced by a more acceptable equivalent.

Of the many substitutes that have been derived to replace Euclid's parallel postulate, the one most commonly used is that made well known in modern times by the Scottish physicist and mathematician John Playfair (1748–1819), although this particular alternative had been used by others and had even been stated as early as the fifth century by Proclus. It is the substitute most frequently encountered in present-day high-school geometry texts: *Through a given point not on a given line can be drawn only one line parallel to the given line.*[5] Some other proposed alternatives for the parallel postulate are (1) *There exists at least one triangle having the sum of its three angles equal to two right angles.* (2) *There exists a pair of similar noncongruent triangles.* (3) *There exists a pair of straight lines everywhere equally distant from one another.* (4) *A circle can be passed through any three noncollinear points.* (5) *Through any point within an angle less than 60° there can always be drawn a straight line intersecting both sides of the angle.*

The attempts to derive the parallel postulate as a theorem from the remaining nine "axioms" and "postulates" occupied geometers for over 2000 years and culminated in some of the most far-reaching developments of modern mathematics. Many "proofs" of the postulate were offered, but each was sooner or later shown to rest upon a tacit assumption equivalent to the postulate itself.

The first really scientific investigation of the parallel postulate was not printed until 1773, by the Italian Jesuit priest Girolamo Saccheri (1667–1733).

Little is known of Saccheri's life. He was born in San Remo, showed marked precocity as a youngster, completed his novitiate for the Jesuit Order at the age of twenty-three, and then spent the rest of his life filling a succession of university teaching posts. While instructing rhetoric, philosophy, and theology at a Jesuit College in Milan, Saccheri read Euclid's *Elements* and became

[5] Proposition I 27 guarantees the existence of at least one parallel.

enamored with the powerful method of *reductio ad absurdum*. Later, while teaching philosophy at Turin, Saccheri published his *Logica demonstrativa*, in which the chief innovation is the application of the method of *reductio ad absurdum* to the treatment of formal logic. Some years after, while a professor of mathematics at the University of Pavia, it occurred to Saccheri to apply his favorite method of *reductio ad absurdum* to a study of Euclid's parallel postulate, and he received permission to print a little book entitled *Euclides ab omni naevo vindicatus* (Euclid Freed of Every Flaw), which appeared in Milan in 1733, only a few months before his death.

In his work on the parallel postulate, Saccheri accepts the first twenty-eight propositions of Euclid's *Elements*, which, as we noted above, do not require the parallel postulate for their proof. With the aid of these theorems, he then proceeds to study a quadrilateral $ABCD$ (see Figure 114) in which angles A and B are right angles and sides AD and BC are equal. By drawing the diagonals AC and BD and then using simple congruence theorems (which are found among Euclid's first twenty-eight propositions), Saccheri easily shows, as can any high-school geometry student, that angles D and C are equal. There are, then, three possibilities: angles D and C are equal acute angles, equal right angles, or equal obtuse angles. These three possibilities are referred to by Saccheri as the **hypothesis of the acute angle,** the **hypothesis of the right angle,** and the **hypothesis of the obtuse angle.** The plan of the work is to show that the assumption of either the hypothesis of the acute angle or the hypothesis of the obtuse angle leads to a contradiction; then, by *reductio ad absurdum,* the hypothesis of the right angle must hold. This hypothesis, Saccheri shows, implies the parallel postulate. Tacitly assuming the infinitude of the straight line, Saccheri readily eliminates the hypothesis of the obtuse angle, but the case of the hypothesis of the acute angle proves to be much more difficult. After obtaining many of the now classical theorems of so-called non-Euclidean geometry, Saccheri lamely forces into his development an unconvincing contradiction involving hazy notions about infinite elements. Had he not been so eager to exhibit a contradiction here, but rather had admitted his inability to find one, Saccheri would today unquestionably be credited with the discovery of non-Euclidean geometry. His work was little regarded by his contemporaries and was soon forgotten,[6] and it was not until 1889 that it was resurrected by his countryman, Eugenio Beltrami (1835–1900).

Thirty-three years after Saccheri's publication, Johann Heinrich Lambert (1728–1777) of Switzerland wrote a similar investigation entitled *Die Theorie der Parallellinien,* which, however, was not published until after his death. Lambert chose a quadrilateral containing three right angles (half of a Saccheri quadrilateral) as his fundamental figure and considered three hypotheses according as the fourth angle is acute, right, or obtuse. He went considerably

---

[6] There is an alternative explanation, involving an unpleasant insinuation of suppression, that has been offered to account for the long neglect of Saccheri's masterpiece. See, for example, E. T. Bell, *The Magic of Numbers*, Chapter 25.

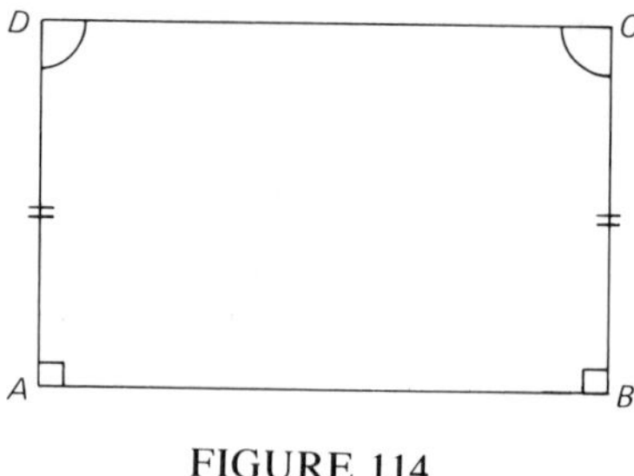

FIGURE 114

beyond Saccheri in deducing propositions under the hypotheses of the acute and obtuse angles. Thus, with Saccheri, he showed that in the three hypotheses the sum of the angles of a triangle is less than, equal to, or greater than two right angles, respectively, and then, in addition, that the deficiency below two right angles in the hypothesis of the acute angle, or the excess above two right angles in the hypothesis of the obtuse angle, is proportional to the area of the triangle. He observed the resemblance of the geometry following from the hypothesis of the obtuse angle to spherical geometry, where the area of a triangle is proportional to its spherical excess, and conjectured that the geometry following from the hypothesis of the acute angle could perhaps be verified on a sphere of imaginary radius. The hypothesis of the obtuse angle was eliminated by making the same tacit assumption as had Saccheri, but his conclusions with regard to the hypothesis of the acute angle were indefinite and unsatisfactory.

Adrien-Marie Legendre (1752–1833), the eminent eighteenth-century French analyst, began anew and considered three hypotheses according as the sum of the angles of a triangle is less than, equal to, or greater than two right angles. Tacitly assuming the infinitude of a straight line, he was able to eliminate the third hypothesis, but, although he made several attempts, he could not dispose of the first hypothesis. These various endeavors appeared in the successive editions of his widely adopted *Éléments de géométrie,* and in this way, he did much to popularize the parallel-postulate problem.

It is no wonder that no contradiction was found under the hypothesis of the acute angle, for it is now known that the geometry developed from a collection of axioms comprising a basic set plus the acute angle hypothesis is as consistent as the Euclidean geometry developed from the same basic set plus the hypothesis of the right angle; that is, the parallel postulate is independent of the remaining postulates and therefore cannot be deduced from them. The first to suspect this fact were Gauss of Germany, Janos Bolyai (1802–1860) of Hungary, and Nicolai Ivanovitch Lobachevsky (1793–1856) of Russia. These men approached the subject through the Playfair form of the parallel postulate by considering the three possibilities: Through a given point can be drawn *more than one,* or *just one,* or *no* line parallel to a given line. These situations are equivalent, respectively, to the hypotheses of the acute, the right, and the obtuse angle. Again, assuming the infinitude of a straight line, the third case was easily eliminated. Suspecting, in time, a consistent geometry under the first possibility, each of these three mathematicians independently carried out extensive geometric and trigonometric developments of the hypothesis of the acute angle.

It is likely that Gauss was the first to reach penetrating conclusions concerning the hypothesis of the acute angle, but since throughout his life he failed to publish anything on the matter, the honor of discovering this particular non-Euclidean geometry must be shared with Bolyai and Lobachevsky. Bolyai published his findings in 1832 in an appendix to a mathematical work of his father. Later it was learned that Lobachevsky had published similar findings as early as 1829–1830, but, because of language barriers and the slowness with which information of new discoveries traveled in those days, Lobachevsky's work did not become known in western Europe for some years. There seems little point in discussing here the intricate, and probably unfounded, theories explaining how various of these men might have obtained and appropriated information of the findings of some other. There was considerable suspicion and incrimination of plagiarism at the time.

Janos (or Johann) Bolyai was a Hungarian officer in the Austrian army, and the son of Farkas (or Wolfgang) Bolyai, a provincial mathematics teacher and long-time personal friend of Gauss. The younger Bolyai undoubtedly received considerable stimulus for his study of the parallel postulate from his father, who had earlier shown an interest in the problem. As early as 1823, Janos Bolyai began to understand the real nature of the problem that faced him, and a letter written during that year to his father shows the enthusiasm he held for his work. In this letter, he discloses a resolution to publish a tract on the theory of parallels as soon as he can find the time and opportunity to put the material in order, and exclaims, "Out of nothing I have created a strange new universe." The father urged that the proposed tract be published as an appendix to his own large two-volume semiphilosophical work on elementary mathe-

NICOLAI LOBACHEVSKY
(New York Public Library Collection)

matics. The expansion and arrangement of ideas proceeded more slowly than Janos had anticipated, but finally, in 1829, he submitted the finished manuscript to his father, and three years later, in 1832, the tract appeared as a twenty-six page appendix to the first volume of his father's work.[7] Janos Bolyai never published anything further, although he did leave behind a great pile of manuscript pages. His chief interest was in what he called "the absolute science of space," by which he meant the collection of those propositions that are independent of the parallel postulate and that consequently hold in both the Euclidean geometry and the new geometry.

Nicolai Ivanovitch Lobachevsky spent the greater part of his life at the University of Kasan, first as a student, later as a professor of mathematics, and finally as rector. His earliest paper on non-Euclidean geometry was published in 1829 and 1830 in the *Kasan Bulletin,* two to three years before Bolyai's work appeared in print. This memoir attracted only slight attention in Russia, and, because it was written in Russian, practically no attention elsewhere. Lobachevsky followed this initial effort with other presentations. For example, in the hope of reaching a wider group of readers, he published, in 1840, a little book written in German entitled *Geometrische Untersuchungen zur Theorie der Parallellinien* (Geometrical Researches on the Theory of Parallels),[8] and then still later, in 1855, a year before his death and after he had become blind, he published in French a final and more condensed treatment entitled *Pangéométrie* (Pangeometry).[9] So slowly did information of new discoveries spread in those days that Gauss probably did not hear of Lobachevsky's work until the appearance of the German publication in 1840, and Janos Bolyai was unaware of it until 1848. Lobachevsky himself did not live to see his work accorded any wide recognition, but the non-Euclidean geometry that he developed is nowadays frequently referred to as **Lobachevskian geometry.**

The actual independence of the parallel postulate from the other postulates of Euclidean geometry was not unquestionably established until consistency proofs of the hypothesis of the acute angle were furnished. These were not long in coming and were supplied by Beltrami, Arthur Cayley, Felix Klein, Henri Poincaré, and others. The method was to set up a model in Euclidean geometry so that the abstract development of the hypothesis of the acute angle could be given a concrete interpretation in a part of Euclidean space. Then any inconsistency in the non-Euclidean geometry would imply a corresponding inconsistency in Euclidean geometry (see Problem Study 13.11).

In 1854, Georg Friedrich Bernhard Riemann (1826–1866) showed that if the infinitude of a straight line be discarded and merely its boundlessness be

---

[7] For a translation of this appendix, see R. Bonola, *Non-Euclidean Geometry,* or D. E. Smith, *A Source Book in Mathematics,* pp. 375–388. Both texts are listed in the Bibliography at the end of this chapter.

[8] For a translation, see R. Bonola, *Non-Euclidean Geometry,* which is listed in the Bibliography at the end of the chapter.

[9] For a translation, see D. E. Smith, *A Source Book in Mathematics,* pp. 360–374, which is listed in the Bibliography at the end of the chapter.

assumed, then, with some other slight adjustments of the remaining postulates, another consistent non-Euclidean geometry can be developed from the hypothesis of the obtuse angle. The three geometries, that of Bolyai and Lobachevsky, that of Euclid, and that of Riemann were given, by Klein in 1871, the names **hyperbolic geometry, parabolic geometry,** and **elliptic geometry.**

13–9 The Liberation of Geometry[10]

The immediate consequence of the discovery of an internally consistent non-Euclidean geometry was, of course, the final settlement of the ages-old problem of the parallel postulate. The parallel postulate was shown to be independent of the other assumptions of Euclidean geometry and, therefore, could not be deduced as a theorem from those other assumptions.

A much more far-reaching consequence than the settlement of the parallel postulate problem was the liberation of geometry from its traditional mold. A deep-rooted and centuries-old conviction that there could be only the one possible geometry was shattered, and the way was opened for the creation of many different systems of geometry. The postulates of geometry became, for the mathematician, mere hypotheses whose physical truth or falsity need not concern him; the mathematician may take his postulates to suit his pleasure, just so long as they are consistent with one another. A postulate, as employed by the mathematician, was seen to have nothing to do with the characteristics of "self-evidence" or "truth" that had been assigned to postulates since the days of the ancient Greeks. With the possibility of inventing purely "artificial" geometries, it became apparent that physical space must be viewed as an empirical concept derived from our external experiences and that the postulates of a geometry designed to describe physical space are simply expressions of this experience, like the laws of a physical science. Euclid's parallel postulate, for example, insofar as it tries to interpret actual space, appears to have the same type of validity as Galileo's law of falling bodies; that is, they are both laws of observation that are capable of verification within the limits of experimental error.

This point of view, that geometry, when applied to actual space, is an experimental science, is in striking contrast to Emmanuel Kant's (1724–1804) theory of space, which dominated philosophical thinking at the time of the discovery of the Lobachevskian geometry. The Kantian theory claimed that space is a framework already existing in the human mind, and that the postulates of Euclidean geometry are *a priori* judgments imposed on the human mind, and that without these postulates, no consistent reasoning about space can be possible. That this viewpoint is untenable was incontestably demon-

---

[10] The material of this section has been adapted from the fuller treatment given in Chapter 3 of Howard Eves, *Foundations and Fundamental Concepts of Mathematics*. 3rd ed. Boston: PWS-KENT Publishing Company, 1990.

strated by the creation of the Lobachevskian geometry. The Kantian theory was so prevalent at the time that to entertain a contrary viewpoint labeled one something of a crackpot. It was a desire to avoid the carping of "the Boeotians" that withheld Gauss from publishing his advanced views of non-Euclidean geometry.

The creation of the Lobachevskian geometry not only liberated geometry but had a similar effect on mathematics as a whole. Mathematics emerged as an arbitrary creation of the human mind and not as something essentially dictated to us of necessity by the world in which we live. The matter is very neatly put in the following words of E. T. Bell:

> In precisely the same way that a novelist invents characters, dialogues, and situations of which he is both author and master, the mathematician devises at will the postulates upon which he bases his mathematical systems. Both the novelist and the mathematician may be conditioned by their environments in the choice and treatment of their material; but neither is compelled by an extrahuman, eternal necessity to create certain characters or invent certain systems.[11]

The creation of the non-Euclidean geometry, by puncturing a traditional belief and breaking a centuries-long habit of thought, dealt a severe blow to the *absolute truth* viewpoint of mathematics. In the words of Georg Cantor, "the essence of mathematics lies in its freedom."

13–10 The Emergence of Algebraic Structure

Ordinary addition and multiplication performed on the set of positive integers are binary operations; to each ordered pair of positive integers a and b are assigned unique positive integers c and d, called, respectively, the *sum* of a and b and the *product* of a and b, and denoted by the symbols

$$c = a + b, \qquad d = a \times b.$$

These two binary operations of addition and multiplication performed on the set of positive integers possess certain basic properties. For example, if a, b, c denote arbitrary positive integers, we have

1. $a + b = b + a$, the so-called *commutative law of addition.*
2. $a \times b = b \times a$, the *commutative law of multiplication.*
3. $(a + b) + c = a + (b + c)$, the *associative law of addition.*
4. $(a \times b) \times c = a \times (b \times c)$, the *associative law of multiplication.*
5. $a \times (b + c) = (a \times b) + (a \times c)$, the *distributive law of multiplication over addition.*

[11] E. T. Bell, *The Development of Mathematics*, p. 330.

In the early nineteenth century, algebra was considered simply symbolized arithmetic.[12] In other words, instead of working with specific numbers, as we do in arithmetic, in algebra we employ letters that represent these numbers. The above five properties, then, are statements that always hold in the algebra of positive integers. But since the statements are symbolic, it is conceivable that they might be applicable to some set of elements other than the positive integers, provided we supply appropriate definitions for the two binary operations involved. This is indeed the case (see, for instance, the examples given in Problem Study 13.13).

It follows that the five basic properties of positive integers just listed may also be regarded as properties of other entirely different systems of elements. The consequences of the preceding five properties constitute an algebra applicable to the positive integers, but it is also evident that the consequences of the five properties constitute an algebra applicable to other systems—that is, there is a common *algebraic structure* (the five basic properties and their consequences) attached to many different systems. The five basic properties may be regarded as postulates for a particular type of algebraic structure, and any theorem formally implied by these postulates would be applicable to any interpretation satisfying the five basic properties. Considered from this view, then, algebra is severed from its tie to arithmetic, and an algebra becomes a purely formal hypothetico-deductive study.

The earliest glimmerings of this modern view of algebra appeared about 1830 in England, with the work of George Peacock (1791–1858), a Cambridge graduate and teacher, and later Dean of Ely. Peacock was one of the first to study seriously the fundamental principles of algebra, and in 1830 he published his *Treatise on Algebra,* in which he attempted to give algebra a logical treatment comparable to that of Euclid's *Elements,* thus winning for himself the title of "the Euclid of algebra." He made a distinction between what he called "arithmetic algebra" and "symbolic algebra." The former was regarded by Peacock as the study that results from the use of symbols to denote ordinary positive decimal numbers, together with signs for operations, like addition and subtraction, to which these numbers may be subjected. Now, in "arithmetic algebra," certain operations are limited by their applicability. In a subtraction, $a - b$, for example, we must have $a > b$. Peacock's "symbolic algebra," on the other hand, adopts the operations of "arithmetic algebra" but ignores their restrictions. Thus, subtraction in "symbolic algebra" differs from the same operation in "arithmetic algebra" in that it is to be regarded as always applicable. The justification of this extension of the rules of "arithmetic algebra" to "symbolic algebra" was called, by Peacock, the **principle of the permanence of equivalent forms.** Peacock's "symbolic algebra" is a universal "arithmetic algebra" whose operations are determined by those of "arithmetic algebra," so

[12] This is still the view of algebra as taught in the high schools and frequently in the freshman year at college.

far as the two algebras proceed in common, and by the principle of the permanence of equivalent forms in all other cases.

The principle of permanence of equivalent forms was regarded as a powerful concept in mathematics, and it played an historical role in such matters as the early development of the arithmetic of the complex number system and the extension of the laws of exponents from positive integral exponents to exponents of a more general kind. In the theory of exponents, for example, if a is a positive rational number and n is a positive integer, then a^n is, by definition, the product of n a's. From this definition, it readily follows that, for any two positive integers m and n, $a^m a^n = a^{m+n}$. By the principle of permanence of equivalent forms, Peacock affirmed that in "symbolic algebra," $a^m a^n = a^{m+n}$, no matter what might be the nature of the base a or of the exponents m and n. The hazy principle of permanence of equivalent forms has today been scrapped, but we are still often guided, when attempting to extend a definition, to formulate the more general definition in such a way that some property of the old definition will be preserved.

British contemporaries of Peacock advanced his studies and pushed the notion of algebra closer to the modern concept of the subject. Thus, Duncan Farquharson Gregory (1813–1844) published a paper in 1840 in which the commutative and distributive laws in algebra were clearly brought out. Further advances in an understanding of the foundations of algebra were made by Augustus De Morgan (1806–1871), another member of the British school of algebraists. In the somewhat groping work of the British school, one can trace the emergence of the idea of algebraic structure and the preparation for the postulational program in the development of algebra. Soon the ideas of the British school spread to continental Europe, where in 1867 they were considered with great thoroughness by the German historian of mathematics Hermann Hankel (1839–1873). Even before Hankel's treatment appeared, however, the Irish mathematician William Rowan Hamilton (1805–1865) and the German mathematician Hermann Günther Grassmann (1809–1877) had published results that were of a far-reaching character, results that led to the liberation of algebra in much the same way that the discoveries of Lobachevsky and Bolyai led to the liberation of geometry, and that opened the floodgates of modern abstract algebra. This remarkable work of Hamilton and Grassmann will be considered in the next section.

13–11 The Liberation of Algebra

Geometry, we have seen, remained shackled to Euclid's version of the subject until Lobachevsky and Bolyai, in 1829 and 1832, liberated it from its bonds by creating an equally consistent geometry in which one of Euclid's postulates fails to hold. With this accomplishment, the former conviction that there could be only the one possible geometry was shattered, and the way was opened for the creation of many new geometries.

The same sort of story can be told of algebra. It seemed inconceivable, in the early nineteenth century, that there could exist an algebra different from the common algebra of arithmetic. To attempt, for example, the construction of a consistent algebra in which the commutative law of multiplication fails to hold, not only probably did not occur to anyone of the time, but had it occurred would surely have been dismissed as a purely ridiculous idea; after all, how could one possibly have a logical algebra in which $a \times b$ is not equal to $b \times a$? Such was the feeling about algebra when, in 1843, William Rowan Hamilton was forced, by physical considerations, to invent an algebra in which the commutative law of multiplication does not hold. The radical step of abandoning the commutative law did not come easily to Hamilton; it dawned on him only after years of cogitation on a particular problem.

It would take us too far afield to go into the physical motivation that lay behind Hamilton's creation. Perhaps the best approach, for our purposes, is through Hamilton's elegant treatment of complex numbers as real number pairs.[13] The mathematicians of his time, like most college freshman mathematics students of today, regarded a complex number as a strange hybrid of the form $a + bi$, where a and b are real numbers and i is some kind of a nonreal number such that $i^2 = -1$, and where addition and multiplication of these numbers are to be accomplished by treating each complex number as a linear polynomial in i, replacing i^2, wherever it might occur, by -1. In this way, one finds, for addition,

$$(a + bi) + (c + di) = (a + c) + (b + d)i,$$

and, for multiplication,

$$(a + bi)(c + di) = ac + adi + bci + bdi^2 = (ac - bd) + (ad + bc)i.$$

If these results should be taken as definitions for the addition and multiplication of pairs of complex numbers, it is not difficult to show that addition and multiplication are commutative and associative, and that multiplication is distributive over addition.

Now, since a complex number $a + bi$ is completely determined by the two real numbers a and b, it occurred to Hamilton to represent the complex number simply and nonmystically by the ordered real number pair (a,b). He defined two such number pairs (a,b) and (c,d) to be equal if and only if $a = c$ and $b = d$. Addition and multiplication of such number pairs he defined (to agree with the above results) to be

$$(a,b) + (c,d) = (a + c, b + d) \quad \text{and} \quad (a,b)(c,d) = (ac - bd, ad + bc).$$

With these definitions, it is easy to show that addition and multiplication of the ordered real number pairs are commutative and associative, and that multiplication is distributive over addition, if one assumes, of course, that these laws hold for the ordinary addition and multiplication of real numbers.

---

[13] Communicated by Hamilton, in 1833, to the Royal Irish Academy.

It is to be noted that the real number system is *embedded* in the complex number system. By this statement is meant that if each real number r is identified with the corresponding number pair $(r,0)$, then the correspondence is preserved under addition and multiplication of complex numbers, for we have

$$(a,0) + (b,0) = (a + b,0) \quad \text{and} \quad (a,0)(b,0) = (ab,0).$$

In practice, a complex number of the form $(r,0)$ can be replaced by its corresponding real number r.

To obtain the older form of a complex number from Hamilton's form, we note that any complex number (a,b) can be written as

$$(a,b) = (a,0) + (0,b) = (a,0) + (b,0)(0,1) = a + bi,$$

where $(0,1)$ is represented by the symbol i, and $(a,0)$ and $(b,0)$ are identified with the real numbers a and b. Finally, we see that

$$i^2 = (0,1)(0,1) = (-1,0) = -1.$$

The former mystical aura surrounding complex numbers has been removed, for there is nothing mystical about an ordered pair of real numbers. This was a great achievement on the part of Hamilton.

The complex number system is a very convenient number system for the study of vectors and rotations in the plane.[14] Hamilton attempted to devise an analogous system of numbers for the study of vectors and rotations in three-dimensional space. In his researches, he was led to the consideration, not of ordered real number pairs (a,b) having the real numbers embedded within them, but of ordered real number quadruples (a,b,c,d) having both the real and the complex numbers embedded within them. In other words, defining two such quadruples (a,b,c,d) and (e,f,g,h) to be equal if and only if $a = e$, $b = f$, $c = g$, $d = h$. Hamilton found it necessary to define an addition and a multiplication of ordered real number quadruples such that, among other restrictions, he would have

$$\begin{aligned}(a,0,0,0) + (b,0,0,0) &= (a + b,0,0,0),\\ (a,0,0,0)(b,0,0,0) &= (ab,0,0,0),\\ (a,b,0,0) + (c,d,0,0) &= (a + c,b + d,0,0),\\ (a,b,0,0)(c,d,0,0) &= (ac - bd,ad + bc,0,0).\end{aligned}$$

---

[14] This convenience results from the fact that when a complex number $z = a + bi$ is considered as representing the point Z having rectangular Cartesian coordinates (a,b), then the complex number z may also be regarded as representing the vector OZ, where O is the origin of coordinates.

Calling such ordered real number quadruples, (real) **quaternions,** Hamilton found that, for his various purposes, he had to formulate the following definitions for addition and multiplication of his quaternions:

$$(a,b,c,d) + (e,f,g,h) = (a + e,\ b + f,\ c + g,\ d + h),$$

$$(a,b,c,d)(e,f,g,h) = (ae - bf - cg - dh,\ af + be + ch - dg,\ ag + ce + df - bh,\ ah + bg + de - cf).$$

It can be shown, with these definitions, that the real numbers and the complex numbers are embedded among the quaternions, and that if we identify the quaternion $(m,0,0,0)$ with the real number m, then

$$m(a,b,c,d) = (a,b,c,d)m = (ma,mb,mc,md).$$

It can also be shown that addition of quaternions is commutative and associative, and that multiplication of quaternions is associative and distributive over addition. But the commutative law for multiplication fails to hold. To see this, consider, in particular, the two quaternions (0,1,0,0) and (0,0,1,0). One finds that

$$(0,1,0,0)(0,0,1,0) = (0,0,0,1),$$

while

$$(0,0,1,0)(0,1,0,0) = (0,0,0,-1) = -(0,0,0,1);$$

that is, the commutative law for multiplication is broken. In fact, if we represent by the symbols 1, i, j, k, respectively, the **quaternionic units** (1,0,0,0), (0,1,0,0), (0,0,1,0), (0,0,0,1), we can verify that the following multiplication table prevails; that is, the desired product is found in the box common to the row headed by the first factor and the column headed by the second factor:

$\times$	1	i	j	k
1	1	i	j	k
i	i	-1	k	$-j$
j	j	$-k$	-1	i
k	k	j	$-i$	-1

Hamilton told the story that the idea of abandoning the commutative law of multiplication came to him in a flash, after fifteen years of fruitless meditation, while he was walking with his wife along the Royal Canal near Dublin just

Here as he walked by
on the 16th of October 1843
Sir William Rowan Hamilton
in a flash of genius discovered
the fundamental formula for
quaternion multiplication
$i^2 = j^2 = k^2 = ijk = -1$
& cut it in a stone of this bridge

before dusk. He was so struck by the unorthodoxy of the idea that he took out his penknife and scratched the gist of the above multiplication table into one of the stones of Broughm Bridge. Today a tablet embedded in the stone of the bridge tells the story (see accompanying figure). Thus, one of the great moments in mathematics is commemorated for us.

We can write the quaternion (a,b,c,d) in the form $a + bi + cj + dk$. When two quaternions are written in this form, they may be multiplied like polynomials in i, j, k, and then the resulting product put into the same form by means of the above multiplication table.

In the year 1844, Hermann Günther Grassmann published the first edition of his remarkable *Ausdehnungslehre*, in which were developed classes of algebras of much greater generality than Hamilton's quaternion algebra. Instead of considering just ordered sets of quadruples of real numbers, Grassmann considered ordered sets of n real numbers. To each such set $(x_1, x_2, \ldots, x_n)$ Grassmann associated a hypercomplex number of the form $x_1e_1 + x_2e_2 + \cdots + x_ne_n$, where $e_1, e_2, \ldots, e_n$ are the fundamental units of his algebra. Two such hypercomplex numbers are added and multiplied like polynomials in $e_1, e_2, \ldots, e_n$. The addition of two such numbers yields, then, a number of the same kind. To make the product of two such numbers a number of the same kind requires the construction of a multiplication table for the units $e_1, \ldots, e_n$ similar to Hamilton's multiplication table for his units 1, i, j, k. Here one has considerable freedom, and different algebras can be created by making different multiplication tables. The multiplication table is governed by the application of the algebra to be made and by the laws of algebra that one wishes to preserve.

Before closing this section, let us consider one more noncommutative algebra—the matric algebra devised by the English mathematician Arthur Cayley (1821–1895) in 1857. Matrices arose with Cayley in connection with linear transformations of the type

$$x' = ax + by,$$

$$y' = cx + dy,$$

where a, b, c, d are real numbers, and which may be thought of as mapping the point (x,y) onto the point (x',y'). Clearly, the preceding transformation is completely determined by the four coefficients a, b, c, d, so the transformation can be symbolized by the square array

$$\begin{bmatrix} a & b \\ c & d \end{bmatrix},$$

which we shall call a **(square) matrix (of order** 2). Because two transformations of the kind under consideration are identical if and only if they possess the same coefficients, we define two matrices

$$\begin{bmatrix} a & b \\ c & d \end{bmatrix} \text{ and } \begin{bmatrix} e & f \\ g & h \end{bmatrix}$$

to be equal if and only if $a = e$, $b = f$, $c = g$, $d = h$. If the transformation given above is followed by the transformation

$$x'' = ex' + fy',$$

$$y'' = gx' + hy',$$

the result can be shown, by elementary algebra, to be the transformation

$$x'' = (ea + fc)x + (eb + fd)y,$$

$$y'' = (ga + hc)x + (gb + hd)y.$$

This leads to the following definition for the product of two matrices:

$$\begin{bmatrix} e & f \\ g & h \end{bmatrix}\begin{bmatrix} a & b \\ c & d \end{bmatrix} = \begin{bmatrix} ea + fc & eb + fd \\ ga + hc & gb + hd \end{bmatrix}.$$

Addition of matrices is defined by

$$\begin{bmatrix} a & b \\ c & d \end{bmatrix} + \begin{bmatrix} e & f \\ g & h \end{bmatrix} = \begin{bmatrix} a + e & b + f \\ c + g & d + h \end{bmatrix},$$

and, if m is any real number, we define

$$m\begin{bmatrix} a & b \\ c & d \end{bmatrix} = \begin{bmatrix} a & b \\ c & d \end{bmatrix}m = \begin{bmatrix} ma & mb \\ mc & md \end{bmatrix}.$$

In the resulting algebra of matrices, it may be shown that addition is both commutative and associative and that multiplication is associative and distributive over addition. But multiplication is not commutative, as is shown by the simple example:

$$\begin{bmatrix} 1 & 0 \\ 0 & 0 \end{bmatrix}\begin{bmatrix} 0 & 1 \\ 0 & 1 \end{bmatrix} = \begin{bmatrix} 0 & 1 \\ 0 & 0 \end{bmatrix}, \qquad \begin{bmatrix} 0 & 1 \\ 0 & 1 \end{bmatrix}\begin{bmatrix} 1 & 0 \\ 0 & 0 \end{bmatrix} = \begin{bmatrix} 0 & 0 \\ 0 & 0 \end{bmatrix}.$$

By developing algebras satisfying structural laws different from those obeyed by common algebra, Hamilton, Grassmann, and Cayley opened the floodgates of modern abstract algebra. Indeed, by weakening or deleting various postulates of common algebra, or by replacing one or more of the postulates by others, which are consistent with the remaining postulates, an enormous variety of systems can be studied. These systems include groupoids, quasigroups, loops, semigroups, monoids, groups, rings, integral domains, lattices, division rings, Boolean rings, Boolean algebras, fields, vector spaces, Jordan algebras, and Lie algebras, the last two being examples of nonassociative algebras. It is probably correct to say that mathematicians have, to date, studied well over 200 such algebraic structures. Most of this work belongs to the twentieth century and reflects the spirit of generalization and abstraction so prevalent in mathematics today. Abstract algebra has become the vocabulary of much of present-day mathematics and has been dubbed "the skeleton key of mathematics."

13–12 Hamilton, Grassmann, Boole, and De Morgan

William Rowan Hamilton, by all odds Ireland's greatest claim to fame in the field of mathematics, was born in Dublin in 1805 and, except for short visits elsewhere, spent his whole life there. He was early orphaned, but even before that, when only a year old, his upbringing was entrusted to an uncle who gave the boy a strenuous but lopsided education with a strong emphasis on languages. William proved to be a prodigy, and when he reached the age of thirteen, he was fluently acquainted with as many foreign languages as he was years old. He developed a fondness for the classics and, with no real success, indulged in what was to become a lifelong desire—the writing of poetry. He became an intimate friend and mutual admirer of the great poet William Wordsworth.

It was not until Hamilton was fifteen that his interests changed and he became excited about mathematics. The change was brought about by his meeting Zerah Colburn, the American lightning calculator, who, though only a youngster himself, gave a demonstration of his powers at an exhibition in Dublin. Shortly after, Hamilton chanced upon a copy of Newton's *Arithmetica universalis*. This he avidly read and then mastered analytic geometry and calculus. Next he read the four volumes of the *Principia* and proceeded to the great mathematical works of the continent. Reading Laplace's *Mécanique céleste*, he uncovered a mathematical error and in 1823 wrote a paper on it that attracted considerable attention. The following year, he entered Trinity College, Dublin.

WILLIAM ROWAN HAMILTON
(Granger Collection)

Hamilton's career at the university was unique, for in 1828, when he was only twenty-two years old and still an undergraduate, the electors unanimously appointed him Royal Astronomer of Ireland, Director of the Dunsink Observatory, and Professor of Astronomy at the University. Shortly after, on mathematical theory alone, he predicted conical refraction in biaxial crystals, which was then dramatically confirmed experimentally by physicists. In 1833, he presented to the Irish Academy his significant paper in which the algebra of complex numbers appears as an algebra of ordered pairs of real numbers (see Section 13–10). He was knighted in 1835.

Following his 1833 paper, Hamilton thought off and on for a long period of years on algebras of ordered triples and quadruples of real numbers, but was always stymied on the matter of how to define multiplication so as to preserve the familiar laws of that operation while at the same time making the operation fit his physical investigations. Finally, in a flash of intuition in 1843 (as described in Section 13–10), it occurred to Hamilton that he was demanding too much and that he had to sacrifice the commutative law, and the algebra of quaternions, the first noncommutative algebra, was suddenly born.

During the remaining twenty-some years of his life, Hamilton expended most of his time and energy in developing his quaternions, which he felt would be of revolutionary significance in mathematical physics. His great work, *Treatise on Quaternions,* appeared in 1853, after which he devoted himself to preparing an enlarged *Elements of Quaternions,* but he died in Dublin in 1865, essentially from alcoholism and a generally rundown condition brought on by a very unhappy married life, before the work was quite completed. The subject of quaternions won a number of staunch supporters, such as Peter Guthrie Tait (1831–1901) of the University of Edinburgh, Alexander Macfarlane (1851–1913), also of the University of Edinburgh but later of the University of Texas and Lehigh University, and Charles Jasper Joly (1864–1906), Hamilton's successor at the Dunsink Observatory. But in time, the more supple vector analysis of the American physicist and mathematician, Josiah Willard Gibbs (1839–1903) of Yale University, and the more general treatment of ordered n-tuples of Hermann Günther Grassmann, tended to relegate the theory of quaternions to little more than a highly interesting museum piece. It is true quaternions were somewhat revived in 1927 as the "spin variables" in Wolfgang Pauli's (1900–1958) quantum theory, and it could be that the future may give quaternions a new lease on life. No matter what, the great importance of quaternions in the history of mathematics lies in the fact that their creation by Hamilton in 1843 liberated algebra from its traditional ties to the arithmetic of real numbers, and accordingly opened the floodgates of modern abstract algebra.

In addition to his work on quaternions, Hamilton wrote on optics, dynamics, the solution of equations of the fifth degree, fluctuating functions, the hodograph curve of a moving particle,[15] and the numerical solution of differential equations.

Hamilton's name is encountered by students of physics in the so-called *Hamiltonian function* and in the *Hamilton-Jacobi differential equations* of dynamics. In matrix theory, there is the *Hamilton-Cayley theorem, equation,* and *polynomial;* in mathematical recreations, one encounters the *Hamiltonian game* played on a regular dodecahedron (see Problem Study 13.24).

It is perhaps pleasing to Americans to recall that in the sad final years of Hamilton's illness and marital strife, the newly founded National Academy of Sciences of the United States elected him as its first foreign associate. Another rare honor and compliment accorded Hamilton occurred when, in 1845, he attended the second Cambridge meeting of the British Association; he was lodged for a week in the sacred rooms of Trinity College in which tradition asserts that Isaac Newton composed his *Principia*.

Sir William Rowan Hamilton is not to be confused with his contemporary, Sir William Hamilton (1788–1856), the noted philosopher of Edinburgh. The latter inherited his title; the former earned his.

Grassmann was born in Stettin, Germany, in 1809, and died there in 1877. He was a man of very broad intellectual interests. He was not only a teacher of mathematics, but of religion, physics, chemistry, German, Latin, history, and geography. He wrote on physics and composed school texts for the study of German, Latin, and mathematics. He was a copublisher of a political weekly in the stormy years of 1848 and 1849. He was interested in music, and in the 1860s, he was an opera critic for a daily newspaper. He prepared a philological treatise on German plants, edited a missionary paper, investigated phonetic laws, wrote a dictionary to the Rig-Veda and translated the Rig-Veda in verse, harmonized folk songs in three voices, composed his great treatise *Ausdehnungslehre,* and raised nine of his eleven children.

---

[15] The extremity of the vector drawn from a fixed point and equal to the velocity vector of a moving point traces a curve called the **hodograph** of the moving point.

HERMAN GÜNTHER GRASSMANN
(David Eugene Smith Collection, Rare Book and Manuscript Library, Columbia University)

It was in the year 1844 that Grassmann published the first edition of his remarkable *Ausdehnungslehre* (Calculus of Extension). Unfortunately, the poor exposition and the obscure presentation caused the work to remain practically unknown to his contemporaries. A second reformulation, put out in 1862, proved scarcely more successful. Discouraged with the reception given his work, Grassmann gave up mathematics for the study of Sanskrit language and literature, a field in which he contributed a number of brilliant papers.

Grassmann spent his entire life in his native city of Stettin, except for the years from 1834 and 1836, when he taught mathematics in an industrial school in Berlin, having succeeded Jacob Steiner to the post. His teaching was entirely at the secondary level, though he had hoped to secure a university position. His father was a teacher of mathematics and physics in the gymnasium at Stettin. His son Hermann Grassmann (born 1859) also became a mathematician. His father wrote two books on mathematics, and the son wrote a treatise on projective geometry.

The *Ausdehnungslehre* has very broad application, with (as was discussed in Section 13–10) no limit to the number of dimensions. In more recent times the wonderful richness and generality of Grassmann's work has become appreciated, and Grassmann's methods have generally been followed, especially on the European continent and in America, in preference to Hamilton's.

We now make brief mention of the two British mathematicians George Boole and Augustus De Morgan, who, among other things, continued the scientific treatment of the fundamental principles of algebra initiated by Hamilton and Grassmann.

George Boole was born in Lincoln, England, in 1815. His father was a struggling lower-class tradesman, so Boole had only a common school education, but he managed to teach himself Greek and Latin. Later, while working as an elementary-school teacher, he learned mathematics by reading the works of Laplace and Lagrange, studied foreign languages, and, through his friend De Morgan, became interested in formal logic. In 1847, Boole published a pamphlet entitled *The Mathematical Analysis of Logic,* which De Morgan praised as epoch making. In his work, Boole maintained that the essential character of mathematics lies in its form rather than in its content; mathematics is not (as some dictionaries today still assert) merely "the science of measurement and number," but, more broadly, any study consisting of symbols along with precise rules of operation upon those symbols, the rules being subject only to the requirement of inner consistency. Two years later, Boole was appointed professor of mathematics at the newly founded Queen's College in Cork, Ireland. In 1854, Boole expanded and clarified his earlier work of 1847 into a book entitled *Investigation of the Laws of Thought,* in which he established both formal logic and a new algebra—the algebra of sets known today as **Boolean algebra.** In more recent times, Boolean algebra has found a number of applications, such as to the theory of electric switching circuits. In 1859 Boole published his *Treatise on Differential Equations,* and then, in 1860, his *Treatise on the Calculus of Finite Differences.* The latter book has remained a standard work in its subject right into present times. Boole died in Cork in 1864.

Augustus De Morgan, whose name appears in several places elsewhere in our book, was born (blind in one eye) in 1806 in Madras, where his father was

AUGUSTUS DE MORGAN
(David Eugene Smith Collection, Rare Book and Manuscript Library, Columbia University)

associated with the East India Company. He was educated at Trinity College, Cambridge, graduating as fourth wrangler, and in 1828 became a professor at the then newly established University of London (later renamed University College), where, through his works and his students, he exercised a wide influence in English mathematics. He was well read in the philosophy and the history of mathematics, and wrote works on the foundations of algebra, differential calculus, logic, and the theory of probability. He was a highly lucid expositor. His witty and amusing book, *A Budget of Paradoxes,* still makes entertaining reading. He continued Boole's work on the algebra of sets, enunciating the principle of duality of set theory, of which the so-called **De Morgan laws** are an illustration: If A and B are subsets of a universal set, then the complement of the union of A and B is the intersection of the complements of A and B, and the complement of the intersection of A and B is the union of complements of A and B (in symbols: $(A \cup B)' = A' \cap B'$ and $(A \cap B)' = A' \cup B'$, where prime denotes complement). Like Boole, De Morgan regarded mathematics as an abstract study of symbols subjected to sets of symbolic operations. De Morgan was an outspoken champion of academic freedom and of religious tolerance. He performed beautifully on the flute and was always jovial company, and he was a confirmed lover of big-city life. He had a fondness for puzzles and conundrums, and when asked either his age or his year of birth would reply, "I was x years old in the year x^2." He died in London in 1871.

13–13 Cayley, Sylvester, and Hermite

The major part of this section is devoted to two brilliant English mathematicians, Arthur Cayley and James Joseph Sylvester, who greatly stimulated one another, frequently researched on the same mathematical problems, created much new mathematics, and, yet, were opposites in temperament, style, and outlook.

Arthur Cayley was born in 1821 at Richmond, in Surrey, and was educated at Trinity College, Cambridge, graduating in 1842 as senior wrangler in the mathematical tripos and in the same year placing first in the even more difficult test for the Smith's prize. For a period of several years, he studied and practiced law, always being careful not to let his legal practice prevent him from working on mathematics. While a student of the bar, he went to Dublin and attended Hamilton's lectures on quaternions. When the Sadlerian professorship was established at Cambridge in 1863, Cayley was offered the chair, which he accepted, thus giving up a lucrative future in the legal profession for the modest provision of an academic life. But then he could devote *all* of his time to mathematics.

Cayley ranks as the third most prolific writer of mathematics in the history of the subject, being surpassed only by Euler and Cauchy. He began publishing while still an undergraduate student at Cambridge, put out between 200 and 300 papers during his years of legal practice, and continued his prolific publication the rest of his long life. The massive *Collected Mathematical Papers* of Cayley

ARTHUR CAYLEY
(Library of Congress)

contains 966 papers and fills thirteen large quarto volumes averaging about 600 pages per volume. There is scarcely an area in pure mathematics that has not been touched and enriched by the genius of Cayley. We have already, in Section 13–10, considered his work in matric algebra. He made pioneering contributions to analytic geometry, transformation theory, the theory of determinants, higher-dimensional geometry, partition theory, the theory of curves and surfaces, the study of binary and ternary forms, and the theory of Abelian, theta, and elliptic functions. But perhaps his most important work was his creation and development of invariant theory. Germs of this theory can be found in the writing of Lagrange, Gauss, and, in particular, of Boole. The basic problem of invariant theory is to find those functions of the coefficients of a given algebraic equation that, when the variables of the equation are subjected to a general linear transformation, remain unchanged except for a factor involving only the coefficients of the transformation. Sylvester became interested in the same field of study, and the two men, both living in London at the time, poured out new discoveries in rapid succession.

Cayley's mathematical style reflects his legal training, for his papers are severe, direct, methodical, and clear. He possessed a phenomenal memory and seemed never to forget anything he had once seen or read. He also possessed a singularly serene, even, and gentle temperament. He has been called "the mathematicians' mathematician."

Cayley developed an unusual avidity for novel reading. He read novels while traveling, while waiting for meetings to start, and at any odd moments that presented themselves. During his life, he read thousands of novels, not only in English, but also in Greek, French, German, and Italian. He took great delight in painting, especially in water colors, and he exhibited a marked talent as a water colorist. He was also an ardent student of botany and nature in general.

Cayley was, in the true British tradition, an amateur mountain climber, and he made frequent trips to the Continent for long walks and mountain scaling. A story is told that he claimed the reason he undertook mountain climbing was that, although he found the ascent arduous and tiring, the grand feeling of exhilaration he attained when he conquered the peak was like that he experienced when he solved a difficult mathematics problem or completed an intricate mathematical theory, and it was easier for him to attain the desired feeling by climbing the mountain.

Cayley died in 1895. Writing in the *Comptes rendus* shortly after, Charles Hermite said: "The mathematical talent of Cayley was characterized by clearness and extreme elegance of analytical form; it was reinforced by an incomparable capacity for work which has caused the distinguished scholar to be compared with Cauchy."

James Joseph Sylvester was born in London in 1814 as the youngest of several children. The surname of the family was originally Joseph, but the eldest son migrated to America where, for some reason not now known, he assumed the new surname Sylvester, which was then adopted by the rest of the family. The American brother was an actuary, and suggested to the Directors of the Lotteries Contractors of the United States that they submit a difficult problem in arrangements that was bothering them to his younger brother James, then only sixteen years old. James' complete and satisfying solution of the problem caused the Directors to award the young mathematician a prize of $500.

JAMES JOSEPH SYLVESTER
(David Eugene Smith Collection, Rare Book and Manuscript Library, Columbia University)

In 1831, James entered St. John's College, Cambridge, and six years later emerged as second wrangler. From 1838 to 1840, he served as professor of natural philosophy at the University of London, and then, in 1841, accepted a professorship in mathematics at the University of Virginia in America, a position from which he resigned after only a few months because of a quarrel he got into with two of his students. Returning to England, he worked as an actuary and was called to the bar in 1850. It was in 1846 that he became associated with Arthur Cayley.

From 1855 to 1870, Sylvester was a professor of mathematics at the Royal Military Academy at Woolwich. In 1876 he returned to America, as a professor of mathematics at the Johns Hopkins University in Baltimore, and there spent seven very happy and highly productive years, becoming the founding editor of the *American Journal of Mathematics* in 1878. During his tenure at Johns Hopkins, he invited Cayley to the university for a series of lectures on Abelian functions; Sylvester himself attended the lectures. In 1884, Sylvester accepted the Savilian chair in geometry at Oxford University. He died in London in 1897, when he was eighty-three years old.

Sylvester's earliest mathematical papers were on Fresnel's optical theory and Sturm's theorem. Then, stimulated by Cayley, he began making important contributions to modern algebra. He wrote papers on elimination theory, transformation theory, canonical forms, determinants, the calculus of forms, partition theory, the theory of invariants, Tchebycheff's method concerning the number of primes within certain limits, latent roots of matrices, the theory of equations, multiple algebra, the theory of numbers, linkage machines, probability theory, and reciprocants. He contributed extensively to mathematical terminology, coining so many new names that he has become known as "the Adam of mathematics."

As remarked earlier, Cayley and Sylvester were antitheses of one another in temperament, style, and outlook. Whereas Cayley was always serene and unruffled, Sylvester was often irritable and prone to show his temper. Cayley's teaching was methodical and prepared; Sylvester's teaching was rambling and off-the-cuff. Cayley wrote severely and to the point; Sylvester wrote discursively with occasional rapturous outbursts. Cayley's lectures were finished pieces; Sylvester often created mathematics in the lecture room. Cayley had a phenomenal memory; Sylvester often could not even recall some of his own findings. Cayley read the mathematical accomplishments of others; Sylvester found it boring to read what others had done. Cayley admired Euclid's *Elements;* Sylvester despised the work. Cayley, although tough and wiry, was slight of build; Sylvester was stocky, muscular, and broad shouldered.

Sylvester had a lifelong interest in poetry and amused himself by writing verses. One evening, at the Peabody Institute in Baltimore, he read his Rosalind poem, which consists of 400 lines all rhyming with the heroine's name "Rosalind." So as not to interrupt the poem, he first spent an hour and a half reading his explanatory footnotes, many of which led to further extemporaneous elaborations. Then, to the remnant of his audience that was left, he read the poem itself. In 1870, he published a curious little booklet entitled *The Laws of Verse,* of which he had a high regard.

Sylvester also had an interest in music and was an amateur singer with a fine voice, having at one time taken singing lessons from the famous French composer Charles François Gounod. On occasion he entertained at workingmen's gatherings with his songs, and it was said that he was prouder of his high C in singing than he was of his contributions to invariant theory in mathematics. In a footnote to his paper, "On Newton's rule for the discovery of imaginary roots," he exclaims: "May not Music be described as the Mathematic of sense, Mathematic as Music of the reason? the soul of each the same! Thus the musician *feels* Mathematic, the mathematician *thinks* Music."

It may be interesting to point out that the most distinguished of the private pupils who studied mathematics under Sylvester in the hard days of his early life was a young woman named Florence Nightingale, later to become world famous as a reformer of hospital nursing.

Many of the beautiful discoveries of Cayley and Sylvester appear in the admirable treatises of George Salmon (1819–1904), provost of Trinity College, Dublin, and one of the finest writers of advanced mathematics texts of his time.

Much of the work of Cayley and Sylvester was continued and expanded by the talented French mathematician Charles Hermite, who made outstanding contributions in both algebra and analysis. Hermite was born at Dieuze in Lorraine in 1822, and after a fitful education, first at the Louis-le-Grand lycée and then briefly at the École Polytechnique, secured, in 1848, the position of

CHARLES HERMITE
(David Eugene Smith Collection, Rare Book and Manuscript Library, Columbia University)

admission's examiner and quiz master at the École Polytechnique. He later served as a professor at the École Polytechnique and the Sorbonne, remaining at the latter institution until his retirement in 1897. He died in Paris in 1901.

Although not a prolific writer, most of Hermite's papers deal with questions of great importance, and his methods exhibit high originality and broad applicability. Even while at Louis-le-Grand, Hermite had two papers, one of quite exceptional quality, accepted by the *Nouvelles annales de mathématiques,* a journal founded in 1842 and devoted to the interests of students in the higher schools. His mentor, Professor Louis Paul Émile Richard, felt compelled to confide to Hermite's father that Charles was "a young Lagrange." Hermite's researches were confined to algebra and analysis. He wrote on the theory of numbers, matrices, algebraic continued fractions, invariants and covariants, quantics, evectants, definite integrals, the theory of equations, elliptic functions, Abelian functions, and the theory of functions. In the last field, he was the foremost French writer of his day. Hermite's collected works, edited by Émile Picard, occupy four volumes.

The two fundamental mathematical results due to Hermite that are of most popular interest are his solution in 1858 of the general quintic equation by means of elliptic functions, and his proof in 1873 of the transcendence of the number e. Hermite's success with the quintic equation later led to the fact that a root of the general equation of degree n can be represented in terms of the coefficients of the equation by means of Fuchsian functions, and the method he employed to prove that e is transcendental was employed by Lindemann in 1882 to prove that π also is transcendental.

Hermite was born with a deformity of his right leg and was lame all his life, requiring a cane to get about. One benefit of this infirmity was that it successfully barred Hermite from any kind of military service. One disadvantage was that after one year at the École Polytechnique, he was dropped from further study because the authorities claimed that his lame leg rendered him unfit for any of the positions open to successful students of the school. Despite his lameness and early difficulties in securing a suitable position, Hermite uniformly maintained the sweetest of dispositions, causing him to be loved by all who knew him. A number of mathematicians have exhibited great generosity to younger men struggling for recognition; Hermite is regarded as unquestionably the finest character of this sort in the entire history of mathematics. In 1856, following a severe illness, he was converted by Cauchy from a tolerant agnostic to a Roman Catholic.

The question of mathematical existence is a highly controversial issue. For example, do mathematical entities and their properties already exist in a sort of timeless twilight land of their own, and we, wandering about in that land, accidentally discover them? In this twilight land, the medians of a triangle are, and always have been, concurrent in a point trisecting each median, and someone, probably in ancient times, wandering about in his mind in the twilight land, came upon this already existing property of the medians of a triangle. In the twilight land, many other remarkable properties of geometrical figures have always existed, but no one has yet stumbled upon them, and may not for years, if ever. In the twilight land, the natural numbers and their host of pretty properties already exist, and always have, but these properties will become existent in the real land of man only when someone wandering about in the twilight land comes upon them.

Pythagoras entertained this idea of mathematical existence, as have many mathematicians after him. Hermite was a confirmed believer in the twilight land of mathematical existence. To him, numbers and all their beautiful properties have always had an existence of their own, and occasionally some mathematical Columbus chances upon one of these already existing properties and then announces his *discovery* to the world.

13–14 Academies, Societies, and Periodicals

The great increase that occurred in scientific and mathematical activity at a time when no periodicals existed led to the formation of a number of discussion circles with regular times of meeting. Some of these groups finally crystallized into academies, the first of which was established in Naples in 1560, followed by the Accademia dei lincei in Rome in 1603. Then, following the northward swing of mathematical activity in the seventeenth century, the Royal Society was founded in London in 1662 and the French Academy in Paris in 1666. These academies constituted centers where scholarly papers could be presented and discussed.

The need for periodicals for the prompt dissemination of new scientific and mathematical findings was increasingly felt, until today the extent of such literature has become enormous. One count claims that prior to 1700 there were only seventeen periodicals containing mathematical articles, the first of these having appeared in 1665. In the eighteenth century, 210 such periodicals appeared, and in the nineteenth century, the number of new journals of this sort reached 950. Many of these, however, often contained little relating to pure mathematics. Perhaps the oldest of the current journals devoted chiefly or entirely to advanced mathematics is the French *Journal de l'École Polytechnique,* launched in 1794. A number of more elementary mathematics journals were started earlier, but many of these aimed to entertain the subscriber with puzzles and problems, rather than to advance mathematical knowledge. Some of our current high-grade mathematical periodicals were started during the first half of the nineteenth century. Foremost among these are the German journal entitled *Journal für die reine und angewandte Mathematik,* first published in 1826 by A. L. Crelle, and the French journal entitled *Journal de mathématiques pures et appliquées,* which appeared in 1836 under the editorship of J. Liouville. These two journals are frequently called *Crelle's Journal* and *Liouville's Journal,* after the names of their founders. In England, the *Cambridge Mathematical Journal* was founded in 1839, became the *Cambridge and Dublin Mathematical Journal* from 1846 to 1854, and in 1855 took the title of *Quarterly Journal of Pure and Applied Mathematics*. The *American Journal of Mathematics* was established in 1878 under the editorship of J. J. Sylvester.

The earliest permanent periodicals devoted to the interests of teachers of mathematics, rather than to mathematical research, are the *Archiv der Mathematik und Physik,* founded in 1841, and the *Nouvelles annales de mathématiques,* founded a year later.

In the second half of the nineteenth century, there was a powerful development that increased the number of the high-quality mathematics journals. This was the formation of a number of large mathematical societies having regular periodicals as their official organs. The earliest of these societies was the London Mathematical Society, organized in 1865, which immediately began to publish its *Proceedings*. This society has become the national mathematical society of England. Seven years later, the Société Mathématique de France was established in Paris, and its official journal is known as its *Bulletin*. In Italy, in 1884, the mathematical society Circolo Matematico di Palermo was organized, and three years later it began to publish its *Rendiconti*. About this time, the Edinburgh Mathematical Society was founded in Scotland, and has since maintained its *Proceedings*. The American Mathematical Society was organized, under a different name, in 1888, and began to issue its *Bulletin,* then later, in 1900, its *Transactions,* and more recently, in 1950, its *Proceedings*. Germany was the last of the leading mathematical countries to organize a national mathematical society, but in 1890 the Deutsche Mathematiker-Vereinigung was organized, which, in 1892, began the publication of its *Jahresbericht*. This last journal carried a number of extensive reports on modern developments in different fields of mathematics, such a report sometimes running into many hundreds of pages. These reports may be regarded as forerunners of the later large encyclopedias of mathematics. The excellent mathematics journals of the Soviet Union, although of later origin, are not to be ignored.

Today almost every country has its mathematical society, and many have additional associations devoted to various levels of mathematical instruction. These societies and associations have become potent factors in the organization and development of research activity in mathematics and in the improvement of methods of teaching the subject. In general, each of these societies and associations sponsors the publication of at least one periodical.

With the great increase in mathematical specialization in the twentieth century, a large number of new mathematics journals have appeared that are devoted to highly limited areas of the subject. Very valuable to researchers is the journal *Mathematical Reviews,* organized by a number of mathematical groups located both in the United States and abroad. This journal appeared in 1940 and contains abstracts and reviews of the current mathematical literature of the world.

Problem Studies

13.1 The Fundamental Theorem of Algebra

Employing the procedure Gauss used in his first proof of the fundamental theorem of algebra, show that

(a) $z^2 - 4i = 0$ has a complex root.
(b) $z^2 + 2iz + i = 0$ has a complex root.

13.2 Basic Properties of Congruence

In the first chapter of *Disquisitiones arithmeticae,* Gauss gives the following definition and notation (here somewhat condensed): Two integers a and b are said to be **congruent modulo n** (where n is a positive integer), symbolized by

$$a \equiv b \pmod{n},$$

if and only if n divides the difference $a - b$. Gauss then goes on to develop the algebra of the congruence relation, which has much in common with the algebra of the ordinary equality relation, but also has many important differences. If n is a fixed positive integer and a, b, c, d are arbitrary integers, show that:

(a) $a \equiv a \pmod{n}$ (the reflexive property).
(b) If $a \equiv b \pmod{n}$, then $b \equiv a \pmod{n}$ (the symmetric property).
(c) If $a \equiv b \pmod{n}$ and $b \equiv c \pmod{n}$, then $a \equiv c \pmod{n}$ (the transitive property).
(d) If $a \equiv b \pmod{n}$ and $c \equiv d \pmod{n}$, then $a + c \equiv b + d \pmod{n}$ and $ac \equiv bd \pmod{n}$.
(e) If $a \equiv b \pmod{n}$, then $a + c \equiv b + c \pmod{n}$ and $ac \equiv bc \pmod{n}$.
(f) If $a \equiv b \pmod{n}$, then $a^k \equiv b^k \pmod{n}$ for any positive integer k.
(g) If $ca \equiv cb \pmod{n}$, then $a \equiv b \pmod{n/d}$, where d is the greatest common divisor of c and n.
(h) If $ca \equiv cb \pmod{n}$ and c and n are relatively prime, then $a \equiv b \pmod{n}$.
(i) If $ca \equiv cb \pmod{p}$, where p is a prime number that does not divide c, then $a \equiv b \pmod{p}$.
(j) If $ab \equiv 0 \pmod{n}$, and if a and b are relatively prime, then either $a \equiv 0 \pmod{n}$ or $b \equiv 0 \pmod{n}$.
(k) If a is prime to n, then the linear congruence $ax \equiv b \pmod{n}$ has only one positive solution x not exceeding n.

13.3 Gauss and Numbers

(a) Using essentially the method of the schoolboy Gauss, find the sum of n terms of an arithmetic progression of initial term a and final term l.
(b) Taking 0 as the first triangular number, express each of the natural numbers from 1 to 100 as a sum of 3 triangular numbers.
(c) Show, from the quadratic reciprocity law, that if p and q are distinct odd primes, then $(q|p) = -(p|q)$ if $p \equiv q \equiv 3 \pmod{4}$.

13.4 Fourier Series

It can be shown, assuming that the trigonometric series of Section 13–2 can be integrated term by term from $-\pi$ to π, that *if* a function $f(x)$ can be represented by such a series, then the coefficients in the series are given by

$$a_n = \frac{1}{\pi}\int_{-\pi}^{\pi} f(x) \cos nx \, dx, \qquad (n \geqq 0)$$
$$b_n = \frac{1}{\pi}\int_{-\pi}^{\pi} f(x) \sin nx \, dx.$$

(a) Show that $\int_{-\pi}^{\pi} \sin nx \, dx = \int_{-\pi}^{\pi} \cos nx \, dx = 0$ when $n \neq 0$.

(b) Show that the Fourier series for the function $f(x)$ defined by

$$f(x) = 2, \qquad -\pi < x < 0,$$
$$f(x) = 1, \qquad 0 < x < \pi,$$

is

$$f(x) = \frac{3}{2} - \frac{2}{\pi}\left(\sin x + \frac{1}{3}\sin 3x + \frac{1}{5}\sin 5x + \cdots\right).$$

(c) Setting $x = \pi/2$ in the Fourier series of (b), obtain the relation

$$\frac{\pi}{4} = 1 - \frac{1}{3} + \frac{1}{5} - \frac{1}{7} + \cdots.$$

(d) Show that the function of (b) can be expressed over the given range by the single equation

$$f(x) = \frac{3}{2} - \frac{x}{2|x|}.$$

13.5 Cauchy and Infinite Series

(a) Establish the convergence or divergence of the following series, using the Cauchy ratio test.
1. $1 + 1/2! + 1/3! + \cdots$.
2. $1/5 - 2/5^2 + 3/5^3 - \cdots$.
3. $1 + 2^2/2! + 3^3/3! + \cdots$.

(b) Establish the convergence or divergence of the following series, using the Cauchy root test.
1. $|\sin \alpha|/2 + |\sin 2\alpha|/2^2 + |\sin 3\alpha|/2^3 + \cdots$.
2. $2|\sin \alpha| + 2^2|\sin 2\alpha| + 2^3|\sin 3\alpha| + \cdots$.

(c) Establish the convergence or divergence of the following series, using the Cauchy integral test.
1. $1/e + 2/e^2 + 3/e^3 + \cdots$.
2. $1/(2 \ln 2) + 1/(3 \ln 3) + 1/(4 \ln 4) + \cdots$.

13.6 Group Theory

A **group** is a nonempty set G of elements over which a binary operation $*$ is defined satisfying the following postulates:

G1: For all a, b, c in G, $(a * b) * c = a * (b * c)$.

G2: There exists an element i of G such that, for all a in G, $a * i = a$. (The element i is called an **identity element** of the group.)

G3: For each element a of G there exists an element a^{-1} of G such that $a * a^{-1} = i$. (The element a^{-1} is called an **inverse element** of a.)

Establish the following theorems about a group.

(a) If a, b, c are in G and $a * c = b * c$, then $a = b$.
(b) For all a in G, $i * a = a * i$.
(c) A group has a unique identity element.
(d) For each a of G, $a^{-1} * a = a * a^{-1}$.
(e) If a, b, c are in G and $c * a = c * b$, then $a = b$.
(f) Each element of a group has a unique inverse element.
(g) If a is in G, then $(a^{-1})^{-1} = a$.
(h) If a and b are in G, then there exist elements x and y of G such that $a * x = b$ and $y * a = b$.

13.7 Examples of Groups

Show that each of the following systems is a group.

(a) The set of all integers under ordinary addition.
(b) The set of all nonzero rational numbers under ordinary multiplication.
(c) The set of all translations

$$T\colon \begin{aligned} x' &= x + h, \\ y' &= y + k, \end{aligned}$$

where h and k are real numbers, of the Cartesian plane, with $T_2 * T_1$ denoting the result of performing first translation T_1 and then translation T_2.
(d) The 4 numbers $1, -1, i, -i$ (where $i^2 = -1$) under ordinary multiplication.
(e) The 4 integers, 1, 2, 3, 4 under multiplication modulo 5.
(f) The 6 expressions

$$r, \qquad 1/r, \qquad 1 - r, \qquad 1/(1 - r), \qquad (r - 1)/r, \qquad r/(r - 1),$$

with $a * b$ denoting the result of substituting the expression b in place of r in the expression a. (This group is known as the **cross ratio group.**)

13.8 Abelian Groups

A group satisfying the further postulate

G4: If a and b are in G, then $a * b = b * a$

is called a **commutative,** or an **Abelian, group.** Which of the groups in Problem Study 13.7 are Abelian?

13.9 Saccheri Quadrilaterals

A **Saccheri quadrilateral** is a quadrilateral $ABCD$ in which the sides AD and BC are equal and the angles at A and B are right angles. Side AB is known as the **base,** the opposite side, DC, as the **summit,** and the angles at D and C as the **summit angles.** Prove, by simple congruence theorems (which do not require the parallel postulate), the following relations:

(a) The summit angles of a Saccheri quadrilateral are equal.

(b) The line joining the midpoints of the base and summit of a Saccheri quadrilateral is perpendicular to both of them.

(c) If perpendiculars are drawn from the extremities of the base of a triangle upon the line passing through the midpoints of the 2 sides, a Saccheri quadrilateral is formed.

(d) The line joining the midpoints of the equal sides of a Saccheri quadrilateral is perpendicular to the line joining the midpoints of the base and summit.

13.10 The Hypothesis of the Acute Angle

The *hypothesis of the acute angle* assumes that the equal summit angles of a Saccheri quadrilateral are acute, or that the fourth angle of a Lambert quadrilateral is acute. In the following, we shall assume the hypothesis of the acute angle.

(a) Let ABC be any right triangle and let M be the midpoint of the hypotenuse AB. At A, construct angle BAD = angle ABC. From M, draw MP perpendicular to CB. On AD, mark off $AQ = PB$ and draw MQ. Prove triangles AQM and BPM congruent, thus showing that angle AQM is a right angle and points Q, M, P are collinear. Then $ACPQ$ is a Lambert quadrilateral with acute angle at A. Now show that, *under the hypothesis of the acute angle, the sum of the angles of any right triangle is less than two right angles.*

(b) Let angle A of triangle ABC be not smaller than either angle B or angle C. Draw the altitude through A and show, by (a), that, *under the hypothesis of the acute angle, the sum of the angles of any triangle is less than two right angles.* The difference between 2 right angles and the sum of the angles of a triangle is known as the **defect** of the triangle.

(c) Consider 2 triangles, ABC and $A'B'C'$, in which corresponding angles are equal. If $A'B' = AB$, then these triangles are congruent. Suppose $A'B' < AB$. On AB, mark off $AD = A'B'$, and on AC mark off $AE = A'C'$. Then triangles ADE and $A'B'C'$ are congruent. Show that E cannot fall on C, since then angle BCA would be greater than angle DEA. Show also that E cannot fall on AC produced, since then DE would cut BC in a point F and the sum of the angles of triangle FCE would exceed 2 right angles. Therefore, E lies between A and C, and $BCED$ is a convex quadrilateral. Show that the sum of the angles of this quadrilateral is equal to 4 right angles. But this is impossible under the hypothesis of the acute angle. It thus follows that we cannot have

$A'B' < AB$ and that, *under the hypothesis of the acute angle, two triangles are congruent if the three angles of one are equal to the three angles of the other.* In other words, in hyperbolic geometry, similar figures of different sizes do not exist.

(d) A line segment joining a vertex of a triangle to a point on the opposite side is called a **transversal.** A transversal divides a triangle into 2 subtriangles, each of which may be similarly subdivided, and so on. Show that if a triangle is partitioned by transversals into a finite number of subtriangles, the defect of the original triangle is equal to the sum of the defects of the triangles in the partition.

13.11 A Euclidean Model for Hyperbolic Geometry

Take a fixed circle, Σ, in the Euclidean plane and interpret the hyperbolic plane as the interior of Σ, a "point" of the hyperbolic plane as a Euclidean point within Σ, and a "line" of the hyperbolic plane as that part of a Euclidean line which is contained within Σ. Verify, in this model, the following statements:

(a) Two "points" determine 1 and only 1 "line."

(b) Two distinct "lines" intersect in at most one "point."

(c) Given a "line" l and a "point" P not on l. Through P can be passed indefinitely many "lines" not meeting "line" l.

(d) Let the Euclidean line determined by the 2 "points" P and Q intersect Σ in S and T, in the order S, P, Q, T. Then we interpret the hyperbolic "distance" from P to Q as $\log [(QS)(PT)/(PS)(QT)]$. If P, Q, R are 3 "points" on a "line," show that

$$\text{"distance" } PQ + \text{"distance" } QR = \text{"distance" } PR.$$

(e) Let "point" P be fixed and let "point" Q move along a fixed "line" through P toward T. Show that "distance" $PQ \to \infty$.

This model was devised by Felix Klein. With the above interpretations, along with a suitable interpretation of "angle" between 2 "lines," it can be shown that all of the postulates for Euclidean plane geometry, except the parallel postulate, are true statements in the geometry of the model. We have seen, in (c), that the Euclidean parallel postulate is not such a statement, but that the Lobachevskian parallel postulate holds instead. The model thus proves that the Euclidean parallel postulate cannot be deduced from the other postulates of Euclidean geometry, for if it were implied by the other postulates, it would have to be a true statement in the geometry of the model.

13.12 Non-Euclidean Geometry and Physical Space

Because of the apparently inextricable entanglement of space and matter, it may be impossible to determine by astronomical methods whether physical space is Euclidean or non-Euclidean. Since all measurements involve both physical and geometrical assumptions, an observed result can be explained in many different ways by merely making suitable compensatory changes in our assumed qualities of space and matter. For example, it is quite possible that a discrepancy observed in the angle-sum of a triangle could be explained by preserving the assumptions of Euclidean geometry but at the same time modifying some physical law, such as some law of optics. And again, the absence of any such discrepancy might be compatible with the assumptions of a non-Euclidean geometry, together with some suitable adjustments in our assumptions about matter. On these grounds, Henri Poincaré maintained the impropriety of asking which geometry is the true one. To clarify this viewpoint, Poincaré devised an imaginary universe Σ occupying the interior of a sphere of radius R in which he assumed the following physical laws to hold:

1. At any point P of Σ, the absolute temperature T is given by $T = k(R^2 - r^2)$, where r is the distance of P from the center of Σ and k is a constant.
2. The linear dimensions of a material body vary directly with the absolute temperature of the body's locality.
3. All material bodies in Σ immediately assume the temperatures of their localities.

(a) Show that it is possible for an inhabitant of Σ to be quite unaware of the preceding 3 physical laws holding in his universe.

(b) Show that an inhabitant of Σ would feel that his universe is infinite in extent on the grounds that he would never reach a boundary after taking a finite number N of steps, no matter how large N may be chosen.

(c) Show that geodesics in Σ are curves bending toward the center of Σ. As a matter of fact, it can be shown that the geodesic through 2 points A and B of Σ is the arc of a circle through A and B that cuts the bounding sphere orthogonally.

(d) Let us impose 1 further physical law on the universe Σ by supposing that light travels along the geodesics of Σ. This condition can be physically realized by filling Σ with a gas having the proper index of refraction at each point of Σ. Show, now, that the geodesics of Σ will "look straight" to an inhabitant of Σ.

(e) Show that in the geometry of geodesics in Σ, the Lobachevskian parallel postulate holds, so that an inhabitant of Σ would believe that he lives in a non-Euclidean world. Here we have a piece of ordinary, and supposedly Euclidean, space, that because of different physical laws, appears to be non-Euclidean.

13.13 Systems with a Common Algebraic Structure

Show that each of the following sets with accompanying definitions of + and × satisfies the 5 basic properties given at the start of Section 13–9.

(a) The set of all even positive integers, with + and × denoting usual addition and multiplication.

(b) The set of all rational numbers, with $+$ and $\times$ denoting usual addition and multiplication.
(c) The set of all real numbers, with $+$ and $\times$ denoting usual addition and multiplication.
(d) The set of all real numbers of the form $m + n\sqrt{2}$, where m and n are integers, with $+$ and $\times$ denoting usual addition and multiplication.
(e) The set of **Gaussian integers** (complex numbers $m + in$, where m and n are ordinary integers and $i = \sqrt{-1}$), with $+$ and $\times$ denoting usual addition and multiplication.
(f) The set of all ordered pairs of integers, where $(a,b) + (c,d) = (a + c, b + d)$ and $(a,b) \times (c,d) = (ac,bd)$.
(g) The set of all ordered pairs of integers, where $(a,b) + (c,d) = (ad + bc, bd)$ and $(a,b) \times (c,d) = (ac,bd)$.
(h) The set of all ordered pairs of integers, where $(a,b) + (c,d) = (a + c, b + d)$ and $(a,b) \times (c,d) = (ac - bd, ad + bc)$.
(i) The set of all real polynomials in the real variable x, with $+$ and $\times$ denoting the ordinary addition and multiplication of polynomials.
(j) The set of all real-valued continuous functions of the variable x defined on the closed interval $0 \leqq x \leqq 1$, with $+$ and $\times$ denoting ordinary addition and multiplication of such functions.
(k) The set consisting of just 2 elements m and n, where we define

$$m + m = m, \qquad m \times m = m,$$

$$m + n = n + m = n, \qquad m \times n = n \times m = m,$$

$$n + n = m, \qquad n \times n = n.$$

(l) The set of all point sets of the plane, with $a + b$ denoting the union of sets a and b, and $a \times b$ denoting the intersection of sets a and b. As a special point set of the plane, we introduce an ideal set, the **null set,** which has no points in it.

13.14 Algebraic Laws

Reduce the left member of each of the following equalities to the right member by using successively an associative, commutative, or distributive law. Following custom, multiplication is here sometimes indicated by a raised dot $(\cdot)$ and sometimes by mere juxtaposition of the factors.

(a) $5(6 + 3) = 3 \cdot 5 + 5 \cdot 6$.
(b) $5(6 \cdot 3) = (3 \cdot 5)6$.
(c) $4 \cdot 6 + 5 \cdot 4 = 4(5 + 6)$.
(d) $a[b + (c + d)] = (ab + ac) + ad$.
(e) $a[b(cd)] = (bc)(ad)$.
(f) $a[b(cd)] = (cd)(ab)$.
(g) $(ad + ca) + ab = a[(b + c) + d]$.
(h) $a + [b + (c + d)] = [(a + b) + c] + d$.

13.15 More on Algebraic Laws

Determine whether the following binary operations $*$ and $|$, defined for positive integers, obey the commutative and associative laws, and whether the operation $|$ is distributive over the operation $*$.

(a) $a * b = a + 2b$, $a \mid b = 2ab$.
(b) $a * b = a + b^2$, $a \mid b = ab^2$.
(c) $a * b = a^2 + b^2$, $a \mid b = a^2b^2$.
(d) $a * b = a^b$, $a \mid b = b$.

13.16 Complex Numbers as Ordered Pairs of Real Numbers

In Hamilton's treatment of complex numbers as ordered pairs of real numbers, show that

(a) Addition is commutative and associative.
(b) Multiplication is commutative and associative.
(c) Multiplication is distributive over addition.
(d) $(a,0) + (b,0) = (a + b,0)$.
(e) $(a,0)(b,0) = (ab,0)$.
(f) $(0,b) = (b,0)(0,1)$.
(g) $(0,1)(0,1) = (-1,0)$.

13.17 Quaternions

(a) Add the 2 quaternions $(1,0,-2,3)$ and $(1,1,2,-2)$.
(b) Multiply, in both orders, the 2 quaternions $(1,0,-2,3)$ and $(1,1,2,-2)$.
(c) Show that addition of quaternions is commutative and associative.
(d) Show that multiplication of quaternions is associative and distributive over addition.
(e) Show that the real numbers are embedded within the quaternions.
(f) Show that the complex numbers are embedded within the quaternions.
(g) Multiply the 2 quaternions $a + bi + cj + dk$ and $e + fi + gi + hk$ like polynomials in i, j, k, and, by means of the multiplication table for the quaternionic units, check into the defined product of the 2 quaternions.

13.18 Matrices

(a) If

$$x' = ax + by, \qquad x'' = ex' + fy',$$
$$y' = cs + dy, \qquad y'' = gx' + hy',$$

show that

$$x'' = (ea + fc)x + (eb + fd)y,$$
$$y'' = (ga + hc)x + (gb + hd)y.$$

(b) Given the matrices

$$A = \begin{bmatrix} 2 & -3 \\ 4 & 1 \end{bmatrix}, \qquad B = \begin{bmatrix} -2 & 2 \\ 0 & 3 \end{bmatrix},$$

calculate $A + B$, AB, BA, and A^2.

(c) Show that addition of matrices is commutative and associative.

(d) Show that multiplication of matrices is associative and distributive over addition.

(e) Show that in matric algebra the matrix $\begin{bmatrix} 1 & 0 \\ 0 & 1 \end{bmatrix}$ plays the role of unity, and the matrix $\begin{bmatrix} 0 & 0 \\ 0 & 0 \end{bmatrix}$ plays the role of zero.

(f) Show that

$$\begin{bmatrix} 0 & 1 \\ 0 & 1 \end{bmatrix}\begin{bmatrix} 1 & 0 \\ 0 & 0 \end{bmatrix} = \begin{bmatrix} 0 & 0 \\ 0 & 0 \end{bmatrix}$$

and that

$$\begin{bmatrix} 1 & 0 \\ 0 & 0 \end{bmatrix}\begin{bmatrix} 0 & 1 \\ 0 & 1 \end{bmatrix} = \begin{bmatrix} 1 & 0 \\ 0 & 0 \end{bmatrix}\begin{bmatrix} 0 & 1 \\ 1 & 0 \end{bmatrix}.$$

What 2 familiar laws or ordinary algebra are broken here?

(g) Show that the matrix $\begin{bmatrix} 0 & 1 \\ 0 & 0 \end{bmatrix}$ has no square root.

(h) Show that for any real number k,

$$\begin{bmatrix} k & 1 + k \\ 1 - k & -k \end{bmatrix}^2 = \begin{bmatrix} 1 & 0 \\ 0 & 1 \end{bmatrix},$$

whence the matrix $\begin{bmatrix} 1 & 0 \\ 0 & 1 \end{bmatrix}$ has an infinite number of square roots.

(i) Show that we may define complex numbers as matrices of the form

$$\begin{bmatrix} a & b \\ -b & a \end{bmatrix},$$

where a and b are real, subject to the usual definitions of addition and multiplication of matrices.

(j) Show that we may define real quaternions as matrices of the form

$$\begin{bmatrix} a + bi & c + di \\ -c + di & a - bi \end{bmatrix},$$

where a, b, c, d are real and $i^2 = -1$, subject to the usual definitions of addition and multiplication of matrices.

13.19 Jordan and Lie Algebras

A (special) **Jordan algebra,** which is used in quantum mechanics, has square matrices for elements, with equality and addition defined as in Cayley's matric algebra, but with the product of 2 matrices A and B defined by $A * B = (AB + BA)/2$, where AB stands for Cayley's product of the 2 matrices A and B. Although multiplication in this algebra is nonassociative, it is obviously commutative. A **Lie algebra** differs from the above Jordan algebra in that the product of 2 matrices A and B is defined by $A \circ B = AB - BA$, where again AB denotes the Cayley product of the matrices A and B. In this algebra, multiplication is neither associative nor commutative.

(a) Taking

$$A = \begin{bmatrix} 1 & 0 \\ -1 & 0 \end{bmatrix}, \quad B = \begin{bmatrix} 1 & 1 \\ -1 & 1 \end{bmatrix}, \quad C = \begin{bmatrix} 1 & 1 \\ 0 & 1 \end{bmatrix}$$

as elements of a Jordan algebra, calculate $A + B$, $A * B$, $B * A$, $A* (B * C)$, and $(A * B) * C$.

(b) Taking A, B, C of (a) as elements of a Lie algebra, calculate $A + B$, $A \circ B$, $B \circ A$, $A \circ (B \circ C)$, and $(A \circ B) \circ C$.

(c) Show that the following relations hold in a Jordan algebra.

J1: $A * B = B * A$,
J2: $(kA) * B = A * (kB) = k(A * B)$, k an arbitrary number,
J3: $A * (B + C) = (A * B) + (A * C)$,
J4: $(B + C) * A = (B * A) + (C * A)$,
J5: $A * (B * A^2) = (A * B) * A^2$, where $A^2 = A * A = AA$.

The name *Jordan algebra* was introduced by A. A. Albert in 1946, inasmuch as the study of these algebras was initiated in 1933 by the physicist Pascual Jordan, one of the founders of modern quantum mechanics. Relation J5 is a special associative law of Jordan algebras.

(d) Show that the following relations hold in a Lie algebra.

L1: $A \circ B = -(B \circ A)$,
L2: $(kA) \circ B = B \circ (kA) = k(A \circ B)$, k an arbitrary number,
L3: $A \circ (B + C) = (A \circ B) + (A \circ C)$,
L4: $(B + C) \circ A = (B \circ A) + (C \circ A)$,
L5: $A \circ (B \circ C) + B \circ (C \circ A) + C \circ (A \circ B) = O$.

Lie algebras are named after the Norwegian mathematician Marius Sophus Lie (1842–1899), who did inaugural work in the study of continuous groups. Relation L5 is known as the **Jacobi identity** of Lie algebras.

(e) Show that

$$A \circ (B * B) = 2[(A \circ B) * B],$$

$$A \circ (B \circ C) = 4[(A * B) * C - (A * C) * B],$$

$$AB = (A * B) + (A \circ B)/2.$$

(f) The **transpose A′** of a square matrix A is the matrix whose successive rows are the successive columns of A. A matrix A is said to be **skew-symmetric** if $A = -A'$. Show that

L6: If A and B are skew-symmetric, then $A \circ B$ is skew-symmetric.

A beautiful theorem about skew-symmetric matrices was established by Jacobi in 1827. He showed that the determinant of a skew-symmetric matrix of odd order equals zero.

13.20 Vectors

Hamilton's quaternions and, to some extent, Grassmann's calculus of extension were devised by their creators as mathematical tools for the exploration of physical space. These tools proved to be too complicated for quick mastery and easy application, but from them emerged the much more easily learned and more easily applied subject of vector analysis. This work was due principally to the American physicist Josiah Willard Gibbs (1839–1903) and is encountered by every student of elementary physics. In elementary physics, a vector is graphically regarded as a directed line segment, or arrow, and the following definitions of equality, addition, and multiplication of these vectors are made:

1. Two vectors a and b are equal if and only if they have the same length and the same direction.
2. Let a and b be any 2 vectors. Through a point in space, draw vectors a' and b' equal, respectively, to vectors a and b, and complete the parallelogram determined by a' and b'. Then the *sum*, $a + b$, of vectors a and b is a vector whose length and direction are those of the diagonal running from the common origin of a' and b' to the fourth vertex of the parallelogram.
3. Let a and b be any 2 vectors. By the **vector product,** $a \times b$, of these 2 vectors is meant a vector whose length is numerically equal to the area of the parallelogram in definition (2), and whose direction is that of the progress of an ordinary screw when placed perpendicular to both a' and b' and twisted through the angle of not more than 180° that will carry vector a' into vector b'.

(a) Show that vector addition is commutative and associative.

(b) Show that vector multiplication is noncommutative and nonassociative.

(c) Show that vector multiplication is distributive over vector addition.

As a native of New Haven, Gibbs studied mathematics and physics at Yale University, receiving a doctor's degree in physics in 1863. He then studied mathematics and physics further at Paris, Berlin, and Heidelberg. In 1871, he was appointed professor of mathematical physics at Yale. As a highly original physicist, he contributed notably to mathematical physics. His *Vector Analysis* appeared in 1881 and again in 1884. In 1902, he published his *Elementary Principles of Statistical Mechanics*. Every student of harmonic analysis encounters the curious *Gibb's phenomenon* of Fourier series.

13.21 An Interesting Algebra

Consider the set of all ordered real number pairs and define the following:

1. $(a,b) = (c,d)$ if and only if $a = c$ and $b = d$.
2. $(a,b) + (c,d) = (a + c,\ b + d)$.
3. $(a,b)(c,d) = (0,ac)$.
4. $k(a,b) = (ka,kb)$.

(a) Show that multiplication is commutative, associative, and distributive over addition.

(b) Show that the product of 3 or more factors is always equal to (0,0).

(c) Construct a multiplication table for the units $u = (1,0)$ and $v = (0,1)$.

13.22 A Point Algebra

Let capital letters $P, Q, R, \ldots$ denote points of the plane. Define *addition* of points P and Q by $P + Q = R$, where triangle PQR is a counterclockwise equilateral triangle.

(a) Show that the addition of points of the plane is noncommutative and nonassociative.

(b) Show that if $P + Q = R$, then $Q + R = P$.

(c) Establish the following identities:

1. $(P + (P + (P + (P + (P + (P + Q)))))) = Q$.
2. $P + (P + (P + Q)) = (Q + P) + (P + Q)$.
3. $(P + Q) + R = (P + (Q + R)) + Q$.

13.23 An Infinite Non-Abelian Group

(a) Show that the set of all 2 by 2 matrices

$$\begin{bmatrix} a & b \\ c & d \end{bmatrix},$$

where a, b, c, d are rational numbers such that $ad - bc \neq 0$, constitutes a group under Cayley matrix multiplication.

(b) Calculate the inverse A^{-1} of matrix $A = \begin{bmatrix} 2 & 1 \\ 3 & -1 \end{bmatrix}$, and show that the product AA^{-1} is the identity matrix.

13.24 The Hamiltonian Game

The Hamiltonian Game consists of determining a route alone the edges of a regular dodecahedron that will pass once and only once through each vertex of the dodecahedron. The game was invented by Sir William Rowan Hamilton, who denoted the vertices of the dodecahedron by letters standing for various towns. Hamilton proposed a number of problems connected with his game.

1. The first problem is to go "all round the world"; that is, starting from some given town, visit every other town once and only once and return to the initial town, where the order of the first $n \leqq 5$ towns may be

prescribed. Hamilton presented a solution of this problem at the 1857 meeting of the British Association at Dublin.

2. Another problem suggested by Hamilton is that of starting at some first given town, visiting certain specific towns in assigned order, then going on to every other town once and only once and ending the journey at some second given town.

The student is invited to read up on the theory of the Hamiltonian game in, say, *Mathematical Recreations and Essays,* by W. W. Rouse Ball and revised by H. S. M. Coxeter.

Essay Topics

13/1 The greatest mathematician of the nineteenth century.

13/2 Stories and anecdotes about Gauss.

13/3 The Wessel-Argand-Gauss plane.

13/4 Cauchy's influence on present-day undergraduate college mathematics.

13/5 Elliptic functions—what they are and why they are so named.

13/6 Instances of independent and nearly simultaneous discovery in geometry in the first half of the nineteenth century.

13/7 The two Sir William Hamiltons.

13/8 De Morgan as one of the most quoted mathematicians.

13/9 The principle of the permanence of equivalent forms.

13/10 The Analytical Society.

13/11 Sylvester in America.

13/12 Music and mathematics.

13/13 Some nineteenth-century mathematical prodigies.

13/14 Some nineteenth-century mathematicians who died early.

13/15 Some nineteenth-century mathematicians frequently associated in pairs, and reasons why.

13/16 The calculating prodigies Colburn, Bidder, and Dase.

13/17 Algebra in the nineteenth century and the oft expressed view that most great mathematical discoveries are made by young people.

13/18 Cambridge mathematicians of the nineteenth century.

13/19 Sophie Germain (1776–1831).

13/20 Niels Abel (1802–1829).

13/21 Anecdotes about Lobachevsky and Janos Bolyai.

13/22 George Green (1793–1841), the miller mathematician.

13/23 Adolphe Quetelet (1796–1874).

13/24 Florence Nightingale (1820–1910) and mathematics.

13/25 The Galois story.

Bibliography

ALTSHILLER-COURT, NATHAN *Modern Pure Solid Geometry.* 2nd ed. New York: Chelsea, 1964.

BALL, W. W. R., and H. S. M. COXETER *Mathematical Recreations and Essays.* 12th ed. Toronto: University of Toronto Press, 1974. Reprinted by Dover Publications, New York.

BELL, E. T. *Men of Mathematics.* New York: Simon and Schuster, 1937. Reprinted by the Mathematical Association of America.

BOLYAI, JOHN *The science of absolute space.* Translated by G. B. Halsted, 1895. In *Non-Euclidean Geometry.* New York: Dover Publications, 1955.

BOLZANO, BERNHARD *Paradoxes of the Infinite.* Translated by D. A. Steele. London: Routledge and Kegan Paul, 1950.

BONOLA, ROBERTO *Non-Euclidean Geometry.* Translated by H. S. Carslaw. New York: Dover Publications, 1955.

BOYER, C. B. *The History of the Calculus and Its Conceptual Development.* New York: Dover Publications, 1949.

BUCCIARRELLI, LOUIS, and NANCY DWORSKY *Sophie Germain: An Essay in the History of the Theory of Elasticity.* Dordrecht, Holland: D. Reidel, 1980.

BÜHLER, W. K. *Gauss, A Biographic Study.* New York: Springer-Verlag, 1981.

BURTON, D. M. *Elementary Number Theory.* Revised edition. Boston: Allyn and Bacon, 1980. Available from Wm. C. Brown Company, Dubuque, Iowa.

CAYLEY, ARTHUR *Collected Mathematical Papers.* 14 vols. Cambridge: 1889–1898.

CROWE, M. J. *A History of Vector Analysis: The Evolution of the Idea of a Vectorial System.* Notre Dame, Ind.: University of Notre Dame Press, 1967.

DE MORGAN, AUGUSTUS *A Budget of Paradoxes.* 2 vols. New York: Dover Publications, 1954.

DE MORGAN, SOPHIA ELIZABETH *Memoir of A. D. M. by His Wife Sophie Elizabeth De Morgan, with Selections from His Letters.* London: 1882.

DUNNINGTON, G. W. *Carl Friedrich Gauss, Titan of Science: A Study of His Life and Work.* New York: Hafner, 1955.

EVES, HOWARD *A Survey of Geometry,* vol. 2. Boston: Allyn and Bacon, 1965.

——— *Elementary Matrix Theory.* New York: Dover Publications, 1980.

——— *Foundations and Fundamental Concepts of Mathematics.* 3rd ed. Boston: PWS-KENT Publishing Company, 1990.

FORDER, H. G. *The Calculus of Extension.* New York: Cambridge University Press, 1941.

FOURIER, J. B. J. *The Analytical Theory of Heat.* New York: Dover Publications, 1955.

GALOS, E. B. *Foundations of Euclidean and Non-Euclidean Geometry.* New York: Holt, Rinehart and Winston, 1968.

GANS, DAVID *An Introduction to Non-Euclidean Geometry.* New York: Academic Press, 1973.

GAUSS, C. F. *Inaugural Lecture on Astronomy and Papers on the Foundations of Mathematics.* Translated by G. W. Dunnington. Baton Rouge, La.: Louisiana State University, 1937.

——— *Theory of the Motion of Heavenly Bodies.* New York: Dover Publications, 1963.

——— *General Investigation of Curved Surfaces*. Translated by Adam Hiltebeitel and James Morehead. New York: Raven Press, 1965.

——— *Disquisitiones arithmeticae*. English translation by A. A. Clarke. New Haven, Conn.: Yale University Press, 1966.

GIBBS, J. W., and E. B. WILSON *Vector Analysis*. New Haven, Conn.: Yale University Press, 1901.

GRABINER, J. V. *The Origins of Cauchy's Rigorous Calculus*. Cambridge, Mass.: M.I.T. Press, 1981.

GRATTAN-GUINESS, I., in collaboration with J. R. RAVETZ *Joseph Fourier, 1768–1830*. Cambridge, Mass.: M.I.T. Press, 1972.

GRAVES, R. P. *Life of Sir William Rowan Hamilton*. 3 vols. Dublin: Hodges, Figgis, 1882.

GRAY, JEREMY *Ideas of Space: Euclidean, Non-Euclidean, and Relativistic*. Oxford: Clarendon Press, 1979.

GREENBERG, MARVIN *Euclidean and Non-Euclidean Geometry: Development and History*. San Francisco: W. H. Freeman, 1974.

HALL, TORD *Carl Friedrich Gauss*. Translated by A. Froderberg. Cambridge, Mass.: M.I.T. Press, 1970.

HALSTED, G. B., ed. *Girolamo Saccheri's Euclides Vindicatus*. New York: Chelsea, 1986.

HERIVEL, J. *Joseph Fourier, The Man and the Physicist*. Oxford: Clarendon Press, 1975.

HOYLE, FRED *Ten Faces of the Universe* (Chap. 1). San Francisco: W. H. Freeman, 1977.

INFELD, LEOPOLD, *Whom the Gods Love: The Story of Evariste Galois*. New York: McGraw-Hill, 1948.

KAGAN, V. N. *Lobachevsky and His Contribution to Science*. Moscow: Foreign Languages Publishing House, 1957.

KLEIN, FELIX *Development of Mathematics in the 19th Century*. Translated by M. Ackerman. Brookline, Mass.: Mathematical Science Press, 1979.

LANGER, R. E. *Fourier Series, the Genesis and Evolution of a Theory*. Oberlin, Ohio: The Mathematical Association of America, 1947.

LOBACHEVSKY, NICHOLAS *"Geometrical researches on the theory of parallels."* Translated by G. B. Halsted, 1891. In R. Bonola, *Non-Euclidean Geometry*. New York: Dover Publications, 1955.

MACFARLANE, ALEXANDER *Lectures on Ten British Mathematicians of the Nineteenth Century*. New York: John Wiley, 1916.

MACHALE, DESMOND *George Boole: His Life and Work*. Boole Press, 1985.

MARTIN, GEORGE *The Foundations of Geometry and the Non-Euclidean Plane*. New York: Intext Educational Publishers, 1975.

MESCHKOWSKI, HERBERT *Ways of Thought of Great Mathematicians*. San Francisco: Holden-Day, 1964.

——— *Evolution of Mathematical Thought*. San Francisco: Holden-Day, 1965.

MERZ, J. T. *A History of European Thought in the Nineteenth Century*. New York: Dover Publications, 1965.

MIDONICK, HENRIETTA O. *The Treasury of Mathematics: A Collection of Source Material in Mathematics Edited and Presented with Introductory Biographical and Historical Sketches*. New York: Philosophical Library, 1965.

MUIR, JANE *Of Men and Numbers: The Story of the Great Mathematicians*. New York: Dodd, Mead, 1961.

MUIR, THOMAS *The Theory of Determinants in the Historical Order of Development.* 4 vols. New York: Dover Publications, 1960.

NAVY, LUBOS *Origins of Modern Algebra.* Groningen: Noordhoff, 1973.

O'DONNELL, SEAN *William Rowan Hamilton: Portrait of a Prodigy.* Boole Press, 1983.

ORE, OYSTEIN *Niels Henrik Abel, Mathematician Extraordinary.* Minneapolis, Minn.: University of Minnesota Press, 1957.

PEACOCK, GEORGE *Treatise on Algebra.* 2 vols., 1840–1845. New York: *Scripta Mathematica,* 1940.

PRASAD, GANESH *Some Great Mathematicians of the Nineteenth Century, Their Lives and Their Works.* 2 vols. Benares: Benares Mathematical Society, 1933–1934.

Quaternion Centenary Celebration. Dublin: Proceedings of the Royal Irish Academy, vol. 50, 1945. (Among the articles is "The Dublin Mathematical School in the first half of the nineteenth century," by A. J. McConnell.)

SARTON, GEORGE *The Study of the History of Mathematics.* New York: Dover Publications, 1957.

SCHAAF, W. L. *Carl Friedrich Gauss.* New York: Watts, 1964.

SIMONS, L. G. *Bibliography of Early American Textbooks on Algebra.* New York: *Scripta Mathematica,* 1936.

SMITH, D. E. *Source Book in Mathematics.* New York: Dover Publications, 1958.

———, and JEKUTHIEL GINSBURG *A History of Mathematics in America Before 1900.* Chicago: Open Court, 1934.

SOMMERVILLE, D. M. Y. *Bibliography of Non-Euclidean Geometry.* London: Harrison, 1911.

STIGLER, S. M. *The History of Statistics: The Measurement of Uncertainty Before 1900.* Cambridge, Mass.: Belknap Press, 1986.

TAYLOR, E. G. R. *The Mathematical Practitioners of Hanoverian England.* New York: Cambridge University Press, 1966.

TURNBULL, H. W. *The Great Mathematicians.* New York: New York University Press, 1961.

VAN DER WAERDON, B. L. *A History of Algebra from al-Khwarizmi to Emmy Noether.* New York: Springer-Verlag, 1985.

WHEELER, L. P. *Josiah Willard Gibbs: The History of a Great Mind.* New Haven Conn.: Yale University Press, 1962.

WOLFE, H. E. *Introduction to Non-Euclidean Geometry.* New York: Holt, Rinehart and Winston, 1945.

WUSSING, HANS *The Genesis of the Abstract Group Concept.* Cambridge, Mass.: M.I.T. Press, 1984.

YOUNG, J. W. A., ed. *Monographs on Topics of Modern Mathematics Relevant to the Elementary Field.* New York: Dover Publications, 1955.

Chapter 14

THE LATER NINETEENTH CENTURY AND THE ARITHMETIZATION OF ANALYSIS

14–1 Sequel to Euclid

Until modern times, it had been thought that the Greeks had pretty well exhausted the elementary synthetic geometry of the triangle and the circle. Such proved to be far from the case, for the nineteenth century witnessed an astonishing reopening of this study. It now seems that this field of investigation must be unlimited, for an enormous number of papers have appeared, and are continuing to appear, concerned with the synthetic examination of the triangle and associated points, lines, and circles. Much of the material has been extended to the tetrahedron and its associated points, planes, lines, and spheres. It would be too great a task here to enter into any sort of a detailed history of this rich and extensive subject. Many of the special points, lines, circles, planes, and spheres have been named after original or subsequent investigators. Among these names are Gergonne, Nagel, Feuerbach, Hart, Casey, Brocard, Lemoine, Tucker, Neuberg, Simson, McCay, Euler, Gauss, Bodenmiller, Fuhrmann, Schoute, Spieker, Taylor, Droz-Farny, Morley, Miquel, Hagge, Peaucellier, Steiner, Tarry, and many others.

Beyond a few isolated earlier discoveries, such as Commandino's theorem of 1565 (see Problem Study 14.2) and Ceva's theorem of 1678 (see Problem Study 9.10), little that was new and significant was discovered in the synthetic geometry of the triangle and the tetrahedron until the nineteenth century. It is true that Euler discovered the Euler line of a triangle (see Problem Study 14.1) in 1765, but this proof was analytic; the first synthetic proof was given by L. N. M. Carnot in his *Géométrie de position* of 1803. A number of the important associated elements of a triangle, such as the nine-point circle and the miscredited Brocard points, were discovered in the first half of the nineteenth century, but it was in the second half of the nineteenth century that the subject really blossomed and grew in a prodigious manner. Most of these discoveries emanated from France, Germany, and England. Today contributions to the field continue, coming now from almost all parts of the world.

Large portions of the above material have been summarized and organized in numerous recent texts bearing the title *modern,* or *college, geometry*. It is not too much to say that a course in this material is very desirable for every prospective teacher of high-school geometry. The material is definitely elementary, but not easy, and is extremely fascinating.

14–2 Impossibility of Solving the Three Famous Problems with Euclidean Tools

It was not until the nineteenth century that the three famous problems of antiquity were finally shown to be impossible of solution with Euclidean tools. Proofs of this fact can now be found in many of the present-day textbooks dealing with the theory of equations, where it is shown that needed criteria for constructibility are essentially algebraic in nature. In particular, the following two theorems are established:[1]

1. *The magnitude of any length constructible with Euclidean tools from a given unit length is an algebraic number.*
2. *From a given unit length it is impossible to construct with Euclidean tools a segment the magnitude of whose length is a root of a cubic equation with rational coefficients but with no rational root.*

The quadrature problem is disposed of by the first theorem. For if we take the radius of the given circle as our unit of length, the side of the sought equivalent square is $\sqrt{\pi}$. Thus, if the problem were possible with Euclidean tools, we could construct from the unit segment another segment of length $\sqrt{\pi}$. This is impossible, however, since π, and hence $\sqrt{\pi}$, was shown by Lindemann in 1882 to be nonalgebraic.

The second theorem disposes of the other two problems. Thus, in the duplication problem, take for our unit of length the edge of the given cube and let x denote the edge of the sought cube. Then we must have $x^3 = 2$. If the problem is solvable with Euclidean tools, we could construct from the unit segment another segment of length x. But this is impossible, since $x^3 = 2$ is a cubic equation with rational coefficients but without any rational root.[2]

We may prove that the *general* angle cannot be trisected with Euclidean tools by showing that some *particular* angle cannot be so trisected. Now, from trigonometry, we have the identity

$$\cos\theta = 4\cos^3\left(\frac{\theta}{3}\right) - 3\cos\left(\frac{\theta}{3}\right).$$

Taking $\theta = 60°$ and setting $x = \cos(\theta/3)$, this becomes

$$8x^3 - 6x - 1 = 0.$$

---

[1] See, for example, Howard Eves, *A Survey of Geometry*, vol. 2, pp. 30–38.

[2] It will be recalled that if a polynomial equation

$$a_0x^n + a_1x^{n-1} + \cdots + a_n = 0,$$

with integral coefficients $a_0, a_1, \ldots, a_n$, has a reduced root a/b, then a is a factor of a_n and b is a factor of a_0. Thus, any rational roots of $x^3 - 2 = 0$ are among $1, -1, 2, -2$. Since by direct testing none of these numbers satisfies the equation, the equation has no rational roots.

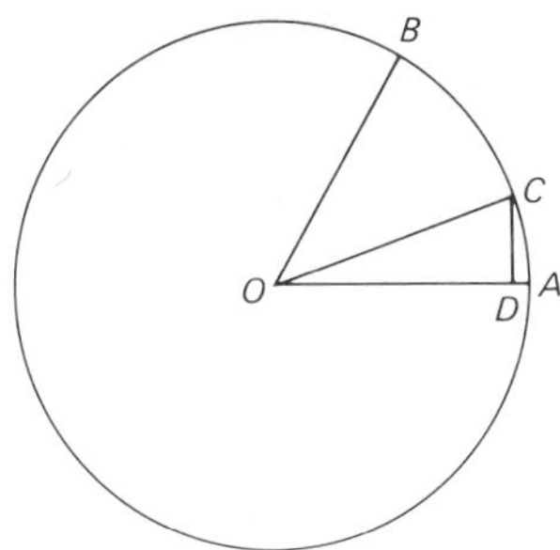

FIGURE 115

Let OA be a given unit segment. Describe the circle with center O and radius OA, and with A as center and AO as radius draw an arc to cut the circle in B (see Figure 115). Then angle $BOA = 60°$. Let trisector OC, which makes angle $COA = 20°$, cut the circle in C, and let D be the foot of the perpendicular from C on OA. Then $OD = \cos 20° = x$. It follows that if a 60° angle can be trisected with Euclidean tools, in other words if OC can be drawn with these tools, we can construct from a unit segment OA another segment of length x. This is impossible, however, by the second theorem, since the above cubic equation has rational coefficients but no rational root.

It should be noted that we have not proved that *no* angle can be trisected with Euclidean tools, but only that *not all* angles can be so trisected. The truth of the matter is that 90° and an infinite number of other angles can be trisected by the use of Euclidean tools.

14–3 Compasses or Straightedge Alone[3]

The eighteenth-century Italian geometer and poet, Lorenzo Mascheroni (1750–1800) made the surprising discovery that all Euclidean constructions, insofar as the given and required elements are points, can be made with the compasses alone, and that the straightedge is thus a redundant tool. Of course, straight lines cannot be drawn with the compasses, but any straight line arrived at in a Euclidean construction can be determined by the compasses alone by finding two points of the line. This discovery appeared in 1797 in Mascheroni's *Geometria del compasso*.

Since in a Euclidean construction new points are found from old points by (1) finding an intersection of two circles, (2) finding an intersection of a straight line and a circle, or (3) finding the intersection of two straight lines, all Mascheroni had to do was to show how, with compasses alone, problems (2) and (3) can be solved, where for a straight line we are given two points of the line.

---

[3] For a fuller treatment of the material of this section, along with proofs, see, for example, Howard Eves, *A Survey of Geometry*, vol. 1, Chapter 4.

Shortly before 1928, a student of the Danish mathematician J. Hjelmslev (1873–1950), while browsing in a bookstore in Copenhagen, came across a copy of an old book, *Euclides danicus,* published in 1672 by an obscure writer named Georg Mohr (1640–1697). Upon examining the book, Hjelmslev was surprised to find that it contained Mascheroni's discovery, with a proof, arrived at 125 years before Mascheroni's publication. In 1890, the Viennese geometer, August Adler (1863–1923), published a new proof of Mascheroni's results, using the inversion transformation.

Inspired by Mascheroni's discovery, the French mathematician Jean Victor Poncelet (1788–1867) considered constructions with straightedge alone. Now not all Euclidean constructions can be achieved with only the straightedge, but, curiously enough, in the presence of one circle and its center drawn on the plane of construction, all Euclidean constructions can be carried out with straightedge alone. This remarkable theorem was conceived by Poncelet in 1822 and then later, in 1833, fully developed by the Swiss-German geometry genius Jacob Steiner (1796–1863). Here it is necessary to show that, in the presence of a circle and its center, constructions (1) and (2) can be solved with straightedge alone, where now a circle is considered as given by its center and a point on its circumference.

It was about 980 that the Arabian mathematician Abû'l-Wefâ (940–998) proposed using the straightedge along with **rusty compasses**—that is, with compasses of a fixed opening. In view of the Poncelet-Steiner theorem, we need, in fact, use the compasses only once, after which the compasses may be discarded. In 1904, the Italian Francesco Severi went still further, and showed that all that is needed is an arc, no matter how small, of one circle, and its center, in order to accomplish all Euclidean constructions with straightedge alone. It has also been shown, by Adler and others, that any Euclidean construction can be carried out with a two-edged straightedge, whether the edges are parallel or not. There are many intriguing construction theorems of this sort, the proofs of which require considerable ingenuity.

Recently,[4] it was shown that the Georg Mohr mentioned above was the author of an anonymously published booklet entitled *Compendium Euclidis curiosi,* which appeared in 1673 and which in effect shows that all the constructions of Euclid's *Elements* are possible with straightedge and rusty compasses.

Lorenzo Mascheroni was born at Castagneta, Italy, in 1750. He began the study of mathematics late in life, having first been interested in the humanities. He taught Greek and poetry at the school in his native town and then at Pavia. He took holy orders and became an abbot. After teaching the humanities, Mascheroni became interested in geometry, and was elected professor of mathematics at Pavia. He wrote on physics, calculus, and the proposed metric system, and published annotations to Euler's *Integral Calculus*. As a friend of

---

[4] See A. E. Hallerberg, "The geometry of the fixed-compass," *The Mathematics Teacher* (April) 1959: 230–244, and A. E. Hallerberg, "Georg Mohr and *Euclidis curiosi,*" *The Mathematics Teacher* (February) 1960: 127–132.

Napoleon, who admired mathematics and was himself an amateur geometer, he interested the general in compass constructions (see Problem Study 14.8(d)). Mascheroni died in Paris in 1800.

Biographical data on Poncelet and Steiner appear in Section 14–4.

The problem of finding the "best" Euclidean solution to a required construction has also been considered, and a science of *geometrography* was developed by Émile Lemoine (1840–1912) for quantitatively comparing one construction with another. To this end, Lemoine considered the following five operations:

S_1: To make the straightedge pass through one given point.
S_2: To rule a straight line.
C_1: To make one compass leg coincide with a given point.
C_2: To make one compass leg coincide with any point of a given locus.
C_3: To describe a circle.

If the preceding operations are performed m_1, m_2, n_1, n_2, n_3 times in a construction, then $m_1S_1 + m_2S_2 + n_1C_1 + n_2C_2 + n_3C_3$ is regarded as the **symbol** of the construction. The total number of operations, $m_1 + m_2 + n_1 + n_2 + n_3$, is called the **simplicity** of the construction, and the total number of coincidences, $m_1 + n_1 + n_2$, is called the **exactitude** of the construction. The total number of loci drawn is $m_2 + n_3$, the difference between the simplicity and the exactitude of the construction. The symbol for drawing the straight line through points A and B is $2S_1 + S_2$, and that for drawing the circle with center C and radius AB is $3C_1 + C_3$.

Émile Lemoine served as editor of *l'Intermédiare des mathématiciens*. He presented his geometrographical proposals (1888–1889, 1892, 1893) at the International Mathematical Congress held in connection with the Chicago World

LORENZO MASCHERONI
(David Smith Collection)

Fair of 1893. His name appears in geometry also in connection with the so-called *Lemoine* (or *symmedian*) *point* of a triangle (see Problem Study 14.5), and the *Lemoine line, Lemoine circle,* and *second Lemoine* (or *cosine*) *circle* of a triangle. In three space, there is a *Lemoine tetrahedron,* and the *Lemoine point* and *Lemoine plane* of a tetrahedron.

14–4 Projective Geometry

Quite apart from the discovery of non-Euclidean geometry, the field of geometry made enormous strides in the nineteenth century. There was, as has already been pointed out, an extensive development of the surprisingly rich material constituting a sequel to Euclid. In the present section, it will be seen that projective geometry also made very impressive and highly fruitful gains. Section 14–5 is devoted to the remarkable nineteenth-century expansion of the methods of analytic geometry, and Section 14–7 to an examination of the extraordinary growth of differential geometry in that century.

Although Desargues, Monge, and Carnot had initiated the study of projective geometry, its truly independent development was launched in the nineteenth century by Jean Victor Poncelet. Poncelet was born at Metz in 1788, attended the lycée there, and then from 1807 to 1810, enrolled at the École Polytechnique, where he studied under Monge. In 1812, following a stint as a student at the military academy at Metz, he entered the army as a lieutenant of engineers, and served in Napoleon's fateful Russian campaign. Left for dead on the battlefield of Krasnoi during the French retreat from Moscow, Poncelet was taken prisoner of war and, after a forced march of nearly five months, was placed in confinement at Saratoff on the Volga River. There, with no books at hand, he planned his great *Traité des propriétés projectives des figures,* which, subsequent to his release and return to Metz late in 1814, he put into form and published in Paris in 1822. Poncelet's later life was devoted to military duties interspersed with writing on mechanics, hydraulics, infinite series, and geometry. He published a treatise on applied mechanics (1826), an interesting memoir on water mills (also 1826), a report on the English machinery and tools displayed at the London International Exhibition of 1851, a two-volume expansion of his earlier work of 1822 (1862, 1865), and numerous geometry articles in the pages of *Crelle's Journal.* Of rugged health all his long life, Poncelet was always conscientious, efficient, and dependable in his military assignments, and he retained his creative abilities in mathematics almost to the time of his death. He died in Paris in 1867 at the age of seventy-nine.

Poncelet's *Traité des propriétés projectives des figures* is a geometric milestone. It gave tremendous impetus to the study of projective geometry and inaugurated the so-called "great period" in the history of the subject. There followed into the field a host of mathematicians, among whom were Gergonne, Brianchon, Chasles, Plücker, Steiner, Staudt, Reye, and Cremona—great names in the history of geometry, and in the history of projective geometry in particular.

JEAN VICTOR PONCELET
(Culver Service)

We shall here restrict ourselves to the consideration of only two of the mathematical tools utilized by Poncelet in his development of projective geometry—the *principle of duality* and the *principle of continuity*.

In plane projective geometry, there is a remarkable symmetry between points and lines, when the ideal elements at infinity are utilized, such that if in a true proposition about "points" and "lines" we should interchange the roles played by these words, and perhaps smooth out the language, we obtain another true proposition about "lines" and "points." As a simple example, consider the following two propositions related in this way:

> *Any two distinct points determine one and only one line on which they both lie.*
>
> *Any two distinct lines determine one and only one point through which they both pass.*

This symmetry, which results in the pairing of the propositions of plane projective geometry, is a far-reaching principle known as the **principle of duality.** Once the principle of duality is established, then the proof of one proposition of a dual pair carries with it the proof of the other. Let us dualize Pascal's theorem. We first restate Pascal's theorem in a form that is perhaps more easily dualized.

> *The six vertices of a hexagon lie on a conic if and only if the points of intersection of the three pairs of opposite sides lie on a line.*

Dualizing this we obtain:

> *The six sides of a hexagon are tangent to a conic if and only if the lines joining the three pairs of opposite vertices intersect in a point.*

This theorem was first published by Charles Julien Brianchon (1785–1864), when a student at the École Polytechnique in Paris, in 1806, nearly 200 years after Pascal had stated his theorem.

There are several ways in which the principle of duality may be established. It is possible to give a set of postulates for projective geometry that are themselves arranged in dual pairs. It follows that the dual of any theorem derived from such a set of postulates may be authenticated by simply dualizing the steps in the proof of the original theorem. The principle may also be established analytically once the concepts of "coordinates" of a line and "equation" of a point are formulated (see Section 14–5). Finally the student who is familiar with the elementary notion of poles and polars with respect to some base conic will realize that, under the correspondence between poles and polars so set up, to each figure consisting of lines and points is associated a dual figure consisting of points and lines. It was in this last way that the principle of duality was first elaborated by Gergonne and Poncelet. The term *pole* had been introduced in 1810 by the French mathematician F. J. Servois (1767–1847), and the corresponding term *polar* by Gergonne (1771–1859) three years later.

It is interesting to point out that principles of duality have been established in several other branches of mathematics, such as solid projective geometry, Boolean algebra, the theory of trigonometric identities, spherical geometry, partially ordered sets, and the calculus of propositions.

Poncelet's other mathematical tool, the *principle of continuity,* may be explained by the following example. Consider the situation of two circles intersecting in the real points A and B. A student of elementary geometry can easily prove that the locus of a point P having equal powers with respect to the two circles is the line AB. This property, having been established, must be provable by the method of analytic geometry. But the method of analytic geometry would take no cognizance of whether the points A and B of intersection of the two circles are real or imaginary. Hence, the chain of equations that proves the proposition in the case in which A and B are real at the same time proves the proposition when A and B are imaginary. It follows that when our two circles do not intersect, the locus of a point P having equal powers with respect to the two circles still is a straight line. This method of reasoning, where from a proof of a theorem for a real situation one obtains the theorem for an imaginary situation, was called by Poncelet the **principle of continuity** of geometry. In projective geometry, there are many instances in which a proposition that can be established for the case of a real projection can be extended by the principle of continuity to the case of an imaginary projection.

Poncelet's principle of continuity met resistance from a number of geometers, and many of Poncelet's articles in *Crelle's Journal* are devoted to defending and illustrating the principle.

Many of Poncelet's ideas in projective geometry were further developed by the Swiss geometer Jacob Steiner, one of the greatest synthetic geometers the world has ever known. Steiner was born at Utzensdorf in 1796 and did not learn to write until he was fourteen. At seventeen, he became a pupil of Johann Heinrich Pestalozzi (1746–1827), the famous Swiss educator, who instilled in the boy a love for mathematics. Later, in 1818, Steiner matriculated at Heidelberg, where he quickly exhibited his ability in mathematics. In 1821, he started giving private lessons in mathematics in Berlin, and soon was appointed a teacher in the Gewerbeakademi. His name became well known through his articles published in the newly founded *Crelle's Journal;* he and Abel were leading contributors to the journal. In 1834, through the influence of Jacobi, Crelle, and von Humbolt, a chair was founded for him at the University of Berlin, where he remained for the rest of his teaching career. His final years were spent in poor health in Switzerland. He died at Bern in 1863.

Described as "the greatest geometrician since the time of Apollonius," Steiner possessed incredible power in the synthetic treatment of geometry. He became a prolific contributor in the field and wrote a number of treatises of the highest rank. It is said that he loathed the analytical method in geometry, regarding it as a crutch for the geometrically feeble-minded. He created new geometry at such a prodigious rate that often he had no time to record his proofs, with the result that many of his findings remained for years as riddles to those seeking demonstrations. His *Systematische Entwicklungen,* published in 1832, immediately made his reputation. This work contains a complete discussion of reciprocation, the principle of duality, homothetic ranges and pencils, harmonic division, and the projective geometry of the conic sections based upon the highly fruitful definition of a conic as the locus of the points of intersection of corresponding lines of two homographic pencils with distinct vertices. He contributed to the study of the n-gon in space, the theory of curves and surfaces, pedal curves, roulettes, and the twenty-seven straight lines on a surface of the third order. He attacked by synthetic geometry problems in maxima and minima that in the hands of others required the paraphernalia of the calculus of variations. His name is met in many places in geometry, as in

JACOB STEINER
(David Smith Collection)

the Steiner solution and generalization of the Malfatti problem, Steiner chains, Steiner's porism, and the Steiner points of the mystic hexagram configuration.

In the treatments of projective geometry by Poncelet and Steiner, many projective concepts are based upon metrical properties. Projective geometry was finally completely freed of any metrical basis by Karl Georg Christian von Staudt in his *Geometrie der Lage* of 1847. Staudt was born at Rothenburg in 1798, held the chair of mathematics at Erlangen, and died at Erlangen in 1867.

The analytical side of projective geometry made spectacular gains in the work of Augustus Ferdinand Möbius (1790–1868), Michel Chasles (1793–1880), and, particularly, Julius Plücker (1801–1868). Plücker became as famous a champion of analytic geometry as Steiner did of synthetic geometry. This work will be considered in more detail in Section 14–5. Michel Chasles was also an outstanding synthetic geometer, and his *Aperçu historique sur l'origine et le développement des méthodes en géométrie* (1837) is still a standard work on the history of geometry. Chasles became professor of geometry and mathematics at the École Polytechnique in 1841 and professor of geometry in the faculty of sciences in 1846. He received the Copley medal of the Royal Society for his *Traité des sections coniques,* which was published in Paris in 1865.

Later it was shown how, by the adoption of a suitable projective definition of a metric, we can study metric geometry in the framework of projective geometry, and how, by the adjunction of an invariant conic to a projective geometry in the plane, we can obtain the classical non-Euclidean geometries. In the late nineteenth and early twentieth centuries, projective geometry received a number of postulational treatments, and finite projective geometries were discovered. It was shown that, by gradually adding and altering postulates, one can move from projective geometry to Euclidean geometry, encountering a number of other important geometries on the way.

14–5 Analytic Geometry

There are plane coordinate systems other than the rectangular and oblique Cartesian systems. As a matter of fact, one can invent coordinate systems rather easily. All one needs is an appropriate frame of reference along with some accompanying rules telling us how to locate a point in the plane by means of an ordered set of numbers referred to the frame of reference. Thus, for the rectangular Cartesian system, the frame of reference consists of two perpendicular axes, each carrying a scale, and we are all familiar with the rules telling us how to locate a point with respect to this frame by the ordered pair of real numbers representing the signed distances of the point from the two axes. The Cartesian systems are much the commonest systems in use and have been developed enormously. Much terminology, like our classification of curves into linear, quadratic, cubic, and so forth, stems from our use of this system. Some curves, however, such as many spirals, have intractable equations when referred to a Cartesian frame, whereas they enjoy relatively simple equations when referred to some other skillfully designed coordinate system. Particularly

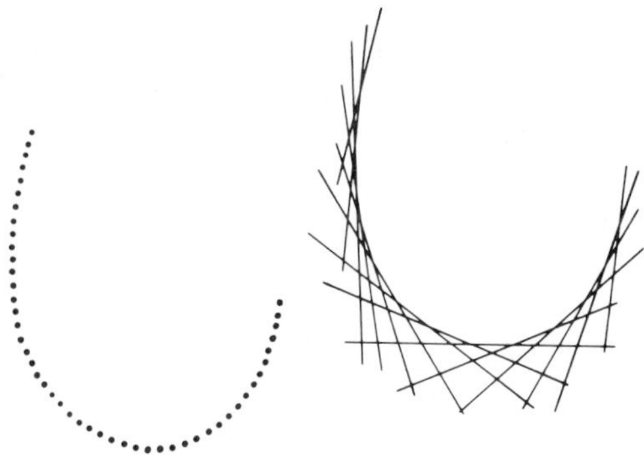

FIGURE 116

useful in the case of spirals is the polar coordinate system, where, it will be recalled, the frame of reference is an infinite ray and where a point is located by a pair of real numbers, one of which represents a distance and the other an angle. The idea of polar coordinates seems to have been introduced in 1691 by Jakob Bernoulli (1654–1705).[5] Further coordinate systems were little investigated until toward the close of the eighteenth century, when geometers were led to break away from the Cartesian systems in situations in which the peculiar necessities of a problem indicated some other algebraic apparatus as more suitable. After all, coordinates were made for geometry and not geometry for coordinates.

An interesting development in coordinate systems was inaugurated by the Prussian geometer Julius Plücker in 1829, when he noted that our fundamental element need not be the point, but can be any geometric entity. Thus, if we choose the straight line as our fundamental element, we might locate any straight line not passing through the origin of a given rectangular Cartesian frame of reference by recording, say, the x and y intercepts of the given line. Plücker actually chose the negative reciprocals of these intercepts as the location numbers of the line and considerably exploited the analytic geometry of these so-called **line coordinates.** A point now, instead of having coordinates, possesses a linear equation namely, the equation satisfied by the coordinates of all the lines passing through the point (see Problem Study 14.15). The double interpretation of a pair of coordinates as either point coordinates or line coordinates and of a linear equation as either the equation of a line or the equation of a point furnishes the basis of Plücker's analytical proof of the principle of duality of projective geometry. A curve may be regarded either as the locus of its points or as the envelope of its tangents (see Figure 116). If, instead of points or straight lines, we should choose circles as fundamental elements, then we would require an ordered triple of numbers to determine one of our elements

[5] See, however, C. B. Boyer, "Newton as an originator of polar coordinates," *The American Mathematical Monthly* (February) 1949: 73–78.

completely. On a rectangular Cartesian frame of reference, for example, we might take the two Cartesian coordinates of the circle's center along with the circle's radius. Ideas such as these led to considerable generalization and the development of a dimension theory. The **dimensionality** of a manifold of fundamental elements was considered as the number of independent coordinates needed to locate each fundamental element. According to this concept, the plane is two dimensional in points, and also in lines, but is three dimensional in circles. It can be shown that the plane is five dimensional if the totality of all conic sections in the plane should be chosen as the manifold of fundamental elements. Dimension theory has, of course, developed far beyond this elementary concept and is today a subject of considerable extent and depth.

Although Descartes had mentioned solid analytic geometry, he did not elaborate it. Others, like the younger Frans van Schooten, La Hire, and Johann Bernoulli, suggested our familiar solid analytic geometry, but it was not until 1700 that the subject was first systematically developed, by Antoine Parent (1666–1716) in a paper presented to the French Academy. A. C. Clairaut, in 1731, was the first to write analytically on nonplanar curves in space. Euler later advanced the whole subject well beyond its elementary stages. These initial workers chose the point as fundamental element. Although space is three dimensional in points, it may be shown that it is four dimensional in lines and also in spheres. It is three dimensional, however, in planes (see Problem Study 14.16).

While synthetic geometers were making easy and spectacular gains, the analytic geometers were bogged down in a morass of algebraic calculations. If analytic geometry was to compete successfully with synthetic geometry, it had to develop new and improved procedures. With great zeal, some of the protagonists of coordinate methods entered the lists in defense of analytic geometry, and the subject commenced its golden period. Foremost among the contributors of improved procedures in analytic geometry was Julius Plücker, who, in a sequence of articles and texts, devised methods that showed that analytic geometry, when properly employed, need concede nothing in elegance and simplicity to synthetic geometry.

Plücker was born at Elberfeld in 1801 and was educated at Bonn, Berlin, and Heidelberg, with a short period of study in Paris, where he attended lectures of Monge and his pupils. Between 1826 and 1836, he held teaching positions successively at Bonn, Berlin, and Halle. In 1836, he returned to the University of Bonn as a professor of mathematics, a position that in 1847 he exchanged for a professorship in physics there. He died at Bonn in 1868.

Plücker's two-volume *Analytisch-geometrische Entwicklungen* was published in 1828 and 1831. In the first volume of this work, the **method of abridged notation,** though employed earlier by Gabriel Lamé and Étienne Bobillier is given its first extensive treatment. The idea of abridged notation lies in representing long expressions by single letters and in the fundamental principle: If $\alpha(x,y) = 0$ and $\beta(x,y) = 0$ are two curves, then $u\alpha + v\beta = 0$, where u and v are any constants or functions of x and y, is a curve passing through the points of intersection of the curves $\alpha = 0$ and $\beta = 0$. Such seemingly algebraically

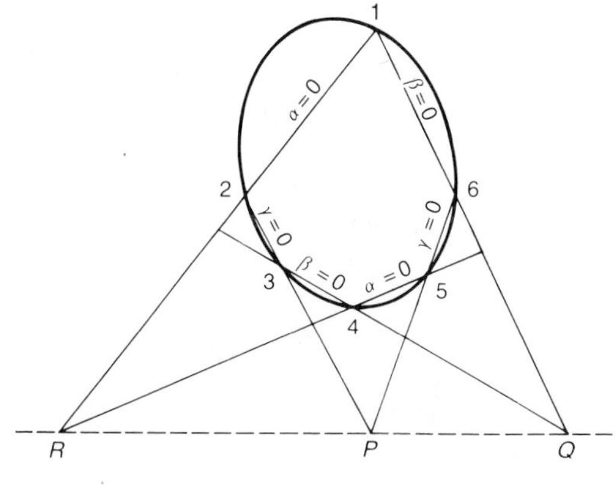

FIGURE 117

complex theorems as Desargues' two-triangle theorem and Pascal's mystic hexagram theorem can be given remarkably neat and brief proofs with the aid of abridged notation. Consider, for example, Pascal's mystic hexagram theorem: *If the six points* 1, 2, 3, 4, 5, 6 *lie on a conic, then the three points* P, Q, R *of intersection of the three pairs of lines* 56 *and* 23, 16 *and* 34, 12 *and* 45 *are collinear.*

Let $\alpha = 0, \beta = 0, \gamma = 0, \alpha' = 0, \beta' = 0, \gamma' = 0$ be equations (see Figure 117) of the lines 12, 34, 56, 45, 61, 23. Consider the cubic curve

$$\alpha\beta\gamma + k\alpha'\beta'\gamma' = 0.$$

Regardless of the value of k, this cubic passes through the nine points 1, 2, 3, 4, 5, 6, P, Q, R. Take another point 7 on the conic and determine k so that the cubic will also pass through point 7. Now a cubic and a conic can intersect in at most $(3)(2) = 6$ points, unless the conic is a part of the cubic, the rest of which is some straight line. This, then, must be the case, and the remaining three points P, Q, R must lie on a straight line.

In the second volume of *Analytisch-geometrische Entwicklungen,* occurs a presentation of homogeneous coordinates of points in the plane. Here (Cartesian) homogeneous coordinates of a point P having Cartesian coordinates (X,Y) are defined as any ordered *triple* (x,y,t) such that $X = x/t$ and $Y = y/t$. It follows that the triples (x,y,t) and (kx,ky,kt) represent the same point. The name *homogeneous* arises from the fact that when one converts the equation $f(X,Y) = 0$ of an algebraic curve in Cartesian coordinates to the form $f(x/t, y/t) = 0$, all the terms in the new equation become of the same degree in the new variables. But, more important, in homogeneous coordinates a triple $(x,y,0)$, which has no counterpart in the Cartesian system, represents a "point at infinity," and the ideal points at infinity of Kepler, Desargues, and Poncelet now receive representation in a coordinate system. The equation $t = 0$ is then the equation of the ideal line at infinity. It follows that homogeneous coordinates

JULIUS PLÜCKER
(David Eugene Smith Collection, Rare Book and Manuscript Library, Columbia University)

furnish a perfect tool for the analytical exploration of projective geometry, which requires both the finite and the infinite points of the plane.

Plücker's book *System der analytischen Geometrie* of 1835 contains a complete classification of cubic curves based upon the nature of their points at infinity, and his *Theorie der algebraischen Curven* of 1839 gives an enumeration of curves of the fourth order and his famous four equations connecting the singularities of algebraic curves. These equations are

$$m = n(n-1) - 2\delta - 3\kappa, \qquad n = m(m-1) - 2\tau - 3\iota,$$

$$\iota = 3n(n-2) - 6\delta - 8\kappa, \qquad \kappa = 3m(m-2) - 6\tau - 8\iota,$$

where m is the class of the curve (the degree of the equation of the curve when expressed in line coordinates), n the order of the curve (the degree of the equation of the curve when expressed in point coordinates), δ the number of nodes, κ the number of cusps, ι the number of points of inflection, and τ the number of bitangents.

For about twenty years following his appointment as professor of physics, Plücker largely devoted himself to researches in spectrum analysis, magnetism, and the Fresnel wave surface. Later in life, he returned to his first love, mathematics, and developed the four-dimensional geometry of lines in space, along with his theory of "complexes" and "congruences" of lines in space.

In a more detailed treatment of the remarkable growth of analytic geometry in the nineteenth century, more than just passing notice would be given to

Joseph Diaz Gergonne (1771–1859, artillery officer, editor, and professor of mathematics), Augustus Ferdinand Möbius (1790–1868, professor at Leipzig), Gabriel Lamé (1795–1870, engineer and professor of mathematics), Étienne Bobillier (1798–1840, professor of mechanics), Ludwig Otto Hesse (1811–1874, professor at Königsberg), Rudolph Friedrich Alfred Clebsch (1833–1872, professor at Königsberg and later at Göttingen), George-Henri Halphen (1844–1889, of Rouen and examinateur at the École Polytechnique in Paris), and others.

A discussion of the application of analytic geometry to the study of a hyperspace which is n-dimensional ($n > 3$) in points is reserved for the following section.

14–6 *N*-Dimensional Geometry

The first nebulous notions of a hyperspace that is n dimensional ($n > 3$) in points are lost in the dimness of the past and were confused by metaphysical considerations. The first published paper dealing explicitly with higher-dimensional point geometry was written by Arthur Cayley (1821–1895) in 1843, following which the subject received the attention of the three British mathematicians Cayley, J. J. Sylvester (1814–1897), and W. K. Clifford (1845–1879). The simultaneous pioneering work in higher-dimensional geometry done by H. G. Grassmann (1809–1877) and Ludwig Schläfli (1814–1895) on the continent failed for some time to attract any attention. In fact, the bulk of Schläfli's work was not published until several years after his death, and by that time Victor Schlegel (1843–1905) and others in Germany had made the subject well known. Higher-dimensional projective geometry was developed almost entirely by the Italian school of geometers, although it was Clifford who inaugurated this study in 1878.

Quite independently of the work described above regarding the beginning of higher-dimensional point geometry, we find the arithmetic aspect of the subject gradually emerging from applications of analysis where an analytical treatment can easily be extended from two or three variables to arbitrarily many variables. Thus, George Green (1793–1841), in 1833, reduced the problem of the mutual attraction of two ellipsoidal masses to analysis and then solved the problem for any number of variables, saying, "It is no longer confined as it were to the three dimensions of space." Other writers made similar generalizations to arbitrarily many variables, and it was but a step further to apply the terminology of geometry to many of the forms and processes of algebra and analysis. This procedure was clearly stated by Cauchy, in 1847, in a paper on analytical loci when he said, "We shall call a set of n variables an analytical point, an equation or system of equations an analytical locus," and so on. Beyond any doubt, the most important early expression of this analytical viewpoint of higher-dimensional geometry is found in Riemann's great probationary lecture of 1854, but it was not published until 1866. It was in this lecture that Riemann built up his notion of n-dimensional manifolds and their measure relations, and throughout the discussion, he held geometrical conceptions and imagery before the mind.

The number of papers and works devoted to higher-dimensional geometry increased greatly after 1870. In 1911, D. M. Y. Sommerville published his *Bibliography on Non-Euclidean Geometry, Including the Theory of Parallels, the Foundations of Geometry, and Space of n Dimensions*. In this bibliography, there appear 1832 references to n-dimensional geometry, of which about one-third are Italian, one-third German, and the rest mostly French, English, and Dutch.

One studies n-dimensional geometry analytically by introducing appropriate concepts into arithmetic space of n dimensions. **Arithmetic space of n dimensions** is the set of all ordered n-tuples $x = (x_1, x_2, \ldots, x_n)$ of real numbers, and each such n-tuple is called a **point** of the space. Relations among these points are defined by formulas analogous to the formulas holding for the corresponding relations among points in, say, Cartesian point spaces of two and three dimensions. Thus, since the distance between the two points (x_1, x_2) and (y_1, y_2) in a two-dimensional rectangular Cartesian system is given by

$$[(x_1 - y_1)^2 + (x_2 - y_2)^2]^{1/2},$$

and the distance between the two points (x_1, x_2, x_3) and (y_1, y_2, y_3) in a three-dimensional rectangular Cartesian system is given by

$$[(x_1 - y_1)^2 + (x_2 - y_2)^2 + (x_3 - y_3)^2]^{1/2},$$

we *define* the **distance** between the two points $x = (x_1, \ldots, x_n)$ and $y = (y_1, \ldots, y_n)$ in arithmetic n-dimensional space to be

$$[(x_1 - y_1)^2 + \cdots + (x_n - y_n)^2]^{1/2}.$$

Similarly, we define an **n-dimensional sphere** of radius r and with center at the point $(a_1, \ldots, a_n)$ to be the collection of all points $x = (x_1, \ldots, x_n)$ such that

$$(x_1 - a_1)^2 + \cdots + (x_n - a_n)^2 = r^2.$$

We define a pair of points to be a **line segment,** and we define any ordered n-tuple of numbers of the form

$$(k(y_1 - x_1), \ldots, k(y_n - x_n)), \qquad k \neq 0,$$

to be **direction numbers** of the line segment xy determined by the points x and y. The **cosine of the angle** θ between the two line segments xy and uv is defined as

$$\cos\theta = \frac{(y_1 - x_1)(v_1 - u_1) + \cdots + (y_n - x_n)(v_n - u_n)}{d(x,y)\, d(u,v)},$$

where $d(x,y)$ is the distance between the points x and y, and $d(u,v)$ is the distance between the points u and v. The two segments are said to be **perpendicular** if and only if the cosine of the angle between them is zero. A transformation of the form

$$y_i = a_i + x_i, \qquad i = 1, \ldots, n,$$

mapping the point x onto the point y, is called a **translation.** Other point transformations of the space onto itself can be similarly defined. It is easy to formulate a definition of an *n-dimensional conicoid* and then to study the pole, polar, and other properties of these conicoids. An n-dimensional geometry of this sort can be regarded as a purely algebraic study that employs geometric terminology.

Higher-dimensional geometries do not lack application in other areas of study. Indeed, it was actually certain needs of physicists and statisticians that largely accounted for much of the expansion and development of the subject. For example, it is quite generally known today, even among laymen, that relativity theory uses the idea of a four-dimensional space. But an easier example may be described here, by showing how the mathematical treatment of the kinetic theory of gases has come to employ higher-dimensional geometry. Consider a closed vessel containing a gas, and suppose the gas is composed of m molecules. These molecules are moving about within the vessel, and any particular one of them is, at a given instant, at a point (x,y,z) of ordinary space, and has, at that instant, certain velocity components u,v,w along the coordinate axes. Only if we know all six numbers x,y,z,u,v,w do we know where the molecule is at the given instant, and the direction and rate of its movement. The m molecules of gas in the vessel thus depend upon $6m$ coordinates. At any instant, these $6m$ coordinates have definite values that define the *state* of the gas at that instant. Now these $6m$ values determine a point in a point space of $6m$ dimensions, and there is a one-to-one correspondence between such points and possible states of the gas. As the state of the gas varies, owing to the motion of the molecules, the corresponding point generates a path, or locus, in the space of $6m$ dimensions. It follows that the behavior, or history, of the gas is geometrically represented by this locus.

14–7 Differential Geometry

Differential geometry is the study of properties of curves and surfaces, and their generalizations, by means of the calculus. For the most part, differential geometry investigates curves and surfaces in the immediate neighborhoods of any of their points. This aspect of differential geometry is known as **local differential geometry,** or **differential geometry in the small.** However, sometimes properties of the total structure of a geometrical figure are implied by certain local properties of the figure that hold at every point of the figure. This leads to what is known as **integral geometry,** or **global differential geometry,** or **differential geometry in the large.**

Although one can find geometrical theorems deduced from a study of evanescent figures in Archimedes' determination of areas and volumes, Apollonius' treatment of normals to conic sections, and later in Cavalieri's method of indivisibles and Huygens' beautiful work on curvature and evolutes, it is probably quite correct to say that differential geometry, at least in its modern form, started in the early part of the eighteenth century with applications of the differential and integral calculus to analytic geometry. The first real stimulus to the subject, however, beyond planar situations, was furnished by Gaspard Monge (1746–1818), who can be considered as the father of the differential geometry of curves and surfaces of space.

Monge was an outstanding teacher, and his lectures at the École Polytechnique of Paris inspired a host of younger men to enter the field of differential geometry. Among these were J. B. Meusnier (1754–1793), E. L. Malus (1775–1812), C. Dupin (1784–1873), and O. Rodrigues (1794–1851), all of whom have important theorems in differential geometry named after them. For example, a theorem of Meusnier states the following: *If PT is a tangent line to a given surface S at a given point P on S, then the circle of curvature at P of a variable section of S through PT is the circle in which the plane of the section cuts the sphere with center C_n and radius r_n, where C_n and r_n are the center and radius of the circle of curvature at P of the normal section of S through the tangent PT*. One of Dupin's theorems states that: *The sum of the normal curvatures, at a point P on a surface S, in any two perpendicular directions is constant*. Dupin is also responsible for the *Dupin indicatrix,* a device that furnishes much information about the nature of a surface *S* at a point *P* of *S*.

Monge and his students formed the start of the great French school of differential geometers, which later included such names as Augustin Louis Cauchy (1789–1857); B. de Saint-Venant (1796–1886), who, among other things, in 1845 supplied the name *binormal* in connection with the local trihedron of a point of a space curve; F. Frenet (1816–1888) and J. A. Serret (1819–1885), who were responsible for the *Frenet-Serret formulas,* which are so important and central in an analytical study of space curves; V. Puiseux (1820–1883); and J. Bertrand (1822–1900), whose name is attached to pairs of space curves for which the principal normals of each are principal normals of the other.

Cauchy's work in differential geometry marks the close of the first period in the history of the subject. The second period was inaugurated by Carl Friedrich Gauss (1777–1855), who introduced the singularly fruitful method of studying the differential geometry of curves and surfaces by means of parametric representations of these objects. We now encounter the names of G. Mainardi (1800–1879) and D. Codazzi (1824–1875), after whom important equations in the subject are named, the blind Belgian physicist J. Plateau (1801–1883); C. G. J. Jacobi (1804–1851); O. Bonnet (1819–1892); E. B. Christoffel (1829–1901); E. Beltrami (1835–1900); J. D. Darboux (1842–1917), after whom a special vector associated with each point of a space curve has been named, and who, among other things, completed work done by Dupin in connection with **triply orthogonal families of surfaces,** wherein each family is orthogonal to the other two families; and the names of many others who contributed to the classical theory of curves and surfaces in space.

The third great period in the history of differential geometry was initiated by George Bernhard Riemann (1826–1866). Here we find an assertion of that tendency of mathematics of modern times to strive for the greatest possible generalization. Ordinary familiar three space is left far behind, and the study concentrates on such things as m-dimensional manifolds immersed in n-dimensional space. Two things were found necessary for this further development: an improved notation and a procedure dependent on the nature of the manifold and not the particular coordinate system employed. The *tensor calculus* was accordingly devised, and this general subject was developed by such mathematicians as G. Ricci-Curbastro (1853–1925), T. Levi-Civita (1873–1941), and A. Einstein (1879–1955). Generalized differential geometries, known as *Riemannian geometries,* were explored intensively, and these in turn led to *non-Riemannian* (and other) *geometries*. Present-day research in differential geometry bears little resemblance to the classical study with its strong ties to the concrete.

A surface can be looked at in two ways: as the boundary of a solid body or as a detached two-dimensional film. The former is the way a construction engineer might regard the surface, and the latter the way a surveyor might regard it. The first viewpoint leads one to search out the properties of the surface that relate it to its surrounding space, and the second viewpoint leads one to search out the properties of the surface that are independent of its surrounding space. Properties of the first kind are called **relative properties** of the surface, and their study is called the **extrinsic geometry** of the surface; properties of the second kind are called **absolute properties** of the surface, and their study is called the **intrinsic geometry** of the surface. It is interesting that the two great early contributors to the differential geometry of surfaces, Monge and Gauss, respectively, saw a surface primarily as the boundary of a solid and primarily as a detached two-dimensional film. Monge is noted, among other things, for his work as a construction engineer of military fortifications, and Gauss is noted, among other things, for his work in geodesy and geodetic surveying.

It was Gauss who, in his great work *Disquisitiones generales circa superficies curvas* of 1827, introduced the important concept of curvature of a surface S at a point P on S. Consider the sections of S made by planes containing the normal to S at a point P on S. Of these sections, there is one having a maximum curvature k at P, and one having a minimum curvature k' at P. These two sections are generally at right angles to one another, and their curvatures at P are called the **principal curvatures** of S at P. The product $K = kk'$ is called the **Gaussian,** or **total, curvature** of the surface S at P. If the two principal curvatures are of the same sense, then K is positive; if the two principal curvatures are of opposite sense, then K is negative; if at least one principal curvature is zero, then K is zero. Gauss discovered the remarkable fact that *if a surface is* *bent* (without stretching, creasing, or tearing), *the total curvature of the surface at each point remains unaltered.* Two surfaces that can be bent so as to coincide are said to be **applicable** to each other, and have the same intrinsic geometry; thus, a plane and a circular cylinder have the same intrinsic geometry, but they certainly do not look alike in space. It should be kept in mind that we are here concerned with local differential geometry and not global differential geometry. A plane and a circular cylinder have the same *local* intrinsic geometry, but obviously not the same *global* intrinsic geometry.

One of Gauss' most remarkable discoveries about surfaces is that the total curvature K is an absolute property of the surface. This, at first encounter, seems incredible, for the total curvature of a surface at a point on the surface is equal to the product of the two principal normal curvatures of the surface at the point. But the normal curvatures at a point are relative properties of the surface! The statement that the total curvature K of a surface is an absolute property of the surface is known as **Gauss' therema egregium.**

Gauss also showed that if we have on a surface a triangle bounded by geodesics (that is, curves of shortest length joining pairs of points on the surface), and if the angles of the triangle are a_1, a_2, a_3, then

$$\int\int_A K\, dA = a_1 + a_2 + a_3 - \pi,$$

where A is the area of the triangle. If the surface has constant total curvature K, then

$$a_1 + a_2 + a_3 - \pi = KA,$$

and the difference between the sum of the angles of the triangle and π is positive, zero, or negative according as $K > 0$, $= 0$, or < 0, and the excess when $K > 0$, or the deficiency when $K < 0$, is proportional to the area of the triangle. It follows that the intrinsic geometry of the geodesics of a surface of constant nonzero total curvature is non-Euclidean, whereas that of a surface of constant zero total curvature is Euclidean.

It was in 1831 that Sophie Germain (1776–1831) introduced the concept of **mean curvature** $M = (k + k')/2$ of a surface at a point P of the surface. Of particular interest are surfaces for which M is everywhere equal to zero; such surfaces are called **minimal surfaces.** It follows that at any point of a minimal surface, the two principal normal curvatures are equal in magnitude but opposite in sign. Minimal surfaces derive their name from the fact that they can be characterized as the surfaces of least area among all surfaces bounded by a given closed space curve. They are illustrated by the shapes assumed by the soap films that result when closed loops of wire of any shape are dipped into a soap solution; the surface tension of the films minimizes the surface areas of the films. The problem of determining the minimal surface through a given closed-space curve was first proposed by Lagrange, but became known as the **problem of Plateau,** since the blind physicist Joseph Plateau was the first to conceive the soap-film method of "seeing" these surfaces. It is interesting that we can characterize minimal surfaces either by a property in the small or by a property in the large. A complete mathematical solution of the problem of Plateau was given in 1931 by the American mathematician Jesse Douglas (1897–1965), when he was only thirty-four years of age, for which he received the Bôcher prize and one of the two first awarded (1936) Fields medals.

14–8 Felix Klein and the Erlanger Programm

In 1872, upon appointment, at the young age of twenty-three, to a full professorship on the Philosophical Faculty and Senate of the University of Erlangen, Felix Klein (1849–1925) presented, according to custom, both an oral speech introducing himself to his new faculty associates and a written paper exhibiting research interests in his field of mathematics. The speech was aimed at a wide university audience and expressed Klein's pedagogical view of the unity of all knowledge and the ideal that a complete education should not be neglected because of special studies. The written work, which was passed out at the time of the speech, was designed for his departmental colleagues. Thus, the two parts of Klein's inauguration exhibit both his deep interest in pedagogical matters and his serious commitment to mathematical research.

The written paper, based upon work by himself and Sophus Lie (1842–1899) in group theory, set forth a remarkable definition of "a geometry" that served to codify essentially all the existing geometries of the time and pointed the way to new and fruitful avenues of geometrical research. It has become known as the **Erlanger Programm,** and it appeared right at the time when group theory was invading almost every domain of mathematics, and some mathematicians were beginning to feel that all mathematics is nothing but some aspect of group theory. This *programm* can be regarded as perhaps Klein's single most important mathematical accomplishment.

Klein's application of groups to geometry depends upon the concept of a **transformation of a set S onto itself,** by which is simply meant a correspondence under which each element of S corresponds to a unique element of S, and each element of S is the correspondent of a unique element of S. By the **product,** T_2T_1, of two transformations T_1 and T_2 of a set S of elements onto itself, we mean the resultant transformation obtained by first performing transformation T_1 and then transformation T_2. If T is a transformation of a set S onto itself, which carries each element a of S into a corresponding element b of S, then the transformation that reverses transformation T, by carrying each element b of S back into its original element a of S, is called the **inverse transformation** of transformation T and is denoted by T^{-1}. The transformation that carries each element of S into itself is called the **identity transformation** on set S, and is denoted by I. The following can now easily be established: *A set* Γ *of transformations of a set S onto itself constitutes a group* (in the technical sense of abstract algebra—see Problem Study 13.6) *under multiplication of transformations if* (1) *the product of any two transformations of the set* Γ *is in the set* Γ, (2) *the inverse of any transformation of the set* Γ *is in the set* Γ. Such a group of transformations is briefly referred to as a **transformation group.**

We are now ready to give Felix Klein's famous definition of a geometry: *A* **geometry** *is the study of those properties of a set S that remain invariant when the elements of set S are subjected to the transformations of some transformation group* Γ. The geometry may be conveniently denoted by the symbol $G(S,\Gamma)$.

To illustrate Klein's definition of a geometry, let S be the set of all points of an ordinary plane, and consider the set Γ of all transformations of S compounded from translations, rotations, and reflections in lines. Since the product of any two such transformations and the inverse of any such transformation are also such transformations, it follows that Γ is a transformation group. The resulting geometry is ordinary **plane Euclidean metric geometry.** Since such properties as length, area, congruence, parallelism, perpendicularity, similarity of figures, collinearity of points, and concurrency of lines are invariant under the group Γ, these properties are studied in plane Euclidean metric geometry. If, now, Γ is enlarged by including, together with the translations, rotations, and reflections in lines, the homothety transformations (in which each point P is carried into a point P' such that $AP = k \cdot AP'$, where A is some fixed point, k is some fixed positive constant, and A, P, P' are collinear), we obtain **plane similarity,** or **plane equiform, geometry.** Under this enlarged group, such properties as length, area, and congruence no longer remain invariant and hence are no longer subjects of study, but parallelism, perpendicularity, similarity of figures, collinearity of points, and concurrency of lines are still invariant properties, and hence do constitute subject matter for study in this geometry. Considered from Klein's point of view, plane projective geometry is the study of those properties of the points of a *projective* plane that remain invariant when the points are subjected to the group of so-called projective transformations. Of the previously mentioned properties, only collinearity of points and concurrency of lines still remain invariant. An important invariant under this group of transformations is the cross ratio of four collinear points; this invariant plays an important role in the study of projective geometry. The plane non-Euclidean metric geometries, considered in earlier chapters, can be thought of as the study of those properties of the points of a *non-Euclidean* plane that remain invariant under the group of transformations compounded from translations, rotations, and reflections in lines.

In all of the above geometries, the fundamental elements upon which the transformations of some transformation group are made to act are points; hence the above geometries are all examples of so-called **point geometries.** There are, as was pointed out in Section 14–5, geometries in which entities other than points are chosen for fundamental elements. Thus geometers have studied line geometries, circle geometries, sphere geometries, and various other geometries. In building up a geometry one is at liberty to choose, first of all, the fundamental element of the geometry (point, line, circle, etc.); next, the manifold or space of these elements (plane of points, ordinary space of points, spherical surface of points, plane of lines, pencil of circles, etc.); and finally, the group of transformations to which the fundamental elements are to be subjected. The construction of a new geometry becomes, in this way, a rather simple matter.

Another interesting feature is the way in which some geometries embrace others. Thus, since the transformation group of plane Euclidean metric geometry is a subgroup of the transformation group of plane equiform geometry, it follows that any theorem holding in the latter geometry must hold in the former. From this point of view, it can be shown that projective geometry lies within each of the former, and we have a sort of sequence of nesting geometries. Until recent times, the transformation group of projective geometry contained as subgroups the transformation groups of practically all other geometries that had been studied. This is essentially what Cayley meant when he remarked that "projective geometry contains all geometry." Actually, so far as the theorems of the geometries are concerned, it is the other way around—the theorems of projective geometry are contained among the theorems of each of the other geometries.

For almost fifty years, the Klein synthesis and codification of geometries remained essentially valid. But shortly after the turn of the century, bodies of mathematical propositions, which mathematicians felt should be called geometries, came to light; these bodies of propositions could not be fitted into this codification, and a new point of view upon the matter was developed, based upon the idea of abstract space with a superimposed structure that may or may not be definable in terms of some transformation group. We shall examine this new point of view in Section 15–3, merely remarking here that some of these new geometries have found application in the modern theory of physical space that is incorporated in Einstein's general theory of relativity. The Kleinian concept is still highly useful where it applies, and we may call a geometry that fits Klein's definition as given above, a **Kleinian geometry.** Partially successful efforts were made in the twentieth century, particularly by Oswald Veblen (1880–1960) and Élie Cartan (1869–1951), to extend and generalize Klein's definition so as to include geometries that lie outside Klein's original Programm.

Felix Klein was born in Düsseldorf in 1849. He studied at Bonn, Göttingen, and Berlin, and served as assistant to Julius Plücker at Bonn. His first professorial position was at the University of Erlangen (1872–1875), where his inaugural paper set forth the geometric program described above. He then taught at Munich, Leipzig University (1880–1886), and Göttingen University (1886–1913), officiating as department head at the latter institution. He was editor of the *Mathematische Annalen* and founder of the great mathematical *Encyklopädie*. He was a lucid expositor, an inspiring teacher, and a gifted lecturer. He died at Göttingen in 1925.

During Klein's tenure as department head at Göttingen University, that institution became a mecca for mathematics students from all over the world. A remarkable number of top-flight mathematicians studied at the university or served there as worthy successors to Gauss, Dirichlet, and Riemann, making the Göttingen school of mathematics one of the most famous of modern times.

FELIX KLEIN
(David Smith Collection)

Among these mathematicians were David Hilbert (1862–1943, the greatest mathematician of recent times), Edmund Landau (1877–1938, a famous number theorist), Hermann Minkowski (1864–1909, born in Russia and creator of geometric number theory), Wilhelm Ackermann (1896–1962, co-worker with Hilbert in mathematical logic), Constantin Carathéodory (1873–1950, a Greek mathematician who achieved fame in function theory), Ernst Zermelo (1871–1953, of Zermelo's postulate fame), Carl Runge (1856–1927, known to students of differential equations through the Runge-Kutta method), Emmy Noether (1882–1935, renowned algebraist), Richard Dedekind (1831–1916, of Dedekind-cut fame), Max Dehn (1878–1952, the first mathematician to solve one of Hilbert's twenty-three Paris problems), Hermann Weyl (1885–1955, specially known for his work in the foundations and philosophy of mathematics), and many, many others.

Among the many students who studied under Felix Klein at Göttingen was the Englishwoman Grace Emily Chisholm, who became Klein's "favorite pupil." At that time, women were not admitted into graduate schools in England, and Miss Chisholm went to Göttingen for her graduate work in mathematics. In 1895, she became the first female to receive a German doctorate through the regular examination process, and the following year she married the English mathematician William Henry Young.

The first comprehensive textbook on set theory and its applications to function theory, *The Theory of Sets of Points*, appeared in England in 1906 and was written by William Henry Young (1863–1942) and his wife Grace Chisholm Young (1868–1944). The Youngs published two other mathematics books and over 200 papers. Their son Laurence C. Young became a noted mathematician.

The great Göttingen school remained a potent force in world mathematics until it was all but destroyed by Adolph Hitler (1889–1945) and the rising Nazi tide. The totalitarian government and racial oppression caused a migration of eminent scholars to other parts of the world, of which the United States was perhaps the foremost beneficiary. There resulted a marked growth of mathematical accomplishment in America during the first half of the twentieth century. Similar, if lesser, migrations of scholars occurred in earlier times, as when Pythagoras went to Crotona and when scholars fled from the ancient University of Alexandria during its latter hectic days.

14–9 The Arithmetization of Analysis

A third profoundly significant mathematical event took place in the nineteenth century, in addition to the liberation of geometry and the liberation of algebra. This third event occurred in the field of analysis, was slow in materializing, and became known as the *arithmetization of analysis*.

When the theory of a mathematical operation is only poorly understood, there is the danger that the operation will be applied in a blindly formal and perhaps illogical manner. The performer, not aware of possible limitations upon the operation, is likely to use the operation in instances in which it does not necessarily apply. Instructors of mathematics see mistakes of this sort made by their students almost every day. Thus, one student of elementary algebra, firmly convinced that $a^0 = 1$ for all real numbers a, will set $0^0 = 1$, whereas another such student will assume that the equation $ax = b$ always has exactly one real solution for each pair of given real values a and b. Again, a student of trigonometry may think that the formula

$$\sqrt{1 - \sin^2 x} = \cos x$$

holds for all real x. A student of the calculus, not aware of improper integrals, may get an incorrect result by apparently correctly applying the rules of formal integration, or he or she may arrive at a paradoxical result by applying to a certain convergent infinite series some rule that holds only for absolutely convergent infinite series. This is essentially what happened in analysis during the century following the invention of the calculus. Attracted by the powerful applicability of the subject, and lacking a real understanding of the foundations upon which the subject must rest, mathematicians manipulated analytical processes in an almost blind manner, often being guided only by a native intuition of what was felt must be valid. A gradual accumulation of absurdities was bound to result, until, as a natural reaction to the pell-mell employment of intuitionism and formalism,[6] some conscientious mathematicians felt bound to attempt the difficult task of establishing a rigorous foundation under the subject.

---

[6] The terms *formalism* and *intuitionism* in this section should not be confused with the special meanings given to these terms in present-day discussions of the philosophies of mathematics. We shall encounter these philosophical connotations in the final chapter of the book.

The first suggestion of a real remedy for the unsatisfactory state of the foundations of analysis came from Jean-le-Rond d'Alembert (1717–1783), who very correctly observed in 1754 that a theory of limits was needed, although a sound development of this theory was not forthcoming until 1821. The earliest mathematician of the first rank actually to attempt a rigorization of the calculus was the Italian-French mathematician Joseph Louis Lagrange (1736–1813). The attempt, based upon representing a function by a Taylor's series expansion, was far from successful, for it ignored necessary matters of convergence and divergence. It was published in 1797 in Lagrange's monumental work, *Théorie des fonctions analytiques*. Lagrange was perhaps the leading mathematician of the eighteenth century, and his work had a deep influence on later mathematical research; with Lagrange's work, the long and difficult task of banishing intuitionism and formalism from analysis had begun.

In the nineteenth century, the superstructure of analysis continued to rise, but on ever-deepening foundations. A debt is undoubtedly owed to Carl Friedrich Gauss, for Gauss, more than any other mathematician of his time, broke from intuitive ideas and set new high standards of mathematical rigor. Also, in a treatment of hypergeometric series made by Gauss in 1812, we encounter what is generally regarded as the first really adequate consideration of the convergence of an infinite series.

A great stride was made in 1821, when the French mathematician Augustin-Louis Cauchy (1789–1857) successfully executed d'Alembert's suggestion by developing an acceptable theory of limits and then defining continuity, differentiability, and the definite integral in terms of the limit concept. It is essentially these definitions that we find in the more carefully written of today's elementary textbooks on the calculus. The limit concept is certainly indispensable for the development of analysis, for convergence and divergence of infinite series also depend upon this concept. Cauchy's rigor inspired other mathematicians to join the effort to rid analysis of formalism and intuitionism.

The demand for an even deeper understanding of the foundations of analysis was strikingly brought out in 1874 with the publicizing of an example, due to the German mathematician Karl Weierstrass, of a continuous function having no derivative, or, what is the same thing, a continuous curve possessing no tangent at any of its points. Georg Bernhard Riemann produced a function that is continuous for all irrational values of the variable but discontinuous for all rational values. Examples such as these seemed to contradict human intuition and made it increasingly apparent that Cauchy had not struck the true bottom of the difficulties in the way of a sound foundation of analysis. The theory of limits had been built upon a simple intuitive notion of the real number system. Indeed, the real number system was taken more or less for granted, as it still is in most of our elementary calculus texts. It became clear that the theory of limits, continuity, and differentiability depend upon more recondite properties of the real number system than had been supposed. Accordingly, Weierstrass advocated a program wherein the real number system itself should first be rigorized, then all the basic concepts of analysis should be derived from this number system. This remarkable program, known as the **arithmetization of**

analysis, proved to be difficult and intricate, but was ultimately realized by Weierstrass and his followers, so that today all of analysis can be logically derived from a postulate set characterizing the real number system.

Mathematicians have gone considerably beyond the establishment of the real number system as the foundation of analysis. Euclidean geometry, through its analytical interpretation, can also be made to rest upon the real number system, and mathematicians have shown that most branches of geometry are consistent if Euclidean geometry is consistent. Again, since the real number system, or some part of it, can serve for interpreting so many branches of algebra, it appears that the consistency of a good deal of algebra can also be made to depend upon that of the real number system. In fact, today it can be stated that essentially all of existing mathematics is consistent if the real number system is consistent. Herein lies the tremendous importance of the real number system for the foundations of mathematics.

Since the great bulk of existing mathematics can be made to rest on the real number system, one naturally wonders if the foundations can be pushed even deeper. In the late nineteenth century, with the work of Richard Dedekind (1831–1916), Georg Cantor (1845–1918), and Giuseppe Peano (1858–1932), these foundations were established in the much simpler and more basic system of natural numbers. That is, these men showed how the real number system, and thence the great bulk of mathematics, can be derived from a postulate set for the natural number system. Then, in the early twentieth century, it was shown that the natural numbers can be defined in terms of concepts of set theory, and thus that the great bulk of mathematics can be made to rest on a platform in set theory. Logicians, led by Bertrand Russell (1872–1970) and Alfred North Whitehead (1861–1947), have endeavored to push the foundations even deeper, by deriving the theory of sets from a foundation in the calculus of propositions of logic, though not all mathematicians feel this step has been successfully executed.

14–10 Weierstrass and Riemann

It is generally thought that a potential mathematician of the first rank, in order to succeed in his field, must start serious mathematical studies at an early age and must not be dulled by an inordinate amount of elementary teaching. Karl Theodor Wilhelm Weierstrass, who was born in Ostenfelde in 1815, is an outstanding exception to these two general rules. A misdirected youth spent in studying the law and finance gave Weierstrass a late start in mathematics, and it was not until he was forty that he finally emancipated himself from secondary teaching by obtaining an instructorship at the University of Berlin, and another eight years passed before, in 1864, he was awarded a full professorship at the university and could finally devote all his time to advanced mathematics. Weierstrass never regretted the years he spent in elementary teaching, and he later carried over his remarkable pedagogical abilities into his university work, becoming probably the greatest teacher of advanced mathematics that the world has yet known.

KARL WEIERSTRASS
(David Smith Collection)

Weierstrass wrote a number of early papers on hyperelliptic integrals, Abelian functions, and algebraic differential equations, but his widest known contribution to mathematics is his construction of the theory of complex functions by means of power series. This, in a sense, was an extension to the complex plane of the idea earlier attempted by Lagrange, but Weierstrass carried it through with absolute rigor. Weierstrass showed particular interest in entire functions and in functions defined by infinite products. He discovered uniform convergence and, as we have seen above, started the so-called arithmetization of analysis, or the reduction of the principles of analysis to real number concepts. A large number of his mathematical findings became possessions of the mathematical world, not through publication by him, but through notes taken of his lectures. He was very generous in allowing students and others to carry out, and receive credit for, investigations of many of his mathematical gems. As an illustration, somewhat in point, it was in his lectures of 1861 that he first discussed his example of a continuous nondifferentiable function, which was finally published in 1874 by Paul du Bois-Reymond (1831–1889). As stated earlier, Bolzano had already given such a function.

In algebra, Weierstrass was perhaps the first to give a so-called postulational definition of a determinant. He defined the determinant of a square matrix A as a polynomial in the elements of A, which is homogeneous and linear in the elements of each row of A, which merely changes sign when two rows of A are permuted, and which reduces to 1 when A is the corresponding identity matrix. He also contributed to the theory of bilinear and quadratic forms, and, along with J. J. Sylvester (1814–1897) and H. J. S. Smith (1826–1883), created the theory of elementary divisors of λ matrices.

Weierstrass was a very influential teacher, and his meticulously prepared lectures established an ideal for many future mathematicians; "Weierstrassian rigor'' became synonymous with ''extremely careful reasoning.'' Weierstrass was ''the mathematical conscience par excellence,'' and he became known as ''the father of modern analysis.'' He died in Berlin in 1897, just one hundred years after the first publication, in 1797 by Lagrange, of an attempt to rigorize the calculus.

Along with this rigorization of mathematics, there appeared a tendency toward abstract generalization, a process that has become very pronounced in present-day mathematics. Perhaps the German mathematician Georg Friedrich Bernhard Riemann influenced this feature of modern mathematics more than any other nineteenth-century mathematician. He certainly wielded a profound influence on a number of branches of mathematics, particularly geometry and function theory, and few mathematicians have bequeathed to their successors a richer legacy of ideas for further development.

Riemann was born in 1826 in a small village in Hanover, the son of a Lutheran pastor. In manner, he was always shy; in health, he was always frail. In spite of the very modest circumstances of his father, Riemann managed to secure a good education, first at the University of Berlin and then at the University of Göttingen. He took his doctoral degree at the latter institution with a brilliant thesis in the field of complex-function theory. In this thesis, one finds the so-called *Cauchy-Riemann differential equations* (known, however, before Riemann's time) that guarantee the analyticity of a function of a complex variable, and the highly fruitful concept of a *Riemann surface,* which introduced topological considerations into analysis. Riemann clarified the concept of integrability by the definition of what we now know as the *Riemann integral,* which led, in the twentieth century, to the more general *Lebesgue integral,* and thence to further generalizations of the integral.

GEORG RIEMANN
(David Smith Collection)

In 1854, Riemann became Privatdocent (official but unpaid lecturer) at Göttingen, and for this privilege presented his famous probationary lecture on the hypotheses that lie at the foundations of geometry. This has been considered the richest paper of comparable size ever presented in the history of mathematics; in it appears a broad generalization of space and geometry. Riemann's point of departure was the formula for the distance between two infinitesimally close points. In Euclidean geometry, this *metric* is given by

$$ds^2 = dx^2 + dy^2 + dz^2.$$

Riemann pointed out that many other distance formulas can be used, each such metric determining the properties of the resulting space and geometry. A space with a metric of the form

$$\begin{aligned} ds^2 = {} & g_{11}dx^2 + g_{12}dx\,dy + g_{13}dx\,dz \\ & + g_{21}dy\,dx + g_{22}dy^2 + g_{23}dy\,dz \\ & + g_{31}dz\,dx + g_{32}dz\,dy + g_{33}dz^2, \end{aligned}$$

where the g's are constants or functions of x, y, and z, is now known as a **Riemannian space,** and the geometry of such a space is called a **Riemannian geometry.** Euclidean space is the very special case in which $g_{11} = g_{22} = g_{33} = 1$ and all the other g's are zero. Later, Albert Einstein and others found Riemann's broad concept of space and geometry the mathematical milieu needed for general relativity theory. Riemann himself contributed in a number of directions to theoretical physics; he was the first, for example, to give a mathematical treatment of shock waves.

Famous in mathematical literature are the so-called **Riemann zeta function** and associated **Riemann hypothesis.** The latter is a celebrated unproved conjecture that is to classical analysis what *Fermat's* last "theorem" is to number theory. Euler had pointed out connections between the theory of prime numbers and the series

$$1/1^s + 1/2^s + 1/3^s + \cdots + 1/n^s + \cdots,$$

where s is an integer. Riemann studied the same series for s a complex number $\sigma + i\tau$. The sum of the series defines a function $\zeta(s)$, which has come to be known as Riemann's zeta function. Riemann, around 1859, conjectured that all the imaginary zeros of the zeta function have their real part $\sigma = 1/2$. In 1914, the English number theorist Sir Godfrey Harold Hardy (1877–1947) succeeded in showing that $\zeta(s)$ has an infinity of zeros with $\sigma = 1/2$. Even though it is now over a century old, the original Riemann conjecture is still unresolved. Hilbert selected the resolution of the Riemann hypothesis as one of his famous twenty-three Paris problems.

In 1857, Riemann was appointed assistant professor at Göttingen, and then, in 1859, full professor, succeeding Dirichlet in the chair once occupied by

Gauss. Riemann died of tuberculosis in 1866, when only forty years of age, in northern Italy, where he had gone to seek an improvement in his health.

14–11 Cantor, Kronecker, and Poincaré

This section will be devoted to a brief consideration of Georg Cantor and Henri Poincaré, two mathematicians with life spans astride the nineteenth and twentieth centuries, and who exerted a considerable influence on much of the mathematics of present times. It is also natural to insert a few words about Leopold Kronecker, the harsh and relentless critic of Cantor's mathematics of the infinite.

Georg Ferdinand Ludwig Philip Cantor was born of Danish parents in St. Petersburg, Russia, in 1845, and moved with his parents to Frankfurt, Germany, in 1856. Cantor's father was a Jew converted to Protestantism and his mother had been born a Catholic. The son took a deep interest in medieval theology and its intricate arguments on the continuous and the infinite. As a consequence, he gave up his father's suggestion of preparing for a career in engineering for concentrating on philosophy, physics, and mathematics. He studied at Zurich, Göttingen, and Berlin (where he came under the influence of Weierstrass and where he took his doctorate in 1867). He then spent a long teaching career at the University of Halle from 1869 until 1905. He died in a mental hospital in Halle in 1918.

Cantor's early interests were in number theory, indeterminate equations, and trigonometric series. The subtle theory of trigonometric series seems to have inspired him to look into the foundations of analysis. He produced his beautiful treatment of irrational numbers, which utilizes convergent sequences

GEORG CANTOR
(David Smith Collection)

of rational numbers and differs radically from the geometrically inspired treatment of Dedekind, and commenced in 1874 his revolutionary work on set theory and the theory of the infinite. With this latter work, Cantor created a whole new field of mathematical research. In his papers, he developed a theory of transfinite numbers, based on a mathematical treatment of the actual infinite, and created an arithmetic of transfinite numbers analogous to the arithmetic of finite numbers. Some of this matter is amplified in Section 15–4.

Cantor was deeply religious, and his work which in a sense is a continuation of the arguments connected with the paradoxes of Zeno, reflects his sympathetic respect for medieval scholastic speculation on the nature of the infinite. His views met considerable opposition, chiefly from Leopold Kronecker of the University of Berlin, and it was Kronecker who steadfastly opposed Cantor's efforts toward securing a teaching post at the University of Berlin. Today, Cantor's set theory has penetrated into almost every branch of mathematics, and it has proved to be of particular importance in topology and in the foundations of real function theory. There are logical difficulties, and paradoxes have appeared. The twentieth-century controversy between the formalists led by Hilbert, and the intuitionists, led by Brouwer, is essentially a continuation of the controversy between Cantor and Kronecker. We look deeper into these matters in the next chapter.

Kronecker was born at Liegnitz, near Breslau, in 1823, and had Kummer as a teacher at the gymnasium of his native town. He next studied at the University of Berlin under Jacobi, Steiner, and Dirichlet, and then at the University of Bonn, again under Kummer. Following his studies, he engaged in business for the eleven years from 1844 to 1855, and as a gifted financier amassed considerable personal wealth. In 1855, he moved to Berlin and began

LEOPOLD KRONECKER
(David Smith Collection)

teaching at the University there in 1861. Kummer had also moved to Berlin, and Kummer, Weierstrass, and Kronecker constituted a strong mathematical trio there. Kronecker specialized in the theory of equations, elliptic functions, and algebraic number theory. As a finitist, he condemned the work of Cantor, regarding it as theology and not as mathematics. Believing that all of mathematics must be based by finite methods upon the whole numbers, he was a nineteenth-century Pythagorean. He once made the toast, "Die ganze Zahlen hat Gott gemacht, alles andere ist Menschenwerk" (God made the whole numbers, all the rest is the work of man). He died in Berlin in 1891.

Jules Henri Poincaré, generally acknowledged to be the outstanding mathematician of his age, was born in Nancy, France, in 1854. He was a first cousin of Raymond Poincaré, the eminent statesman and president of the French republic during World War I. After graduating from the École Polytechnique in 1875, Henri took a degree in mining engineering at the École des Mines in 1879, and in that same year also earned a doctorate in science from the University of Paris. Upon graduating from the École des Mines, he was appointed to a teaching post at Caen University, but two years later moved to the University of Paris, where he held several professorships in mathematics and science until his death in 1912.

Poincaré has been described as the last of the universalists in the field of mathematics. It is certainly true that he commanded and enriched an astonishing range of subjects. At the Sorbonne, he brilliantly lectured each year on a different topic in pure or applied mathematics, many of these lectures shortly after appearing in print. He was a prolific writer, producing more than thirty books and 500 technical papers. He was also one of the ablest popularizers of mathematics and science. His inexpensive paperback expositions were avidly bought and widely read by people in all walks of life; they are masterpieces that, for lucidity of communication and engaging style, have never been excelled, and they have been translated into many foreign languages. In fact, so

HENRI POINCARÉ
(David Smith Collection)

great was the literary excellence of Poincaré's popular writing that he was awarded the highest honor that can be conferred on a French writer—he was elected a member of the literary section of the French *Institut*.

Poincaré never cared to remain in one field for very long, but preferred to jump nimbly from area to area. He was described by one of his contemporaries as "a conqueror, not a colonist." His doctoral dissertation on differential equations concerned itself with existence theorems. This work led him to develop the theory of automorphic functions, and, in particular, the so-called zeta-Fuchsian functions, which Poincaré showed can be used to solve second-order linear differential equations with algebraic coefficients. Like Laplace, Poincaré contributed notably to the subject of probability. He also anticipated the twentieth-century interest in topology, and his name is found today in the *Poincaré groups* of combinatorial topology. We have already, in Section 13–7 and Problem Study 13.12, seen Poincaré's interest in non-Euclidean geometry. In applied mathematics, this versatile genius contributed to such diverse subjects as optics, electricity, telegraphy, capillarity, elasticity, thermodynamics, potential theory, quantum theory, theory of relativity, and cosmogony.

All his life, Poincaré was physically awkward, nearsighted, and absent-minded, but he possessed almost complete retention and instant recall of anything he had ever read. He worked his mathematics in his head while restlessly pacing about, and when it was completely thought through, he committed it to paper rapidly and with essentially no rewriting or erasures. In contrast to his hasty and extensive production, one recalls the meticulously prepared productions of Gauss, and Gauss' motto: "Few, but ripe."

There are stories of Poincaré's lack of manual dexterity. It was said of him that he was ambidextrous—that is, he could perform equally badly with either hand. He had no ability whatever in drawing, and he earned a flat zero in the subject in school. At the end of the school year, his classmates jokingly organized a public exhibition of his artistic masterpieces. They carefully labeled each item in Greek—"This is a house," "This is a horse," and so on.

It may well be that Poincaré will be the last person of whom it can in a reasonable sense be claimed that *all* of mathematics was his province. Mathematics has grown at such an incredible rate in modern times that it is believed quite impossible for anyone ever again to achieve such a distinction.

14–12 Sonja Kovalevsky, Emmy Noether, and Charlotte Scott

Sophia Korvin-Krukovsky, later known as Sonja Kovalevsky, was born into a family of Russian nobility in Moscow in 1850. When she was seventeen, she went to St. Petersberg and studied calculus with a teacher of the naval school there. Barred, because of her sex, from pursuing advanced studies in a Russian university, she contracted a nominal marriage with the sympathetic Vladimir Kovalevsky (later to become a noted paleontologist) to be free of parental objections to studying abroad. The marriage took place in 1868, and in the following spring, the pair went to Heidelberg.

At Heidelberg, Kovalevsky attended the mathematics lectures of Leo Königsberger (1837–1921) and du Bois-Reymond (1831–1889) and the physics lectures of Kirchhoff (1824–1887) and Helmholz (1821–1894). Königsberger had earlier studied under Karl Weierstrass of the University of Berlin, and his enthusiastic reports of his mentor instilled in Kovalevsky a desire to study under the great teacher. She arrived in Berlin in 1870, but found the University adamant in its exclusion of women students. She accordingly approached Weierstrass directly, who, upon receiving a strong recommendation from Königsberger, accepted her as a private student. Kovalevsky soon became Weierstrass' favorite pupil, and he repeated his university lectures to her. She won Weierstrass' admiration and studied under the master for four years (1870–1874), during which time she not only covered the university course in mathematics, but also wrote three important papers, one on the theory of partial differential equations, one on the reduction of Abelian integrals of the third kind, and one supplementing Laplace's research on the form of Saturn's rings.

In 1874, Sonja Kovalevsky was awarded, *in absentia,* the degree of Doctor of Philosophy by Göttingen University and, because of the outstanding quality of a submitted paper on partial differential equations, was excused from taking the oral examination. In 1888, when thirty-eight years old, she achieved her greatest success when the French Academy awarded her the prestigious Prix Bordin for her memoir "On the Problem of the Rotation of a Solid Body about a Fixed Point." Of the fifteen papers submitted for the prize, Kovalevsky's was judged the best; it was considered so exceptional that the prize was raised from 3000 francs to 5000 francs.

SONJA KOVALEVSKY
(David Smith Collection)

From 1884 until her death in 1891, Kovalevsky served as a professor of higher mathematics at Stockholm University. Her motto was: "Say what you know, do what you must, come what may."

There is an oft-told story about an early influence, other than her mathematically inclined father and uncle, that attracted Kovalevsky to mathematics when she was only a child. It seems that one of the children's rooms of her home was temporarily papered with sheets of calculus lecture notes dating from her father's student days. These sheets fascinated her, and she spent hours trying to decipher them and to put them in proper order.

Amalie Emmy Noether, one of the most outstanding mathematicians in the field of abstract algebra, was born in Erlangen, Germany, in 1882. Although she was born in the late nineteenth century, she did her work in the first half of the twentieth century. Her father, Max Noether (1844–1921), was a distinguished mathematician at the University of Erlangen. Max Noether was an algebraist, as was Paul Gordan (1837–1912), who also was associated with the university and was a close friend of the Noether family. It is no wonder that Emmy Noether, who studied at the University, also became an algebraist. She wrote her doctoral thesis, "On Complete Systems of Invariants for Ternary Biquadratic Forms," under Gordan in 1907. When Gordan retired in 1910, he was followed one year later by Ernst Fischer (1875–1959), another algebraist with particular interests in the theory of elimination and the theory of invariants. His influence on Noether was great, and under his direction, her preoccupation

AMALIE EMMY NOETHER
(Bryn Mawr College Archives)

passed from the algorithmic aspect of Gordan's work to the abstract axiomatic approach of Hilbert.

After leaving Erlangen, Emmy Noether studied at Göttingen, where she passed her habilitation examination in 1919, after overcoming objections of some of the faculty who were opposed to women lecturers. "What will our soldiers think" they queried, "when they return to the University and find that they are expected to learn at the feet of a woman?" David Hilbert was very annoyed at the question, and responded, "Meine Herren, I do not see that the sex of the candidate is an argument against her admission as a Privatdozent. After all, the Senate is not a bathhouse." In 1922, she became extraordinary professor at Göttingen, a position she held until 1933, when, under the excesses of the German national revolution, she, as well as many others, was prohibited from academic participation. She thereupon left Germany to accept a professorship at Bryn Mawr College in Pennsylvania and to become a member of the Institute for Advanced Study at Princeton. Her short time in America was perhaps her happiest and most productive period. She died in 1935, at the age of fifty-three and at the height of her creative powers.

Although Noether was a poor lecturer and lacked pedagogical skill, she managed to inspire a surprising number of students who also left marks in the field of abstract algebra. Her studies on abstract rings and ideal theory have been particularly important in the development of modern algebra.

In the ceremonies following her death, Emmy Noether received a glowing tribute from Albert Einstein. Someone once described her as the daughter of Max Noether. To this Edmund Landau replied: "Max Noether was the father of Emmy Noether. Emmy is the origin of coordinates in the Noether family." Hermann Weyl characterized her as "warm, like a loaf of bread." A centenary celebration of Emmy Noether's birth was held at Bryn Mawr College in 1982.

It is natural when considering Emmy Noether to recall her eminent predecessor at Bryn Mawr College, Charlotte Angas Scott (1858–1931). Charlotte Scott became, in 1885, the first British woman to receive a doctorate (in any field): it was a D.Sc. in mathematics, granted by the University of London, and for which she passed the qualifying examinations at First-Class Level. She had spent nine years at Cambridge University, but that university did not grant degrees to women until 1948.

In addition to being a research mathematician (she published over twenty articles in the research journals of her day), Scott was a superb teacher who insisted on the highest academic standards. Her research study lay chiefly in the geometry of curves. She wrote three books, of which her *An Introductory Account of Certain Modern Ideas and Methods in Plane Analytic Geometry*, published in 1894, was an inspiring masterpiece.[7]

From 1899 until 1926, Scott served as coeditor of the *American Journal of Mathematics* that had been founded by J. J. Sylvester (1814–1897) in 1878 when

[7] Reprinted by Chelsea in 1961. See the Bibliography at the end of the chapter.

CHARLOTTE ANGAS SCOTT
(Bryn Mawr College Archives)

he was department head at Johns Hopkins. Scott also played an active role in the founding of the New York Mathematical Society, which, in 1894, was reorganized into the American Mathematical Society.[8]

The seven mythical daughters of Atlas have become enshrined in the northern sky as the seven principal stars in the Pleiades cluster. In imitation, the seven mathematicians (Hypatia, Maria Gaetana Agnesi, Sophie Germain, Mary Fairfax Somerville, Sonja Kovalevsky, Grace Chisholm Young, and Amalie Emmy Noether) have become known as The Mathematical Pleiades, or The Seven Daughters of Mathesis. Not only were these women competent mathematicians, but they have inspired and enabled other women to enter mathematics. The sex barrier in mathematics of the nineteenth and early twentieth centuries was broken, and universities became open to the attendance and academic recognition of women and to the acceptance of women on their faculties.

There was formed in America in 1971 the Association for Women in Mathematics (open to both female and male members), founded to put male and female mathematicians on an equal footing. The male population has no inherent superiority in mathematical thinking or creativity, and there is today a rapid increase in the number of women among the topflight practitioners and creators of mathematics.

---

[8] For an excellent account of Charlotte Scott, see Patricia C. Kenschaft, "Charlotte Angas Scott, 1858–1931," *The College Mathematics Journal* (March) 1987, 98–110.

14–13 The Prime Numbers

The prime numbers have enjoyed a long history, running from the days of the ancient Greeks up into the present. Since some of the most important discoveries about the primes were made in the nineteenth century, this seems a fitting place to discuss these interesting numbers.

The fundamental theorem of arithmetic says that the prime numbers are building bricks from which all other integers are multiplicatively made. Accordingly, the prime numbers have received much study, and considerable efforts have been spent trying to determine the nature of their distribution in the sequence of positive integers. The chief results obtained in antiquity are Euclid's proof of the infinitude of the primes and Eratosthenes' sieve for finding all primes below a given integer n.

From the sieve of Eratosthenes can be obtained a cumbersome formula that will determine the number of primes below n when the primes below $\sqrt{n}$ are known. This formula was considerably improved in 1870 by Ernst Meissel, who succeeded in showing that the number of primes below 10^8 is 5,761,455. The Danish mathematician Bertelsen continued these computations and announced, in 1893, that the number of primes below 10^9 is 50,847,478. In 1959, the American mathematician D. H. Lehmer showed that this last result is incorrect and that it should read 50,847,534; he also showed that the number of primes below 10^{10} is 455,052,511.

No practicable procedure is yet known for testing large numbers for primality, and the effort spent on testing certain special numbers has been enormous. For more than seventy-five years, the largest number actually verified as a prime was the thirty-nine-digit number

$$2^{127} - 1 = 170{,}141{,}183{,}460{,}469{,}231{,}731{,}687{,}303{,}715{,}884{,}105{,}727,$$

given by the French mathematician Anatole Lucas (1842–1891) in 1876. In 1952, the EDSAC machine, in Cambridge, England, established primality of the much larger (seventy-nine-digit) number

$$180(2^{127} - 1)^2 + 1.$$

Since then other digital computers have shown the primality of the enormous numbers $2^n - 1$ for $n = 521$, 607, 1279, 2203, 2281, 3217, 4253, 4423, 9689, 9941, 11213, 19937, 21701, 23209, 86243, 132049, and 216091.

A dream of number theorists is the finding of a function $f(n)$ that, for positive integral n, will yield only prime numbers, the sequence of primes so obtained containing infinitely many different primes. Thus

$$f(n) = n^2 - n + 41$$

yields primes for all such $n < 41$, but $f(41) = (41)^2$, a composite number. The quadratic polynomial $f(n) = n^2 - 79n + 1601$ yields primes for all $n < 80$.

Polynomial functions can be obtained that will successively yield as many primes as desired, but no such functions can be found that will always yield primes. It was about 1640 that Pierre de Fermat conjectured that $f(n) = 2^{2^n} + 1$ is prime for all non-negative integral n, but this, as we have pointed out in Section 10–3, is incorrect. An interesting recent result along these lines is the proof, by W. H. Mills in 1947, of the existence of a real number A, such that the largest integer not exceeding A^{3^n} is a prime for every positive integer n. Nothing was shown about the actual value, nor even the rough magnitude, of the real number A.

A remarkable generalization of Euclid's theorem on the infinitude of the primes was established by Lejeune-Dirichlet (1805–1859), who succeeded in showing that every arithmetic sequence,

$$a,\ a + d,\ a + 2d,\ a + 3d,\ \ldots,$$

in which a and d are relatively prime, contains an infinitude of primes. The proof of this result is far from simple.

Perhaps the most amazing result yet found concerning the distribution of the primes is the so-called **prime number theorem.** Suppose we let A_n denote the number of primes below n. The prime number theorem then says that $(A_n \log_e n)/n$ approaches 1 as n becomes larger and larger. In other words, A_n/n, called the **density** of the primes among the first n integers, is approximated by $1/\log_e n$, the approximation improving as n increases. This theorem was conjectured by Gauss from an examination of a large table of primes and was independently proved in 1896 by the French and Belgian mathematicians J. Hadamard and C. J. de la Vallée Poussin.

Extensive factor tables are valuable in researches on prime numbers. Such a table for all numbers up to 24,000 was published by J. H. Rahn (1622–1676) in 1659, as an appendix to a book on algebra. In 1668, John Pell (1611–1685) of England extended this table up to 100,000. As a result of appeals by the German mathematician J. H. Lambert (1728–1777), an extensive and ill-fated table was computed by a Viennese schoolmaster named Anton Felkel (born 1740). The first volume of Felkel's computations, giving factors of numbers up to 408,000, was published in 1776 at the expense of the Austrian imperial treasury. There were very few subscribers to the volume, however, so the treasury recalled almost the entire edition and converted the paper into cartridges to be used in a war for killing Turks! In the nineteenth century, the combined efforts of Chernac, Burckhardt, Crelle, Glaisher, and the lightning calculator Dase led to a table covering all numbers up to 10,000,000 and published in ten volumes. The greatest achievement of this sort, however, is the table calculated by J. P. Kulik (1773–1863), of the University of Prague. His as yet unpublished manuscript is the result of a twenty-year hobby, and covers all numbers up to 100,000,000. The best available factor table is that of the American mathematician D. N. Lehmer[9] (1867–1938). It is a cleverly prepared one-volume table covering num-

[9] Father of D. H. Lehmer. D. N. Lehmer has pointed out that Kulik's table contains errors.

bers up to 10,000,000. With the advent of the modern electronic computers, testing for primality and constructing tables of special primes have been greatly enhanced. For example, in the November 1980 issue of *Crux mathematicorum* appears a table of all 93 five-digit and all 668 seven-digit palindromic primes (a palindromic number is a number that reads the same in both directions, as the palindromic prime 3417143). The calculation was done on a PDP-11/45 at the University of Waterloo, and the computer time was slightly more than one minute. A particularly attractive nine-digit palindromic prime is the number 345676543, given by Léo Sauvé, editor of the above journal, who stated that there are 5172 nine-digit palindromic primes.

There are many unproved conjectures regarding prime numbers. One of these is to the effect that there are infinitely many pairs of **twin primes,** or primes of the form p and $p + 2$, like 3 and 5, 11 and 13, and 29 and 31. Another is the conjecture made by Christian Goldbach (1690–1764) in 1742 in a letter to Euler. Goldbach had observed that every even integer, except 2, seemed representable as the sum of two primes. Thus, $4 = 2 + 2$, $6 = 3 + 3$, $8 = 5 + 3, \ldots,$ $16 = 13 + 3$, $18 = 11 + 7, \ldots, 48 = 29 + 19, \ldots, 100 = 97 + 3$, and so forth. Progress on this problem was not made until 1931, when the Russian mathematician L. G. Schnirelmann (1905–1935) showed that every positive integer can be represented as the sum of not more than 300,000 primes! Somewhat later the Russian mathematician I. M. Vinogradoff (born 1891) showed that there exists a positive integer N such that any integer $n > N$ can be expressed as the sum of at most four primes, but the proof in no way permits us to appraise the size of N. The Goldbach conjecture has been verified for numbers up through 100 million.

The following questions (in which n represents a positive integer) about primes have never been answered: Are there infinitely many primes of the form $n^2 + 1$? Is there always a prime between n^2 and $(n + 1)^2$? Is any n from some point onwards either a square or the sum of a prime and a square? Are there infinitely many **Fermat primes** (primes of the form $2^{2^n} + 1$)?

Problem Studies

14.1 The Feuerbach Configuration

Prominent in the modern elementary geometry of the triangle is the *nine-point circle*. In a given triangle $A_1A_2A_3$, of circumcenter O and orthocenter (intersection of the three altitudes) H, let O_1, O_2, O_3 be the midpoints of the sides, H_1, H_2, H_3 the feet of the 3 altitudes, and C_1, C_2, C_3 the midpoints of the segments HA_1, HA_2, HA_3. Then the 9 points, $O_1, O_2, O_3, H_1, H_2, H_3, C_1, C_2, C_3$, lie on a circle, known as the **nine-point circle** of the given triangle. This circle, due to later misplaced credit for earliest discovery, is sometimes referred to as **Euler's circle.** In Germany, it is called **Feuerbach's circle,** because Karl Wilhelm Feuerbach (1800–1834) published a pamphlet in which he not only arrived at the nine-point circle, but also proved that it is tangent to the inscribed and the 3 escribed circles of the given triangle. This last fact is known as **Feuerbach's theorem** and is justly regarded as one of the more elegant theorems in the modern geometry of the triangle. The 4 points of contact of the nine-point circle with the inscribed and escribed circles are known as the **Feuerbach points** of the triangle and have received considerable study. The center, F, of the nine-point circle is at the midpoint of OH. The centroid (intersection of the 3 medians of the triangle), G, also lies on OH such that $HG = 2(GO)$. The line of collinearity of O, F, G, H is known as the **Euler line** of the given triangle. Let H_2H_3, H_3H_1, H_1H_2 intersect the opposite sides A_2A_3, A_3A_1, A_1A_2, in P_1, P_2, P_3. Then P_1, P_2, P_3 lie on a line known as the **polar axis** of triangle $A_1A_2A_3$, and the polar axis is perpendicular to the Euler line. If the nine-point circle and the circumcircle intersect, then the polar axis is the line of the common chord of these 2 circles, and the circle on HG as diameter, the so-called **orthocentroidal circle** of the given triangle, also passes through the same points of intersection.

Draw a large and carefully constructed figure of an obtuse triangle with its centroid, orthocenter, circumcenter, incenter, 3 excenters, Euler line, polar axis, nine-point circle, Feuerbach points, circumcircle, and orthocentroidal circle.

14.2 Commandino's Theorem

Federigo Commandino (1509–1575) published in 1565 one of the first theorems, after the time of the Greeks, in the geometry of the tetrahedron. The theorem concerns the medians of the tetrahedron, in which a **median** is defined as the line segment joining a vertex of the tetrahedron to the centroid of the opposite face. Commandino's theorem states: The 4 medians of a tetrahedron are concurrent in a point that quadrisects each median.

(a) Prove Commandino's theorem analytically.

(b) Prove Commandino's theorem synthetically.

(c) Prove that the plane determined by the centroids of 3 faces of a tetrahedron is parallel to the fourth face of the tetrahedron.

The tetrahedron formed by the planes passing through the vertices of a given tetrahedron and parallel to the respective opposite faces is called the **anticomplementary tetrahedron** of the given tetrahedron.

(d) Prove that the vertices of a tetrahedron are the centroids of the faces of the anticomplementary tetrahedron.

(e) Prove that an edge of the anticomplementary tetrahedron is trisected by the 2 faces of the given tetrahedron that meet that edge.

14.3 The Altitudes of a Tetrahedron

The 3 altitudes of a triangle are concurrent. Are the 4 altitudes of a tetrahedron concurrent?

14.4 Space Analogs

State theorems in three space that are analogs of the following theorems in the plane.

(a) The bisectors of the angles of a triangle are concurrent at the center of the inscribed circle of the triangle.

(b) The area of a circle is equal to the area of a triangle the base of which has the same length as the circumference of the circle and the altitude of which is equal to the radius of the circle.
(c) The foot of the altitude of an isosceles triangle is the midpoint of the base.

14.5 Isogonal Elements

Two lines through the vertex of an angle and symmetrical with respect to the bisector of the angle are called **isogonal conjugate lines** of the angle. There is an attractive theorem about triangles that states that if 3 lines through the vertices of a triangle are concurrent, then the 3 isogonal conjugate lines through the vertices of the triangle are also concurrent. The 2 points of concurrency are called a pair of **isogonal conjugate points** of the triangle. The 6 feet of the perpendiculars dropped from a pair of isogonal conjugate points on the sides of a triangle lie on a circle whose center is the midpoint of the line segment joining the pair of isogonal conjugate points.

(a) Draw a figure illustrating the above facts.
(b) Prove that the orthocenter and the circumcenter of a triangle constitute a pair of isogonal conjugate points.
(c) Try to find 3-space analogs of the definitions and theorems stated at the start of this Problem Study.

The isogonal conjugate point of the centroid of a triangle is called the **symmedian point,** or **Lemoine point,** of the triangle. This point was first given by Émile Lemoine (1840–1912) in 1873, in a paper read before the Association Française pour l'Avancement des Sciences and that can be claimed to have seriously started the modern study of the geometry of the triangle.

14.6 Impossible Constructions

(a) Establish the identity: $\cos\theta = 4\cos^3(\theta/3) - 3\cos(\theta/3)$.
(b) Show that it is impossible with Euclidean tools to construct a regular polygon of 9 sides.
(c) Show that it is impossible with Euclidean tools to construct an angle of 1°.
(d) Show that it is impossible with Euclidean tools to construct a regular polygon of 7 sides.
(e) Show that it is impossible with Euclidean tools to trisect an angle whose cosine is 2/3.
(f) Given a segment s, show that it is impossible with Euclidean tools to construct segments m and n such that $s:m = m:n = n:2s$.
(g) Show that it is impossible with Euclidean tools to construct the radius of a sphere whose volume is the sum of the volumes of 2 arbitrary spheres whose radii are given.
(h) Show that it is impossible with Euclidean tools to construct a line segment whose length equals the circumference of a given circle.
(i) Given an angle AOB and a point P within the angle. The line through P cutting OA and OB in C and D so that $CE = PD$, where E is the foot of

the perpendicular from O on CD, is known as **Philon's line** for angle AOB and point P. It can be shown that Philon's line is the minimum chord CD that can be drawn through P. Show that in general it is impossible to construct with Euclidean tools Philon's line for a given angle and a given point.

14.7 Some Approximate Constructions

(a) For an approximate construction of a regular heptagon inscribed in a given circle, take for a side of the heptagon the apothem of the regular inscribed hexagon. How good an approximation is this?

(b) To trisect a given central angle of a circle, someone suggests trisecting the chord of the arc cut off by the angle and then connecting these trisection points with the center of the circle. Show that this leads to a poor approximation for large obtuse angles.

(c) Study the accuracy of the following procedure for approximately trisecting an angle; it was given by Kopf in 1919 and then later improved by O. Perron and M. d'Ocagne. Let the given angle AOB be taken as a central angle in a circle of diameter BOC. Find D, the midpoint of OC, then P on OC produced such that $CP = OC$. At D, erect a perpendicular to cut the circle in E, then mark off between C and D the point F such that $DF = (DE)/3$. With F as center and FB as radius, describe an arc to cut CA produced in A'. Then angle $A'PB$ is approximately equal to 1/3 of angle AOB.

(d) Study the accuracy of the following procedure for approximately trisecting an angle; it was given by M. d'Ocagne in 1934 and is surprisingly accurate for small angles. Let the given angle AOB be taken as a central angle of a circle of diameter BOC. Let D be the midpoint of OC and M the midpoint of arc AB. Then angle MDB is approximately equal to 1/3 of angle AOB.

14.8 Mascheroni Construction Theorem

Let us designate the circle with center at point C and passing through the point A by the symbol $C(A)$, and the circle with center at point C and with radius equal to the segment AB by the symbol $C(AB)$. Prove the following chain of constructions and show that they establish the Mascheroni construction theorem: *Any Euclidean construction, insofar as the given and required elements are points, may be accomplished with the Euclidean compasses alone.* The constructions are recorded in a tabular form in which the upper line indicates what is to be drawn, while the lower line indicates the new points that are thus constructed.

(a) To construct with Euclidean compasses the circle $C(AB)$.

$C(A), A(C)$	$M(B), N(B)$	$C(X)$
M, N	X	

(*Note:* This construction shows that the Euclidean and modern compasses are equivalent tools.)

(b) To construct with modern compasses the intersection of $C(D)$ with the line determined by points A and B.

CASE 1 C *not on* AB.

$A(C), B(C)$	$C(D), C_1(CD)$
C_1	X, Y

CASE 2 C *on* AB.

$A(D), C(D)$	$C(DD_1), D(C)$	$C(DD_1), D_1(C)$	$F(D_1), F_1(D)$	$F(CM), C(D)$
D_1	F, fourth vertex of parallelogram CD_1DF	F_1, fourth vertex of parallelogram CDD_1F_1	M	X, Y

(c) To construct with modern compasses the point of intersection of the lines determined by the pairs of points A, B and C, D.

$A(C)$, $B(C)$	$A(D)$, $B(D)$	$C(DD_1)$, $D_1(CD)$	$C_1(G)$, $G(D_1)$	$C_1(C)$, $G(CE)$	$C(F)$, $C_1(CF)$
C_1	D_1	G, collinear with C, C_1	E, either intersection	F, collinear with C_1, E	X

(d) On page 268 of Cajori's *A History of Mathematics* we read: "Napoleon proposed to the French mathematicians the problem, to divide the circumference of a circle into 4 equal parts by the compasses only. Mascheroni does this by applying the radius 3 times to the circumference; he obtains the arcs AB, BC, CD; then AD is a diameter; the rest is obvious." Complete the "obvious" part of the construction.

14.9 Constructions with Straightedge and Rusty Compasses

Solve, with straightedge and rusty compasses, the following first 14 constructions found in Mohr's *Compendium Euclidis curiosi:*

1. To divide a given line segment into 2 equal parts.
2. To erect a perpendicular to a line from a given point in the line.
3. To construct an equilateral triangle on a given side.
4. To erect a perpendicular to a line from a given point off the line.
5. Through a given point to draw a line parallel to a given line.
6. To add 2 given line segments.
7. To subtract a shorter segment from a given segment.
8. Upon the end of a given line segment to place a given segment perpendicularly.

9. To divide a line segment into any number of equal parts.
10. Given 2 line segments, to find the third proportional.
11. Given 3 line segments, to find the fourth proportional.
12. To find the mean proportional of 2 given segments.
13. To change a given rectangle into a square.
14. To draw a triangle, given the 3 sides.

14.10 Lemoine's Geometrography

Find the *symbol, simplicity,* and *exactitude* for the following familiar constructions of a line through a given point A and parallel to a given line MN.

(a) Through A draw any line to cut MN in B. With any radius r, draw the circle $B(r)$ to cut MB in C and AB in D. Draw circle $A(r)$ to cut AB in E. Draw circle $E(CD)$ to cut circle $A(r)$ in X. Draw AX, obtaining the required parallel.

(b) With any suitable point D as center, draw circle $D(A)$ to cut MN in B and C. Draw circle $C(AB)$ to cut circle $D(A)$ in X. Draw AX.

(c) With any suitable radius r, draw the circle $A(r)$ to cut MN in B. Draw circle $B(r)$ to cut MN in C. Draw circle $C(r)$ to cut circle $A(r)$ in X. Draw AX.

Find the *symbol, simplicity,* and *exactitude* for the following constructions of a perpendicular to a given line m at a given point P on m.

(d) With P as center and with any convenient radius, draw a circle to cut m in A and B. With A and B as centers and with any suitable radius, draw arcs to intersect in Q. Draw PQ, the required perpendicular.

(e) With any convenient point not on m as center, draw the circle through P to cut m again in Q, and draw diameter QR of this circle. Draw PR, the required perpendicular.

14.11 Principle of Duality

(a) Dualize 9.12 (a). (See Problem Study 9.12.)

(b) Given 5 lines, find on any 1 of them the point of contact of the conic touching the 5 lines.

(c) Given 4 tangents to a conic and the point of contact of any 1 of them, construct further tangents to the conic.

(d) Dualize 9.12 (d).

(e) Dualize 9.12 (e).

(f) Given 3 tangents to a conic and the points of contact of 2 of them, construct the point of contact of the third.

(g) Dualize Desargues' 2-triangle theorem.

14.12 A Self-Dual Postulate set for Projective Geometry

(a) Show that the following postulate set for projective geometry, given by Karl Menger in 1945, is self-dual.

P1: *There is 1 and only 1 line on every 2 distinct points, and 1 and only 1 point on every two distinct lines.*

P2: *There exist 2 points and 2 lines such that each of the points is on just 1 of the lines and each of the lines is on just 1 of the points.*

P3: *There exist 2 points and 2 lines, the points not on the lines, such that the point on the two lines is on the line on the 2 points.*

(b) Verify the 3 postulates of (a) for the 7 "points," designated by the letters A, B, C, D, E, F, G, and the 7 "lines," denoted by the trios (AFB), (BDC), (CEA), (AGD), (BGE), (CGF), (DEF). This example establishes the existence of *finite* projective geometries (that is, projective geometries containing only a finite number of points and lines).

14.13 Principle of Duality of Trigonometry

If, in a trigonometric equation, each trigonometric function that appears is replaced by its cofunction, the new equation obtained is called the **dual** of the original equation. Establish the following **principle of duality** of trigonometry: If a trigonometric equation involving a single angle is an identity, then its dual is also an identity.

14.14 Coordinate Systems

Let us designate as a **bipolar coordinate system,** one for which the frame of reference is a horizontal line segment AB of length a, with respect to which a point P of the plane is located by recording as coordinates the counterclockwise angle $\alpha = \measuredangle BAP$ and the clockwise angle $\beta = \measuredangle ABP$ (see Figure 118).

(a) Find the bipolar equation of (1) the perpendicular bisector of AB, and (2) an arc of a circle having AB as chord.

(b) Find the equations of transformation connecting the bipolar coordinate system with the rectangular Cartesian coordinate system having x-axis along AB and origin at the midpoint of AB.

(c) Identify the curves (1) $\cot\alpha \cot\beta = k$, (2) $\cot\alpha/\cot\beta = k$, and (3) $\cot\alpha + \cot\beta = k$, where k is a constant.

(d) Find rectangular Cartesian equations of the following curves given by polar equations: (1) lemniscate of Bernoulli, $r^2 = a^2 \cos 2\theta$, (2) cardioid, $r = a(1 - \cos\theta)$, (3) spiral of Archimedes, $r = a\theta$, (4) equiangular spiral, $r = e^{a\theta}$, (5) hyperbolic spiral, $r\theta = a$, (6) 4-leaved rose, $r = a \sin 2\theta$.

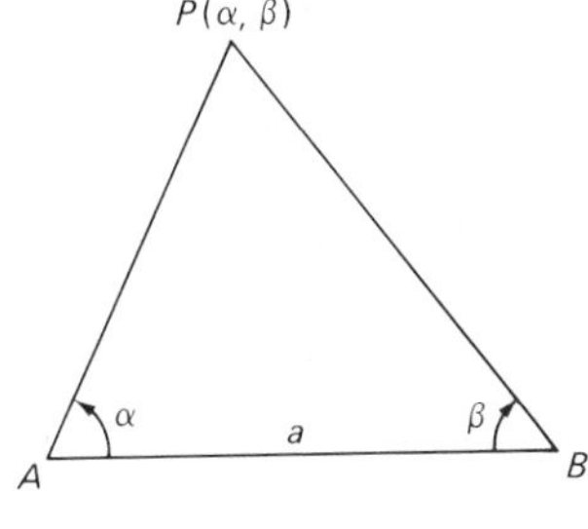

FIGURE 118

(e) Describe the latitude and longitude coordinate system on a spherical surface.

(f) A natural extension to space of the polar coordinate system of the plane consists in fixing an origin O and then taking as coordinates of a point P the length r of the radius vector OP and the latitude ϕ and longitude θ of P for the sphere having center O and radius OP. These coordinates are known as **spherical coordinates.** Find equations connecting the spherical coordinates (r,ϕ,θ) of a point P and rectangular Cartesian coordinates (x,y,z) of the point. Essentially such relations are found in the works of Lagrange (1736–1813).

(g) Design a coordinate system to locate points on (1) a circular cylindrical surface, and (2) a torus.

14.15 Line Coordinates

(a) Show that on a rectangular Cartesian frame of reference we can use the slope and the y-intercept of a line as its coordinates, or the length of the perpendicular upon the line from the origin and the angle which that perpendicular makes with the x-axis.

(b) The negative reciprocals, u and v, of the x- and y-intercepts of a line are known as the line's **Plücker** coordinates. Find the Plücker coordinates of the lines whose Cartesian equations are $5x + 3y - 6 = 0$ and $ax + by + 1 = 0$. Write the Cartesian equation of the line having Plücker coordinates (1,3).

(c) Show that the Plücker coordinates, u, v, of all lines passing through the point with Cartesian coordinates (2,3), satisfy the linear equation $2u + 3v + 1 = 0$. This equation is taken as the Plücker equation of the point (2,3). What are the Cartesian coordinates of the points whose Plücker equations are $5u + 3v - 6 = 0$ and $au + bv + 1 = 0$? Write the Plücker equation of the point having Cartesian coordinates (1,3).

14.16 Dimensionality

(a) Show that the plane is 4-dimensional in directed line segments.

(b) What is the dimensionality of the plane in directed line segments of a given length?

(c) Show that space is 4-dimensional in lines.

(d) Show that space is 3-dimensional in planes.

(e) Show that space is 4-dimensional in spheres.

What is the dimensionality of the manifold of

(f) Lines cutting across 2 skew lines?

(g) Lines through a point in space?

(h) Planes through a point in space?

(i) Circles in space through a fixed point?

(j) Spheres in space through a fixed point?

(k) All circles on a given sphere?

(l) All circles in space?

(m) All circles whose planes pass through a fixed line of space?

(n) All lines tangent to a given sphere?
(o) All planes tangent to a given sphere?

14.17 Abridged Notation

Establish the following theorems.

(a) If $\alpha = 0$ and $\beta = 0$ are the normal (perpendicular) equations of 2 distinct straight lines not passing through the origin, and if $m\alpha + n\beta = 0$, where m and n are constants, is a line through their point of intersection, then m/n is minus the ratio of the signed distance of a point on the line $m\alpha + n\beta = 0$ from line $\beta = 0$ to its signed distance from line $\alpha = 0$.

(b) If $\alpha = 0$ and $\beta = 0$ are the normal equations of 2 given nonparallel straight lines not passing through the origin, then $\alpha - \beta = 0$ and $\alpha + \beta = 0$ are the bisectors of the angles formed by the 2 given lines, the first being the bisector of the angle that contains the origin.

(c) If $\alpha = 0$ and $\beta = 0$ are normal equations of 2 nonparallel lines not passing through the origin, then $m\alpha + n\beta = 0$ and $n\alpha + m\beta = 0$, where m and n are constants, are isogonal lines for the angles formed by the 2 original lines.

(d) Let $\alpha = 0$, $\beta = 0$, $\gamma = 0$ be equations of the sides of a triangle. Then the 3 cevians $m\beta - n\gamma = 0$, $r\gamma - s\alpha = 0$, and $u\alpha - v\beta = 0$ are concurrent if and only if $mru = nsv$.

(e) If $\alpha = 0$, $\beta = 0$, $\gamma = 0$ are equations of the sides of a triangle, any 3 concurrent cevians can be written as $r\beta - s\gamma = 0$, $s\gamma - t\alpha = 0$, $t\alpha - r\beta = 0$.

(f) The bisectors of the angles of a triangle are concurrent.

(g) The altitudes of a triangle are concurrent.

(h) The medians of a triangle are concurrent.

(i) If 3 cevians of a triangle are concurrent, then so also are their 3 isogonal cevians.

(j) The locus of a point that moves so that the product of its distances from 1 pair of opposite sides of a quadrilateral is proportional to the product of its distances from the other pair of opposite sides is a conic passing through the vertices of the quadrilateral.

(k) The locus of a point that moves so that the product of its distances from 2 lines is proportional to the square of its distance from a third line is a conic tangent to the first 2 lines at the points where they are cut by the third line.

14.18 Homogeneous Coordinates

(a) Write the corresponding homogeneous Cartesian coordinates of the points (2,3), (−1,0), (0,7).

(b) Write homogeneous Cartesian coordinates of the points at infinity on the lines $x = y$, $3x + 2y - 7 = 0$, $ax + by + c = 0$.

(c) Write the corresponding nonhomogeneous Cartesian coordinates of the points (7,3,−4), (1,1,1), (0,−2,2).

(d) Write a nonhomogeneous Cartesian equation of a line passing through the ideal point (1,−2,0).
(e) Write a corresponding homogeneous Cartesian equation of the circle

$$x^2 + y^2 + 2fy + 2gx + c = 0.$$

(f) Show that every circle passes through the 2 imaginary ideal points $(1,-i,0)$ and $(1,i,0)$. These points are called the **circular points at infinity.**
(g) Show that any real conic passing through the 2 circular points at infinity is a circle.

14.19 Plücker's Numbers

Using Plücker's equations connecting the singular points of an algebraic curve, show that:

(a) Every conic is of class 2.
(b) The cubic curve $y = x^3$ is of class 3 and has a cusp at infinity.
(c) The cubic curve $y^2 = x^3$ is of class 6 and has an inflection at infinity.
(d) $\iota - \kappa = 3(m - n)$.

14.20 N-Dimensional Geometry

(a) How might one define, analytically, the *straight line* in hyperspace determined by the 2 points $(x_1, \ldots, x_n)$ and $(y_1, \ldots, y_n)$?
(b) How might one define *direction cosines* of the straight line in (a)?
(c) How might one define a point to *lie between* the 2 points in (a)?
(d) How might one define the *midpoint* of the line segment determined by the 2 points of (a)?
(e) Justify the definition given in Section 14–6 of the cosine of the angle between 2 line segments in hyperspace; that is, show that $0 \leqq |\cos\theta| \leqq 1$.
(f) If x, y, z are any 3 points in hyperspace, and if $d(x,y)$ denotes the distance between points x and y, show that
   1. $d(x,y) \geqq 0$,
   2. $d(x,y) = 0$ if and only if $x = y$,
   3. $d(x,y) = d(y,x)$,
   4. $d(x,z) \leqq d(x,y) + d(y,z)$.

14.21 Gaussian Curvature

(a) Is there a quadric surface whose total curvature is everywhere positive? Everywhere negative? Everywhere zero? In some places positive and in others negative?
(b) Show that a pair of applicable surfaces have their points in 1-to-1 correspondence such that at pairs of corresponding points the total curvatures of the 2 surfaces are equal.
(c) Show that when 1 surface is bent into another surface, the geodesics of the first surface go into geodesics of the second surface.

(d) Show that a sphere of radius r has constant positive total curvature equal to $1/r^2$.
(e) Show that a plane has constant zero total curvature.
(f) Show that a cylindrical surface has constant zero total curvature. Is a cylindrical surface applicable to a plane?
(g) Show that if a surface is applicable upon itself in all positions, its total curvature must be constant.
(h) Show that the only surfaces upon which free mobility of figures is possible are those of constant total curvature.
(i) Show that a sphere is not applicable to a plane. (This is why, in terrestrial map-making, some sort of distortion in the map is necessary.)

14.22 The Tractoid

The graph of $y = k \cosh (x/k)$ is the **catenary,** the form assumed by a perfectly flexible inextensible chain of uniform density hanging from 2 supports not in the same vertical line. Let this catenary (see Figure 119) cut the y-axis in A; let P be any point on the curve, and let F be the foot of the ordinate through P; let the tangent to the curve at P cut the x-axis in T, and let Q be the foot of the perpendicular from F on PT.

(a) With simple calculus, show that QF is constant and equal to k.
(b) With the aid of integral calculus, show that QP is equal to the length of the arc AP.
(c) Show that if a string AP is unwound from the catenary, the tracing end A will describe a curve AQ having the property that the length of the tangent QF is constant and equal to k. In other words, the locus of Q, which is an involute of the catenary, is a tractrix.
(d) It can be shown that for a surface of revolution the principal curvatures (see Section 14–7, p. 557) at a point Q on the surface are the curvature of the meridian through Q and the curvature of the section through Q that is normal to the meridian through Q. If the normal to the surface at Q meets the axis of revolution of the surface in T, then the latter

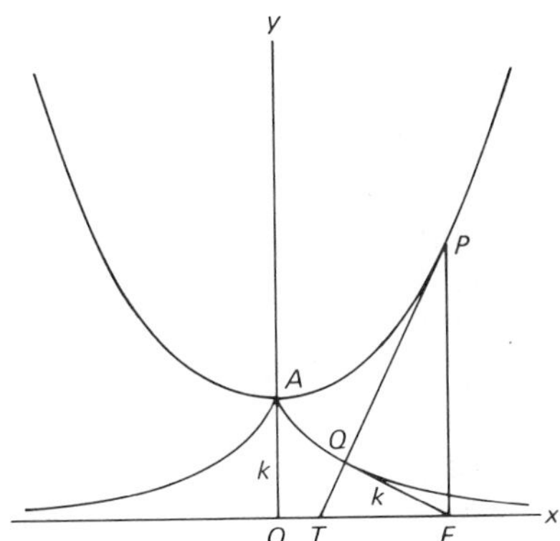

FIGURE 119

curvature is known to be equal to $1/QT$. Show that the principal curvatures at Q of the tractoid obtained by revolving the tractrix of (c) about the x-axis are given by $1/QP$ and $1/QT$.

(e) Show that the curvature (see Problem Study 14.21) of the tractoid of (d) is constant, and everywhere equal to $-1/k^2$.

14.23 The Erlanger Programm

(a) Show that the set of all projective transformations of the projective plane onto itself that carry a fixed line of the plane (call it the **line at infinity**) onto itself constitutes a transformation group. (The geometry associated with this group is known as **plane affine geometry.**)

(b) Show that the set of all projective transformations of the projective plane onto itself that carry a fixed line of the plane onto itself and a fixed point of the plane not on the fixed line into itself constitutes a transformation group. (The geometry associated with this group is known as **plane centro-affine geometry.**)

(c) Show that we have the following nesting of geometries:

$$\{\text{Euclidean metric, equiform, centro-affine, affine, projective}\},$$

wherein the transformation group of any one of the geometries is a subgroup of the transformation group of any one of the following geometries of the sequence.

(d) Show that the set of all projective transformations of the projective plane onto itself that carry a given circle S of the plane onto itself and the interior of S onto itself constitutes a transformation group. (With appropriate definitions of distance and angular measurement, it can be shown that the geometry associated with this transformation group is equivalent to plane Lobachevskian metric geometry.)

14.24 Mysticism and Absurdity in the Early Calculus

(a) One of the ablest criticisms of the faulty foundation of the early calculus came from the eminent metaphysician Bishop George Berkeley (1685–1753), who insisted that the development of the calculus by Newton involved the logical fallacy of a *shift in the hypothesis*. Point out the shift in the hypothesis in Newton's following determination of the derivative (or fluxion, as he called it) of x^3. We here paraphrase Newton's treatment as given in his *Quadrature of Curves* of 1704:

In the same time that x, by growing, becomes $x + o$, the power x^3 becomes $(x + o)^3$, or

$$x^3 + 3x^2o + 3xo^2 + o^3,$$

and the growths, or increments,

$$o \text{ and } 3x^2o + 3xo^2 + o^3$$

are to each other as

$$1 \text{ to } 3x^2 + 3xo + o^2.$$

Now let the increments vanish, and their last proportion will be 1 to $3x^2$, whence the rate of change of x^3 with respect to x is $3x^2$.

(b) Explain Bishop Berkeley's sarcastic description of derivatives as "ghosts of departed quantities."

(c) Discuss the following postulate made by Johann Bernoulli to sanction operations like that illustrated in (a) above: "A quantity which is increased or decreased by an infinitely small quantity is neither increased nor decreased."

14.25 Early Difficulties with Infinite Series

Seventeenth-century and eighteenth-century mathematicians had little understanding of infinite series. They often applied, to such series, operations that hold for *finite* series but apply to *infinite* series only under certain restrictions. Not being aware of the restrictions, the result was that paradoxes arose in work with infinite series.

(a) A bothersome series in the early days of the calculus was the alternating series

$$1 - 1 + 1 - 1 + 1 - 1 + \cdots,$$

and much discussion arose as to the sum S that should be assigned to this series. Show that the grouping

$$(1 - 1) + (1 - 1) + (1 - 1) + \cdots$$

leads to $S = 0$, and that the grouping

$$1 - (1 - 1) - (1 - 1) - (1 - 1) - \cdots$$

leads to $S = 1$. Some argued that since the sums 0 and 1 are equally probable, the correct sum of the series is the average value 1/2. Show that this value, too, can be obtained in a purely formal manner by the grouping

$$1 - (1 - 1 + 1 - 1 + 1 - 1 + \cdots).$$

(b) The binomial expansion

$$(a + b)^n = a^n + C(n,1)a^{n-1}b + C(n,2)a^{n-2}b^2 + C(n,3)a^{n-3}b^3 + \cdots,$$

where

$$C(n,r) = \frac{n(n-1)(n-2)\cdots(n-r+1)}{(1)(2)(3)\cdots(r)},$$

holds only under certain restrictions. That is, the series on the right converges to the expression on the left only under certain restrictions on a, b, and n. Not knowing these restrictions, and applying the expansion as though universally true, can lead to paradoxes. Obtain such a paradox (as did Euler) by formally applying the binomial expansion to $(1 - 2)^{-1}$.

(c) By dividing $1 - x$ into x and $x - 1$ into x, and then adding the results, obtain the ridiculous result found by Euler:

$$\cdots + \frac{1}{x^2} + \frac{1}{x} + 1 + x + x^2 + \cdots = 0$$

for all x different from 0 and 1.

(d) Explain the following paradox. Let S denote the sum of the *convergent* series

$$\frac{1}{(1)(3)} + \frac{1}{(3)(5)} + \frac{1}{(5)(7)} + \cdots.$$

Then

$$\begin{aligned} S &= \left(\frac{1}{1} - \frac{2}{3}\right) + \left(\frac{2}{3} - \frac{3}{5}\right) + \left(\frac{3}{5} - \frac{4}{7}\right) + \cdots \\ &= 1 - \frac{2}{3} + \frac{2}{3} - \frac{3}{5} + \frac{3}{5} - \frac{4}{7} + \cdots = 1, \end{aligned}$$

since all terms after the first cancel out. Again

$$\begin{aligned} S &= \frac{\left(\frac{1}{1} - \frac{1}{3}\right)}{2} + \frac{\left(\frac{1}{3} - \frac{1}{5}\right)}{2} + \frac{\left(\frac{1}{5} - \frac{1}{7}\right)}{2} + \cdots \\ &= \frac{1}{2} - \frac{1}{6} + \frac{1}{6} - \frac{1}{10} + \frac{1}{10} - \frac{1}{14} + \cdots = \frac{1}{2}, \end{aligned}$$

since all terms after the first cancel out. It follows that $1 = 1/2$.

14.26 Some Paradoxes in Elementary Algebra

When the theory of a mathematical operation is only poorly understood, there is the danger that the operation will be applied in a blindly formal and perhaps illogical manner. The performer, not aware of possible limitations upon the operation, is likely to use the operation in instances in which it does not necessarily apply. This is essentially what happened in analysis during the century following the invention of the calculus, with the result that a gradual accumulation of absurdities resulted. The present Problem Study illustrates how such absurdities can arise in elementary algebra when certain algebraic operations are performed without realization of limitations upon those operations.

(a) Explain the following paradox:

Certainly

$$3 > 2.$$

Multiplying both sides by log (1/2), we find

$$3 \log \left(\frac{1}{2}\right) > 2 \log \left(\frac{1}{2}\right)$$

or

$$\log \left(\frac{1}{2}\right)^3 > \log \left(\frac{1}{2}\right)^2,$$

whence

$$\left(\frac{1}{2}\right)^3 > \left(\frac{1}{2}\right)^2 \text{ or } \frac{1}{8} > \frac{1}{4}.$$

(b) Explain the following paradox:

Clearly $(-1)^2 = (+1)^2$. Taking the logarithm of each side, we have $\log (-1)^2 = \log (1)^2$. Therefore, $2 \log (-1) = 2 \log 1$, or $-1 = 1$.

(c) Most students of elementary algebra will agree to the following theorem: *If 2 fractions are equal and have equal numerators, then they also have equal denominators*. Now consider the following problem. We wish to solve the equation

$$\frac{x + 5}{x - 7} - 5 = \frac{4x - 40}{13 - x}.$$

Combining the terms on the left side, we find

$$\frac{(x + 5) - 5(x - 7)}{x - 7} = \frac{4x - 40}{13 - x}$$

or

$$\frac{4x - 40}{7 - x} = \frac{4x - 40}{13 - x}.$$

By the above theorem, it follows that $7 - x = 13 - x$, or, upon adding x to both sides, that $7 = 13$. What is wrong?

(d) Find the fallacy in the following proof by mathematical induction: $P(n)$: *All numbers in a set of n numbers are equal to one another.*

1. $P(1)$ is obviously true.
2. Suppose k is a natural number for which $P(k)$ is true. Let $a_1, a_2, \ldots, a_k, a_{k+1}$ be any set of $k + 1$ numbers. Then, by the supposition, $a_1 = a_2 = \cdots = a_k$ and $a_2 = \cdots = a_k = a_{k+1}$. Therefore $a_1 = a_2 = \cdots = a_k = a_{k+1}$, and $P(k + 1)$ is true.

It follows that $P(n)$ is true for all natural numbers n.

(e) Find the fallacy in the following proof by mathematical induction: $P(n)$: *If a and b are any two natural numbers such that max(a,b) = n, then a = b.* [*Note:* By max(a,b), when $a \neq b$, is meant the larger of the 2 numbers a and b. By max(a,a) is meant the number a. Thus, max(5,7) = 7, max(8,2) = 8, max(4,4) = 4.]

1. $P(1)$ is obviously true.
2. Suppose k is a natural number for which $P(k)$ is true. Let a and b be any 2 natural numbers such that $\max(a,b) = k + 1$, and consider $\alpha = a - 1, \beta = b - 1$. Then $\max(\alpha,\beta) = k$, whence, by the supposition, $\alpha = \beta$. Therefore, $a = b$ and $P(k + 1)$ is true.

It follows that $P(n)$ is true for all natural numbers n.

(f) Explain the concluding 3 paradoxes involving square-root radicals:

1. Since $\sqrt{a}\sqrt{b} = \sqrt{ab}$, we have

$$\sqrt{-1}\sqrt{-1} = \sqrt{(-1)(-1)} = \sqrt{1} = 1.$$

But, by definition, $\sqrt{-1}\sqrt{-1} = -1$. Hence $-1 = +1$.

2. We have, successively,

$$\sqrt{-1} = \sqrt{-1},$$

$$\sqrt{\frac{1}{-1}} = \sqrt{\frac{-1}{1}},$$

$$\frac{\sqrt{1}}{\sqrt{-1}} = \frac{\sqrt{-1}}{\sqrt{1}},$$

$$\sqrt{1}\sqrt{1} = \sqrt{-1}\sqrt{-1},$$

$$1 = -1.$$

3. Consider the following identity, which holds for all values of x and y:

$$\sqrt{x - y} = i\sqrt{y - x}.$$

Setting $x = a$, $y = b$, where $a \neq b$, we find

$$\sqrt{a - b} = i\sqrt{b - a}.$$

Now setting $x = b$, $y = a$, we find

$$\sqrt{b - a} = i\sqrt{a - b}.$$

Multiplying the last 2 equations, member by member, we get

$$\sqrt{a - b}\,\sqrt{b - a} = i^2\sqrt{b - a}\,\sqrt{a - b}.$$

Dividing both sides by $\sqrt{a - b}\,\sqrt{b - a}$, we finally get

$$1 = i^2, \quad \text{or} \quad 1 = -1.$$

14.27 Some Paradoxes in Calculus

(a) By standard procedure we find

$$\int_{-1}^{1} \frac{dx}{x^2} = \left[-\frac{1}{x}\right]_{-1}^{1} = -1 - 1 = -2.$$

The function $y = 1/x^2$ is never negative, however; hence, the above "evaluation" cannot be correct.

(b) Let e denote the eccentricity of the ellipse $x^2/a^2 + y^2/b^2 = 1$. It is well known that the length r of the radius vector drawn from the left-hand focus of the ellipse to any point $P(x,y)$ on the curve is given by $r = a + ex$. Now $dr/dx = e$. Since there are no values of x for which dr/dx vanishes, it follows that r has no maximum or minimum. But the only closed curve for which the radius vector has no maximum or minimum is a circle. It follows that every ellipse is a circle.

(c) Consider the isosceles triangle ABC of Figure 120, in which base $AB = 12$ and altitude $CD = 3$. Surely there is a point P on CD such that

$$S = PC + PA + PB$$

is a minimum. Let us try to locate this point P. Denote DP by x. Then $PC = 3 - x$ and $PA = PB = (x^2 + 36)^{1/2}$. Therefore

$$S = 3 - x + 2(x^2 + 36)^{1/2},$$

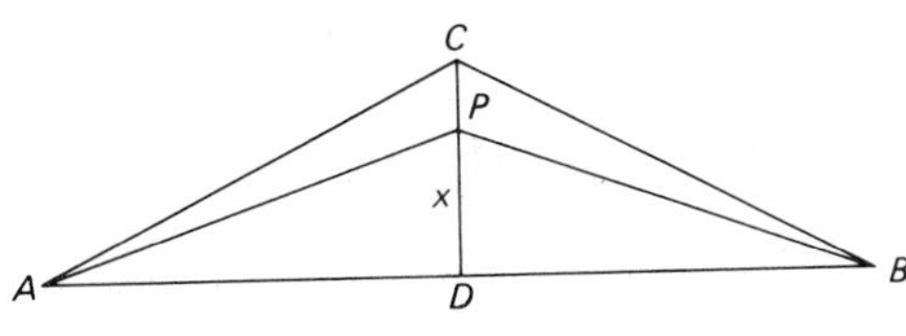

FIGURE 120

and

$$\frac{dS}{dx} = -1 + 2x(x^2 + 36)^{-1/2}.$$

Setting $dS/dx = 0$, we find $x = 2\sqrt{3} > 3$, and P lies outside the triangle on DC produced. Hence, there is no point on the segment CD for which S is a minimum.

(d) Consider the integral

$$I = \int \sin x \cos x \, dx.$$

Then we have

$$I = \int \sin x(\cos x \, dx) = \int \sin x \, d(\sin x) = \frac{\sin^2 x}{2}.$$

Also

$$I = \int \cos x(\sin x \, dx) = -\int \cos x \, d(\cos x) = -\frac{\cos^2 x}{2}.$$

Therefore

$$\sin^2 x = -\cos^2 x,$$

or

$$\sin^2 x + \cos^2 x = 0.$$

But, for any x,

$$\sin^2 x + \cos^2 x = 1.$$

(e) Since

$$\int \frac{dx}{x} = \int \frac{-dx}{-x},$$

we have $\log x = \log(-x)$ or $x = -x$, whence $1 = -1$.

14.28 A Continuous Curve Having No Tangents

It is well known that a continuous curve can be defined geometrically as the limit of a sequence of polygonal curves, and this process has been used by a number of mathematicians to produce a continuous curve that has no tangent

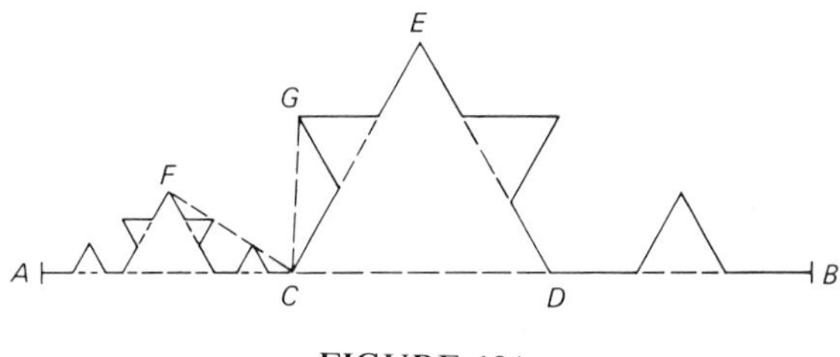

FIGURE 121

or half tangent at any of its points. We here consider such a curve created by the Swedish mathematician Helge von Koch (1870–1924).

Divide the horizontal line segment AB (see Figure 121) into 3 equal parts by the points C and D; on the middle part, CD, construct an equilateral triangle CED on the left side of the directed segment AB, and then efface the open segment CD. Now effect the same construction on each of the directed segments AC, CE, ED, DB. Repeat the construction indefinitely. The limit approached by the figure is the **Koch curve.**

(a) Considering a tangent to a curve at a point P of the curve as the limiting position, if such exists, of a secant line through P and a neighboring point Q of the curve as Q moves along the curve into coincidence with P, show that the Koch curve of Figure 121 has no tangent at the point C.

(b) Show that the Koch curve is infinite in length.

(c) On each side of an equilateral triangle construct, exterior to the triangle, a Koch curve. The resulting closed curve is sometimes called the **snowflake curve.** Show that the snowflake curve is a simple closed continuous curve of infinite length bounding a finite area.

(d) Let T_1 be a horizontal planar equilateral triangular region. Divide T_1 into 4 congruent parts by joining the midpoints of the sides of T_1. On the central piece, construct a regular tetrahedron lying above the plane of T_1; efface the central piece of T_1; denote the resulting surface by T_2. Describe a continuation of this process that will yield a continuous tangentless surface in 3-space.

14.29 Algebraic and Transcendental Numbers

A complex number is said to be **algebraic** if it is a root of some polynomial having rational coefficients; otherwise, it is said to be **transcendental.** It was F. Lindemann (1852–1939) who first proved, in 1882, that π is transcendental.

(a) Show that every rational number is an algebraic number and hence that every real transcendental number is irrational.

(b) Is every irrational number a transcendental number?

(c) Is the imaginary unit i algebraic or transcendental?

(d) Using Lindemann's result, show that $\pi/2$ is transcendental.

(e) Using Lindemann's result, show that $\pi + 1$ is transcendental.

(f) Using Lindemann's result, show that $\sqrt{\pi}$ is transcendental.

(g) Generalize (d), (e), and (f).

(h) Show that an algebraic number is a root of a polynomial having *integral* coefficients.

14.30 Bounds

A real number a is called an upper bound of a nonempty set M of real numbers if for each number m of M we have $m \leqq a$, and a is called a least upper bound of M if $a < b$ whenever b is any other upper bound of M. A basic and important property of the real number system asserts that *if a nonempty set of real numbers has an upper bound, then it has a least upper bound.*

(a) Give a definition of a lower bound and of a greatest lower bound of a nonempty set of real numbers.

(b) Prove that a nonempty set of real numbers can have at most one least upper bound and at most one greatest lower bound.

(c) Give an example of a nonempty set M of real numbers that has the following:

1. Both an upper and a lower bound.
2. An upper bound but no lower bound.
3. A lower bound but no upper bound.
4. Neither an upper nor a lower bound.
5. A least upper bound that is in the set M.
6. A least upper bound that is not in the set M.

(d) Prove that if a nonempty set M of real numbers has a lower bound, then it has a greatest lower bound.

(e) Let M be a nonempty set of real numbers, and let t be any fixed positive real number. Let N be the set of numbers of the form tx, where x is in M. Show that if b is the least upper bound of M, then tb is the least upper bound of N.

(f) Let M and N be two nonempty sets of real numbers having a and b, respectively, as least upper bounds. Let P be the set of all numbers of the form $x + y$, where x is in M and y is in N. Show that $a + b$ is the least upper bound of P.

(g) Let M be the set of real numbers

$$x_n = (-1)^n(2 - 4/2^n), \qquad n = 1, 2, \ldots .$$

Find the least upper bound and greatest lower bound of M. Do the same for the set of numbers

$$y_n = (-1)^n + 1/n, \qquad n = 1, 2, \ldots .$$

(h) If we are restricted to just the rational numbers, does the existence of an upper bound of a nonempty set M necessarily imply the existence of a least upper bound of M?

(i) If we are restricted to only the nonzero real numbers, does the existence of an upper bound of a nonempty set M necessarily imply the existence of a least upper bound of M?

14.31 Prime Numbers

(a) Find, by the sieve of Eratosthenes, all the primes below 500.

(b) Prove that a positive integer p is prime if it has no prime factor not exceeding the greatest integer whose square does not exceed p. This theorem says that, in the elimination process of the sieve of Eratosthenes, we may stop as soon as we reach a prime $p > \sqrt{n}$, for the cancellation of every pth number from p will merely be a repetition of cancellations already effected. Thus, in finding the primes less than 500, we may stop after crossing off every nineteenth number from 19, since the next prime, 23, is greater than $\sqrt{500}$.

(c) Compute $(A_n \log_e n)/n$ for $n = 500$, 10^8, and 10^9.

(d) Prove that there can always be found n consecutive composite integers, however great n may be.

(e) How many pairs of twin primes are there less than 100?

(f) Express each even positive integer less than 100, other than 2, as the sum of 2 primes.

(g) Show that the formulas $2 + \sin^2(n\pi/2)$, $3(\cos 2n\pi)$, and $3(n^0)$ yield prime numbers for all positive integral values of n.

(h) Show that $n^2 + n + 17$ is prime for all integral n from 1 to 16 and that $2n^2 + 29$ is prime for all integral n from 1 to 28.

(i) Show that 11 is the only palindromic prime containing an even number of digits.

(j) Find all 15 three-digit palindromic primes.
(It is not known if there are infinitely many palindromic primes.)

Essay Topics

14/1 Peaucellier's cell and Hart's contraparallelogram.

14/2 Linkages.

14/3 Augustus Ferdinand Möbius (1790–1868).

14/4 Karl Feuerbach (1800–1834).

14/5 William Kingdon Clifford (1845–1879).

14/6 Charles Lutwidge Dodgson (1832–1898).

14/7 Giuseppe Peano (1858–1932).

14/8 Pole-polar theory and the principle of duality.

14/9 A self-dual postulate set for plane projective geometry.

14/10 A principle of duality in the study of spherical triangles.

14/11 The Malfatti problem.

14/12 The tesseract, or hypercube.

14/13 The intrinsic versus the extrinsic geometry of surfaces.

14/14 Klein's years as chairman of the mathematics department at Göttingen.

14/15 Geometry as the theory of invariants and algebra as the theory of structure.

14/16 Riemann's probationary lecture of 1854.

14/17 Poincaré as a popular writer.

14/18 An analytic geometry without coordinates or a frame of reference.

14/19 The importance of existence theorems.

14/20 Developments in analysis in the nineteenth century motivated by internal factors in mathematics.

14/21 Some nineteenth-century mathematicians who were outstanding in both research and teaching.

14/22 Jacob Steiner and Julius Plücker contrasted.

14/23 The Cantor-Kronecker dispute.

14/24 Why there have been so few eminent women in mathematics.

Bibliography

ALTSHILLER-COURT, NATHAN *College Geometry, An Introduction to the Modern Geometry of the Triangle and the Circle*. 2d ed. Revised and enlarged, containing historical and bibliographical notes. New York: Barnes & Noble, 1952.

——— *Modern Pure Solid Geometry*. 2d ed., containing historical and bibliographical notes. New York: Chelsea, 1964.

BARON, MARGARET E. *The Origins of the Infinitesimal Calculus*. Oxford: Pergamon Press, 1969.

BELL, E. T. *Men of Mathematics*. New York: Simon and Schuster, 1937.

BIRKHOFF, GARRETT, ed. *A Source Book in Classical Analysis*. Cambridge, Mass.: Harvard University Press, 1973.

BOYER, C. B. *The History of the Calculus and Its Conceptual Development*. New York: Dover Publications, 1949.

——— *History of Analytic Geometry*. New York: *Scripta Mathematica,* 1956.

BREWER, J. W., and MARTHA SMITH *Emmy Noether: A Tribute to Her Life and Work*. New York: Marcel Dekker, 1981.

BURRILL, C. W. *Foundations of Real Numbers*. New York: McGraw-Hill, 1967.

COHEN, L. W., and GERTRUDE EHRLICH *The Structure of the Real Number System*. Princeton, N.J.: Van Nostrand Reinhold, 1963.

COOKE, ROGER *The Mathematics of Sonya Kovalevskaya*. New York: Springer-Verlag, 1984.

COOLIDGE, J. L *A Treatise on the Circle and the Sphere*. Oxford: The Clarendon Press, 1916.

——— *A History of Geometrical Methods*. Oxford: The Clarendon Press, 1940.

DEDEKIND, RICHARD *Essays on the Theory of Numbers*. Translated by W. W. Beman. Chicago: Open Court, 1901. Reprinted by Dover Publications, New York, 1963.

DICK, AUGUSTE *Emmy Noether 1882–1935*. Basel, Switzerland: Birkhauser Verlad, 1970.

DICKSON, L. E. *History of the Theory of Numbers*. 3 vols. New York: Chelsea, 1962.

——— "Construction with ruler and compasses." In *Monographs on Topics of Modern Mathematics*. Edited by J. W. A. Young. New York: Longmans, Green, 1924.

DODGE, C. W. *Numbers and Mathematics*. 2d ed. Boston: Prindle, Weber & Schmidt, 1975.

EVES, HOWARD *A Survey of Geometry*. 2 vols. Boston: Allyn and Bacon, 1972 and 1965.

——— *Foundations and Fundamental Concepts of Mathematics*. 3rd ed. Boston: PWS-KENT Publishing Company, 1990.

GRATTAN-GUINESS, IVOR *The Development of the Foundations of Mathematical Analysis from Euler to Riemann*. Cambridge, Mass.: M.I.T. Press, 1970.

HANCOCK, HARRIS *Foundations of the Theory of Algebraic Numbers*. 2 vols. New York: Dover Publications, 1931 and 1932.

——— *Development of the Minkowski Geometry of Numbers*. 2 vols. New York: Dover Publications, 1939.

HUDSON, H. P. *Ruler and Compasses*. Reprinted in *Squaring the Circle, and Other Monographs*. New York: Chelsea, 1953.

KAZARINOFF, N. C. *Ruler and the Round*. Boston: Prindle, Weber & Schmidt, 1970.

KEMPE, A. B. *How to Draw a Straight Line; A Lecture on Linkages*. Reprinted in *Squaring the Circle, and Other Monographs*. New York: Chelsea, 1953.

KENNEDY, D. H. *Little Sparrow: A Portrait of Sophia Kovalevsky*. Athens, Ohio: Ohio University Press, 1983.

KENNEDY, HUBERT, Jr. *Peano: Life and Works of Giuseppe Peano*. Dordrecht, Holland: D. Reidel, 1980.

———, ed. *Selected Works of Giuseppe Peano*. Toronto: University of Toronto Press, 1973.

KLEIN, FELIX *Elementary Mathematics from an Advanced Standpoint*. 2 vols. Translated by E. R. Hedrick and C. A. Noble. New York: Dover Publications, 1939 and 1945.

——— *Famous Problems of Elementary Geometry*. Translated by W. W. Beman and D. E. Smith. Reprinted in *Famous Problems, and Other Monographs*. New York: Chelsea, 1955.

KOBLITZ, ANN *A Convergence of Lives. Sofia Kovaleskaia: Scientist, Writer, Revolutionary*. Boston: Birkhauser, 1983.

KOSTOVKII, A. N. *Geometrical Constructions Using Compasses Only*. Translated by Halina Moss. New York: Blaisdell, 1961.

KOVALEVSKAYA, S. *A Russian Childhood*. Translated by B. Stillman. Berlin: Springer-Verlag, 1978.

LANDAU, EDMUND *Foundations of Analysis*. Translated by F. Steinhardt. New York: Chelsea, 1951.

LANG, SERGE *Algebraic Numbers*. Reading, Mass.: Addison-Wesley, 1964.

MESCHOWSKI, HERBERT *Ways of Thought of Great Mathematicians*. San Francisco: Holden-Day, 1964.

——— *Evolution of Mathematical Thought*. San Francisco: Holden-Day, 1965.

MOORE, GREGORY *Zermelo's Axiom of Choice: Its Origins, Development and Influence*. New York: Springer-Verlag, 1982.

MUIR, JANE *Of Men and Numbers, The Story of the Great Mathematicians*. New York: Dodd, Mead, 1961.

NIVEN, IVAN *Irrational Numbers*. Carus Mathematical Monograph No. 11. New York: John Wiley, 1956.

POINCARÉ, HENRI *The Foundations of Science*. Translated by G. B. Halsted. Lancaster, Pa.: The Science Press, 1946.

POLLARD, HARRY *The Theory of Algebraic Numbers*. Carus Mathematical Monograph No. 9. New York: John Wiley, 1950.

PRASAD, GANESH *Some Great Mathematicians of the Nineteenth Century*. 2 vols. Benares: Benares Mathematical Society, 1933 and 1934.

PUKERT, WALTER, and H. J. ILGAUDS *Georg Cantor, 1845–1918*. Boston: Birkhauser, 1981.

SANDHEIMER, ERNST, and ALAN ROGERSON *Numbers and Infinity, A Historical Account of Mathematical Concepts*. New York: Cambridge University Press, 1981.

SIEGEL, C. L. *Transcendental Numbers*. Princeton, N.J.: Princeton University Press, 1949.

SMITH, D. E. *A Source Book in Mathematics*. New York: McGraw-Hill, 1929.

SMOGORZHEVSKII, A. S. *The Ruler in Geometrical Construction*. Translated by Halina Moss. New York: Blaisdell, 1961.

STEINER, JACOB *Geometrical Constructions with a Ruler*. Translated by M. E. Stark. Edited by R. C. Archibald. New York: *Scripta Mathematica*, 1950.

TEMPLE, GEORGE *100 Years of Mathematics, A Personal Viewpoint*. New York: Springer-Verlag, 1981.

THURSTON, H. A. *The Number-System*. New York: Interscience, 1956.

TURNBULL, H. W. *The Great Mathematicians*. New York: New York University Press, 1961.

WAISMANN, FRIEDRICH *Introduction to Mathematical Thinking: The Formation of Concepts in Modern Mathematics*. Translated by T. J. Benac. New York: Frederick Ungar, 1951.

YATES, R. C. *Geometrical Tools*. St. Louis: Educational Publishers, 1949.

——— *The Trisection Problem*. Ann Arbor, Mich.: Edward Brothers, 1947.

YOUNG, J. W. A., ed. *Monographs on Topics of Modern Mathematics Relevant to the Elementary Field*. New York: Dover Publications, 1955.

Cultural Connection 

THE ATOM AND THE SPINNING WHEEL

The Twentieth Century
(to accompany Chapter Fifteen)

Special problems confront anyone writing twentieth-century history. The twentieth century, unlike past epochs, is not yet over; consequently, we cannot know with certainty how it will look when complete. In the nine preceeding Cultural Connections, we have endeavored to identify each of the considered historical periods with a central theme. In prehistoric times, people lived as nomadic hunters and gatherers. We saw the Agricultural Revolution begin in ancient China, Egypt, and the Middle East at the dawn of recorded history. Democracy developed in Classical Greece, and grand empires arose in Rome and China. India and Arabia witnessed the birth of dynamic new religions that shaped the character of civilization there. With the Middle Ages came the fall of the Roman Empire and the emergence of a new, feudal, European culture. After A.D. 1500, European society expanded into other continents in the Age of Exploration. The eighteenth century witnessed the rise of the middle class, and in the nineteenth century, that middle class engineered the Industrial Revolution.

It is not possible to identify the twentieth century with such a single theme. If we look at history as a vast tapestry, we would see that the part depicting the present era has been only partially woven, and we could not yet identify the picture that will eventually take shape there. We can, however, examine the threads from which the tapestry is woven, and from these threads, and from the part of the panorama that is complete, guess what the finished picture will look like.

We can see much in the portion of the tapestry already woven. The great imperial powers of the nineteenth century fought a bloody "war to end all wars," World War I (1914–1918), which did not end all wars, but left the old industrial empires spent and broken. The Russian Revolution (1917) overthrew a centuries-old monarchy, replacing it with the world's first socialist state. Nationalists, meanwhile, established new regimes in Poland, Yugoslavia, Czechoslovakia, and Hungary in 1920. Fascists, self-righteous ultranationalist zealots, seized control of Germany, Spain, Italy, and Japan during a Great Depression in the 1930s. In the Holocaust, a vicious program of genocide against innocent people, European fascists imprisoned and brutally murdered millions of Jewish, homosexual, and other minority citizens. The German,

Italian, and Japanese fascists, giddy with power, embarked on a frenzy of conquests and wars that led to their ultimate defeat in World War II (1939–1945) at the hands of Great Britain, the Soviet Union, the United States (along with other nations), and underground bands of antifascist rebel partisans. The aftermath of World War II left the United States and the Soviet Union as the dominant world powers, their respective spheres of influence dividing the globe into western and eastern blocs.

The end of World War II also marked the beginning of the gradual disintegration of the old nineteenth-century colonial and economic empires. Dozens of newly independent nations came into being in Africa, Asia, the Pacific, and elsewhere, ranging in size from gigantic India and Indonesia (the former Dutch East Indies) to tiny Nauru and Grenada. These former colonies suffered from great disadvantages. Their former imperial masters had viewed them only as sources of raw materials and had been uninterested in promoting industrialization there. Poor, overpopulated, unindustrialized, and undereducated, these "Third-World" countries struggled against illiteracy, famine, and disease. They often went into debt, and their poverty made them ripe for revolutionary violence. Still, some Third-World nations made remarkable progress in improving the living standards of their peoples and achieving political stability. We can cite Saudi Arabia, Egypt, and China as relatively successful counterpoints to more turbulent places like Vietnam, Zaire, Nigeria, and Uganda, although we must acknowledge that, at least in the case of Vietnam, instability was in part imposed from without. Due to their numbers, Third-World nations have, as a bloc, been influential at the United Nations, a global assembly formed in the wake of World War II.

Thus we glimpse at part of the incomplete picture of the tapestry of the twentieth century. If we look at the threads that make up the weave of the tapestry, crisscrossing each other in vertical and horizontal contrasts, we can pick out two competing tendencies: the mechanistic and the organic. As historian and philosopher Carolyn Merchant has suggested in her book *The Death of Nature* (1980), there are two essential ways of looking at the world, both of which date back to Classical Greece. One view, the mechanistic, holds that nature and civilization work like machines, made up of component parts over which humankind has control and with which we must continually tinker. The other view, the organic, understands the world as a living whole, of which humankind is only a part, and that lies in a delicate, natural balance. Both world views are ancient; both are compatible with science; both are with us still.

In the twentieth century, for many, the atom came to symbolize the mechanistic world view. Harnessed to both destructive and constructive ends, atomic power represented both humanity's ultimate domination of nature and its potential for self-destruction. For Third-World countries, the atom was a constant reminder of the primacy of the superpowers, the United States and the Soviet Union, and of the Third World's secondary status. Atomic power was the result of the twentieth-century merger of pure science with technology, a merger that created an unprecedented demand for the fruits of science within government and business, and, consequently, thousands of jobs for scientists. Whether the atom will lead us to a mechanistic utopia, to atomic war, or to a hopelessly polluted environment we cannot yet say.

In the Third World, Indian leader Mohandas Gandhi (1869–1948) proposed the spinning wheel as the symbol of the organic philosophy. A simple machine, the spinning wheel is powered by human hands, not electricity, and to Gandhi represented a harmony between humanity and nature. Its round shape reminded Gandhi of the spherical earth; its symmetry bespoke the symbolic unity of humankind. Although the atom appears to have dominated the twentieth century, we can distinguish the influence of the spinning wheel as well: the break-up of the nineteenth century colonial empires; the civil rights movements in the United States and Africa since 1955; the environmental and women's movements in Europe, America, and Asia; the antinuclear crusade and the call for "appropriate technology"; the religious fundamentalism in the United States and Iran—all tend to see the world in nonmechanistic terms.

Neither can we completely separate the two ideas into easily identifiable opposing camps. No one represented the mutual interdependence of the atom and the spinning wheel better than the twentieth century's greatest scientist, Albert Einstein (1879–1955). Einstein worked in mechanics, yet was a compassionate humanist. He recognized the mechanical nature of the universe, but also, in his theory of relativity, understood it as a glorious interlocking whole. He helped harness the power of the atom but was also wise enough to warn the world of the perils of its misuse.

The twentieth century still has several years to go before it is over, and it will be years after that before we can evaluate it objectively. In describing a hero of an earlier time, English poet Alfred, Lord Tennyson, wrote, "much is taken, much abides." Future generations, when considering the twentieth century, will decide what to take, what will remain, and whether or not those of us alive today have lived up to the proud words found in the Charter of the United Nations: "to promote social progress and better standards of life in larger freedom, and for those ends to practice tolerance and live together in peace with one another as good neighbors. . . ."

Chapter 15

INTO THE TWENTIETH CENTURY

15–1 Logical Shortcomings of Euclid's "Elements"

Much of the mathematical work of the twentieth century has been devoted to examining the logical foundations and structure of the subject. This in turn led to the creation of **axiomatics,** or the study of postulate sets and their properties. Many of the basic concepts of mathematics underwent remarkable evolution and generalization, and such deeply fundamental subjects as set theory, abstract algebra, and topology were extensively developed. General set theory led to some profound and disturbing paradoxes that required urgent treatment. Logic itself, as the apparatus used in mathematics to obtain conclusions from accepted hypotheses, was carefully scrutinized, and *mathematical logic* came into being. The ties between logic and philosophy led to various important present-day schools of philosophy of mathematics. The computer revolution of the twentieth century also deeply affected a number of the branches of mathematics. All in all, the old view of "the tree of mathematics" became obsolete. Curiously enough, like so much of mathematics, most of these modern considerations trace their origins back to the work of the ancient Greeks and, in particular, to the great *Elements* of Euclid.

It would be very remarkable indeed if Euclid's *Elements,* being such an early and colossal attempt at the postulational method of presentation, should be free of logical blemishes. The searchlight of subsequent criticism has revealed many defects in the logical structure of the work. Perhaps the gravest of these defects are numerous tacit assumptions made by Euclid, assumptions not granted by his postulates. Thus, although Postulate P2 asserts that a straight line may be produced indefinitely,[1] it does not necessarily imply that a straight line is infinite, but merely that it is endless, or boundless. The arc of a great circle joining two points on a sphere may be produced indefinitely along the great circle, making the prolonged arc endless, but certainly not infinite. The great German mathematician Riemann, in his famous probationary lecture *Über die Hypothesen welche der Geometrie zu Grunde liegen,* of 1854, distinguished between the boundlessness and the infinitude of straight lines. There are numerous occasions, for instance in the proof of Proposition I 16, where Euclid unconsciously assumed the infinitude of straight lines. Again, Euclid tacitly assumed, in his proof of Proposition I 21 for example, that if a straight

[1] See Section 5–7 for statements of Euclid's axioms and postulates.

line enters a triangle at a vertex it must, if sufficiently produced, intersect the opposite side. Moritz Pasch (1843–1930) recognized the necessity of a postulate to take care of this situation. Another oversight of Euclid's geometry is the assumption of the existence of points of intersection of certain lines and circles. Thus, in Proposition I 1, it is assumed that circles with centers at the ends of a line segment and having the line segment as a common radius intersect, and do not, somehow or other, slip through each other with no common point. Some sort of continuity postulate, such as one later furnished by R. Dedekind, is needed to assure us of the existence of such a point of intersection. Also, Postulate P1 guarantees the existence of at least one straight line joining two points A and B, but does not assure us that there cannot be more than one such joining line. Euclid frequently assumed there is a unique line joining two distinct points. Objections can also be raised to the principle of superposition, used by Euclid, with apparent reluctance, to establish some of his early congruence theorems, although these objections can partially be met by Axiom A4.

Not only is Euclid's work marred by numerous tacit assumptions, but some of his preliminary definitions are also open to criticism. Euclid made an attempt to define all the technical terms of his discourse. Now, actually, it is as impossible to define *explicitly* all of the technical terms of a discourse as it is to prove all of the statements of the discourse, for a technical term must be defined by means of other technical terms, and these other terms by means of still others, and so on. In order to get started, and to avoid circularity of definition, one is forced to set down at the very start of the discourse a collection of primitive, or basic, technical terms whose meanings are not to be questioned. All subsequent technical terms of the discourse must be defined, ultimately, by means of these initial primitive ones. The postulates of the discourse are, then, in final analysis, assumed statements about the primitive terms. From this point of view, the primitive terms may be regarded as defined *implicitly,* in the sense that they are any things or concepts that satisfy the postulates, and this implicit definition is the only kind of definition that the primitive terms can receive.

In Euclid's development of geometry, the terms *point* and *line,* for example, could well have been included in a set of primitive terms for the discourse. At any rate, Euclid's definition of a point as "that which has no part" and a line as "length without breadth" are easily seen to be circular and therefore, from a logical viewpoint, woefully inadequate. One distinction between the Greek conception and the modern conception of the axiomatic method lies in this matter of primitive terms; in the Greek conception, there is no listing of the primitive terms. The excuse for the Greeks is that to them geometry was not just an abstract study, but an attempted logical analysis of idealized physical space. Points and lines were, to the Greeks, idealizations of very small particles and very thin threads. It is this idealization that Euclid attempted to express in some of his initial definitions. There are still other differences between the Greek and the modern views of the axiomatic method.

It was not until the end of the nineteenth century and the early part of the twentieth century, after the foundations of geometry had been subjected to an intensive study, that satisfactory postulate sets were supplied for Euclidean plane and solid geometry. Prominent among such sets are those of M. Pasch, G. Peano, M. Pieri, D. Hilbert, O. Veblen, E. V. Huntington, G. D. Birkhoff, and L. M. Blumenthal. Hilbert's set contains twenty-one postulates and has *point, straight line, plane, on, congruent,* and *between* as primitive terms; Pieri's set contains twenty postulates and has *point* and *motion* as primitive terms; Veblen's set contains sixteen postulates and has *point* and *order* as primitive terms; Huntington's set contains twenty-three postulates and has *sphere* and *inclusion* as primitive terms.

Since about the middle of the twentieth century, a number of authors and writing groups have attempted the task of producing textual materials for the high school geometry class wherein geometry is developed rigorously from a postulational base. In these attempts, usually either the Hilbert postulate set or the Birkhoff postulate set (often somewhat altered and/or augmented) is adopted.

15–2 Axiomatics[2]

It was largely the modern search for a logically acceptable postulate set of Euclidean geometry and the revelation furnished by the discovery of equally consistent non-Euclidean geometries that led to the development of axiomatics, or the study of postulate sets and their properties.

One of the pitfalls of working with a deductive system is too great a familiarity with the subject matter of the system. It is this pitfall that accounts for most of the blemishes in Euclid's *Elements*. In order to escape this pitfall, it is advisable to replace the primitive or undefined terms of the discourse by symbols, like x, y, z, and so forth. Then the postulates of the discourse become statements about these symbols and are thus devoid of concrete meaning; conclusions, therefore, are obtained upon a strictly logical basis without the intrusion of intuitive factors. The study of axiomatics considers properties of such sets of postulates.

Clearly, we cannot take as a postulate set *any* set of statements about the primitive terms. There are certain required and certain desired properties that a postulate set should possess. It is essential, for example, that the postulates be **consistent**—that is, that no contradictions can be deduced from the set.

The most successful method so far invented for establishing consistency of a postulate set is the method of models. A model of a postulate set is obtained if we can assign meanings to the primitive terms of the set that convert the postulates into true statements about the assigned concepts. There are two types of models—**concrete models** and **ideal models.** A model is said to be *concrete* if the meanings assigned to the primitive terms are objects and rela-

[2] For a fuller treatment of axiomatics see, for example, Chapter 6 of Howard Eves and C. V. Newsom, *An Introduction to the Foundations and Fundamental Concepts of Mathematics*. Revised edition. New York: Holt, Rinehart and Winston, 1965.

tions adapted from the real world, whereas a model is said to be *ideal* if the meanings assigned to the primitive terms are objects and relations adapted from some other postulational development.

When a concrete model has been exhibited, we feel that we have established the *absolute* consistency of our postulate system, for if contradictory theorems are implied by our postulates, then corresponding contradictory statements would hold in our concrete model. But contradictions in the real world we believe are impossible.

It is not always possible to set up a concrete model of a given postulate set. Thus, if the postulate set contains an infinite number of primitive elements, a concrete model would certainly be impossible, for the real world does not contain an infinite number of objects. In such instances, we attempt to set up an ideal model by assigning to the primitive terms of the postulate system A, say, concepts of some other postulate system B, in such a manner that the interpretations of the postulates of system A are logical consequences of the postulate system B. But now our test of consistency of the postulate set A can no longer claim to be an absolute test, but only a *relative* test. All we can say is that postulate set A is consistent if postulate set B is consistent, and we have reduced the consistency of system A to that of another system B.

Whether a postulate set might be consistent without our being able to establish the fact is one of the interesting open questions of axiomatics. Studies upon consistency have led to several disturbing and controversial results for those concerned with the foundations of mathematical knowledge. Proof of consistency by the method of models is an indirect process. It is conceivable that absolute consistency may be established by a direct procedure that endeavors to show that by following the rules of deductive inference no two theorems can be arrived at, from a given postulate set, that will contradict each other. In recent years, Hilbert considered, with only partial success, such a direct method.

A set of postulates is said to be **independent** if no postulate of the set is implied by the other postulates of the set. To show that any particular postulate of the set is independent, one must devise an interpretation of the primitive terms that falsifies the concerned postulate but that verifies each of the remaining postulates. It we are successful in finding such an interpretation, then the concerned postulate cannot be a logical consequence of the other postulates, for if it were a logical consequence of the other postulates, then the interpretation that converts all the other postulates into true propositions would also have to convert it into a true proposition. A test, along these lines, of the independence of an entire set of postulates can apparently be a lengthy business, for if there are n postulates in the set, n separate tests (one for each postulate) will have to be formulated. It was the matter of independence that was so important in connection with the non-Euclidean geometry.

A given body of material may be deducible from more than one postulate set. All that is required of two postulate sets $P^{(1)}$ and $P^{(2)}$, in order that they lead to the same development, is that the primitive terms in each be definable by means of the primitive terms of the other, and the postulates of each be deducible from the postulates of the other. Two such postulate sets are said to be **equivalent.** The notion of equivalent postulate sets arose in trying to find substitutes for Euclid's parallel postulate.

There are other properties of postulate sets studied in axiomatics besides those of consistency, independence, and equivalence. The subject is closely connected with symbolic logic and with the philosophy of mathematics. There have been, and are at present, many contributors to this field. Prominent among such contributors are Hilbert, Peano, Pieri, Veblen, Huntington, Russell, Whitehead, Gödel, and many others.

15–3 The Evolution of Some Basic Concepts

Following the development of the theory of sets by Georg Cantor toward the end of the nineteenth century, interest in that theory developed rapidly until today virtually every field of mathematics has felt its impact. Notions of space and the geometry of a space, for example, have been completely revolutionized by the theory of sets. Also, the basic concepts in analysis, such as those of limit, function, continuity, derivative, and integral, are now most aptly described in terms of set-theory ideas. Most important, however, has been the opportunity for new mathematical developments undreamed of fifty years ago. Thus, in companionship with the new appreciation of postulational procedures in mathematics, abstract spaces have been born, general theories of dimension and measure have been created, and the branch of mathematics called *topology* has undergone a spectacular growth. In short, under the influence of set theory, a considerable unification of traditional mathematics has occurred, and new mathematics has been created at an explosive rate.

To illustrate the historical evolution of basic mathematical concepts, let us first consider notions of space and the geometry of a space. These concepts have undergone marked changes since the days of the ancient Greeks. For the Greeks there was only one space and one geometry; these were absolute concepts. The space was not thought of as a collection of points, but rather as a realm, or locus, in which objects could be freely moved about and compared with one another. From this point of view, the basic relation in geometry was that of congruence or superposability.

With the advent of analytic geometry in the seventeenth century, space came to be regarded as a collection of points, and with the creation of the classical non-Euclidean geometries in the nineteenth century, mathematicians accepted the situation that there is more than one geometry. But space was still regarded as a locus in which figures can be compared with one another. The central idea became that of a group of congruent transformations of space onto itself, and a geometry came to be regarded as the study of those properties of configurations of points that remain unchanged when the enclosing space is subjected to these transformations. We have seen in Section 14–8 how this point of view was expanded by Felix Klein in his *Erlanger Programm* of 1872. In the *Erlanger Programm*, a geometry was defined as the invariant theory of a transformation group. This concept synthesized and generalized all earlier concepts of geometry and supplied a singularly neat classification of a large number of important geometries.

At the end of the nineteenth century, with the development of the idea of a branch of mathematics as an abstract body of theorems deduced from a set of postulates, each geometry became, from this point of view, a particular branch of mathematics. Postulate sets for a large variety of geometries were studied, but the *Erlanger Programm* was in no way upset, for a geometry could be regarded as a branch of mathematics that is the invariant theory of a transformation group.

In 1906, however, Maurice Fréchet (1878–1973) inaugurated the study of abstract spaces (see Problem Study 15.15), and very general geometries came into being that no longer necessarily fit into the neat Kleinian classification. A space became merely a set of objects, usually called *points,* together with a set of relations in which these points are involved, and a geometry became simply the theory of such a space. The set of relations to which the points are subjected is called the **structure** of the space, and this structure may or may not be explainable in terms of the invariant theory of a transformation group. Thus, through set theory, geometry received a further generalization. Although abstract spaces were first formally introduced in 1906, the idea of a geometry as the study of a set of points with some superimposed structure was really already contained in remarks made by Riemann in his great lecture of 1854. It is interesting that some of these new geometries have found valuable application in the Einstein theory of relativity, and in other developments of modern physics.

The concept of function, like the notions of space and geometry, has undergone a marked evolution, and a student of mathematics encounters various refinements of this evolution as his studies progress from the elementary courses of high school into the more advanced and sophisticated courses of the college postgraduate level.

The history of the term *function* furnishes another interesting example of the tendency of mathematicians to generalize and extend their concepts. The word *function,* in its Latin equivalent, seems to have been introduced by Leibniz in 1694, at first as a term to denote any quantity connected with a curve, such as the coordinates of a point on the curve, the slope of the curve, the radius of curvature of the curve, and so on. Johann Bernoulli, by 1718, had come to regard a function as any expression made up of a variable and some constants, and Euler, somewhat later, regarded a function as any equation or formula involving variables and constants. This latter idea is the notion of a function formed by most students of elementary mathematics courses. The Euler concept remained unchanged until Joseph Fourier (1768–1830) was led, in his investigations of heat flow, to consider so-called trigonometric series. These series involve a more general type of relationship between variables than had previously been studied, and, in an attempt to furnish a definition of function broad enough to encompass such relationships, Lejeune Dirichlet (1805–1859) arrived at the following formulation: A **variable** is a symbol that represents any one of a set of numbers; if two variables x and y are so related that whenever a value is assigned to x there is automatically assigned, by some rule or correspondence, a value to y, then we say y is a (single-valued) **function** of x. The variable x, to which values are assigned at will, is called the **independent variable,** and the variable y, whose values depend upon those of x, is called the **dependent variable.** The permissible values that x may assume constitute the **domain of definition** of the function, and the values taken on by y constitute the **range of values** of the function.

The student of mathematics used to meet the Dirichlet definition of function in his introductory course in calculus. The definition is a very broad one and does not imply anything regarding the possibility of expressing the relationship between x and y by some kind of analytic expression; it stresses the basic idea of a relationship between two sets of numbers.

Set theory has extended the concept of function to embrace relationships between any two sets of elements, be the elements numbers or anything else. Thus, in set theory, a **function f** is defined to be any set of ordered pairs of elements such that if $(a_1,b_1) \in f$, $(a_2,b_2) \in f$, and $a_1 = a_2$, then $b_1 = b_2$. The set A of all first elements of the ordered pairs is called the **domain (of definition)** of the function, and the set B of all second elements of the ordered pairs is called the **range (of values)** of the function. A functional relationship is thus nothing but a special kind of subset of the Cartesian product set $A \times B$. A **one-to-one correspondence** is, in its turn, a special kind of function, namely, a function f such that if $(a_1,b_1) \in f$, $(a_2,b_2) \in f$, and $b_1 = b_2$, then $a_1 = a_2$. If, for a functional relationship f, $(a,b) \in f$, we write $b = f(a)$.

The notion of function pervades much of mathematics, and since the early part of the present century, various influential mathematicians have advocated the employment of this concept as the unifying and central principle in the organization of elementary mathematics courses. The concept seems to form a natural and effective guide for the selection and development of textual material. There is no doubt of the value of a mathematics student's early acquaintance with the function concept.

15–4 Transfinite Numbers

The modern mathematical theory of sets is one of the most remarkable creations of the human mind. Because of the unusual boldness of some of the ideas found in its study, and because of some of the singular methods of proof to which it has given rise, the theory of sets is indescribably fascinating. Above this, the theory has assumed tremendous importance for almost the whole of mathematics. It has enormously enriched, clarified, extended, and generalized many domains of mathematics, and its role in the study of the foundations of mathematics is very basic. It also forms one of the connecting links between mathematics on the one hand and philosophy and logic on the other.

Two sets are said to be **equivalent** if and only if they can be placed in one-to-one correspondence. Two sets that are equivalent are said to have the same

cardinal number. The cardinal numbers of finite sets may be identified with the natural numbers. The cardinal numbers of infinite sets are known as **transfinite numbers,** and their theory was first developed by Georg Cantor in a remarkable series of articles beginning in 1874, and published, for the most part, in the German mathematics journals *Mathematische Annalen* and *Journal für Mathematik.* Prior to Cantor's study mathematicians accepted only one infinity, denoted by some symbol like ∞, and this symbol was employed indiscriminately to indicate the "number" of elements in such sets as the set of all natural numbers and the set of all real numbers. With Cantor's work, a whole new outlook was introduced, and a scale and arithmetic of infinities was achieved.

The basic principle that equivalent sets are to bear the same cardinal number presents us with many interesting and intriguing situations when the sets under consideration are infinite sets. Galileo Galilei observed as early as the latter part of the sixteenth century that, by the correspondence $n \leftrightarrow 2n$, the set of all positive integers can be placed in one-to-one correspondence with the set of all even positive integers. Hence, the same cardinal number should be assigned to each of these sets, and, from this point of view, we must say that there are as many even positive integers as there are positive integers in all. It is observed at once that the Euclidean postulate that states that the whole is greater than a part cannot be tolerated when cardinal numbers of infinite sets are under consideration. In fact, Dedekind, in about 1888, actually *defined* an **infinite set** to be one that is equivalent to some proper subset of itself.

We shall designate the cardinal number of the set of all natural numbers by d and describe any set having this cardinal number as being **denumerable.**[3] It follows that a set S is denumerable if and only if its elements can be written as an unending sequence $\{s_1, s_2, s_3, \ldots\}$. Since it is easily shown that any infinite set contains a denumerable subset, it follows that d is the "smallest" transfinite number.

Cantor, in one of his earliest papers on set theory, proved the denumerability of two important sets that scarcely seem at first glance to possess this property.

The first set is the set of all rational numbers. This set has the important property of being **dense.** By this is meant that between any two distinct rational numbers there exists another rational number—in fact, infinitely many other rational numbers. For example, between 0 and 1 lie the rational numbers

$$1/2,\ 2/3,\ 3/4,\ 4/5,\ 5/6,\ \ldots,\ n/(n+1),\ \ldots;$$

between 0 and 1/2 lie the rational numbers

$$1/3,\ 2/5,\ 3/7,\ 4/9,\ 5/11,\ \ldots,\ n/(2n+1),\ \ldots;$$

[3] Cantor designated the cardinal number by the Hebrew letter aleph with the subscript zero, that is, by $\aleph_0$.

between 0 and 1/4 lie the rational numbers

$$1/5,\ 2/9,\ 3/13,\ 4/17,\ 5/21,\ \ldots,\ n/(4n+1),\ \ldots;$$

and so on. Because of this property, one might well expect the transfinite number of the set of all rational numbers to be greater than d.[4] Cantor showed that this is *not* the case, and that, on the contrary, the set of all rational numbers is denumerable. His proof is interesting and runs as follows.

THEOREM 1: *The set of all rational numbers is denumerable.*
Consider the array

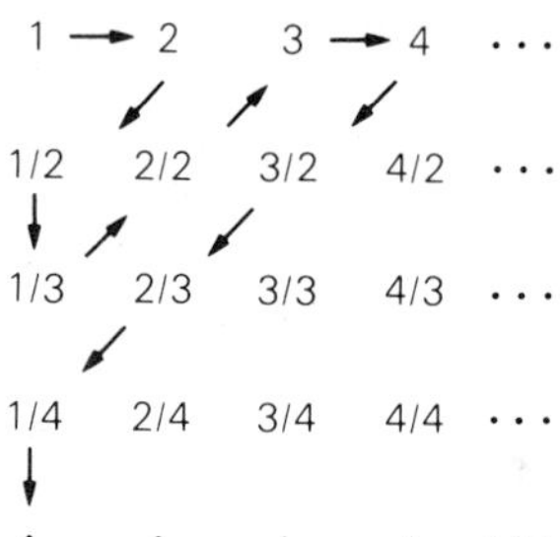

in which the first row contains, in order of magnitude, all the natural numbers (that is, all positive fractions with denominator 1), the second row contains, in order of magnitude, all the positive fractions with denominator 2, the third row contains, in order of magnitude, all the positive fractions with denominator 3, etc. Obviously, every positive rational number appears in this array, and if we list the numbers in the order of succession indicated by the arrows, omitting numbers that have already appeared, we obtain an unending sequence

$$1,\ 2,\ 1/2,\ 1/3,\ 3,\ 4,\ 3/2,\ 2/3,\ 1/4,\ \ldots$$

in which each positive rational number appears once and only once. Denote this sequence by $\{r_1, r_2, r_3, \ldots\}$. Then the sequence $\{0, -r_1, r_1, -r_2, r_2, \ldots\}$ contains the set of all rational numbers, and the denumerability of this set is established.

The second set considered by Cantor is a seemingly much more extensive set of numbers than the set of rational numbers. We first make the following definition.

DEFINITION 1: A complex number is said to be **algebraic** if it is a root of some polynomial

$$f(x) = a_0x^n + a_1x^{n-1} + \cdots + a_{n-1}x + a_n,$$

[4] The cardinal number of a set A is said to be *greater than* the cardinal number of a set B if and only if B is equivalent to a proper subset of A, but A is equivalent to no proper subset of B.

where $a_0 \neq 0$ and all the a_k's are integers. A complex number that is not algebraic is said to be **transcendental.**

It is quite clear that the algebraic numbers include, among others, all rational numbers and all roots of such numbers. Accordingly, the following theorem is somewhat astonishing:

THEOREM 2: *The set of all algebraic numbers is denumerable.*

Let $f(x)$ be a polynomial of the kind described in Definition 1, where, without loss of generality, we may suppose $a_0 > 0$. Consider the so-called **height** of the polynomial, defined by

$$h = n + a_0 + |a_1| + |a_2| + \cdots + |a_{n-1}| + |a_n|.$$

Obviously h is an integer $\geqq 1$, and there are plainly only a finite number of polynomials of a given height h, and therefore only a finite number of algebraic numbers arising from polynomials of a given height h. We may now list (theoretically speaking) all the algebraic numbers, refraining from repeating any number already listed, by first taking those arising from polynomials of height 1, then those arising from polynomials of height 2, then those arising from polynomials of height 3, and so on. We thus see that the set of all algebraic numbers can be listed in an unending sequence, whence the set is denumerable.

In view of the preceding two theorems, there remains the possibility that all infinite sets are denumerable. That this is not so was shown by Cantor in a striking proof of the following significant theorem:

THEOREM 3: *The set of all real numbers in the interval* $0 < x < 1$ *is nondenumerable.*

The proof is indirect and employs an unusual method known as the **Cantor diagonal process.** Let us, then, assume the set to be denumerable. Then we may list the numbers of the set in a sequence $\{p_1, p_2, p_3, \ldots\}$. Each of these numbers p_i can be written uniquely as a nonterminating decimal fraction; in this connection, it is useful to recall that every rational number may be written as a "repeating decimal"; a number such as 0.3, for example, can be written as 0.29999. . . . We can then display the sequence in the following array,

$$\begin{aligned} p_1 &= 0.a_{11}a_{12}a_{13}\cdots \\ p_2 &= 0.a_{21}a_{22}a_{23}\cdots \\ p_3 &= 0.a_{31}a_{32}a_{33}\cdots \\ &\cdots\cdots\cdots\cdots\cdots \end{aligned}$$

where each symbol a_{ij} represents some one of the digits 0, 1, 2, 3, 4, 5, 6, 7, 8, 9. Now, in spite of any care that has been taken to list all the real numbers between 0 and 1, there is a number that could not have been listed. Such a number is $0.b_1b_2b_3\ldots$, where, say, $b_k = 7$ if $a_{kk} \neq 7$ and $b_k = 3$ if $a_{kk} = 7$, for $k = 1, 2, 3, \ldots, n, \ldots$. This number clearly lies between 0 and 1, and it must differ from each number p_i, for it differs from p_1 in at least the first decimal place, from p_2 in at least the second decimal place, from p_3 in at least the third decimal place, and so on. Thus, the original assumption that all the real numbers between 0 and 1 can be listed in a sequence is untenable, and the set must therefore be nondenumerable.

Cantor deduced the following remarkable consequence of Theorems 2 and 3:

THEOREM 4: *Transcendental numbers exist.*
Since, by Theorem 3, the set of all real numbers between 0 and 1 is nondenumerable, it is easily demonstrated that the set of all complex numbers is also nondenumerable. By Theorem 2, however, the set of all algebraic numbers is denumerable. It follows that there must exist complex numbers that are not algebraic, and the theorem is established.

Not all mathematicians are willing to accept the above proof of Theorem 4. The acceptability or nonacceptability of the proof hinges on what one believes mathematical existence to be, and there are some mathematicians who feel that mathematical existence is established only when one of the objects whose existence is in question is actually constructed and exhibited. Now the above proof does not establish the existence of transcendental numbers by producing a specific example of such a number. There are many existence proofs in mathematics of this nonconstructive sort, where existence is presumably established by merely showing that the assumption of nonexistence leads to a contradiction. Most proofs of the Fundamental Theorem of Algebra, for example, are formulated along such lines.

Because of the dissatisfaction of some mathematicians with nonconstructive existence proofs, a good deal of effort has been made to replace such proofs by those that actually yield one of the objects concerned.

The proof of the existence of transcendental numbers and the proof that some particular number is transcendental are two quite different matters, the latter often being a very difficult problem. It was Hermite who, in 1873, proved that the number e, the base for natural logarithms, is transcendental, and Lindemann, in 1882, who first established the transcendentality of the number π. Unfortunately, it is inconvenient for us to prove these interesting facts here. The difficulty of identifying a particular given number as algebraic or transcendental is illustrated by the fact that it is not yet known whether the number π^π is algebraic or transcendental. A recent gain along these lines was the establishment of the transcendental character of any number of the form a^b, where a is an algebraic number different from 0 or 1, and b is any irrational algebraic number. This result, achieved in 1934 by Alexsander Osipovich Gelfond (1906–1968), and now known as Gelfond's theorem, was a culmination of an almost thirty-year effort to prove that the so-called **Hilbert number,** $2^{\sqrt{2}}$, is transcendental.

Since the set of all real numbers in the interval $0 < x < 1$ is nondenumerable, the transfinite number of this set is greater than d. We shall denote it by c, and shall refer to it as the **cardinal number of the continuum.** It is generally believed that c is the next transfinite number after d—that is, that there is no set having a cardinal number greater than d but less than c. This belief is known as the **continuum hypothesis,** but, in spite of strenuous efforts, no proof has been found to establish it. Many consequences of the hypothesis have been deduced. In about 1940, the Austrian logician Kurt Gödel (1906–1978) succeeded in showing that the continuum hypothesis is consistent with a famous postulate set of set theory provided these postulates themselves are consistent. Gödel conjectured that the denial of the continuum hypothesis is also consistent with the postulates of set theory. This conjecture was established, in 1963, by Paul J. Cohen (born 1934) of Stanford University, thus proving that the continuum hypothesis is independent of the postulates of set theory, and hence can never be deduced from those postulates. The situation is analogous to that of the parallel postulate in Euclidean geometry.

It has been shown that the set of all single-valued functions $f(x)$ defined over the interval $0 < x < 1$ has a cardinal number greater than c, but whether this cardinal number is the next after c is not known. Cantor's theory provides for an infinite sequence of transfinite numbers, and there are demonstrations that purport to show that an unlimited number of cardinal numbers greater than that of the continuum actually exist.

15–5 Topology

Topology started as a branch of geometry, but during the second quarter of the twentieth century it underwent such generalization and became involved with so many other branches of mathematics that it is now perhaps more properly considered, along with geometry, algebra, and analysis, as a fundamental division of mathematics. Today, topology may be roughly defined as the mathematical study of continuity. In this section we shall restrict ourselves to some of those aspects of the subject that reflect its geometric origin. From this point of view, topology may be regarded as the study of those properties of geometric figures that remain invariant under so-called **topological transformations;** that is, under single-valued continuous mappings possessing single-valued continuous inverses. By a geometric figure, we mean a point set in three-dimensional (or higher-dimensional) space; a single-valued continuous mapping is one that, given a Cartesian coordinate system in the space, can be represented by single-valued continuous functions of the coordinates.

Since the set of all topological transformations of a geometric figure constitute a transformation group, topology can, from our viewpoint, be considered as a Kleinian geometry, and hence codified within Klein's *Erlanger Programm*. Those properties of a geometric figure that remain invariant under topological transformations of the figure are called **topological properties** of the figure, and two figures that can be topologically transformed into one another are said to be **homeomorphic,** or **topologically equivalent.**

The mapping functions of a topological transformation need not be defined over the whole of the space in which the geometric figure is imbedded, but may be defined over just the point set making up the geometric figure. We can, then, regard the **intrinsic** topological properties of the figure as those that remain invariant under *all* topological transformations of the figure, and the **extrinsic** topological properties of the figure as those that remain invariant only under topological transformations of the whole space that contains the figure. The intrinsic topological properties of the figure are those that are independent of the imbedding space, whereas the extrinsic topological properties are those that depend upon the imbedding space, and we are reminded of the similar situation in Section 14–7 in connection with the differential geometry of surfaces in three-space.

Topology, as a self-connected study, scarcely predates the mid-nineteenth century, but one can find some earlier isolated topological investigations. Toward the end of the seventeenth century, Leibniz used the term **geometria situs** to describe a sort of qualitative mathematics that today would be thought of as topology, and he predicted important studies in this field, but his prediction was slow in materializing. An early discovered topological property of a simple closed polyhedral surface is the relation $v - e + f = 2$, where v, e, f denote the number of vertices, edges, and faces, respectively, of the polyhedral surface. This relation was adumbrated by Descartes in 1640, and the first proof of the formula was given by Euler in 1752. Euler had earlier, in 1736, considered some topology of linear graphs in his treatment of the Königsberg bridge problem (see Problem Study 12.8). Gauss made several contributions to topology. Of the several proofs that he furnished of the fundamental theorem of algebra, two are explicitly topological. His first proof of this theorem employs topological techniques and was given in his doctoral dissertation in 1799 when he was twenty-two years old. Later, Gauss briefly considered the theory of knots, which today is an important subject in topology. About 1850, Francis Guthrie conjectured the four-color problem,[5] which was later taken up by Augustus De Morgan, Arthur Cayley, and others. At this time, the subject of topology was known as **analysis situs.** The term *topology* was introduced by J. B. Listing (1808–1882), one of Gauss' students, in 1847, in the title, *Vorstudien zur Topologie,* of the first book devoted to the subject. The German word *Topologie* was later anglicized to *topology* by Professor Solomon Lefschetz (1884–1972) of Princeton University. G. R. Kirchhoff (1824–1887), another of Gauss' students, in 1847 employed the topology of linear graphs in his study of electrical networks. Of all of Gauss' students, however, the one who contributed by far the most to topology was Bernhard Riemann, who, in his doctoral thesis of 1851, introduced topological concepts into the study of complex-function theory. The chief stimulus to topology furnished by Riemann was his notion of **Riemann surface,** a topological device for rendering multiple-valued complex functions into single-valued functions. Also of importance in topology is Riemann's probationary lecture of 1854 concerning the hypotheses that lie at the foundations of geometry. This lecture furnished the breakthrough to higher dimensions, and

[5] That any map on a plane or a sphere can be colored with at most four colors.

the term and concept of *manifold* were introduced here. About 1865, A. F. Möbius (1790–1868) wrote a paper in which a polyhedral surface was viewed simply as a collection of joined polygons. This introduced the concept of two complexes into topology. In his systematic development of two complexes, Möbius was led to the one-sided and one-edged surface now referred to as a *Möbius strip*. In 1873, J. C. Maxwell (1831–1879) used the topological theory of connectivity in his study of electromagnetic fields. Others, such as H. Helmholtz (1821–1894) and Lord Kelvin (William Thomson, 1824–1907), can be added to the list of physicists who applied topological ideas with success. Henri Poincaré (1854–1912) ranks high among the early contributors to topology. A paper of his, written in 1895 and entitled *Analysis situs,* is the first significant paper devoted wholly to topology. It was in that paper that the important homology theory of n dimensions was introduced. It was also Poincaré who introduced the Betti groups into topology. With Poincaré's work, the subject of topology was well under way, and an increasing number of mathematicians entered the field. Especially important names in topology since Poincaré are O. Veblen (1880–1960), J. W. Alexander (1888–1971), S. Lefschetz (1884–1972), L. E. J. Brouwer (1881–1966), and M. Fréchet (1878–1973).

The notion of a geometric figure as made up of a finite set of joined fundamental pieces, emphasized by Möbius, Riemann, and Poincaré, gradually gave way to the Cantorian concept of an arbitrary set of points, and it then was recognized that any collection of things—be it a set of numbers, algebraic entities, functions, or nonmathematical objects—can constitute a topological space in some sense or other. This latter, and very general, viewpoint of topology has become known as **set topology,** whereas studies more intimately connected with the earlier viewpoint have become known as **combinatorial,** or **algebraic topology.** The classical formulation of set topology was given by Felix Hausdorff (1868–1942) in his *Grundzüge der Mengenlehre* of 1914. Here we find a systematic exposition of the subject, in which the nature of the fundamental elements is of no consequence. In the latter part of the work we find a development of the topological spaces today known as *Hausdorff spaces* (see Problem Study 15.27).

15–6 Mathematical Logic

A mathematical theory results from the interplay of two factors, a set of postulates and a logic. The set of postulates constitutes the basis from which the theory starts, and a logic constitutes the rules by which such a basis may be expanded into a body of theorems. Clearly both factors are important, and accordingly each factor has been carefully examined and studied. The study of the first factor forms the subject of axiomatics, which we have already considered in Section 15–2; in this section we look into the second of the two factors.

Although the ancient Greeks considerably developed formal logic, and Aristotle (384–322 B.C.) systematized the material, this early work was all carried out with the use of ordinary language. Mathematicians of today have found it a well nigh hopeless task to discuss modern considerations of logic in a similar way. A symbolic language has become necessary in order to achieve the required exact scientific treatment of the subject. Because of the presence of such symbolism, the resulting treatment is known as **symbolic,** or **mathematical, logic.** In symbolic logic, the various relations among propositions, classes, and so forth, are represented by formulas whose meanings are free from the ambiguities so common to ordinary language. It becomes possible to develop the subject from a set of initial formulas in accordance with certain clearly prescribed rules of formal transformation, much like the development of a piece of common algebra. Also, and again as in the development of a piece of common algebra, the advantages of the symbolic language over ordinary language, insofar as compactness and ease of comprehension are concerned, are very great.

Leibniz is regarded as the first to consider seriously the desirability of a symbolic logic. One of his earliest works was an essay, *De arte combinatoria,* published in 1666, in which he indicated his belief in the possibility of a universal scientific language, expressed in an economical and workable symbolism for guidance in the reasoning process. Returning to these ideas between the years 1679 and 1690, Leibniz made considerable headway toward the creation of a symbolic logic, and he formulated a number of the concepts that are so important in modern studies.

An important renewal of interest in symbolic logic took place in 1847 when George Boole (1815–1864) published his little pamphlet entitled *The Mathematical Analysis of Logic, Being an Essay towards a Calculus of Deductive Reasoning*. Another paper followed in 1848, and finally, in 1854, Boole gave a notable exposition of his ideas in the work, *An Investigation into the Laws of Thought, on Which Are Founded the Mathematical Theories of Logic and Probability*.

Augustus De Morgan (1806–1871) was a contemporary of Boole, and his treatise on *Formal Logic; or, the Calculus of Inference, Necessary and Probable,* published in 1847, in some ways went considerably beyond Boole. Later De Morgan also made extended studies of the hitherto neglected logic of relations.

In the United States, outstanding work in the field was contributed by Charles Sanders Peirce (1839–1914), son of the distinguished Harvard mathematician, Benjamin Peirce (1809–1880). Peirce rediscovered some of the principles enunciated by his predecessors. It is unfortunate that his work appeared somewhat out of the stream of normal development; only in comparatively recent times has the merit of much of Peirce's thought been properly appreciated.

The notions of Boole were given a remarkable completeness in the massive treatise by Ernst Schröder (1841–1902) entitled *Vorlesungen über die Algebra der Logic,* published during the period between 1890 and 1895. In fact, modern logicians are inclined to characterize symbolic logic in the Boolean tradition by the term **Boole-Schröder algebra.** Considerable work is still being done in Boolean algebra, and many papers upon the subject are to be found in present-day research journals.

A still more modern approach to symbolic logic originated with the work of the German logician Gottlob Frege (1848–1925) during the period between 1879 and 1903, and with studies of Giuseppe Peano (1858–1932). Peano's work was motivated by a desire to express all mathematics in terms of a logical calculus, and Frege's work stemmed from the need of a sounder foundation for mathematics. Frege's *Begriffsschrift* appeared in 1879, and his historically important *Grundgesetze der Arithmetik* in the period between 1893 and 1903; the *Formulaire de mathématiques* of Peano and his co-workers began its appearance in 1894. The work started by Frege and Peano led directly to the very influential and monumental *Principia mathematica* (1910–1913) of Alfred North Whitehead (1861–1947) and Bertand Russell (1872–1970). The basic idea of this work is the identification of much of mathematics with logic by the deduction of the natural number system, and hence of the great bulk of existing mathematics, from a set of premises or postulates for logic itself. In the period between 1934 and 1939 appeared the comprehensive *Grundlagen der Mathematik* of David Hilbert (1862–1943) and Paul Bernays (1888–1977). This work, based upon a series of papers and university lectures given by Hilbert, attempts to build up mathematics by the use of symbolic logic in a new way that renders possible the establishment of the consistency of mathematics.

At the present time, elaborate studies in the field of symbolic logic are being pursued by many mathematicians, chiefly as a result of the impetus given to the work by the publication of the *Principia mathematica*. A periodical, known as the *Journal of Symbolic Logic*, was established in 1935 to publicize the writings of this group.

An interesting analogy (if it is not pushed too far) exists between the parallelogram law of forces and the axiomatic method. By the parallelogram law, two component forces are combined into a single resultant force. Different resultant forces are obtained by varying one or both of the component forces, although it is possible to obtain the same resultant force by taking different pairs of initial component forces. Now, just as the resultant force is determined by the two initial component forces, so (see Figure 122) is a mathematical theory determined by a set of postulates and a logic; that is, the set of statements constituting a mathematical theory results from the interplay of an initial

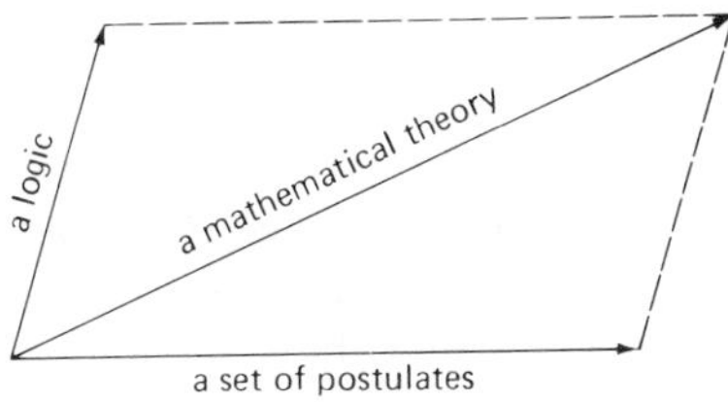

FIGURE 122

set of statements, called the postulates, and another initial set of statements, called the logic or the rules of procedure. For some time, mathematicians have been aware of the variability of the first set of initial statements (namely, the postulates) but until recent times the second set of initial statements (namely, the logic) was universally thought to be fixed, absolute, and immutable. Indeed, this is still the prevailing view among most people, for it seems quite inconceivable, except to the very few students of the subject, that there can be any alternative to the laws of logic stated by Aristotle in the fourth century B.C. The general feeling is that these laws are in some way attributes of the structure of the universe and that they are inherent in the very nature of human reasoning. As with many other absolutes of the past, this one, too, has toppled, but only as late as 1921. The modern viewpoint can hardly be more neatly put than in the following words of the outstanding American logician, Alonzo Church.

> We do not attach any character of uniqueness or absolute truth to any particular system of logic. The entities of formal logic are abstractions, invented because of their use in describing and systematizing facts of experience or observation, and their properties, determined in rough outline by this intended use, depend for their exact character on the arbitrary choice of the inventor. We may draw the analogy of a three dimensional geometry used in describing physical space, a case for which we believe, the presence of such a situation is more commonly recognized. The entities of the geometry are clearly of abstract character, numbering as they do planes without thickness and points which cover no area in the plane, point sets containing an infinitude of points, lines of infinite length, and other things which cannot be reproduced in any physical experiment. Nevertheless the geometry can be applied to physical space in such a way that an extremely useful correspondence is set up between the theorems of the geometry and observable facts about material bodies in space. In building the geometry, the proposed application to physical space serves as a rough guide in determining what properties the abstract entities shall have, but does not assign these properties completely. Consequently there may be, and actually are, more than one geometry whose use is feasible in describing physical space. Similarly, there exist, undoubtedly, more than one formal system whose use as a logic is feasible, and of these systems one may be more pleasing or more convenient than another, but it cannot be said that one is right and the other wrong.

It will be recalled that new geometries first came about through the denial of Euclid's parallel postulate and that new algebras first came about through the denial of the commutative law of multiplication. In similar fashion, the new so-called "many-valued logics" first came about by denying Aristotle's law of the excluded middle. According to this law, the disjunctive proposition "p or not-p" is a tautology, and a proposition p in Aristolelian logic is always either true or false. Because a proposition may possess any of the two possible truth values—namely, truth or falsity—this logic is known as a two-valued logic. In 1921, in a short two-page paper, J. Lukasiewicz considered a three-valued logic, or a logic in which a proposition p may possess any one of three possible truth values. Very shortly after, and independently of Lukasiewicz' work, E. L. Post considered m-valued logics, in which a proposition p may possess any one of m possible truth values, where m is any integer greater than 1. If m exceeds 2, the logic is said to be **many-valued.** Another study of m-valued logics was given in 1930 by Lukasiewicz and A. Tarski. Then, in 1932, the m-valued truth systems were extended by H. Reichenbach to an infinite-valued logic, in which a proposition p may assume any one of infinitely many possible values.[6]

Not all new logics are of the type just discussed. Thus, A. Heyting has developed a symbolic two-valued logic to serve the intuitionist school of mathematicians; it differs from Aristotelian logic in that it does not universally accept the law of the excluded middle or the law of double negation. Like the many-valued logics, then, this special-purpose logic exhibits differences from Aristotelian laws. Such logics are known as **non-Aristotelian logics.**

Like the non-Euclidean geometries, the non-Aristotelian logics have proved not to be barren of application. Reichenbach actually devised his infinite-valued logic to serve as a basis for the mathematical theory of probability. In 1933, F. Zwicky observed that many-valued logics can be applied to the quantum theory of modern physics. Many of the details of such an application have been supplied by Garrett Birkhoff, J. von Neumann, and H. Reichenbach. The part that non-Aristotelian logics may play in the future development of mathematics is uncertain but intriguing to contemplate; the application of Heyting's symbolic logic to intuitionistic mathematics indicates that the new logics may be mathematically valuable. In the next section, we point out a possible use of these logics in the resolution of a modern crisis in the foundations of mathematics.

From the above discussion there emerges a remarkable principle of discovery and advancement—namely, the constructive doubting of a traditional belief. When Einstein was asked how he came to invent the theory of relativity he replied, "By challenging an axiom." Lobachevsky and Bolyai challenged Euclid's axiom of parallels; Hamilton and Cayley challenged the axiom that multiplication is commutative; Lukasiewicz and Post challenged Aristotle's axiom of the excluded middle. Similarly, in the field of science, Copernicus challenged the axiom that the earth is the center of the solar system; Galileo challenged the axiom that the heavier body falls the faster; Einstein challenged the axiom that of two distinct instants one must precede the other. This constructive challeng-

[6] As a matter of historical interest, in 1936 K. Michalski discovered that the three-valued logics had actually been anticipated as early as the fourteenth century by the medieval schoolman, William of Occam. The possibility of a three-valued logic had also been considered by the philosopher Hegel and, in 1896, by Hugh MacColl. These speculations, however, had little effect on subsequent thought and so cannot be considered as decisive contributions.

ing of axioms has become one of the commoner ways of making advances in mathematics, and it undoubtedly lies at the heart of Georg Cantor's famous aphorism: "The essence of mathematics lies in its freedom."

15–7 Antinomies of Set Theory

A study of the history of mathematics from Greek antiquity to the present reveals that the foundations of mathematics have undergone three profoundly disturbing crises wherein, in each instance, some sizable portion of mathematics that had been thought established became suspect and in urgent need of revision.

The first crisis in the foundations of mathematics arose in the fifth century B.C.; indeed, such a crisis could not have occurred much earlier, for, as we have seen, mathematics as a deductive study originated not earlier than the sixth century B.C., perhaps with Thales, Pythagoras, and their pupils. The crisis was precipitated by the unexpected discovery that not all geometrical magnitudes of the same kind are commensurable with one another; it was shown, for example, that the diagonal and side of a square contain no common unit of measure. Since the Pythagorean development of magnitudes was built upon the firm intuitive belief that all like magnitudes are commensurable, the discovery that like magnitudes may be incommensurable proved to be highly devastating. For instance, the entire Pythagorean theory of proportion with all of its consequences had to be scrapped as unsound. The resolution of this first crisis in the foundations of mathematics was neither easily nor quickly realized. It was finally achieved about 370 B.C. by the brilliant Eudoxus, whose revised theory of magnitude and proportion is one of the great mathematical masterpieces of all time. Eudoxus' remarkable treatment of incommensurables may be found in the fifth book of Euclid's *Elements;* it coincides essentially with the modern exposition of irrational numbers that was given by Richard Dedekind in 1872. We considered this first crisis in the foundations of mathematics in Section 3–5, and its resolution by Eudoxus in Section 5–5. It is quite possible that this crisis is largely responsible for the subsequent formulation and adoption of the axiomatic method in mathematics.

The second crisis in the foundations of mathematics followed the invention of the calculus by Newton and Leibniz in the late seventeenth century.[7] We have seen how the successors of these men, intoxicated by the power and applicability of the new tool, failed to consider sufficiently the solidity of the base upon which the subject was founded, so that instead of having demonstrations justify results, results were used to justify demonstrations. With the passage of time, contradictions and paradoxes arose in increasing numbers, and a serious crisis in the foundations of mathematics became evident. It was realized more and more that the edifice of analysis was being built upon sand, and

[7] Forewarnings of this crisis can be seen in the renowned paradoxes of Zeno of about 450 B.C.

finally, in the early nineteenth century, Cauchy took the first steps toward resolving the crisis by replacing the hazy method of infinitesimals by the precise method of limits. With the subsequent so-called arithmetization of analysis by Weierstrass and his followers, it was felt that the second crisis in the foundations of mathematics had been overcome, and that the whole structure of mathematics had been redeemed and placed upon an unimpeachable base. The origin and resolution of this second crisis in the foundations of mathematics constituted the subject matter of Section 14–9.

The third crisis in the foundations of mathematics materialized with shocking suddenness in 1897, and, though now well over three-quarters of a century old, is still not resolved to the satisfaction of all concerned. The crisis was brought about by the discovery of paradoxes or antinomies in the fringe of Cantor's general theory of sets. Since so much of mathematics is permeated with set concepts and, for that matter, can actually be made to rest upon set theory as a foundation, the discovery of paradoxes in set theory naturally cast into doubt the validity of the whole foundational structure of mathematics.

In 1897, the Italian mathematician Burali-Forti brought to light the first publicized paradox of set theory. As originally conceived and stated by Burali-Forti, the paradox involves technical terms and ideas that, in our limited treatment, we lack space to develop. The essence of the paradox can be given, however, by a nontechnical description of a very similar paradox found by Cantor two years later. In his theory of sets, Cantor had succeeded in proving that for any given transfinite number there is always a greater transfinite number, so that just as there is no greatest natural number, there also is no greatest transfinite number. Now consider the set whose members are all possible sets. Surely no set can have more members than this set of all sets. But if this is the case, how can there be a transfinite number greater than the transfinite number of this set?

Whereas the Burali-Forti and Cantor paradoxes involve results of set theory, Bertrand Russell discovered in 1902 a paradox depending on nothing more than just the concept of set itself. Before describing the Russell paradox, we note that sets either are members of themselves or are not members of themselves. Thus, the set of all abstract ideas is itself an abstract idea, but the set of all men is not a man. Again, the set of all sets is itself a set, but the set of all stars is not a star. Let us represent the set of all sets that are members of themselves by M, and the set of all sets that are not members of themselves by N. We now ask ourselves whether set N is or is not a member of itself. If N is a member of itself, then N is a member of M and not of N, and N is not a member of itself. On the other hand, if N is not a member of itself, then N is a member of N and not of M, and N is a member of itself. The paradox lies in the fact that in either case we are led to a contradiction.

A more compact and less wordy presentation of the Russell paradox may be given as follows. Let X denote *any* set. Then, by the definition of N,

$$(X \in N) \leftrightarrow (X \notin X).$$

Now take X to be N, and we have the contradiction

$$(N \in N) \leftrightarrow (N \notin N).$$

This paradox was communicated by Russell to Frege just after the latter had completed the last volume of his great two-volume treatise on the foundations of arithmetic. Frege acknowledged the communication at the end of his treatise by the following pathetic and remarkably restrained sentences. "A scientist can hardly meet with anything more undesirable than to have the foundation give way just as the work is finished. In this position I was put by a letter from Mr. Bertrand Russell as the work was nearly through the press." Thus terminated the labor of a dozen or more years.

The Russell paradox has been popularized in many forms. One of the best known of these forms was given by Russell himself in 1919 and concerns the plight of the barber of a certain village who has enunciated the principle that he shaves all those persons and only those persons of the village who do not shave themselves. The paradoxical nature of this situation is realized when we try to answer the question, "Does the barber shave himself?" If he does shave himself, then he shouldn't according to his principle; if he doesn't shave himself, then he should according to his principle.

Since the discovery of the above contradictions within Cantor's theory of sets, additional paradoxes have been produced in abundance. These modern paradoxes of set theory are related to several ancient paradoxes of logic. For example, Eubulides, of the fourth century B.C., is credited with making the remark, "This statement I am now making is false." If Eubulides' statement is true, then, by what it says, the statement must be false. On the other hand, if Eubulides' statement is false, then it follows that his statement must be true. Thus, Eubulides' statement can be neither true nor false without entailing a contradiction. Still older than the Eubulides paradox may be the unauthenticated Epimenides paradox. Epimenides, who himself was a Cretan philosopher of the sixth century B.C., is claimed to have made the remark, "Cretans are always liars." A simple analysis of this remark easily reveals that it, too, is self-contradictory.

The existence of paradoxes in set theory, like those described above, clearly indicates that something is wrong. Since their discovery, a great deal of literature on the subject has appeared, and numerous attempts at a solution have been offered.

So far as mathematics is concerned, there seems to be an easy way out. One has merely to reconstruct set theory on an axiomatic basis sufficiently restrictive to exclude the known antinomies. The first such attempt was made by Zermelo in 1908, and subsequent refinements have been made by Fraenkel (1922, 1925), Skolem (1922, 1929), von Neumann (1924, 1928), Bernays (1937–1948), and others. But such a procedure has been criticized as merely avoiding the paradoxes; certainly it does not explain them. Moreover, this procedure carries no guarantee that other kinds of paradoxes will not crop up in the future.

There is another procedure that apparently both explains and avoids the known paradoxes. If examined carefully, it will be seen that each of the paradoxes considered above involves a set S and a member m of S whose definition depends upon S. Such a definition is said to be **impredicative,** and impredicative definitions are, in a sense, circular. Consider, for instance, Russell's barber paradox. Let us designate the barber by m and the set of all members of the barber's village by S. Then m is defined impredicatively as "that member of S who shaves all those members and only those members of S who do not shave themselves." The circular nature of this definition is evident—the definition of the barber involves the members of the village and the barber himself is a member of the village.

Poincaré considered the cause of the antinomies to lie in impredicative definitions, and Russell expressed the same view in his Vicious Circle Principle: *No set S is allowed to contain members m definable only in terms of S, or members m involving or presupposing S*. This principle amounts to a restriction on the concept of set. Cantor had attempted to give the concept of set a very general meaning by stating: *By a set S we are to understand any collection into a whole of definite and separate objects m of our intuition or our thought; these objects m are called the elements of S*. The theory of sets constructed on Cantor's general concept of set leads, as we have seen, to contradictions, but if the notion of set is restricted by the Vicious Circle Principle, the resulting theory avoids the known antinomies. The outlawing of impredicative definitions would appear, then, to be a solution to the known paradoxes of set theory. There is, however, one serious objection to this solution; namely, there are parts of mathematics that mathematicians are very reluctant to discard that contain impredicative definitions.

An example of an impredicative definition in mathematics is that of the least upper bound of a given nonempty set of real numbers—the least upper bound of the given set is the smallest member of the set of all upper bounds of the given set. There are many similar instances of impredicative definitions in mathematics, though many of them can be circumvented. In 1918, Hermann Weyl undertook to find out how much of analysis can be constructed genetically from the natural number system without the use of impredicative definitions. Although he succeeded in obtaining a considerable part of analysis, he was unable to derive the important theorem that every nonempty set of real numbers having an upper bound has a least upper bound.

Other attempts to solve the paradoxes of set theory look for the trouble in logic, and it must be admitted that the discovery of the paradoxes in the general theory of sets has brought about a thorough investigation of the foundations of logic. Very intriguing is the suggestion that the way out of the difficulties of the paradoxes may be through the use of a three-valued logic. For example, in the Russell paradox given above, we saw that the statement, "N is a member of itself," can be neither true nor false. Here a third possibility would be helpful. Denoting truth quality of a proposition by T, false quality by F, and a *third* quality, which is neither T nor F, by ? (meaning, perhaps, *undecidable*), the situation would be saved if we could simply classify the statement as ?.

There have arisen three main philosophies, or schools of thought, concerning the foundations of mathematics—the so-called logistic, intuitionist, and formalist schools. Naturally, any modern philosophy of the foundations of mathematics must, somehow or other, cope with the present crisis in the foundations of mathematics. In the next section, we very briefly consider these three schools of thought and point out how each proposes to deal with the antinomies of general set theory.

15–8 Philosophies of Mathematics

A philosophy may be regarded as an explanation that attempts to make some kind of sense out of the natural disorder of a set of experiences. From this point of view, it is possible to have a philosophy of almost anything—a philosophy of art, of life, of religion, of education, of society, of history, of science, of mathematics, even of philosophy itself. A philosophy amounts to a process of refining and ordering experiences and values; it seeks relations among things that are normally felt to be disparate and finds important differences between things normally considered as the same; it is the description of a theory concerning the nature of something. In particular, a philosophy of mathematics essentially amounts to an attempted reconstruction in which the chaotic mass of mathematical knowledge accumulated over the ages is given a certain sense or order. Clearly, a philosophy is a function of time, and a particular philosophy may become outdated or have to be altered in the light of additional experiences. We are here concerned with only contemporary philosophies of mathematics—philosophies that take account of the recent advances in mathematics and of the current crisis in the subject.

There are three principal present-day philosophies of mathematics, each of which has attracted a sizable group of adherents and developed a large body of associated literature. These are referred to as the logistic school, of which Russell and Whitehead are the chief expositors; the intuitionist school, led by Brouwer; and the formalist school, developed principally by Hilbert. There are, of course, present-day philosophies of mathematics other than these three. There are some independent philosophies and some that constitute various mixtures of the principal three, but these other points of view have not been so widely cultivated, or do not comprise a reconstruction of mathematics of similar extent.

LOGICISM: The logistic thesis is that mathematics is a branch of logic. Rather than being just a tool of mathematics, logic becomes the progenitor of mathematics. All mathematical concepts are to be formulated in terms of logical concepts, and all theorems of mathematics are to be developed as theorems of logic; the distinction between mathematics and logic becomes merely one of practical convenience.

The notion of logic as a science containing the principles and ideas underlying all other sciences dates back at least as far as Leibniz (1666). The actual reduction of mathematical concepts to logical concepts was engaged in by

Dedekind (1888) and Frege (1884–1903), and the statement of mathematical theorems by means of a logical symbolism was undertaken by Peano (1889–1908). These men, then, are forerunners of the logistic school, which received its definitive expression in the monumental *Principia mathematica* of Whitehead and Russell (1910–1913). This great and complex work purports to be a detailed reduction of the whole of mathematics to logic. Subsequent modifications and refinements of the program have been supplied by Wittgenstein (1922), Chwistek (1924–1925), Ramsey (1926), Langford (1927), Carnap (1931), Quine (1940), and others.

The logistic thesis arises naturally from the effort to push back the foundations of mathematics to as deep a level as possible. We have seen how these foundations were established in the real number system and then how they were pushed back from the real number system to the natural number system, and thence into set theory. Since the theory of classes is an essential part of logic, the idea of reducing mathematics to logic certainly suggests itself. The logistic thesis is thus an attempted synthesization suggested by an important trend in the history of the application of the axiomatic method.

The *Principia mathematica* starts with "primitive ideas" and "primitive propositions," corresponding to the "undefined terms" and "postulates" of a formal abstract development. These primitive ideas and propositions are not to be subjected to interpretation but are restricted to intuitive concepts of logic; they are to be regarded as, or at least are to be accepted as, plausible descriptions and hypotheses concerning the real world. In short, a concrete rather than an abstract point of view prevails, and consequently no attempt is made to prove the consistency of the primitive propositions. The aim of *Principia mathematica* is to develop mathematical concepts and theorems from these primitive ideas and propositions, starting with a calculus of propositions, proceeding up through the theory of classes and relations to the establishment of the natural number system, and thence to all mathematics derivable from the natural number system. In this development, the natural numbers emerge with the unique meanings that we ordinarily assign to them and are not nonuniquely defined as *any things* that satisfy a certain set of abstract postulates.

To avoid the contradictions of set theory, *Principia mathematica* employs a "theory of types." Somewhat oversimply described, such a theory sets up a hierarchy of levels of elements. The primary elements constitute those of type 0; classes of elements of type 0 constitute those of type 1; classes of elements of type 1 constitute those of type 2; and so on. In applying the theory of types, one follows the rule that all the elements of any class must be of the same type. Adherence to this rule precludes impredicative definitions and thus avoids the paradoxes of set theory. As originally presented in *Principia mathematica,* hierarchies within hierarchies appeared, leading to the so-called "ramified" theory of types. In order to obtain the impredicative definitions needed to establish analysis, an "axiom of reducibility" had to be introduced. The nonprimitive and arbitrary character of this axiom drew forth severe criticism, and much of the subsequent refinement of the logistic program lies in attempts to devise some method of avoiding the disliked axiom of reducibility.

Whether or not the logistic thesis has been established seems to be a matter of opinion. Although some accept the program as satisfactory, others have found many objections to it. For one thing, the logistic thesis can be questioned on the ground that the systematic development of logic (as of any organized study) presupposes mathematical ideas in its formulation, such as the fundamental idea of iteration that must be used, for example, in describing the theory of types or the idea of deduction from given premises.

Alfred North Whitehead was born at Ramsgate, England, in 1861, and was educated at Sherborne School and Trinity College, Cambridge. He lectured on mathematics at Trinity College from 1885 to 1911, and then on applied mathematics and mechanics at University College of the University of London. He was a professor of mathematics at the Imperial College of Science and Technology at the University of London from 1914 to 1924, after which he went to the United States as a professor of philosophy at Harvard University, a post that he held until his retirement in 1936. He died at Cambridge, Massachusetts in 1947. Like his most distinguished student, Bertrand Russell, Whitehead viewed philosophy from the standpoint of mathematics, and together the two men wrote their epochal *Principia mathematica* in the years 1910 to 1913. Whitehead published a number of notably lucid works on mathematics and philosophy.

Bertrand Arthur William Russell, descendant of an aristocratic family, was born near Trelleck, Wales, in 1872. The winner of an open scholarship at Trinity College, Cambridge, he took high honors in mathematics and philosophy, and studied under Whitehead. In addition to lecturing, largely at universities in the United States, he wrote over forty books on mathematics, logic, philosophy, sociology, and education. He received many awards, such as both

BERTRAND RUSSELL
(New York Public Library Collection)

the Sylvester and de Morgan medals of the Royal Society (1934), the Order of Merit (1940), and the Nobel Prize for Literature (1950). His outspoken views often embroiled him in controversies. During World War I, he was dismissed from Cambridge University and imprisoned for four months because of his pacifist views and his opposition to conscription. In the early 1960s, he led pacifist moves to ban nuclear weapons and was again briefly imprisoned. A man of remarkable mind and ability, he died in 1970, mentally alert to the end, at the advanced age of ninety-eight.

INTUITIONISM: The intuitionist thesis is that mathematics is to be built solely by finite constructive methods on the intuitively given sequence of natural numbers. According to this view, then, at the very base of mathematics lies a primitive intuition, allied, no doubt, to our temporal sense of before and after, that allows us to conceive a single object, then one more, then one more, and so on endlessly. In this way, we obtain unending sequences, the best known of which is the sequence of natural numbers. From this intuitive base of the sequence of natural numbers, any other mathematical object must be built in a purely constructive manner, employing a finite number of steps or operations. In the intuitionist thesis, we have the genetical development of mathematics pushed to its extreme.

The intuitionist school (as a school) originated about 1908 with the Dutch mathematician L. E. J. Brouwer, although one finds some of the intuitionist ideas uttered earlier by such men as Kronecker (in the 1880s) and Poincaré (1902–1906). The school has gradually strengthened with the passage of time, has won over some eminent present-day mathematicians, and has exerted a tremendous influence on all thinking concerning the foundations of mathematics.

Some of the consequences of the intuitionist thesis are little short of revolutionary; thus, the insistence on constructive methods leads to a conception of mathematical existence not shared by all practicing mathematicians. For the intuitionists, an entity whose existence is to be proved must be shown to be constructible in a finite number of steps; it is not sufficient to show that the assumption of the entity's nonexistence leads to a contradiction. This means that many existence proofs found in current mathematics are not acceptable to the intuitionists.

An important instance of the intuitionists' insistence upon constructive procedures is in the theory of sets. For the intuitionists, a set cannot be thought of as a ready-made collection, but must be considered as a law by means of which the elements of the set can be constructed in a step-by-step fashion. This concept of set rules out the possibility of such contradictory sets as "the set of all sets."

There is another remarkable consequence of the intuitionists' insistence upon finite constructibility, and this is the denial of the universal acceptance of the law of the excluded middle. Consider, for example, the number x, which is defined to be $(-1)^k$, where k is the number of the first decimal place in the decimal expansion of π where the sequence of consecutive digits 123456789 begins, and, if no such k exists, $x = 0$. Now, although the number x is well-defined, we cannot at the moment, under the intuitionists' restrictions, say that the proposition "$x = 0$" is either true or false. This proposition can be said to be true only when a proof of it has been constructed in a finite number of steps, and it can be said to be false only when a proof of this situation has been constructed in a finite number of steps. Until one or the other of these proofs is constructed, the proposition is neither true nor false, and the law of the excluded middle is inapplicable. If, however, k is further restricted to be less than a billion, say, then it is perfectly correct to say that the proposition is now either true or false, for, with k less than a billion, the truth or falseness can certainly be established in a finite number of steps.

Thus, for the intuitionists, the law of the excluded middle holds for finite sets but should not be employed when dealing with infinite sets. This state of affairs is blamed by Brouwer on the sociological development of logic. The laws of logic emerged at a time in man's evolution when he had a good language for dealing with finite sets of phenomena; he then later made the mistake of applying these laws to the infinite sets of mathematics, with the result that antinomies arose.

In the *Principia mathematica,* the law of the excluded middle and the law of contradiction are equivalent. For the intuitionists, this situation no longer prevails, and it is an interesting problem to try, if possible, to set up the logical apparatus to which intuitionist ideas lead us. This was done in 1930 by A. Heyting, who succeeded in developing an intuitionist symbolic logic. Intuitionist mathematics thus produces its own type of logic, and mathematical logic, as a consequence, is a branch of mathematics.

There is the final important question: How much of existing mathematics can be built within the intuitionistic restrictions? If all of it can be so rebuilt, without too great an increase in the labor required, then the present problem of the foundations of mathematics would appear to be solved. Now the intuitionists have succeeded in rebuilding large parts of present-day mathematics, including a theory of the continuum and a set theory, but there is a great deal that is still wanting. So far, intuitionist mathematics has turned out to be considerably less powerful than classical mathematics, and in many ways it is much more complicated to develop. This is the fault found with the intuitionist approach—too much that is dear to most mathematicians is sacrificed. This situation may not exist forever, because there remains the possibility of an intuitionist reconstruction of classical mathematics carried out in a different and more successful way. Meanwhile, in spite of present objections raised against the intuitionist thesis, it is generally conceded that its methods do not lead to contradictions.

Besides being the leader and untiring advocate of the intuitionist view of mathematics, Brouwer left his mark in other areas of the subject. He is regarded as one of the founders of modern topology, and is particularly known for his **invariance theorem** and his **fixed-point theorem.** The former asserts that the dimensionality of a Cartesian n-dimensional number-manifold is a topological invariant, and the latter that every continuous mapping of an n-dimensional sphere onto itself has at least one fixed point.

Brouwer was born in 1881, spent the major part of his professional life at the University of Amsterdam, and died in 1966. He was a ruthless fighter for his beliefs. As the editor of the *Mathematische Annalen* in charge of acceptance or rejection of submitted papers, he opened his attack on the free use of *reductio ad absurdum* by refusing all papers that applied the law of the excluded middle to propositions whose truth or falsity could not be decided in a finite number of steps. The editorial board of the journal met the crisis by resigning and then re-electing themselves, minus Brouwer. The Dutch government was so indignant over this snub of their leading mathematician that they created a rival mathematics journal, with Brouwer in charge.

The intuitionists' ranks were greatly strengthened when Hermann Weyl joined the group. Weyl was born in 1885 near Hamburg. He entered Göttingen University when he was eighteen, became one of Hilbert's most gifted students, and remained there (except for one year at Munich) until he was called to Zurich in 1913, where he met Einstein. In 1930, he was invited to Göttingen as Hilbert's successor. He remained at Göttingen for only three years, after which he resigned because of the dismissal of so many of his colleagues by the Nazis. In 1933, he accepted an offer of permanent membership at the newly founded Institute for Advanced Study at Princeton. During his last years, he spent half of each year at Princeton and the other half at Zurich. He died suddenly in 1955.

FORMALISM: The formalist thesis is that mathematics is concerned with formal symbolic systems. In fact, mathematics is regarded as a collection of such abstract developments, in which the terms are mere symbols and the statements are formulas involving these symbols; the ultimate base of mathematics does not lie in logic but only in a collection of prelogical marks or symbols and in a set of operations with these marks. Since, from this point of view, mathematics is devoid of concrete content and contains only ideal symbolic elements, the establishment of the consistency of the various branches of mathematics becomes an important and necessary part of the formalist program. Without such an accompanying consistency proof, the whole study is essentially senseless. In the formalist thesis, we have the axiomatic development of mathematics pushed to its extreme.

The formalist school was founded by David Hilbert after completing his postulational study of geometry. In his *Grundlagen der Geometrie* (1899), Hilbert had sharpened the mathematical method from the material axiomatics of Euclid to the formal axiomatics of the present day. The formalist point of view was developed later by Hilbert to meet the crisis caused by the paradoxes of set theory and the challenge to classical mathematics caused by intuitionistic criticism. Although Hilbert talked in formalistic terms as early as 1904, not until after 1920 did he and his collaborators, Bernays, Ackermann, von Neumann, and others, seriously start work on what is now known as the formalist program.

The success or failure of Hilbert's program to save classical mathematics hinges upon the solution of the consistency problem. Freedom from contradiction is guaranteed only by consistency proofs, and the older consistency proofs based upon interpretations and models usually merely shift the question of consistency from one domain of mathematics to another. In other words, a consistency proof by the method of models is only relative. Hilbert, therefore, conceived a new direct approach to the consistency problem. Much as one may prove, by the rules of a game, that certain situations cannot occur within the game, Hilbert hoped to prove, by a suitable set of rules of procedure for obtaining acceptable formulas from the basic symbols, that a contradictory formula can never occur. In logical notation, a contradictory formula is any formula of the type "*F* and not-*F*," where *F* is some accepted formula of the system. If one can show that no such contradictory formula is possible, then one has established the consistency of the system.

The development of the preceding ideas of a direct test for consistency in mathematics is called, by Hilbert, the **proof theory.** Hilbert and Bernays planned to give a detailed exposition (and application to all classical mathematics) of the proof theory in their great *Grundlagen der Mathematik,* which may be considered as the "Principia mathematica" of the formalist school. The *Grundlagen der Mathematik* was finally published in two volumes, Volume I in 1934 and Volume II in 1939, but, as the work was being written, unforeseen difficulties arose, and it was not possible to complete the proof theory. For certain elementary systems, proofs of consistency were carried out, which illustrated what Hilbert would like to have done for all classical mathematics, but, for the system *in toto,* the problem of consistency remained refractory.

As a matter of fact, the Hilbert program, at least in the form originally envisioned by Hilbert, appears to be doomed to failure; this truth was brought out by Kurt Gödel in 1931, actually before the publication of the *Grundlagen* had taken place. Gödel showed, by unimpeachable methods acceptable to the

DAVID HILBERT
(David Smith Collection)

followers of any one of the three principal schools of the philosophy of mathematics, that it is impossible for a sufficiently rich formalized deductive system, such as Hilbert's system for all classical mathematics, to prove consistency of the system by methods belonging to the system. This remarkable result is a consequence of an even more fundamental one; Gödel proved the incompleteness of Hilbert's system—that is, he established the existence within the system of "undecidable" problems, of which consistency of the system is one. These theorems of Gödel are too difficult to consider in their technical details here. They are certainly among the most remarkable in all mathematics, and they reveal an unforeseen limitation in the methods of formal mathematics. They show "that the formal systems known to be adequate for the derivation of mathematics are unsafe in the sense that their consistency cannot be demonstrated by finitary methods formalized within the system, whereas any system known to be safe in this sense is inadequate."[8]

David Hilbert was born in Königsberg in 1862 and received his Ph.D. from the university there in 1885. He taught at the University of Königsberg, first as Privatdozent (1886–1892) and then as professor (1893–1894). In 1895, he became a professor at Göttingen, a post that he held to his retirement in 1930. He died in Göttingen in 1943.

Hilbert was a broad mathematician and made highly important contributions in many areas, usually neatly completing each area before passing on to the next. These areas include the theory of algebraic invariants (1885–1892); the theory of algebraic numbers (1893–1899); the foundations of geometry, which initiated his work in axiomatics (1898–1899); the Dirichlet problem and the calculus of variations (1900–1905); integral equations, including spectral theory and the concept of Hilbert space (to 1912); followed by contributions in mathematical physics to the kinetic theory of gases and the theory of relativity; and, finally, his critical studies of the foundations of mathematics and mathematical logic. His stimulating lectures attracted students from all parts of the world. He was a powerhouse at the University of Göttingen, and, with a galaxy of great colleagues, he made Göttingen a Mecca for mathematicians until the destructive political events of the 1930s. Hilbert received many honors and became editor of *Mathematische Annalen* in 1902. At the International Mathematical Congress in Paris in 1900, he proposed twenty-three significant unsolved mathematical problems, subsequent work upon which has greatly enriched mathematics.

---

[8] F. De Sua, "Consistency and completeness: A résumé," *The American Mathematical Monthly* 63 (1956): 295–305. Here we also find the following interesting remark: "Suppose we loosely define a *religion* as any discipline whose foundations rest on an element of faith, irrespective of any element of reason which may be present. Quantum mechanics for example would be a religion under this definition. But mathematics would hold the unique position of being the only branch of theology possessing a rigorous demonstration of the fact that it should be so classified." See also, Howard Eves, *Foundations and Fundamental Concepts of Mathematics*. 3rd ed. (Appendix A.9) Boston: PWS-KENT Publishing Company, 1990.

15–9 Computers

A very important twentieth-century achievement in the area of mathematics was the development of the simple mechanical calculating aids of earlier days into the remarkable and amazing large-scale electronic computing devices of today. Particularly revolutionary was the concept of incorporating into the machines a program of instruction as well as a set of operating data. We devote this section to a brief history of the calculating aids that culminated in these latest marvels.

Beyond the computational aid given to man by nature in the form of his ten fingers (still used in school classrooms) and the highly efficient and inexpensive abacus of ancient origin (still used in many parts of the world), the invention of the first calculating machine is attributed to Blaise Pascal, who, in 1642, devised an adding machine to assist his father in the auditing of the government accounts at Rouen. The instrument was able to handle numbers not exceeding six digits. It contained a sequence of engaging dials, each marked from 0 to 9, so designed that when one dial of the sequence turned from 9 to 0 the preceding dial of the sequence automatically turned one unit. Thus the "carrying" process of addition was mechanically accomplished. Pascal manufactured over fifty machines, some of which are still preserved in the Conservatoire des Arts et Métiers at Paris. It is interesting that Pascal has also been credited with the invention of the one-wheeled wheelbarrow as we know it today.

Later in the century, Leibniz (1671) in Germany and Sir Samuel Morland (1673) in England invented machines that multiplied. Similar attempts were made by a number of others, but most of these machines proved to be slow and impractical. In 1820, Thomas de Colmar, although not familiar with Leibniz' work, transformed a Leibniz type of machine into one that could perform subtractions and divisions. This machine proved to be the prototype of almost all commercial machines built before 1875, and of many developed since that time. In 1875, the American Frank Stephen Baldwin was granted a patent for the first practical calculating machine that could perform the four fundamental

One of Pascal's arithmetic machines, devised by him in 1642.

Part of Babbage's difference engine.

operations of arithmetic without any resetting of the machine. In 1878, Willgodt Theophile Odhner, a Swede, was granted a United States patent on a machine very similar in design to that of Baldwin. The electrically operated desk calculators, such as those of Friden, Marchant, and Monroe, of the first half of the twentieth century, have essentially the same basic construction as the Baldwin machine.

About 1812, the English mathematician Charles Babbage (1792–1871) began to consider the construction of a machine to aid in the calculation of mathematical tables. He resigned the Lucasian professorship at Cambridge in order to devote all his energies to the construction of his machine. In 1823, after investing and losing his own personal fortune in the venture, he secured financial aid from the British government and set to work to make a *difference engine* capable of employing twenty-six significant figures and of computing and printing successive differences out to the sixth order. But Babbage's work did not progress satisfactorily, and ten years later the governmental aid was withdrawn. Babbage thereupon abandoned his difference engine and commenced work on a more ambitious machine which he called his *analytic engine*, which was intended to execute completely automatically a whole series of arithmetic operations assigned to it at the start by the operator. This machine, also, was never completed, largely because the necessary precision tools were not as yet made.

The first direct descendant of the Babbage analytic engine is the great *IBM Automatic Sequence Controlled Calculator* (the ASCC), completed at Harvard University in 1944 as a joint enterprise by the University and the International Business Machines Corporation under contract for the Navy Department. The machine is fifty-one feet long, eight feet high, with two panels six feet long, and weighs about five tons. An improved second model of the ASCC was made for use, beginning in 1948, at the Naval Proving Ground, Dahlgren, Virginia. Another descendent of Babbage's effort is the *Electronic Numerical Integrator and Computer* (the ENIAC), a multipurpose electronic computer completed in 1945 at the University of Pennsylvania under contract with the Ballistic Research Laboratory of the Army's Aberdeen Proving Ground. This machine requires a thirty-by-fifty foot room, contains 19,000 vacuum tubes, and weighs about thirty tons; it may now be found in the Smithsonian Institution in Washington, D.C. These amazing high-speed computing machines, along with similar projects, like the *Selective Sequence Electronic Calculator* (SSEC) of the International Business Machines Corporation, the *Electronic Discrete Variable Calculator* (EDVAC) of the University of Pennsylvania, the MANIAC of the

CHARLES BABBAGE
(David Smith Collection)

Institute for Advanced Study at Princeton, the *Universal Automatic Computer* (UNIVAC) of the Bureau of Standards, and the various *differential analyzers,* presaged machines of even more fantastic accomplishment. Every few years, a new generation of machines seems to eclipse in speed, reliability, and memory those of the preceding generation. The following table of comparisons of calculations of π performed on electronic computers illustrates the rapid increase in computational speed that took place in the short span of years from 1949 to 1961.

Author	Machine	Date	Decimal Places	Time
Reitwiesner	ENIAC	1949	2037	70 hours
Nicholson and Jeenel	NORC	1954	3089	13 minutes
Felton	Pegasus	1958	10000	33 hours
Genuys	IBM 704	1958	10000	100 minutes
Genuys	IBM 704	1959	16167	4.3 hours
Shanks and Wrench	IBM 7090	1961	100265	8.7 hours

The increase in computational speed in more recent times has been astonishing, as witness the calculation by D. H. Bailey on a Cray-2 supercomputer in 1986 of π to 29,360,000 decimal places in 28 hours. In addition to increase in speed, computers have successively become lighter and more compact. These latter features are largely due to the progression over the years from vacuum tubes to transistors to microchips.

Most of the early computers were designed to solve military problems, but today they are also being designed for business, engineering, government, and other purposes. From luxury tools they have become vital and necessary instruments of modern development. Because of this, numerical analysis has received a tremendous stimulus in recent times and has become a subject of ever-growing importance. It is becoming common for secondary schools to offer introductory courses in computer science with their own minicomputers. University and college Departments of Mathematics are increasingly becoming Departments of Mathematics and Computer Science. Babbage's dream has come true!

Unfortunately, there is a developing feeling, not only among the general public but also among young students of mathematics, that from now on any mathematical problem will be resolved by a sufficiently sophisticated electronic machine, and that all mathematics of today is computer oriented. Teachers of mathematics must combat this disease of "computeritis," and should point out that the machines are merely extraordinarily fast and efficient calculators, and are invaluable in mathematics only in those problems where extensive computing or enumeration can be utilized.

Nonetheless, in their areas of applicability, the machines have secured some remarkable mathematical victories. For example, the recent accomplishments described in Section 3–3 concerning amicable and perfect numbers, and in Section 14–13 concerning prime numbers, would have been virtually impossible without the assistance of a computer. These machines have proved valuable not only in certain parts of number theory, but in other mathematical studies, such as group theory, finite geometries, and recreational mathematics. In the last field, for example, in 1958 Dana S. Scott instructed the MANIAC digital computer to search out all solutions to the problem of putting all twelve pentominos[9] together to form an eight-by-eight square with a two-by-two hole in the middle. After operating for about $3\frac{1}{2}$ hours, the machine produced a complete list of sixty-five distinct solutions, wherein no solution can be obtained from another by rotations and reflections. Similarly, the enumeration and construction of all 880 distinct normal magic squares of order four were achieved with a computer, and it is not difficult to program the corresponding problem for normal magic squares of order five, or higher.

A very spectacular mathematical triumph of the computer was the resolution, in 1976, of the famous four-color conjecture of topology; this conjecture asserts that any map on a plane or a sphere needs at most four colors to color it so that no two countries sharing a common boundary will have the same color. The conjecture emerged about 1850, and an enormous amount of effort has since been spent trying to establish or disestablish it, yielding many partial or allied results but always leaving the conjecture itself unsettled. Then, in the summer of 1976, Kenneth Appel and Wolfgang Haken of the University of Illinois, established the conjecture by an immensely intricate computer-based analysis. The proof contains several hundred pages of complex detail and subsumes over a thousand hours of computer calculation. The method of proof involves an examination of 1936 reducible configurations, each requiring a search of up to half a million logical options to verify reducibility. This last phase of the work occupied six months and was finally completed in June, 1976. Final checking took most of the month of July, and the results were communicated to the *Bulletin of the American Mathematical Society* on July 26, 1976.

The Appel-Haken solution is unquestionably an astounding accomplishment, but a solution based on computerized analyses of close to 2000 cases with a total of something like a billion logical options seems, to many mathematicians, far indeed from elegant mathematics. Certainly on at least an equal footing with a solution to a problem is the elegance of the solution itself. Although a second, and considerably less complex, computer proof of the four-color conjecture was given in the following year, 1977, by F. Allaire, the existence and possible necessity of such treatments of mathematical problems have raised philosophical questions as to just what should be allowed to constitute a *proof* of a mathematical proposition.

Very useful to students, businessmen, and engineers are the pocket-size calculators now available for less than $10 and becoming more sophisticated each year. It was in 1971 that Bowmar Instrument Corporation introduced the first of these calculators into the consumer market; it was a three-by-five inch model selling for $249. Within a year and a half, nearly a dozen firms were selling pocket calculators in the stores. The fierce competition drove the low-

---

[9] A **pentomino** is a planar arrangement of five unit squares joined along their edges.

est-priced machines to below \$100, and another year later the price was under \$50. As early as 1974, annual sales of pocket calculators topped \$10,000,000. With increasing cheapness and new battery designs, the calculators slimmed down roughly to the size and thickness of a credit card, and today they are among the largest selling consumer products with an annual sale in the billions of dollars. These little machines can handle numbers of about eight digits, possess a memory, and instantly perform any arithmetic operation, and also, in some models, trigonometric calculations. They are becoming widely used in the schools and colleges, and specially designed calculator-coordinated courses and textbooks (in such subjects as trigonometry and calculus) are becoming common in the colleges. At the centennial convention of the American Mathematical Society in 1988, the HP 28 S (with original list price of \$235) was introduced. It graphs functions, does differentiation of functions and both definite and indefinite integration, manipulates algebraic functions, solves equations and systems of linear equations, operates with complex numbers and vectors as well as real numbers, and does fairly sophisticated programming with a 16K memory.

Any discussion of modern computers would not be complete without at least a brief mention of the great Hungarian mathematician John von Neumann, for it was he who was most responsible for initiating the first fully electronic calculator and for the concept of a stored program digital computer. His studies of the human brain and of logic proved useful in his researches on the development of the computer.

Von Neumann was born in Budapest in 1903 and was soon recognized as a scientific prodigy. He took his doctorate in Budapest in 1926, migrated to America in 1930, and in 1933 became a permanent member of the Institute for Advanced Study at Princeton. He already had an international reputation for his contributions to operator theory, quantum theory, and game theory. He did much to determine the direction of a great deal of twentieth-century mathematics. His work was remarkably bold and original. During World War II he engaged in scientific and administrative work related to the hydrogen and atomic bombs and to long-range weather forecasting. He died of cancer in 1957.

15–10 The New Math and Bourbaki

Two characteristics of twentieth-century mathematics are an emphasis on abstraction and an increasing concern with the analysis of broad underlying structures and patterns. By mid-century, these characterizing features were noted by those interested in the instruction of school mathematics, and it was felt by some that the features should be incorporated into the teaching of mathematics in the schools. Accordingly, competent and enthusiastic writing groups were formed to revamp and "modernize" the school offerings in mathematics, and the so-called *new math* came into being.

Since abstract ideas of mathematics can often most neatly and concisely be expressed in terms of set concepts and set notation, and since set theory had become recognized as a foundation of mathematics, the new math starts with an elementary introduction to set theory, and then continues with a persistent use of set ideas and set notation. The new math also stresses, as does twentieth-century mathematics, the underlying structures of the subject. Thus, in elementary algebra, much more attention is given than formerly to the basic structures and laws of algebra, such as the commutative, associative, distributive, and other laws. As often happens with new ideas, there was a tendency on the part of some enthusiasts to go overboard and to apply the tenets of the new approach even to situations where they did not lead to clarification or simplification, and some pedagogues expressed a concern that in the endeavor to stress the *why* in mathematics, the *how* was becoming short-changed. There seems little doubt, however, that a saner application of the basic ideas of the new math is probably here to stay.

Since 1939, a comprehensive set of volumes in mathematics, purposely reflecting, at the more advanced level, the tendencies of twentieth-century mathematics, has been appearing in France under the alleged authorship of a Nicolas Bourbaki. Bourbaki first appeared in connection with some notes, reviews, and other papers published in the *Comptes rendus* of the French Academy of Sciences, and elsewhere. Then began the piecemeal construction of Bourbaki's major treatise. The purpose of the major treatise was explained in a paper that appeared translated into English and published in 1950 in *The American Mathematical Monthly* under the title "The architecture of mathematics." A footnote to this paper reads: "Professor N. Bourbaki, formerly of the Royal Poldavian Academy, now residing in Nancy, France, is the author of a comprehensive treatise of modern mathematics, in course of publication under the title *Éléments de Mathématique* (Herman et Cie, Paris, 1939–), of which ten volumes have appeared so far." More than thirty volumes appeared by 1970.

Nicolas Bourbaki's name is Greek, his nationality is French, and he must be ranked as one of the most influential mathematicians of our century. His works are much read and quoted. He has enthusiastic supporters and scathing critics, and, most curious of all, he does not exist.

Nicolas Bourbaki is a collective pseudonym employed by an informal group of mathematicians. Though the members of the organization have taken no oath of secrecy, it has amused most of them to be somewhat cryptic about themselves. Nevertheless, their names are largely an open secret to most mathematicians. It is believed that among the original members were C. Chevalley, J. Delsarte, J. Dieudonné, and A. Weil. The membership has varied over the years, sometimes involving as many as twenty mathematicians. The only rule of the group is to have no rules except compulsory retirement from membership at the age of fifty. The work of the group is based upon the unprovable metaphysical belief that for each mathematical question there is, among the many possible ways of dealing with it, a best, or optimal, way. Although the founders of the Bourbaki group have purposely shrouded the origin of the name Nicolas Bourbaki in mystery, there are a couple of legends that endeavor to explain the choice.

A colorful officer, General Charles Denis Sauter Bourbaki, achieved some fame in the Franco-Prussian War. In 1862, when he was forty-six, he was offered the throne of Greece, which he declined. In a disastrous military campaign in 1871, he was forced to retreat into Switzerland, where he was interned and tried to shoot himself. Apparently his attempt at suicide failed, for he lived to the good age of eighty-three. There is said to be a statue of him in Nancy, France, and this might be the connection between him and the later group of mathematicians, for several of the group were at one time or another associated with the University of Nancy. This explanation still leaves the derivation of the "Nicolas" part of the name unresolved.

Another legend concerning the origin of the name Bourbaki is based upon a story that, around 1930, entering students at the École Normale Supérieure, where so many French mathematicians received their training, were exposed to a lecture by a distinguished visitor named Nicolas Bourbaki, who in reality was merely a disguised amateur actor or perhaps an upperclassman, skilled in seemingly plausible mathematical double talk.

The Bourbaki conception of present-day mathematics, or at least Jean Dieudonné's conception, is that mathematics today is like a ball of many tangled strands of yarn (see Figure 123), where those strands in the center of the ball react tightly upon one another in a nearly unpredictable manner. In this tangle of yarn, there are strands, and ends of strands, that issue outward in various directions and that have no intimate connection with anything within. The Bourbaki method is to snip off these free strands and to concentrate only on the tight core of the ball from which all the rest unravels. The tight core contains the basic structures and the fundamental processes or tools of mathematics—those parts of mathematics that have graduated from strategems to methods, and have attained a considerable degree of fixedness. It is only this part of mathematics that Bourbaki attempts to arrange logically and to shape into a coherent and easily applied theory. It follows that much of mathematics is purposely left outside the province of the Bourbaki group.

15–11 The Tree of Mathematics

It became popular some years ago to picture mathematics in the form of a tree, usually a great oak tree. The roots of the tree were labeled with such titles as *algebra, plane geometry, trigonometry, analytic geometry,* and *irrational num-*

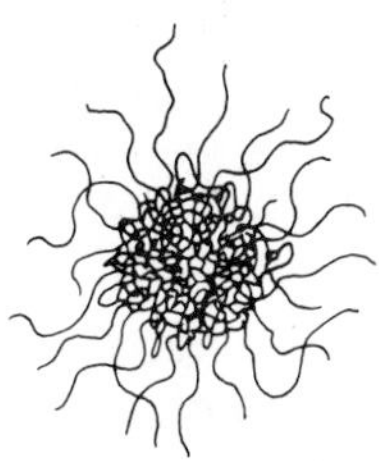

FIGURE 123

bers. From these roots rose the powerful trunk of the tree, on which was printed *calculus*. Then, from the top of the trunk, numerous branches issued and subdivided into smaller branches. These branches bore such titles as *complex variables, real variables, calculus of variations, probability,* and so on, through the various "branches" of higher mathematics.

The purpose of the tree of mathematics was to point out to the student not only how mathematics had historically grown, but also the trail the student should follow in pursuing a study of the subject. Thus, in the schools and perhaps the freshman year at college, students should occupy themselves with the fundamental subjects forming the roots of the tree. Then, early in the college career, through a specially heavy program, the calculus should be thoroughly mastered. After this is accomplished, the student can then ascend those advanced branches of the subject that he or she may wish to pursue.

The pedagogical principle advocated by the tree of mathematics is probably a sound one, for it is based on the famous law pithily stated by biologists in the form: "Ontogeny recapitulates phylogeny," which simply means that, in general, "The individual repeats the development of the group." That is, at least in rough outline, a student learns a subject pretty much in the order in which the subject developed over the ages. As a specific example, consider geometry. The earliest geometry may be called **subconscious geometry,** which originated in simple observations stemming from human ability to recognize physical form and to compare shapes and sizes. Geometry then became **scientific,** or **experimental, geometry,** and this phase of the subject arose when human intelligence was able to extract from a set of concrete geometrical relationships a general abstract relationship (a geometrical law) containing the former as particular cases. In early chapters of this book, we have noted how the bulk of pre-Hellenic geometry was of this experimental kind. Later, actually in the Greek period, geometry advanced to a higher stage and became **demonstrative geometry.** The basic pedagogical principle here under consideration claims, then, that geometry should first be presented to young children in its subconscious form, probably through simple art work and simple observations of nature. Then, somewhat later, this subconscious basis is evolved into scientific geometry, wherein the pupils induce a considerable array of geometric facts through experimentation with compasses and straightedge, with ruler and protractor, and with scissors and paste. Still later, when the student has become sufficiently sophisticated, geometry can be presented in its demonstrative, or deductive, form, and the advantages and disadvantages of the earlier inductive processes can be pointed out.

So we have no quarrel with the pedagogical principle advocated by the tree of mathematics. But what about the tree itself? Does it still present a reasonably true picture of modern mathematics? We think not. A tree of mathematics is clearly a function of time. The oak tree described above certainly could not, for example, have been the tree of mathematics during the great Alexandrian period. The oak tree does represent fairly well the situation in mathematics in the eighteenth century and a good part of the nineteenth century, for in those years the chief mathematical endeavors were the development, extension, and application of the calculus. But with the enormous growth of mathematics in the twentieth century, the general picture of mathematics as given by the oak tree no longer holds. It is perhaps quite correct to say that today the larger part of mathematics has no, or very little, connection with the calculus and its extensions. Consider, for example, the vast areas covered by abstract algebra, finite mathematics, set theory, combinatorics, mathematical logic, axiomatics, nonanalytical number theory, postulational studies of geometry, finite geometries, and on and on.

We must redraw the tree of mathematics if it is to represent the mathematics of today. Fortunately, there is an ideal tree for this new representation—the banyan tree. A banyan tree is a many-trunked tree, ever growing newer and newer trunks. Thus, from a branch of a banyan tree, a threadlike growth extends itself downward until it reaches the ground. There is takes root, and over the succeeding years, the thread becomes thicker and stronger, and, in time, itself becomes a trunk with many branches, each dropping its threadlike growths to the ground.

There are some banyan trees in the world having many scores of trunks, which together cover city blocks in area. Like the great oak tree, these trees are both beautiful and long lived; it is claimed that the banyan tree in India, against which Buddha rested while meditating, is still living and growing. In the banyan tree, then, we have a worthy and more accurate tree of mathematics for today. Over future years, newer trunks will emerge, and some of the older trunks may atrophy and die away. Different students can select different trunks of the tree to ascend, each student first studying the foundations covered by the roots of his chosen trunk. All these trunks, of course, are connected overhead by the intricate branch system of the tree. The calculus trunk is still alive and doing well, but there is also, for example, a linear algebra trunk, a mathematical logic trunk, and others.

Mathematics has become so extensive that today one can be a very productive and creative mathematician and yet have scarcely any knowledge of the calculus and its extensions. We who teach mathematics in the colleges today are probably doing a disservice to some of our mathematics students by insisting that *all* students must first ascend the calculus trunk of the tree of mathematics. In spite of the great fascination and beauty of calculus, not all students of mathematics find it their "cup of tea." By forcing *all* students up the calculus trunk, we may well be killing off some potentially able mathematicians of the noncalculus fields. In short, it is perhaps time to adjust our mathematical pedagogy to fit a tree of mathematics that better reflects the recent historical development of the subject.

15–12 What's Ahead?

There is no crystal ball that reveals the future lines of mathematical development, nor, in view of failures of past efforts, does there seem to be much wisdom in attempting to predict those lines of development. History has shown that areas of mathematics that were particularly alive have suddenly died off, and areas that seemed exhausted have suddenly revived and become highly productive again. Mathematicians have too often witnessed the creation of brand new and completely unanticipated developments in their subject, such as, for example, the recent concepts of category theory, fractals, and catastrophe theory, undreamed of only a few years ago. Who, in the early 1900s, could have foreseen the present fantastic development of the electronic calculators and computers?

Nevertheless, there do seem to be a few points upon which it appears safe to prognosticate, namely:

1. The tremendous and incredible twentieth-century development of computers will for some time continue into the future, leading to a speed of calculation and to applications that at present are scarcely imaginable.
2. Having finally won emancipation in the field of mathematics, women in mathematics will become increasingly more prevalent and important.[10]
3. The cleavage between pure and practical mathematics will continue to become ever more blurred. On the one hand, as G. H. Hardy once pointed out, pure mathematics is the really practical mathematics, because what is useful above all in mathematics is technique, and technique is acquired in pure mathematics. On the other hand, as illustrated by the ultimate application of the ancient Greek study of conic sections to modern celestial mechanics, *all* mathematics is practical mathematics—application is sometimes merely a matter of time.

What do mathematicians themselves feel about the future of their subject? Most feel that the well of mathematics is infinitely deep and that they will be able to continue indefinitely to draw from it. They point out that the sustaining life of mathematics lies in its current supply of unsolved problems. Mathematicians will never give up attempts to solve such problems, and it is in these attempts that new developments in their subject arise. As Julian Lowell Coolidge once remarked, a nice thing about mathematics is that it never solves a problem without creating new ones.

On the other hand, not all mathematicians have shared, or do now share, the optimistic view; a number have expressed fear that the well of mathematics is drying up. No less a creative mathematician than Lagrange imparted to d'Alembert his feelings that "mathematics is beginning to decline," and more than a few present-day mathematicians have expressed a similar fear that the modern drive in mathematics toward more and ever more abstraction is sounding the death knell of the subject.

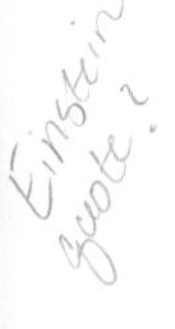

There is an even gloomier concern about the possible future of mathematics. Some feel that mathematics is becoming a Frankenstein monster that will ultimately kill itself. Mathematics plays a cardinal role in our nuclear age. It is largely mathematics that has been responsible for the development of the atomic bomb and other globally destructive weapons, and it seems axiomatic

[10] This prognostication is well buttressed by Edna E. Kramer in the final chapter of her elegant book *The Nature and Growth of Modern Mathematics*.

that any mathematics that *can* be used for destructive purposes *will* be used for those purposes. In connection with this deeper concern about mathematics, witness Norbert Wiener's famous letter, written after the fateful dropping of the bombs on Hiroshima and Nagasaki in 1945, in which he decried the custom of mathematicians freely sharing their knowledge and findings, and witness the pangs of conscience suffered by Albert Einstein and others over the part their mathematics played in certain aspects of the nuclear age. More recently, at the joint meeting of the American Mathematical Society and the Mathematical Association of America held in San Antonio in January 1987, a group of mathematicians called upon their colleagues to refrain from participation in the S. D. I. (Strategic Defense Initiative) "star wars" program.

A growing number of mathematicians feel that we now have two contrasting areas of mathematics, a safe area and an unsafe one, and these mathematicians, on conscientious grounds, try in their research to stay clear of the unsafe area. These mathematicians are bothered by such questions as the following: Can we in the future expect massive migrations of mathematicians, like that in ancient times from the University of Alexandria during the hectic days of the dissolution of the Egyptian empire, and that in modern times from Germany during the repressive days of the Nazi upsurge? Is the world heading toward another Dark Ages, but this time one of perhaps global extent, brought on by a nuclear war or by nuclear pollution? Is the present relationship between mathematics and the military morally defensible?

Let us hope that a sane outcome will prevail, that the grand subject of mathematics will continue to flourish indefinitely, and, paraphrasing words of Carl Gustav Jacobi, that it will continue to ennoble the human mind and spirit.

Problem Studies

15.1 Tacit Assumptions Made By Euclid

Look up (in T. L. Heath's *The Thirteen Books of Euclid's Elements,* for example) the statements and proofs of Propositions I 1, I 16, and I 21, and show that:

(a) In Proposition I 1 Euclid tacitly assumed that 2 circles having centers at the ends of a line segment and having the line segment as a common radius intersect one another.

(b) In Proposition I 16 Euclid tacitly assumed the infinitude of straight lines.

(c) In Proposition I 21 Euclid tacitly assumed that if a straight line enters a triangle at a vertex it must, if sufficiently produced, intersect the opposite side.

15.2 Three Geometrical Paradoxes

If an assumption tacitly made in a deductive development should involve a misconception, its introduction may lead not only to a proposition that does not follow from the postulates of the deductive system, but to one that may actually contradict some previously established proposition of the system. From this point of view, criticize the following 3 geometrical paradoxes:

(a) *To prove that any triangle is isosceles.*

Let ABC be any triangle (see Figure 124). Draw the bisector of $\sphericalangle C$ and the perpendicular bisector of side AB. From their point of intersection E, drop perpendiculars EF and EG on AC and BC, respectively, and draw EA and EB. Now right triangles CFE and CGE are congruent, since each has CE as hypotenuse and since $\sphericalangle FCE = \sphericalangle GCE$; therefore, $CF = CG$. Again, right triangles EFA and EGB are congruent, since leg EF of one equals leg EG of the other (any point E on the bisector of an angle C is equidistant from the sides of the angle) and since hypotenuse EA of one equals hypotenuse EB of the other (any point E on the perpendicular bisector of a line segment AB is equidistant from the extremities of that line segment); therefore, $FA = GB$. It now follows that $CF + FA = CG + GB$, or $CA = CB$, and the triangle is isosceles.

(b) *To prove that a right angle is equal to an obtuse angle.*

Let $ABCD$ be any rectangle (see Figure 125). Draw BE outside the rectangle and equal in length to BC, and hence to AD. Draw the perpendicular bisectors of DE and AB; since they are perpendicular to nonparallel lines, they must intersect in a point P. Draw AP, BP, DP, EP. Then $PA = PB$ and $PD = PE$ (any point on the perpendicular bisector of a line segment is equidistant from the extremities of the line segment). Also, by construction, $AD = BE$. Therefore, triangles APD and BPE are congruent, since the 3 sides of one are equal to the three sides of the other. Hence, $\sphericalangle DAP = \sphericalangle EBP$. But $\sphericalangle BAP = \sphericalangle ABP$, since these angles are base angles of the isosceles triangle APB. By subtraction it now follows that right angle DAG = obtuse angle EBA.

(c) *To prove that there are two perpendiculars from a point to a line.*

Let any 2 circles intersect in A and B (see Figure 126). Draw the diameters AC and AD, and let the join of C and D cut the respective

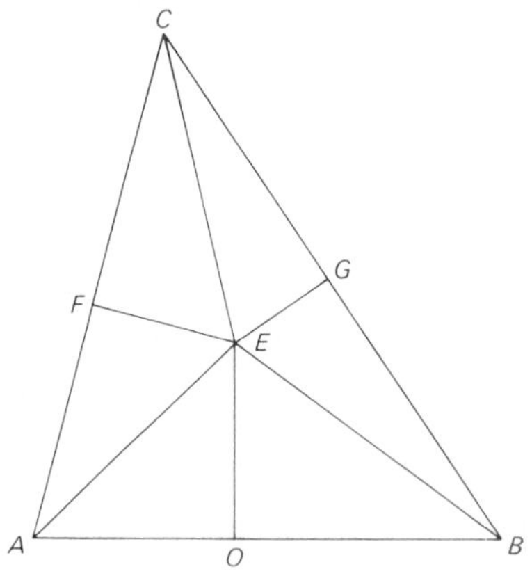

FIGURE 124

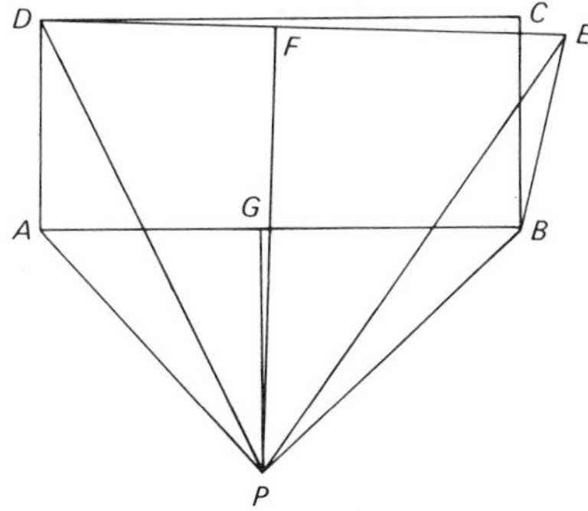

FIGURE 125

circles in M and N. Then angles AMC and AND are right angles, since each is inscribed in a semicircle. Hence, AM and AN are 2 perpendiculars to CD.

15.3 Dedekind's Continuity Postulate

To guarantee the existence of certain points of intersection (of line with circle and circle with circle) Richard Dedekind (1831–1916) introduced into geometry the following continuity postulate: *If all points of a straight line fall into two classes, such that every point of the first class lies to the left of every point of the second class, then there exists one and only one point which produces this division of all points into two classes, that is, this severing of the straight line into two portions.*

(a) Complete the details of the following indicated proof of the theorem: *The straight line segment joining a point A inside a circle to a point B outside the circle has a point in common with the circle.*

Let O be the center and r the radius of the given circle (see Figure 127), and let C be the foot of the perpendicular from O on the line determined by A and B. The points of the segment AB can be divided into 2 classes: those points P for which $OP < r$ and those points Q for

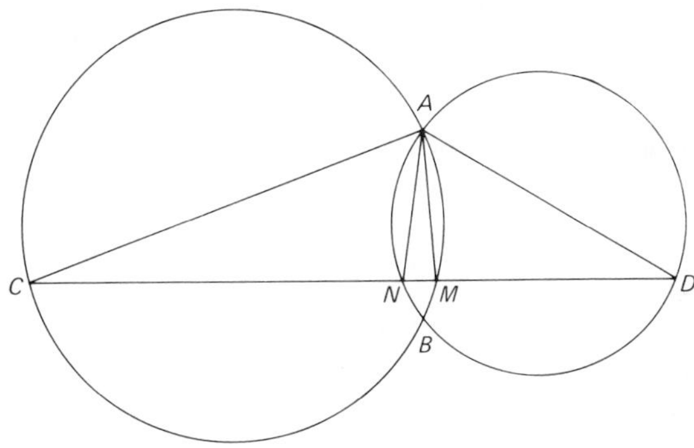

FIGURE 126

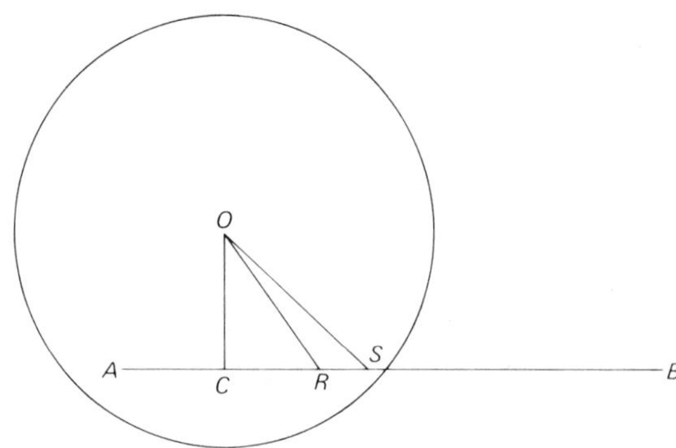

FIGURE 127

which $OQ \geqq r$. It can be shown that, in every case, $CP < CQ$; hence, by Dedekind's Postulate, there exists a point R of AB such that all points that precede it belong to 1 class and all that follow it belong to the other class. Now $OR \nless r$, for otherwise we could choose S on AB, between R and B, such that $RS < r - OR$. But, since $OS < OR + RS$, this would imply the absurdity that $OS < r$. Similarly, it can be shown that $OR \ngtr r$. Hence, we must have $OR = r$, and the theorem is established.

(b) How might Dedekind's Postulate be extended to cover angles?

(c) How might Dedekind's Postulate be extended to cover circular arcs?

15.4 A Coordinate Interpretation of Euclid's Postulates

Let us, for convenience, restate Euclid's first 3 postulates in the following equivalent forms:

1. *Any two distinct points determine a straight line.*
2. *A straight line is unbounded.*
3. *There exists a circle having any given point as center and passing through any second given point.*

Show that Euclid's postulates, partially restated above, hold if the points of the plane are restricted to those whose rectangular Cartesian coordinates for some fixed frame of reference are rational numbers. Show, however, that under this restriction a circle and a line through its center need not intersect each other.

15.5 A Spherical Interpretation of Euclid's Postulates

Show that Euclid's postulates (as partially restated in Problem Study 15.4) hold if we interpret the plane as the surface of a sphere, straight lines as great circles on the sphere, and the points as points on the sphere. Show, however, that in this interpretation the following are true:

(a) Parallel lines do not exist.

(b) All perpendiculars to a given line erected on 1 side of the line intersect in a point.

(c) It is possible to have 2 distinct lines joining the same 2 points.

(d) The sum of the angles of a triangle exceeds 2 right angles.
(e) There exist triangles having all 3 angles right angles.
(f) An exterior angle of a triangle is not always greater than either of the 2 remote interior angles.
(g) The sum of 2 sides of a triangle can be less than the third side.
(h) A triangle with a pair of equal angles may have the sides opposite them unequal.
(i) The greatest side of a triangle does not necessarily lie opposite the greatest angle of the triangle.

15.6 Pasch's Postulate

In 1882, Moritz Pasch formulated the following postulate: *Let* A, B, C, *be three points not lying in the same straight line, and let* m *be a straight line lying in the plane of* ABC *and not passing through any of the points* A, B, C. *Then, if the line* m *passes through a point of segment* AB, *it will also pass through a point of the segment* BC *or a point of the segment* AC. This postulate is 1 of those assumptions classified by modern geometers as a *postulate of order,* and it assists in bringing out the idea of "betweenness."

(a) Prove, as a consequence of Pasch's Postulate, that *if a line enters a triangle at a vertex, it must cut the opposite side.*
(b) Show that Pasch's Postulate does not always hold for a spherical triangle cut by a great circle.

15.7 An Abstract Mathematical System

Consider a set K of undefined elements, which we shall denote by lower case letters, and let R denote an undefined dyadic relation that may or may not hold between a given pair of elements of K. If element a of K is related to element b of K by the R relation we shall write $R(a,b)$. We now assume the following 4 postulates concerning the elements of K and the dyadic relation R.

P1: *If* a *and* b *are any 2 distinct elements of* K, *then we have either* $R(a,b)$ *or* $R(b,a)$.

P2: *If* a *and* b *are any 2 elements of* K *such that we have* $R(a,b)$, *then* a *and* b *are distinct elements.*

P3: *If* a, b, c *are any 3 elements of* K *such that we have* $R(a,b)$ *and* $R(b,c)$, *then we have* $R(a,c)$. (In other words, the R relation is transitive.)

P4: K *consists of exactly 4 distinct elements.*

Deduce the following 7 theorems from the preceding 4 postulates:

T1: *If we have* $R(a,b)$ *then we do not have* $R(b,a)$. (In other words, the R relation is not symmetric.)

T2: *If we have* $R(a,b)$ *and if* c *is in* K, *then we have either* $R(a,c)$ *or* $R(c,b)$.

T3: *There is at least 1 element of* K *not* R*-related to any element of* K. (This is an *existence* theorem.)

T4: *There is at most 1 element of* K *not* R*-related to any element of* K. (This is a *uniqueness* theorem.)

Definition 1: *If we have $R(b,a)$, then we shall say we have $D(a,b)$.*

T5: *If we have $D(a,b)$ and $D(b,c)$, then we have $D(a,c)$.*

Definition 2: *If we have $R(a,b)$ and there is no element c such that we also have $R(a,c)$ and $R(c,b)$, then we shall say we have $F(a,b)$.*

T6: *If we have $F(a,c)$ and $F(b,c)$, then a is identical with b.*

T7: *If we have $F(a,b)$ and $F(b,c)$, then we do not have $F(a,c)$.*

Definition 3: *If we have $F(a,b)$ and $F(b,c)$, then we shall say we have $G(a,c)$.*

15.8 Axiomatics

(a) Establish the consistency of the postulate set of Problem Study 15.7 by means of each of the following interpretations:

1. Let K consist of a man, his father, his father's father, and his father's father's father, and let $R(a,b)$ mean "a is an ancestor of b."
2. Let K consist of 4 distinct points on a horizontal line, and let $R(a,b)$ mean "a is to the left of b."
3. Let K consist of the 4 integers 1, 2, 3, 4, and let $R(a,b)$ mean "$a < b$."

The postulates of this set are those for *sequential relation among four elements*. Any R that interprets the postulates is called a **sequential relation,** and the elements of K are said to form a **sequence.** The interpretations suggested above furnish 3 applications of the abstract branch of mathematics developed in Problem Study 15.7.

(b) Write out the statements of the theorems and definitions in Problem Study 15.7 for each of the interpretations of (a).

(c) Establish the independence of the postulate set of Problem Study 15.7 by means of the following 4 partial interpretations:

1. Let K consist of 2 brothers, their father, and their father's father, and let $R(a,b)$ mean "a is an ancestor of b." This establishes the independence of Postulate P1.
2. Let K consist of the 4 integers 1, 2, 3, 4, and let $R(a,b)$ mean "$a \leqq b$." This establishes the independence of Postulate P2.
3. Let K consist of the 4 integers 1, 2, 3, 4, and let $R(a,b)$ mean "$a \neq b$." This establishes the independence of Postulate P3.
4. Let K consist of the 5 integers 1, 2, 3, 4, 5, and let $R(a,b)$ mean "$a < b$." This establishes the independence of Postulate P4.

(d) Show that P1, T1, P3, P4 constitute a postulate set equivalent to P1, P2, P3, P4.

15.9 Associated Hypothetical Propositions

(a) Prove the proposition: *If a triangle is isosceles, then the bisectors of its base angles are equal.*

(b) State the *converse* of the proposition of (a). (This converse, which is somewhat troublesome to establish, has become known as the *Steiner-Lehmus problem.*)

(c) State the *opposite* (or *inverse*) of the proposition of (a).
(d) If a proposition of the form *If A then B* is true, does it necessarily follow that its converse is true? Its opposite?
(e) Show that if a proposition of the form *If A then B*, and its opposite, are both true, then the converse is also true.
(f) State the propositions that must be true if A is a necessary condition for B; a sufficient condition for B; a necessary and sufficient condition for B. (If A is both necessary and sufficient for B, then A is called a **criterion** for B.)

15.10 Intuition versus Proof

Answer the following questions intuitively, and then check your answers by calculation:

(a) A car travels from P to Q at the rate of 40 miles per hour and then returns from Q to P at the rate of 60 miles per hour. What is the average rate for the round trip?
(b) A can do a job in 4 days, and B can do it in 6 days. How long will it take A and B together to do the job?
(c) A man sells half of his apples at 3 for 17 cents and then sells the other half at 5 for 17 cents. At what rate should he sell all of his apples in order to make the same income?
(d) If a ball of yarn 4 inches in diameter costs 20 cents, how much should you pay for a ball of yarn 6 inches in diameter?
(e) Two jobs have the same starting salary of \$6000 per year and the same maximum salary of \$12,000 per year. One job offers an annual raise of \$800 and the other offers a semiannual raise of \$200. Which is the better-paying job?
(f) Each bacterium in a certain culture divides into 2 bacteria once a minute. If there are 20 million bacteria present at the end of 1 hour, when were there exactly 10 million bacteria present?
(g) Is a salary of 1 cent for the first half month, 2 cents for the second half month, 4 cents for the third half month, 8 cents for the fourth half month, and so on until the year is used up, a good or poor total salary for the year?
(h) A clock strikes six in 5 seconds. How long will it take to strike twelve?
(i) A bottle and a cork together cost \$1.10. If the bottle costs a dollar more than the cork, how much does the cork cost?
(j) Suppose that in 1 glass there is a certain quantity of a liquid A, and in a second glass an equal quantity of another liquid B. A spoonful of liquid A is taken from the first glass and put into the second glass, then a spoonful of the mixture from the second glass is put back into the first glass. Is there now more or less liquid A in the second glass than there is liquid B in the first glass?
(k) Suppose that a large sheet of paper one one-thousandth of an inch thick is torn in half and the 2 pieces put together, 1 on top of the other. These are then torn in half, and the 4 pieces put together in a pile. If

this process of tearing in half and piling is done 50 times, will the final pile of paper be more or less than a mile high?

(l) Is a discount of 15 percent on the selling price of an article the same as a discount of 10 percent on the selling price followed by a discount of 5 percent on the reduced price?

(m) Four-fourths exceeds three-fourths by what fractional part?

(n) A boy wants the arithmetical average of his 8 grades. He averages the first 4 grades, then the last 4 grades, and then finds the average of these averages. Is this correct?

15.11 A Miniature Mathematical System

Consider the following postulate set:

P1: *Every abba is a collection of dabbas.*

P2: *There exist at least two dabbas.*

P3: *If p and q are two dabbas, then there exists one and only one abba containing both p and q.*

P4: *If L is an abba, then there exists a dabba not in L.*

P5: *If L is an abba, and p is a dabba not in L, then there exists one and only one abba containing p and not containing any dabba that is in L.*

(a) What are the primitive terms in this postulate set?

(b) Show that the postulate set is absolutely consistent.

(c) Establish the independence of Postulates P3 and P5.

(d) Deduce the following theorems from the following postulate set:

1. *Every dabba is contained in at least 2 abbas.*
2. *Every abba contains at least 2 dabbas.*
3. *There exist at least 4 distinct dabbas.*
4. *There exist at least 6 distinct abbas.*

15.12 A Set of Inconsistent Statements

If p, q, r represent propositions, show that the following set of 4 statements is inconsistent:

1. If q is true, then r is false.
2. If q is false, then p is true.
3. r is true.
4. p is false.

15.13 A Postulate Set Related to Relativity Theory

Let S be a set of elements and F a dyadic relation satisfying the following postulates:

P1: *If a and b are elements of S and if b F a, then* $a \not F b$. (Here $b\ F\ a$ means that element b is F-related to element a.)

P2: *If a is an element of S, then there is at least 1 element b of S such that b F a.*

P3: *If a is an element of S, then there is at least 1 element b of S such that a F b.*

P4: *If a, b, c are elements of S such that b F a and c F b, then c F a.*

P5: *If a and b are elements of S such that b F a, then there is at least one element c of S such that c F a and b F c.*

(a) Show that the statement, "If a is an element of S, then there is at least 1 element b of S, distinct from a, such that $b \not F a$ and $a \not F b$," is consistent with the above postulates. (This set of postulates, augmented by the above statement, has been used in relativity theory, where the elements of S are interpreted as *instants of time* and F as meaning "follows.")

(b) Rewrite the above postulates and the statement of (a) in terms of the interpretation mentioned in (a).

15.14 Bees and Hives

Consider the following postulate set, in which *bee* and *hive* are primitive terms:

P1: *Every hive is a collection of bees.*

P2: *Any 2 distinct hives have 1 and only 1 bee in common.*

P3: *Every bee belongs to 2 and only 2 hives.*

P4: *There are exactly 4 hives.*

(a) Show that this set of postulates is absolutely consistent.

(b) Show that Postulates P2, P3, and P4 are independent.

(c) Deduce the following theorems from the given postulate set:

T1: *There are exactly 6 bees.*

T2: *There are exactly three bees in each hive.*

T3: *For each bee, there is exactly one other bee not in the same hive with it.*

15.15 Metric Space

In 1906, Maurice Fréchet introduced the concept of a *metric space*. A **metric space** is a set M of elements, called **points,** together with a real number $d(x,y)$, called the **distance function** or **metric** of the space, associated with each ordered pair of points x and y of M, satisfying the following 4 postulates:

M1: $d(x,y) \geqq 0$.

M2: $d(x,y) = 0$ *if and only if* $x = y$.

M3: $d(x,y) = d(y,x)$.

M4: $d(x,y) \leqq d(x,z) + d(z,y)$, *where* x, y, z *are any 3, not necessarily distinct, points of* M. (This is referred to as the **triangle inequality.**)

(a) Show that the set M of all real numbers x, along with $d(x_1,x_2) = |x_1 - x_2|$, is a metric space.

(b) Show that the set M of all ordered pairs $p = (x,y)$ of real numbers, along with

$$d(p_1,p_2) = [(x_1 - x_2)^2 + (y_1 - y_2)^2]^{1/2},$$

where $p_1 = (x_1,y_1)$ and $p_2 = (x_2,y_2)$, is a metric space.

(c) Show that the set M of all ordered pairs $p = (x,y)$ of real numbers, along with

$$d(p_1p_2) = |x_2 - x_1| + |y_2 - y_1|,$$

where $p_1 = (x_1, y_1)$ and $p_2 = (x_2, y_2)$, is a metric space. (By plotting on a Cartesian plane, one readily sees why this metric space is sometimes referred to as **taxicab space.**)

(d) Show that the set M of all ordered pairs $p = (x,y)$ of real numbers, along with

$$d(p_1, p_2) = \max(|x_2 - x_1|, |y_2 - y_1|),$$

where $p_1 = (x_1, y_1)$ and $p_2 = (x_2, y_2)$, is a metric space.

(e) Show that Postulates **M1, M3,** and **M4** of a metric space may be replaced by the single postulate **M1′** : $d(x,y) \leqq d(y,z) + d(z,x)$, *where* x, y, z *are any 3, not necessarily distinct, points of* M.

(f) Show that any set M of elements can be made into a metric space by setting $d(x,y) = 1$ if $x \neq y$, and $d(x,y) = 0$ if $x = y$.

(g) Show that if $d(x,y)$ is a metric for a set M, then we may also use as a metric for M:

1. $k\, d(x,y)$, where k is a positive real number.
2. $[d(x,y)]^{1/2}$.
3. $d(x,y)/[1 + d(x,y)]$. Show that here all distances are less than 1.

(h) Let c be a point of a metric space and let r be a positive real number. Define the set of all points x *of* M such that $d(c,x) = r$ is a **circle,** with **center** c and **radius** r, in the metric space. Describe the appearance of a circle in the Cartesian representations of the metric spaces of (a), (b), (c), and (d).

15.16 Equivalent Segments

(a) Whether an endpoint A (or B) of a line segment AB is to be considered as belonging or not belonging to the segment will be indicated by using a bracket or a parenthesis, respectively, about the letter A (or B). Using this notation, show that the segments $[AB]$, $(AB]$, $[AB)$, (AB), considered as sets of points, are equivalent to one another.

(b) Show that the set of points composing any finite segment and the set composing any infinite segment are equivalent to each other.

15.17 Some Denumerable and Nondenumerable Sets

(a) Prove that the union of a finite number of denumerable sets is a denumerable set.

(b) Prove that the union of a denumerable number of denumerable sets is a denumerable set.

(c) Show that the set of all irrational numbers is nondenumerable.

(d) Show that the set of all transcendental numbers is nondenumerable.

15.18 Polynomials of Heights 1, 2, 3, 4, and 5

(a) Show that 1 is the only polynomial of height 1.

(b) Show that x and 2 are the only polynomials of height 2.

(c) Show that x^2, $2x$, $x + 1$, $x - 1$, and 3 are the only polynomials of height 3, and they yield the distinct algebraic numbers 0, 1, -1.

(d) Form all possible polynomials of height 4 and show that the only new real algebraic numbers contributed are -2, $-1/2$, $1/2$, 2.

(3) Show that polynomials of height 5 contribute 12 more real algebraic numbers.

15.19 The Measure of a Denumerable Set of Points

(a) Complete the details of the following proof that the set of all points on a line segment AB is nondenumerable:

Take the length of AB to be one unit, and assume that the points on AB constitute a denumerable set. The points on AB can then be arranged in a sequence $\{P_1, P_2, P_3, \ldots\}$. Enclose point P_1 in an interval of length $1/10$, point P_2 in an interval of length $1/10^2$, point P_3 in an interval of length $1/10^3$, and so on. It follows that the unit interval AB is entirely covered by an infinite sequence of possibly overlapping subintervals of lengths $1/10$, $1/10^2$, $1/10^3$, ... But the sum of the lengths of these subintervals is

$$1/10 + 1/10^2 + 1/10^3 + \cdots = 1/9 < 1.$$

(b) By choosing the subintervals in (a) to be of lengths $\varepsilon/10$, $\varepsilon/10^2$, $\varepsilon/10^3$, . . . , where ε is an arbitrarily small positive number, show that a denumerable set of points can be covered by a set of intervals the sum of whose lengths can be made as small as we please. (Using the terminology of measure theory, we say that a denumerable set of points has **zero measure.**)

15.20 Transfinite Numbers and Dimension Theory

Let E_1 denote the set of all points on the segment $(0,1]$, and let E_2 denote the set of all points in the unit square $0 < x,y \leqq 1$. A point Z of E_1 may be designated by an unending decimal $z = 0.z_1z_2z_3\ldots$ lying between 0 and 1, and a point P of E_2 may be designated by an ordered pair of unending decimals

$$(x = 0.x_1x_2x_3\ldots, y = 0.y_1y_2y_3\ldots),$$

each decimal lying between 0 and 1. Suppose we let each z_i, x_i, y_i in these representations denote either a nonzero digit or a nonzero digit preceded by a possible block of zeroes. For example, if $z = 0.73028007\ldots$, then $z_1 = 7$, $z_2 = 3$, $z_3 = 02$, $z_4 = 8$, $z_5 = 007$, ... Show that a 1-to-1 correspondence may be set up between the points of E_1 and those of E_2 by associating with the point $0.z_1z_2z_3\ldots$ of E_1 the point

$$(0.z_1z_3z_5\ldots, 0.z_2z_4z_6\ldots)$$

of E_2, and with the point

$$(0.x_1x_2x_3 \ldots, 0.y_1y_2y_3 \ldots)$$

of E_2 the point $0.x_1y_1x_2y_2x_3y_3 \ldots$ of E_1. Thus, show that the set of all points in a unit square has the transfinite number c. (This shows that the dimension of a manifold cannot be distinguished by the transfinite number of the manifold.)

15.21 Circles and Lines

(a) Show that if a circle has a center with at least 1 irrational coordinate, then there are at most 2 points on the circle with rational coordinates.
(b) Show that if a circle has a center with at least 1 transcendental coordinate, then there are at most 2 points on the circle with algebraic coordinates.
(c) Is it possible for a straight line or a circle in the Cartesian plane to contain only points having rational coordinates? Algebraic coordinates?
(d) Show that any infinite set of mutually external closed intervals on a straight line is denumerable.
(e) Show that any infinite set of mutually external circles lying in a plane is denumerable.

15.22 Homeomorphic Surfaces

Two surfaces are **homeomorphic,** or **topologically equivalent,** if we can pass from one to the other by a process of stretching, shrinking, and bending (without tearing or welding), and with cutting, if desired, as long as we ultimately rejoin the 2 lips of each cut in the same fashion in which they were joined before cutting.

(a) Arrange the surfaces formed by the 26 letters of alphabet soup into topologically equivalent classes.
(b) Show that a regular tetrahedron with its edges replaced by material wires yields a surface that is homeomorphic to the surface of a sphere with 3 tea cup handles attached to it.
(c) Explain the facetious remark: "A topologist can't tell the difference between a doughnut and his coffee cup."

15.23 Sides and Edges

(a) The surface formed by a band of paper twisted through 180° and then whose ends are glued together is called a **Möbius strip.** Show that a Möbius strip is 1-sided and has 1 unknotted edge.
(b) Construct a surface that is 1-sided and has 1 knotted edge.
(c) Construct a surface that is 2-sided and has 1 knotted edge.
(d) Construct a surface that is 2-sided and has 1 unknotted edge.

15.24 Paradromic Rings

Discuss the procedure of the witch doctor who gave advice to couples wondering if they should get married. If he wished to prophesy a future breakup in the proposed marriage, he would split an untwisted band; if he wished to prophesy that the couple would quarrel but still stay together, he would split a band having a full twist; if he wished to prophesy a perfect marriage, he would split a Möbius strip.

15.25 Polyhedral Surfaces

(a) Calculate $v - e + f$ for each of the regular polyhedral surfaces. (It can be shown that $v - e + f = 2$ for all polyhedral surfaces homeomorphic to a sphere.)

(b) Give examples of simple closed polyhedral surfaces with 6 edges and with 8 edges, and show that there is none with 7 edges.

(c) Show, from the relation $v - e + f = 2$, that there cannot be more than 5 regular polyhedra.

15.26 Faces and Vertices of Polyhedral Surfaces

Consider a simple closed polyhedral surface P of v vertices, e edges, and f faces. Let f_n denote the number of faces having n edges, and let v_n denote the number of vertices from which n edges issue.

(a) Show that

1. $f = f_3 + f_4 + \cdots$,
2. $v = v_3 + v_4 + \cdots$,
3. $2e = 3f_3 + 4f_4 + 5f_5 + \cdots$,
4. $2e = 3v_3 + 4v_4 + 5v_5 + \cdots$.

Now show, from the relation $v - e + f = 2$, that

5. $2(v_3 + v_4 + \cdots) = 4 + f_3 + 2f_4 + 3f_5 + 4f_6 + \cdots$.

Similarly show that

6. $2(f_3 + f_4 + \cdots) = 4 + v_3 + 2v_4 + 3v_5 + 4v_6 + \cdots$.

Doubling (6) and adding (5) obtain

7. $3f_3 + 2f_4 + f_5 = 12 + 2v_4 + 4v_5 + \cdots + f_7 + 2f_8 + \cdots$.

(b) From (7) of (a) deduce the following:

1. There is no P each of whose faces has more than 5 edges.
2. If P has no triangular or quadrilateral faces, then at least 12 faces of P are pentagonal.
3. If P has no triangular or pentagonal faces, then at least 6 faces of P are quadrilateral.
4. If P has no quadrilateral or pentagonal faces, then at least 4 faces of P are triangular.

(c) P is said to be **trihedral** if precisely 3 edges issue from each vertex. Show that:

1. If P is trihedral and has only pentagonal and hexagonal faces, then the number of pentagonal faces is 12.

2. If P is trihedral and has only quadrilateral and hexagonal faces, then the number of quadrilateral faces is 6.
3. If P is trihedral and has only triangular and hexagonal faces, then the number of triangular faces is 4.

15.27 Hausdorff Space

In 1914, Felix Hausdorff developed an abstract topological space that has since become known as a **Hausdorff space.** Such a space is a set H of elements, called **points,** together with subsets of H, called **neighborhoods,** satisfying the following 4 postulates:

H1: *For each point x of H there exists at least one neighborhood N_x containing x.*

H2: *For any two neighborhoods N_x and N'_x of x, there exists a third neighborhood N''_x contained in both N_x and N'_x.*

H3: *If y is a point of N_x, there is a neighborhood N_y of y such that N_y is contained in N_x.*

H4: *If x and y are distant points of H, there exists an N_x and an N_y that have no point in common.*

(a) Show that the set of all points on a straight line can be made into a Hausdorff space by selecting for neighborhoods of a point x the open segments having x as midpoint. (The arithmetic counterpart of this Hausdorff space is important in the study of analysis.)

(b) Show that the set of all points in the plane can be made into a Hausdorff space by selecting for neighborhoods of a point P the interiors of circles having P as center.

(c) Show that the set of all points in the plane can be made into a Hausdorff space by selecting for neighborhoods of a point P the interiors of squares centered at P and having sides parallel to 2 given perpendicular lines of the plane.

(d) Show that any set of points can be made into a Hausdorff space if we select for neighborhoods the points themselves.

(e) Show that any metric space can be made into a Hausdorff space if we select for neighborhoods the interiors of "circles." (See Problem Study 15.15.)

Definition: A point x of a Hausdorff space H is called a **limit point** of a subset S of H, provided every neighborhood of x contains at least one point of S distinct from x.

(f) Prove that any neighborhood N_x of a limit point x of a subset S of a Hausdorff space H contains an infinite number of points of S.

15.28 Allied Propositions

Related to the proposition "If p, then q" are the following three propositions:

1. The *converse,* "If q, then p."
2. The *inverse,* "If not-p, then not-q."
3. The *contrapositive,* "If not-q, then not-p."

Show that:

(a) The converse of a true implication is not necessarily true.
(b) The inverse of a true implication is not necessarily true.
(c) The contrapositive of a true implication is true.
(d) The contrapositive of an implication is the converse of the inverse of the implication.
(e) Is the inverse of the converse of an implication the same as the converse of the inverse of the implication?

15.29 Three-Valued Logics

(a) Show that there are 256 different ways of developing a truth table for conjunction in a 3-valued logic, where we assume "p and q" is true when and only when both p and q are true.
(b) Show that there are 12 different ways of developing a truth table for negation in a 3-valued logic, where we assume that when p is true, not-p must fail to be true, and when p is false, not-p must fail to be false.
(c) Assuming, as is the case in the customary 2-valued logic, that all the other logical connectives can be formed by defining these other connectives in terms of conjunction and negation, show that there are altogether 3072 possible 3-valued logics.
(d) How many possible m-valued logics are there analogous to the 3072 possible 3-valued logics?

15.30 The Russell Paradox

Consider the following popularizations of the Russell paradox:

(a) Every municipality of a certain country must have a mayor, and no 2 municipalities may have the same mayor. Some mayors do not reside in the municipalities they govern. A law is passed compelling nonresident mayors to reside by themselves in a certain special area A. There are so many nonresident mayors that A is proclaimed a municipality. Where shall the mayor of A reside?
(b) An adjective in the English language is said to be **autological** if it applies to itself; otherwise, the adjective is said to be **heterological.** Thus, the adjectives "short," "English," and "polysyllabic" all apply to themselves and hence are autological, whereas the adjectives "long," "French," and "monosyllabic" do not apply to themselves and hence are heterological. Now is the adjective "heterological" autological or heterological?
(c) Suppose a librarian compiles, for inclusion in his library, a bibliography of all those bibliographies in his library that do not list themselves.

15.31 A Paradox

Examine the following paradox. Every natural number can be expressed in simple English, without the use of numerical symbols; thus, 5 can be expressed as "five," or as "half of ten," or as "the second odd prime," or as "the positive square root of twenty-five," and so on. Now consider the expression, "the least natural number not expressible in fewer than twenty-three syllables." This expression expresses in twenty-two syllables a natural number that cannot be expressed in fewer than twenty-three syllables.

15.32 Some Dilemmas and Some Questions

(a) A crocodile, which has stolen a child, promises the child's father to return the child provided the father guesses whether the child will be returned or not. What should the crocodile do if the father guesses that the child will not be returned?

(b) An explorer has been captured by cannibals who offer the explorer the opportunity to make a statement under the condition that if it is true he will be boiled and if it is false he will be roasted. What should the cannibals do if the explorer states, "I will be roasted."?

(c) Is the statement, "Every general statement has its exceptions," self contradictory?

(d) What would happen if an irresistible force should collide with an immovable body?

(e) If Zeus can do anything, can he make a stone that he cannot lift?

15.33 Recreational Mathematics

(a) Construct all 12 pentominos and empirically find at least 1 of the 65 ways of putting them together to form an 8×8 square with a 2×2 hole in the middle.

(b) Place 8 queens on a chessboard so that no queen can take any other. (This problem was originally proposed by Franz Nauck in 1850. There are 12 fundamental solutions, that is, such that no 2 can be obtained from one another by rotations or reflections.)

Essay Topics

15/1 Bertrand Russell (1872–1970).
15/2 Stories and anecdotes about David Hilbert.
15/3 Hermann Minkowski (1864–1909).
15/4 Hardy and Littlewood.
15/5 Albert Einstein (1879–1955).
15/6 Anna Johnson Pell Wheeler (1883–1966).
15/7 Srinivasa Ramanujan (1887–1920).
15/8 Norbert Wiener (1894–1964).
15/9 Properties of axiom systems.
15/10 A postulate set for Boolean algebra.
15/11 The principle of duality of Boolean algebra.
15/12 Recreational aspects of the Möbius strip.

15/13 Homeomorphic surfaces.
15/14 Impredicative definitions.
15/15 Gödel's theorems.
15/16 Computerized art.
15/17 Computer stamps from the Netherlands.
15/18 The fourth dimension in landscape planning.
15/19 The Polish school of mathematics.
15/20 The psychology of mathematical creation.
15/21 Is mathematics invented or discovered?
15/22 Esthetics of mathematics and mathematics of esthetics.
15/23 Moral obligations of the mathematician.
15/24 Evolution of viewpoints as to what geometry is.
15/25 Stories and anecdotes about Bourbaki.
15/26 Aftermath of the new math.
15/27 Lessons from the tree of mathematics.
15/28 Why study the history of mathematics?
15/29 The history of mathematics as a teaching tool.
15/30 Instances of history of mathematics in the high-school classroom.
15/31 Adolph Hitler's "contribution" to mathematics.
15/32 The Institute for Advanced Study at Princeton.

Bibliography

ALEXANDROFF, PAUL *Elementary Concepts of Topology*. Translated by A. N. Obolensky. New York: Frederick Ungar, 1965.

AMBROSE, ALICE, and MORRIS LAZEROWITZ *Fundamentals of Symbolic Logic*. New York: Holt, Rinehart and Winston, 1948.

APOSTLE, H. G. *Aristotle's Philosophy of Mathematics*. Chicago: University of Chicago Press, 1952.

AUGARTEN, STAN *Bit by Bit, An Illustrated History of Computers*. New York: Picknor & Fields, 1986.

BARKER, S. F. *Philosophy of Mathematics*. Englewood Cliffs, N.J.: Prentice-Hall, 1964.

BEGLE, E. G., ed. *The Role of Axiomatics in Problem Solving in Mathematics*. Boston: Ginn, 1966.

BENACERRAF, PAUL, and HILARY PUTNAM, eds. *Philosophy of Mathematics: Selected Readings*. Englewood Cliffs, N.J.: Prentice-Hall, 1964.

BERKELEY, E. C. *Giant Brains; Or, Machines that Think*. New York: John Wiley, 1949.

BERNAYS, PAUL *Axiomatic Set Theory*. Amsterdam, Holland: North-Holland, 1958.

BERNSTEIN, JEREMY *The Analytical Engine: Computers—Past, Present and Future*. New York: Random House, 1963.

BETH, E. W. *The Foundations of Mathematics*. Amsterdam, Holland: North-Holland, 1959.

——— *Mathematical Thought: An Introduction to the Philosophy of Mathematics*. New York: Gordon & Breach Science Publishers, 1965.

BIGGS, N. L., E. K. LLOYD, and R. J. WILSON *Graph Theory, 1736–1936*. Oxford: Clarendon Press, 1976.

BIRKHOFF, G. D., and RALPH BEATLEY *Basic Geometry*. Chicago: Scott, Foresman, 1940; New York: Chelsea, 1959.

BLACK, MAX *The Nature of Mathematics: A Critical Survey*. London: Routledge & Kegan Paul, 1965.

——— *Critical Thinking: An Introduction to Logic and Scientific Method*. 2d ed. Englewood Cliffs, N.J.: Prentice-Hall, 1952.

BLANCHÉ, ROBERT *Axiomatics*. Translated by G. B. Kleene. London: Routledge & Kegan Paul, 1962.

BLUMENTHAL, L. M. *A Modern View of Geometry*. San Francisco: W. H. Freeman, 1961.

BOCHENSKI, J. M. *A Precis of Mathematical Logic*. Translated by Otto Bird. Dordrecht, Holland: D. Reidel, 1959.

——— *A History of Formal Logic*. Translated by Ivor Thomas. Notre Dame, Ind.: University of Notre Dame Press, 1961.

BOOLE, GEORGE *An Investigation of the Laws of Thought*. New York: Dover Publications, 1951.

BRADIS, V. M., V. L. MINKOVSKII, and A. K. KHARCHEVA *Lapses in Mathematical Reasoning*. New York: Macmillan, 1959.

BREUER, JOS. H. *Introduction to the Theory of Sets*. Translated by H. F. Fehr. Englewood Cliffs, N.J.: Prentice-Hall, 1958.

BROWDER, FELIX, ed. *Mathematical Developments Arising from Hilbert's Problems* (*Proceedings of Symposia in Pure Mathematics,* vol. 28). Providence, R.I.: American Mathematical Society, 1976.

BYNUM, T. W. *Gottlob Frege: Conceptual Notations and Related Articles*. Oxford: Oxford University Press, 1972.

CANTOR, GEORG *Contributions to the Founding of the Theory of Transfinite Numbers*. Translated by P. E. B. Jourdain. La Salle, Ill.: Open Court, 1952.

CARMICHAEL, R. D. *The Logic of Discovery*. Chicago: Open Court, 1930.

CARNAP, RUDOLF *The Logical Syntax of Language*. New York: Brace & World, 1937.

CARTWRIGHT, M. L. *The Mathematical Mind*. New York: Oxford University Press, 1955.

CHAPMAN, F. M., and PAUL HENLE *The Fundamentals of Logic*. New York: Charles Scribner, 1933.

CHURCH, ALONZO *Introduction to Mathematical Logic* (*Part 1*), Annals of Mathematical Studies, No. 13. Princeton, N.J.: Princeton University Press, 1944.

COHEN M. R., and ERNEST NAGEL *Introduction to Logic and Scientific Method*. Harcourt, Brace & World, 1934.

COHEN, PAUL *Set Theory and the Continuum Hypothesis*. New York: W. A. Benjamin, 1966.

COOLEY, J. C. *A Primer of Formal Logic*. New York: Macmillan, 1942.

COURANT, RICHARD, and HERBERT ROBBINS *What is Mathematics? An Elementary Approach to Ideas and Methods*. New York: Oxford University Press, 1941.

CURRY, H. B. *Outlines of a Formalist Philosophy of Mathematics*. Amsterdam, Holland: North-Holland, 1951.

DAVIS, MARTIN *Computability & Unsolvability*. New York: McGraw-Hill, 1958.

DELACHET, ANDRÉ *Contemporary Geometry*. Translated by H. G. Bergmann. New York: Dover Publications, 1962.

DUBBEY, J. M. *Development of Modern Mathematics*. London: Butterworth, 1970.

DUMMETT, MICHAEL *Frege: Philosophy of Language*. New York: Harper & Row, 1973.

——— *Elements of Intuitionism*. Oxford: Oxford University Press, 1977.

ENDERTON, HERBERT *Elements of Set Theory*. New York: Academic Press, 1977.

ENRIQUES, FEDERIGO *The Historic Development of Logic*. Translated by J. Rosenthal. New York: Holt, Rinehart and Winston, 1929.

EVES, HOWARD *A Survey of Geometry,* vol. 2. Boston: Allyn and Bacon, 1965.

——— *Foundations and Fundamental Concepts of Mathematics*. 3rd ed. Boston: PWS-KENT Publishing Company, 1990.

EXNER, R. M., and M. F. ROSSKOPF *Logic in Elementary Mathematics*. New York: McGraw-Hill, 1959.

FANG, J *Hilbert: Towards a Philosophy of Modern Mathematics*. Hauppauge, N.Y.: Paideia Press, 1970.

FEARNSIDE, W. W., and W. B. HOLTHER *Fallacy, the Counterfeit of Argument*. Englewood Cliffs, N.J.: Prentice-Hall, 1959.

FÉLIX, LUCIENNE *The Modern Aspect of Mathematics*. Translated by J. H. and F. H. Hlavaty. New York: Basic Books, 1960.

FORDER, H. G. *The Foundations of Euclidean Geometry*. New York: Cambridge University Press, 1927.

FRAENKEL, ABRAHAM *Abstract Set Theory*. 3d revised edition. Amsterdam, Holland: North-Holland, 1966.

——— *Set Theory and Logic*. Reading, Mass.: Addison-Wesley, 1966.

———, and Y. BAR-HILLEL *Foundations of Set Theory*. Amsterdam, Holland: North-Holland, 1958.

FRÉCHET, MAURICE, and KY FAN *Initiation to Combinatorial Topology*. Translated and augmented with notes by Howard Eves. Boston: Prindle, Weber & Schmidt, 1967.

FREGE, GOTTLOB *The Foundations of Arithmetic*. Translated by J. L. Austing. Evanston, Ill.: Northwestern University Press, 1968.

GALILEO GALILEI *Dialogue Concerning Two New Sciences*. Translated by H. Crew and A. deSalvio. New York: Dover Publications, 1951.

GARDNER, MARTIN *Logic Machines and Diagrams*. New York: McGraw-Hill, 1958.

GÖDEL, KURT *On Undecidable Propositions of Formal Mathematical Systems*. Princeton, N.J.: Princeton University Press, 1934.

——— *Consistency of the Axiom of Choice and of the Generalized Continuum Hypothesis with the Axioms of Set Theory*. Revised edition. Princeton, N.J.: Princeton University Press, 1951.

——— *On Formally Undecidable Propositions of Principia Mathematica and Related Systems*. New York: Basic Books, 1962.

GOLDSTINE, H. H. *The Computer from Pascal to Von Neumann*. Princeton, N.J.: Princeton University Press, 1972.

GOODSTEIN, R. L. *Essays in the Philosophy of Mathematics*. Leicester, England: Leicester University Press, 1965.

GRADSHTEIN, I. S. *Direct and Converse Theorems*. Translation by T. Boddington. New York: Macmillan, 1963.

HADAMARD, JACQUES *An Essay on the Psychology of Invention in the Mathematical Field*. Princeton, N.J.: Princeton University Press, 1945.

HALMOS, PAUL *Naive Set Theory*. Princeton, N.J.: D. Van Nostrand, 1960.

HALSTED, G. B. *Rational Geometry*. New York: John Wiley, 1904.

HARDY, G. H. *Bertrand Russell and Trinity*. Cambridge: Cambridge University Press, 1970.

——— *A Mathematician's Apology*. New York: Cambridge University Press, 1941.

HATCHER, WILLIAM *Foundations of Mathematics*. Philadelphia: W. B. Saunders, 1968.

HAUSDORFF, FELIX *Mengenlehre*. New York: Dover Publications, 1944.

——— *Grundzüge der Mengenlehre*. New York: Chelsea, 1949.

HEATH, T. L. *The Thirteen Books of Euclid's Elements*. 2d ed., 3 vols. New York: Dover Publications, 1966.

HEYTING, A. *Intuitionism: An Introduction*. Amsterdam, Holland: North-Holland Publishing Company, 1956.

HILBERT, DAVID *The Foundations of Geometry*. 10th ed. Revised and enlarged by Paul Bernays. Translated by Leo Unger. Chicago: Open Court, 1971.

———, and WILHELM ACKERMANN *Principles of Mathematical Logic*. Translated by L. M. Hammond, et al. New York: Chelsea, 1950.

HYMAN, ANTHONY *Charles Babbage, Pioneer of the Computer*. Princeton, N.J.: Princeton University Press, 1982.

INFELD, LEOPOLD *Albert Einstein: His Work and Its Influence on Our World*. New York: Charles Scribner's, 1950.

JAMES, GLENN, ed. *The Tree of Mathematics*. Pacoma, Calif.: The Digest Press, 1957.

JOHNSON, P. E. *A History of Set Theory*. Boston: Prindle, Weber & Schmidt, 1972.

KAMKE, E. *Theory of Sets,* Translated by F. Bagemihl. New York: Dover Publications, 1950.

KATTSOFF, LOUIS *A Philosophy of Mathematics*. Freeport, N.Y.: Books for Libraries Press, 1969.

KELLY, J. L. *General Topology*. Princeton, N.J.: Van Nostrand Reinhold, 1955.

KENELLY, J. W. *Informal Logic*. Boston: Allyn and Bacon, 1967.

KEYSER, C. J. *Mathematical Philosophy: A Study of Fate and Freedom*. New York: E. P. Dutton, 1922.

——— *Mathematics and the Question of Cosmic Mind, with Other Essays*. New York: *Scripta Mathematica,* 1935.

——— *Thinking about Thinking*. New York: *Scripta Mathematica,* 1942.

KILMISTER, C. W. *Language, Logic and Mathematics*. New York: Barnes & Noble, 1967.

KING, AMY, and C. B. READ *Pathways to Probability*. New York: Holt, Rinehart and Winston, 1963.

KLEENE, S. C. *Introduction to Metamathematics*. Princeton, N.J.: Van Nostrand Reinhold, 1952.

——— *Mathematical Logic*. New York: John Wiley, 1967.

KLINE, MORRIS *Mathematics: The Loss of Certainty*. New York: Oxford University Press, 1979.

KNEALE, WILLIAM and MARTHA *The Development of Logic*. New York: Oxford University Press, 1962.

KNEEBONE, G. T. *Mathematical Logic and the Foundations of Mathematics.* Princeton, N.J.: Van Nostrand Reinhold, 1963.

KÖRNER, STEPHAN *The Philosophy of Mathematics: An Introduction.* New York: Harper & Row, 1962.

KURATOWSKI, K., and A. FRAENKEL *Axiomatic Set Theory.* Amsterdam, Holland: North-Holland, 1968.

LANGER, S. K. *An Introduction to Symbolic Logic.* 2d revised edition. New York: Dover Publications, 1953.

LASALLE, J. P., and SOLOMON LEFSCHETZ, eds. *Recent Soviet Contributions to Mathematics.* New York: Macmillan, 1962.

LEIBNIZ, G. W. *Logical Papers.* Edited and translated by G. A. R. PARKINSON. New York: Oxford University Press, 1966.

LE LIONNAIS, F., ed. *Great Currents of Mathematical Thought.* 2 vols. Translated by R. A. Hall and H. G. Bergmann. New York: Dover Publications, 1971.

LEVY, AZRIEL *Basic Set Theory.* Berlin: Springer-Verlag, 1979.

LUCHINS, A. S. and E. H. *Logical Foundations of Mathematics for Behavioral Scientists.* New York: Holt, Rinehart and Winston, 1965.

LUKASIEWICZ, JAN *Elements of Mathematical Logic.* Translated by O. Wojtasiewicz. New York: Macmillan, 1963.

MACH, ERNST *Space and Geometry.* Translated by T. J. McCormack. Chicago: Open Court, 1943.

MANHEIM, J. H. *The Genesis of Point Set Topology.* New York: Macmillan, 1964.

MAOR, ELI *To Infinity and Beyond: A Cultural History of Infinity.* Boston: Birkhäuser, 1987.

MAZIARZ, E. A. *The Philosophy of Mathematics.* New York: Philosophical Library, 1950.

MESCHKOWSKI, HERBERT *Ways of Thought of Great Mathematicians.* Translated by John Dyer-Bennet. San Francisco: Holden-Day, 1964.

——— *Evolution of Mathematical Thought.* Translated by J. H. Gayl. San Francisco: Holden-Day, 1965.

MONNA, A. F. *Methods, Concepts and Ideas in Mathematics: Aspects of an Evolution.* CWI Tract V. 23. Math Centrum, 1986.

MOORE, GREGORY *Zermelo's Axiom of Choice: Its Origins, Development and Influence.* New York: Springer-Verlag, 1982.

MORRISON, PHILIP and EMILY *Charles Babbage and His Calculating Engines (selected writings of Charles Babbage and others).* New York: Dover Publications, 1961.

MOSTOWSKI, ANDRZEJ *Thirty Years of Foundational Studies.* New York: Barnes & Noble, 1966.

NAGEL, ERNEST, and J. R. NEWMAN *Gödel's Proof.* New York: New York University Press, 1958.

NEWMAN, M. H. A. *Elements of the Topology of Plane Sets of Points.* New York: Cambridge University Press, 1939.

NEWSOM, C. V. *Mathematical Discourses: The Heart of Mathematical Science.* Englewood Cliffs, N.J.: Prentice-Hall, 1964.

NICOD, JEAN *Foundations of Geometry and Induction.* New York: The Humanities Press, 1950.

POINCARÉ, HENRI *The Foundations of Science.* Translated by G. B. Halsted. Lancaster, Pa.: The Science Press, 1913.

POLYA, GEORGE *How to Solve It: A New Aspect of Mathematical Method*. Princeton, N.J.: Princeton University Press, 1945.

——— *Induction and Analogy in Mathematics*. Princeton, N.J.: Princeton University Press, 1954.

——— *Patterns of Plausible Inference*. Princeton, N.J.: Princeton University Press, 1954.

——— *Mathematical Discovery*. 2 vols. New York: John Wiley, 1962 and 1965.

PRASAD, GANESH *Mathematical Research in the Last Twenty Years*. Berlin: Walter de Gruyter, 1923.

QUINE, W. V. *Mathematical Logic*. New York: W. W. Norton, 1940.

——— *Elementary Logic*. Revised edition. Cambridge, Mass.: Harvard University Press, 1980.

——— *Methods of Logic*. New York: Holt, Rinehart and Winston, 1950.

RAMSEY, F. P. *The Foundations of Mathematics and Other Logical Essays*. New York: The Humanities Press, 1950.

RASHEVSKY, N. *Looking at History Through Mathematics*. Cambridge, Mass.: The M.I.T. Press, 1968.

REICHENBACH, HANS *The Theory of Probability: An Inquiry into the Logical and Mathematical Foundations of the Calculus of Probability*. Berkeley, Calif.: University of California Press, 1949.

REID, CONSTANCE. *Hilbert*. New York: Springer-Verlag, 1970.

——— *Courant in Göttingen and New York. The Story of an Improbable Mathematician*. New York: Springer-Verlag, 1976.

RESCHER, NICHOLAS *Hypothetical Reasoning*. Amsterdam, Holland: North-Holland, 1964.

ROBB, A. A. *A Theory of Time and Space*. New York: Cambridge University Press, 1914.

ROBINSON, G. DE B. *The Foundations of Geometry*. 2d ed. Toronto: University of Toronto Press, 1946.

ROSENBLOOM, P. C. *The Elements of Mathematical Logic*. New York: Dover Publications, 1950.

ROSSER, J. B. *Logic for Mathematicians*. New York: McGraw-Hill, 1953.

———, and A. R. TURQUETTE *Many-valued Logics*. Amsterdam, Holland: North-Holland, 1951.

RUSSELL, BERTRAND *Introduction to Mathematical Philosophy*. 2d ed. New York: Macmillan, 1924.

——— *Mysticism and Logic*. New York: W. W. Norton, 1929.

——— *Principles of Mathematics*. 2d ed. New York: W. W. Norton, 1937.

——— *An Essay on the Foundations of Geometry*. New York: Dover Publications, 1956.

——— *The Autobiography of Bertrand Russell,* 3 vols. London: George Allen and Unwin, Ltd., 1967–1969.

SCHAAF, W. L. *Mathematics, Our Great Heritage: Essays on the Nature and Cultural Significance of Mathematics*. Revised edition. New York: Collier Books, 1963.

SCHOLZ, HEINRICH *Concise History of Logic*. Translated by K. F. Leidecker. New York: Philosophical Library, 1961.

SIERPÍNSKI, WACLAW *Introduction to General Topology*. Translated by C. C. Krieger. Toronto: University of Toronto Press, 1934.

——— *Cardinal and Ordinal Numbers*. 2d ed. Warsaw: Polish Scientific Publications, 1965.

SINGH, JAGJIT *Great Ideas of Modern Mathematics: Their Nature and Use*. New York: Dover Publications, 1959.

STABLER, E. R. *An Introduction to Mathematical Thought*. Reading, Mass.: Addison-Wesley, 1953.

STEIN, DOROTHY *Ada: A Life and a Legacy*. Cambridge, Mass.: The M.I.T. Press, 1985.

STIBITZ, G. R., and J. A. LARRIVEE *Mathematics and Computers*. New York: McGraw-Hill, 1957.

STOLL, R. R. *Sets, Logic and Axiomatic Theories*. San Francisco: W. H. Freeman, 1961.

STYAZHKIN, N. I. *History of Mathematical Logic from Leibniz to Peano*. Cambridge, Mass.: The M.I.T. Press, 1969.

SUPPES, PATRICK *Axiomatic Set Theory*. Princeton, N.J.: Van Nostrand Reinhold, 1960.

TARSKI, ALFRED *Introduction to Logic and to the Methodology of Deductive Sciences*. Translated by O. Helmer. New York: Oxford University Press, 1954.

TEMPLE, GEORGE *100 Years of Mathematics, A Personal Viewpoint*. New York: Springer-Verlag, 1981.

VAN HEIJENOORT, JEAN *From Frege to Gödel*. Cambridge, Mass.: Harvard University Press, 1967.

VON NEUMANN, JOHN *The Computer and the Brain*. New Haven: Yale University Press, 1959.

WAISMANN, FRIEDRICH *Introduction to Mathematical Thinking*. Translated by T. J. Benac. New York: Frederick Ungar, 1951.

WANG, HOA *A Survey of Mathematical Logic*. Amsterdam, Holland: North-Holland, 1963.

WEDBERG, ANDERS *Plato's Philosophy of Mathematics*. Stockholm, Sweden: Almqvist and Wiksell, 1955.

WEYL, HERMANN *Philosophy of Mathematics and Natural Science*. Revised and augmented English edition, based on translation by O. Helmer. Princeton, N.J.: Princeton University Press, 1949.

WHITEHEAD, A. N., and B. RUSSELL *Principia Mathematica*. 2d ed., 3 vols. Cambridge: Cambridge University Press, 1965.

WIENER, NORBERT *I Am a Mathematician: The Later Life of a Prodigy*. Garden City, N.Y.: Doubleday, 1956.

WILDER, R. L. *Introduction to the Foundations of Mathematics*. 2d ed. New York: John Wiley, 1965.

——— *The Evolution of Mathematical Concepts: A Historical Approach*. New York: John Wiley, 1968.

——— *Mathematics as a Cultural System*. New York: Pergamon Press, 1981.

WOLFF, PETER *Breakthroughs in Mathematics*. New York: New American Library, 1963.

WOODGER, J. H. *The Axiomatic Method in Biology*. New York: Cambridge University Press, 1937.

YOUNG, J. W. *Lectures on Fundamental Concepts of Algebra and Geometry*. New York: Macmillan, 1936.

YOUNG, J. W. A., ed. *Monographs on Topics of Modern Mathematics Relevant to the Elementary Field*. New York: Dover Publications, 1955.

ZUCKERMAN, MARTIN *Sets and Transfinite Numbers*. New York: Macmillan, 1974.

GENERAL BIBLIOGRAPHY

ALBERS, D. J., G. L. ALEXANDERSON, and CONSTANCE REID *International Mathematical Congresses: An Illustrated History 1893–1986*. New York: Springer-Verlag, 1986.

———, and G. L. ALEXANDERSON, eds. *Mathematical People: Profiles and Interviews*. Boston: Birkhauser, 1987.

ARCHIBALD, R. C. "Outline of the history of mathematics." Herbert Ellsworth Slaught Memorial Paper No. 2. Buffalo, N.Y.: The Mathematical Association of America, 1949.

BALL, W. W. R. *A Primer of the History of Mathematics*. 4th ed. New York: Macmillan, 1895.

——— *A Short Account of the History of Mathematics*. 5th ed. New York: Macmillan, 1912.

BELL, E. T. *The Development of Mathematics*. 2d ed. New York: McGraw-Hill, 1945.

——— *Mathematics, Queen and Servant of Science*. New York: McGraw-Hill, 1951. Reprinted by the Mathematical Association of America, Washington, D.C., 1987.

BOCHNER, SALOMON *The Role of Mathematics in the Rise of Science*. Princeton, N.J.: Princeton University Press, 1966.

BOYER, C. B. *A History of Mathematics*. 2d ed. New York: John Wiley. Revised by U.C. Merzbach, 1991.

BURTON, D. M. *The History of Mathematics, An Introduction*. Dubuque, Iowa: Wm. C. Brown Company Publishers, 1983.

BOTTAZZINI, UMBERTO *The Higher Calculus: A History of Real and Complex Analysis from Euler to Weierstrass*. Translated by W. Van Egmond. New York: Springer-Verlag, 1986.

CAJORI, FLORIAN *The Teaching and History of Mathematics in the United States*. Washington, D.C.: Government Printing Office, 1890. Reprinted by Scholarly Press, 1974.

——— *A History of Mathematics*. 4th ed. New York: Macmillan, 1924. Reprinted by Chelsea, 1985.

——— *A History of Elementary Mathematics*. 4th ed. New York: Macmillan, 1924. Reprinted by Chelsea, 1985.

——— *A History of Mathematical Notations*. 2 vols. Chicago: Open Court, 1929.

CALINGER, RONALD, ed. *Classics of Mathematics*. Oak Park, Ill.: Moore Publishing Co., 1982.

CAMPBELL, D. M. *The Whole Craft of Number*. Boston: Prindle, Weber & Schmidt, 1976.

———, and J. C. HIGGINS, eds. *Mathematics: People, Problems, Results*. 3 vols. Pacific Grove, Calif.: Wadsworth & Brooks/Cole, 1984.

CARRUCCIO, ETTORE *Mathematics and Logic in History and in Contemporary Thought*. Translated by Isabel Quigly. Chicago: Adline, 1964.

DAUBEN, J. W. *The History of Mathematics from Antiquity to the Present: A Selected Bibliography*. New York: Gadland Publishing, 1985.

DAVID, PHILIP, and REUBEN HERSH *The Mathematical Experience*. Boston: Birkhauser, 1981.

DEDRON, P., and J. ITARD *Mathematics and Mathematicians*. 2 vols. Translated by J. V. Field. London: London Transworld Publications, 1973.

DÖRRIE, HEINRICH *100 Great Problems of Elementary Mathematics: Their History and Solution*. Translated by D. Antin. New York: Dover Publications, 1965.

DUBIN, J. R. *Mathematics: Its Spirit and Evolution*. Boston: Allyn and Bacon, 1973.

EVES, HOWARD *Great Moments in Mathematics (Before 1650)*. Washington, D.C.: The Mathematical Association of America, 1981.

——— *Great Moments in Mathematics (After 1650)*. Washington, D.C.: The Mathematical Association of America, 1982.

——— *Foundations and Fundamental Concepts of Mathematics*. 3rd ed. Boston: PWS-KENT Publishing Company, 1990.

FAUVEL, JOHN, and JEREMY GRAY, eds. *The History of Mathematics: A Reader*. London: Macmillan, 1987.

FINK, CARL *A Brief History of Mathematics*. Translated by W. W. Beman and D. E. Smith. Chicago: Open Court, 1900.

FREEBURY, H. A. *A History of Mathematics*. New York: Macmillan, 1961.

GILLISPIE, C. C., ed. *Dictionary of Scientific Biography*. 16 vols. New York: Charles Scribner, 1970–1980.

GITTLEMAN, ARTHUR *History of Mathematics*. Columbus, Ohio: Charles E. Merrill, 1975.

GRATTAN-GUINESS, I., ed. *From the Calculus to Set Theory, 1630–1910*. London: Duckworth, 1980.

GRINSTEIN, L. S., and P. J. CAMPBELL, eds. *Women of Mathematics: A Bibliographical Sourcebook*. Westport, Conn.: Greenwood Press, 1987.

HOFMANN, J. E. *The History of Mathematics*. New York: Philosophical Library, 1957.

——— *Classical Mathematics, A Concise History of the Classical Era in Mathematics*. New York: Philosophical Library, 1959.

HOGBEN, L. T. *Mathematics for the Millions*. New York: W. W. Norton, 1937.

HOOPER, ALFRED *Makers of Mathematics*. New York: Random House, 1948.

HOWSON, GEOFFREY *A History of Mathematics Education in England*. Cambridge: Cambridge University Press, 1982.

ITÔ, KIYOSI, ed. *Encyclopedic Dictionary of Mathematics*. 2d ed., 4 vol. Cambridge, Mass.: M.I.T. Press, 1987.

JAMES, GLENN, and R. C. JAMES *Mathematical Dictionary*. 2d ed. Princeton, N.J.: Van Nostrand Reinhold, 1959.

KITCHER, PHILIP *The Nature of Mathematical Knowledge*. New York: Oxford University Press, 1983.

KLINE, MORRIS *Mathematics in Western Culture*. New York: Oxford University Press, 1953.

——— *Mathematics and the Physical World*. New York: Thomas Y. Crowell, 1959.

——— *Mathematics, a Cultural Approach*. Reading, Mass.: Addison-Wesley, 1962.

——— *Mathematical Thought from Ancient to Modern Times*. New York: Oxford University Press, 1972.

——— *Mathematics and the Search for Knowledge*. New York: Oxford University Press, 1985.

KRAMER, E. E. *The Main Stream of Mathematics*. Greenwich, Conn.: Fawcett Publications, 1964.

——— *The Nature and Growth of Modern Mathematics*. New York: Hawthorn Books, 1970.

KUZAWA, SISTER MARY GRACE *Modern Mathematics, The Genesis of a School in Poland*. New Haven, Conn.: College & University Press, 1968.

LARRETT, DENHAM *The Story of Mathematics*. New York: Greenberg Publishers, 1926.

LE LIONNAIS, F., ed. *Great Currents of Mathematical Thought*. Translated by S. Hatfield. Freeport, N.Y.: Books for Libraries Press, 1970.

MAY, K. O. *Bibliography and Research Manual of the History of Mathematics*. Toronto: University of Toronto Press, 1973.

MESCHKOWSKI, HERBERT *The Ways of Thought of Great Mathematicians*. San Francisco: Holden-Day, 1964.

MIDONICK, H. O. *The Treasury of Mathematics*. New York: Philosophical Library, 1965.

MORGAN, BRYAN *Men and Discoveries in Mathematics*. London: John Murray, 1972.

MORITZ, R. E. *On Mathematics and Mathematicians*. New York: Dover Publications, 1958.

NEWMAN, JAMES, ed. *The World of Mathematics*. 4 vol. New York: Simon and Schuster, 1956.

OSEN, L. M. *Women in Mathematics*. Cambridge, Mass.: M.I.T. Press, 1974.

PEARSON, E. S. *The History of Statistics in the 17th and 18th Centuries*. New York: Macmillan, 1978.

PERI, TERI *Math Equals: Biographies of Women Mathematicians + Related Activities*. Reading, Mass.: Addison-Wesley, 1978.

PHILLIPS, E. R., ed. *Studies in the History of Mathematics,* MAA Studies in Mathematics, vol. 26. Washington, D.C.: Mathematical Association of America, 1987.

PLEDGE, H. T. *Science Since 1500: A Short History of Mathematics, Physics, Chemistry, and Biology*. New York: Harper Brothers, 1959.

SANFORD, VERA *A Short History of Mathematics*. Boston: Houghton Mifflin, 1930.

SARTON, GEORGE *The Study of the History of Mathematics*. New York: Dover Publications, 1954.

SCOTT, J. F. *A History of Mathematics from Antiquity to the Beginning of the Nineteenth Century*. London: Taylor & Francis, 1958.

SHARLAU, W., and H. OPALKA *From Fermat to Minkowski: Lectures on the Theory of Numbers and Its Historical Development*. New York: Springer-Verlag, 1985.

SMITH, D. E. *Mathematics*. Boston: Marshall Jones, 1923.

——— *History of Mathematics*. 2 vols. Boston: Ginn & Company, 1923–25. Reprinted by Dover Publications, 1958.

——— *A Source Book in Mathematics*. New York: McGraw Hill, 1929. Reprinted by Dover Publications, 1959.

———, and JEKUTHIEL GINSBURG *A History of Mathematics in America Before 1900*. Carus Mathematical Monograph No. 5. Chicago: Open Court, 1934. Reprinted by Arno Press, New York, 1980.

SMITH, S. B. *The Great Mental Calculators*. New York: Columbia University Press, 1983.
STRUIK, D. J. *A Concise History of Mathematics*. Revised edition. New York: Dover Publications, 1987.
——— *A Source Book in Mathematics, 1200–1800*. Cambridge, Mass.: Harvard University Press, 1969. Reprinted by Princeton University Press, 1986.
TARWATER, D., ed. *The Bicentennial Tribute to American Mathematics, 1776–1976*. Washington, D.C.: The Mathematical Association of America, 1977.
TIETZE, HEINRICH *Famous Problems of Mathematics*. New York: Graylock, 1965.
TURNBULL, H. W. *The Great Mathematicians*. New York: New York University Press, 1969.
WEIL, ANDRÉ *Number Theory: An Approach Through History from Hammurabi to Legendre*. Boston: Birkhauser, 1984.
WILLERDING, MARGARET *Mathematical Concepts, A Historical Approach*. Boston: Prindle, Weber & Schmidt, 1967.
YOUNG, L. C. *Mathematicians and Their Times*. Amsterdam, Holland: North-Holland, 1981.

Journal literature on the history of mathematics is vast. For an excellent beginning, one may consult the following list:

READ, C. B. "The history of mathematics—a bibliography of articles in English appearing in six periodicals." *School Science and Mathematics,* February 1966: 147–59. This is a bibliography of over 1000 articles devoted to the history of mathematics and appearing prior to September 15, 1965 in the following six journals: *The American Mathematical Monthly, The Mathematical Gazette* (New Series), *The Mathematics Teacher, National Mathematics Magazine* (volumes 1–8 were published as *Mathematics News Letter;* starting with volume 21, the title became *Mathematics Magazine*), *Scripta Mathematica,* and *School Science and Mathematics*. The articles are classified into some thirty convenient categories.

Of particular value to the researcher and the serious worker in the history of mathematics is the international journal *Historia Mathematica* (volume 1 appeared in May, 1974). It is published by Academic Press in Orlando, Florida.

Highly useful for any teacher of mathematics is *Historical Topics for the Mathematics Classroom* (31st yearbook), National Council of Teachers of Mathematics, Washington, D.C., 1969.

For stories and anecdotes about mathematics and mathematicians, see the following collections:

EVES, HOWARD *In Mathematical Circles*. 2 vols. Boston: Prindle, Weber & Schmidt, 1969.
——— *Mathematical Circles Revisited*. Boston: Prindle, Weber & Schmidt, 1971.
——— *Mathematical Circles Squared*. Boston: Prindle, Weber & Schmidt, 1972.
——— *Mathematical Circles Adieu*. Boston: Prindle, Weber & Schmidt, 1977.
——— *Return to Mathematical Circles*. Boston: PWS and KENT, 1988.

There are some films and videotapes involving the history of mathematics. Many of these may be found in the following catalogue:

SCHNEIDER, D. I. *An Annotated Bibliography of Films and Videotapes for College Mathematics*. Washington, D.C.: The Mathematical Association of America, 1980.

A *Mathematical Sciences Calendar* is published yearly by Rome Press, Inc., Crabtree Valley Station, Box 31451, Raleigh, N.C. 27632.

Excellent and copious is *A Calendar of Mathematical Dates,* composed by V. F. Rickey. This work is computer generated for continual updating and is available from the author, Department of Mathematics and Statistics, Bowling Green State University, Bowling Green, Ohio 43403.

A CHRONOLOGICAL TABLE[1]

It has been estimated that the sun originated about 5 trillion years ago, the earth about 5 billion years ago, and man about 2 million years ago.

−50000	Evidence of counting.
−25000	Primitive geometric art.
−6000	Approximate date of the Ishango bone.
−4700	Possible beginning of Babylonian calendar.
−4228	Hypothetical origin of Egyptian calendar.
−3500	Writing in use; potter's wheel.
−3100	Approximate date of a royal Egyptian mace in a museum at Oxford.
−3000	Discovery of bronze; wheeled vehicles in use.
−2900	Great pyramid of Gizeh erected.
−2400	Babylonian tablets of Ur; positional notation in Mesopotamia.
−2200	Date of many mathematical tablets found at Nippur; mythical date of the *lo-shu,* the oldest known example of a magic square.
−1850	Moscow, or Golenishev, papyrus (twenty-five numerical problems, "greatest Egyptian pyramid"); oldest extant astronomical instrument.
−1750	Rule of Hammurabi; Plimpton 322 dates somewhere from −1900 to −1600.
−1700	Stonehenge in England (?).
−1650	Rhind, or Ahmes, papyrus (85 numerical problems).
−1600	Approximate date of many of the Babylonian tablets in the Yale collection.
−1500	Largest existing obelisk; oldest extant Egyptian sundial.
−1350	Phoenician alphabet; iron discovered; water clocks; date of later mathematical tablets found at Nippur; Rollin papyrus (elaborate bread problems).
−1200	Trojan War.
−1167	Harris papyrus (list of temple wealth).

[1] A minus sign before a date indicates that the date is B.C. Many of the dates are approximate.

−1105 Possible date of the *Chóu-peï,* oldest Chinese mathematical work.
−776 First Olympiad.
−753 Rome founded.
−740 Works of Homer (?).
−650 Papyrus introduced into Greece by this date.
−600 Thales (beginning of demonstrative geometry).
−540 Pythagoras (geometry, arithmetic, music).
−516 Inscription executed on the Behistun rock at the command of Darius the Great.
−500 Possible date of the *Śulvasūtras* (religious writings showing acquaintance with Pythagorean numbers and with geometric constructions); appearance of Chinese rod numerals.
−480 Battle of Thermopylae.
−461 Beginning of Age of Pericles.
−460 Parmenides (sphericity of the earth).
−450 Zeno (paradoxes of motion).
−440 Hippocrates of Chios (reduction of the duplication problem, lunes, arrangement of the propositions of geometry in a scientific fashion); Anaxagoras (geometry).
−430 Antiphon (method of exhaustion).
−429 Plague at Athens.
−425 Hippias of Elis (trisection with quadratrix); Theodorus of Cyrene (irrational numbers); Socrates.
−410 Democritus (atomistic theory).
−404 Athens finally defeated by Sparta.
−400 Archytas (leader of Pythagorean school at Tarentum, applications of mathematics to mechanics).
−399 Death of Socrates.
−380 Plato (mathematics in the training of the mind, Plato's Academy).
−375 Theaetetus (incommensurables, regular solids).
−370 Eudoxus (incommensurables, method of exhaustion, astronomy).
−350 Menaechmus (conics); Dinostratus (quadrature with quadratrix, brother of Menaechmus); Xenocrates (history of geometry); Thymaridas (solution of systems of simple equations).
−340 Aristotle (systematizer of deductive logic).
−336 Alexander the Great began his reign.
−335 Eudemus (history of mathematics).
−332 Alexandria founded.
−323 Alexander the Great died.
−320 Aristaeus (conics, regular solids).
−306 Ptolemy I (Soter) of Egypt.

−300 Euclid (*Elements,* perfect numbers, optics, data).

−280 Aristarchus (Copernican system).

−260 Conon (astronomy, spiral of Archimedes); Dositheus (recipient of several papers by Archimedes).

−250 Stone columns erected by King Aśoka containing earliest preserved examples of our present number symbols.

−240 Nicomedes (trisection with conchoid).

−230 Eratosthenes (prime number sieve, size of the earth).

−225 Apollonius (conic sections, plane loci, tangencies, circle of Apollonius); Archimedes (greatest mathematician of antiquity, measurement of the circle and the sphere, computation of π, area of a parabolic segment, spiral of Archimedes, infinite series, method of equilibrium, mechanics, hydrostatics).

−213 Burning of the books in China.

−210 Great Chinese Wall begun.

−196 Rosetta Stone engraved.

−180 Hypsicles (astronomy, number theory); Diocles (duplication with cissoid).

−140 Hipparchus (trigonometry, astronomy, star catalogue).

−100 Probable date of carvings on the walls of a cave near Poona.

−75 Cicero found the tomb of Archimedes.

−50 Sun-tzï (indeterminate equations).

−44 Death of Julius Caesar.

75 Possible date of Heron (machines, plane and solid mensuration, root extraction, surveying).

100 Nicomachus (number theory); Menelaus (spherical trigonometry); Theodosius (geometry, astronomy); *Arithmetic in Nine Sections;* Plutarch.

150 Ptolemy (trigonometry, table of chords, planetary theory, star catalogue, geodesy, *Almagest*).

200 Probable date of inscriptions carved in the caves at Nasik.

250 Probable date of Diophantus (number theory, syncopation of algebra).

265 Wang Fan (astronomy, π = 142/45); Liu Hui (commentary on the *Arithmetic in Nine Sections*).

300 Pappus (*Mathematical Collection,* commentaries, isoperimetry, projective invariance of cross ratio, Castillon-Cramer problem, arbelos theorem, generalization of Pythagorean theorem, centroid theorems, Pappus' theorem).

320 Iamblichus (number theory).

390 Theon of Alexandria (commentator, edited Euclid's *Elements*).

410 Hypatia of Alexandria (commentator, first woman mentioned in the history of mathematics, daughter of Theon of Alexandria).

460 Proclus (commentator).

476 Birth of Āryabhata; fall of Rome.

480 Tsu Ch'ung-chih's approximation of π as 355/113.

500 Metrodorus and the *Greek Anthology*.

505 Varāhamihira (Hindu astronomy).

510 Boethius (writings on geometry and arithmetic became standard texts in the monastic schools); Āryabhata the Elder (astronomy and arithmetic).

529 School at Athens closed.

530 Simplicius (commentator).

560 Eutocius (commentator).

622 Flight of Mohammed from Mecca.

625 Wang Hs'iao-t'ung (cubic equations).

628 Brahmagupta (algebra, cyclic quadrilaterals).

641 Last library at Alexandria burned.

710 Bede (calendar, finger reckoning).

711 Saracens invade Spain.

766 Brahmagupta's works brought to Baghdad.

775 Alcuin called to the court of Charlemagne; Hindu works translated into Arabic.

790 Harun al-Rashid (caliph patron of learning).

820 Mohammed ibn Mûsâ al-Khowârizmî (wrote influential treatise on algebra and a book on the Hindu numerals, astronomy, "algebra," "algorithm"); al-Mâmûn (caliph patron of learning).

850 Mahāvira (arithmetic, algebra).

870 T̲âbit ibn Qorra (translator of Greek works, conics, algebra, magic squares, amicable numbers).

871 Alfred the Great began his reign.

900 Abû Kâmil (algebra).

920 Al-Battânî, or Albategnius (astronomy).

950 *Bakhshālī manuscript* (date very uncertain).

980 Abû'l-Wefâ (geometric constructions with compasses of fixed opening, trigonometric tables).

1000 Alhazen (optics, geometric algebra); Gerbert, or Pope Sylvester II (arithmetic, globes).

1020 Al-Karkhî (algebra).

1042 Edward the Confessor became king.

1048 Death of al-Biruni.

1066 Norman Conquest.

1095 First Crusade.

1100 Omar Khayyam (geometric solution of cubic equations, calendar).

1115 Important edition of the *Arithmetic in Nine Sections* printed.

1120 Plato of Tivoli (translator from the Arabic); Adelard of Bath (translator from the Arabic).

1130 Jabir ibn Aflah, or Gerber (trigonometry).

1140 Johannes Hispalensis (translator from the Arabic); Robert of Chester (translator from the Arabic).

1146 Second Crusade.

1150 Gherardo of Cremona (translator from the Arabic); Bhāskara (algebra, indeterminate equations).

1170 Murder of Thomas à Becket.

1202 Fibonacci (arithmetic, algebra, geometry, Fibonacci sequence, *Liber abaci*).

1215 Magna Carta.

1225 Jordanus Nemorarius (algebra).

1250 Sacrobosco (Hindu-Arabic numerals, sphere); Nasîr ed-dîn (trigonometry, parallel postulate); Roger Bacon (eulogized mathematics); Ch'in Kiu-shao (indeterminate equations, symbol for zero, Horner's method); Li Yeh (notation for negative numbers); rise of European universities.

1260 Campanus (translation of Euclid's *Elements,* geometry); Yang Hui (decimal fractions, earliest extant presentation of Pascal's arithmetic triangle); reign of Kublai Kahn began.

1271 Marco Polo began his travels.

1296 Invention of eyeglasses (approximately).

1303 Chu Shï-kié (algebra, numerical solution of equations, Pascal's arithmetic triangle).

1325 Thomas Bradwardine (arithmetic, geometry, star polygons).

1349 Black Death destroyed a large part of the European population.

1360 Nicole Oresme (coordinates, fractional exponents).

1431 Joan of Arc burned.

1435 Ulugh Beg (trigonometric tables).

1450 Nicholas Cusa (geometry, calendar reform); printing from movable type.

1453 Fall of Constantinople.

1460 Georg von Peurbach (arithmetic, astronomy, table of sines).

1470 Regiomontanus, or Johann Müller (trigonometry).

1478 First printed arithmetic, in Treviso, Italy.

1482 First printed edition of Euclid's *Elements.*

1484 Nicolas Chuquet (arithmetic, algebra); Borghi's arithmetic.

1489 Johann Widman (arithmetic, algebra, + and − signs).
1491 Calandri's arithmetic.
1492 Columbus discovered America.
1494 Pacioli (*Sūma,* arithmetic, algebra, double entry bookkeeping).
1498 Execution of Savonarola.
1500 Leonardo da Vinci (optics, geometry).
1506 Scipione del Ferro (cubic equation); Antonio Maria Fior (cubic equation).
1510 Albrecht Dürer (curves, perspective, approximate trisection, patterns for folding the regular polyhedra).
1514 Jakon Köbel (arithmetic).
1517 Protestant Reformation.
1518 Adam Riese (arithmetic).
1521 Luther excommunicated.
1522 Tonstall's arithmetic.
1525 Rudolff (algebra, decimals); Buteo (arithmetic).
1530 Da Coi (cubic equation); Copernicus (trigonometry, planetary theory).
1544 Stifel: *Arithmetica integra.*
1545 Ferrari (quartic equation); Tartaglia (cubic equation, arithmetic, science of artillery); Cardano (algebra: *Ars magna*).
1550 Rhaeticus (tables of trigonometric functions); Scheubel (algebra); Commandino (translator, geometry).
1556 First work on mathematics printed in the New World.
1557 Robert Recorde (arithmetic, algebra, geometry, = sign).
1558 Elizabeth became Queen of England.
1564 Birth of Shakespeare; death of Michelangelo.
1570 Billingsley and Dee (first English translation of the *Elements*).
1572 Bombelli (algebra, irreducible case of cubic equations).
1573 Valentin Otho found early Chinese value of π, namely 355/113.
1575 Xylander, or Wilhelm Holzmann (translator).
1580 François Viète, or Vieta (algebra, geometry, trigonometry, notation, numerical solution of equations, theory of equations, infinite product converging to $2/\pi$).
1583 Clavius (arithmetic, algebra, geometry, calendar).
1584 Assassination of William of Orange.
1588 Drake defeated the Spanish Armada.
1590 Cataldi (continued fractions); Stevin (decimal fractions, compound interest table, statics, hydrostatics).
1593 Adrianus Romanus (value of π, problem of Apollonius).
1595 Pitiscus (trigonometry).

1598 Edict of Nantes.

1600 Thomas Harriot (algebra, symbolism); Jobst Bürgi (logarithms); Galileo (falling bodies, pendulum, projectiles, astronomy, telescopes, cycloid); Shakespeare.

1603 Accademia dei Lincei founded (Rome).

1608 Telescope invented.

1610 Kepler (laws of planetary motion, volumes, star polyhedra, principle of continuity); Ludolf van Ceulen (computation of π).

1612 Bachet de Méziriac (mathematical recreations, edited Diophantus' *Arithmetica*).

1614 Napier (logarithms, rule of circular parts, computing rods).

1619 Savilian professorships at Oxford established.

1620 Gunter (logarithmic scale, Gunter's chain in surveying); Paul Guldin (centroid theorems of Pappus); Snell (geometry, trigonometry, refinement of classical method of computing π, loxodromes); landing of the Pilgrims.

1624 Henry Briggs (common logarithms, tables).

1630 Mersenne (number theory, Mersenne numbers, clearinghouse for mathematical ideas); Oughtred (algebra, symbolism, slide rule, first table of natural logarithms); Mydorge (optics, geometry); Albert Girard (algebra, spherical geometry).

1635 Fermat (number theory, maxima and minima, probability, analytic geometry, Fermat's last "theorem"); Cavalieri (method of indivisibles).

1636 Harvard College founded.

1637 Descartes (analytic geometry, folium, ovals, rule of signs).

1640 Desargues (projective geometry); de Beaune (Cartesian geometry); Torricelli (physics, geometry, isogonic center); Frénicle de Bessy (geometry); Roberval (geometry, tangents, indivisibles); de la Loubère (curves, magic squares).

1643 Louix XIV crowned.

1649 Charles I executed.

1650 Blaise Pascal (conics, cycloid, probability, Pascal triangle, calculating machines); John Wallis (algebra, imaginary numbers, arc length, exponents, symbol for infinity, infinite product converging to $\pi/2$, early integration); Frans van Schooten (edited Descartes and Viète); Grégoire de Saint-Vincent (circle squarer, other quadratures); Wingate (arithmetic); Nicolaus Mercator (trigonometry, astronomy, series computation of logarithms); John Pell (algebra, incorrectly credited with the so-called Pell equation).

1660 Sluze (spirals, points of inflection); Viviani (geometry); Brouncker (first president of Royal Society, rectification of parabola and cycloid, infinite series, continued fractions); the Restoration.

1662 Royal Society founded (London).

1663 Lucasian professorships at Cambridge established.

1666 French Academy founded (Paris).

1670 Barrow (tangents, fundamental theorem of the calculus); James Gregory (optics, binomial theorem, expansion of functions into series, astronomy); Huygens (circle quadrature, probability, evolutes, pendulum clocks, optics); Sir Christopher Wren (architecture, astronomy, physics, rulings on hyperboloid of one sheet, arc length of cycloid).

1671 Giovanni Domenico Cassini (astronomy, Cassinian curves).

1672 Mohr (geometric constructions with limited tools).

1675 Greenwich observatory founded.

1680 Sir Isaac Newton (fluxions, dynamics, hydrostatics, hydrodynamics, gravitation, cubic curves, series, numerical solution of equations, challenge problems); Johann Hudde (theory of equations); Robert Hooke (physics, spring-balance watches); Seki Kōwa (determinants, calculus).

1682 Leibniz (calculus, determinants, multinomial theorem, symbolic logic, notation, calculating machines); *Acta eruditorum* founded.

1685 Kochanski (approximate rectification of circle).

1690 Marquis de l'Hospital (applied calculus, indeterminate forms); Halley (astronomy, mortality tables and life insurance, translator); Jakob (James, Jacques) Bernoulli (isochronous curves, clothoid, logarithmic spiral, probability); de la Hire (curves, magic squares, maps); Tschirnhausen (optics, curves, theory of equations).

1691 Rolle's theorem in the calculus.

1700 Johann (John, Jean) Bernoulli (applied calculus); Giovanni Ceva (geometry); David Gregory (optics, geometry); Parent (solid analytic geometry).

1706 William Jones (first use of π for circle ratio).

1715 Taylor (expansion in series, geometry).

1720 De Moivre (actuarial mathematics, probability, complex numbers, Stirling's formula).

1731 Alexis Clairaut (solid analytic geometry).

1733 Saccheri (forerunner of non-Euclidean geometry).

1734 Bishop Berkeley (attack on the calculus).

1740 Marquise du Châtelet (French translation of Newton's *Principia*); Frederick the Great became King of Prussia.

1743 Maclaurin (higher plane curves, physics).

1748 Agnesi (analytic geometry, witch of Agnesi).

1750 Euler (notation, $e^{i\pi} = -1$, Euler line, $v - e + f = 2$, quartic equation, ϕ-function, beta and gamma functions, applied mathematics); Cramer's rule.

1770 Lambert (non-Euclidean geometry, hyperbolic functions, map projection, irrationality of π).

1776 United States' independence.

1777 Comte du Buffon (calculation of π by probability).

1780 Lagrange [calculus of variations, differential equations, mechanics, numerical solution of equations, attempted rigorization of calculus (1797), theory of numbers].

1789 French Revolution.

1790 Meusnier (surfaces).

1794 École Polytechnique and École Normale founded; Monge (descriptive geometry, differential geometry of surfaces).

1797 Mascheroni (geometry of compasses); Wessel (geometric representation of complex numbers).

1799 Republic of France adopted the metric system of weights and measures; Rosetta Stone unearthed.

1800 Gauss (polygon construction, number theory, differential geometry, non-Euclidean geometry, fundamental theorem of algebra, astronomy, geodesy).

1803 Carnot (modern geometry).

1804 Napoleon made emperor.

1805 Laplace (celestial mechanics, probability, differential equations); Legendre [*Éléments de géométrie* (1794), theory of numbers, elliptic functions, method of least squares, integrals].

1806 Argand (geometrical representation of complex numbers).

1810 Gergonne (geometry, editor of *Annales*).

1815 "The Analytical Society" at Cambridge; Battle of Waterloo.

1816 Germain (theory of elasticity, mean curvature).

1819 Horner (numerical solution of equations).

1820 Poinsot (geometry).

1822 Fourier (mathematical theory of heat, Fourier series); Poncelet (projective geometry, ruler constructions); Feuerbach's theorem.

1824 Thomas Carlyle (English translation of Legendre's *Géométrie*).

1826 *Crelle's Journal;* principle of duality (Poncelet, Plücker, Gergonne); elliptic functions (Abel, Gauss, Jacobi).

1827 Cauchy (rigorization of analysis, functions of a complex variable, infinite series, determinants); Abel (algebra, analysis).

1828 Green (mathematical physics).

1829 Lobachevsky (non-Euclidean geometry); Plücker (higher analytic geometry).

1830 Poisson (mathematical physics, probability); Peacock (algebra); Bolzanno (series); Babbage (computing machines); Jacobi (elliptic functions, determinants).

1831 Somerville (exposition of Laplace's *Mécanique céleste*).
1832 Bolyai (non-Euclidean geometry); Galois (groups, theory of equations).
1834 Steiner (higher synthetic geometry).
1836 *Liouville's Journal.*
1837 Trisection of an angle and duplication of cube proved impossible.
1839 *Cambridge Mathematical Journal,* which in 1855 became *Quarterly Journal of Pure and Applied Mathematics.*
1841 *Archiv der Mathematik und Physik.*
1842 *Nouvelles annales de mathématiques.*
1843 Hamilton (quaternions).
1844 Grassmann (calculus of extension).
1846 Rawlinson deciphered the Behistun rock.
1847 Staudt (freed projective geometry of metrical basis).
1849 Dirichlet (number theory, series).
1850 Mannheim (standardized the modern slide rule).
1852 Chasles (higher geometry, history of geometry).
1854 Riemann (analysis, non-Euclidean geometry, Riemannian geometry); Boole (logic).
1855 Zacharias Dase (lightning calculator).
1857 Cayley (matrices, algebra, higher-dimensional geometry).
1865 London Mathematical Society founded; *Proceedings of the London Mathematical Society.*
1872 Société Mathématique de France founded; Klein's *Erlanger Programm;* Dedekind (irrational numbers).
1873 Hermite proved e transcendental; Brocard (geometry of the triangle).
1874 Georg Cantor (set theory, irrational numbers, transcendental numbers, transfinite numbers).
1877 Sylvester (algebra, invariant theory).
1878 *American Journal of Mathematics.*
1881 Gibbs (vector analysis).
1882 Lindemann (transcendence of π, squaring of circle proved impossible).
1884 Circolo Matematico di Palermo founded.
1887 *Rendiconti.*
1888 Lemoine (geometry of the triangle, geometrography); American Mathematical Society founded (at first under a different name; *Bulletin of the American Mathematical Society*); Kovalevsky (partial differential equations, Abelian integrals, Prix Bordin).
1889 Peano (axioms for the natural numbers).

1890 Weierstrass (arithmetization of analysis); Deutsche Mathematiker-Vereinigung organized.

1892 *Jahresbericht.*

1894 Scott (geometry of curves); *The American Mathematical Monthly.*

1895 Poincaré (*Analysis situs*).

1896 Prime number theorem proved by Hadamard and de la Vallée Poussin.

1899 Hilbert (*Grundlagen der Geometrie,* formalism).

1900 *Transactions of American Mathematical Society.*

1903 Lebesgue integration.

1906 Grace Young (first woman to receive a German doctorate through the regular examination process, set theory); Fréchet (functional analysis, abstract spaces).

1907 Brouwer (intuitionism).

1909 Russell and Whitehead (*Principia mathematica,* logicism).

1914 Start of World War I.

1915 Mathematical Association of America founded.

1916 Einstein (general theory of relativity).

1917 Hardy and Ramanujan (analytical number theory); Russian Revolution.

1922 E. Noether (abstract algebra, rings, ideal theory).

1923 Banach spaces.

1927 Lindberg flew across the Atlantic.

1931 Gödel's theorem.

1933 Hitler became chancellor of Germany; Institute for Advanced Study founded at Princeton.

1934 Gelfond's theorem.

1939 Bourbaki's works commenced.

1941 Pearl Harbor bombed.

1944 IBM Automatic Sequence Controlled Calculator (ASCC).

1945 Electronic Numerical Integrator and Computer (ENIAC); Hiroshima bombed.

1948 Improved ASCC installed at Naval Proving Ground, Dahlgren, Virginia.

1963 P. J. Cohen on the continuum hypothesis; President Kennedy assassinated.

1971 First pocket calculator offered for sale in the consumer market; Association for Women in Mathematics founded.

1976 Four-color conjecture established by K. Appel and W. Haken.

1985 Supercomputers in use.

1987 Bieberbach conjecture established.

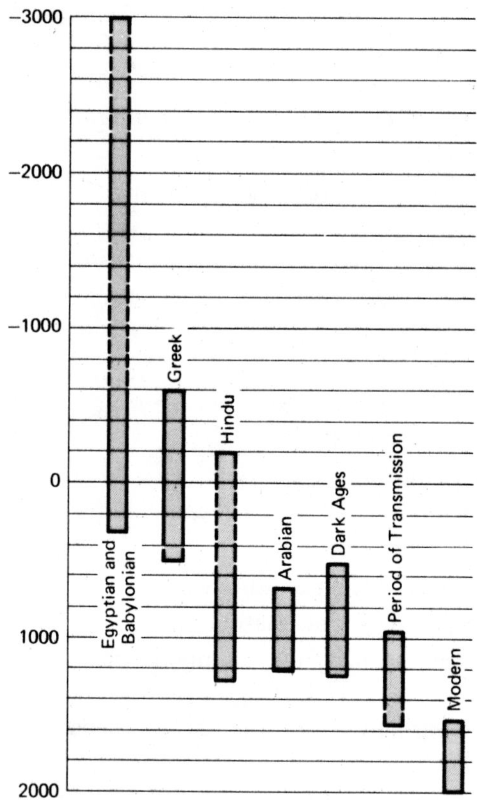

FIGURE 128 **Mathematical periods.**

ANSWERS AND SUGGESTIONS FOR THE SOLUTION OF THE PROBLEM STUDIES

1.1 **(a)** "one man" = 20 (10 fingers plus 10 toes), etc.

(b) If one counts on the fingers of the open hand by folding the fingers down one by one, when 5 is reached all fingers are folded down and the hand has "come to an end," or "died."

(c) The peak-finger is the middle finger, which would denote 3 when counting on one's fingers, starting with the little finger as 1.

(d) Here we have number names that originated from gestures formerly used to express the numbers.

(e) A husband and wife sleep on the same mattress.

(f) This refers to the 9 months of pregnancy.

1.3 **(a)** 27, 3, 2.

(b) 5780 = $\varepsilon'\psi\pi$, 72,803 = $\zeta M\beta'\omega\gamma$, 450,082 = $\mu M\varepsilon M\pi\beta$, 3,257,888 = $\tau M\kappa M\varepsilon M\zeta'\omega\pi\eta$.

1.4 **(d)** 360 = $2(5^3) + 4(5^2) + 2(5)$ = (()))))**,
252 = $2(5)^3) + 2(1)$ = ((//,
78 = $3(5^2) + 3(1)$ =)))///,
33 = $1(5^2) + 1(5) + 3(1)$ =)*///.

(e) 360 = *(* #, 252 = * ##*, 78 =)#), 33 = //).

1.5 **(a)** Note that $ab = [(a - 5) + (b - 5)]10 + (10 - a)(10 - b)$.

1.6 **(b)** Multiply the decimal fraction by b, then the decimal part of this product by b, and so forth.

(c) $(.3012)_4 = 99/128 = .7734375$.

1.8 **(a)** First express to base 10, and then to base 8.

(b) 9, 8, 7.

(c) no, yes, yes, no.

(d) In the first case, we have $79 = b^2 + 4b + 2$.

(e) Denoting the digits by a, b, c we have $49a + 7b + c = 81c + 9b + a$, where a, b, c are less than 7.

(f) We must have $3b^2 + 1 = t^2$, t and b positive integers, $b > 3$.

1.9 **(a)** Express w in the binary scale.

1.10 **(a)** Let t be the tens digit and u the units digit. Following instructions we have

$$2(5t + 7) + u = (10t + u) + 14$$

as the announced final result. The trick is now obvious.

2.1 **(a)** Suppose n is regular. Then

$$\frac{1}{n} = a_0 + \frac{a_1}{60} + \cdots + \frac{a_r}{60^r} = \frac{a_0 60^r + a_1 60^{r-1} + \cdots + a_r}{60^r} = \frac{m}{60^r}, \text{ say.}$$

It follows that $mn = 60^r$ and n can have no prime factors other than those of 60.

(e) 3.

2.2 **(a)** We have $(1.2)^x = 2$, whence $x = (\log 2)/(\log 1.2)$.

2.4 **(a)** We have $x^2 + y^2 = 1000$, $y = 2x/3 - 10$.

(d) 20, 12.

(e) altitude of trapezoid = 24.

(f) 0;18.

(g) Yes.

2.5 **(c)** 31;15.

(d) Denoting the right members of the given equations by a and b, respectively, one finds that $x^8 + a^2x^4 = b^2$.

2.6 **(b)** Set $x = 2y$.

(c) Eliminate x and y, obtaining a cubic equation in z.

(d) Take the cubic in x with unit leading coefficient and subject it to a linear transformation of the type $x = y + m$. Determine m so that the resulting cubic in y lacks the linear term.

2.8 **(b)** Express, in the binary scale, the factor that is successively halved.

2.9 **(c)** Take $p = 1, 3, 9$ in turn.

(e) If $n = 3a$, the other unit fraction is $1/2a$.

(f) If $n = 5a$, the other unit fraction is $1/3a$.

(h) Apply the relation given in (d).

2.10 **(a)** $2/7 = 1/4 + 1/28$.

(b) $2/97 = 1/49 + 1/4753$.

(c) Denote the given fraction by a/b, where $a < b$, and let

$$b/a = x + r/a, \; r < a.$$

Then

$$a/b = 1/(x + r/a), \; 0 < r/a < 1.$$

But

$$1/x > 1/(x + r/a) > 1/(x + 1).$$

2.11 **(b)** Yes.

(c) $5\frac{1}{2}$.

(d) $(35)^2/13$ cubits.

2.12 **(a)** By *fraction,* Ahmes means *unit fraction.* Only the denominators of unit fractions were written.

(c) Let x be the largest share and d the common difference in the arithmetic progression. Then we find $5x - 10d = 100$ and $11x - 46d = 0$.

2.13 **(a)** 256/81, or approximately 3.16.

(c) Consider the right triangle T_1 with legs a and b, and any other triangle T_2 with sides a and b. Place T_2 on T_1 so that one pair of equal sides coincide. Or use the formula $K = (1/2)ab \sin C$.

(d) Draw the diagonal DB and use (c).

(e) $(a + c)(b + d)/4 = [(ad + bc)/2 + (ab + cd)/2]/2$. Now use (d).

(f) The corollary is not correct.

2.14 **(b)** Start with $\sqrt{m} - \sqrt{n} \geqq 0$.

(c) Complete the pyramid of which the frustum is a part, and express the volume of the frustum as the difference between the volumes of the completed and added pyramids.

2.15 **(a)** 3, 4.

(b) 4, 10.

2.16 The 4 right triangles having legs of lengths 3 and 4, along with the small unit square, form a square whose area is 25. It follows that the hypotenuse of a right triangle having legs 3 and 4 is 5. Since a triangle is determined by its 3 sides, it now follows that a 3, 4, 5 triangle is a right triangle.

3.2 **(a)** Show that $2^{mn} - 1$ contains the factor $2^m - 1$.

(b) 8128.

(c) If $a_1, a_2, \ldots, a_n$ represent all the divisors of N, then $N/a_1, N/a_2, \ldots, N/a_n$ also represent all the divisors of N.

(d) The sum of the proper divisors of p^n is $(p^n - 1)/(p - 1)$.

(h) (1) For $n = 7$ we have $2^6(2^7 - 1) = 2^{13} - 2^6 \approx 2^{13}$. Log $2^{13} = 13 \log 2 = 4^+$. Therefore the answer is 4.

(i) The five-link sociable chain is 12496, 14288, 15472, 14536, 14264.

(j) The divisors of 120 are 1, 2, 3, 4, 5, 6, 8, 10, 12, 15, 20, 24, 30, 40, 60, 120.

(k) Yes.

3.3 **(a)** 1, 6, 15, 28.

(b) An oblong number is of the form $a(a + 1)$.

(d) See Figure 129.

(g) $2^{n-1}(2^n - 1) = 2^n(2^n - 1)/2$.

(h) $a = (m - 2)/2$, $b = (4 - m)/2$.

(i) $a = 5/2$, $b = -3/2$.

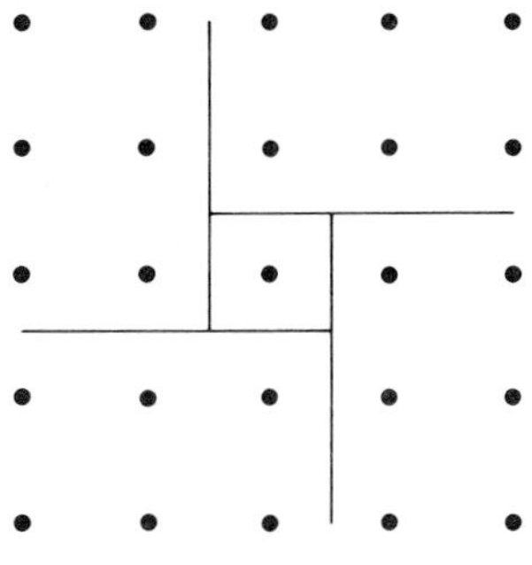

FIGURE 129

3.4 **(a)** Use the fact that $(a - b)^2 \geqq 0$.

(c) Multiply the first equation by b and the second by a, and then eliminate ab/n.

(e) A cube has 8 vertices, 12 edges, and 6 faces.

(f) Set $m = a/(b + c)$, $n = c/(a + b)$. Using the fact that $b = 2ca/(c + a)$, show that $2mn/(m + n) = b/(c + a)$.

3.6 **(c)** If there were an isosceles right triangle with integral sides, then $\sqrt{2}$ would be rational.

(d) If there are positive integers a, b, c $(a \neq 1)$ such that $a^2 + b^2 = c^2$ and $b^2 = ac$, then a, b, c cannot be relatively prime. But if there is a Pythagorean triple in which one integer is a mean proportional between the other two, there must be a primitive Pythagorean triple of this sort.

(g) Show that $(3a + 2c + 1)^2 + (3a + 2c + 2)^2 = (4a + 3c + 2)^2$ if $a^2 + (a + 1)^2 = c^2$.

(h) Use (g).

(i) In the parametric representation of primitive Pythagorean triples, given in Section 2–6, either u or v must be even, whence leg a is a multiple of 4. If u or v is a multiple of 3, then leg a is a multiple of 3. If neither u nor v is a multiple of 3, then u is of the form $3m \pm 1$ and v is of the form $3n \pm 1$, and it follows that $u^2 - v^2$ is a multiple of 3, and therefore leg b is a multiple of 3. If u or v is a multiple of 5, then leg a is a multiple of 5. If neither u nor v is a multiple of 5, then u is of the form $5m \pm 1$ or $5m \pm 2$ and v is of the form $5n \pm 1$ or $5n \pm 2$. If $u = 5m \pm 1$ and $v = 5n \pm 1$, or if $u = 5m \pm 2$ and $v = 5n \pm 2$, then $u^2 - v^2$ is a multiple of 5. If $u = 5m \pm 1$ and $v = 5n \pm 2$, or if $u = 5m \pm 2$ and $v = 5n \pm 1$, then $u^2 + v^2$ is a multiple of 5. It follows that either leg b is a multiple of 5 or hypotenuse c is a multiple of 5.

(j) If n is odd and >2, $(n, (n^2 - 1)/2, (n^2 + 1)/2)$ is a Pythagorean triple. If n is even and >2, $(n, n^2/4 - 1, n^2/4 + 1)$ is a Pythagorean triple.

(k) Since $a^2 = (c - b)(c + b)$ it follows that $b + c$ is a factor of a^2. Therefore $b < a^2$ and $c < a^2$, and the number of combinations of such natural numbers b and c is finite.

3.7 **(a)** If the line should pass through the point (a,b) of the coordinate lattice, we would have $\sqrt{2} = b/a$, a rational number.

(c) Assume $\sqrt{p} = a/b$, where a and b are relatively prime.

(d) Assume $\log_{10} 2 = a/b$, where a and b are integers. Then we must have $10^a = 2^b$, which is impossible.

(f) Let (see Figure 130) AC and BC be commensurable with respect to AP.

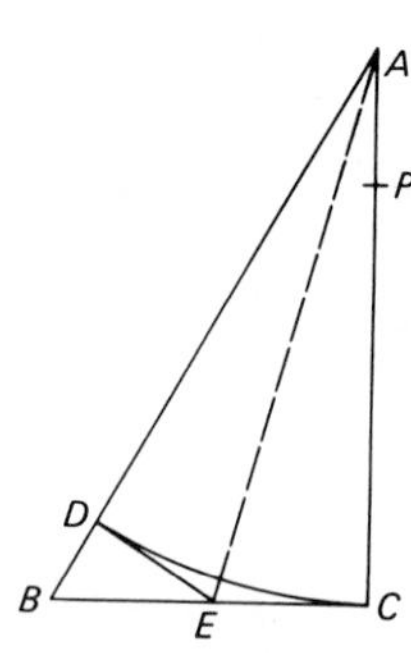

FIGURE 130

Show that then DE and DB are also commensurable with respect to AP, and so on.

3.9 **(b)** ab is the fourth proportional to $1, a, b$.

(c) a/b is the fourth proportional to $b, 1, a$.

(d) $\sqrt{a}$ is a mean proportional between 1 and a.

(g) Construct the mean proportional between a and na.

(h) Use the fact that $a^3 + b^3 = (a + b)(a^2 + b^2 - ab)$.

(i) Use the fact that

$$a(1 + \sqrt{2} + \sqrt{3})^{1/2} = [a(a + a\sqrt{2} + a\sqrt{3})]^{1/2}.$$

(j) Use the fact that $(abcd)^{1/4} = [(ab)^{1/2}(cd)^{1/2}]^{1/2}$.

(k) $60°$.

3.10 **(a)** Obtain $\sqrt{12}$ as in 3.9(d).

(c) Denote the parts by x and $a - x$. Then $x^2 - (a - x)^2 = x(a - x)$, or $x^2 + ax - a^2 = 0$.

(e) Show that $OM + ON = g$ and $(OM)(ON) = h$.

(g) Let A be the point (0,2) and let RS cut the x-axis in L and the tangent to the circle at A in T. We have the following equations:

$$\begin{aligned} &\text{circle:} && x^2 + y(y - 2) = 0, \\ &\text{line } AR\text{:} && 2x + r(y - 2) = 0, \\ &\text{line } AS\text{:} && 2x + s(y - 2) = 0. \end{aligned}$$

Therefore (line AR)(line AS) $-$ 4(circle) $= 0$ yields

$$(y - 2)[2x(r + s) + rs(y - 2) - 4y] = 0,$$

a pair of lines on the intersections of the circle with the lines AR and AS. It follows that the second factor set equal to 0 represents the line RS. Setting $y = 0$, we find $OL = rs/(r + s) = h/g$; setting $y = 2$, we find $AT = 4/(r + s) = 4/g$.

3.11 **(b)** First trisect the diagonal BD by points E and F. Then the broken lines AEC and AFC divide the figure into three equivalent parts. Transform these parts so as to fulfull the conditions by drawing parallels to AC through E and F.

(d) Through B draw BD parallel to MN to cut AC in D. Then, if the required triangle is $AB'C'$, AC' is a mean proportional between AC and AD.

(e) Let ABC be the given triangle. Draw AB' making the given vertex angle with AC and let it cut the parallel to AC through B in B'. Now use (d).

3.12 **(a)** A convex polyhedral angle must contain at least 3 faces, and the sum of its face angles must be less than 360°.

(b) $V = e^3\sqrt{2}/3, \quad A = 2e^2\sqrt{3}$.

3.13 This Problem Study would make a good junior research project for the better prepared student, who might care to consult the mensuration formulas for regular polyhedra as given, for example, in the *CRC Standard Mathematical Tables*.

3.14 **(a)** Denote the longer segment by y and the shorter one by x. Then $x + y : y = y : x$, or $x^2 + xy - y^2 = 0$, or $(x/y)^2 + x/y - 1 = 0$, or $x/y = (\sqrt{5} - 1)/2$.

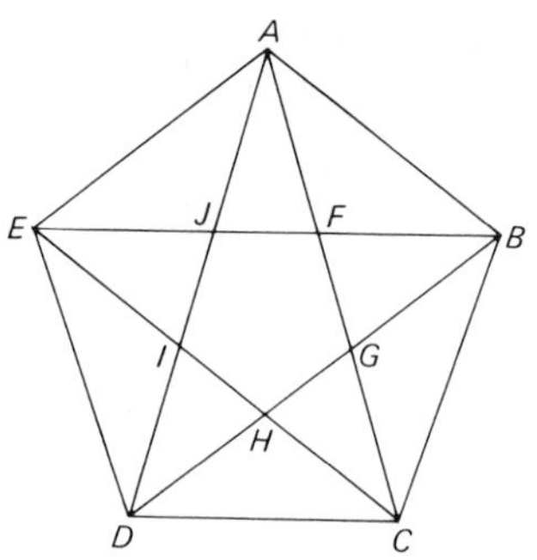

FIGURE 131

(b) In Figure 131, isosceles triangles DAC and DGC are similar. Therefore, $AD:DG = DC:GC$, whence $DB:DG = DG:GB$.

(c) $AG:AH = AG:GB = AB:AG = AB - AG:AG - AH = GB:HG = AH:HG$.

(d) Let HG in Figure 131 be the given side. Draw a right triangle PQR with legs PR and QR equal to HG and $HG/2$, respectively. On PQ produced mark off $QT = QR$. Then $PT = GB = GC = HC$, and so on.

(e) Let DB in Figure 131 be the given diagonal. Draw a right triangle PQR with legs PR and QR equal to $DB/2$ and DB, respectively. On PQ mark off $PT = PR$. Then $TQ = DG = DC$, and so on.

3.16 See *The Mathematical Gardner* (Prindle, Weber & Schmidt, 1980), pp. 276, 277.

4.1 **(b)** Let A be the given point and BC the given line segment. Construct by Proposition 1 an equilateral triangle ABD. Draw circle $B(C)$ and let DB produced cut this circle in G. Now draw circle $D(G)$ to cut DA produced in L. Then AL is the sought segment.

(c) Use Proposition 2 of Book I.

4.2 **(a)** See T. L. Heath, *A Manual of Greek Mathematics,* pp. 155–57.

(b) (1) The equations of the parabolas may be taken as $x^2 = sy$ and $y^2 = 2sx$, where s and $2s$ are the latera recta of the parabolas. (2) The equations of the parabola and hyperbola may be taken as $x^2 = sy$ and $xy = 2s^2$.

4.3 **(a)** Let M be the midpoint of OA and let E be the center of the rectangle $OADB$. Then, by Proposition 6, Book II (see Section 3–6), $(OA')(AA') + (MA)^2 = (MA')^2$. Adding $(ME)^2$ to both sides we find

$$(OA')(AA') + (EA)^2 = (EA')^2.$$

Similarly, $(OB')(BB') + (EB)^2 = (EB')^2$. Therefore

$$(OA')(AA') = (OB')(BB').$$

4.4 **(a)** We have $r = P_1P_2 = AP_1 \tan\theta = 2a \sin\theta \tan\theta$. It follows that $r = 2a(y/r)(y/x)$, or $r^2x = 2ay^2$.

(b) Denote the coordinates of P by (x,y). Then $(AQ)^3/(OA)^3 = y^3/x^3 = y/(2a - x) = RP/RA = OD/OA = n$, where R is the foot of the perpendicular from P on OA.

(c) Let S be the foot of the perpendicular from R on MN, and let T be the midpoint of RS. Draw the circle $S(T)$ to cut TP in U. Then $SCPU$ is a parallelogram. Let TP cut MN in V and the tangent to $S(T)$ at the point Q diametrically opposite to T in W. Triangles SUV and APV are congruent, and $UV = VP$. It is now easy to show that $TP = UW$. Thus P lies on the cissoid of $S(T)$ and QW for the pole T.

4.5 **(a)** The equation of the hyperbola, referred to its asymptotes as coordinate axes, is $xy = ab$, where $(b/2,a/2)$ is the center of the rectangle. The equation of the circumcircle of the rectangle is $x^2 + y^2 - ay - bx = 0$. The point of intersection of the hyperbola and circle, other than the point (b,a), is $(\sqrt[3]{a^2b}, \sqrt[3]{ab^2})$. But $\sqrt[3]{a^2b}$ and $\sqrt[3]{ab^2}$ are the mean proportionals between a and b.

4.6 **(a)** Denote AB by a, AC by b, BC by c, angle ADB by θ. Then, by the law of sines, applied first to triangle BCD and then to triangle ABD, $\sin 30°/\sin\theta = a/c$, $\sin\theta/\sin 120° = a/(b + a)$. Consequently $1/\sqrt{3} = \tan 30° = a^2/c(b + a)$. Squaring both sides and recalling that $c^2 = b^2 - a^2$, we find $2a^3(2a + b) = b^3(2a + b)$, or $b^3 = 2a^3$.

(b) Draw CO and use the fact that an exterior angle of a triangle is equal to the sum of the 2 remote interior angles.

4.7 **(a)** Let R be the foot of the perpendicular from Q on the x-axis and let RQ cut c in S. Then $OQ/RQ = PQ/SQ$.

(b) See 4.6(a).

(c) See 4.6(b).

(d) See 4.4.

4.8 **(a)** Let Q and N be the feet of the perpendiculars from P and M on OA, and let QP cut OM in S. Since P and R are on the hyperbola, we have $(OQ)(QP) = (ON)(NR)$, or $NR = (OQ)(NM)/ON$. Hence $SP = RM$. But, from similar triangles OQS and ONM, $QS = (OQ)(NM)/ON$. It follows that $SRMP$ is a rectangle. If T is the center of this rectangle, $OP = PT = TM$.

(b) Take radius $OA = 1$ and denote angle AOB by 3θ. Take P on arc AB such that angle $AOP = 1/3$ angle AOB, and let Q be the foot of the perpendicular from P on OC. Then $AP = 2\sin\theta/2 = 2\,PQ$.

4.9 **(a)** Use the fact that the sum of the infinite geometric series $1/2 - 1/4 + 1/8 - 1/16 + \cdots$ is $1/3$. For another asymptotic Euclidean solution of the trisection problem, see Problem 4134, *The American Mathematical Monthly* (December) 1945: 587–589.

4.10 **(a)** We have angle $AOP = k\pi/2$ when $OM = k(OA) = k$. Therefore, if we denote the coordinates of P by (x,y), $y = k = x\tan(k\pi/2) = x\tan(\pi y/2)$.

(c) Let the quadratrix cut OA in Q. Then

$$OQ = \lim_{y\to 0} \frac{y}{\tan\dfrac{\pi y}{2}} = \frac{2}{\pi},$$

by l'Hospital's rule in calculus. It is now easy to show that

$$\widehat{AC} : OA = OA : OQ.$$

4.11 **(a)** 3.1414.

(b) 3.14153.

(c) $GB/BA = EF/FA = (DE)^2/(DA)^2 = (DE)^2/[(BA)^2 + (BC)^2]$. Therefore, $GB = 4^2/(7^2 + 8^2) = 16/113 = 0.1415929\ldots$. This leads to 355/113 as an approximation of π.

4.13 **(a)** Let $\alpha = \tan^{-1}(1/5)$ and $\beta = \tan^{-1}(1/239)$. Then show that $4\alpha - \beta = \pi/4$ by showing that $\tan(4\alpha - \beta) = 1$.

(b) Consider a circle of unit radius. Then the side of an inscribed square is given by $\sec\theta$, where $\theta = 45°$. The sum of 2 sides of a regular inscribed 8-gon is given by $\sec\theta \sec\theta/2$; the sum of 4 sides of a regular inscribed 16-gon is given by $\sec\theta\sec\theta/2\sec\theta/4$; and so on. It follows that

$$\sec\theta \sec\frac{\theta}{2}\sec\frac{\theta}{4}\cdots \to \frac{\pi}{2},$$

the length of a quadrant of the circle. Therefore,

$$\frac{2}{\pi} = \cos\theta\cos\frac{\theta}{2}\cos\frac{\theta}{4}\cdots.$$

Now use the fact that $\cos\theta = \sqrt{2}/2$ and $\cos\theta/2 = [(1 + \cos\theta)/2)^{1/2}$, $\cos\theta/4 = [(1 + \cos\theta/2)/2]^{1/2}$, and so on.

(c) Set $x = \sqrt{1/3}$ in Gregory's series.

(f) Let θ denote $\pi/2n$. Then $\sin\theta = s_{2n}/2R$, $\cos\theta = s_n/2s_{2n}$. Now use the fact that $\sin^2\theta + \cos^2\theta = 1$.

(g) Let θ denote $\pi/2n$. Then $\tan 2\theta = S_n/2r$, $\tan\theta = S_{2n}/2r$. Now use the fact that $\tan 2\theta = (2\tan\theta)/(1 - \tan^2\theta)$.

(h) First show that $p_n = 2nR\sin(\pi/n)$, $P_n = 2nR\tan(\pi/n)$.

(i) First show that $a_n = nR^2\sin(\pi/n)\cos(\pi/n)$, $A_n = nR^2\tan(\pi/n)$.

4.14 **(a)** arc $AR = \pi/2$, $AT = 3/2$.

(b) Let M be the foot of the perpendicular from P on OA. Then $PM = \sin\theta$, $OM = \cos\theta$, whence $\tan\phi = \sin\theta/(2 + \cos\theta)$.

(c) Let PS cut the circle again in N. Then, since $ON < SN$, angle $SON = \phi + \varepsilon$, where $\varepsilon > 0$. Therefore angle $ONP = 2\phi + \varepsilon$, and $\theta = 3\phi + \varepsilon$.

4.15 **(a)** The 32nd decimal place in the expansion of π is occupied by 0.

5.1 **(c)** Suppose $a > b$. Then the algorithm may be summarized as follows:

$$\begin{array}{ll} a = q_1 b + r_1 & 0 < r_1 < b \\ b = q_2 r_1 + r_2 & 0 < r_2 < r_1 \\ r_1 = q_3 r_2 + r_3 & 0 < r_3 < r_2 \\ \cdots\cdots & \cdots\cdots \\ r_{n-2} = q_n r_{n-1} + r_n & 0 < r_n < r_{n-1} \\ r_{n-1} = q_{n+1} r_n & \end{array}$$

Now, from the last step, r_n divides r_{n-1}. From next to the last step r_n divides r_{n-2}, since it divides both terms on the right. Similarly r_n divides r_{n-3}. Successively, r_n divides each r_i, and finally a and b.

On the other hand, from the first step, any common divisor c of a and b divides r_1. From the second step, c then divides r_2. Successively, c divides each r_i. Thus c divides r_n.

(d) From the next to the last step in the algorithm, we can express r_n in terms of r_{n-1} and r_{n-2}. From the preceding step, we can then express r_n in terms of r_{n-2} and r_{n-3}. Continuing this way, we finally obtain r_n in terms of a and b.

5.2 **(a)** If p does not divide u, then integers P and Q exist such that $Pp + Qu = 1$, or $Ppv + Quv = v$.

(b) Suppose there are two prime factorizations of the integer n. If p is one of the prime factors in the first factorization it must, by (a), divide one of the factors in the second factorization, that is, coincide with one of the factors.

(c) We note that $273 = (13)(21)$. Find [see 5.1(e)] integers p and q such that $13p + 21q = 1$. Dividing by 273, we then have $p/21 + q/13 = 1/273$. Similarly find integers r and s such that $1/21 = r/3 + s/7$.

5.5 **(c)** For each b_i in (b) may have $a_i + 1$ values.

(f) Since b divides ac, we have $b_i \leqq a_i + c_i$. Also, since a and b are relatively prime, we have $a_i = 0$ or $b_i = 0$. In either case $b_i \leqq c_i$.

(h) Suppose $\sqrt{2} = a/b$, where a and b are positive integers. Then, since $a^2 = 2b^2$, we have $(2a_1, 2a_2, \ldots) = (1 + 2b_1, 2b_2, \ldots)$, whence $2a_1 = 1 + 2b_1$, which is impossible.

5.6 **(c)** Let ABC be the given triangle and let XY, parallel to BC, cut AB in X and AC in Y. Draw BY and CX. Show that $\Delta BXY : \Delta AXY = \Delta CXY : \Delta AXY$. But, by VI 1, $\Delta BXY : \Delta AXY = BX : XA$ and $\Delta CXY : \Delta AXY = CY : YA$.

5.7 **(c)** For [see 5.1(f)] there exist positive integers p and q such that $pr - qs = \pm 1$. Then the difference between the angle subtended at the center of the s-gon by p of its sides and the angle subtended at the center of the r-gon by q of its sides is

$$p\left(\frac{360°}{s}\right) - q\left(\frac{360°}{r}\right) = (pr - qs)\frac{360°}{rs} = \frac{\pm 360°}{rs}.$$

(f) To see how Euclid established this proposition, consult Heath, *The Thirteen Books of Euclid's Elements*. A pretty trigonometric proof can be formulated along the following lines. Let $u = 18°$. Then $\sin 4u = \cos u$ and $\cos 4u = \sin u$. Show that these, respectively, imply

$$-8 \sin^4 u + 4 \sin^2 u = \sin u$$

and

$$8 \sin^4 u - 8 \sin^2 u + 1 = \sin u,$$

from which we obtain

$$-16 \sin^4 u + 12 \sin^2 u = 1.$$

Now, if p and d represent the sides of a regular pentagon and a regular decagon inscribed in a unit circle, show that $p = 2 \sin 2u$ and $d = 2 \sin u$, whence

$$p^2 - d^2 = -16 \sin^4 u + 12 \sin^2 u = 1,$$

which establishes the proposition.

(g) Show that $\tan(180°/17)$ is approximately equal to $3/16$.

5.12 **(c)** $h_c = b \sin A$.
(f) $h_a = t_a \cos [(B - C)/2]$.
(g) $4h_a^2 + (b_a - c_a)^2 = 4m_a^2$.
(h) $b_a - c_a = 2R \sin (B - C)$.
(i) $4R(r_a - r) = (r_a - r)^2 + a^2$. If M and N are the midpoints of side BC and arc BC, then $MN = (r_a - r)/2$; clearly any 2 of R, a, MN determine the third.
(j) $h_a = 2rr_a/(r_a - r)$.

5.13 **(b)** See Problem 3336, *The American Mathematical Monthly,* August 1929.
(c) See Problem E 1447, *The American Mathematical Monthly,* September 1961. The solution given in this reference is a singularly fine application of the method of data.

5.14 **(b)** Let M be the midpoint of BC. The broken line EMA bisects the area. Through M draw MN parallel to AE to cut a side of triangle ABC in N. Then EN is the sought line.
(c) Let a, b, h denote the bases and altitude of the given trapezoid, let c be the sought parallel line, let p be the altitude of the trapezoid with bases a and c, and let q be the altitude of the trapezoid with bases c and b. Then we have $(a + c)p = (c + b)q$, $p + q = h$, $(a + c)2p = (a + b)h$. Eliminating p and q and solving for c we find $c = [(a^2 + b^2)/2]^{1/2}$, the *root-mean-square* of a and b.

6.1 **(a)** $\sec [(29/30)90°] = \sec 87° = 19.11$.

6.2 **(d-1)** The volume of the segment is equal to the volume of a spherical sector minus the volume of a cone. Also, $a^2 = h(2R - h)$.
(d-2) The segment is the difference of 2 segments, each of 1 base, and having, say, altitudes u and v. Then

$$V = \pi R(u^2 - v^2) - \frac{\pi(u^3 - v^3)}{3}$$

$$= \pi h\left[(Ru + Rv) - \frac{u^2 + uv + v^2}{3}\right].$$

But $u^2 + uv + v^2 = h^2 + 3uv$ and also $(2R - u)u = a^2$ and $(2R - v)v = b^2$. Therefore

$$V = \pi h\left(\frac{a^2 + b^2}{2} + \frac{u^2 + v^2}{2} - \frac{h^2}{3} - uv\right)$$

$$= \pi h\left(\frac{a^2 + b^2}{2} + \frac{h^2}{2} + uv - \frac{h^2}{3} - uv\right), \text{ and so on.}$$

(f) Pass the planes perpendicularly through the trisection points of a diameter of the sphere.

6.4 **(a)** $(GC)^2 + (TW)^2 = 4r_1r_2$.

6.5 **(a)** Produce CB to E so that $BE = BA$. Prove triangles MBA and MBE are congruent. For a singularly elegant alternative proof, see Solution I of Problem 466 in *Crux Mathematicorum* (June–July) 1980: 189.

6.7 **(b)** Let A and B be points, and C a straight line. Produce AB to cut line C in S. Now find T on line C such that $(ST)^2 = (SA)(SB)$. In general, there are 2 solutions.

(c) Reflect the given point in a bisector of the angle determined by the 2 given lines.

(d) Reflect the focus F in the line m, obtaining a point F'. Now, by (b), find the centers of the circles passing through F and F' and touching the given directrix.

6.8 **(b)** For problem (1) take A and B on the x-axis and reflections of one another in the origin.

(c) 1. Let the interior and exterior bisectors of angle APB cut AB in M and N. Then M and N are on the required locus and angle MPN is a right angle. 2. Let A and B be the fixed points, P the moving point, and O the midpoint of AB. Add the expressions for $(PA)^2$ and $(PB)^2$ as given by the law of cosines applied to triangles PAO and PBO.

6.9 **(a)** Let $ABCD$ be the cyclic quadrilateral. Find E on diagonal AC such that $\measuredangle ABE = \measuredangle DBC$. From similar triangles ABE and DBC, obtain $(AB)(DC) = (AE)(BD)$. From similar triangles ABD and EBC, obtain $(AD)(BC) = (EC)(BD)$.

(b-1) In (a) take AC as a diameter, $BC = a$, and $CD = b$.

(b-2) In (a) take AB as a diameter, $BD = a$, and $BC = b$.

(b-3) In (a) take AC as a diameter, $BD = t$ and perpendicular to AC.

(d-1) Taking the side of the triangle as 1 unit, apply Ptolemy's theorem to quadrilateral $PACB$.

(d-2) Taking the side of the square as 1 unit, apply Ptolemy's theorem to quadrilaterals $PBCD$ and $PCDA$.

(d-3) Taking the side of the pentagon as 1 unit, apply Ptolemy's theorem to quadrilaterals $PCDE$, $PCDA$, $PBCD$.

(d-4) Taking the side of the hexagon as 1 unit, apply Ptolemy's theorem to quadrilaterals $PBCD$, $PEFA$, $PBCF$, $PCFA$.

6.11 **(b)** Let a ray of light emanating from a point A hit the mirror at point M and reflect toward a point B. If B' is the image of B in the mirrow, then BB' is perpendicularly bisected by the plane of the mirror, and we must have AMB' a straight line.

(c) Apply (b).

6.12 **(b)** Show, from a figure, that $ab = 2rs$ and $a + b = r + s$, and then solve simultaneously.

6.13 **(a)** 120 apples.

(b) 60 years old.

(c) 960 talents.

(d) Each Grace had $4n$ apples, gave away $3n$, and had n left.

6.14 **(a)** 2/5 of a day.

(b) 144/37 hours.

(c) 30.5 minae of gold, 9.5 minae of copper, 14.5 minae of tin, and 5.5 minae of iron.

6.15 **(a)** 84 years old.

(b) 7, 4, 11, 9.

(c) Set $CD = 3x$, $AC = 4x$, $AD = 5x$, $CB = 3y$. Then, since $AB/DB = AC/CD$, we find $AB = 4(y - x)$. By the Pythagorean theorem, we are led to $7y = 32x$. We finally get $AB = 100$, $AD = 35$, $AC = 28$, $BD = 75$, $DC = 21$.

(d) 1806.

6.16 **(a)** $481 = 20^2 + 9^2 = 16^2 + 15^2$.

(b) We have $5 = 2^2 + 1^2$, $13 = 3^2 + 2^2$, $17 = 4^2 + 1^2$. By using the identities of (a) we find

$$(5)(13) = 8^2 + 1^2 = 7^2 + 4^2,$$
$$(5)(17) = 9^2 + 2^2 = 7^2 + 6^2,$$
$$(13)(17) = 14^2 + 5^2 = 11^2 + 10^2.$$

Again, by the identities of (a), we find

$$1105 = 33^2 + 4^2 = 32^2 + 9^2 = 31^2 + 12^2 = 24^2 + 23^2.$$

6.17 **(a)** From the similar triangles DFB and DBO, $FD/DB = DB/OD$. Therefore $FD = (DB)^2/OD = 2(AB)(BC)/(AB + BC)$.

(b) From similar right triangles, $OA/OB = AF/BD = AF/BE = AC/CB = (OC - OA)/(OB - OC)$. Now solve for OC.

(c) Let HA cut BC in R and LM in S, and let LB cut DH in U and MC cut FH in V. Then ▱ $ABDE$ = ▱ $ABUH$ = ▱ $BRSL$, and ▱ $ACFG$ = ▱ $ACVH$ = ▱ $RCMS$.

(e) An analytic solution is easy if we recall that the coordinates of a point dividing the segment joining points (a,b) and (c,d) in the ratio m/n are $(na + mc)/(m + n)$ and $(nb + md)/(m + n)$ and that the coordinates of the centroid of the triangle determined by (a,b), (c,d), (e,f) are

$$(a + c + e)/3 \text{ and } (b + d + f)/3.$$

A synthetic solution is not so easy. One, due to Fuhrmann, is given in R. A. Johnson, *Modern Geometry*, Section 276, p. 175.

6.18 **(a)** $V = 2\pi^2r^2R$, $S = 4\pi^2rR$.

(b) The centroid of the semicircular arc lies on the bisecting radius of the semicircle and at the distance $2r/\pi$, where r is the radius of the semicircle, from the diameter of the semicircle.

(c) The centroid of the semicircular area lies on the bisecting radius of the semicircle and at the distance $4r/3\pi$, where r is the radius of the semicircle, from the diameter of the semicircle.

6.19 **(a)** Let P be the point (x,y). Then, from similar triangles, $x^2/a^2 = (OB)^2/(AB)^2$ and $y^2/b^2 = (OA)^2/(AB)^2$, whence $x^2/a^2 + y^2/b^2 = 1$.

(b) Keuffel and Esser Company has manufactured an ellipsograph based on the trammel construction.

6.20 See Howard Eves, *A Survey of Geometry*, vol. 1, Section 2–3.

6.21 This Problem Study, along with Problem Studies 2.14, 3.4, 4.13(h) and (i), and 6.17(a) and (b), constitutes a good junior research project of average difficulty.

7.1 **(d)** 37/4 dou, 17/4 dou, 11/4 dou.

7.2 **(a)** height = 9.6 ch'ih, width = 2.8 ch'ih.

(b) 12 feet.

7.3 **(a)** The magic constant $= (1 + 2 + 3 + \cdots + n^2)/n$.

(c) Denote the numbers in the magic square by letters and then add together the letters of the middle row, the middle column, and the 2 diagonals.

(d) Use (c) and an indirect argument.

7.4 **(a)** $x = hd/(2h + d)$.
(b) 8 cubits and 10 cubits.
(c) 40.

7.5 **(a)** 8 days.
(b) 18 mangoes.
(c) 8 for a citron and 5 for a wood apple.
(d) 36 camels.

7.6 **(a)** 72 bees.
(b) 20 cubits.
(c) 22/7 yojanas.
(d) 10!, 4!.
(e) 100 arrows.

7.7 **(a)** Suppose $\sqrt{a} = b + \sqrt{c}$. Then $\sqrt{c} = (a - b^2 - c)/2b$.
(b) If $a + \sqrt{b} = c + \sqrt{d}$, then $\sqrt{b} = (c - a) + \sqrt{d}$. Now use (a).

7.8 **(b)** It is easily shown that $x = x_1 + mb$ and $y = y_1 - ma$ constitute a solution. Conversely, assume x and y form a solution. Then $a(x - x_1) = b(y_1 - y)$, or $x - x_1 = mb$ and $y_1 - y = ma$.
(c) Dividing by 7, we find

$$x + 2y + \frac{2}{7}y = 29 + \frac{6}{7}.$$

Therefore there exists an integer z such that

$$\frac{2}{7}y + z = \frac{6}{7},$$

or

$$2y + 7z = 6.$$

This can be solved by inspection to give $z_1 = 0$, $y_1 = 3$. Then $x_1 = 23$. The general solution of the original equation is then, by (b),

$$x = 23 + 16m, \qquad y = 3 - 7m.$$

Since, by requirement, $x > 0$, $y > 0$, we must have $m \geqq -1$ and $m \leqq 0$. The only permissible values for m are 0 and -1. We thus get two solutions

$$x = 23,\ y = 3 \quad \text{and} \quad x = 7,\ y = 10.$$

Or find, as in 5.1(f), p and q such that $7p + 16q = 1$. Then we may take $x_1 = 209p$ and $y_1 = 209q$.
(d) There are the 4 solutions: $x = 124$, $y = 4$; $x = 87$, $y = 27$; $x = 50$, $y = 50$; $x = 13$, $y = 73$.
(e) Let x represent the number of dimes and y the number of quarters. Then we must have $10x + 25y = 500$.
(f) Let x denote the number of fruits in a pile and y the number of fruits each traveler receives. Then we have $63x + 7 = 23y$. The smallest permissible value for x is 5.

7.9 **(a)** Draw the circumdiameter through the vertex through which the altitude passes, and use similar triangles.

(b) Apply (a) to triangles DAB and DCB.

(c) Use the result of (b) along with Ptolemy's relation, $mn = ac + bd$.

(d) Here $\theta = 0°$ and $\cos\theta = 1$. Now use (b) and (c).

7.10 **(b)** Since the quadrilateral has an incircle we have $a + c = b + d = s$. Therefore, $s - a = c$, $s - b = d$, $s - c = a$, $s - d = b$.

(c) In Figure 132, we have

$$a^2 + c^2 = r^2 + s^2 + m^2 + n^2 - 2(rn + sm)\cos\theta,$$
$$b^2 + d^2 = r^2 + s^2 + m^2 + n^2 + 2(sn + rm)\cos\theta.$$

Therefore $a^2 + c^2 = b^2 + d^2$ if and only if $\cos\theta = 0$, $\theta = 90°$.

(d) Use (c).

(e) The consecutive sides of the quadrilateral are 39, 60, 52, 25; the diagonals are 56 and 63; the circumdiameter is 65; the area is 1764.

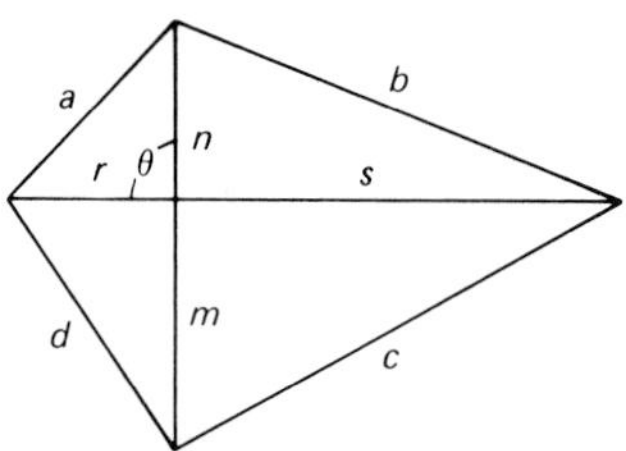

FIGURE 132

7.11 **(c)** See T. L. Heath, *A Manual of Greek Mathematics*, pp. 340–42.

7.12 **(a)** We shall indicate a proof of the theorem for the 4-digit number N having a, b, c, d for its thousands, hundreds, tens, and units digits; the proof is easily generalized. Now

$$N = 1000a + 100b + 10c + d.$$

Let $S = a + b + c + d$. Then

$$N = 999a + 99b + 9c + S = 9(111a + 11b + c) + S,$$

and so on.

(b) Let M and N be any two numbers with excesses e and f. Then there exist integers m and n such that

$$M = 9m + e, \qquad N = 9n + f.$$

Now

$$M + N = 9(m + n) + (e + f),$$

and

$$MN = 9(9mn + ne + mf) + ef,$$

and so on.

(d) Let M be the given number and N that obtained by some permutation of the digits of M. Then, since M and N consist of the same digits, they have [by (a)] the same excess e. Thus, we have

$$M = 9m + e, \qquad N = 9n + e.$$

and

$$M - N = 9(m - n).$$

(e) By (d) the final product must be divisible by 9, whence, by (a), the excess for the sum of the digits in the product must be 0.

(f) Replace 9 by $(b - 1)$.

7.14 **(b)** $x = 2.3696$.

(c) $x = 4.4934$.

7.15 **(a)** Find z such that $b/a = a/z$, then m such that $n/z = a/m$.

(c) The positive roots are 2 and 4; the negative root is -1.

7.16 **(a)** The real roots are given by the abscissas of the points of intersection of the line $ay + bx + c = 0$ with the cubic curve $y = x^3$.

(b) $x = 1.7^+$.

(c) $x = -3.5, 1, 2.5$.

(e) $x = -6, -2, -1$.

7.17 **(a)** Draw any circle Σ on the sphere and mark any 3 points A, B, C on its circumference. On a plane, construct a triangle congruent to triangle ABC, find its circumcircle, and thus obtain the radius of Σ. Construct a right triangle having the radius of Σ as 1 leg and the polar chord of Σ as hypotenuse. It is now easy to find the diameter of the given sphere.

(b) If d is the diameter of the sphere and e the edge of the inscribed cube, then $e = (d\sqrt{3})/3$, whence e is one-third the altitude of an equilateral triangle of side $2d$.

(c) If d is the diameter of the sphere and e the edge of the inscribed regular tetrahedron, then $e = (d\sqrt{6})/3$, whence e is the hypotenuse of a right isosceles triangle with leg equal to the edge of the inscribed cube. See (b).

8.1 **(a)** Let x, y, z denote the number of men, women, and children. Then we must have

$$6x + 4y + z = 200 \quad \text{and} \quad x + y + z = 100,$$

or $5x + 3y = 100$. It follows that y must be a multiple of 5, say, $5n$. Then $x = 20 - 3n$ and $z = 80 - 2n$. One easily finds that the only permissible values for n are 1, 2, 3, 4, 5, 6. The solution given in Alcuin's collection corresponds to $n = 3$—namely, 11 men, 15 women, 74 children.

(b) It is easily shown that each son must receive the same number of entirely empty flasks as full ones. There are many solutions.

(c) Let x denote the required number of leaps. Then $9x - 7x = 150$.
(d) Find 2 solutions. For other problems of this sort see Maurice Kraitchik, *Mathematical Recreations,* pp. 214–22.
(e) How about 5/27 to the mother, 15/27 to the son, and 7/27 to the daughter?
(f) Let the legs, hypotenuse, and area of the triangle be a, b, c, K, $a \geqq b$. Then

$$a^2 + b^2 = c^2, \qquad ab = 2K.$$

Solving for a and b we find

$$a = \frac{\sqrt{c^2 + 4K} + \sqrt{c^2 - 4K}}{2}, \qquad b = \frac{\sqrt{c^2 + 4K} - \sqrt{c^2 - 4K}}{2}.$$

8.2 **(b-1)** Use mathematical induction. Assume the relation true for $n = k$. Then

$$\begin{aligned} u_{k+2}u_k &= (u_{k+1} + u_k)u_k \\ &= u_{k+1}u_k = u_k^2 \\ &= u_{k+1}u_k + u_{k+1}u_{k-1} - (-1)^k \\ &= u_{k+1}(u_k + u_{k-1}) + (-1)^{k+1} \\ &= u_{k+1}^2 + (-1)^{k+1}, \end{aligned}$$

and so on. Or use the expression for u_n given in (b-2).
(b-2) Set $v_n = [(1 + \sqrt{5})^n - (1 - \sqrt{5})^n]/2^n\sqrt{5}$. Show that $v_n + v_{n+1} = v_{n+2}$ and that $v_1 = v_2 = 1$. Then $v_n = u_n$.
(b-3) Use the expression for u_n given in (b-2).
(b-4) Use the relation given in (b-1).

8.3 **(a)** A has 121/17 denarii and B has 167/17 denarii.
(b) 33 days. This may be solved as a problem in *variation.*
(c) Let x represent the value of the estate and y the amount received by each son. Then the first son receives $1 + (x - 1)/7$, and the second receives

$$2 + \frac{x - \left(1 + \dfrac{x-1}{7}\right) - 2}{7}.$$

Equating these we find $x = 36$, $y = 6$, and the number of sons was $36/6 = 6$.

8.4 **(b)** The following is essentially Fibonacci's solution to the problem. Let s denote the original sum and $3x$ the total sum returned. Before each man received a third of the sum returned, the 3 men possessed $s/2 - x$, $s/3 - x$, $s/6 - x$. Since these are the sums possessed after putting back 1/2, 1/3, 1/6 of what they had first taken, the amounts first taken were $2(s/2 - x)$, $(3/2)(s/3 - x)$, $(6/5)(s/6 - x)$, and these amounts added together equal s. Therefore $7s = 47x$, and the problem is indeterminate. Fibonacci took $s = 47$ and $x = 7$. Then the sums taken by the men from the original pile are 33, 13, 1.
(c) 382 apples.

8.6 **(a)** Denote the given angle by y and angle AOF by x. Since OF is equal and parallel to DE, $OFED$ is a parallelogram. It follows that $FE = OD = FO$

and triangle OFE is isosceles. Now angle OFE = angle ODE = angle $OAE = x$. Therefore, summing the 3 angles of triangle OFE, $2(90 - y + x) + x = 180$, or $x = 2y/3$.

(b) Calling the 2 parts x and y, we have $x + y = 10$, $x^2 + y^2 = 58$. Therefore, we take $x = 7$, $y = 3$.

8.8 **(a)** Following is esentially the solution given by Regiomontanus. We are given (see Figure 133) $p = b - c$, h, $q = m - n$. Now $b^2 - m^2 = h^2 = c^2 - n^2$, or $b^2 - c^2 = m^2 - n^2$, or $b + c = qa/p$. Therefore

$$b = \frac{qa + p^2}{2p} \quad \text{and} \quad m = \frac{a + q}{2}.$$

Substituting these expressions in the relation $b^2 - m^2 = h^2$, we obtain a quadratic in the unknown a.

(b) Following is essentially the solution given by Regiomontanus. Here we are given (see Figure 133) a, h, $k = c/b$. Set $2x = m - n$. Then

$$4n^2 = (a - 2x)^2, \qquad 4c^2 = 4h^2 + (a - 2x)^2,$$
$$4m^2 = (a + 2x)^2, \qquad 4b^2 = 4h^2 + (a + 2x)^2.$$

Then

$$k^2[4h^2 + (a + 2x)^2] = 4h^2 + (a - 2x)^2.$$

Solving this quadratic, we obtain x, and then b and c.

The triangle is easily constructed by using a circle of Apollonius. See Problem Study 6.8(b).

(c) On AD produced (see Figure 134) take $DE = bc/a$, the fourth proportional to the given segments a, b, c. Then triangles DCE and BAC are similar and $CA/CE = a/c$. Thus c is located as the intersection of 2 loci, a circle of Apollonius and a circle with center D and radius c.

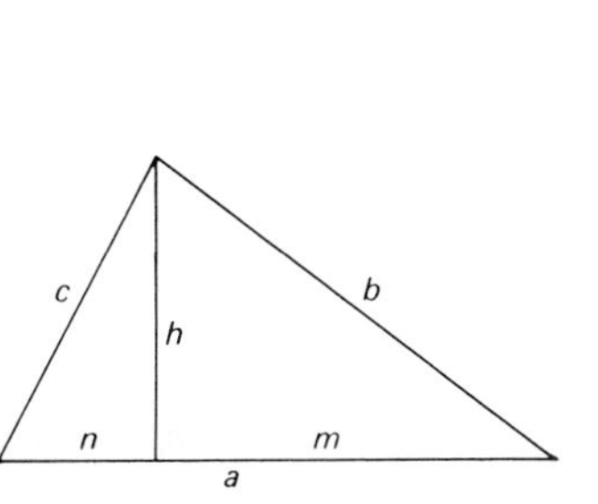

FIGURE 133

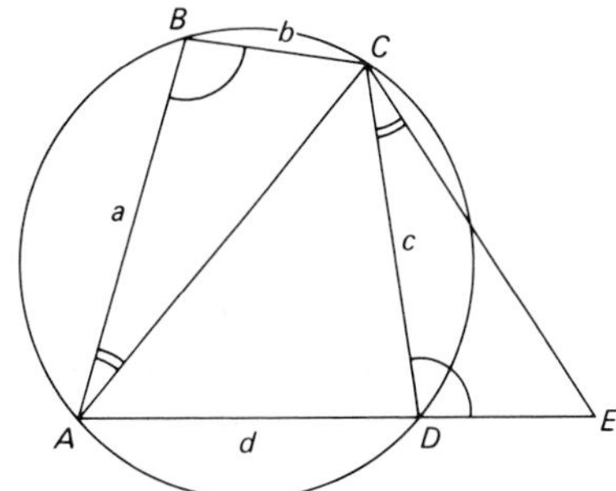

FIGURE 134

8.9 **(a)** $29.

(b) 180/11 days.

(c) The price of each cask is 120 francs and the duty on a cask is 10 francs.

(d) Suppose $a/c < b/d$. Then $ad < bc$, $ac + ad < ac + bc$, $a(c + d) < c(a + b)$, $a/c < (a + b)/(c + d)$, and so on.

8.10 **(a)** Using standard notation we have

$$(rs)^2 = s(s - a)(s - b)(s - c),$$

or

$$16s^2 = s(s - 14)(6)(8),$$

and $s = 21$. The required sides are then $21 - 6 = 15$ and $21 - 8 = 13$. This is not Pacioli's method of solving the problem; his solution is needlessly involved.

8.11 **(b)** 463 7/23.

(f) Profits are proportional to the time the money is in the service of the company as well as to the amount.

(g) Over 16 percent.

8.14 **(b)** $H = (3ac - b^2)/9a^2$, $G = (2b^3 - 9abc + 27a^2d)/27a^3$.

(d) $x = 4$. The other 2 roots are imaginary.

8.15 **(a)** $(3 \pm \sqrt{5})/2$, $(-5 \pm \sqrt{21})/2$.

(c) $y^3 + 15y^2 + 36y = 450$, $y^6 - 6y^4 - 144y^2 = 2736$.

8.16 **(a)** $R\,q \llcorner R\,c \llcorner R\,q\ 68\ p\ 2 \lrcorner m\ R\,c \llcorner R\,q\ 68\ m\ 2 \lrcorner \lrcorner$.

(b) $\sqrt[3]{4 + (-11)^{1/2}} + \sqrt[3]{4 - (-11)^{1/2}}$.

(c) A cub $- B$ 3 in A quad $+ C$ plano 4 in A aequatur D solido 2.

8.17 **(b)** $\cos 5\theta = 16 \cos^5 \theta - 20 \cos^3 \theta + 5 \cos \theta$.

(c) $x = 243$.

(d) $x_2 = (r - qx - px^2 - x^3)/(3x^2 + 2px + q)$.

8.18 **(a)** 10.

(b) 28 beggars, \$2.20.

(c) \$92.

8.19 **(a)** Bombelli's solution runs as follows. Denote the sought square by $DEFG$, where D is on AB and G is on AC. Let the altitude AM of triangle ABC cut DG in N. By Heron's formula, the area of triangle ABC is 84, whence $AM = 12$. Let $DG = 14x$. Then $AN = 12x$. It follows that $12x + 14x = 12$, or $x = 6/13$. The side of the square is then $(14)(6/13) = 84/13$.

(b) $(BP)^2 = (VC)^2 = (AV)(VB)$. Placing a rectangular coordinate framework with origin at V and positive x-axis along VW, and representing the coordinates of P by (x,y), we then have $y^2 = px$.

9.1 **(a)** See almost any text on college algebra or trigonometry.

(b) 1. Set $y = \log_b N$, $z = \log_a N$, $w = \log_a b$. Then $b^y = N$, $a^z = N$, $a^w = b$, whence $a = b^{1/w}$, or $a^z = b^{z/w} = b^y$. Thus, $y = z/w$.

2. Set $y = \log_b N$ and $z = \log_N b$. Then $b^y = N$, $N^z = b$, whence $N = b^{1/z} = b^y$. Thus, $y = 1/z$.

3. Set $y = \log_N b$ and $z = \log_{1/N}(1/b)$. Then $N^y = b$, $(1/N)^z = 1/b$, whence $N = b^{1/z} = b^{1/y}$. Thus $y = z$.

(c) $\log 4.26 = 1/2 + 1/8 + 1/256 + \cdots = 0.6294 \ldots$.

9.2 **(b)** $\cos c = \cos a \cos b$.

(c) (1) $A = 122°39'$, $C = 83°5'$, $b = 109°22'$. (2) $A = 105°36'$, $b = 44°0'$, $c = 78°46'$.

9.3 Popsicle sticks or tongue depressors make excellent rods.

9.5 **(a)** Acceleration is the increase in velocity during a unit period of time.

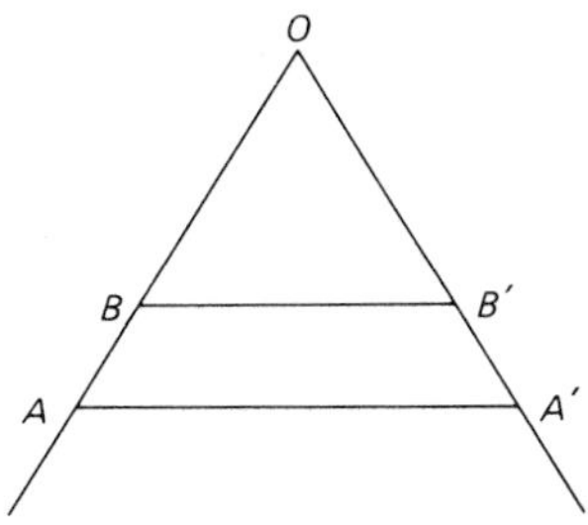

FIGURE 135

9.6 **(a)** Open the compasses so that the given segment AA' stretches between the 100 marks on the 2 simple scales of the compasses (see Figure 135). Then the distance between the two 20 marks is 1/5 of the given segment. How does one solve this problem if the given segment is too long to fit between the legs of the instrument?

(b) Open the compasses so that AA'/OA is the desired ratio of scale. Then BB' is the new length to be associated with the old length OB.

(c) Connect a on one arm to b on the other. Through c on the first arm draw a parallel to the line just drawn, to cut the other arm in the sought fourth proportional.

(d) Open the compasses so that the distance between the 106 marks is equal to 150. Then the distance between the 100 marks represents the amount of the investment a year ago. Perform this operation 5 times to find the required amount.

9.7 **(b)** Show that $(HG)^2 = (HB)^2 = (BF)^2 - (HF)^2 = (HE)^2 - (HF)^2$, and so on.

(c) Two classes that can be placed in one-to-one correspondence are said to be **equivalent** or to have the same **cardinality.** That is, it is possible to ascertain at a party of boys and girls, for example, that the count of boys is the same as that of the girls if each boy has one and only one partner among the girls, and vice versa. The difference between a finite and an infinite class is that an infinite class is equivalent to a part of itself.

9.8 **(c)** 1000 years.

(d) 25 A.U.

(f) 1 hour and 24 minutes.

9.10 **(a)** Choose for π' a plane parallel to the plane determined by point O and line l.

(c) Project line OU to infinity.

(d) Project line LMN to infinity and use the elementary fact that the joins of corresponding vertices of 2 similar and similarly situated triangles are concurrent.

(e) Choose a plane π' parallel to the minor axis of the ellipse, such that the angle θ between π' and the plane of the given ellipse is such that $\cos \theta = b/a$, where a and b are the semimajor and semiminor axes of the ellipse. Now project the ellipse orthogonally onto π'.

(g) Let c be any line through the intersection of a and b (see Figure 136). Let PA cut c in Q and QB cut MP in M'.

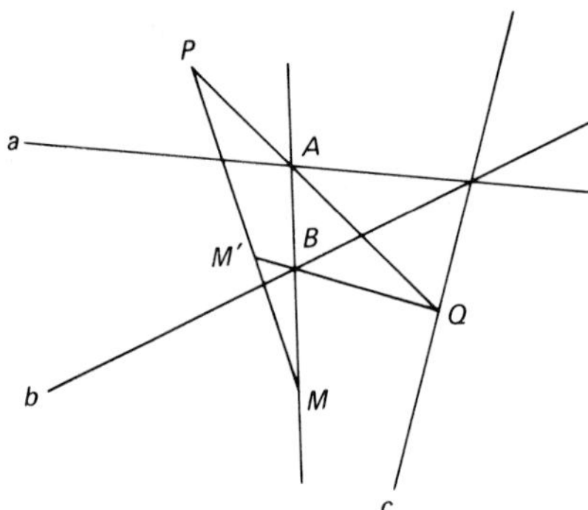

FIGURE 136

9.12 **(a)** Let points 1 and 6 coincide so that the line 16 becomes the tangent to the conic at point 1.

(b) Use (a).

(c) Let 1, 2, 3, 4 be the four points and 45 the tangent at $4 \equiv 5$, and let 12 cut 45 in P. Through 1 draw any line 16 cutting 34 in R, and then draw the Pascal line PR to cut 23 in Q. Then $5Q$ cuts 16 in a point 6 on the conic.

(d) Take $1 \equiv 6$ and $3 \equiv 4$, and then take $2 \equiv 3$ and $5 \equiv 6$.

(e) Take $1 \equiv 2$, $3 \equiv 4$, $5 \equiv 6$.

(f) Use (e).

9.13 **(a)** This follows from the definition of the arithmetic triangle as given in Section 9–9.

(b) By successive applications of (a).

(c) Use mathematical induction and (a).

(d) By (c).

(e) By (a).

(f) By (e).

(g) By (c).

10.1 **(a)** See Figure 137.

(c) By (a) and (b), and 3.9(c).

(d) We have $r + s = g$ and $rs = h$.

(e) We have $-r + s = -g$ and $-rs = -h$.

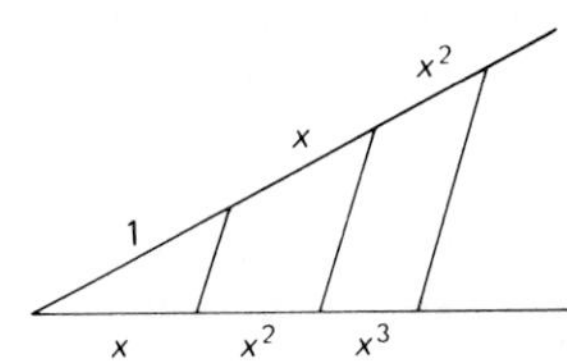

FIGURE 137

10.2 **(a)** $x^3 - 2ax^2 - a^2x + 2a^3 = axy$.

(b) See 10.1(c).

(c) Consider the equations of L_1, L_2, L_3, L_4 in *normal*, or *perpendicular, form*. We then easily see that the equation of the locus is quadratic.

(d) We find $x_2 - x_1 = m$.

10.4 **(b)** $r = (3a \sin\theta \cos\theta)/(\sin^3\theta + \cos^3\theta)$.

(c) $x = 3at/(1 + t^3)$, $y = 3at^2/(1 + t^3)$; loop $(0,\infty)$, lower arm $(-\infty, -1)$, upper arm $(-1,0)$.

(d)
$$y = \pm x\sqrt{\frac{3 - x\sqrt{2}}{3x\sqrt{2} + 3}}.$$

(e) We find $h + m - k^2 = -2$, $k(m - h) = 8$, $mh = -3$. Eliminating m and h, by solving the first 2 relations for h and m in terms of k and then substituting in the third relation, we get

$$k^6 - 4k^4 + 16k^2 - 64 = 0,$$

a cubic in k^2.

10.5 **(a)** $\phi(n)$, for $n = 2, \ldots, 12$, is 1, 2, 2, 4, 2, 6, 4, 6, 4, 10, 4.

(b) The only positive integers not exceeding p^a and not prime to p^a are the p^{a-1} multiples of p,

$$p, 2p, \ldots, p^{a-1}p.$$

(e) Let $n = ab$. Then, if $x^n + y^n = z^n$, we have $(x^a)^b + (y^a)^b = (z^a)^b$.

(f) Suppose the point $(a/b, c/d)$, where a, b, c, d are integers, is on the curve. Then $(ad)^n + (bc)^n = (bd)^n$.

(g) Consider the right triangle whose sides are given by

$$a = 2mn,\ b = m^2 - n^2,\ c = m^2 + n^2.$$

The area of this triangle is

$$A = (1/2)\,ab = mn(m^2 - n^2).$$

Taking $m = x^2$ and $n = y^2$, and setting $x^4 - y^4 = z^2$, we find

$$A = x^2y^2(x^4 - y^4) = x^2y^2z^2.$$

Therefore, if $x^4 - y^4 = z^2$ has a solution in positive integers, x, y, z, there exists an integral-sided right triangle whose area is a square number.

Finally, if $x^4 + v^4 = z^4$, then $z^4 - v^4 = (x^2)^2$.

(h) Suppose $\sqrt{3} = a/b$, where a and b are positive integers. Now

$$\sqrt{3} + 1 = 2/(\sqrt{3} - 1).$$

Replacing the $\sqrt{3}$ on the right side by a/b, we find

$$\sqrt{3} = (3b - a)/(a - b).$$

Since $3/2 < a/b < 2$, it follows that $3b - a$ and $a - b$ are positive integers with $3b - a < a$ and $a - b < b$.

10.6 **(a)** 15 : 1.

(b) 21 : 11.

10.8 **(c)** For, by the definition of a cissoid (see Problem Study 4.4),

$$r = OP = AB.$$

By the law of sines applied to triangle OBC (see Figure 138),

$$\frac{\sin \alpha}{\frac{a\sqrt{2}}{2}} = \frac{\sin \theta}{\frac{a}{2}}$$

whence

$$r = AB = a \cos \alpha = a\sqrt{1 - 2 \sin^2 \theta},$$

and

$$r^2 = a^2 \cos 2\theta.$$

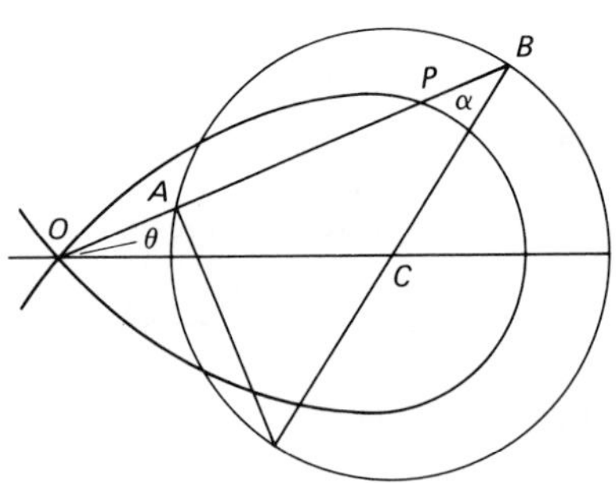

FIGURE 138

10.9 **(b)** Let the number chosen be x. Then

$$x = 3a' + a = 4b' + b = 5c' + c,$$

whence

$$\frac{40a + 45b + 36c}{60} = \frac{2(x - 3a')}{3} + \frac{3(x - 4b')}{4} + \frac{3(x - 5c')}{5}$$

$$= 2x - (2a' + 3b' + 3c') + \frac{x}{60}.$$

(c) In the general case, B ends up with $q(p + 1)$ counters.

10.10 **(a)** In the case of the ellipse, consider a point on the curve as moving away from one focus and toward the other; in the case of the hyperbola, consider a point on the curve as moving either away from or toward both foci. Now in the first case, the sum of the focal radii of the moving point is constant, and in the second case, the difference of the focal radii is constant.

(b) A **lune** is a portion of the surface of a sphere bounded by 2 semicircumferences of great circles; the **angle** of the lune is the angle between the two semicircumferences.

(c) Produce the sides of the given triangle ABC to complete the great circles; let A', B', C' be the points antipodal to A, B, C respectively. Triangles $A'BC$ and $AB'C'$ are symmetrical, and therefore equivalent. It follows that

$$\Delta ABC + \Delta AB'C' = \text{lune } ABA'C.$$

Also

$$\Delta ABC + \Delta AB'C = \text{lune } BAB'C,$$
$$\Delta ABC + \Delta ABC' = \text{lune } CAC'B.$$

Now

$$ABC + AB'C' + AB'C + ABC' = 360 \text{ spherical degrees}$$

and

$$ABA'C + BAB'C + CAC'B = 2(A + B + C) \text{ spherical degrees.}$$

Therefore

$$2\,ABC + 360 \text{ spherical degrees} = 2(A + B + C) \text{ spherical degrees.}$$

(d) Let S be the area of the sphere. Then $A : S = E : 720$. But $S = 4\pi r^2$.

(e) $98\,\pi$ square inches.

10.11 **(a)** $\ln 2 = 0.69315$.

(b) $\ln 3 = 1.09861$.

(c) $\ln 4 = 2 \ln 2 = 1.3863$.

11.1 **(a)** Let M denote the given magnitude and m, taken less than M, any assigned magnitude of the same kind. By the axiom of Archimedes, there exists an

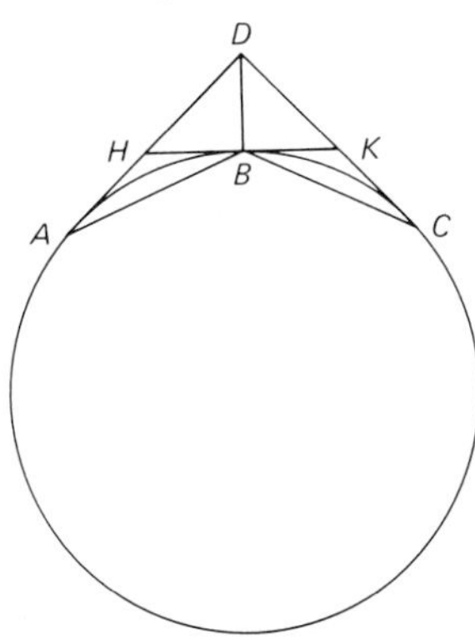

FIGURE 139

integer $n \geqq 2$ such that $nm > M$. Since $n \geqq 2$, it follows that $n/2 \leqq n - 1$. Let M_1 be what remains after we subtract from M a part not less than its half. Then

$$M_1 \leqq \frac{M}{2} < \frac{nm}{2} \leqq (n - 1)m.$$

Continuing this process we finally get $M_{n-1} < m$.

(b) In Figure 139, $HA = HB < HD$. Therefore $\Delta HBD > \Delta HBA$, or $\Delta HKD > (1/2)(ABCD)$.

11.2 **(a)** We have $(OM)(AO) = (OP)(AC)$. Summing we then find

$$(\text{area of segment})(HK) = (\Delta AFC)KC/3.$$

11.3 **(a)** $2\pi rh$.

(b) Consult a calculus text.

(c) $V = 2r^2h/3$, r = radius of cylinder and h = altitude of wedge.

(d) $V = 16r^3/3$.

11.4 **(a)** (1) Consider the triangular prism ABC–$A'B'C'$. Dissect the prism by the planes $B'AC$ and $B'A'C$.

(c) $V = 2r^2h/3$.

(d) $V = \pi h^3/6$.

(e) See (d).

(g) $V = 2\pi^2cr^2$.

(h) $A = \pi a^2$.

(i) Equally spaced chords between 2 sides of a polygon change length uniformly, whereas equally spaced chords in a circle do not.

11.5 **(b)** Let O be any point in the midsection and remove from the prismatoid the pyramids P_U and P_L having O as vertex and having the upper and lower bases, respectively, as bases. Then the volumes of P_U and P_L are given by $hU/6$ and $hL/6$. Now draw face diagonals, if necessary, so that all lateral faces of the prismatoid are triangles, and pass planes through O and the lateral edges, dividing the remaining piece of the prismatoid into a set of pyramids each having O as a vertex and a lateral triangular face of the prismatoid for opposite base. Show that the volume of one of these pyramids is $4hS/6$, where S is the area of the midsection of the prismatoid included in the pyramid.

(c) Any section, being a quadratic function of the distance from one base, is equal to the algebraic sum of a constant section area of a prism, a section area (proportional to the distance from the base) of a wedge, and a section area (proportional to the square of the distance from the base) of a pyramid. Thus the prismatoid is equal to the algebraic sum of the volumes of a parallelepiped, a wedge, and a pyramid. Now apply (a).

(d) Let $A(x) = ax^2 + bx + c$. Show that

$$V = \int_0^h A(x)dx = \frac{h}{6}\left[A(0) + 4A\left(\frac{h}{2}\right) + A(h)\right].$$

11.6 **(b)** Use mathematical induction.

11.8 **(b)** Set $x = y + h$. Then, by (a),

$$f(x) \equiv f(y + h) \equiv f(h) + f'(h)y + \cdots + f^{(n)}(h)\frac{y^n}{n!}.$$

If h is such that $f(h), f'(h), \ldots, f^{(n)}(h)$ are all positive, then the equation $f(y + h) = 0$ in y cannot have a positive root. That is, $f(x) = 0$ has no root greater than h, and h is an upper bound for the roots of $f(x)$.

(c) We have

$$f^{(n-k)}(a + h) \equiv f^{(n-k)}(a) + f^{(n-k+1)}(a)h + \cdots + f^{(n)}(a)\frac{h^k}{k!},$$

which shows that if $f^{(n-k)}(a), f^{(n-k+1)}(a), \ldots, f^{(n)}(a)$ are all positive, and h is also positive, then $f^{(n-k)}(a + h)$ must be positive. Similarly, the other functions are also positive for $x = a + h$.

(d) The greatest root lies between 3 and 4.

11.9 **(a)** Consider the four cases illustrated in Figure 140.

(b) 2.0945514, correct to 7 places.

(c) 4.4934.

(h) See, for example, W. V. Lovitt, *Elementary Theory of Equations,* p. 144.

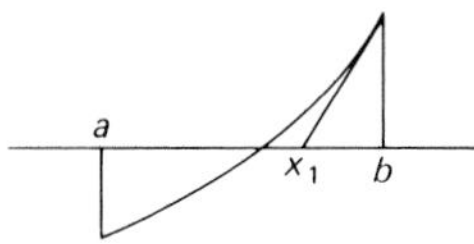

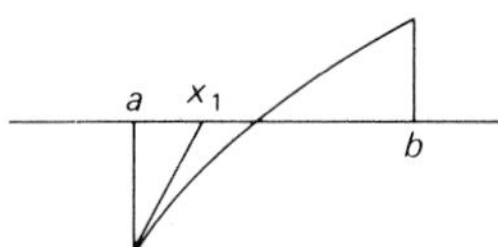

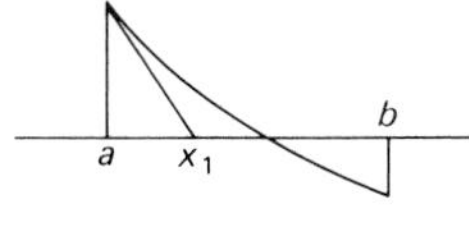

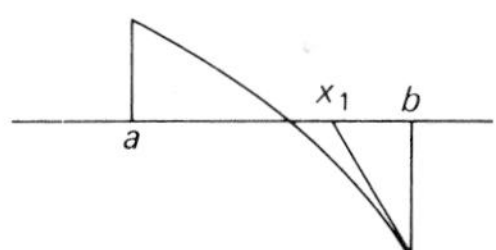

FIGURE 140

12.1 **(a)** $B_1 = 1/6$, $B_2 = 1/30$, $B_3 = 1/42$, $B_4 = 1/30$, $B_5 = 5/66$.

(b) $7709321041217 = 37(208360028141)$.

(c) $B_4 = -1 + 1/2 + 1/3 + 1/5$; $B_8 = 6 + 1/2 + 1/3 + 1/5 + 1/17$.

12.2 **(a)** Use mathematical induction.

(b) $\cos 4x = 8\cos^4 x - 8\cos^2 x + 1$,
$\sin 4x = 4 \sin x \cos x - 8 \sin^3 x \cos x$.

(c)
$$\begin{aligned}(-1 - i)^{15} &= 2^{15/2}(\cos 225° + i \sin 225°)^{15}\\ &= 2^{15/2}(\cos 3375° + i \sin 3375°)\\ &= 2^{15/2}(\cos 135° + i \sin 135°)\\ &= 2^7(-1 + i).\end{aligned}$$

(d) $\cos(n\pi/2) + i \sin(n\pi/2) = [\cos(\pi/2) + i \sin(\pi/2)]^n = i^n$.

(e) ± 1, $\pm(\sqrt{2} + i\sqrt{2})/2$, $\pm i$, $\pm(\sqrt{2} - i\sqrt{2})/2$.

12.3 **(c)** 2.996 heads per toss.
(d) 2 heads per toss.
(e) 3 heads per toss.
(f) The mean rises greatly; the median may rise a little; the mode remains the same.
(g) The mode.
(h) They are all the same.

12.4 **(a)** $\sin z = z - z^3/3! + z^5/5! - z^7/7! + \ldots,$
$\cos z = 1 - z^2/2! + z^4/4! - z^6/6! + \ldots,$
$e^z = 1 + z + z^2/2! + z^3/3! + z^4/4! + \ldots.$

12.7 See, for example, Cadwell, *Topics in Recreational Mathematics,* Chapter 15.

12.8 See, for example, Ball, *Mathematical Recreations and Essays,* 11th ed., pp. 242–54.

12.10 **(b)** We have $du = x\,dy - y\,dx$. For the circle this becomes

$$du = x\,d(1 - x^2)^{1/2} - (1 - x^2)^{1/2}\,dx = -\frac{dx}{(1 - x^2)^{1/2}},$$

whence

$$u = \int_1^x \frac{-dx}{(1 - x^2)^{1/2}} = \cos^{-1} x.$$

For the hyperbola we have

$$du = x\,d(x^2 - 1)^{1/2} - (x^2 - 1)^{1/2}\,dx = \frac{dx}{(x^2 - 1)^{1/2}},$$

whence

$$u = \int_1^x \frac{dx}{(x^2 - 1)^{1/2}} = \ln[x + (x^2 - 1)^{1/2}].$$

12.16 For synthetic treatments of (b), (c), (d), (f) see, respectively, Sections 228, 232, 233, 299 of Altshiller-Court, *Modern Pure Solid Geometry,* 2d ed. New York: Chelsea, 1964. An analytical treatment of the parts of this Problem Study constitutes a good junior research project in solid analytic geometry.

12.17 **(a)** Consider the three cases: (1) C between A and B, (2) B between A and C, (3) A between C and B.
(b) Use (a).
(c) By (b) we have, for the left member,

$$AD(DC - DB) + BD(DA - DC) + CD(DB - DA).$$

(d) Start with $AM = MB$, and then insert an origin at P.
(e) Insert an origin at P.
(f) Set $AA' = OA' - OA = (O'A' - O'O) - OA$, etc.
(g) Insert an origin O and let M and N denote the midpoints of CR and PQ. Then $4OM = 2OR + 2OC = OA + OB + 2OC = OB + OC + 2OQ =$

$2OP + 2OQ = 4ON$. Or, M and N clearly coincide if A, B, C are not collinear; now let C approach collinearity with A and B.

12.18 **(a)** If the sides BC, CA, AB of a triangle ABC meet an n-ic in $P_1, P_2, \ldots, P_n$; $Q_1, Q_2, \ldots, Q_n$; $R_1, R_2, \ldots, R_n$, then

$$(AR_1)(AR_2)\cdots(AR_n)(BP_1)(BP_2)\cdots(BP_n)(CQ_1)(CQ_2)\cdots(CQ_n)$$
$$= (AQ_1)(AQ_2)\cdots(AQ_n)(BR_1)(BR_2)\cdots(BR_n)(CP_1)(CP_2)\cdots(CP_n).$$

(b) If the sides AB, BC, CD, . . . of a polygon intersect a conic in A_1 and A_2, B_1 and B_2, C_1 and C_2, . . . , then

$$(AA_1)(AA_2)(BB_1)(BB_2)(CC_1)(CC_2)\cdots$$
$$= (BA_1)(BA_2)(CB_1)(CB_2)(DC_1)(DC_2)\cdots.$$

(c) Show that under a translation wherein the origin is shifted to the point (x_0,y_0), the coefficients of the terms of highest degree of a polynomial $f(x,y)$ are left unaltered, and the constant term becomes $f(x_0,y_0)$.

(d) Through any point O in the plane of the polygon draw lines parallel to the sides of the polygon. Now apply (c) to each pair of adjacent sides of the polygon.

13.1 **(a)** We have the 2 lines $x \pm y = 0$ and the hyperbola $xy = 2$.

(b) We have the 2 hyperbolas $x^2 - y^2 - 2y = 0$ and $2xy + 2x + 1 = 0$.

13.2 See, for example, D. M. Burton, *Elementary Number Theory,* revised edition, Chapter 4.

13.3 **(a)** $n(a + l)/2$.

(c) Let $p = 4m + 3$, $q = 4n + 3$. Then $P = (p - 1)/2$ and $Q = (q - 1)/2$ are both odd, whence $(-1)^{PQ} = -1$.

13.5 **(a)** (1) convergent, (2) absolutely convergent, (3) divergent.

(b) (1) convergent, (2) divergent.

(c) (1) convergent, (2) divergent.

13.6 **(a)** By G3 there exists c^{-1}. From $a * c = b * c$, we then have $(a * c) * c^{-1} = (b * c) * c^{-1}$, or, by G1, $a * (c * c^{-1}) = b * (c * c^{-1})$. Employing G3, we now have $a * i = b * i$, whence finally, by G2, $a = b$.

(b) By G3 there exists a^{-1}. Hence, applying G1, G3, G2, G3 in turn, we have $(i * a) * a^{-1} = i * (a * a^{-1}) = i * i = i = a * a^{-1}$. By (a) we then have $i * a = a$. But, by G2, we have $a * i = a$. It now follows that $i * a = a * i$.

(c) Let i and j be 2 identity elements for the group. Then, by G2 applied to the identity element j, $i * j = i$. Also, by (b), $i * j = j * i$. But, by G2 applied to the identity element i, $j * i = j$. It now follows that $i = j$.

(d) By G1, G3, and (b), applied in turn, $(a^{-1} * a) * a^{-1} = a^{-1} * (a * a^{-1}) = a^{-1} * i = i * a^{-1}$. Therefore, by (a), $a^{-1} * a = i$. But, by G3, $a * a^{-1} = i$. It now follows that $a^{-1} * a = a * a^{-1}$.

13.8 Those of (a), (b), (c), (d), (e).

13.9 **(a)** Let M be the midpoint of the base AB. Draw DM and CM.

(c) Drop a perpendicular from the vertex of the triangle upon the line joining the midpoints of the 2 sides of the triangle.

13.15 **(a)** $*$ is neither commutative nor associative; $|$ is both commutative and associative; the distributive law holds.

(b) None of the laws holds.

(c) Only the 2 commutative laws hold.
(d) | is associative and the distributive law holds.

13.18 **(f)** There are no divisors of 0; the left cancellation law for multiplication.
(g) Show that $\begin{bmatrix} 0 & 1 \\ 0 & 0 \end{bmatrix} = \begin{bmatrix} a & b \\ c & d \end{bmatrix}^2$ implies:(1) $b(a + d) = 1$, (2) $c(a + d) = 0$, (3) $a^2 + bc = 0$, (4) $cb + d^2 = 0$. From (1) it follows that $a + d \neq 0$. Therefore, from (2), $c = 0$. Hence, from (3) and (4), $a = d = 0$. This contradicts the conclusion that $a + d \neq 0$.

13.19 See, for example, H. Eves, *Elementary Matrix Theory,* Sections 1.7A and 6.7.

13.20 **(c)** This may be shown in several ways, but each is tricky. Look up the proof as given in some textbook on vector analysis.

14.2 **(b)** See for example, Altshiller-Court, *Modern Pure Solid Geometry*. 2d ed. Section 170, p. 57.
(c) See *loc. cit.,* Section 172, p. 58.
(d) See *loc. cit.,* Section 176.1, p. 59.

14.3 Only if each edge of the tetrahedron is orthogonal to its opposite edge. (Such a tetrahedron is called an **orthocentric tetrahedron.**)

14.5 **(c)** Instead of isogonal conjugate lines of a plane angle, consider *isogonal conjugate planes* of a dihedral angle.

14.6 **(a)** $\cos \theta = \cos(2\theta/3 + \theta/3)$.
(b) The central angle of a regular polygon of nine sides is $40° = (2/3)\ 60°$.
(d) Let $7\theta = 360°$. Then $\cos 3\theta = \cos 4\theta$, or, setting $x = \cos \theta$, $8x^3 + 4x^2 - 4x - 1 = 0$.
(f) $m^3 = 2s^3$.
(h) Let c be the circumference of a circle with unit radius. Then $c = 2\pi$.
(i) Take $AOB = 90°$ and let M and N be the feet of the perpendiculars from P on OA and OB. Let R be the center of the rectangle $OMPN$. Now if CD is Philon's line for angle AOB and point P, show that $RE = RP$, and hence that $RD = RC$. We now have Apollonius' solution of the duplication problem (see Problem Study 4.3).

14.7 **(c), (d)** See R. C. Yates, *The Trisection Problem.*

14.8 See Howard Eves, *A Survey of Geometry,* vol. 1, Section 4–4.

14.9 See A. E. Hallerberg, "The geometry of the fixed-compass," *The Mathematics Teacher* (April) 1959: 230–244 and A. E. Hallerberg, "Georg Mohr and Euclidis Curiosi," *The Mathematics Teacher* (February) 1960: 127–132.

14.10 **(a)** Simplicity 13, exactitude 8.
(b) Simplicity 9, exactitude 6.
(c) Simplicity 9, exactitude 5.
(d) Simplicity 9, exactitude 5.
(e) Simplicity 8, exactitude 5.

14.11 **(g)** The theorem is self-dual.

14.14 **(a)** (1) $\alpha = \beta$, (2) $\alpha + \beta = k$.
(b) $x = a(\cot \alpha - \cot \beta)/2(\cot \alpha + \cot \beta)$, $y = a/(\cot \alpha + \cot \beta)$, $\alpha = \cot^{-1}[(a + 2x)/2y]$, $\beta = \cot^{-1}[(a - 2x)/2y]$, where $a = AB$.
(c) (1) An ellipse, (2) a vertical straight line, (3) a straight line.
(d) (1) $(x^2 + y^2)^2 = a^2(x^2 - y^2)$, (2) $x^2 + y^2 + ax = a\sqrt{x^2 + y^2}$.
(f) $x = r \cos \phi \cos \theta$, $y = r \cos \phi \sin \theta$, $z = r \sin \phi$.

14.16 **(f)** 2; **(g)** 2; **(h)** 2; **(i)** 4; **(j)** 3; **(k)** 3; **(l)** 6; **(m)** 4; **(n)** 3; **(o)** 2.

14.17 See, for example, H. Eves, *A Survey of Geometry,* vol. 2, Section 9–2.

14.18 **(b)** $(b,-a,0)$.
(d) $2x + y + k = 0$.
(e) $x^2 + y^2 + 2fyz + 2gxz + cz^2 = 0$.
(g) See, for example, H. Eves, *A Survey of Geometry*, vol. 2, Theorem 10.3.9 (p. 83).

14.20 **(a)** The set of all points z such that

$$z_i = (1 - t)x_i + ty_i,$$

where t is an arbitrary real number.
(b) The ordered set of numbers $(y_i - x_i)/d$, $i = 1, \ldots, n$, where d is the distance between the two given points.
(c) In (a) limit t so that $0 < t < 1$.
(d) The point z such that $z_i = (x_i + y_i)/2$.
(f) (1), (2), and (3) are obvious. To prove (4), first show that the distance between 2 points is invariant under a translation. The validity of (4) will therefore be unaltered if the points x, y, z are transformed by a translation that carries y onto the origin. Then (4) reduces to the inequality

$$[\Sigma(x_i - z_i)^2]^{1/2} \leqq (\Sigma x_i^2)^{1/2} + (\Sigma z_i^2)^{1/2},$$

which may be established by simple algebra.

14.22 **(a)** Use the relations of Problem Study 12.10.
(c) This is an immediate consequence of (a) and (b).
(e) $K = -(1/QP)(1/QT) = -1/(QF)^2 = -1/k^2$.

14.26 **(a)** $\log(1/2) < 0$.
(c) If two fractions are equal and have equal *nonzero* numerators, they also have equal denominators.
(d) Examine step 2 for $k = 2$.
(e) Examine step 2 for $a = 1$ or $b = 1$.

14.27 **(a)** The integral is improper, since the integrand is discontinuous at $x = 0$.
(b) Examine for endpoint maxima and minima.
(c) Examine for endpoint maxima and minima.
(d) Do not forget the constant of integration.

14.28 See, for example, Howard Eves, *A Survey of Geometry*, vol. 2, Section 13.4.

14.29 **(b)** No. For example, $\sqrt{2}$ is algebraic, since it is a root of $x^2 - 2 = 0$.
(c) Algebraic. It is a root of $x^2 + 1 = 0$.
(d) If $\pi/2$ is a root of the polynomial equation $f(x) = 0$, then π is a root of the polynomial equation $f(x/2) = 0$.
(e) If $\pi + 1$ is a root of the polynomial equation $f(x) = 0$, then π is a root of the polynomial equation $f(x + 1) = 0$.
(f) If $\sqrt{\pi}$ is a root of the polynomial equation $f(x) = 0$, then π is a root of the equation $f(\sqrt{x}) = 0$. Solve for $\sqrt{x}$, square both sides, etc.

14.30 **(d)** Consider the set N of all numbers of the form $-x$, where x is in M.
(g) l.u.b. = 2, g.l.b. = -2; l.u.b. = 3/2, g.l.b. = -1.
(h) No.
(i) No.

14.31 **(b)** If p is composite, then $p = ab$, where $a \leqq b$ and, consequently, $a^2 \leqq p$.
(c) For $n = 10^9$ we have $(A_n \log_e n)/n = 1.053 \ldots$.
(d) Consider $(n + 1)! + 2, (n + 1)! + 3, \ldots, (n + 1)! + (n + 1)$.

15.4 Verification of the first 4 postulates presents little difficulty. To verify the fifth postulate, it suffices to show that 2 ordinarily intersecting straight lines, each determined by a pair of restricted points, intersect in a restricted point. This may be accomplished by showing that the equation of a straight line determined by 2 points having rational coordinates has rational coefficients, and that 2 such lines, if they intersect, must intersect in a point having rational coordinates. For the last part of the problem, consider the unit circle with center at the origin, and the line through the origin having slope one.

15.6 **(a)** Let the line enter the triangle through vertex A. Take any point U on the line and lying inside the triangle. Let V be any point on the segment AC, and draw line UV. By Pasch's postulate, UV will (1) cut AB, or (2) cut BC, or (3) pass through B. If UV cuts AB, denote the point of intersection by W and draw WC; now apply Pasch's postulate, in turn, to triangles VWC and BWC. If UV cuts BC, denote the point of intersection by R; now apply Pasch's postulate to triangle VRC. If UV passes through B, apply Pasch's postulate to triangle VBC.

15.7 **T1** Suppose we have both $R(a,b)$ and $R(b,a)$. Then, by P3, we have $R(a,a)$. But this is impossible by P2. Hence, the theorem by *reductio ad absurdum*.

T2 Since $c \neq a$ we have, by P1, either $R(a,c)$ or $R(c,a)$. If we have $R(c,a)$, since we also have $R(a,b)$, we have, by P3, $R(c,b)$. Hence the theorem.

T3 Suppose the theorem is false, and let a be any element of K. Then there exists an element b of K such that we have $R(a,b)$. By P2, $a \neq b$. Thus a and b are distinct elements of K.

By our supposition, there exists an element c of K such that we have $R(b,c)$. By P2, $b \neq c$. By P3, we also have $R(a,c)$. By P2, $a \neq c$. Thus, a, b, c are distinct elements of K.

By our supposition, there exists an element d of K such that we have $R(c,d)$. By P2, $c \neq d$. By P3, we also have $R(b,d)$ and $R(a,d)$. By P2, $b \neq d$, $a \neq d$. Thus, a, b, c, d are distinct elements of K.

By our supposition, there exists an element e of K such that we have $R(d,e)$. By P2, $d \neq e$. By P3, we also have $R(c,e)$, $R(b,e)$, $R(a,e)$. By P2, $c \neq e$, $b \neq e$, $a \neq e$. Thus, a, b, c, d, e are distinct elements of K.

We now have a contradiction of P4. Hence, the theorem by *reductio ad absurdum*.

T4 By T3 there is at least one such element, say a. Let $b \neq a$ be any other element of K. By P1, we have either $R(a,b)$ or $R(b,a)$. But, by hypothesis, we do not have $R(a,b)$. Therefore, we must have $R(b,a)$, and the theorem is proved.

T5 By Definition 1, we have $R(b,a)$ and $R(c,b)$. By P3, we then have $R(c,a)$, or, by Definition 1, we have $D(a,c)$.

T6 Suppose $a \neq b$. Then, by P1, we have either $R(a,b)$ or $R(b,a)$. Suppose we have $R(a,b)$. Since we have $F(b,c)$ we also have, by Definition 2, $R(b,c)$. This is impossible since we are given that we have $F(a,c)$. Suppose we have $R(b,a)$. Since we have $F(a,c)$ we also have, by Definition 2, $R(a,c)$. This is impossible since we are given that we have $F(b,c)$. Thus, in either case we are led to a contradiction of our hypothesis. Hence the theorem by *reductio ad absurdum*.

T7 For, by Definition 2, we have $R(a,b)$ and $R(b,c)$. Hence, by Definition 2, we cannot have $F(a,c)$.

15.8 **(b)** **T1:** *If a is an ancestor of b, then b is not an ancestor of a.*

T2: *If a is an ancestor of b and if c is some third member of K distinct from a and b, then either a is an ancestor of c or c is an ancestor of b.*

T3: *There is some man in K who is not an ancestor of anyone in K.*

T4: *There is only one man in K who is not an ancestor of anyone in K.*

Definition 1: If b is an ancestor of a, we say that a is a *descendant* of b.

T5: *If a is a descendant of b, and b is a descendant of c, then a is a descendant of c.*

Definition 2: If a is an ancestor of b and there is no individual c of K such that a is an ancestor of c and c is an ancestor of b, then we say that a is a *father* of b.

T6: *A man in K has at most one father in K.*

T7: *If a is the father of b and b is the father of c, then a is not the father of c.*

Definition 3: If a is the father of b and b is the father of c, we say that a is a *grandfather* of c.

(d) Since T1 has been deduced from P1, P2, P3, P4, all that remains is to deduce P2 from P1, T1, P3, P4.

15.9 **(b)** The *converse* of "If A then B" is "If B then A."

(c) The *opposite* of "If A then B" is "If not-A then not-B."

15.10 **(a)** 48 miles per hour.

(b) 2.4 days.

(d) $67\frac{1}{2}$ cents.

(e) The second one.

(f) At the end of 59 seconds.

(g) A very good salary.

(h) 11 seconds.

(i) Five cents.

(j) Neither; the amounts are equal.

(k) The final pile will be over 17,000,000 miles high.

(l) No.

(m) One-third.

(n) Yes.

15.13 **(a)** Interpret the elements of S as a set of all rectangular Cartesian frames of reference that are parallel to one another but with no axis of one frame coincident with an axis of another frame, and let bFa mean that the origin of frame b is in the first quadrant of frame a. Or, interpret the elements of S as the set of all ordered pairs of real numbers (m,n), and let $(m,n)F(u,v)$ mean $m > u$ and $n > v$.

15.14 **(a)** Interpret the bees as 6 people, A, B, C, D, E, F, and the 4 hives as the 4 committees (A, B, C), (A, D, E), (B, F, E), and (C, F, D). Or, interpret the bees and the hives as 6 trees and 4 rows of trees, respectively, forming the vertices and sides of a complete quadrilateral.

(b) To show independence of P2, interpret the bees and the hives as 4 trees and 4 rows of trees forming the vertices and sides of a square. To show independence of P3, interpret the bees as 4 trees located at the vertices and the foot of an altitude of an equilateral triangle, and the hives as the 4 rows of trees along the sides and the altitude of the triangle. To show independence of P4, interpret the bees and the hives as 3 trees and 3 rows forming the vertices and sides of a triangle.

	a	b	c	d
a		1	2	3
b	1		4	5
c	2	4		6
d	3	5	6	

FIGURE 141

(c) Denote the 4 hives by a, b, c, d and denote the bees by natural numbers 1, 2, 3, ... The postulates lead, of necessity, to the schema of Figure 141, wherein the natural number in any box indicates the unique bee common to the 2 hives given by the headings of the row and column containing the box. All three theorems are now apparent from the schema.

15.15 **(e)** By **M1**′ we have $d(x,y) \leqq d(y,z) + d(z,x)$ and, by interchanging x and y, $d(y,x) \leqq d(x,z) + d(z,y)$. Setting $z = x$ in the first of these inequalities, and $z = y$ in the second one, we find (recalling **M2**), $d(x,y) \leqq d(y,x) \leqq d(x,y)$. It follows that $d(x,y) = d(y,x)$.

In $d(x,z) \leqq d(z,y) + d(y,x)$ set $z = x$. Then, since $0 = d(x,x)$ by **M2,** $0 \leqq d(x,y) + d(y,x) = 2d(x,y)$, by the above. Hence $d(x,y) \geqq 0$, and so on.

(g-3) It is only the verification of the triangle inequality that presents any difficulty. Denote $d(y,z)$, $d(z,x)$, $d(x,y)$ by a, b, c, respectively. Then we have

$$\frac{b}{1+b} = \frac{1}{\dfrac{1}{b}+1} \leqq \frac{1}{\dfrac{1}{c+a}+1} = \frac{c+a}{1+c+a}$$
$$= \frac{c}{1+c+a} + \frac{a}{1+c+a} \leqq \frac{c}{1+c} + \frac{a}{1+a}.$$

(h) For (c), a *circle* is a square with center at c and having its diagonals equal to $2r$ in length and lying parallel to the coordinate axes.

15.16 **(a)** Let M_1 be the midpoint of AB, M_2 the midpoint of M_1B, M_3 the midpoint of M_2B, and so on. Denote by E the set of all points on $[AB]$ with the exception of points $A, B, M_1, M_2, M_3, \ldots$. Then we have

$$\begin{aligned}[AB] &= E, A, B, M_1, M_2, M_3, \ldots, \\ (AB] &= E, B, M_1, M_2, M_3, \ldots, \\ [AB) &= E, A, M_1, M_2, M_3, \ldots, \\ (AB) &= E, M_1, M_2, M_3, \ldots .\end{aligned}$$

It is now apparent how we may put the points of any one of the 4 segments in one-to-one correspondence with the points of any other one of the 4 segments.

(b) Start with Figure 142.

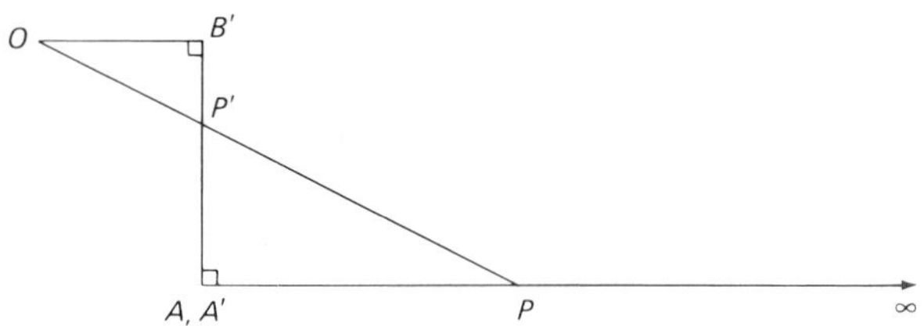

FIGURE 142

15.17 **(b)** Use the idea employed in the proof of Theorem 1 of Section 15–4.

(c) Use an indirect argument along with (a) and Theorem 1 of Section 15–4.

(d) Use an indirect argument along with (a) and Theorem 2 of Section 15–4.

15.21 **(a)** See Problem E832, *The American Mathematical Monthly* 56 (1949): 407.

(c) No, for there are c points on a straight line or a circle, and there are only d rational numbers and d algebraic numbers.

(d) Take a number axis on the given straight line. In each interval, choose a point with rational coordinates. These points are all distinct, and therefore in one-to-one correspondence with the intervals, and they constitute an infinite subset of the denumerable set of all rational numbers.

15.23 **(b)** The surface formed by a band of paper twisted through 540° and then whose ends are glued together.

(c) See Figure 143. This surface was discovered by F. Frankl and L. S. Pontryagin in 1930.

(d) A circular disc.

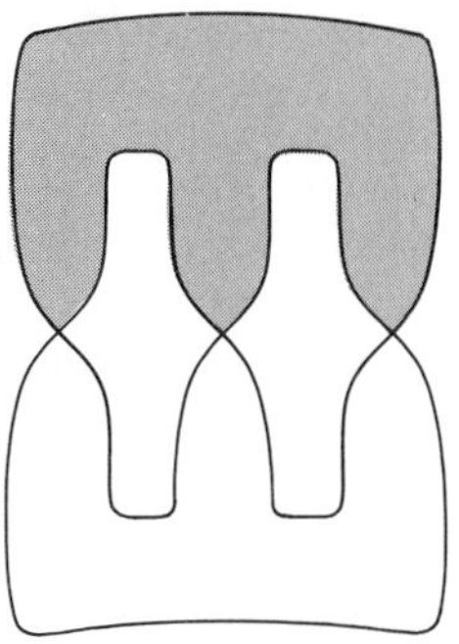

FIGURE 143

15.25 **(b)** A tetrahedron has six edges, and a pyramid with a square base has eight edges. Suppose there exists a seven-edged simple closed polyhedron. Concentrate on any particular face of the polyhedron and suppose that face has n edges. Since at least three edges issue from each vertex of this face, we see that $2n \leqq 7$, or $n < 4$. It follows that all faces of the polyhedron must be triangles, whence $3f = 2e = 14$. But this is impossible, since f is an integer.

15.26 **(a)** Relations (1) and (2) are obvious. Relations (3) and (4) follow since each edge belongs to precisely two faces, and each edge issues from precisely two vertices. To obtain (5) note that $v - e + f = 2$, or $2v + 2f = 4 + 2e$.

Substituting (1), (2), (3) we then find

$$2(v_3 + v_4 + \ldots) + 2(f_3 + f_4 + \ldots) = 4 + 3f_3 + 4f_4 + 5f_5 + \ldots,$$

or

$$2(v_3 + v_4 + \ldots) = 4 + f_3 + 2f_4 + 3f_5 + 4f_6 + \ldots.$$

To obtain (6) we similarly substitute (1), (2), (3) in $2v + 2f = 4 + 2e$. Doubling (6) and adding (5) yields

$$4(f_3 + f_4 + \ldots) + 2(v_3 + v_4 + \ldots) = 8 + (2v_3 + 4v_4 + 6v_5 + 8v_6 + \ldots) + 4 + (f_3 + 2f_4 + 3f_5 + 4f_6 + \ldots)$$

or

$$3f_3 + 2f_4 + f_5 = 12 + (2v_4 + 4v_5 + 6v_6 + \ldots) + (f_7 + 2f_8 + 3f_9 + \ldots),$$

which is (7).

(b) These are easy consequences of relation (7) of (a).

(c) For (1), relation (7) of (a) reduces to $f_5 = 12$; for (2), it reduces to $2f_4 = 12$, or $f_4 = 6$; for (3), it reduces to $3f_3 = 12$, or $f_3 = 4$.

15.27 **(e)** We omit the obvious verification of **H1** and **H2.**

To verify **H3,** let $d(x,y) < r$ and set $R = r - d(x,y) > 0$. The triangle inequality states that $d(x,y') \leqq d(x,y) + d(y,y') = (r - R) + d(y,y') < r$, if $d(y,y') < R$. Denoting the interior of the circle with center c and radius r by $S(c,r)$, we now have $S(y,R)$ is contained in $S(x,r)$.

To verify **H4,** let x be distinct from y and set $r = d(x,y) > 0$. Then it is easy to show that $S(x,r/3)$ and $S(y,r/3)$ have no common point.

(f) Since x is a limit point of S, any neighborhood N_x of x contains a point y_1 of S, where $y_1 \neq x$. By **H4,** there then exist disjoint neighborhoods N_{y_1} and N'_x of y_1 and x. Again, by **H2,** there exists a neighborhood N''_x of x contained in both N_x and N'_x. It follows that y_1 is not in N''_x. But, since x is a limit point of S, N''_x, and hence N_x, contains a point y_2 of S, where y_2 is distinct from both x and y_1. Continuing in this way, we find that N_x contains an infinite sequence of distinct points $y_1, y_2, \ldots$ of S, and the theorem is established.

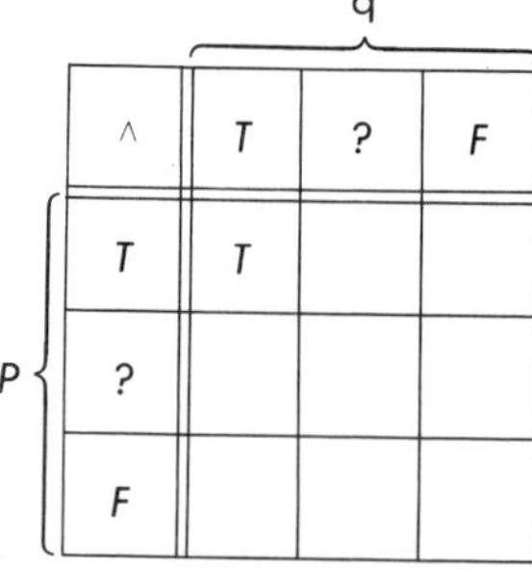

FIGURE 144

p	not-p
T	
?	
F	

FIGURE 145

15.29 **(a)** Denote the 3 possible truth values of a proposition by T (true), F (false), ? (otherwise). We may construct the truth table for conjunction as indicated in Figure 144, where, by our agreement concerning the meaning of "p and q," the top left box in the table must contain a T, and no other box in the table is allowed to contain a T. Since there are 8 remaining boxes and each may be filled in either of 2 possible ways, (namely with either an F or a?) there are altogether $2^8 = 256$ possible ways of filling in the 8 boxes.

(b) The truth table for negation may be constructed as in Figure 145, in which there are 2 ways of filling in the top box under not-p (namely, with F or ?), 3 ways of filling in the middle box under not-p (namely with T, F, or ?), and 2 ways of filling in the bottom box under not-p (namely with T or ?).

(c) $(256)(12) = 3072$.

(d) $m^{m-2}(m - 1)^{m^2+1}$.

15.33 **(a)** See, for example, Howard Eves, *Mathematical Circles Squared*, Boston: Prindle, Weber & Schmidt, 1972, pp. 53–55.

(b) See, for example, Ball-Coxeter, *Mathematical Recreations and Essays*. New York: Macmillan, 1939, pp. 165–170.

INDEX

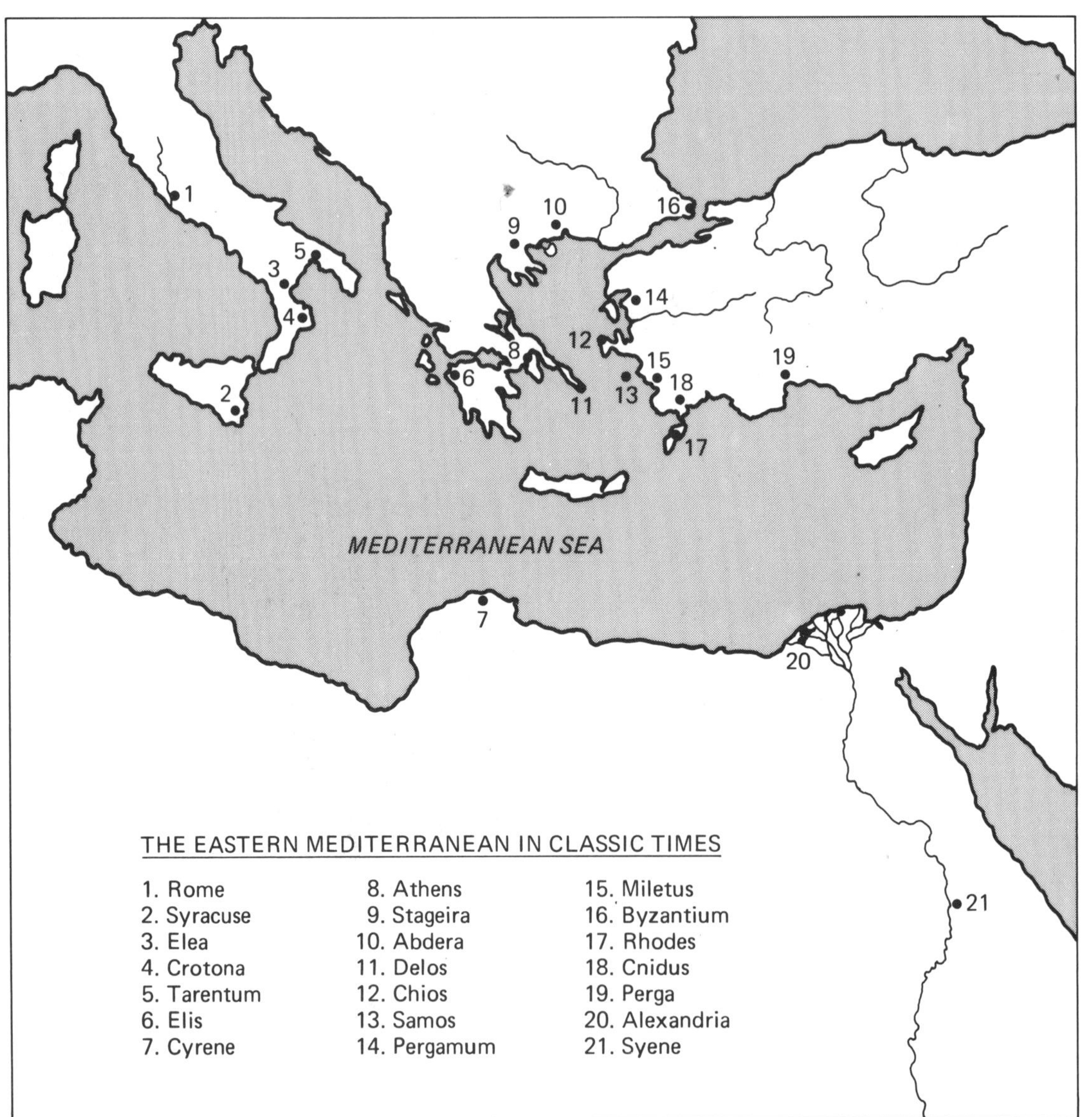

MEDITERRANEAN SEA
THE EASTERN MEDITERRANEAN IN CLASSIC TIMES
1. Rome
2. Syracuse
3. Elea
4. Crotona
5. Tarentum
6. Elis
7. Cyrene
8. Athens
9. Stageira
10. Abdera
11. Delos
12. Chios
13. Samos
14. Pergamum
15. Miletus
16. Byzantium
17. Rhodes
18. Cnidus
19. Perga
20. Alexandria
21. Syene